Values of some physical and chemical constants (in SI units and based on the ^{12}C scale)

Avogadro's number	$\mathfrak{N} = 6.0222 \times 10^{23}$ mol^{-1}
Velocity of light	$c = 2.997925 \times 10^{8}$ m s^{-1}
Mass of electron	$m = 0.91096 \times 10^{-30}$ kg
Electronic charge	$e = 0.16022 \times 10^{-18}$ C
Faraday	$\mathfrak{F} = \mathfrak{N}e = 96{,}490$ C mol^{-1}
Planck's constant	$h = 0.66262 \times 10^{-33}$ J s^{-1}
Boltzmann constant	$k = 1.3806 \times 10^{-23}$ J deg^{-1}
Gas constant	$R = \mathfrak{N}k = 8.3143$ J deg^{-1} mol^{-1}
	$= 0.08206$ liter atm deg^{-1} mol^{-1}
Atmospheric pressure	1 atm $= 101{,}325$ N m^{-2}
Standard molar gas volume	22.415 liters
Absolute zero	$-273.15°$C
Permittivity	$4\pi\epsilon_0 = 1.11264 \times 10^{-10}$ C^2 N^{-1} m^{-2}

Energy conversion factors

	J molecule^{-1}	kJ mol^{-1}	erg molecule^{-1}	kcal mol^{-1}	eV
1 J molecule^{-1}	1	0.16603×10^{-20}	1×10^{7}	1.4395×10^{20}	6.2420×10^{18}
1 kJ mol^{-1}	6.0229×10^{20}	1	0.16603×10^{-13}	0.23900	0.010363
1 erg molecule^{-1}	1×10^{-7}	6.0229×10^{-13}	1	1.4395×10^{13}	6.2420×10^{11}
1 kcal mol^{-1}	0.69468×10^{-20}	4.1840	0.69468×10^{-13}	1	0.043361
1 eV	0.16022×10^{-18}	96.4905	0.16022×10^{-11}	23.0618	1

PHYSICAL CHEMISTRY

PHYSICAL CHEMISTRY

Third Edition

Gordon M. Barrow

McGRAW-HILL BOOK COMPANY
New York
St. Louis
San Francisco
Düsseldorf
Johannesburg
Kuala Lumpur
London
Mexico
Montreal
New Delhi
Panama
Rio de Janeiro
Singapore
Sydney
Toronto

PHYSICAL CHEMISTRY

1 2 3 4 5 6 7 8 9 0 HDMM 7 9 8 7 6 5 4 3

This book was set in News Gothic by York Graphic Services, Inc.
The editors were William P. Orr and Laura Warner; the designer was Merrill Haber; and the production supervisor was Joe Campanella. The drawings were done by York Graphic Services, Inc., and Vantage Art, Inc.
The printer was Halliday Litho Corporation; the binder, The Maple Press Company.

Library of Congress Cataloging in Publication Data
Barrow, Gordon M
Physical chemistry.

Includes bibliographies.
1. Chemistry, Physical and theoretical.
QD453.2.B37 1973 541 72–10320
ISBN 0–07–003823–6

CONTENTS

PREFACE **XV**

1

PROPERTIES OF GASES **1**

1-1 The Dependence of the Volume of a Gas on Pressure: Boyle's Law 2
1-2 The Volume-Temperature Behavior of Ideal Gases 4
1-3 The *PVT* Behavior of Ideal Gases 5
1-4 The Gas Constant: Energy Units 9
1-5 Some Properties of Gas Mixtures 10
1-6 The Nonideal Behavior of Gases 12
1-7 Condensation of Gases, the Critical Point, and the Law of Corresponding States 15
1-8 The Ideal-gas Thermometer 18
1-9 Graham's Law of Effusion 19
1-10 The Viscosity of Gases 20

2

THE KINETIC-MOLECULAR THEORY OF GASES **26**

2-1 The Kinetic-Molecular Gas Model 26
2-2 The Pressure of a Gas 27
2-3 Kinetic Energies and Temperature 30
2-4 Numerical Values for Molecular Energies and Molecular Speeds 32
2-5 Distribution of Molecular Velocities in One Dimension 34
2-6 Distribution of Molecular Velocities in Three Dimensions 37
*2-7 Effusion and Molecular Beams 41

*Sections preceded by an asterisk may be omitted without loss of continuity.

2-8 The Mean Free Path, Collision Diameter, and Collision Number 42
2-9 The Kinetic Theory of Gas Viscosity 45
2-10 Numerical Values of Collision Properties 48
2-11 Theory of Nonideal Behavior—van der Waals' Equation 50
2-12 Van der Waals' Equation and the Critical Point 53
2-13 Van der Waals' Equation and the Law of Corresponding States 55

3

INTRODUCTION TO THE MECHANICS OF ATOMS AND MOLECULES **60**

3-1 The Nature of Light 61
3-2 Atomic Spectra 63
3-3 The Wave Nature of Particles 66
3-4 Wave Mechanics and the Schrödinger Equation 69
3-5 A One-dimensional Illustration of the Schrödinger Equation: A Particle in a One-dimensional Square Potential Well 71
3-6 A Three-dimensional Illustration of the Schrödinger Equation: The Cubic Potential-well Problem 75
3-7 The Use of Angular Momentum to Impose Quantum Restrictions 77

4

AN INTRODUCTION TO MOLECULAR ENERGIES **83**

4-1 Categories of Thermal Energies of Molecules of Gases 83
4-2 The Translational Energies of a Molecule of an Ideal Gas 86
4-3 The Allowed Rotational Energies of a Molecule of a Gas 87
4-4 The Vibrational Energies of Gas-phase Molecules 89
4-5 Summary and a Comment on Electronic States of Molecules 92

5

THE ENERGIES OF COLLECTIONS OF MOLECULES: THE MOLECULAR APPROACH **94**

5-1 The Boltzmann Distribution 94
*5-2 Derivation of the Boltzmann Distribution Expression 96
*5-3 The Thermal Energy due to the Translational Motion of the Molecules of a Gas and the Deduction That $\beta = 1/kT$ 99
5-4 The Partition Function 101
5-5 The Three-dimensional Translational Energy of a Mole of Gas 103
5-6 The Thermal Energy due to the Rotational Motion of the Molecules of a Gas 105
5-7 The Thermal Energy due to the Vibrational Motion of Molecules 106
5-8 Thermal Energies 109

6

THE FIRST LAW OF THERMODYNAMICS **112**

6-1 The Nature of Thermodynamics 112
6-2 Measurement and Relation of Thermal and Mechanical Energy 113

*Sections preceded by an asterisk may be omitted without loss of continuity.

6-3 The First Law of Thermodynamics 115
6-4 Determination of ΔE: Reversible and Irreversible Processes 118
6-5 $\Delta E_{\text{mech res}}$ for Expansions and Contractions of the System 119
6-6 The Enthalpy Function 122
6-7 Some Properties of State Functions 124
6-8 The Expansion of an Ideal Gas: The Dependence of Internal Energy and Enthalpy of an Ideal Gas on Pressure 127
6-9 The Dependence of E and H on Temperature: The Heat Capacities C_p and C_v of Ideal Gases 129
6-10 The Expansion of Real Gases: The Joule-Thomson Coefficient 130
6-11 Adiabatic Expansions of Ideal Gases 132
6-12 A Molecular Interpretation of E and H of Ideal Gases 134
6-13 Molecular Interpretation of C_v and C_p 136

7

THERMOCHEMISTRY **143**

7-1 Measurements of Heats of Reaction 143
7-2 Internal-energy and Enthalpy Changes in Chemical Reactions 145
7-3 Relation between ΔE and ΔH 146
7-4 Thermochemical Equations 147
7-5 Indirect Determination of Heats of Reaction 148
7-6 Standard Heats of Formation 149
7-7 Standard Heats of Formation of Ions in Aqueous Solutions 151
7-8 Temperature Dependence of Heats of Reaction 153
7-9 Heats of Reaction and the Molecular Model 157
7-10 Bond Enthalpies and Bond Energies 159

8

ENTROPY AND THE SECOND AND THIRD LAWS OF THERMODYNAMICS **166**

8-1 General Statements of the Second Law of Thermodynamics 167
8-2 Entropy and Another Statement of the Second Law of Thermodynamics 168
8-3 The Carnot Cycle 174
8-4 The Efficiency of the Transformation of Heat into Work 178
8-5 Entropy Is a State Function 179
8-6 The Unattainability of Absolute Zero 181
8-7 Entropy and the Third Law of Thermodynamics 184

9

FREE ENERGY AND CHEMICAL EQUILIBRIA **190**

9-1 A Convenient Measure of the Driving Force of a Reaction: The Free Energy 190
9-2 Some Free-energy Calculations 193
9-3 Standard Free Energies 196
9-4 The Dependence of Free Energy on Pressure and Temperature 197
9-5 The Pressure Dependence of the Free Energy of Nonideal Gases: The Fugacity 198
9-6 The Standard State for Nonideal Gases: The Activity and Activity Coefficient 203

*9-7 The Dependence of Entropy, Enthalpy, and Heat Capacity on Pressure 205
9-8 Quantitative Relation of ΔG and the Equilibrium Constant 206
9-9 Equilibrium Constants for Systems of Real Gases 210
9-10 Temperature Dependence of the Free Energy of a Reaction and the Equilibrium Constant 211

10

ENTROPY, FREE ENERGY, AND CHEMICAL EQUILIBRIA: THE MOLECULAR APPROACH **219**

10-1 The Molecular Interpretation of Entropy 219
10-2 The Translational Entropy of an Ideal Gas 222
10-3 The Rotational Entropy of the Molecules of an Ideal Gas 224
10-4 The Vibrational Entropy of the Molecules of an Ideal Gas 227
10-5 The Molecular Interpretation of Free Energy and Equilibria 228
10-6 Molecular Interpretation of the Third Law 235

11

INTRODUCTION TO THE THEORY OF CHEMICAL BONDING **238**

11-1 The Solution of the Schrödinger Equation for the Hydrogen Atom 239
11-2 The Hydrogen-atom Wave Functions 242
11-3 Experimental Characterization of Atomic States 248
11-4 The Basis for Descriptions of Many-electron Atoms: Central Forces, Electron Orbitals, and the Pauli Exclusion Principle 251
11-5 Electron Configurations and the Periodic Table 254
11-6 Size and Energy of Atomic Orbitals and Atoms: Approximate Methods, Self-consistent Field Results 255
11-7 The Beginnings of Bonding Theory 262
11-8 The Ionic Bond 263
11-9 Introduction to the Covalent Bond 265
11-10 The Hydrogen-molecule Ion 267
11-11 The Hydrogen Molecule: The Electron-pair Bond 270
11-12 The Nature of Chemical Bonds: The Virial Theorem 272

12

THE NATURE OF THE BONDING IN CHEMICAL COMPOUNDS **277**

12-1 Bonding in Homonuclear Diatomic Molecules 277
12-2 Heteronuclear Bonds and the Ionic Character of Bonds 281
12-3 Electronegativities 285
12-4 The Valence Bond: Hybridization 288
12-5 π Bonding 293
12-6 Resonance in Conjugated Systems: The Valence-bond Description 296
12-7 The Molecular Orbital Approach to Aromatic and Conjugated Molecules 299
12-8 Bonding with d Orbitals 301
12-9 Bonding in Coordination Compounds 301

*Sections preceded by an asterisk may be omitted without loss of continuity.

13

EXPERIMENTAL STUDY OF MOLECULAR STRUCTURE: SPECTROSCOPIC METHODS **309**

13-1 Rotational Spectra 310
13-2 Vibrational Spectra 316
13-3 Rotation-Vibration Spectra 321
13-4 Raman Spectroscopy 325
13-5 Electronic Spectra 326
13-6 Electronic Energies of Polyatomic Molecules 329
13-7 The Energy Levels of Nuclei in Magnetic Fields 332
13-8 *nmr* Spectroscopy 335
13-9 Chemical Shifts and Nuclear Magnetic Interactions 336
13-10 Electron-spin-resonance Spectroscopy 341

14

EXPERIMENTAL STUDIES OF MOLECULAR STRUCTURE: DIFFRACTION METHODS **347**

14-1 The Interference Phenomenon 348
14-2 The Wave Nature and Scattering of a Beam of Electrons 352
14-3 The Wierl Equation 354
*14-4 The Radial-distribution Method 359
14-5 Covalent Radii 361
14-6 Crystal Shapes 364
14-7 Symmetry Elements and Symmetry Operations 364
14-8 Lattices and Unit Cells 368
14-9 Crystal Planes 370
14-10 X-rays and X-ray Diffraction 372
14-11 The Determination of the Lattice Type and Unit-cell Dimensions 375
14-12 The Structure Factor 377
14-13 The Fourier Synthesis 382
14-14 Ionic Radii 384
14-15 Van der Waals Radii 386

15

EXPERIMENTAL STUDIES OF THE ELECTRICAL AND MAGNETIC PROPERTIES OF MOLECULES **390**

15-1 Dipole Moments of Molecules 391
15-2 Some Basic Electrostatic Ideas 393
15-3 Electrostatics for Dielectric Media 395
15-4 The Molecular Basis for Dielectric Behavior 397
15-5 Determination of the Dipole Moment and the Molecular Polarizability 401
15-6 Dipole Moments and Ionic Character 403
15-7 Bond Moments 406
15-8 Determination of Magnetic Molecular Properties 408
15-9 Molecular Interpretation of Diamagnetism 411
15-10 Molecular Interpretation of Paramagnetism 412

*Sections preceded by an asterisk may be omitted without loss of continuity.

15-11 Magnetic Results for Molecules 415
15-12 Magnetic Results for Coordination Compounds 416

16

THE NATURE OF CHEMICAL REACTIONS: RATES AND MECHANISM **419**

16-1 Measurement of the Rates of Chemical Reactions 421
16-2 Introduction to Rate Equations 422
16-3 The Fitting of Rate Data to First- and Second-order Rate Equations 424
16-4 Enzyme Kinetics 432
16-5 Introduction to Elementary Processes in the Gas Phase 435
16-6 The Rate of Gas-phase Collisions between Molecules of Different Types 436
16-7 Nature, Lifetime, and Reactions of Gas-phase Free-radical Intermediates 439
16-8 Elementary Reactions in Liquid Solution: Diffusion-controlled Reactions 440
16-9 Lifetime of Liquid-phase Intermediates: The Cage Effect 442
16-10 Ionic Intermediates in Liquid-phase Reactions 443
16-11 Reaction Mechanisms and Rate Laws: The Stationary-state Method 446
16-12 A Mechanism for Enzyme-catalyzed Reactions 450

17

THE NATURE OF ELEMENTARY REACTIONS **456**

17-1 Temperature Dependence of the Rates of Chemical Reactions 457
17-2 Introduction to Theories of Elementary-reaction Processes 460
17-3 The Collision Theory 462
17-4 The Transition-state Theory 467
*17-5 Comparison of the Results of the Collision and the Transition-state Theories 471
17-6 Application of the Transition-state Theory to Reactions in Solution 472
17-7 The Entropy of Activation 473
17-8 Light Absorption 475
17-9 The Primary Process 477
17-10 Secondary Processes 478
17-11 Some Photochemical Reactions 480
17-12 Flash Photolysis 482
17-13 Mass-spectroscopic Results 483
17-14 High-energy Radiation 486

18

CRYSTALS **490**

18-1 Crystal Forces and Crystal Types 490
18-2 Cohesive Energy of Ionic Crystals: Thermodynamic Determination 494
18-3 Calculation of the Crystal Energies of Ionic Crystals 495
18-4 The Heat Capacity of Crystals 500
18-5 An Introduction to the Study of Metallic Crystals 504

*Sections preceded by an asterisk may be omitted without loss of continuity.

19

LIQUIDS **512**

19-1 The Heat of Vaporization and Intermolecular Forces 512
19-2 Entropy of Vaporization, Trouton's Rule, and Free-volume Theories of the Liquid State 519
19-3 The Heat Capacity of Liquids 524
19-4 Surface Tension 526
19-5 Surface Tension and Vapor Pressure of Small Droplets 529
19-6 Structure of Liquids 531
19-7 The Viscosity of Liquids 532
19-8 Theory of Viscosity 536

20

PHASE EQUILIBRIA **539**

20-1 Pressure-Temperature Diagrams for One-component Systems 539
20-2 Qualitative Thermodynamic Interpretation of Phase Equilibria of One-component Systems 544
20-3 Quantitative Treatment of Phase Equilibria: The Clausius-Clapeyron Equation 546
20-4 The Number of Phases 549
20-5 The Number of Components 549
20-6 The Number of Degrees of Freedom 551
20-7 The Phase Rule for One-component Systems 552
20-8 The Phase Rule 552
20-9 Two-component Liquid Systems 554
20-10 Two-component Solid-Liquid Systems: Formation of a Eutectic Mixture 557
20-11 Two-component Solid-Liquid Systems: Compound Formation 560
20-12 Two-component Solid-Liquid System: Miscible Solids 562
20-13 Three-component Systems 562
20-14 Liquid-Vapor Equilibria of Solutions 566
20-15 Vapor-pressure Diagrams Showing Liquid and Vapor Compositions 569
20-16 Boiling-point–composition Diagrams 571
20-17 Distillation 572
20-18 Distillation of Immiscible Liquids 575

21

THE THERMODYNAMIC TREATMENT OF MULTICOMPONENT SYSTEMS **580**

21-1 Introduction to Solutions: The Thermodynamics of Ideal Solutions 580
21-2 Thermodynamic Properties of Real Solutions 584
21-3 Properties of the Components of Real Solutions: Partial Molal Quantities 587
21-4 The Free Energies of the Components of a Solution: Solvents 591
21-5 The Free Energies of the Components of a Solution: Solutes 594
21-6 Solute Activities from Solvent Properties: An Application of the Gibbs-Duhem Equation 597
21-7 The Dependence of the Free Energy of the Solvent of an Ideal Solution on Temperature, Pressure, and Composition 599
21-8 Vapor-pressure Lowering 601

21-9 The Boiling-point Elevation 602
21-10 The Freezing-point Depression 605
21-11 Osmotic Pressure 608
21-12 Osmotic-pressure Determination of Molecular Masses 612

22

THE NATURE OF ELECTROLYTES IN SOLUTION **617**

22-1 Electrical Conductivity of Solutions 618
22-2 Equivalent Conductance 620
22-3 The Arrhenius Theory of Dissociations 623
22-4 Colligative Properties of Aqueous Solutions of Electrolytes 625
22-5 Dissociation Equilibria 627
22-6 Electrolysis and the Electrode Process 629
22-7 Transference Numbers 630
22-8 Ionic Conductances 635
22-9 Ionic Mobilities 636
22-10 Some Applications of Conductance Measurements 639
22-11 The Role of the Solvent: Dielectric Effect 642
22-12 The Role of the Solvent: Solvation Energies of Ions 644
*22-13 The Debye-Hückel Theory of Interionic Interactions 647
22-14 Interpretation of the Strong Electrolyte Conductance Results 652

23

THE ELECTROMOTIVE FORCE OF CHEMICAL CELLS AND THERMODYNAMICS OF ELECTROLYTES **659**

23-1 Types of Electrodes 659
23-2 Electrochemical Cells, Emfs, and Cell reactions 663
23-3 Free-energy Changes for Cell Reactions 665
23-4 Standard Emfs and Electrode Potentials 666
23-5 The Concentration and Activity Dependence of the Emf 670
23-6 Activities from Emf Measurements 672
23-7 The Debye-Hückel Theory of Ionic Activity Coefficients 675
23-8 Equilibrium Constants and Solubility Products from Emf Data 678
23-9 Electrode-concentration Cells 679
23-10 Electrolyte-concentration Cells 680
23-11 Electrolyte-concentration Cells with Liquid Junction 682
23-12 The Salt Bridge 684
23-13 The Glass Electrode 685
23-14 pH Definition and Measurement 686
23-15 Activity Coefficients from the Dissociation of a Weak Electrolyte 688
23-16 Activity Coefficients from Solubility Measurements 690
23-17 Activity Coefficients in More Concentrated Solutions 694
23-18 Thermodynamic Data from Emf Measurements 697
23-19 The Effect of Electrostatic Interactions on Reaction Rates 698

*Sections preceded by an asterisk may be omitted without loss of continuity.

24

ADSORPTION AND HETEROGENEOUS CATALYSIS **703**

24-1 Liquid Films on Liquids 704
24-2 Classifications of Adsorptions of Gases on Solids 708
24-3 Heat of Adsorption 709
24-4 The Adsorption Isotherm 710
24-5 The Langmuir Adsorption Isotherm 711
24-6 Determination of Surface Areas 714
24-7 Adsorption from Solution 716
24-8 The Nature of the Adsorbed State 716
24-9 Importance of the Preparation of the Surface 719
24-10 Some Experimental Methods and Results 720
24-11 Kinetics of Heterogeneous Decompositions 721

25

MACROMOLECULES **727**

25-1 Types and Sizes of Particles 728
25-2 Synthetic Polymers 730
25-3 Proteins 732
25-4 Nucleic Acids 736
25-5 The Polysaccharides 737
25-6 The Polyisoprenes 739
25-7 Molecular Masses of Polymers 740
25-8 Osmotic-pressure Determinations of Molecular Masses 741
25-9 Diffusion 741
25-10 Sedimentation and the Ultracentrifuge 744
25-11 Viscosity 747
*25-12 Light Scattering 749
25-13 Electrokinetic Effects 755
25-14 The Donnan Membrane Equilibrium and Dialysis 759
25-15 The Structure of Proteins 761
25-16 The Structure of Nucleic Acids 767
25-17 Crystallinity of High Polymers 769
25-18 Electron Microscopy 771

Appendix 1 Evaluation of Integrals of the Type $\int_0^\infty x^n e^{-ax^2}\,dx$ 774
Appendix 2 Stirling's Approximation 775
Appendix 3 The Method of Lagrange Multipliers 776
Appendix 4 Replacing a Sum by an Integral 778
Appendix 5 Thermodynamic Properties of Substances at 1 Atm Pressure and 25°C in the Physical State Indicated 779
Appendix 6 Thermodynamic Properties of Substances in Aqueous Solution at Unit Activity and 25°C 783
Appendix 7 SI Units 785

INDEX **789**

*Sections preceded by an asterisk may be omitted without loss of continuity.

PREFACE

One of the two time-dependent phenomena leading to a textbook revision, such as this third edition, is the development of the field itself. Physical chemistry has now clearly outgrown its original limitation to the study and organization of the macroscopic chemical world. It now includes, on an equal footing, the study of the atomic and molecular world. The heart and enduring character of the subject is the bringing together of these worlds. The student of the subject acquires the conviction that, on a broad front, knowledge of these two worlds can be coupled and that remarkable insights into both worlds result. This was the view of physical chemistry presented in the first edition of this book, and it remains the guiding attitude for this edition. It is in the extension of our knowledge of the molecular world, and thus our increased ability to develop encounters with the macroscopic world, that updating of the content has been made.

The second motivation for a revision is the added insight that an author receives—from students, other teachers, and reviewers, and from very general experience. On this basis a large part of the second-edition material has been rewritten and reorganized. While the goals and scope are not changed, I hope that the clarity of presentation has been improved and the unity of the various aspects of the subject will be more readily appreciated.

I have continued, in this revision, to try to resolve the conflict between the use of developments that stay close to the physical approach, and are at all stages understandable on this basis, and the introduction of concise mathematical treatments that provide elegant paths to the desired result. As in the second edition, I have attempted to get the best of both approaches. Thus, where additional mathematical formalism appears, it is because I now think the treatment that is carried out reveals more clearly the physical nature of the relation that is established.

An easily noticed change throughout the text is the use of SI units. Although it is hard for a chemist to give up his long-remembered quantities—that R is 1.987 calories per degree mole, for example—the student will probably find SI units easier to deal with. In any case, the trend is toward SI units in most branches of science, both in the United States and abroad. I have, however, been unwilling at this stage to give up the pressure unit of atmospheres, which serves as the basis of standard states, or the angstrom, which will likely be used for some time for the wavelength of light and for scaled molecular models. Also, the treatment of magnetic properties has not been changed, since conversion to SI units involves more than a unit change and would lead too far from current and past chemical practice.

One of my delights in the development of the revision has been the receipt of detailed and incisive reviews on many parts of the manuscript. I particularly would like to acknowledge the very great help that James W. Richardson of Purdue University has given me, as he did when the previous edition was being prepared. I am also greatly indebted to DeWitt Coffey, Jr., San Diego State College; Benjamin P. Dailey, Columbia University; A. James Diefenderger, Lehigh University; Eugene Hamori, University of Delaware; and Alan S. Rodgers, Texas A & M University. Their efforts have, in many places, put me on the right track, and I am sure the instructors and students who use this textbook would share my appreciation if they could recognize their contributions.

GORDON M. BARROW
Carmel Valley
California

1

PROPERTIES OF GASES

The study of the nature of gases provides an ideal introduction to physical chemistry. This study, undertaken in the first two chapters, has three clearly recognizable aspects. One is the organization of the experimental results that are obtained from studies of the world around us, the macroscopic world, into general statements, or *laws*. Second is the development of a molecular model, i.e., a study of the microscopic, or molecular, world. Finally, these aspects are brought together to give a molecular-level interpretation of the observed macroscopic phenomena.

A considerable appreciation of the world of molecules can be obtained from the second and third steps, and this comes about without recourse to the more elaborate, and more powerful, theories and experiments that will be encountered later in this text. The deduction of some of the innermost details of the molecular world from the simple experimental results of Chap. 1 and the equally simple theory of Chap. 2 should be appreciated as an elegant accomplishment of science.

Seldom are the experimental and theoretical aspects of a study so neatly separated as they are here. A clear illustration is provided of how these two aspects of scientific study go hand in hand to lead to a more profound interpretation of our physical world. The division of the subject into its empirical and theoretical aspects, it must be admitted, ignores the historical sequence of events. However, most of the results reported in this chapter predate the theoretical deductions of the following chapter. For reference one can recall that the molecular view of matter was born with the nineteenth century and became quite mature and respectable by the end of that century. The earlier dates

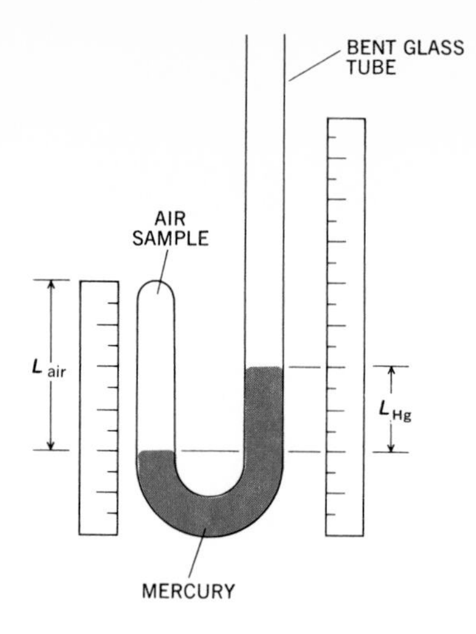

FIGURE 1-1
Apparatus for the measurement of the relation between the pressure and the volume of a sample of air.

attached to some of the empirical studies should emphasize the fact that these studies were indeed purely empirical and were not appreciably guided by any existing theory.

1-1 The Dependence of the Volume of a Gas on Pressure: Boyle's Law

As early as the year 1660 Robert Boyle performed a series of experiments in which he determined the effect of pressure on the volume of a given amount of air with the apparatus illustrated in Fig. 1-1. A little mercury was added through the open end of the tube to seal off a quantity of air in the closed end. The volume of the enclosed air, for various amounts of mercury added through the open end, could then be measured. Table 1-1 shows some of the results that Boyle obtained.

Qualitatively, it is immediately obvious that, as the pressure on the air increases, the volume of the air decreases. Such data prompt one

TABLE 1-1 Data of Boyle on the pressure-volume relation for air

L_{air} (arbitrary units)	L_{Hg} (inches)	$L_{Hg} + 29\frac{1}{8}$ (inches)	$(L_{Hg} + 29\frac{1}{8})L_{air}$
12	0	$29\frac{2}{16}$	349
10	$6\frac{3}{16}$	$35\frac{5}{16}$	353
8	$15\frac{1}{16}$	$44\frac{3}{16}$	353
6	$29\frac{11}{16}$	$58\frac{13}{16}$	353
4	$58\frac{2}{16}$	$87\frac{4}{16}$	349
3	$88\frac{7}{16}$	$117\frac{9}{16}$	353

to go further and see if there is a simple quantitative relation between the pressure P and the volume V. One tries the relation

$$V \propto \frac{1}{P} \quad \text{or} \quad V = \frac{\text{const}}{P} \quad \text{or} \quad PV = \text{const} \qquad [1]$$

The data are easily compared with the final form of this relation, and included in Table 1-1 is the calculated product of the effective length of the column of mercury and the length of the air column. (The units used are not pertinent since it is only the constancy of the result that is of interest.) Within experimental error a constant value is obtained, and so Boyle was able to conclude that the volume of air varies inversely as the pressure. Later experiments showed that this relation required the temperature to be maintained constant and, furthermore, that many gases, as well as air, conformed quite closely to this behavior. Boyle's law can now be written as follows: *The volume of a given quantity of a gas varies inversely as the pressure, the temperature remaining constant.*

Processes which are performed at constant temperature are said to be *isothermal*. The pressure-volume data obtained at constant temperature in demonstrating Boyle's law are frequently exhibited on a plot of P versus V. The hyperbolic curve obtained, as in Fig. 1-2, at any given temperature, is an example of what is called an *isotherm*.

According to Boyle's law, the pressure and volume of a given amount of gas, at a fixed temperature, vary so that the product PV always has the same value. Sometimes one deals with an isothermal process which takes the gas from the initial values P_1 and V_1 to some new values P_2 and V_2. Since the product of P and V is constant, one can write a frequently convenient form of Boyle's law as

$$P_1V_1 = P_2V_2 \qquad [2]$$

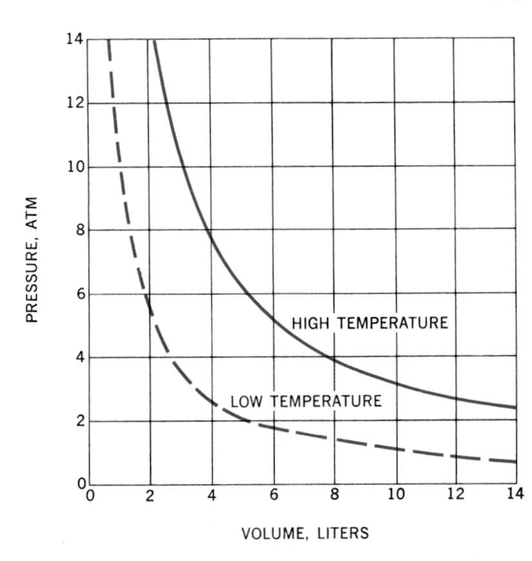

FIGURE 1-2
Isotherms for a gas obeying Boyle's law.

Later it will be shown that more accurate or more extensive measurements reveal that gases do not, in fact, behave exactly in accordance with Boyle's law. It is convenient, to begin with, to ignore these additional complications of gas behavior and to restrict our attention to what is known as *ideal behavior,* a term implying a simply described and generally followed behavior.

The simplicity of Boyle's law and its frequent presentation in elementary chemistry courses should not lead to the view that this is the "expected" behavior. For liquids and solids, by contrast, no simple relation exists between V and P. The fact, for example, that doubling the pressure on any of a wide variety of gases reduces the volume of the gas sample to half its original value is a rather remarkable result that the theory of the next chapter must explain.

1-2 The Volume-Temperature Behavior of Ideal Gases

More than a century elapsed before the counterpart of Boyle's law, a relation between the temperature and gas volume, was discovered. The reason for this long interval stems from the difficulty of the concept of temperature as compared with that of pressure. Although qualitative differences between hot and cold can be readily recognized, the means for making quantitative measurements of the "degree of heat" are not so easily devised. Toward the end of the eighteenth century, however, the use of the expansion of a liquid in a glass tube, i.e., a modern thermometer, was generally accepted as a satisfactory method for measuring temperature. On the continent of Europe, furthermore, some agreement had been reached to choose the freezing point of water as 0° on the temperature scale and the boiling point as 100°. The existence of thermometers, with agreed-upon scales, allowed investigations to be made of the variation of the volume of a gas with the temperature.

The early work of Charles in 1787, and then further work by Gay-Lussac in 1808, showed that, if the pressure is kept constant, the volume of a sample of gas varies linearly with the temperature in a manner indicated by the solid lines of Fig. 1-3.

The extrapolation to low temperatures of curves such as those of Fig. 1-3 is revealing. One finds that all the curves extrapolate to $V = 0$ at a temperature of about -273°C. In view of this common behavior, it is often more convenient to measure temperatures from this point, i.e., from $t = -273$°C, rather than from the zero of the Celsius scale. If the size of the degree is kept the same as in the Celsius scale but the zero is shifted, the ***absolute Kelvin temperature scale*** is obtained. Using the best modern value for absolute zero, temperatures T on this scale are related to Celsius scale temperatures t by

$$T = t + 273.15 \qquad [3]$$

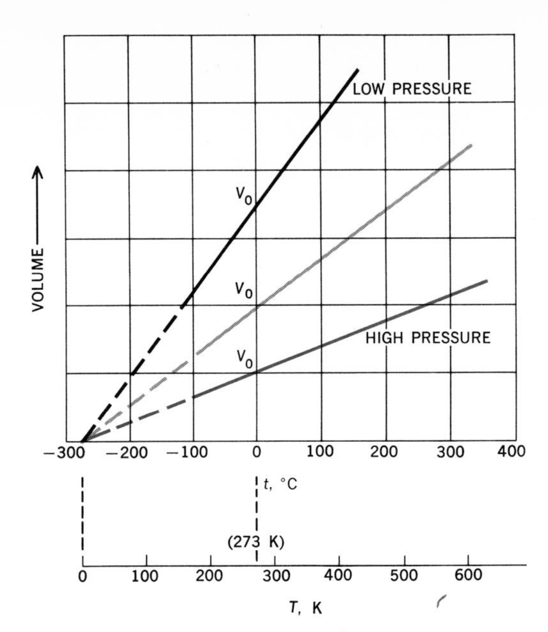

FIGURE 1-3
Variation of the volume of a sample of gas at different constant pressures as a function of the temperature according to Gay-Lussac's law. All slopes are equal to $V_0/273$.

A temperature on this absolute scale is denoted by the symbol K, after Lord Kelvin. (Later in the chapter, after we have looked more carefully into the PVT behavior of real gases, we shall consider in more detail the basis and method of setting up such a temperature scale.)

With the introduction of this absolute temperature scale, shown below the Celsius scale in Fig. 1-3, a relation, paralleling the PV relation of Boyle, can be written for V and T. The linear curves of Fig. 1-3, all extrapolating to $V = 0$ at $T = 0$, allow us to write

$$V \propto T \qquad \text{or} \qquad V = \text{const } T \qquad \text{or} \qquad \frac{V}{T} = \text{const} \tag{4}$$

This result is Gay-Lussac's law, also sometimes referred to as Charles' law: *The volume of a given mass of gas varies directly as the absolute temperature if the pressure remains constant.* Like Boyle's law, this relation is approximately followed by many gases, and obedience of a gas to Gay-Lussac's law constitutes another feature of ideal-gas behavior.

1-3 The PVT Behavior of Ideal Gases

The gas laws of Boyle and Gay-Lussac separately describe the dependence of the volume of a gas sample on the pressure and temperature. The dependence on both variables can be illustrated by the surface of Fig. 1-4. Any cross section perpendicular to the T axis, i.e., a section for constant temperature, would show a Boyle's law hyperbolic curve. Any section perpendicular to the P axis, i.e., a section for constant

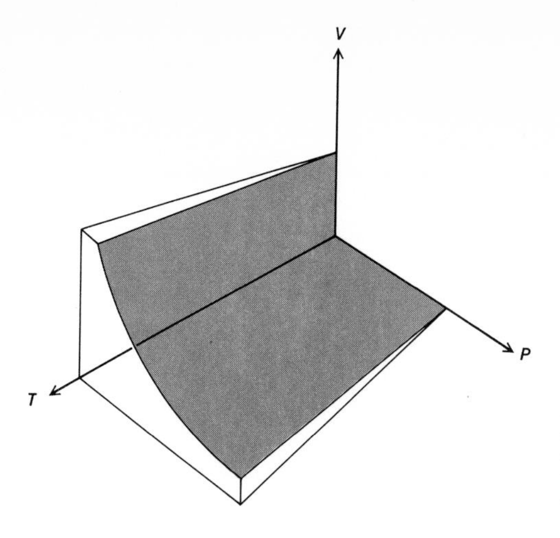

FIGURE 1-4
A section of the surface depicting the dependence of the volume of a gas sample on pressure and temperature.

pressure, would show the linear temperature-volume relation of Gay-Lussac's law.

The surface of Fig. 1-4 requires, for its mathematical description, an expression of the form

$$V = f(P,T) \quad [5]$$

where $f(P,T)$ implies some function of the variables P and T.

The proportionality to which Eq. [5] corresponds must give, as special cases, the proportionalities of Boyle and Gay-Lussac, and on this basis we can write

$$V \propto \frac{T}{P} \quad [6]$$

For a gas sample whose volume is specified at some temperature and pressure, the equality

$$V = \text{const}\,\frac{T}{P} \quad [7]$$

can be written to describe the surface for that sample on a display like Fig. 1-4.

Laboratory studies often lead us to use Eq. [7] to calculate the volume of a gas sample at pressures and temperatures other than those existing under the experimental conditions. A dramatic natural illustration of gases subjected to various temperatures and pressures is also found in nature and is provided by the earth's atmosphere. The variation in temperature with altitude is shown in Fig. 1-5, and this variation and that of pressure are shown in tabular form in Table 1-2.

To calculate changes in gas volume from Eq. [7], or to make estimates from a figure like Fig. 1-4, the gas sample must be specified by a statement of its volume at some indicated temperature and pressure. It is, however, often convenient to treat the gas sample in

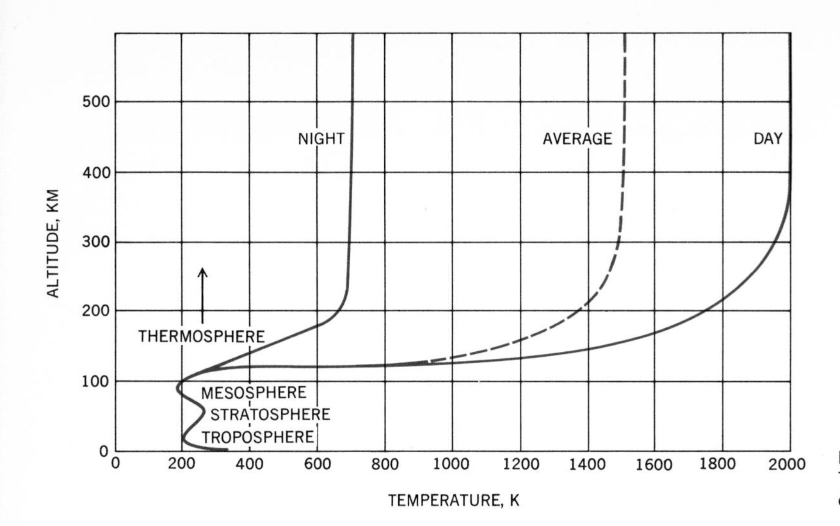

FIGURE 1-5
Temperature trends in the earth's atmosphere.

terms of the *number of moles n* of the gas. When the amount of gas in a gas sample is measured in this way, a gas-law expression that is applicable to all gases that behave ideally can be set up.

The recognition of the molecular nature of matter and the introduction of the mole as a measure of an amount of gas does, however, take us outside the empirical studies of this chapter and anticipates the subject of the following chapter. It is this anticipation, however, that allows us to use the generalizing statement: One mole of any ideal gas occupies the same volume at the same temperature and pressure,

TABLE 1-2 Data for a "model atmosphere"
(Implying an atmosphere in which the variations with day and night, location, and so forth, are avoided and the data form a self-consistent set)

Altitude (km)	Temperature (K)	Pressure (atm)	Density ($kg\ m^{-3} = g\ liter^{-1}$)
0	288	1.00	1.22
20	217	5.4×10^{-2}	8.9×10^{-2}
40	261	3.0×10^{-3}	4.0×10^{-3}
60	254	2.5×10^{-4}	3.5×10^{-4}
80	166	9.9×10^{-6}	2.1×10^{-5}
100	199	2.1×10^{-7}	3.7×10^{-7}
200	1404	1.6×10^{-9}	3.7×10^{-10}
300	1423	2.5×10^{-10}	4.7×10^{-11}
400	1480	5.6×10^{-11}	9.0×10^{-12}
500	1576	1.6×10^{-11}	2.2×10^{-12}
600	1691	5.2×10^{-12}	6.6×10^{-13}
700	1812	2.0×10^{-12}	2.3×10^{-13}

and the *mole* is defined so that the volume occupied by 1 mol is 22.414 liters at a temperature of 273.15 K and a pressure of 1 atm.

Now the proportionality of Eq. [6] can be written to include the volume dependence on the amount of gas, as measured by the number of moles n, as

$$V \propto n\frac{T}{P} \qquad [8]$$

Then, with the generalization stated above, and with R written for the proportionality constant, the equality that can be written is

$$V = R\,n\frac{T}{P}$$

or

$$PV = nRT \qquad [9]$$

Furthermore, R will have a single value, applicable to all gases that behave ideally.

This result is useful in making gas-volume calculations and is, moreover, a summary of the empirical laws of Boyle and Gay-Lussac and the hypothesis of Avogadro. The principal goal of the theory of the nature of gases, to be developed in the next chapter, will be the theoretical derivation of this important result.

A numerical value for the gas constant can be obtained from the result that at 1 atm and 0°C, 1 mol of a gas that behaves ideally occupies 22.414 liters. Substitution of these data in the gas-law expression [9] gives

$$R = \frac{(1\text{ atm})(22.414\text{ liters})}{(1\text{ mol})(273.15\text{ deg})} = 0.082056\text{ liter atm/deg mol}$$

A rearrangement of the important $PV = nRT$ relationship that is sometimes informative depends on recognizing that n, the number of moles, is given by the mass m of the sample divided by the molar mass M, that is, m/M, and that the density can be expressed as m/V. Then we can rearrange $PV = nRT$ first to $PV = (m/M)RT$, then to $PM = (m/V)RT$, and finally to $M = (m/V)RT/P$, or

$$M = d\left(\frac{RT}{P}\right) \qquad [10]$$

Thus the molar mass of the gas is given in terms of its density at some pressure and temperature.

Again we can turn to the atmosphere of the earth for an interesting application of this relation. Data for the density, temperature, and pressure can be used to calculate the average mass of a mole at various altitudes. For example, a value of M of 29.0 g is obtained from the data of Table 1-2 for sea-level air, a value that also could have been deduced from the proportions of nitrogen and oxygen. As the thermosphere is entered, however, the calculated values begin to fall off. At 700 km, for example, M is deduced to be 17.0 g, and this can be related to the occurrence of appreciable amounts of atomic oxygen. At still greater altitudes the composition, and thus the mean mass per mole, continues to change as the dominant species become atomic oxygen, then helium, and, at several thousand kilometers, atomic hydrogen.

1-4 The Gas Constant: Energy Units

The gas constant not only enters into PVT calculations but, as will frequently be seen in succeeding chapters, also plays a very important role in all phenomena involving the energies of molecular systems. This aspect of the gas constant is less surprising when one sees that R involves the dimensions of work, or energy.

If pressure is written as force per unit area and the volume as area times length, one sees that dimensionally

$$\text{Pressure} \times \text{volume} = \frac{\text{force}}{\text{area}} \times \text{area} \times \text{length} = \text{force} \times \text{length}$$

The dimensions of force times length are those of energy, which you recall can be based on the force times the distance through which the force acts. It follows that R has the dimensions of energy per degree per mole.

A numerical value for R involving an energy unit will now be obtained. In so doing we will begin the introduction of the consistent set of units that are referred to as *SI* (for Système International d'Unités). These you will recognize are mks units, i.e., the meter-kilogram-second system generally used by physicists. Appendix 7 summarizes some of the basic and derived units of the SI system and their relation to other units with which chemical quantities have been reported.

Chemists have often used the cgs, i.e., centimeter-gram-second, system and, moreover, have added special units outside this system. Two of these, the pressure unit of the atmosphere and the volume unit of the liter, or the related unit of the milliliter (ml), equal to 10^{-3} liter, have already been introduced here. These units are so convenient in chemical studies and are in such common use that we will continue to report pressures and volumes in these nonsystematic units. To work from them to pressures and volumes expressed in SI units you will have to use the conversions

$$1 \text{ liter} = 10^{-3} \text{ m}^3$$

and

$$1 \text{ atm} = 101{,}325 \text{ N m}^{-2}$$

Now R in the SI energy units of joules, *joule* being the special name for the SI unit newton meter, can be evaluated as

$$\begin{aligned} R &= 0.082056 \text{ liter atm deg}^{-1} \text{ mol}^{-1} \\ &= 0.082056 \frac{\text{liter atm}}{\text{deg mol}} \times \frac{1 \text{ m}}{10^3 \text{ liters}} \times \frac{101{,}325 \text{ N m}^{-2}}{1 \text{ atm}} \\ &= 8.3143 \text{ J deg}^{-1} \text{ mol}^{-1} \end{aligned} \qquad [11]$$

The chemical literature generally has treated R with the energy units of *calories* (1 calorie = 4.1840 joules), and in these units R takes on the value 1.9872 cal deg^{-1} mol^{-1}. This conversion factor, 4.1840 J cal^{-1}, will in fact be needed often if energies reported in the chemical literature are to be converted to joules.

1-5 Some Properties of Gas Mixtures

The atmospheric example of Sec. 1-3 shows that the gas-law expressions, such as $PV = nRT$, can be applied to gases that are mixtures of different components or are pure, single-component gases. In the former case it is often necessary to relate the properties of the gas mixture to those of its components. The basis for this relation for ideal gases that form an ideal-gas mixture is an empirical generalization, known as *Dalton's law of partial pressures*. This follows from experiments in which the pressures exerted by given amounts of gases put separately into a container are measured and compared with the pressure obtained when the same amounts of the gases are placed in the container together. The results of such experiments show that *the total pressure exerted by a mixture of gases is equal to the sum of the pressures which each component would exert if placed separately into the container*. This law, like those of Boyle and Gay-Lussac, is followed quite closely by mixtures of many gases. But deviations are observed in such cases as where the components tend to react with one another.

This law makes it profitable to introduce the term *partial pressure* to denote the pressure exerted by one component of a gaseous mixture. The total pressure P is then the sum of the partial pressures P_i of the components; i.e.,

$$P = P_1 + P_2 + P_3 + \cdots \qquad [12]$$

Dalton's law, furthermore, allows each of the partial pressures to be treated as if each of the components were occupying the container separately. The ideal-gas law can then be applied to each component to give

$$P = n_1\frac{RT}{V} + n_2\frac{RT}{V} + n_3\frac{RT}{V} + \cdots$$

$$= \frac{RT}{V}\sum_i n_i$$

$$= n\frac{RT}{V} \qquad [13]$$

where n is the total number of moles of the gas mixture in the volume V.

The final result, Eq. [13], shows that when the expression $PV = nRT$ is used for mixtures of gases, the number of moles n is simply the sum of the numbers of moles of the individual components. The fact that both Boyle's and Gay-Lussac's laws are independent of the type of molecules that make up the gas anticipates this result.

In dealing with gas mixtures, it is frequently necessary to be able to express the fraction which one component contributes to the total mixture. Two of the most convenient ways of doing this are the use of the *pressure fraction* and the *mole fraction*. The pressure fraction of the ith component is defined as P_i/P and can be seen to be identical, for ideal systems, with the mole fraction x_i, defined as n_i/n, by writing

$$\frac{P_i}{P} = \frac{n_i(RT/V)}{n(RT/V)} = \frac{n_i}{n} = x_i \qquad [14]$$

Thus, from information on the partial pressures of the components of a gas mixture, the relative number of moles of the components in a given sample can be deduced.

Furthermore, a value of the average molecular mass can be obtained. For example, the three principal components of dry sea-level air are nitrogen, oxygen, and argon. They exert partial pressures of 0.781, 0.209, and 0.009 atm, respectively. The mole fractions of each are thus 0.781, 0.209, and 0.009, and the average molar mass M is

$$M = 0.781(28.0) + 0.209(32.0) + 0.009(39.9)$$
$$= 28.9$$

You can note also that you can deal with the fraction of the total volume that can be ascribed to each component, each being treated as being subject to the total pressure. Then you would write

$$V = \frac{nRT}{P} \qquad \text{and} \qquad V_i = \frac{n_iRT}{P}$$

to give

$$\frac{V_i}{V} = \frac{n_i}{n}$$

A noteworthy property of units such as pressure, volume, and mole fractions is that the sum of such fractions over all the components of the system is unity; thus

$$\frac{n_1}{n} + \frac{n_2}{n} + \frac{n_3}{n} + \cdots = \sum_i x_i = 1$$

and

$$\sum_i \frac{P_i}{P} = 1 \qquad \text{and} \qquad \sum_i \frac{V_i}{V} = 1 \tag{15}$$

If one can refrain from anticipating the explanation of gas properties in terms of molecular theory, the properties of gas mixtures embodied in Dalton's law seem quite remarkable. If gases are thought of as nothing more than homogeneous fluids, it is not at all obvious that they should obey such a simple law as that of Dalton. The independent behavior of the components of a gas mixture was, in fact, one of the results that stimulated the ideas of the molecular nature of matter.

1-6 The Nonideal Behavior of Gases

The PVT behavior of gases has so far been presumed to follow Boyle's and Gay-Lussac's laws and, with the mole concept, to lead to the result $PV = nRT$. When measurements are extended to higher pressures, or even when very accurate measurements are made at ordinary pressures, it is found that deviations from these laws do exist. Behavior in accordance with the ideal-gas law $PV = nRT$ is elegantly simple and is at least approximately obeyed by almost all gases, and, as has been mentioned, such behavior is said to be *ideal,* or *perfect.*

An actual gas exhibits, to some extent, deviations from the ideal-gas law, and when these deviations are recognized, the gas is said to behave as a *real, nonideal,* or *imperfect* gas.

The very accurate data for a few gases at relatively low pressures shown in Fig. 1-6 indicate that ideal behavior, which requires the product PV to be independent of P, is not strictly followed. To represent such behavior analytically, the expression $PV = nRT$ must be modified. To do so it is convenient to specify 1 mol of a gas and to indicate this stipulation by writing a small capital v. Then behavior such as shown in Fig. 1-6 can be described by

$$P\mathrm{v} = RT + b'P \qquad \text{or} \qquad P\mathrm{v} = RT(1 + bP) \tag{16}$$

where b' and b are characteristic of the gas and, moreover, are functions of temperature.

The equations one writes to describe the PVT behavior of gases,

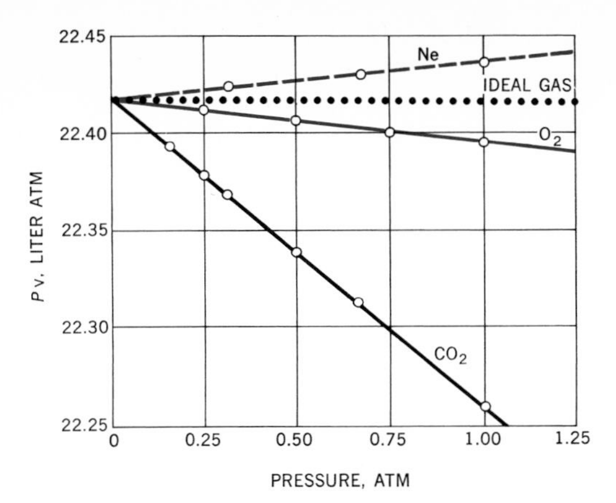

FIGURE 1-6
Accurate data for the product Pv for 1 mol of the gas as a function of pressure, at 0°C. (*Adapted from L. P. Hammett, "Introduction to the Study of Physical Chemistry," McGraw-Hill Book Company, New York, 1952.*)

or other states of matter, are known as *equations of state*. The equation $PV = nRT$ or $P\mathrm{v} = RT$ is the equation of state for an ideal gas. Equations [16] are the first of a number of examples of equations of state that we shall encounter in our studies of real gases. Various equations are used, and this variety results from the highly individualistic behavior of real gases, to the extent that they deviate from ideal behavior, and the need for equations that deal with various pressure and temperature ranges and that are useful for various types of calculations.

Data at pressures higher than 1 atm, as shown in Fig. 1-7, indicate that a more flexible expression than Eqs. [16] is required, and a frequently used empirical equation of state, known as a *virial equation,* is written. Thus, for 1 mol,

$$P\mathrm{v} = RT + BP + CP^2 + \cdots \qquad [17]$$

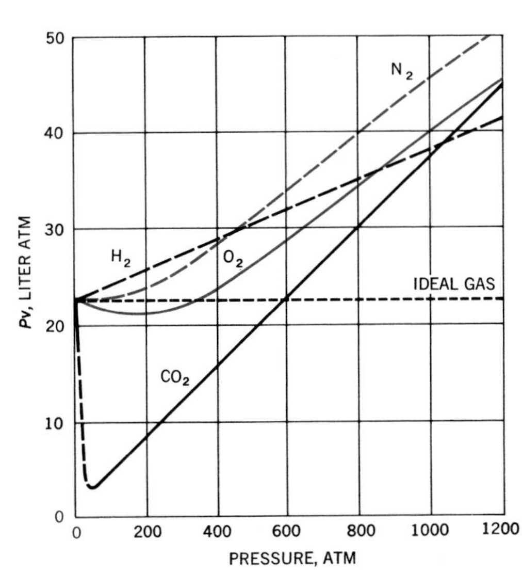

FIGURE 1-7
The product Pv versus pressure for 1 mol of gas at 0°C. The dashed section of the CO_2 curve indicates the liquid state.

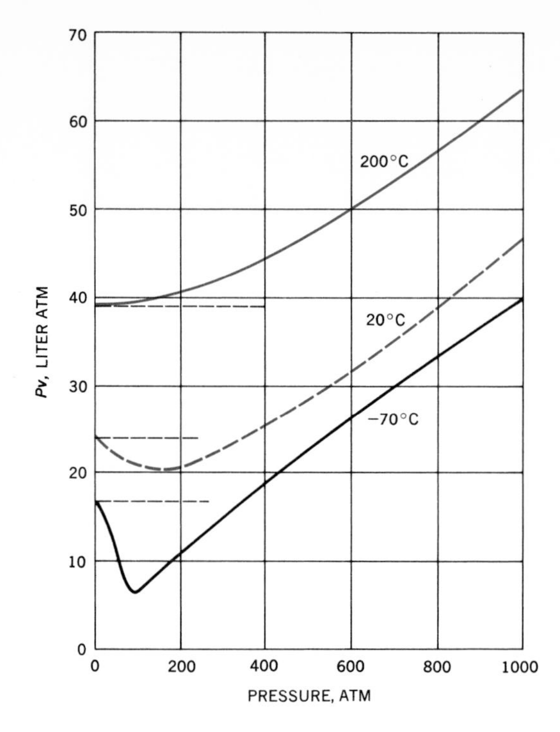

FIGURE 1-8
$P\text{v}$ versus P curves for methane at three different temperatures. [*H. M. Kvalnes and V. L. Gaddy, J. Am. Chem. Soc.*, **53**:*394* (*1931*).]

where B, C, . . . are called *virial coefficients* and again are functions of the nature of the gas and the temperature. Other equations of state that are suggested by the nature of the molecular properties will be dealt with in the following chapter.

The deviations from ideality as a function of temperature can be shown by a display like that of Fig. 1-8. It is often more convenient, however, to plot the ratio $P\text{v}/RT$ versus pressure. For ideal behavior this ratio will be unity for all pressures and temperatures. For real gases, however, some deviation from unity will occur.

The quantity $P\text{v}/RT$ is sufficiently convenient in the discussion of the nonideality of gases so that it is called the *compressibility factor* and is given the symbol Z; that is,

$$Z = \frac{P\text{v}}{RT} \qquad [18]$$

Ideal behavior requires Z to have the value unity at all pressures and temperatures, and the gas imperfection is immediately apparent as the difference between the observed value of Z and unity. To illustrate this function, the behavior of methane at a number of temperatures is shown in Fig. 1-9.

The gas behavior described so far does not account for all the results that can be obtained when the PVT relations of gases are studied. If such studies are conducted at low enough temperatures and

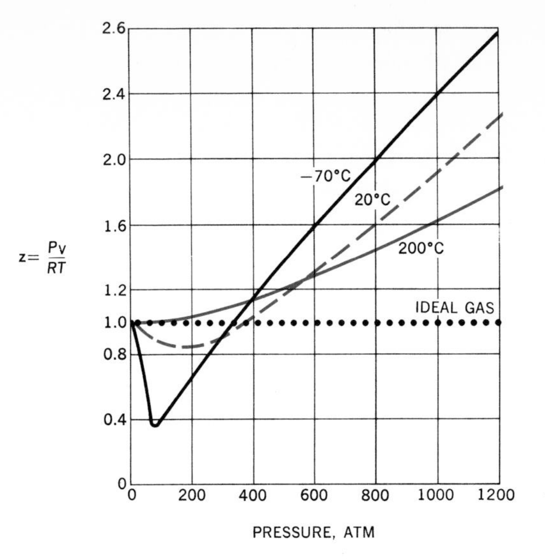

FIGURE 1-9
The compressibility factor of methane as a function of pressure at three different temperatures. [*H. M. Kvalnes and V. L. Gaddy, J. Am. Chem. Soc.*, **53**:*394 (1931)*.]

high enough pressures, for example, all gases will condense to form a liquid phase. This extreme aspect of nonideal-gas behavior must now be investigated.

1-7 Condensation of Gases, the Critical Point, and the Law of Corresponding States

A set of isotherms which extend into the region where condensation occurs is shown for CO_2 in Fig. 1-10. The data for this figure came from the pioneering work of Andrews in 1869 on the behavior of gases. The higher-temperature isotherms show only slight deviations from the

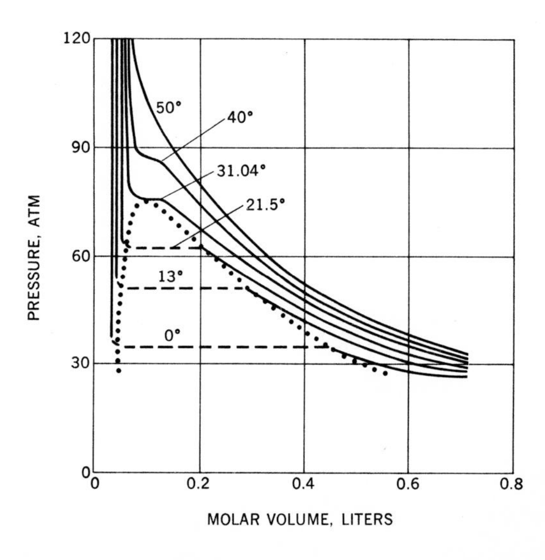

FIGURE 1-10
Isotherms of CO_2 near the critical point. (*From E. D. Eastman and G. K. Rollefson, "Physical Chemistry," McGraw-Hill Book Company, New York, 1947.*)

hyperbolic curves expected for an ideal gas. Lower-temperature isotherms also conform somewhat, at the low-pressure large-volume end, to ideal behavior. As the pressure is increased at such temperatures, the volume decreases approximately according to Boyle's law, until a point on the dotted line of Fig. 1-10 is reached. At this point the gas begins to condense to a liquid. Now the volume decreases as the gas is continually converted to a liquid, the pressure staying constant at the equilibrium vapor pressure for that temperature. When the left limit marked off by the dotted line is reached, the entire gas has been condensed and further application of pressure results in only a minor decrease in volume, as shown by the steep section at the left end of the isotherm.

The region beneath the dotted curve is seen to represent situations in which liquid and vapor coexist. Curves to the right and above this region delineate the system considered to consist of a gas; to the left, considered to correspond to a liquid.

Of particular interest in the study of the nonideal behavior of gaseous systems are the unique isotherms, such as that touching the top of the dotted curve of Fig. 1-10. This isotherm is called the *critical isotherm,* and its temperature, the *critical temperature,* is seen to be the highest temperature at which the gas can be condensed to a liquid. The point at which this isotherm shows its horizontal point of inflection is called the *critical point,* and the pressure and volume per mole at this point are known as the *critical pressure* and *critical volume.* Some data for the critical point are shown in Table 1-3.

The fact that the deviations from ideal behavior shown by real gases appear to depend on the difference between the conditions of the gas and those of the critical point suggests that it might be convenient to

TABLE 1-3 Values of P, v, and T at the critical point

Gas	P_c (atm)	v_c (liters mol^{-1})	T_c (K)
H_2	12.8	0.0650	33.3
He	2.26	0.0576	5.3
CH_4	45.6	0.0988	190.2
NH_3	111.5	0.0724	405.6
H_2O	217.7	0.0450	647.2
CO	35.0	0.0900	134.0
N_2	33.5	0.0900	126.1
O_2	49.7	0.0744	153.4
CH_3OH	78.5	0.1177	513.1
Ar	48.0	0.0771	150.7
CO_2	73.0	0.0957	304.3
n-C_5H_{12}	33.0	0.3102	470.3
C_6H_6	47.9	0.2564	561.6

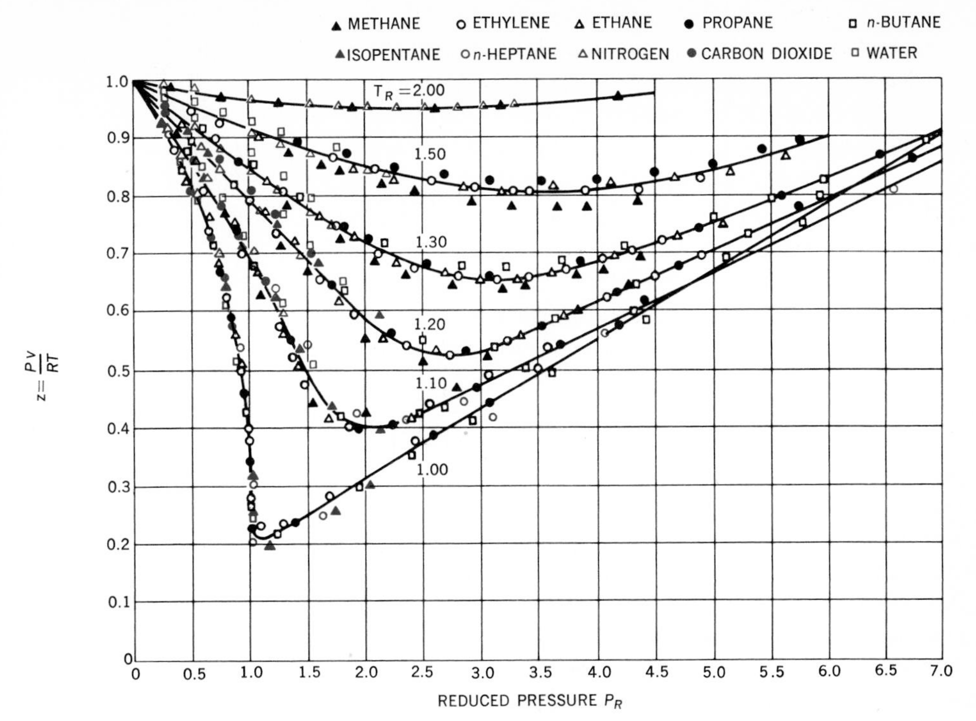

FIGURE 1-11
The compressibility as a function of the reduced pressure at various reduced temperatures. [*From Gouq-Jen Su, Ind. Eng. Chem.*, **38**:*803* (*1946*).]

introduce new variables that relate P, V, and T to the value of these variables at the critical point. To do this, we define the *reduced variables* P_R, v_R, and T_R in terms of the critical constants P_C, v_C, and T_C as

$$P_R = \frac{P}{P_C} \qquad v_R = \frac{v}{v_C} \qquad T_R = \frac{T}{T_C} \tag{19}$$

If the critical constants are known, it is possible of course to treat the behavior of a gas in terms of the reduced variables just as easily as in terms of the ordinary variables.

One can investigate the usefulness of these reduced variables by plotting the compressibility factor $Z = Pv/RT$ for a gas as a function of the reduced pressure. When this is done for a number of different gases at various reduced temperatures, as in Fig. 1-11, the result obtained is equivalent to the statement that all gases deviate from ideal behavior in a way that depends only on the reduced pressure and temperature. This statement constitutes the *law of corresponding states*. The name expresses the fact that gases in states with the same values of the reduced variables will deviate from ideality to nearly the same extent and are said to be in corresponding states.

The law of corresponding states introduces a considerable simplification into the treatment of nonideal gases. The uniform behavior of such gases in terms of the reduced variables should not, however, obscure the fact that critical data, which are characteristic of the detailed nature of the molecules of each gas, are implicit in the reduced variables. No "ideal" generalization such as $PV = nRT$ is possible when the behavior of gases is studied accurately or over a wide range of pressures and temperatures.

1-8 The Ideal-gas Thermometer

The behavior of real gases, shown in Figs. 1-6 to 1-9, for example, is sufficient to allow us to look more closely into the way that the temperature scale and the value for the absolute zero of temperature are deduced.

Experiments such as those discussed in Sec. 1-2 that led to Gay-Lussac's law show, really, only that the thermal expansion of gases at constant pressure is proportional to the expansion of mercury, or some other liquid, in a glass tube. This way of stating the results of Sec. 1-2 emphasizes the arbitrary, and not entirely satisfactory, nature of the temperature-measuring devices that were assumed when Gay-Lussac's law was being deduced. Not all liquids expand in the same way as the temperature increases, and water, for instance, even contracts with increasing temperature in the range 0 to 4°C. Clearly, it would be much more satisfactory to construct a temperature-measurement device that does not depend on the behavior of any one compound.

Let us begin by defining a temperature scale T such that the value of the temperature is proportional to the volume V of a given amount of an ideal gas at some fixed pressure. Then the ratio of the temperatures before and after the addition of an amount of heat is given by the ratio of the initial and final volumes of the ideal-gas sample; i.e.,

$$\frac{T_2}{T_1} = \frac{V_2}{V_1} \qquad [20]$$

To remove the requirement that an ideal gas be used, the ratio of the volumes of a real gas at the low-pressure limit where ideal behavior is approached could be used. The volumes then would become awkwardly large, approaching infinity. It is more practical to employ the ratio of the products PV as P approaches the limit of zero pressure. We thus have

$$\frac{T_2}{T_1} = \frac{\lim_{P \to 0} (PV)_2}{\lim_{P \to 0} (PV)_1} \qquad [21]$$

Rather than deal with such ratios, it is more convenient to choose a reference condition and to assign some numerical value of the temperature to that condition. Then actual values rather than ratios can be used to describe the temperature of other degrees of heat. For a volume V_T corresponding to a temperature T, we then can write

$$T = T_{\text{ref}} \frac{\lim_{P \to 0} (PV)_T}{\lim_{P \to 0} (PV)_{T,\text{ref}}} \qquad [22]$$

For experimental reasons the *triple point* of water, the temperature at which liquid water, water vapor, and ice exist in equilibrium, is chosen as the reference condition. The assignment of the value 273.1600° to this point gives temperatures very nearly equal to those based on the scale established on the ice point and boiling point of water. (The ice point is, in fact, 0.010° lower than the triple point, and thus the assignment for the triple point leads to a value of 273.150° for the ice point.) When this point is used, the ideal-gas temperature scale becomes

$$T = 273.1600 \frac{\lim_{P \to 0} (PV)_T}{\lim_{P \to 0} (PV)_{\text{tp}}} \qquad [23]$$

This approach to the definition of temperature is now generally agreed upon, and in incorporating the value of $T = 0$ when $PV = 0$ into the definition, it has the merit of requiring the arbitrary assignment of a temperature value to a single reference point.

1-9 Graham's Law of Effusion

Let us now turn from the PVT behavior of gases to two properties of a different type.

The process by which a gas moves from a higher to a lower pressure through a porous wall or a very small diameter tube is known as *diffusion*. If the process consists of molecular rather than bulk flow through an orifice, the word *effusion* is used. The rate with which a gas effuses, under given conditions, is a property characteristic of the gas. Since it is rather difficult, both experimentally and theoretically, to deal with the absolute rates of effusion of gases through an orifice of well-defined dimensions, attention is usually confined to the relative rates of effusion of gases.

Measurements of the effusion ratios of a number of gases were made by Graham in 1829. He found that, at a constant temperature and at a constant-pressure drop, the rates of effusion of various gases were inversely proportional to the square roots of the densities of the gases. This relationship resulted when the rates of effusion were measured in terms of the volume of gas, at a particular temperature and pressure, that effused per unit time. If the effusion rate is denoted by v and the density by d, this result, for gases 1 and 2, is written

$$\frac{v_1}{v_2} = \sqrt{\frac{d_2}{d_1}} \qquad [24]$$

An alternative and frequently convenient form of this law can be obtained by recalling the relation between gas density and molecular mass given by Eq. [10]. The density ratio of two gases, at the same pressure and temperature, is seen to be equal to the ratio of the molar masses of the two gases. With this result Graham's law becomes

$$\frac{v_1}{v_2} = \sqrt{\frac{M_2}{M_1}} \qquad [25]$$

This effusion law makes itself evident, for example, in the fact that a system which is satisfactorily leakproof to air, molar mass about 29, may fail to hold gases like hydrogen, molar mass 2, or helium, molar mass 4.

Graham's law provides yet another property of gases for which the theory of the nature of gases must account.

1-10 The Viscosity of Gases

When a fluid flows through a pipe, tube, or trough, flow occurs only as a result of the application of a driving force to the fluid. The resistance to flow which this force overcomes is dependent on the viscosity of the fluid.

A quantitative definition of the viscosity can be made by considering the flow of a fluid near the bottom of a rectangular container as shown in Fig. 1-12. A gas or liquid flowing in a tube or trough forms a very thin stationary layer in contact with the walls of the container. The force required to make the fluid flow results from having to push the fluid relative to this stationary layer. The flow can be understood in terms of a force required to move a layer of fluid relative to another layer. This force is proportional to the areas A of the layers and to the difference in velocity v that is maintained between the layers and is inversely proportional to the distance l between the layers. The

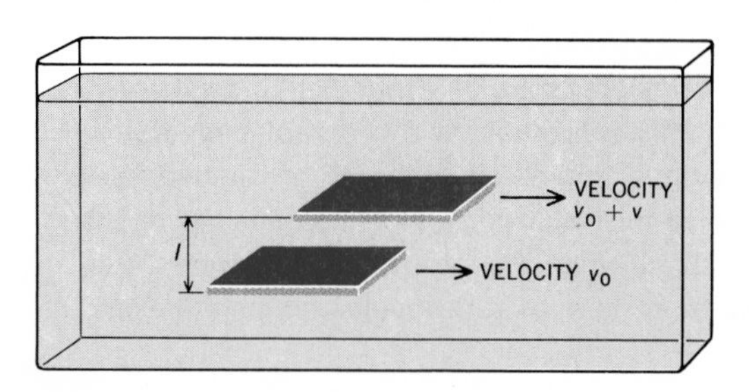

FIGURE 1-12 The relative motion of two layers of a fluid. The viscosity of the fluid requires a force to be applied to the upper layer to keep it moving relative to the lower layer.

coefficient of viscosity, or simply the viscosity, is introduced as a proportionality factor, and the equation

$$\text{Force} = \eta \frac{Av}{l} \qquad [26]$$

can be written. The viscosity can be thought of as the force required to make a layer of unit area move with a unit velocity greater than that of another layer a unit distance away. Thick liquids, such as molasses, have high viscosities; thin liquids, such as gasoline, have low viscosities. Gases have relatively much lower, but not zero, viscosities.

In practice, one often measures viscosity from the rate of flow through a cylindrical tube. Again, the fluid forms a stationary layer along the wall, and a force is required to make the fluid flow through the tube. By integrating the force required to move the annular layers of the fluid relative to this layer, as will be done when the viscosity of liquids is studied in Chap. 19, Eq. [26] can be extended to give the rate of flow through a cylindrical tube as a function of the viscosity η, the radius of the tube R, and the pressure difference $P_2 - P_1$ applied over the tube of length l. The result that will be obtained is

$$\text{Rate of flow} = \frac{\pi(P_1 - P_2)R^4}{8l\eta} \qquad [27]$$

where the rate of flow is measured as the volume of gas, measured at 1 atm, passing through the tube per second. Measurement of all the quantities other than η in Eq. [27] allows this quantity to be determined.

Table 1-4 shows viscosity data for some gases. When all the quantities in Eq. [27] are given in cgs units, the viscosity η is obtained in cgs units, which are given the name *poise.* This unit, not in the SI system, is that with which viscosity data have generally been reported. In systematic calculations the conversion 1 poise $= 10^{-1}$ kg m^{-1} s^{-1} must be used. In the following chapter use will be made of the measured viscosities to obtain information on the size of the molecules and other features of their behavior.

TABLE 1-4
Gas viscosities at 25°C

Gas	*Viscosity (poise)**
N_2	1.78×10^{-4}
O_2	2.08
H_2	0.90
Ar	2.27
H_2O	0.98
CO	1.76
CO_2	1.50
HI	1.72
He	1.97
Hg	2.50

*The unit of poise is related to SI units by 1 poise $= 10^{-1}$ kg m^{-1}s^{-1}.
SOURCE: From S. Dushman, "Scientific Foundations of Vacuum Technique," John Wiley & Sons, Inc., New York, 1949.

Problems

1 A gas occupies a volume of 0.25 liter at a pressure of 12 atm. What volume will it occupy, at the same temperature, at a pressure of 0.075 atm?

2 What volume will an ideal gas occupy at a temperature of 0°C if its volume at 100°C is 3.64 liters, the pressure remaining constant?

3 Prepare plots of P versus V at 25 and 300°C for a sample of a gas obeying Boyle's and Gay-Lussac's laws and having a volume of 0.1 liter at 25°C and 1 atm.

4 Plot the volume of 1 mol of an ideal gas as a function of the absolute temperature over the range 0 to 400 K at pressures of 0.2, 1, and 5 atm.

5 A gas is collected at 24°C and 0.89 atm pressure in a bulb of volume 0.763 liter. What would be the volume of the gas at standard temperature and pressure 0°C and 1 atm?

6 Sketch a surface showing pressure, as the variable along the vertical axis, as a function of the volume and temperature of a gas sample. What would be the mathematical expression for this surface?

7 Calculate the volume that would be occupied by a gas sample of constant composition that is 1 liter at the sea-level conditions of 1 atm and 25°C at the altitudes of 10, 50, 100, 200, and 500 km.

8 Draw graphs to illustrate the variation of pressure and density of the atmosphere with altitude. Use the data of Table 1-2. (*Hint:* Consider the use of logarithms to produce convenient graphs.)

9 Draw a line on the V versus T and P surface, like that of Fig. 1-4, to suggest the changes that occur as a sample of gas is moved from sea level to an altitude of 700 km.

10 What is the concentration in moles per liter and in molecules per liter of an ideal gas at 25°C and (*a*) 1 atm pressure, (*b*) 10^{-9} atm, which is a typical "vacuum" reached in the laboratory with a mercury-vapor pump?

11 The density of gas which behaves ideally is 2.76 g liter^{-1} at a pressure of 2 atm and a temperature of 25°C. What is the molecular mass?

12 A $\frac{1}{2}$-liter bulb weighs 38.7340 g when evacuated and 39.3135 g when filled with air at 1 atm pressure and 24°C. Assuming that air behaves as an ideal gas at this pressure, calculate the effective molecular mass of air.

Ans. 28.2 g.

13 In 1894 W. Ramsey, following up studies on the density of nitrogen, removed, by various absorption procedures, the oxygen, nitrogen, carbon dioxide, and water vapor from a sample of air. Similar separations, but by diffusion processes, were performed at the same time by Lord Rayleigh. They were left with a small amount of gas which had a density of 1.63 g liter^{-1} at 25°C and 1 atm. What element had they discovered?

14 Calculate, from the data of Table 1-2, the molar mass of air at an altitude of 400 km.

15 At an altitude of 1000 km, the principal components of air are helium and atomic oxygen. The average molecular mass at that altitude, averaged over day and night, is 9 g mol^{-1}. What are the corresponding mole fractions of helium and atomic oxygen?

16 The following data have been obtained for the density of CO_2 as a function of pressure at 10°C:

P (atm)	0.68	2.72	8.14
d (g liter^{-1})	1.29	5.25	16.32

By a suitable graphical extrapolation based on the expression $M = (d/P)RT$, obtain the molecular weight of CO_2.

17 Using T_0 and P_0 to represent 0°C and 1 atm, obtain a rearranged form of the expression

$$PV = nRT$$

which shows explicitly that the volume of 1 mol of gas at 0°C and 1 atm is 22.414 liters and that Boyle's and Gay-Lussac's laws allow the calculation of the volume at other pressures and temperatures.

18 What volume of a 0.964-M hydrochloric acid solution would be required to neutralize an ammonium hydroxide solution made when the ammonia in a 2-liter gas bulb at a pressure of 0.98 atm and temperature of 23°C is dissolved in 250 ml of water?

19 The density of the vapor in equilibrium with solid NH_4Cl was found, in an experiment by W. H. Rodebush and J. C. Michalek, reported in *J. Am. Chem. Soc.*, **51**:748 (1929), to be 0.1373 g liter^{-1} at a temperature of 596.9 K and a pressure of 0.253 atm. From these data deduce the nature of the vapor of ammonium chloride under these conditions.

20 Into a gas bulb of 2.83 liters are introduced 0.174 g of H_2 and 1.365 g of N_2, which can be assumed to behave ideally. The temperature is 0°C. What are the partial pressures of H_2 and N_2, and what is the total gas pressure? What are the mole fractions of each gas? What are the pressure fractions? *Ans.* $P_{H_2}/P = n_{H_2}/n = 0.639$.

21 A synthetic sample of air can be made, except for the minor components, by mixing 79 ml of N_2 with 21 ml of O_2, both measured at 1 atm and 25°C. What volume of air would be obtained if this synthetic sample were compressed to 5.37 atm, the temperature being 25°C? What are the mass fractions, pressure fractions, and mole fractions of the two components? Calculate the effective molecular mass to which this composition corresponds, and compare with the value obtained in Prob. 12.

22 The value of 1 atm pressure in SI units can be calculated by recognizing that this pressure is equal to the force per unit area exerted by a 760-mm, or 0.760-m, column of mercury. Using this approach and the fact that the density of mercury is 13.6 g ml^{-1}, or 13,600 kg m^{-3}, and the acceleration due to gravity is 980 cm s^{-2}, or 9.80 m s^{-2}, verify, to three significant figures, the conversion factor given in Sec. 1-4.

23 The virial equation that has been given for 1 mol of methane at 20°C is

$$\frac{P\text{v}}{RT} = 1 - 2.024 \times 10^{-3}P + 3.72 \times 10^{-6}P^2 + 4.26 \times 10^{-12}P^4$$

a Show graphically, up to pressures of a few atmospheres, that the slope of the curve for methane, like those of Fig. 1-6, is essentially constant.
b Calculate the pressure for the minimum in this curve of $P\text{v}/RT$ versus P, and verify that it is a minimum. What is the value of $P\text{v}/RT$ at this point compared with that of an ideal gas?
c At what two pressures is $P\text{v}/RT = 1$?
d Plot the compressibility factor $P\text{v}/RT$ versus pressure up to 500 atm.

24 The variation in the volume with temperature and with pressure is often stated, for gases, liquids, and solids, in terms of the coefficient of expansion $\alpha = 1/V(\partial V/\partial T)_P$ and the compressibility $\beta = -1/V(\partial V/\partial P)_T$.
a Calculate α and β from the expression $PV = nRT$ for an ideal gas at 0°C and 1 atm pressure.
b Calculate α and β for an ideal gas at 0°C and 100 atm pressure.
c At what pressure would an ideal gas have a compressibility equal to that of a typical liquid, for which $\beta = 10^{-5}$ atm^{-1}?

25 At 100°C and 1 atm pressure the density of water vapor is 0.0005970 g ml^{-1}.
a What is the molar volume, and how does this compare with the ideal-gas value? *Ans.* v (observed) = 30.18; v (ideal) = 30.62 liters.
b What is the compressibility factor Z? *Ans.* $Z = 0.986$.

26 For the isotherm of argon at −50°C the following data have been obtained:

P (atm)	8.99	17.65	26.01	34.10	41.92	49.50	56.86	64.02
v *(liters)*	2.000	1.000	0.667	0.500	0.400	0.333	0.286	0.250

The critical temperature and pressure are 151 K and 48 atm. Plot the compressibility versus the reduced pressure for this temperature, and compare with the curves of Fig. 1-11.

27 Show by reference to Fig. 1-10 that one can carry a gas from the left of the dashed liquid-vapor equilibrium region to the right of that region without encountering an observable phase change. Deduce, therefore, for what systems the terms gas and liquid are really meaningful.

28 Attempt to sketch the P versus V isotherms for 1 mol of water in the temperature range 25 to 400°C. Use the following data, which are found in the handbooks, as guides, and indicate what parts of the sketch are determined by these data. (The range of data is such that linear P and V scales are awkward. You might want to resort to logarithmic scales.)

- **a** The critical point has $t_c = 374$°C, $P_c = 218$ atm, and the critical density $d_c = 0.4$ g ml^{-1}.
- **b** The normal boiling point of water is 100°C.
- **c** The equilibrium vapor at the normal boiling point behaves nearly ideally.
- **d** The vapor pressure of water at 25°C is 0.03 atm, and the vapor then behaves ideally.
- **e** The density of liquid water is 1 g ml^{-1} and is not very sensitive to temperature and pressure.

29 The curve for $T_R = 1.00$ of Fig. 1-11 has a vertical section. Explain why, in view of the horizontal section of the critical curve of Fig. 1-10, this is to be expected.

30 Suppose it had been agreed that the normal freezing point of water would have a temperature of 100 on an absolute scale. What then would be the temperature of the normal boiling point of water?

31 A tube with a porous wall allows 0.53 liter of N_2 to escape per minute from a pressure of 1 atm to an evacuated chamber. What will be the amount escaping under the same conditions for He, CCl_4 vapor, and UF_6?

32 In 1846 Graham reported the following results for the time taken, relative to air, for given volumes of various gases to effuse:

Gas	Air	O_2	CO	CH_4	CO_2
Time	1.000	1.053	0.987	0.765	1.218

How well do these data substantiate Graham's law of effusion?

References

NEVILLE, R. G.: The Discovery of Boyle's Law, *J. Chem. Educ.*, **39**:356 (1962). The circumstances, apparatus, and results of Boyle's pressure-volume experiments.

ROGERS, E. M.: "Physics for the Enquiring Mind," pp. 416–424, Princeton University Press, Princeton, N.J., 1960. An entertaining but thought-provoking discussion of temperature and temperature-measuring devices.

PARTINGTON, J. S.: "An Advanced Treatise on Physical Chemistry," vol. 1, pp. 551ff, Longmans, Green & Co., Ltd., London, 1949. A very informative and rather complete summary of studies of gases, including information on all aspects of gases dealt with in this text. Many references to original work.

RESNICK, R., and D. HALLIDAY: "Physics for Students of Science and Engineering," vol. 1, p. 450, John Wiley & Sons, Inc., New York, 1960. A discussion of temperature and temperature scales.

MIDDLETON, W. E. KNOWLES: "A History of the Thermometer," The Johns Hopkins Press, Baltimore, 1966. An attractively illustrated account of the early developments of thermometers and temperature scales.

HARDY, JAMES D. (ed.): "Temperature, Its Measurement and Control in Science and Industry," Reinhold Publishing Corporation, New York, vol. 1, 1941, vol. 2, 1955, vol. 3, 1962. A clear indication of the magnitude of the importance and ubiquity of temperature-dependent phenomena and temperature-measuring techniques.

LEWIS, G. N., and M. RANDALL: "Thermodynamics," 2d ed. (rev. by K. S. Pitzer and L. Brewer), p. 31, McGraw-Hill Book Company, New York, 1961. Mention, in a book that will be referred to in later chapters, of the present agreement on a temperature scale.

International Critical Tables, vol. 3, p. 3, McGraw-Hill Book Company, New York, 1928. A collection of data on the *PVT* behavior of gases.

OTT, J. B., J. R. COATES, and H. T. HALL, JR.: Comparisons of Equations of State in Effectively Describing *PVT* Relations, *J. Chem. Educ.,* **48:**515 (1971).

2

THE KINETIC-MOLECULAR THEORY OF GASES

2-1 The Kinetic-Molecular Gas Model

In the empirical study of the physical behavior of gases in Chap. 1, no attention was given to the natural questions: Why is it that a gas obeys Boyle's law? Gay-Lussac's law? Why does it have the viscosity it has? And so forth. In this chapter an attempt will be made to understand gases so that some questions of this type can be answered. The kinetic-molecular theory is not, however, primarily introduced to provide an explanation of the gas laws. It is the quantitative look into the molecular world provided by this theory that is our principal interest.

It is not possible to deduce the nature of gases directly from the measured properties. These data must be used in a roundabout manner. The procedure is to guess at the underlying characteristics of gases and on this basis to deduce their physical properties. A comparison of the deduced properties with those observed allows the usefulness of the guesses to be estimated. A body of assumptions, such as that concerning gases, is called a *model.*

The gas laws and the properties of gases described in Chap. 1 can be understood through a model according to which gases are composed of a large number of small particles, called *molecules,* that move about and collide with one another and with the walls of the container. The complex mass of chemical knowledge that led to thinking in terms of molecules need not be investigated here. It is enough to recognize that the conception of chemical compounds as being composed of particles evolved gradually and that during the 1800s the concept of atoms and molecules became generally accepted. This idea was applied primarily to chemical studies, and proved valuable in explaining the compositions and transformations of chemical substances. Such applications of atomic theory do not, however, lead to information on the size, shape, or properties of the individual molecules.

This molecular concept provided the basis, however, on which the behavior of gases could be studied. In this application, known as the *kinetic-molecular theory of gases,* much information on the properties of individual molecules appeared. The work of Boltzmann, Maxwell, and Clausius during the late 1800s was primarily responsible for the development of the theory.

Here some of the derivations of gas properties by means of the kinetic-molecular theory will be followed through, occasionally simplified so that the principal features of the theory are not obscured by mathematical operations. Some of these simplifications will be removed in later treatments so that the route to a detailed analysis of the motions of the molecules of a gas will be apparent.

The kinetic-molecular model for a gas is described by the following statements:

1 A gas is made up of a large number of particles, or molecules, that are small in comparison with both the distances between them and the size of the container.

2 The molecules are in continuous random motion.

3 Collisions between the molecules and between the molecules and the walls of the container are perfectly elastic; i.e., none of the translational energy is lost by conversion into internal energy at a collision.

The first step with this model is to show that it does lead to the observed properties of gases.

2-2 The Pressure of a Gas

The pressure exerted by N molecules, each of mass m, that are contained in a cubic container of side l can be calculated on the basis of the kinetic-molecular model. The outward pressure exerted by these molecules is the result of their collisions with the walls of the container. To maintain a fixed volume in equilibrium, then, an inward-directed external pressure must be applied through the agency of the walls of the container.

To begin with, only one of the N molecules will be considered. Let its velocity, i.e., its speed and direction, be $\mathbf{u}$. Boldface type is used to indicate a *vector* quantity. The symbol u will indicate the magnitude of the velocity, i.e., the speed. The velocity can be resolved into the

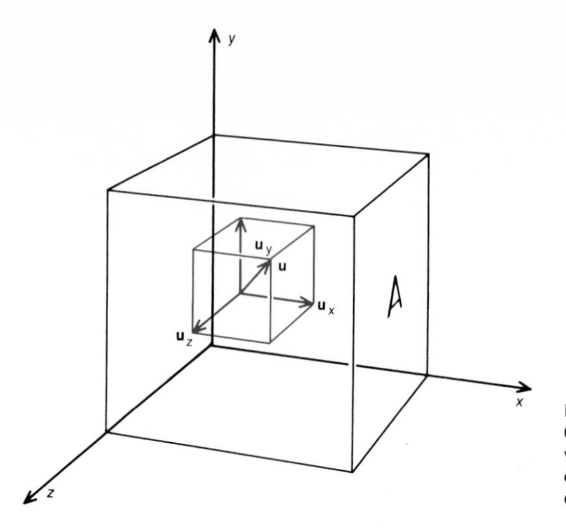

FIGURE 2-1
Coordinate and molecular velocity and velocity components for the derivation of gas pressure.

components $\mathbf{u}_x$, $\mathbf{u}_y$, and $\mathbf{u}_z$, which are perpendicular to the walls of the container as drawn in Fig. 2-1. (Note that, for example, $\mathbf{u}_x$ implies direction information, and this can be given by allowing it to take on positive or negative values to indicate motion in the positive or negative x direction. On the other hand, u_x, the speed along the x direction, is the magnitude of $\mathbf{u}_x$, and can have only positive values.)

The effect of the x component of velocity of the one molecule will now be considered. As a result of this velocity component in the x direction, the molecule will collide with one end of the container, which is perpendicular to the x axis, bounce back, and subsequently collide with the opposite end. It is the total effect of such impacts with the container walls that, according to the kinetic-molecular theory, produces the gas pressure.

The force exerted when a molecule collides with the wall of a container can be calculated from Newton's second law of motion, which states that the rate of change of the momentum, in a given direction, of a particle is equal to the force acting on the particle in that direction. The momentum with which the molecule approaches wall A of Fig. 2-1 is $m\mathbf{u}_x$. After the collision the molecule moves away from the wall, with $\mathbf{u}_y$ and $\mathbf{u}_z$ unchanged but with the sign of $\mathbf{u}_x$, and thus of $m\mathbf{u}_x$, reversed as shown in Fig. 2-2. Thus at impact the wall exerts a force that produces a momentum change in the direction perpendicular to the wall of $2m\mathbf{u}_x$.

The number of such momentum changes per second at wall A is the number of collisions per second which the molecule makes with wall A. Since the molecule travels a distance u_x along the x direction in 1 s, and since the distance traveled between collisions with side A is $2l$, the number of collisions per second with A is $u_x/2l$.

The rate of change of momentum, i.e., the change of momentum per second, is therefore

$$2m\mathbf{u}_x \frac{u_x}{2l} = \mathbf{F}_x \qquad [1]$$

This is the force that side A exerts on the one molecule that is being considered. Since the pressure is the force per unit area, the pressure exerted by side A is

$$\mathbf{P}_x = \frac{\mathbf{F}_x}{l^2} = \frac{2m\mathbf{u}_x(u_x/2)}{l^3} = \frac{2m\mathbf{u}_x(u_x/2)}{V} \qquad [2]$$

where $V = l^3$ is the volume of the container. Now we can recognize that the pressure is the same for all walls of the container and, moreover, acts for all walls perpendicular to the wall. Thus we can discard the restriction "by side A" and the direction—indicating vector notation. We thus have

$$P = \frac{mu_x^2}{V} \qquad [3]$$

Now let us consider N molecules, instead of just one, to be in the cubic container. Our model claims that these molecules act independently of one another and that each molecule will make a contribution to the pressure according to a term like Eq. [3]. (If they interact to the extent of occasionally colliding, the fact that momentum in the component directions must be conserved leads to the same result as obtained for completely independent particles.) Summation of the pressure contributions of the N molecules can be indicated as

$$P = \sum_i \frac{(mu_x^2)_i}{V} \qquad [4]$$

We can consider all molecules to have the same mass, but the model assumes that the molecules move in various directions with various

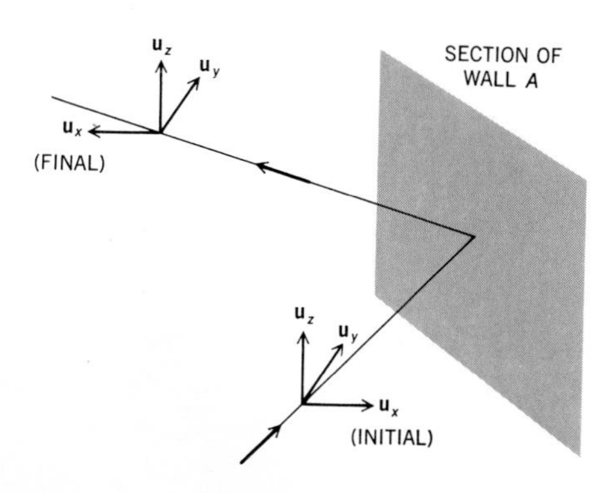

FIGURE 2-2
The reversal of $\mathbf{u}_x$ as a result of a collision with wall A.

speeds, and thus u_x and u_x^2 will be different for each molecule. If the average value of u_x^2 for the N molecules of the gas is denoted by $\overline{u_x^2}$, Eq. [4] can be developed to show the pressure required to confine N molecules to a volume V to be

$$P = \frac{m}{V} \sum_i (u_x^2)_i$$

$$= \frac{m}{V} N\overline{u_x^2} = \frac{Nm\overline{u_x^2}}{V} \qquad [5]$$

More convenient, however, is a relation between the pressure due to the N molecules and an average of the magnitude of the velocities of the molecules. The relation between the square of the component velocities, with which we have been dealing, and the magnitude of the square of the velocity itself is, as will be recalled from the more familiar resolution along two perpendicular directions,

$$u^2 = u_x^2 + u_y^2 + u_z^2 \qquad [6]$$

For a large number of molecules moving in random directions

$$\overline{u_x^2} = \overline{u_y^2} = \overline{u_z^2} \qquad [7]$$

and therefore

$$\overline{u^2} = 3\overline{u_x^2} \qquad [8]$$

(Note that u^2 and $\overline{u^2}$ are scalar, not vector, quantities; i.e., they indicate magnitudes, and not directions, of the molecular velocities.) Substitution of Eq. [8] in Eq. [5] gives the desired relation between P and $\overline{u^2}$; that is,

$$P = \frac{\frac{1}{3}Nm\overline{u^2}}{V} \quad \text{or} \quad PV = \tfrac{1}{3}Nm\overline{u^2} \qquad [9]$$

This important equation is, in a way, the end result of the present derivation. It is as far as one can go in explaining the basis for the pressure of a gas from the kinetic-molecular postulates of Sec. 2-1. This result cannot, however, be compared with the empirical gas laws; but with the additional considerations of the following section, this comparison can be made.

2-3 Kinetic Energies and Temperature

Frequently, as in the case under discussion, it is convenient to show explicitly the kinetic energy rather than the molecular speed. The average kinetic energy of one molecule of a gas is written $\overline{\text{ke}}$. This quantity is related to the average squared molecular speed by

$$\overline{\text{ke}} = \tfrac{1}{2}m\overline{u^2} \qquad [10]$$

Equation [9] can thus be developed from $PV = \frac{2}{3}N(\frac{1}{2}m\overline{u^2})$ to

$$PV = \tfrac{2}{3}N\overline{\text{ke}} \qquad [11]$$

The empirical results of Chap. 1 dealt with molar quantities of gases. The present theoretical results involve the properties of individual molecules, such as $\overline{\text{ke}}$. These can be brought together through Avogadro's number $\mathfrak{N}$, which relates the number of molecules N to the number of moles n by the relation

$$N = n\mathfrak{N} \qquad [12]$$

Equation [8] can now be written

$$PV = \tfrac{2}{3}n(\mathfrak{N}\overline{\text{ke}}) \qquad [13]$$

Furthermore, we introduce the term KE to denote the kinetic energy of an Avogadro's number of molecules and rewrite Eq. [11]

$$PV = \tfrac{2}{3}n\text{KE} \qquad [14]$$

At this stage it is necessary to recall the empirical result

$$PV = nRT$$

and to compare the kinetic-molecular-theory result, Eq. [14], with this expression. Historically, the apparent discrepancy between the two relations presented no problem. Heat and temperature had often been associated, in a qualitative way, with the idea of agitation and motion of the particles of the hot material. It was a relatively straightforward matter, therefore, to make the theoretical deduction agree with the experimental law by putting $\frac{2}{3}$KE equal to RT, or

$$\text{KE} = \tfrac{3}{2}RT \qquad [15]$$

Thus, if the translational kinetic energy of an Avogadro's number of molecules, i.e., 1 mol, has the value $\frac{3}{2}RT$, then the ideal-gas laws, as embodied in $PV = nRT$, are derivable from the postulates of the kinetic-molecular theory.

Some of the empirical results given in Chap. 1 have now been derived. The expression $PV = \frac{2}{3}n\text{KE}$, together with the postulate $\text{KE} = \frac{3}{2}RT$, in effect reproduces Boyle's and Gay-Lussac's laws. Furthermore, the derived result $PV = nRT$ holds, to the extent that the postulates of the kinetic-molecular theory are followed, for any gas; i.e., it is independent of the molecular mass or any other property characteristic of the molecules of the gas. This expression can hold for two different gases at the same temperature and pressure only if equal volumes of the different gases contain the same number of moles or molecules. Thus Avogadro's hypothesis is derived. Dalton's law follows directly from the original postulates since, on the assumption that molecules are noninteracting and occupy no appreciable volume, one set of gas molecules will have no effect on another set. The kinetic-molecular derivation for a mixture of gases would therefore lead to a total pressure P that is the sum of the pressures of each component, i.e.,

$$P = P_1 + P_2 \cdots$$

2-4 Numerical Values for Molecular Energies and Molecular Speeds

The intention in this chapter is to reveal some of the properties, in particular energies and speeds, of the molecules of which a gas is composed. It has been shown so far that the qualitative postulates of the kinetic-molecular theory are a sufficiently accurate description of the molecular world to lead to the ideal-gas laws. More quantitative information is obtained from the result that the kinetic energy of an Avogadro's number of molecules is $\frac{3}{2}RT$.

The value of 8.3143 J deg^{-1} mol^{-1} obtained in Sec. 1-4 for R yields for the translational-motion contribution to the energy of 1 mol of any ideal gas at 25°C the result

$$\tfrac{3}{2}RT = \tfrac{3}{2}(8.31)(298) = 2480 \text{ J mol}^{-1}$$

The average kinetic energy of one molecule of the gas can be calculated as

$$\overline{\text{ke}} = \frac{\text{KE}}{\mathfrak{N}} = \frac{3}{2}\frac{R}{\mathfrak{N}}T \qquad [16]$$

Since much of our subsequent work will be concerned with the energies of individual molecules and atoms, it is useful to introduce a new constant k, called *Boltzmann's constant,* as

$$k = \frac{R}{\mathfrak{N}}$$
$$= 1.3806 \times 10^{-23} \text{ J deg}^{-1} \qquad [17]$$

Boltzmann's constant is therefore the gas constant per molecule, and the average kinetic energy of one molecule is

$$\overline{\text{ke}} = \tfrac{3}{2}kT \qquad [18]$$

The average kinetic energy of a gas molecule at 25°C is

$$\overline{\text{ke}} = \tfrac{3}{2}(1.380 \times 10^{-23})(298) = 4.11 \times 10^{-21} \text{ J}$$

Although the values of these kinetic energies are very important and will become progressively more meaningful, they are at first difficult to appreciate. It is therefore worthwhile to consider a related and more readily visualized molecular property, the speeds with which molecules travel.

The kinetic energy of an Avogadro's number of molecules can be written

$$KE = \mathfrak{N}(\tfrac{1}{2}m\overline{u^2}) = \tfrac{1}{2}M\overline{u^2} \qquad [19]$$

where M is the molar mass. A molecular-speed term is obtained by combining this result with the kinetic-molecular-theory postulate KE $= \frac{3}{2}RT$ to get $\overline{u^2} = 3RT/M$, or

$$\sqrt{\overline{u^2}} = \sqrt{\frac{3RT}{M}} \qquad [20]$$

The cumbersome term $\sqrt{\overline{u^2}}$ is known as the *root-mean-square* (rms) speed. It is necessary to note that this term implies that the magnitude of each of the molecular velocities is squared; then the average value of the squared terms is taken; and finally the square root of this average is determined. This procedure leads to a quantity which is different from a simple average speed, but different only by about 10 percent, as will be seen in Sec. 2-6. For the present, the values of $\sqrt{\overline{u^2}}$ which are deduced will be taken as indicative of average molecular speeds.

For N_2 at 25°C, for example, M has the SI unit value 0.02802 kg, and we obtain

$$\begin{aligned}\sqrt{\overline{u^2}} &= \sqrt{\frac{3(8.3143)(298.15)}{0.02802}} \\ &= 515 \text{ m s}^{-1} \\ &= 1150 \text{ mph}\end{aligned} \qquad [21]$$

Table 2-1 shows additional results for a few simple molecules. Notice that, since the average kinetic energy at a given temperature is the same for all molecules regardless of their mass, light molecules have greater speeds than heavy molecules.

TABLE 2-1 Average speeds of gas molecules [equal to 0.921 $\sqrt{u^2}$ at 25°C (298 K) and 1000°C (1273 K)]

	25°C		1000°C	
Gas	m s^{-1}	mph	m s^{-1}	mph
H_2	1770	3960	3660	8180
He	1260	2820	2600	5830
H_2O	590	1320	1220	2730
N_2	470	1060	970	2190
O_2	440	990	910	2050
CO_2	380	840	780	1740
Cl_2	300	670	620	1380
HI	220	490	450	1010
Hg	180	400	370	830

You should also notice that just as molecular speeds can be interpreted in terms of components in three perpendicular directions by

$$u^2 = u_x^2 + u_y^2 + u_z^2$$

so also can the average of the average kinetic energy. That is,

$$\tfrac{1}{2}m\overline{u^2} = \tfrac{1}{2}m\overline{u_x^2} + \tfrac{1}{2}m\overline{u_y^2} + \tfrac{1}{2}m\overline{u_z^2}$$

or

$$\overline{\text{ke}} = (\overline{\text{ke}})_x + (\overline{\text{ke}})_y + (\overline{\text{ke}})_z \qquad [22]$$

It follows, since these average component energies are equal, that $\overline{\text{ke}} = \frac{3}{2}kT$ leads to

$$(\overline{\text{ke}})_x = (\overline{\text{ke}})_y = (\overline{\text{ke}})_z = \tfrac{1}{2}kT \qquad [23]$$

The three perpendicular directions in which the speeds and energies have been resolved are called *degrees of freedom.* We thus can say that *the average translational energy of a molecule per degree of freedom is* $\frac{1}{2}kT$. You will see that this statement is a far-reaching and important guide in the studies of molecular energies.

2-5 Distribution of Molecular Velocities in One Dimension

Having considered and tabulated some average molecular speeds, it is now appropriate to investigate in more detail the molecular speeds that contribute to the average values already worked out.

The basic relation for handling questions regarding the number of molecules that have various speeds, or energies, is *Boltzmann's distribution.* The deduction of this important relationship is best done after the quantum rules that will be seen to govern molecular behavior are studied. Here the results of the derivation, which will be given in Chap. 5, will be anticipated, and the Boltzmann distribution will be used to obtain the desired information on the distribution of molecular speeds.

According to the model on which the kinetic-molecular theory is based, the molecules of a gas are moving with a variety of speeds and directions, i.e., with various velocities. These velocities can be pictured on a diagram, like that of Fig. 2-3, where each point represents, by its distance from the origin, the magnitude of the velocity, i.e., the speed of a particle, and, by its direction from the origin, the direction in which the particle is moving. It helps to clarify the diagram, even if it is really not necessary, to add the vector velocity arrows.

Since gases behave similarly in all directions, i.e., they are *isotropic,* a diagram like that of Fig. 2-3, for a large enough number of molecules, must be the same in all directions. The nature of the variation in the

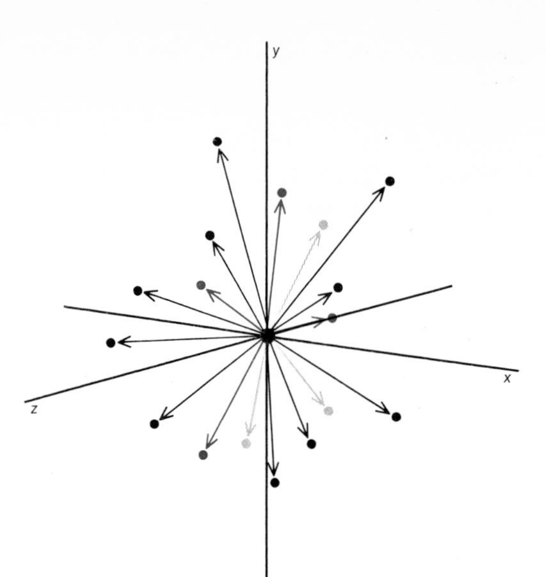

FIGURE 2-3
The velocities of molecules. Each magnitude and direction is represented by the length and direction of the arrow.

density of the velocity points as one goes out from the origin is the distribution of molecular speeds, which will be investigated in this section and in Sec. 2-6.

Here we will investigate the distribution of the molecular speeds along a particular direction, say the x direction. In terms of Fig. 2-4, this requires us to find the ratio between the number of points dN lying in the volume elements shown in Fig. 2-4 and the total number of velocity points N. That is, we must find the fraction dN/N of the velocity points in the speed interval u_x to $u_x + du_x$. According to the Boltzmann distribution expression, this fraction is proportional to an exponential term whose exponent is the ratio of the corresponding kinetic energy, $\frac{1}{2}mu_x^2$ to kT. Explicitly,

$$\frac{dN}{N} = Ae^{-(1/2)mu_x^2/kT}\,du_x \qquad [24]$$

where A is a proportionality constant. This constant can be evaluated by recognizing that integration of the right side of Eq. [24] over all

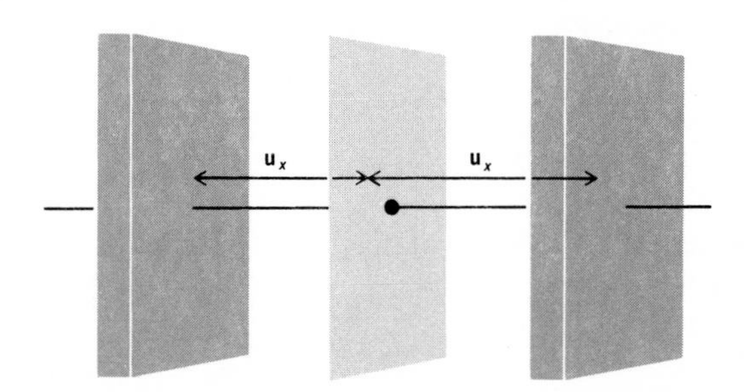

FIGURE 2-4
The two volume elements that together contain all the molecules moving with speed between u_x and $u_x + du_x$.

possible values of u_x, that is, from $u_x = 0$ to $u_x = \infty$, must account for all the velocity points. Thus we can write

$$\int_0^\infty Ae^{-(1/2)mu_x^2/kT}\,du_x = 1 \qquad [25]$$

so that the proportionality constant A is given by

$$A = \frac{1}{\int_0^\infty e^{-(1/2)mu_x^2/kT}\,du_x} \qquad [26]$$

The value of the integral is seen from Appendix 1 to be $\frac{1}{2}\sqrt{\pi}$, and we obtain

$$A = \sqrt{\frac{2m}{\pi kT}} \qquad [27]$$

Finally, the equation for the distribution of speeds along one direction for a sample of N molecules can be written

$$\frac{dN/N}{du_x} = \sqrt{\frac{2m}{\pi kT}}e^{-(1/2)mu_x^2/kT} \qquad [28]$$

Graphs of this one-dimensional distribution function are shown for two temperatures for the example of nitrogen in Fig. 2-5*a*.

In addition to the speed scale, a kinetic-energy scale can be attached

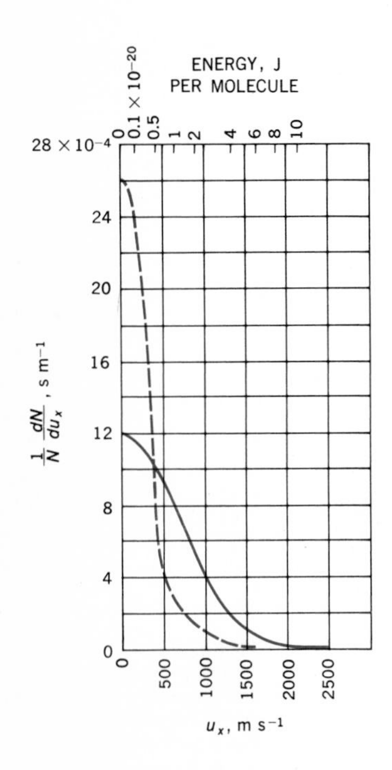

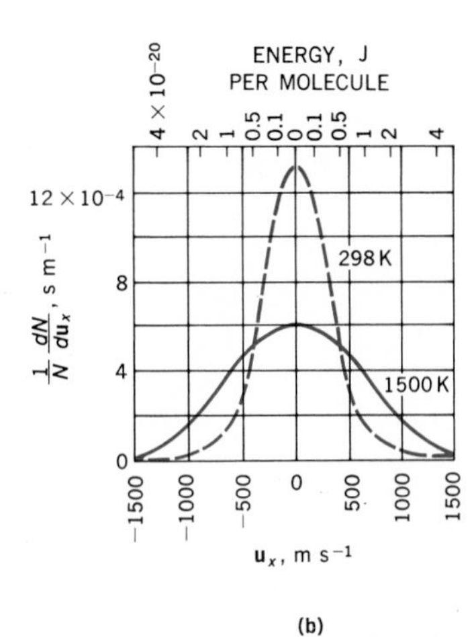

FIGURE 2-5
The distribution of (*a*) speeds and (*b*) velocities in the x direction for N_2 molecules at 298 and 1500 K.

to Fig. 2-5. This scale is shown along the top of the figure. Qualitatively, from the figure, or more accurately by replotting on a linear energy scale, it can easily be verified that the total kinetic energy for this one-dimensional motion is indeed $\frac{1}{2}RT$ mol^{-1}.

The x-component *velocity* distribution, i.e., the distribution of both magnitudes and directions, can also easily be shown. The distribution in the $+x$ direction will be the same as that in the $-x$ direction, and thus velocity-distribution curves can be drawn from the speed-distribution curves of Fig. 2-5a by dividing the ordinate values by 2 and including curves for distribution over positive and negative ranges of velocities, as is done in Fig. 2-5b. The corresponding analytical expression for such x-velocity-distribution curves is

$$\frac{dN/N}{d\mathbf{u}_x} = \sqrt{\frac{m}{2\pi kT}}\,e^{-(1/2)mu_x^2/kT} \qquad [29]$$

2-6 Distribution of Molecular Velocities in Three Dimensions

Let us now pass over the distribution of molecular speeds in two dimensions and proceed directly to the more important case of the distribution in three-dimensional space. Again the Boltzmann distribution will provide the basis on which the distribution expression is derived. The original derivation of the result that will be obtained was given by James Clerk Maxwell, and the result itself is now known as the *Maxwell-Boltzmann distribution* expression.

In the one-dimensional problem that we have already solved, we dealt with the density of points along any direction, for example, along the direction of the x axis. One can see in Fig. 2-3, although too few points are exhibited, that the number of points contained in a volume element, $d\mathbf{u}_x$ in the figure, falls off as the element is moved away, in either direction, from the origin. In a similar manner, of course, one-dimensional distributions along the y and z, or $\mathbf{u}_y$ or $\mathbf{u}_z$, axes could be dealt with. These one-dimensional distributions can, in fact, be combined to give the fraction of the molecules that have velocity components between $\mathbf{u}_x$ and $\mathbf{u}_x + d\mathbf{u}_x$, $\mathbf{u}_y$ and $\mathbf{u}_y + d\mathbf{u}_y$, and $\mathbf{u}_z$ and $\mathbf{u}_z + d\mathbf{u}_z$. This is equal to the fraction of points that occur in the outlined cubic element of volume in Fig. 2-6. It is given analytically as the product of the fractions of the molecules that lie in the appropriate volume elements perpendicular to the axes. Thus we can write

$$\frac{dN}{N} = \left(\sqrt{\frac{m}{2\pi kT}}\,e^{-(1/2)mu_x^2/kT}\,d\mathbf{u}_x\right)\left(\sqrt{\frac{m}{2\pi kT}}\,e^{-(1/2)mu_y/kT}\,d\mathbf{u}_y\right)\left(\sqrt{\frac{m}{2\pi kT}}\,e^{-(1/2)mu_z/kT}\,d\mathbf{u}_z\right)$$

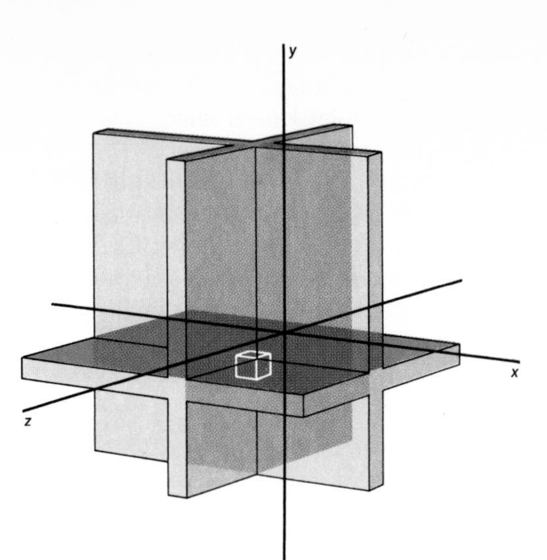

FIGURE 2-6
The volume elements that are combined in the derivation of $(dN/N)\ d\mathbf{u}_x\ d\mathbf{u}_y\ d\mathbf{u}_z$.

or

$$\frac{dN/N}{d\mathbf{u}_x\, d\mathbf{u}_y\, d\mathbf{u}_z} = \left(\frac{m}{2\pi kT}\right)^{3/2} e^{-(1/2)m(u_x{}^2+u_y{}^2+u_z{}^2)/kT} \qquad [30]$$

This result gives the distribution of the molecular velocities in that it expresses the density of points in the volume element of Fig. 2-6.

What is wanted, however, is the density of points within a volume element like that of Fig. 2-7, because all points lying in such an element correspond to the same speed, u. Since this spherical shell has a volume $4\pi u^2\, du$, the number of points in this element is obtained by multiplying

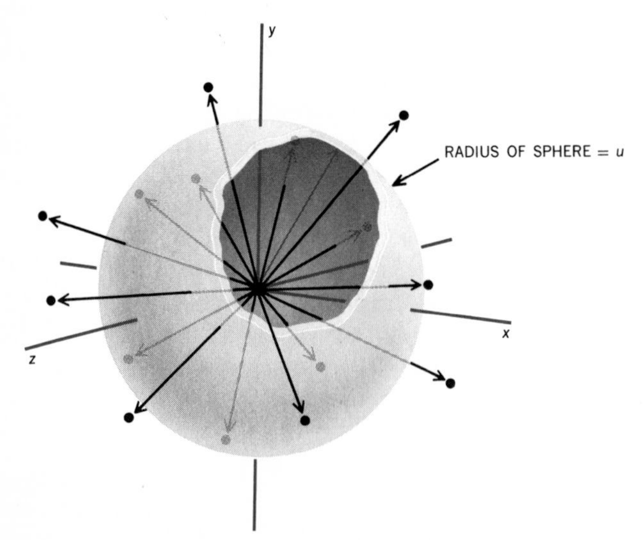

FIGURE 2-7
The molecular velocities and the volume element used to obtain the distribution of the magnitude of the velocities in three dimensions. The volume of the spherical shell is equal to $4\pi u^2\, du$.

the coefficient of the volume element in Eq. [30] by $4\pi u^2\,du$ rather than by $d\mathbf{u}_x\,d\mathbf{u}_y\,d\mathbf{u}_z$. In this way, and with the substitutions $u^2 = u_x{}^2 + u_y{}^2 + u_z{}^2$, one obtains the three-dimensional speed-distribution equation

$$\frac{dN/N}{du} = 4\pi\left(\frac{m}{2\pi kT}\right)^{3/2} e^{-(1/2)mu^2/kT}u^2 \qquad [31]$$

This Maxwell-Boltzmann distribution expression is plotted for N_2 at two temperatures in Fig. 2-8. The average speed has a value that can immediately be seen to be like that previously calculated for the root-mean-square speed. One should also notice that at low temperatures the molecules tend to have speeds bunched in a relatively narrow range. At higher temperatures the distribution is broader and—what is for some purposes very important—the high-speed end of the curve tends to spread out to much higher speeds.

Knowledge of the distribution curve allows the calculation of any desired kind of average. The Maxwell-Boltzmann distribution can, for example, be used to calculate a root-mean-square speed which, by other means, has already been shown, in Sec. 2-4, to have the value

$$\sqrt{\overline{u^2}} = \sqrt{\frac{3kT}{m}} = \sqrt{\frac{3RT}{M}}$$

To obtain such averages from the distribution expression, one multiplies the fraction of molecules that have a particular value of the quantity to be averaged by that value of the quantity, and then sums,

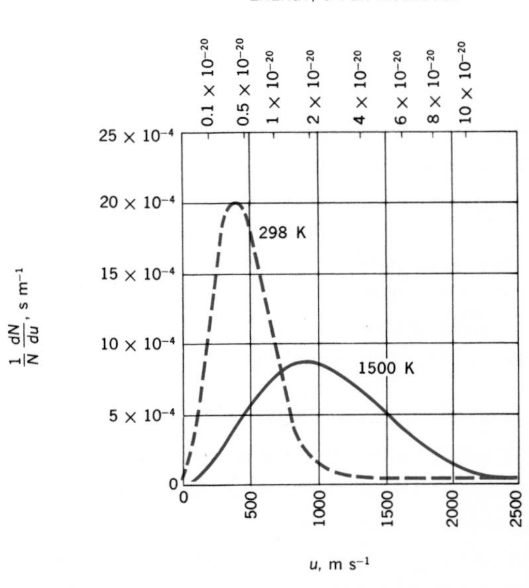

FIGURE 2-8
The distributions of the speeds of N_2 molecules at 298 and 1500 K.

or integrates, over all possible values of the quantity. Thus $\overline{u^2}$ is given by

$$\overline{u^2} = \int_{u=0}^{u=\infty} u^2 \frac{dN}{N}$$

$$= 4\pi \left(\frac{m}{2\pi kT}\right)^{3/2} \int_0^\infty u^4 \, e^{-(1/2)mu^2/kT} \, du$$

One of the integrals of Appendix 1 can then be used to give for $\overline{u^2}$ the result

$$\overline{u^2} = 4\pi \left(\frac{m}{2\pi kT}\right)^{3/2} \left(\frac{2kT}{m}\right)^{5/2} \tfrac{3}{8}\sqrt{\pi} = \frac{3kT}{m} \qquad [32]$$

Finally, $\sqrt{\overline{u^2}} = \sqrt{3kT/m}$, as obtained from the kinetic-molecular treatment of Sec. 2-4.

In a similar way one obtains the average speed as

$$\bar{u} = \int_{u=0}^{u=\infty} u \frac{dN}{N}$$

Substitution of the distribution relation and evaluation of the integral give the result

$$\bar{u} = \sqrt{\frac{8kT}{\pi m}} = \sqrt{\frac{8RT}{\pi M}} \qquad [33]$$

Finally, one sometimes deals with the most probable speed, i.e., the speed corresponding to the maxima of the curves like those of Fig. 2-8. To determine this speed it is only necessary to differentiate the distribution expression, set the result equal to zero, and determine the value of u that this relation implies. In this way, if α denotes the most probable speed,

$$\alpha = \sqrt{\frac{2kT}{m}} = \sqrt{\frac{2RT}{M}} \qquad [34]$$

These three speeds—the rms, the average, and the most probable—are not very different, being in the ratio

$$\sqrt{\overline{u^2}}:\bar{u}:\alpha = 1.00:0.92:0.82 \qquad [35]$$

One or the other of these usually provides sufficient information on the molecular speeds in any given problem. Where a more detailed knowledge of the molecular-speed distribution is required, reference must be made to the distribution expression or to graphs such as Fig. 2-8.

As a consequence of the Boltzmann distribution, we have learned a great deal more about the speeds with which molecules travel than we did by our earlier comparison of the result $P\text{v} = \frac{1}{3}\mathfrak{N}\, m\overline{u^2}$ and $P\text{v} = RT$. While obtaining the detailed distribution of molecular speeds, we were able to verify, by recalculating $\sqrt{\overline{u^2}}$, that the approach of this section is consistent with the simple kinetic-molecular-theory result.

* 2-7 Effusion and Molecular Beams

Most phenomena that we deal with depend on the average three-dimensional properties of the many molecules of a gas sample. The pressure of a gas is an example. But some phenomena depend, in a more detailed way, on the collisions of gas molecules with a surface—as in absorption and heterogeneous catalysis—and some depend on the passage of gas molecules through a pinhole or slit in a container—as in effusion and molecular-beam studies. In such cases, a knowledge of the randomly directed velocities of gas molecules, as has been developed in this section, is not sufficient, and the extensions that are then necessary are now presented.

Molecular beams now provide a valuable technique for studies of the properties and reactions of molecules, and the analysis here can be directed toward an understanding of some of their features. Such beams result when a gas is allowed to pass through a small hole in a sample container into an evacuated region in such a way that molecular collisions are infrequent. Then the molecules travel in straight lines, and a beam of molecules can be selected by appropriate baffles.

Let us first calculate the average speed of the molecules of such a beam, which is taken to be in the positive x direction.

The information necessary for the calculation is given in Fig. 2-5 or in Eq. [29]. Equation [29] gives the fraction of all the molecules with x-component velocities between $\mathbf{u}_x$ and $\mathbf{u}_x + d\mathbf{u}_x$. Thus the required average of only those molecules having a velocity component in the positive x direction is

$$\bar{u}_{x(+)} = \frac{\displaystyle\int_0^\infty \mathbf{u}_x \frac{dn/N}{du_x} d\mathbf{u}_x}{\displaystyle\int_0^\infty \frac{dN/N}{du_x} d\mathbf{u}_x}$$

or with Eq. [29],

$$\bar{u}_{x(+)} = \frac{\sqrt{\dfrac{m}{2\pi kT}} \displaystyle\int_0^\infty u_x e^{-(1/2)mu_x^2/kT}\, d\mathbf{u}_x}{\sqrt{\dfrac{m}{2\pi kT}} \displaystyle\int_0^\infty e^{-(1/2)mu_x^2/kT}\, d\mathbf{u}_x}$$

The denominator, the fraction of molecules having x-component velocities in the positive x direction, is found, with the integrals of Appendix 1, to have the expected value of $\frac{1}{2}$. Evaluation, again using the integrals of Appendix 1, of the numerator then yields

$$\bar{u}_{x(+)} = \sqrt{\frac{2kT}{\pi m}}$$

and with $\bar{u} = \sqrt{8kT/\pi m}$ from Sec. 2-6, we can write

$$\bar{u}_{x(+)} = \tfrac{1}{2}\bar{u} \qquad [36]$$

Since the speed distributions in the x, y, and z directions are independent of one another, the result $\bar{u}_{x(+)} = \frac{1}{2}\bar{u}$ applies both to molecules moving in

* Sections marked with an asterisk and set in reduced type need not be studied to preserve continuity. The regular material is arranged so that it does not depend on any of the developments or concepts introduced in the material set in this way. These optional sections, however, introduce aspects or extensions that will be of interest in the reader's study of physical chemistry. (Problems preceded by an asterisk are based on materials presented in these optional sections.)

random directions in a container and to the directionally selected molecules in a molecular beam.

With this result we can immediately obtain an effusion equation that shows the rate with which molecules will escape from a hole in a container when the gas pressure is low enough and the hole, a "pinhole," is small enough so that the molecules pass through in an individual rather than collective flow manner.

Consider a unit area of a container wall. How many molecules per second will collide with this area or, if the area represents a hole, will pass through it? Let there be N^* molecules per cubic centimeter so that $N^*/2$ are moving in the positive x direction. The number of these that reach the unit surface area per second is equal to the number in the region with unit cross-section area and length $\bar{u}_{x(+)} = \bar{u}/2$. Thus

$$\text{Rate of effusion} = \frac{N^*}{2}\bar{u}_{x(+)} = \tfrac{1}{4}N^*\bar{u}$$

or

$$\text{Rate of effusion} = N^* \sqrt{\frac{kT}{2\pi m}} = N^* \sqrt{\frac{RT}{2\pi M}} \qquad [37]$$

This is the important effusion equation that gives the rate with which the molecules of a gas escape in an effusion process. The result is in agreement with the empirical Graham's law that sets the number of molecules escaping per unit time inversely proportional to the square root of the molecular mass. Nonideality of gases enters, as for other ideal-gas expressions, to upset this mass dependence and the predicted temperature dependence. (The equation is often claimed to apply and to be found to be valid when the conditions for molecular effusion are not met. This results from the fact that diffusion processes that consist of a gas flow involving intermolecular and wall collisions have the same dependence on variables as does the effusion process.)

You should note that Eq. [36] gives the average x-direction speed not only of the molecules escaping and moving out with varying speeds in the y and z directions, but also of those which, by a slit assembly, are selected to have very small speeds in the y and z directions and thus form a molecular beam in the x direction.

2-8 The Mean Free Path, Collision Diameter, and Collision Number

So far we have been able to treat molecules as point particles, i.e., to ignore the size of molecules and the possibility of collisions between molecules. Thus the derivation of the pressure exerted by a gas in Sec. 2-2 was based on the idea that the molecules bounce back and forth between the walls of the container. We shall see that, except for very low gas pressures, a molecule of a gas collides many times with other molecules in traversing the container. It can be shown, however, that since these collisions do not change the net momentum of the colliding molecules in the direction being considered, these collisions in no way affect the derivation of the pressure of a gas. This implies that information about these molecular collisions cannot be obtained from calculations, such as those of Sec. 2-2, of the pressure of a gas.

Section 2-9 will show that, in contrast, the viscosity of a gas is dependent on the collisions of the gas molecules with each other. A kinetic-molecular derivation of the viscosity of a gas will therefore lead to added information about the experiences of the rapidly moving gas molecules. Three questions might come to mind about the collision properties of the molecules of a gas: How far, on the average, does a molecule travel between collisions? How many collisions per second does a molecule experience on the average? And how many collisions per second take place in a given volume of a gas? How little we are at home in the molecular world is impressed on us when we try to guess the answers to these questions.

Before proceeding to the calculation of gas viscosities, we shall show that the answers to all three questions can be related to one molecular property. This property is the diameter of the molecules of the gas. The use of only one quantity, the diameter, to define the size of the molecules means that the simplifying assumption of spherical molecules is being made and, furthermore, that the effective size of the molecules is independent of the energy involved in the molecular collision. Molecules are treated as hard spheres with no mutual attractions.

Let us consider a particular molecule A in Fig. 2-9, moving in the direction indicated. If the speed of the molecule A is $\bar{u}$, it will travel a distance of $\bar{u}$ m in 1 s. Furthermore, if only A is assumed to move and all the other molecules remain stationary, molecule A will collide in 1 s with all the molecules that have their centers within the cylinder of Fig. 2-9. The volume of the cylinder whose radius is equal to the molecular diameter is $\pi d^2\bar{u}$. The number of molecules in the cylinder is $\pi d^2\bar{u}N^*$, where N^* is the number of molecules per cubic meter. The mean free path, i.e., the distance traveled between collisions, is the length of the cylinder, $\bar{u}$, divided by the number of collisions oc-

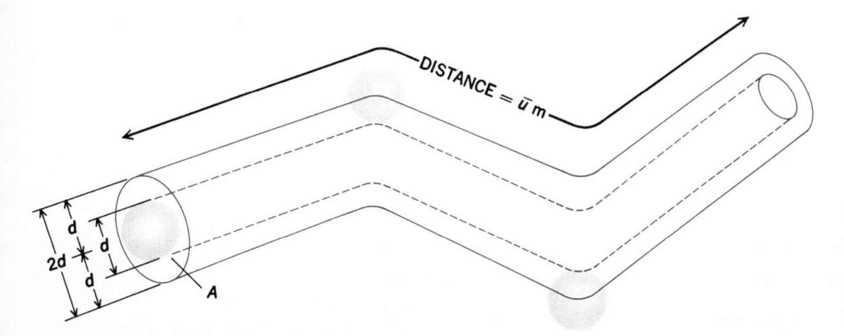

FIGURE 2-9
Path swept out by molecule A in 1 s. Molecules are shown greatly enlarged compared with the distance between them, and under conditions such as STP, many more collisions will occur in 1-s interval.

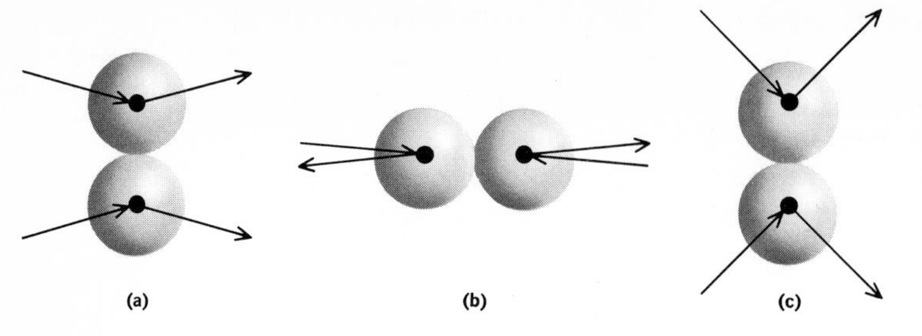

FIGURE 2-10
Types of molecular collisions. The relative velocity in an average collision is $\sqrt{2}\,\bar{u}$.
(a) Relative velocity $\cong 0$.
(b) Relative velocity $= 2\bar{u}$.
(c) Relative velocity $= \sqrt{2}\,\bar{u}$.

curring as the molecule traverses this length. Thus, if L is introduced to denote the mean-free-path length,

$$L = \frac{\bar{u}}{\pi d^2 \bar{u} N^*} = \frac{1}{\pi d^2 N^*} \qquad [38]$$

A more detailed calculation shows that this result is not exactly correct. The assumption that only molecule A moves implies a relative speed of the colliding molecules of $\bar{u}$. In fact, as Fig. 2-10 suggests, if the molecules are all moving with speed $\bar{u}$, all types of collisions will occur, ranging from glancing collisions, where the relative speed may be very small, to head-on collisions, where the relative speed is $2\bar{u}$. In an average collision, it turns out, the molecules move at right angles to each other, and the relative speed is $\sqrt{2}\bar{u}$. A correct result can be obtained in place of Eq. [38] by recognizing that, although molecule A moves a distance $\bar{u}$ in 1 s, it collides with other molecules with a relative speed of $\sqrt{2}\bar{u}$. The mean free path is then

$$L = \frac{1}{\sqrt{2}\,\pi d^2 N^*} \qquad [39]$$

The answer to the first question, as to how far a molecule travels between collisions, has now been shown to be dependent on the number of molecules per unit volume and on the as yet unknown quantity d.

The second problem to be investigated is the number of collisions per second that a molecule makes. This is called the *collision number* and is denoted by Z_1. In relation to the other molecules, the molecule A travels with an effective speed equal to $\sqrt{2}\bar{u}$. The number of collisions per second of this molecule is therefore equal to the number of molecules in a cylinder of radius d and of length $\sqrt{2}\bar{u}$. We thus have

$$\begin{aligned} Z_1 &= (\sqrt{2}\,\bar{u})(\pi d^2)N^* \\ &= \sqrt{2}\,\pi d^2 \bar{u} N^* \end{aligned} \qquad [40]$$

The last of the three problems to be investigated is the number of collisions occurring in a unit volume per unit time. As can be imagined, this quantity is of considerable importance in understanding the rates of chemical reactions. The number of collisions per second per cubic meter is also called the collision number, but is denoted by Z_{11}.

The collision number Z_{11} is closely related to the number Z_1. Since there are N^* molecules per cubic meter and each of these molecules has Z_1 collisions per second, the total number of collisions per second per cubic meter will be $\frac{1}{2}N^*Z_1$. The factor $\frac{1}{2}$ ensures that each collision will not be counted twice. We therefore obtain

$$Z_{11} = \tfrac{1}{2}\sqrt{2}\,\pi d^2\bar{u}(N^*)^2$$

$$= \frac{1}{\sqrt{2}}\pi d^2\bar{u}(N^*)^2 \qquad [41]$$

Both of the collision numbers and the mean free path have now been expressed in equations that involve the molecular diameter d. Since the molecular speeds and the number of molecules per cubic meter of a particular gas can be determined, only molecular diameters need be known in order to evaluate L, Z_1, and Z_{11}. Many methods, as we shall see, are available for determining the size of molecules. For the present, the kinetic-molecular derivation of gas viscosities will be relied upon to yield these values.

2-9 The Kinetic Theory of Gas Viscosity

The molecular theory, in accordance with which molecules move freely about, with large spaces between them, might seem at first to imply a complete absence of viscous forces. The source of viscous drag in gases can be understood, however, by focusing attention on two layers of a gas moving parallel to each other but with different flow rates. Over and above their random thermal motion, the molecules in the faster-moving layer will have a greater velocity component in the direction of flow than will the molecules in the slower layer. But because of their random movement, some of the molecules of the faster layer will move into the slower layer, imparting to it their additional momentum in the direction of flow, and thus tending to speed it up. Likewise, some of the molecules of the slower layer will reach the faster layer and tend to slow it down. The net effect of this exchange of molecules is a tendency toward equalizing the flow rates of the different parts of the gas. The viscous effect is just the difficulty of moving one part of a fluid with respect to another part.

A simplified kinetic-molecular theory of viscosity can be given on this basis. Consider two layers of unit area, separated by a distance equal to the mean free path, of a gas flowing as in Fig. 2-11. The gas flows in the x direction with a velocity v and a velocity gradient dv/dy; that is, the flow rate increases by an amount dv for each increment of distance dy in the y direction. Since the layers under consideration are a mean free path apart, on the average, a molecule leaving one layer will arrive in the other layer, collide, and contribute its greater or lesser momentum in the flow direction to that layer. According to a simple approach, which gives almost the correct result, one-third of

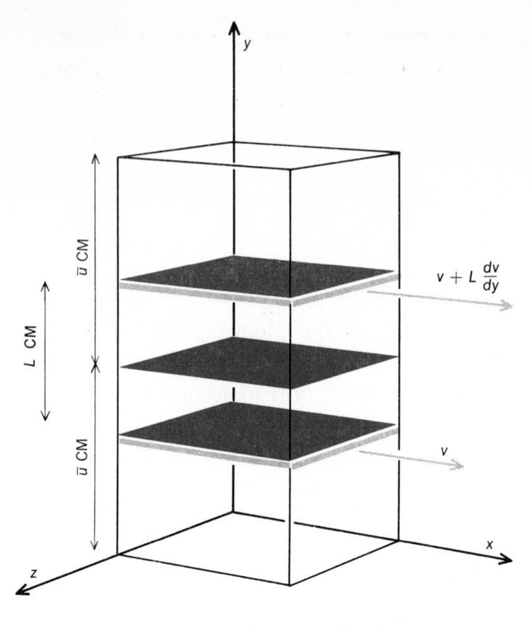

FIGURE 2-11
Two layers of a gas moving in the x direction with a velocity gradient of dv/dy.

the molecules have an x component of velocity, one-third a y component, and one-third a z component. Only the third with a y component are effective in the momentum exchange between the gas layers.

The momentum increment which each transferred molecule adds to or subtracts from the new layer is $mL(dv/dy)$, that is, m times the difference in flow velocity of the two layers. The force between the two layers can again be calculated from the rate of change of momentum. It is now necessary, therefore, to calculate the number of these molecular transfers per second.

The numbers of molecules that move up from the lower layer and down from the upper layer in 1 s are those which are in the lower volume shown and have a y component of velocity upward, and those which are in the upper volume shown and have a y component of velocity downward. These two volumes are both drawn with a length $\bar{u}$ m, so that in 1 s all the molecules with the appropriate direction of flight will have passed the shaded cross section. If there are N^* molecules per cubic meter, there will be $\frac{1}{6}N^*\bar{u}$ molecules in the lower volume, all of which will move into the upper volume in 1 s. A similar number will move down into the lower volume in 1 s. The total interchanges between layers, therefore, is $\frac{1}{3}N^*\bar{u}$ s^{-1}.

The rate of change of momentum is then $(\frac{1}{3}N^*\bar{u})mL(dv/dy)$, and according to Newton's law this is the force exerted by the layers on each other. Thus

$$f = \tfrac{1}{3}N^*\bar{u}mL\frac{dv}{dy} \qquad [42]$$

The coefficient of viscosity η has previously been defined by Eq. [26]

of Chap. 1, which, in differential form and for the unit-layer areas of this derivation, is

$$f = \eta \frac{dv}{dy} \qquad [43]$$

Comparison of these equations gives, for the kinetic-molecular derivation of viscosity, the result

$$\eta = \tfrac{1}{3}N^*\bar{u}mL \qquad [44]$$

A more detailed derivation takes into account the detailed distribution of molecular velocities and leads to the slightly different expression

$$\eta = \tfrac{1}{2}N^*\bar{u}mL \qquad [45]$$

It is this expression that will be used in the deduction of the molecular properties d, L, Z_1, and Z_{11}.

It is now convenient to replace L by means of Eq. [39] so that an expression involving the collision diameter,

$$\eta = \frac{\bar{u}m}{2\sqrt{2}\pi d^2} \qquad [46]$$

is obtained.

This important result permits the calculation of the collision diameter of a gas molecule from measurements of the viscosity of the gas. It is necessary also to have values for the mass of a molecule of the gas and the average speed of the gas molecules. Both these quantities are available, the latter as a result of the derivation of Sec. 2-6, which gave the expression

$$\bar{u} = \sqrt{\frac{8RT}{\pi M}} \qquad [47]$$

Before proceeding to a discussion of the values for the molecular properties which have been introduced in Sec. 2-8, it is interesting to point out an implication of Eq. [46] on the nature of the viscosity of gases. For a given gas, m and d are constants, and $\bar{u}$ varies as the square root of T, according to Eq. [47]. The theoretical derivation makes the prediction, therefore, that the viscosity of a gas should be independent of the pressure and proportional to the square root of the absolute temperature. This rather remarkable result (it seems "reasonable" that the viscosity of a gas should increase as the gas is compressed and becomes more dense) was one of the few theoretical deductions to be made before the experimental measurements had been performed. Maxwell's prediction of this behavior and its subsequent experimental verification provided one of the most dramatic triumphs of the kinetic-molecular theory. (It should be pointed out, however, that at higher pressures the nonideal behavior of gases seriously interferes with these deduced relations.)

2-10 Numerical Values of Collision Properties

As an example, consider the calculation of the collision properties d, L, Z_1, and Z_{11} for N_2 at 1 atm and 25°C. Table 1-4 gives the viscosity as

$$\eta = 1.78 \times 10^{-4} \text{ poise} = 1.78 \times 10^{-3} \text{ kg m}^{-1} \text{ s}^{-1}$$

The number of molecules in 1 m^3 at 25°C and 1 atm is obtained from the number 6.022×10^{23} in 22.414 liters, or 0.022 414 m^3, at STP, as

$$N^* = \frac{6.022 \times 10^{23}}{0.022414} \times \frac{273.15}{298.15} = 2.461 \times 10^{25} \text{ m}^{-3}$$

The average molecular speed is

$$\bar{u} = \sqrt{\frac{(8)(8.314)(298.15)}{0.02802}} = 0.475 \times 10^3 \text{ m s}^{-1}$$

and the mass of one molecule is

$$m = \frac{0.02802}{6.022 \times 10^{23}} = 4.65 \times 10^{-26} \text{ kg}$$

With these data the collision diameter of the N_2 molecule can be calculated from a rearrangement of Eq. [46] as

$$d = \sqrt{\frac{\bar{u}m}{2\sqrt{2}\pi\eta}}$$
$$= 3.74 \times 10^{-10} \text{ m}$$

With this value of the molecular diameter, Eqs. [39] to [41] can be used to obtain the remaining collision properties of N_2 at the specified conditions. Thus

$$L = \frac{1}{\sqrt{2}\pi d^2 N^*}$$
$$= 6.50 \times 10^{-8} \text{ m}$$
$$Z_1 = \sqrt{2}\pi d^2 \bar{u} N^*$$
$$= 7.31 \times 10^9 \text{ collisions s}^{-1}$$
$$Z_{11} = \frac{1}{\sqrt{2}}\pi d^2 \bar{u} (N^*)^2$$
$$= 8.99 \times 10^{34} \text{ collisions m}^{-3} \text{ s}^{-1}$$

Table 2-2 shows similar results for a few other simple molecules.

The data of Table 2-2 indicate the details which can be obtained about the molecular world from the kinetic-molecular theory. A valuable insight into molecular phenomena is provided by these data, and an effort should be made to become familiar with the order of magnitude of these quantities.

Notice, for example, that molecules are indeed small compared with the size of the region which a gas sample occupies under ordinary conditions. Molecular dimensions, as Table 2-2 shows, are of the order of 10^{-10} m. In spite of the small scale of the molecular world, it is not difficult to become accustomed to thinking in terms of molecular dimensions. One aid is an appropriate unit of length. The *angstrom* (abbreviated Å) is the unit in which molecular dimensions are almost always expressed. It is defined as

$$1 \text{ Å} = 10^{-10} \text{ m}$$

Although the angstrom unit of length is not consistent with the SI system, it has been used so extensively, and is so convenient, for molecular dimensions that we will at times report lengths in angstrom units.

It is important to realize that the diameters of Table 2-2 reflect the particular method by which the size of the molecules was measured. The determination of a collision diameter requires, to begin with, the assumption of a spherical molecule so that its size can be specified by the single variable, the diameter. This single-parameter assumption, moreover, implies that the molecules are being considered to be hard spheres with no mutual attractive forces. The effect of relaxing this restriction will be developed in the study of the rates of chemical reactions in Chap. 17.

The mean free path of gases at 1 atm pressure, as the sample calculation for N_2 showed, though hundreds of times larger than the molecular diameter, is short compared with the size of ordinary containers. The molecules of a gas in such a container will therefore collide many times with one another between the collisions they make with the walls of the container. But the mean free path, for a given gas, is inversely proportional to the number of molecules per unit volume, and thus it is inversely proportional to the pressure. At very low pressures, as occur in the upper atmosphere, mean free paths can be of

TABLE 2-2 Some kinetic-molecular-theory gas properties (at 25°C and 1 atm)

	Collision diameter d		Mean free path L	Collision number Z_1	Collision number Z_{11}
	(m)	(Å)	(m)	(s^{-1})	(m^{-3} s^{-1})
H_2	2.73×10^{-10}	2.73	12.3×10^{-8}	14.4×10^{9}	17.7×10^{34}
He	2.18	2.18	19.0	6.6	8.1
N_2	3.74	3.74	6.50	7.3	9.0
O_2	3.57	3.57	7.14	6.1	7.5
Ar	3.96	3.96	5.80	6.9	8.5
CO_2	4.56	4.56	4.41	8.6	10.6
HI	3.50	3.50	7.46	3.0	3.7

some number of meters rather than the small fractions of meters shown in Table 2-2.

Finally, the collision numbers Z_1 and Z_{11} should be considered. Under the conditions indicated in Table 2-2, the very many collisions per second, very short distance traveled between collisions, and the very high molecular speeds, all indicate the tremendous activity in the molecular world. At low pressures, again, although the molecular speeds are unchanged, the collisions experienced by a molecule in a unit time decrease, and the total collisions occurring in a given volume of gas drops off even more rapidly, a consequence of the inverse-square dependence on the particle density.

2-11 Theory of Nonideal Behavior—van der Waals' Equation

The simple model of the kinetic-molecular theory is satisfactory in that it leads to the derivation of the ideal-gas laws. We have seen, however, that real gases show PVT relations that deviate more or less widely from the ideal laws. A question naturally arises as to the possibility of understanding these deviations by the use of a more elaborate model for a gas than that used previously. This can be done, and in our quest for molecular information it is of interest to investigate what refinements of the previous treatment are necessary.

In 1873, the Dutch chemist van der Waals showed that the addition of two items to the molecular model of Sec. 2-1 could account for much of the deviation of real gases from ideal behavior. He attributed the failure of the derived $PV = nRT$ relation to duplicate the behavior of real gases to the neglect of:

1 The volume occupied by the gas molecules

2 The attractive forces among the molecules

The corrections introduced by the recognition of these two factors will be treated one at a time.

When n mol of a gas is placed in a container of volume V, the volume in which the molecules are free to move is equal to V only if the volume occupied by the molecules themselves is negligible. The presence of molecules of nonvanishing size means that a certain volume, called the *excluded volume,* is not available for the molecules to move in. If the volume excluded by 1 mol of a gas is represented by b, then instead of writing $PV = nRT$, a more appropriate equation would be

$$P(V - nb) = nRT \qquad [48]$$

The excluded volume b is usually treated as a constant which is characteristic of each gas and must be determined empirically so that a good correction to the simple gas-law expression is obtained.

The relation of b to the size of the molecules can be seen by considering Fig. 2-12. The molecules are again assumed to be spherical and to have a diameter d. The volume in which the centers of two molecules cannot move because of each other's presence is indicated by the lightly shaded circle in Fig. 2-12. The radius of this sphere is equal to the molecular diameter. The volume excluded per pair of molecules is $\frac{4}{3}\pi d^3$. We then obtain

$$\text{Actual volume of a molecule} = \tfrac{4}{3}\pi\left(\frac{d}{2}\right)^3$$

$$\text{Excluded volume per molecule} = \tfrac{1}{2}(\tfrac{4}{3}\pi d^3) = 4\left[\tfrac{4}{3}\pi\left(\frac{d}{2}\right)^3\right]$$

It is seen, therefore, that the excluded volume is four times the actual volume of the molecules. Since b is the excluded volume *per mole,* we have

$$b = 4\mathfrak{N}\left[\tfrac{4}{3}\pi\left(\frac{d}{2}\right)^3\right] \qquad [49]$$

where $\mathfrak{N}$ is Avogadro's number.

(You might notice that this result, that the excluded volume is four times the volume of all the molecules, applies only if the volume that is excluded results from the coming together of pairs of molecules. At rather high gas pressures more than one molecule might be adjacent to a given molecule and the excluded volume *per* molecule is then reduced. When many molecules are packed as closely together as possible, the volume that is occupied, and thus excluded to other molecules, is 1.35 times the actual volume of the molecules.)

We might be tempted to make use of our previously determined values of d to calculate b. It is more satisfactory, however, to adjust b so that the derived equation corresponds as well as possible to the

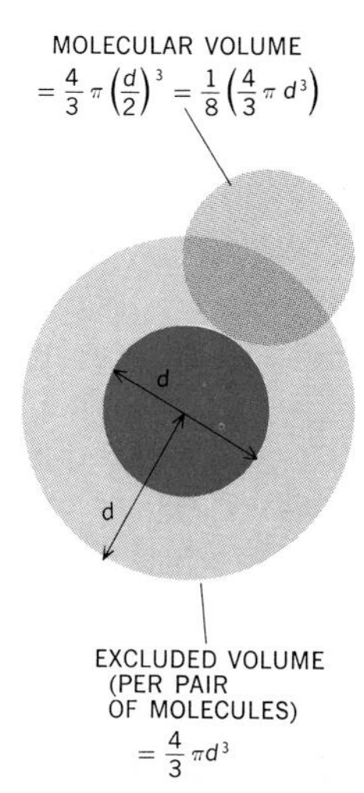

FIGURE 2-12
The excluded volume (light shade) for a pair of molecules according to van der Waals' treatment.

observed PVT data. This procedure, required principally by the difficulties caused by the second correction term, results in van der Waals' equation being *semiempirical.* The form of the derived equation follows from a theoretical treatment, but the numerical values of the constants appearing in the equation are obtained from the experimental PVT data. Semiempirical equations are not uncommon in chemistry and frequently are considerably more satisfactory than completely empirical relations.

The second van der Waals correction term concerns the attractive forces among molecules. That such forces exist is clearly demonstrated by the tendency of all gases to condense at temperatures low enough so that these forces can overcome the kinetic energy of the molecules. That these attractions exist is clear; exact knowledge of their source and quantitative values are much harder to come by. The semiempirical approach, however, requires only that a suitable term representing these attractions be inserted. Its value can be obtained by adjusting it, as is done with b, to give an equation that best fits the PVT data.

The attraction a molecule exerts on its neighbors tends to draw them in toward itself; i.e., the attraction acts with the confining pressure to hold the molecules together. Complete analysis of the consequence of this is difficult; but qualitatively, the effect is that of reducing the independence of each molecule. The pressure exerted by the gas is thus reduced just as it would be if the number of independent molecules decreased. The effect of one molecule in helping to hold the gas together through these forces of attraction is proportional to the number of nearby molecules on which it can act. If there is n mol of gas in a volume V, this number is proportional to n/V, the number of moles per unit volume. Since each of the neighboring molecules is likewise attracting its neighbors, the total pulling together of the gas due to these interactions is proportional to $(n/V)^2$. The gas is confined, therefore, not only by the external pressure P, but also by these intermolecular attractions, which contribute a term proportional to $(n/V)^2$. If the proportionality factor is denoted by a, van der Waals' complete equation becomes

$$\left(P + \frac{an^2}{V^2}\right)(V - nb) = nRT \qquad [50]$$

The success of this equation in fitting the PVT behavior of real gases is judged by choosing values of a and b, different for each gas and for each temperature, to give as good a fit to the observed data as possible. Although perfect agreement of calculated and observed volumes over a wide range of pressure is not obtained, the improvement over the ideal-gas-law expression $PV = nRT$ is very considerable. Figure 2-13 and Table 2-3 indicate the amount of improvement in regions of very nonideal behavior. The success of van der Waals' equation in representing PVT behavior is very much better than would be expected

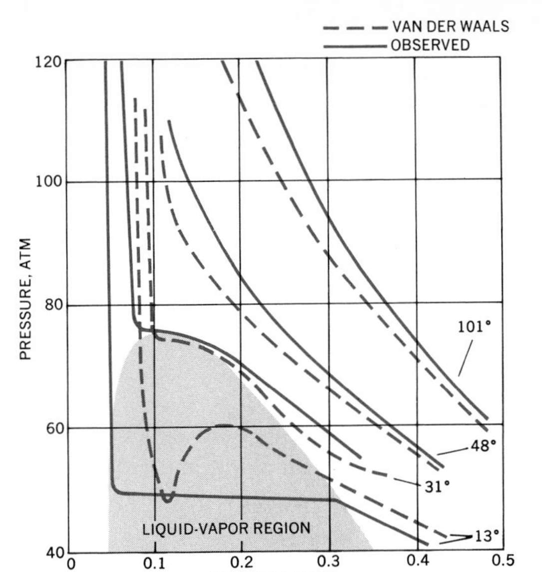

FIGURE 2-13
Comparison of van der Waals' *PV* curves for CO_2 with the observed behavior near the critical point.

for any purely empirical expression with only two adjustable constants. The behavior in the two-phase region, such as shown in Fig. 2-13, however, cannot be followed by van der Waals' equation, and the maxima and minima must be disregarded.

2-12 Van der Waals' Equation and the Critical Point

As Fig. 2-13 shows, van der Waals' equation follows reasonably well the behavior of a gas near the region of liquid-vapor equilibrium. There will, moreover, be one temperature for which van der Waals' equation, with given values of a and b, will show the horizontal point of inflection that is to be identified with the critical isotherm. This identification provides a convenient, but not always the most satisfactory, way of obtaining values for a and b.

TABLE 2-3 Molar volumes of CO_2 at 320 K
Comparison of van der Waals' equation and the ideal-gas law near the critical temperature

	v (liters)		
P (atm)	Observed	van der Waals	Ideal
1	26.2	26.2	26.3
10	2.52	2.53	2.63
40	0.54	0.55	0.66
100	0.098	0.10	0.26

Equation [50] can be rearranged and written for 1 mol to give

$$P = \frac{RT}{\mathrm{v} - b} - \frac{a}{\mathrm{v}^2} \qquad [51]$$

To investigate the horizontal point of inflection on a plot of P versus v, we obtain

$$\frac{dP}{d\mathrm{v}} = \frac{-RT}{(\mathrm{v} - b)^2} + \frac{2a}{\mathrm{v}^3} \qquad [52]$$

and

$$\frac{d^2P}{d\mathrm{v}^2} = \frac{2RT}{(\mathrm{v} - b)^3} - \frac{6a}{\mathrm{v}^4} \qquad [53]$$

At the critical point the first and second derivatives are zero, and the pressure, volume per mole, and temperature can be written P_c, v_c, and T_c. At this point, Eqs. [51] to [53] become, respectively,

$$P_c = \frac{RT_c}{\mathrm{v}_c - b} - \frac{a}{\mathrm{v}_c^2} \qquad [54]$$

$$0 = \frac{-RT_c}{(\mathrm{v}_c - b)^2} + \frac{2a}{\mathrm{v}_c^3} \qquad [55]$$

and

$$0 = \frac{2RT_c}{(\mathrm{v}_c - b)^3} - \frac{6a}{\mathrm{v}_c^4} \qquad [56]$$

These three equations can be solved for a, b, and R in terms of P_c, v_c, and T_c. (The gas constant R, of course, is better known by other means, but it appears in van der Waals' equation as if it were another adjustable constant.) After some manipulation, the following relations are obtained:

$$b = \tfrac{1}{3}\mathrm{v}_c \qquad [57]$$

$$a = 3P_c\mathrm{v}_c^2 \qquad [58]$$

$$R = \frac{8P_c\mathrm{v}_c}{3T_c} \qquad [59]$$

With these results, the critical data of Table 1-3 can be used to obtain values for the van der Waals constants a and b, and calculated values of the parameter R, shown in Table 2-4. Also included are values of the molecular diameter d that are based on the interpretation given to b by Eq. [49]. The values of the empirical constants calculated in this way should result in a good fit of curves, calculated from van der Waals' equation, to the experimental results in the neighborhood of the critical point. If the PVT behavior in some other region is of particular interest, it might be advantageous to adjust the values of a and b to something other than those calculated from Eqs. [57] and [58].

TABLE 2-4 Values for the constants *a*, *b*, and *R* of van der Waals' equation
Calculated from the critical-point data of Table 1-3 and the relations of Eqs. [57] to [59]. Molecular diameter values are obtained by use of Eq. [49].

Gas	a (atm liters^2 mol^{-2})	b (liters mol^{-1})	R (liter atm mol^{-1} deg^{-1})	Molecular diam. d (Å)
H_2	0.162	0.0217	0.067	2.58
He	0.022	0.0192	0.065	2.48
CH_4	1.33	0.0329	0.063	2.97
NH_3	1.75	0.0241	0.053	2.67
H_2O	1.32	0.0150	0.040	2.29
CO	0.85	0.0300	0.063	2.88
N_2	0.81	0.0300	0.064	2.88
O_2	0.82	0.0248	0.064	2.71
CH_3OH	3.26	0.0392	0.048	3.15
Ar	0.86	0.0257	0.065	2.74
CO_2	2.01	0.0319	0.061	2.94
n-C_5H_{12}	9.52	0.1034	0.058	4.35
C_6H_6	9.44	0.0855	0.058	4.08

from p16

The molecular diameters calculated from van der Waals' equation are seen to be in rough agreement with those obtained from viscosity measurements, and this encourages confidence in these numbers as giving some measure of the effective diameters of molecules.

Some discussion of the values of the van der Waals constant a might seem called for. It is advisable, however, to leave the subject of intermolecular forces until it is treated in Chap. 19.

In spite of the success of van der Waals' equation in handling the PVT behavior of real gases, in much of our subsequent work we shall revert to the simple ideal-gas expression $PV = nRT$. At low pressures and not too low temperatures, deviations from this relation are, for many purposes, not significant. Furthermore, the simplicity of the ideal-gas expression and the fact that it can be used for all gases without adjustment of any constants make its use very advantageous.

2-13 Van der Waals' Equation and the Law of Corresponding States

The empirical data plotted in Fig. 1-11 show that, as the law of corresponding states claims, many gases behave in a similar manner when they are treated in terms of the reduced variables P_R, v_R, and T_R. It is interesting to show that van der Waals' equation is consistent with the law of corresponding states in that, when it is written in terms of the reduced variables, no quantities remain to be empirically adjusted for the particular gas to which it is applied.

The constants a, b, and R are related to the critical-point constants by Eqs. [57] to [59]. These relations can be substituted for a, b, and R and used to write van der Waals' equation in a form that shows the critical constants explicitly. Rearrangement of this equation leads to only the terms P/P_c, v/v_c, and T/T_c. Introducing the reduced variables P_R, v_R, and T_R to represent these ratios, we have, for 1 mol of gas,

$$\left(P_R + \frac{3}{\mathrm{v}_R^2}\right)\left(\mathrm{v}_R - \frac{1}{3}\right) = \tfrac{8}{3}T_R \qquad [60]$$

It can now be seen that van der Waals' equation is consistent with the empirical law of corresponding states. This form of van der Waals' equation applies without the evaluation of any additional constants and illustrates the fact that, in terms of the reduced variables, gases behave approximately in a like manner.

Problems

1 Calculate the pressure exerted by 10^{23} gas particles, each of mass 10^{-25} kg, in a container of volume 1 liter. The rms speed is 100 m s^{-1}. What is the total kinetic energy of these particles? What must be the temperature?
Ans. $P = 0.33$ atm; energy $= 50$ J; $T = 24.2$ K.

2 A 1-liter gas bulb contains 1.03×10^{23} H_2 molecules. If the pressure exerted by these molecules is 1 atm, what must the average squared molecular speed be? What must the temperature be?

3 Estimate the number of molecules left in a volume the size of a pinhead, about 10^{-9} m^3, when air is pumped out to give a vacuum of 10^{-9} atm at 25°C.

4 For a gas sample of N molecules that consists of N_1 molecules of mass m_1 and N_2 molecules of mass m_2, follow through a derivation like that of Secs. 2-2 and 2-3, which led to the result $PV = \frac{1}{3}Nm\overline{u^2}$ and $PV = nRT$. What assumptions are necessary for the derivation to give the empirical result $PV = nRT$ regardless of the masses of the molecules of the sample?

5 For 25°C calculate and compare the average kinetic energies, speeds, and momenta of the molecules of He and Hg.

6 Calculate the molecular rms speeds of He atoms at 10, 100, and 1000 K in units of meters per second and miles per hour. What values would be obtained if the pressure were specified to be 10^{-9} atm?

7 The kinetic-molecular theory attributes an average kinetic energy of $\frac{3}{2}kT$ to each particle. What speed would a mist particle of mass 10^{-15} kg have at room temperature according to the kinetic-molecular theory? Compare this value with the molecular speeds of Table 2-1. *Ans.* 3.5×10^{-3} m s^{-1}.

8 How much heat must be added to 3.45×10^{-3} kg of neon in a 10-liter bulb to raise the temperature from 0 to 100°C? By what ratio is the average squared speed changed by this temperature change?

9 How many degrees would the temperature of 1 mol of liquid water be raised by the addition of an amount of energy equal to the translational kinetic energy at 25°C of 1 mol of water vapor? The heat capacity of liquid water is about 18 cal deg^{-1} mol^{-1}, or 75 J deg^{-1} mol^{-1}. *Ans.* 49°C.

10 The following values are given for the speed of sound in air:

Temp. (°*C*)	20	100	500	1000
Speed (*m sec*$^{-1}$)	344	386	553	700

Compare these values with the rms speeds of N_2 molecules at these temperatures.

11 Using the value of d from Table 2-2 for argon, calculate the mean free path, the average number of collisions a molecule experiences per second, and the average number of collisions per cubic meter per second for the molecules of argon at 0°C and 1 atm pressure. What values would be obtained at 1000°C and 1 atm? What values at 0°C and 100 atm?

12 Derive an expression for the mean free path for any gas in terms of the collision diameter, the temperature, and the pressure. Prepare a convenient graph showing the variation of the mean free path with pressure for 0°C and the pressure range 10^{-9} to 1 atm if the gas is nitrogen.

13 Using Eq. [33] for the average speed of a gas, obtain expressions for the collision numbers Z_1 and Z_{11} as functions of d, M, P, and T. For the pressure range 10^{-9} to 1 atm show graphically the variation of Z_1 and Z_{11} with pressure for N_2 at 0°C.

14 The ratio of the number of molecules that have a speed three times the average speed $\bar{u}$ to the number that have the average speed is a guide to the number of fast molecules that are present.
a Calculate this ratio for a gas at 25°C. *Ans.* 3.46×10^{-4}.
b Calculate the ratio of the number of molecules with speeds $3\bar{u}_{25°C}$ to the number with $\bar{u}_{25°C}$ for a gas at 40°C. Note from this calculation how a small temperature increase has a large effect on the number of fast molecules. *Ans.* 5.46×10^{-4}.

15 Plot the one-dimensional and three-dimensional population-versus-speed curves for H_2 molecules at 0°C.

16 Obtain an expression for the distribution of the molecules of a gas throughout x-component translational energies; i.e., obtain an expression for $\frac{1}{N}\frac{dN}{d\epsilon_x}$. $\left(\text{Make use of the derivative relation } \frac{dN}{d\epsilon_x} = \frac{dN}{du_x}\frac{du_x}{d\epsilon_x}.\right)$

17 By graphical means show that the population-energy function of Prob. 16 is consistent with an average translational energy of $\frac{1}{2}kT$ per degree of freedom.

18 From the one-dimensional distribution expression of Prob. 16, determine by integration the average kinetic energy for the motion of a molecule in one dimension.

19 Repeat the distribution and average-value deductions of Prob. 16, but for the total, three-dimensional translational energies of gas-phase molecules.

20 Verify from the distribution expressions that the most probable speed of a molecule is zero in one dimension and is given by Eq. [34] in three dimensions.

21 Verify that integration leads to the average speed given by Eq. [33]. (Consult the list of definite integrals in Appendix 1.)

***22** Obtain an expression for the rate of mass loss as a result of a gas sample leaking by effusion through a pinhole.

* Problems preceded by an asterisk are based on optional material presented in sections marked with an asterisk and set in reduced type.

***23** What expression for the product PV would you obtain if you calculated it from the product of the average number of collisions per unit surface area, using the effusion equation of Sec. 2-7, and the average momentum change per collision, using the average x velocity component of that section? Explain why this result deviates, i.e., is larger or smaller, the way it does from the correct result.

***24** In a study of the effect of a solid catalyst on a gas-phase reaction, a gas mixture at 25°C and at a total pressure of 10^{-8} atm was exposed to the solid catalyst for 15 min. How many molecules and how many moles of the gas mixture came in contact with a surface area of 10^{-4} m^2 of the catalyst during this reaction time? (Ignore the fact that the same molecules may have been responsible for repeated collisions.)

***25** How long would it take 1 mol of (a) hydrogen and (b) xenon to escape by effusion into a vacuum through a pinhole of diameter 10^{-6} m^2 if the pressure and temperature of the gases were 10^{-5} atm and 0°C?

***26** Calculate the fraction of helium molecules that have upward speeds great enough to escape from the earth: (a) for a temperature of 230 K, which is the average temperature up to an altitude of about 100 km, and (b) for a temperature of 2000 K, as occurs in the daytime in the thermosphere. (The escape speed from the earth is 1.12×10^4 m s^{-1}.)

27 Compare the volume of 20 g of H_2O at 100°C and 0.50 atm pressure given by the ideal-gas law with that given by van der Waals' equation.

28 Show that at fairly low pressures, where $PV = nRT$ can be inserted in the van der Waals correction term, van der Waals' equation for 1 mol can be reduced to

$$PV = RT(1 + BP)$$

where

$$B = \frac{b}{RT} - \frac{a}{(RT)^2}$$

Use this approximation to calculate a virial coefficient at 20°C from the van der Waals constants for CH_4. Compare with the value given in Prob. 23 of Chap. 1. *Ans.* $B = -0.00211$ atm^{-1}.

29 Real gases show deviations from ideal behavior that, as is suggested by Figs. 1-8 and 1-9, give PV versus P curves that at first depart either above or below the ideal-gas line. For any gas there is one temperature, known as the *Boyle temperature,* at which the curve begins tangent with the ideal-gas line. Deduce, from van der Waals' equation, an expression for the Boyle temperature. Compare calculated values of the Boyle temperature with the curve shapes of the gas of Fig. 1-9.

30 The compressibility factor of many gases at their critical points is approximately the same, and an average value is 0.28. In what way is van der Waals' equation in line with this empirical result and in what way is it out of line?

References

TABOR, D.: "Gases, Liquids, and Solids," Penguin Books, Inc., Baltimore, 1969. Further treatment of the kinetic theory of ideal and real gases, as well as valuable supplementary material on solids and liquids that is pertinent to the studies of Chaps. 18 and 19.

HILDEBRAND, J. H.: "An Introduction to Kinetic Theory," Reinhold Publishing Corporation, New York, 1963. A brief, very readable discussion of the nature of ideal gases, real gases, and the kinetic-molecular theory.

LOEB, L. B.: "Kinetic Theory of Gases," Dover Publications, Inc., New York, 1961. A more complete treatment of the kinetic-molecular theory so presented and at a level that makes it a very suitable extension to the material of this chapter.

JEANS, J. H.: "The Dynamical Theory of Gases," Dover Publications, Inc., New York, 1954. A comprehensive but rather mathematical treatment of the kinetic-molecular theory of gases.

KENNARD, E. H.: "Kinetic Theory of Gases," McGraw-Hill Book Company, New York, 1938. Similar to the treatise by Jeans in scope and approach.

BRUSH, S. G.: "Kinetic Theory," vols. 1 and 2, Pergamon Press, Oxford, 1965 and 1966. Well-chosen selections, with orienting introductory remarks, of the original contributions to the development of atomic theory. An excellent introduction to the history of the subject.

3

INTRODUCTION TO THE MECHANICS OF ATOMS AND MOLECULES

The kinetic-molecular theory of the properties of gases provides a great deal of information on the nature and behavior of molecules. This information results, so to speak, from an external view of the molecules. Much of the physical chemistry with which we shall deal in later chapters will be more understandable if we here continue our investigation of the molecular world and consider the internal structure of atoms and molecules.

Although we have seen that relatively simple derivations based on the kinetic-molecular model for gases can lead to quite detailed information about individual molecules, any attempt to learn about the internal structure and properties of individual molecules may seem to be an overly bold undertaking. Not only is the small size of the molecule forbidding, but also the fact—as became evident as the actual investigations were pursued—that the electrons and atoms which make up molecules exhibit behavior that is quite outside our ordinary experience. You will see, however, that there are very powerful theoretical and experimental methods that let us probe the inner workings of atoms and molecules. Here, after you have been reminded of the descriptions we use for light and electromagnetic radiation, you will see that detailed and quantitative data that bear on the behavior of the electrons of atoms and molecules come from studies of the radiation emitted from, or absorbed by, atoms. Then the mechanics obeyed by the electrons of atoms and molecules will be introduced, and you will begin to see the nature, and utility, of the *wave,* or *quantum, mechanics* that will open so many doors into the atomic and molecular world.

There is, for chemists, one principal obstacle to an understanding of atoms and molecules in sufficient detail so that the chemical and physical properties of materials can be interpreted: that is the complexity that is introduced by the electrons which surround the nuclei of atoms and molecules and, moreover, provide the "glue" that binds atoms together into molecules. To tackle this problem we will here accept the earlier developments that led to the concept of the "nuclear atom," i.e., the concept that atoms have small heavy nuclei that can be treated as containing neutrons and protons, surrounded by electrons, which, in the neutral atom, are equal in number to the nuclear protons.

3-1 The Nature of Light

Much of our present knowledge of the detailed nature of atoms and molecules comes from experiments in which light or, more generally, radiation and matter interact. Atomic spectroscopy was the first of such experiments to provide the data necessary for advancing the theory of the atom. Before treating these experiments, however, it is necessary to mention the two models of electromagnetic radiation—the wave model and the corpuscular, or particle, model—and the bridge between these models.

With the wave model, electromagnetic radiation is pictured as consisting of electric and magnetic disturbances traveling out from the light source with the speed of light.

Electromagnetic radiation travels through a vacuum with a speed c which has the value

$$c = 2.9979 \times 10^8 \text{ m s}^{-1}$$

The wave-model characteristics, the frequency ν, and the wavelength λ of a particular type of electromagnetic radiation are related, as Fig. 3-1 shows, by the equation

$$\nu = \frac{c}{\lambda} \qquad [1]$$

The categories into which electromagnetic radiation is divided correspond to different ranges of wavelengths or frequencies. The former are often expressed in angstroms (1 Å = 10^{-10} m) as well as in meters.

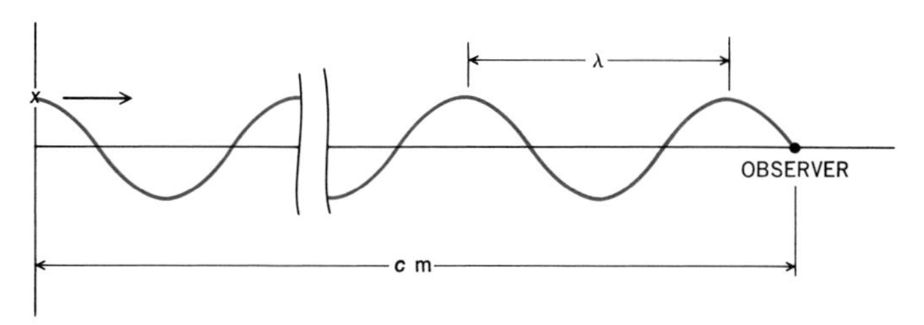

FIGURE 3-1
The basis of the relation $\nu = c/\lambda$.

Thus radiation that is detected by the eye has wavelengths in the range 4000 to 7000 Å. Yellow light, for example, has a wavelength of about 5800 Å, or 0.58×10^{-6} m, and its frequency is therefore

$$\nu = \frac{c}{\lambda} = \frac{3 \times 10^8}{0.58 \times 10^{-6}} = 5.2 \times 10^{14} \text{ Hz} \qquad [2]$$

The description of light as a wave motion successfully explained many of the phenomena which were observed. Diffraction and interference effects, such as the colors produced by an oil film on water, were readily accounted for. This theory of the nature of light, however, never succeeded in completely displacing the "corpuscular theory" originally due to Newton. This second theory views light, or radiation, as consisting of a flow of particles or corpuscles.

The dilemma of these two somewhat satisfactory, but very different, theories of the nature of light persisted for more than two centuries. The dilemma was finally resolved as a result of a bold step, which was to have great consequences in the development of physics and chemistry, taken by Max Planck in 1900. At that time he was attempting to explain the experimental results for the energy emitted by a hot blackbody as a function of the frequency or wavelength of the radiation. All previous attempts to explain the energy distribution that would be emitted by the vibrating particles of the hot body had been unsuccessful. Planck was led to the unprecedented view that the energy of the oscillating particles was "quantized," i.e., that only certain discrete energies were allowed. He further assumed that transitions between these allowed energies could occur and that the energy change, when the oscillator moved to a lower allowed energy state, was related to the frequency of the emitted radiation by the relation

$$\Delta\epsilon = h\nu \qquad [3]$$

where $\Delta\epsilon$ = energy change of oscillator of blackbody
ν = frequency of radiation
h = a constant, known as Planck's constant,[1] having the value 0.6626×10^{-33} J s

On the basis of this relation Planck obtained an expression which correctly gave the energy distribution of the radiation of a hot body.

Einstein recognized in Planck's equation a tying together of the two theories of light. The energy loss of the oscillator can be identified with the energy of a quantum of radiation, a corpuscular-theory concept. This energy, moreover, can be calculated from the frequency of the radiation, a wave-motion concept. Yellow light, for example, previously described as having a frequency of oscillation of 5.2×10^{14} Hz,

[1]In the cgs system h has the value 6.626×10^{-27} or 0.6626×10^{-26} erg s^{-1}. The relation to the SI unit depends on the conversion 1 J = 10^7 ergs.

TABLE 3-1 Some regions of the electromagnetic spectrum

	Typical wavelength λ		Typical frequency ν	Typical energy $\Delta\epsilon$
Description	meters	angstroms	hertz	joules
X-rays	1×10^{-10}	1	3×10^{18}	2×10^{-15}
Ultraviolet	2×10^{-7}	2000	1.5×10^{15}	1×10^{-18}
Visible	5×10^{-7}	5000	0.6×10^{15}	4×10^{-19}
Infrared	1×10^{-5}		3×10^{13}	2×10^{-20}
Radar, or microwave	1×10^{-2}		3×10^{10}	2×10^{-23}
Radio wave	3×10^{3}		1×10^{5}	7×10^{-29}

can now also be characterized as a flow of quanta, or photons, each with an energy of

$$\begin{aligned}\Delta\epsilon &= (0.66 \times 10^{-33})(5.2 \times 10^{14}) \\ &= 3.4 \times 10^{-19}\ \text{J}\end{aligned} \tag{4}$$

Values of the quantum energies of radiation in the various categories are included in Table 3-1.

3-2 Atomic Spectra

When the radiation emitted by the hot atoms of a discharge tube or electric arc is spread out by a prism or grating, as in Fig. 3-2, the radiation can be seen as a function of wavelength, frequency, or quan-

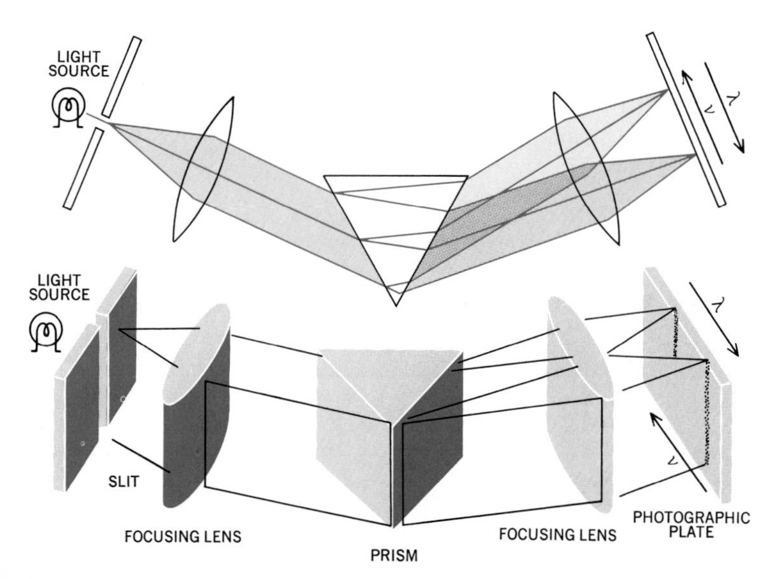

FIGURE 3-2
Optical arrangement of a prism spectrograph.

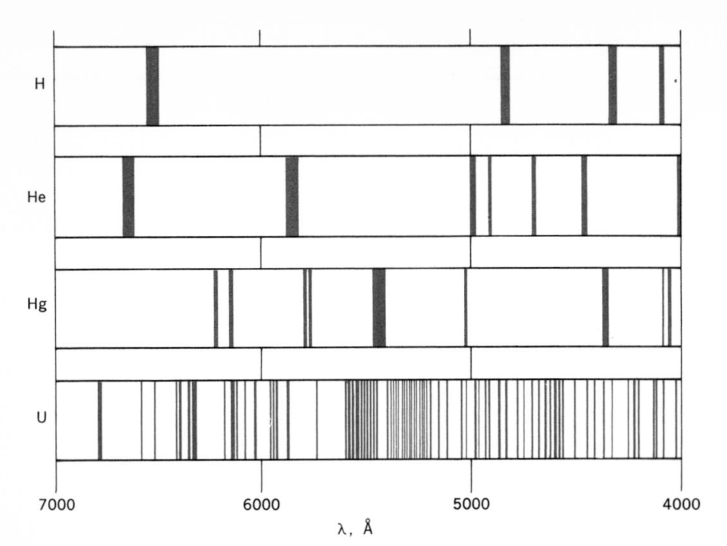

FIGURE 3-3
The spectral lines in the visible region for some atoms of different elements.

tum energy. All spectra obtained in this way show that such radiation consists of many narrow spectral lines and that these occur in usually complex patterns, as in Fig. 3-3, that are characteristic of the element whose atoms are responsible for the emission. (Emission from *solids,* such as that from a tungsten-filament light bulb, is different in that a continuum of radiation is found when it is studied spectroscopically.) Some atoms, such as sodium and especially hydrogen, fortunately gave relatively simple patterns. Although the understanding of the basis for these atomic spectra came about from considerations of the spectra of a number of elements, the theoretical studies of atomic structure were, at this stage, centered on the hydrogen atom; and it is sufficient to consider here the experimental studies of the hydrogen-atom spectrum. The spectrum obtained from the radiation emitted by hydrogen atoms at high temperature is shown in Fig. 3-4.

Considerably before the nature of atoms was understood, it was

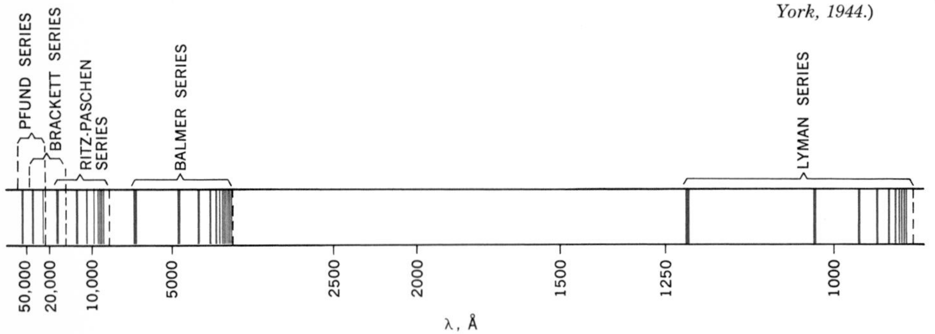

FIGURE 3-4
The hydrogen-atom spectrum with the series of Table 3-2 identified. *(From G. Herzberg, "Atomic Spectra and Atomic Structure," Dover Publications, New York, 1944.)*

recognized that the radiation emitted by a hot, or *excited,* atom presented a body of valuable data on the internal nature of the atom. The actual spectra were of little use, however, until some pattern or relation could be found for the frequencies of the emitted radiation. The initial step in bringing these data to bear on the problem of atomic structure was the completely empirical one of discovering some basic design in the emitted frequencies.

Many unsuccessful attempts were made to explain the observed spectral lines as harmonics or overtones of some set of fundamental frequencies. Gradually, however, a completely empirical approach yielded some understanding of the spectra of the simpler atoms. Finally, in 1885, Balmer showed that the frequencies of some of the observed spectral lines of the hydrogen atom, now known as the *Balmer series,* could be expressed by an empirical equation which we write

$$\nu = 3.2898 \times 10^{15}\left(\frac{1}{2^2} - \frac{1}{n_1^{\,2}}\right) \qquad \text{Hz} \qquad [5]$$

where $n_1 = 3, 4, 5, \ldots$.

This entirely empirical correlation proved a valuable clue, and very shortly it was shown by Rydberg that a more general expression of this type could be written to correlate the frequencies of all the observed spectral lines of the hydrogen atom. The expression, known as a *Rydberg formula,*[1] is

$$\nu = 3.2898 \times 10^{15}\left(\frac{1}{n_2^{\,2}} - \frac{1}{n_1^{\,2}}\right) \qquad \text{Hz} \qquad [6]$$

where n_1 and n_2 are to be assigned integral values 1, 2, 3, . . . such that $n_1 > n_2$. Frequencies calculated by choosing various sets of integers correspond to the observed spectral lines, and these lines can be grouped, as Table 3-2 and Fig. 3-4 show, in series, each series being calculated from a given value of n_2 and named after the discoverer of the series. The Rydberg formula, since it can be used to generate all the spectral lines of the hydrogen atom, provides a concise analytical summary of the spectral data.

It is important to recognize, as was apparent after the acceptance

[1]The frequency of a spectral line can be reported by a related quantity, the *wave number,* usually denoted by $\bar{\nu}$. You recall that $\nu = c/\lambda$, and since c is effectively constant, we could define $\bar{\nu} = 1/\lambda$ and have a measure that is proportional to ν. If SI units were used, you would, for example, see that the Rydberg equation constant would have the value

$$\frac{3.2898 \times 10^{15}\ \text{Hz}}{2.9979 \times 10^{8}\ \text{m s}^{-1}} = 109{,}737 \times 10^{2}\ \text{m}^{-1}$$

The wave-number quantity $\bar{\nu}$ will, however, be seen with the cgs unit of cm^{-1}, and then $\bar{\nu} = 1/\lambda(\text{cm})$, and, for example, the Rydberg number would be 109,737 cm^{-1}.

TABLE 3-2 Hydrogen-atom spectral series and Rydberg integers for the equation $\bar{\nu} = 109{,}677\left(\frac{1}{n_2^2} - \frac{1}{n_1^2}\right)$ cm^{-1}

Series	n_2	n_1	Spectral region
Lyman	1	2, 3, 4, . . .	Ultraviolet
Balmer	2	3, 4, 5, . . .	Visible
Paschen	3	4, 5, 6, . . .	Infrared
Brackett	4	5, 6, 7, . . .	Infrared
Pfund	5	6, 7, 8, . . .	Infrared

of the quantum-wave relation $\Delta\epsilon = h\nu$, that the emission of only specific frequencies of radiation by an atom corresponds to the emission of quanta of discrete energies. This can be emphasized by making use of Planck's relation to write the Rydberg formula more explicitly in terms of energy. The result is

$$\Delta\epsilon = 2.1798 \times 10^{-18}\left(\frac{1}{n_2^2} - \frac{1}{n_1^2}\right) \quad \text{J} \qquad [7]$$

again with n_1 and n_2 selected from the integers 1, 2, 3, . . . with $n_1 > n_2$.

The spectroscopic results show, therefore, that the electron of the hydrogen atom can change its energy only by the definite amounts that are calculated from the Rydberg formula. These amounts can be obtained from the spectrum of hydrogen atoms or can be calculated from the Rydberg expression, which of course is based on the spectral results.

From the energy changes that are determined spectrally, an energy diagram for the hydrogen atom can be constructed. This diagram, and the relation of the spectral series to it, are shown in Fig. 3-5.

The next step is clearly that of finding some description of the way in which electrons behave in atoms and molecules that leads to an electron behavior consistent with the detailed spectral information that can be obtained, as illustrated by Fig. 3-5 for the hydrogen atom.

3-3 The Wave Nature of Particles

The behavior of electrons confined to atoms and molecules cannot be deduced by applying the laws that have been found to govern the behavior of objects moving in ordinary-sized, i.e., macroscopic, regions of space. The first successful response to this difficulty was that of Niels Bohr, who in a 1913 study of the hydrogen atom recognized that an otherwise classical treatment could be made to give results that were in agreement with the spectral studies if he added the restriction that

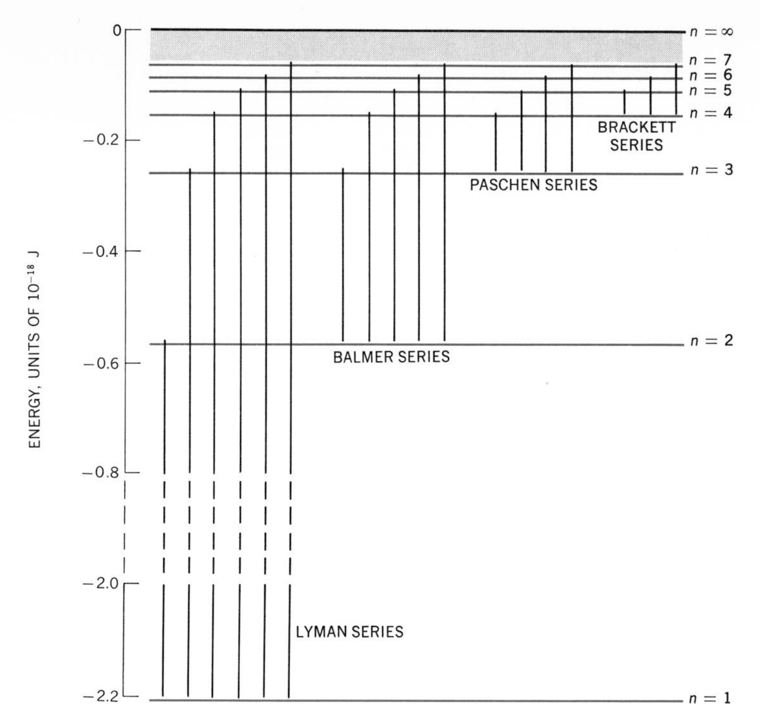

FIGURE 3-5
The hydrogen-atom spectral series on an energy-level diagram.

the electron could move around the nucleus only in ways that gave it certain specified amounts of angular momentum. His recognition of "quantized angular momentum" remains a powerful approach to the mechanics of systems, like atoms, in which particles move about a central point. This restriction can, moreover, be related to another approach that provides a more general way of modifying the macroscopic laws of motion so that they are applicable to particles such as the electrons of atoms and molecules.

The suggestion by Louis de Broglie in 1923 that wave properties could be associated with particles, such as the electron, pointed the way to a very general method for handling atomic and molecular problems.

The wavelength of the wave which is to be associated with a particle is suggested by combining the electromagnetic wave equation $\Delta\epsilon = h\nu$ and the Einstein relation $\Delta\epsilon = mc^2$. These can be combined to give $h\nu = mc^2$. Introduction of the wavelength according to $\lambda = c/\nu$ and rearrangement then gives

$$\lambda = \frac{h}{mc} \qquad [8]$$

This expression, as is seen from its derivation, applies only to electromagnetic radiation. For particles, de Broglie suggested that the wavelength could be deduced from the parallel expression

$$\lambda = \frac{h}{mv} \qquad [9]$$

where v is the velocity of the particle. Thus a wave, with wavelength $\lambda = h/mv$, could be associated with a stream of any particles, each of mass m, moving with velocity v, or more briefly, with momentum mv.

The relation of the wave-nature property of particles to the problem faced by Bohr in assigning certain restrictions to the "orbits" in which he assumed the electron of a hydrogen atom could move is seen in Fig. 3-6. The electron revolving in an orbit will have, according to de Broglie, a wave associated with it, and the wavelength will be related to the mass and velocity of the electron by the relation $\lambda = h/mv$.

For a general orbit and electron velocity, the wave, if drawn around the orbit path, will not fit, and successive cycles will tend to interfere destructively. Constructive interference can occur, as shown in Fig. 3-6a. The relation for constructive interference is that

$$2\pi r = n\lambda \qquad [10]$$

where n is an integer showing the number of waves that fit in the orbit path. Substitution of the de Broglie relation for λ now gives, after rearrangement,

$$mvr = n\frac{h}{2\pi} \qquad n = 1, 2, 3, \ldots \qquad [11]$$

This is the same stipulation, made more arbitrarily by Bohr—the collection of terms mvr being the angular momentum that Bohr recognized as quantized—and it is interesting to see how it comes neatly out of the idea that the allowed orbits are such that the de Broglie wave associated with the electron must form a standing wave.

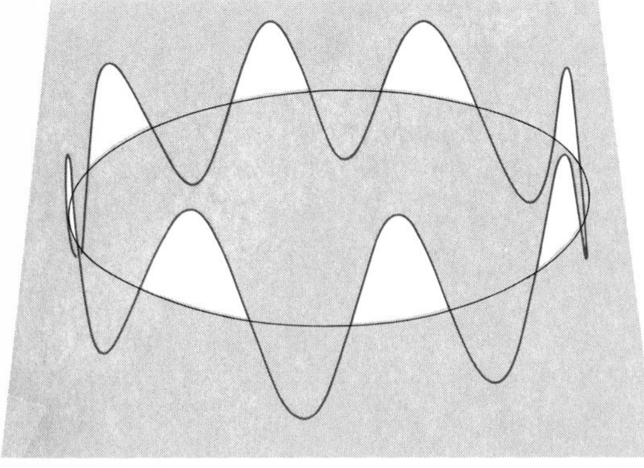

(a)

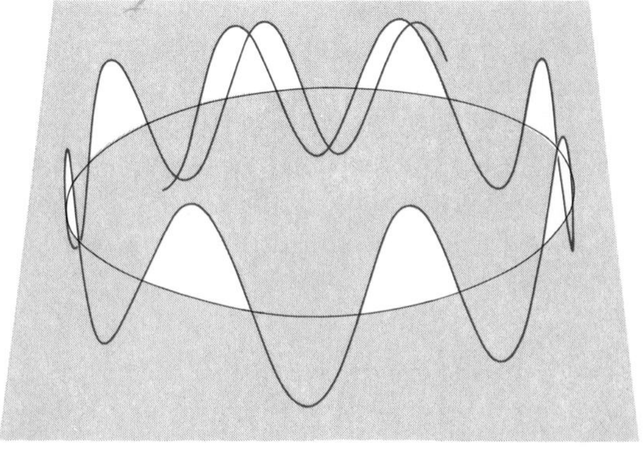

(b)

FIGURE 3-6 The de Broglie wave associated with an electron in a Bohr orbit. In (a) an integral number of waves fit in the orbit; constructive interference then leads to a standing wave. In (b) destructive interference occurs.

The validity of the de Broglie relation was later to be verified more directly when experiments were performed in which a beam of electrons, and later of neutrons, behaved in a wavelike manner and showed diffraction, or interference, effects. Some of these effects will be studied in a later chapter. For the present, the suggestion of de Broglie is important in that it introduces the useful concept of the wave nature of particles.

3-4 Wave Mechanics and the Schrödinger Equation

You have seen, for the special case of electron orbits in the hydrogen atom, that the "fitting in" of de Broglie waves into a region in which a particle is free to move can lead to an expression that shows the restrictions placed on the behavior of the particle. For this approach to be used for the wide range of atomic and molecular problems that are of interest to chemists, we must have a generally applicable procedure for "fitting in" the waves. Such an approach is provided by the *wave equation* introduced by Erwin Schrödinger in 1926, and in different form, independently, by W. Heisenberg in the same year. Schrödinger's method is based on an equation and a set of rules that allow the calculation of the behavior of matter, particularly in systems in which a particle is restricted to regions with the dimensions of atoms and molecules. This method is, in a sense, the exact counterpart of Newton's laws of motion. Newton's laws, which are always presented without any derivation or proof, let one calculate the mechanical behavior of objects of ordinary and even planetary size. Schrödinger's equation is likewise presented without derivation or proof and is of particular interest in chemistry because it is applicable to the behavior of objects down to molecular and atomic, but probably not subnuclear, dimensions. Just as one uses and becomes trusting in $f = ma$, so must one use and, to the extent that seems justifiable, become trusting in the Schrödinger equation.

In a study of an atom, the quantities that might be of prime interest are the energies of the electrons in the atom and the positions of the electrons relative to the nucleus and, possibly, relative to one another. The Schrödinger method yields directly the energies of systems such as the electron and nucleus combination that constitutes an atom. It leads, however, to less specific information on the positions of the particles of the system. Only the probability that the particle, or electron, is at a given position is obtained from wave mechanics. The lack of information on the exact position of a particle at some specific time seems to be a characteristic of atomic dimensions, and not to be a defect in the Schrödinger method.

In studying the hydrogen atom by application of the Schrödinger method we shall see, in a later chapter, that the allowed energies of the electron are obtained and are the same as those deduced by Bohr, and required to account for the hydrogen-atom spectrum. The probability of the electron being at a given distance from the nucleus is also given, and the probability function obtained is related to the orbits of the Bohr theory. To obtain these results, the Schrödinger equation makes use only of the data for the mass of the electron and the potential energy which it experiences as a result of the coulombic attraction to the nucleus.

Just as one practices applying the ordinary laws of motion to simple problems, such as inclined planes and pulley systems, so also is it helpful to practice using the Schrödinger equation on some simple systems of atomic dimensions. Study of the Schrödinger equation is conveniently begun with problems in which the particle under consideration is required to move along one dimension only. For such a particle, the potential energy will be some function of this one dimension, which can be taken along the x axis and can be represented by $U(x)$. The information which will go into the Schrödinger equation in a particular problem will be the nature of $U(x)$ and the mass m of the particle. The information which we attempt to obtain is the allowed values of the energy of the particle and the relative probabilities of the particle being at various positions along the x axis. Solution of the Schrödinger equation will yield a function of x, denoted by $\psi(x)$, or simply ψ, which is called the *wave function* for the particle. It is the square of this function which gives the relative probability of the particle being at various distances along the x axis.

The Schrödinger equation in one dimension is

$$-\frac{h^2}{8\pi^2 m}\frac{d^2\psi}{dx^2} + U(x)\psi = \epsilon\psi \qquad [12]$$

The behavior of a particle is deduced, according to Schrödinger, by finding some function which will solve this differential equation when the appropriate values of $U(x)$ and m have been substituted. Satisfactory solution functions ψ will generally exist only for certain values of ϵ, and these are the allowed energies of the particle. Finally, the probability function ψ^2 is readily obtained from the solution function ψ. (In general, ψ may be either a real or complex function, and to allow for the second possibility, we should write not ψ^2 but $|\psi|^2$, implying that the probability function is obtained by taking the product of it and its complex conjugate. Here we will not carry out any calculations with complex wave functions, and we can therefore use the simple squared function symbol ψ^2.)

The example of the following section will illustrate the necessary inputs to the Schrödinger equation, the treatment that must be given to the equation, and the results that this treatment yields.

3-5 A One-dimensional Illustration of the Schrödinger Equation: A Particle in a One-dimensional Square Potential Well

The one-dimensional square-well potential function of Fig. 3-7 leads to simple Schrödinger equation solutions and yet can be looked upon as a model for a number of molecular problems. Between $x = 0$ and $x = a$ the potential energy has a constant value which can be taken as zero, and elsewhere the potential is infinitely high. An electron in a piece of wire, for example, experiences a potential-energy function which for some purposes can be so represented. Of more chemical interest is the fact that the double bonding, or π, electrons of a conjugated system of double bonds in a molecule behave approximately as though the potential energy which they experience is such a simple square-well function. The present purpose, however, is only to show how the Schrödinger equation is applied to a specific problem and to illustrate the nature of the solutions that are obtained.

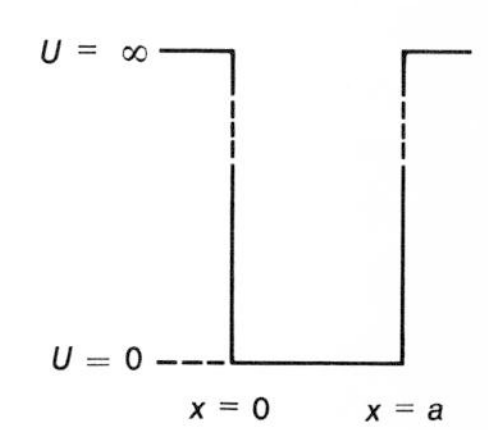

FIGURE 3-7
The square-well potential for the Schrödinger equation example.

Outside of the region between $x = 0$ and $x = a$ the potential energy is infinitely high and there will be zero probability of finding the particle in these regions. (Again the electron in a wire is a helpful analogy. There the difference in energy of an electron in the wire and outside the wire is large, although not infinite.) It follows that, since ψ^2 must be zero for $x < 0$ and $x > a$, so also must ψ be zero in these regions. In the region $0 < x < a$ the potential-energy function is $U(x) = 0$, and the Schrödinger equation in this region is

$$-\frac{h^2}{8\pi^2 m}\frac{d^2\psi}{dx^2} = \epsilon\psi \qquad [13]$$

It is now necessary to find well-behaved solutions for this equation. To be well behaved and avoid a discontinuity at $x = 0$ and $x = a$, the function ψ in the region between 0 and a must be such that it equals zero at $x = 0$ and $x = a$. Functions which solve the differential equation and also satisfy these boundary conditions can be seen by inspection to be

$$\psi = A \sin\frac{n\pi x}{a} \qquad \text{where } n = 1, 2, 3, \ldots \qquad [14]$$

and A is some constant factor. The expression $n\pi x/a$ has been arranged, as can be checked, so that the function goes to zero at $x = 0$ and at $x = a$ for any integral value of n. That the function satisfies the Schrödinger equation can be tested by substitution in Eq. [13] to give

$$\text{Left side} = -\frac{h^2}{8\pi^2 m}\left(-\frac{n^2\pi^2}{a^2}\right)A \sin\frac{n\pi x}{a}$$

$$= \frac{n^2h^2}{8ma^2}\left(A \sin\frac{n\pi x}{a}\right) \qquad [15]$$

$$\text{Right side} = \epsilon\left(A \sin \frac{n\pi x}{a}\right) \qquad [16]$$

The left and right sides of Eq. [13] are equal, and the expression

$$\psi = A \sin \frac{n\pi x}{a}$$

is therefore a solution if

$$\epsilon = \frac{n^2h^2}{8ma^2} \qquad n = 1, 2, 3, \ldots \qquad [17]$$

No really different solution can be found, and no energies other than these will result. (The value $n = 0$ in Eq. [14], it should be mentioned, provides a solution to Eq. [13] but gives a wave function that is everywhere zero. This leads to a zero probability of a particle being anywhere in the box and is therefore unacceptable.) The allowed energies ϵ, which are represented in Fig. 3-8, are seen to be quantized as a result of the quite natural introduction of the integers in the solutions of the Schrödinger equation. A similar situation occurs generally in atomic and molecular problems. The quantum phenomena, which were so arbitrarily introduced in the Bohr theory, are seen to result much more naturally in Schrödinger's approach.

The functions ψ and ψ^2 are shown in Fig. 3-8, alongside the corresponding energy level, for the first few states. The ψ^2 functions show the relative probability of the particle being at various positions when the quantum number has some particular value. If one assigns a value to A such that the total probability of the particle being between $x = 0$ and $x = a$ is unity, the wave functions are said to be *normalized.*

The probability curves of Fig. 3-8, which are typical of those obtained in other such problems, are by no means understandable in terms of the behavior of ordinary-sized objects. The presence of positions at which the probability of finding the particle is zero is most striking. Similarly, a particle in a "box" of atomic or molecular dimensions can have only certain allowed energies, and is not even permitted to have zero energy. Only the success of the Schrödinger method in treating a number of problems where the solutions can be tested against experiment makes us put up with such strange results.

This "particle-in-a-box" problem should be appreciated as being typical of those encountered in applying the Schrödinger equation to problems of chemical interest. In general, three dimensions will be involved and the potential-energy function will be somewhat more complicated. The procedure, however, will consist in writing the Schrödinger equation with the particular potential function and particle mass and then looking for suitable solution functions. This process will generally lead to only certain allowed energies. An application in which numerical values are introduced will illustrate this further.

Something of the quantum restrictions that are imposed on electrons that are confined to atoms or molecules can be seen by considering the simpler problem of an electron confined to a one-dimensional square-well potential with, say, a width of 3 Å, or 3×10^{-10} m. The necessary numerical values for this problem are

$$h = 0.6626 \times 10^{-33} \text{ J}$$
$$m = 0.9109 \times 10^{-30} \text{ kg}$$

and

$$a = 3 \times 10^{-10} \text{ m}$$

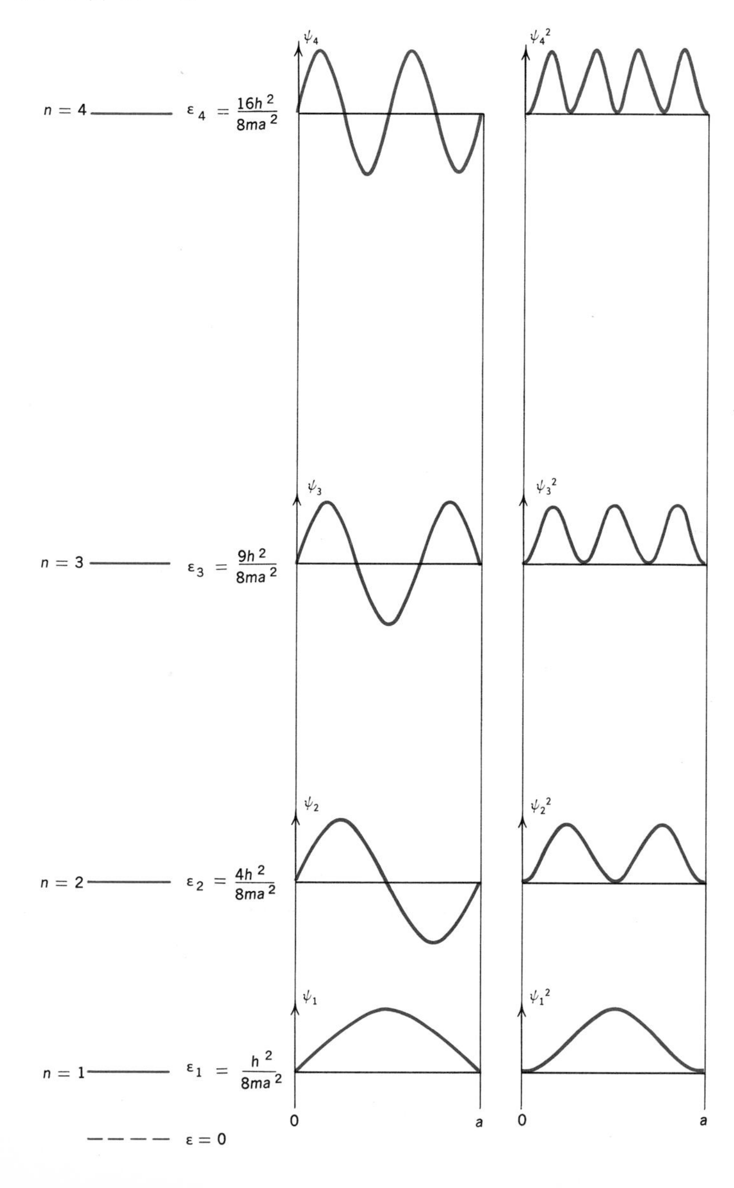

FIGURE 3-8
The energy levels, the wave functions ψ, and the probability distribution functions ψ^2 for the first few Schrödinger equation solutions for the potential of Fig. 3-7.

These values give the result

$$\epsilon = \frac{n^2h^2}{8ma^2} = n^2(6.7 \times 10^{-19}) \quad \text{J} \tag{18}$$

The energy-level pattern will be like that of Fig. 3-8, and as the next chapter will show, the electron will normally occupy the lowest allowed energy level.

The emission of a quantum of radiation as the electron of an excited atom or molecule goes from the first excited $n = 2$ level to the $n = 1$ level, or the absorption of a quantum for the $n = 1$ to $n = 2$ transition, corresponds to processes that might be observed spectroscopically. The energy involved in such a transition is calculated as

$$\Delta\epsilon = 6.7 \times 10^{-19}(2^2 - 1^2) = 2.0 \times 10^{-18} \text{ J} \tag{19}$$

The wavelength of radiation emitted or absorbed in such a transition is obtained with the help of Planck's equation $\Delta\epsilon = h\nu$ and the relation $\lambda = c/\nu$. The energy change would lead, therefore, to radiation with a wavelength of about 10^{-7} m, or 1000 Å. This corresponds to ultraviolet radiation, and it is a fact that atoms and simple molecules do generally absorb radiation in this region. In this regard, the simple square-well model is satisfactory. The finer details of the energy-level pattern for an atom or molecule, and even the qualitative behavior at higher energies, of course require a three-dimensional treatment and a better approximation to the potential energy.

Finally, it is informative to obtain the energies of the allowed levels for this electron confined to a region like that of an atom or molecule in the energy units used in the kinetic-molecular theory. The energy for an Avogadro's number of electrons, if the simple square-well model of this section is assumed, is given by

$$\begin{aligned} \Delta E &= n^2(6.7 \times 10^{-19})(6.02 \times 10^{23}) \\ &= n^2(400{,}000) \quad \text{J mol}^{-1} \\ &= n^2(400) \quad \text{kJ mol}^{-1} \end{aligned} \tag{20}$$

where n is the quantum number of the excitable electron. From this result it is apparent that the $n = 2$ level is of very high energy compared with the $n = 1$ level in terms of the available room-temperature thermal energy of $\frac{1}{2}RT$ per degree of freedom. A general and very useful conclusion that is indicated by the present calculation is that *the allowed energy levels for electrons in atoms and molecules are usually very widely spaced compared with ordinary thermal energies.*

3-6 A Three-dimensional Illustration of the Schrödinger Equation: The Cubic Potential-well Problem

The Schrödinger equation can be applied with little added complexity to the three-dimensional problem of a particle in a cubic container, and this application will reveal some additional feature of the behavior of particles restricted to regions of atomic and molecular dimensions.

For any three-dimensional problem the potential energy will in general be a function of three coordinates, and for a cubic potential box the cartesian coordinates of Fig. 3-9 are convenient. Again, the potential energy can be taken as zero everywhere inside the box, but becomes infinite at the walls. The differential equation that must be solved is now the Schrödinger equation in three dimensions, which is

$$-\frac{h^2}{8\pi^2 m}\left(\frac{\partial^2\psi}{\partial x^2}+\frac{\partial^2\psi}{\partial y^2}+\frac{\partial^2\psi}{\partial z^2}\right)+U(x,y,z)\psi=\epsilon\psi \qquad [21]$$

where it must be expected that the solution function ψ will depend on the three coordinates x, y, and z. For such differential equations it is often profitable to see whether or not they can be separated into parts, each part involving only one of the three coordinates. For the cubic container of Fig. 3-9, the potential energy can, for example, be written $U(x) + U(y) + U(z)$, where each function is like the one-dimensional potential-energy well of Fig. 3-7. Now one can try the substitution

$$\psi(x,y,z) = \varphi(x)\varphi(y)\varphi(z) \qquad [22]$$

and see whether or not such a separated function simplifies Eq. [21]. Substitution of Eq. [22] into Eq. [21] gives

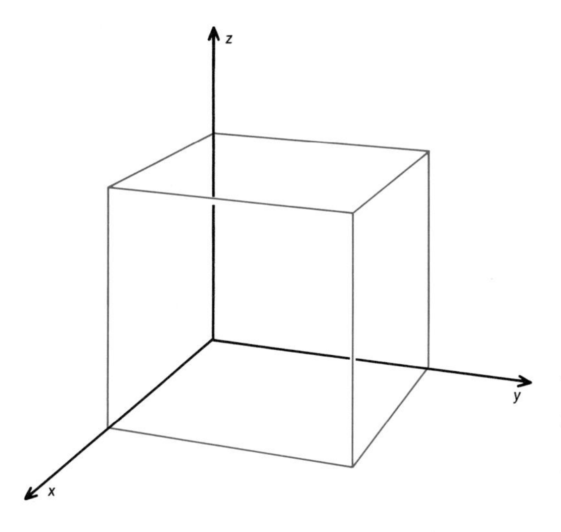

FIGURE 3-9
The three-dimensional square-well potential-energy function. The potential U is infinitely high except within the box, where it can be assigned the value zero.

$$-\frac{h^2}{8\pi^2 m}\left[\varphi(y)\varphi(z)\frac{d^2\varphi(x)}{dx^2} + \varphi(x)\varphi(z)\frac{d^2\varphi(y)}{dy^2} + \varphi(x)\varphi(y)\frac{d^2\varphi(z)}{dz^2}\right]$$
$$+ [U(x) + U(y) + U(z)]\varphi(x)\varphi(y)\varphi(z) = \epsilon[\varphi(x)\varphi(y)\varphi(z)] \quad [23]$$

Division by $\varphi(x)\varphi(y)\varphi(z)$ gives

$$-\frac{h^2}{8\pi^2 m}\left[\frac{1}{\varphi(x)}\frac{d^2\varphi(x)}{dx^2} + \frac{1}{\varphi(y)}\frac{d^2\varphi(y)}{dy^2} + \frac{1}{\varphi(z)}\frac{d^2\varphi(z)}{dz^2}\right]$$
$$+ [U(x) + U(y) + U(z)] = \epsilon \quad [24]$$

For the equation to be satisfied for all values of x, y, and z, each term on the left must equal a component of ϵ, and we can write

$$\epsilon = \epsilon_x + \epsilon_y + \epsilon_z \quad [25]$$

and the Schrödinger equation of Eq. [24] can be broken down to three identical equations of the type

$$-\frac{h^2}{8\pi^2 m}\frac{1}{\varphi(x)}\frac{d^2\varphi(x)}{dx^2} + U(x) = \epsilon_x$$

or

$$-\frac{h^2}{8\pi^2 m}\frac{d^2\varphi(x)}{dx^2} + U(x)\varphi(x) = \epsilon_x\varphi(x) \quad [26]$$

These equations are identical with that written for the one-dimensional problem. The solution to the three-dimensional square-box problem is therefore

$$\psi = \varphi(x)\varphi(y)\varphi(z)$$

with

$$\begin{aligned} \varphi(x) &= A \sin\frac{n_x\pi x}{a} & \epsilon_x &= \frac{n_x^2h^2}{8ma^2} & n_x &= 1, 2, 3, \ldots \\ \varphi(y) &= A \sin\frac{n_y\pi y}{a} & \epsilon_y &= \frac{n_y^2h^2}{8ma^2} & n_y &= 1, 2, 3, \ldots \\ \varphi(z) &= A \sin\frac{n_z\pi z}{a} & \epsilon_z &= \frac{n_z^2h^2}{8ma^2} & n_z &= 1, 2, 3, \ldots \end{aligned} \quad [27]$$

One might try to visualize the probability distribution ψ^2 which is obtained from this wave function, but a graphical representation will be seen to be rather difficult. One can proceed by repeating the curves of Fig. 3-8 to show the separate factors of the wave-function components $\varphi(x)$, $\varphi(y)$, and $\varphi(z)$, and the squares of these, and leaving it to the viewer to imagine a three-dimensional graphical display of the product. (To depict the combined result, one must show the values of ψ or of ψ^2 throughout the region of the cubic box, and this, at best, leads to qualitative displays such as that of Fig. 3-10.)

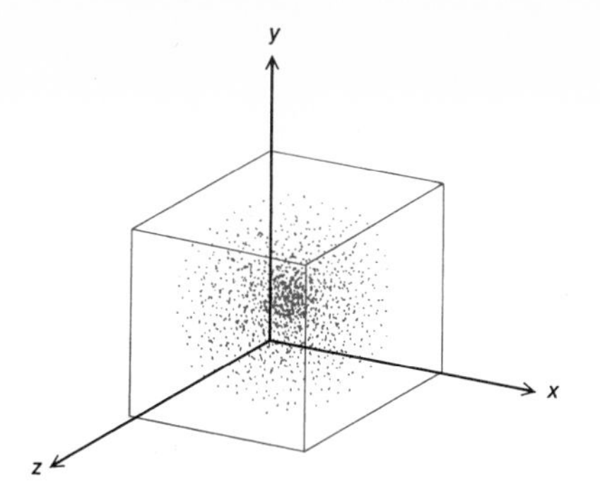

FIGURE 3-10
The function ψ^2 for the lowest energy state, $n_x = n_y = n_z = 1$, for a particle in a cubic box. The density of the dots is intended to be proportional to the value of the function.

The allowed energy levels for a particle in this three-dimensional cubic potential-energy box are

$$\epsilon = \epsilon_x + \epsilon_y + \epsilon_z = (n_x^2 + n_y^2 + n_z^2)\frac{h^2}{8ma^2} \qquad [28]$$

The pattern of energies of the allowed states differs from that for the one-dimensional case, Fig. 3-8, in two ways. First, as Fig. 3-11 shows, the pattern is less regular, a result that follows from the reduced regularity of the sum of the squares of the three integers compared with the squares of integers themselves. Second, some of the energies correspond to more than one state. Thus, although the energy $3(h^2/8ma^2)$ corresponds to only the state specified by $n_x = 1$, $n_y = 1$, $n_z = 1$, the energy $6(h^2/8ma^2)$ corresponds to the three states $n_x = 2$, $n_y = 1$, $n_z = 1$; $n_x = 1$, $n_y = 2$, $n_z = 1$; and $n_x = 1$, $n_y = 1$, $n_z = 2$. Such states, all of which have the same energy, are said to be *degenerate*.

This is as far as we will pursue here this wave-mechanical method for deducing properties of the states that are allowed to confined particles. Applications to the properties of electrons in atoms and molecules will be pursued in Chap. 11, when the nature of the chemical bonding is studied. Quite different uses will also be made in Chap. 4, where the properties of the states allowed to the molecules of a gas, not just the electrons of the molecules, but the molecules themselves, will be sought.

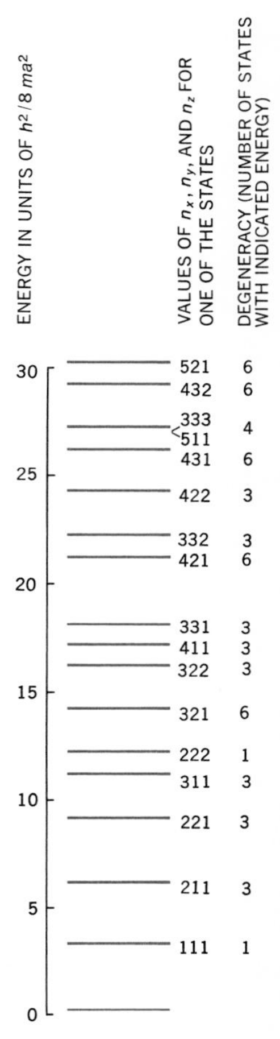

FIGURE 3-11
The energy-level pattern formed by the lower energy states for a particle in a cubic box, according to Eq. [28].

3-7 The Use of Angular Momentum to Impose Quantum Restrictions

When a particle is confined to a region of space, the wave-mechanics equation of Schrödinger, or in simple cases, the pictorial fitting in of de Broglie waves, allows the possible behaviors of the particle to be deduced. In the examples we have dealt with, the aspects of behavior of particular interest have been the particle energy and position. Had we applied this approach, as we will in Sec. 11-1 to a hydrogen atom, or to any particle or collection of particles that is free to rotate, we might also have considered the allowed speeds of rotation, or better, the allowed angular momenta. These could be deduced from the Schrödinger wave functions, but as Bohr anticipated, for systems that can freely rotate, postulates regarding the allowed angular momenta can be used as an alternative approach to the possible properties of the system.

Before seeing the implications of the restrictions that are placed on angular momentum, some of the quantities, and relations, that are used for rotating systems will be introduced. We can do this in terms of a particle rotating about a fixed point, as shown in Fig. 3-12. The speed of rotation can be described in terms of the particle's velocity v, or the number of revolutions per second, i.e., the frequency ν. Since the particle travels a distance v m s^{-1} and the distance for one revolution is $2\pi r$ m, we have the relation

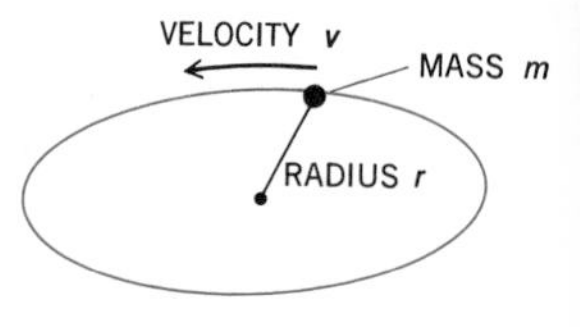

FIGURE 3-12
A particle revolving about a fixed point.

$$\nu = \frac{v}{2\pi r} \qquad [29]$$

Furthermore, angular velocity is also expressed in terms of radians per second, denoted by ω. Since there are 2π rad in a revolution,

$$\omega = \frac{v}{r} \qquad [30]$$

The kinetic energy of a particle, $\frac{1}{2}mv^2$, can be rearranged to a form more convenient for rotary systems by rewriting this as

$$\epsilon = \frac{1}{2}mr^2\left(\frac{v}{r}\right)^2 \qquad [31]$$

Then, if I, the *moment of inertia,* is introduced for the often-occurring collection of terms mr^2, and if ω is inserted for v/r, we have

$$\epsilon = \frac{1}{2}I\omega^2 \qquad [32]$$

You can notice here that in expressions for rotary systems, as compared with those for linear systems, I plays the role of m, and ω plays the role of v.

Finally, we return to the angular momentum, which is defined as $I\omega$. One also can see that

$$I\omega = mr^2\frac{v}{r} = mvr \qquad [33]$$

The expressions in terms of I and ω can, with suitable interpretations of I, be carried over to any rotating system, and quantum restrictions expressed in terms of these quantities are widely applicable.

The basic postulates that give quantum restrictions in line with observed behavior are:

1 The total angular momentum $I\omega$ can have only the values

$$I\omega = \sqrt{l(l+1)}\,\frac{h}{2} \qquad l = 0, 1, 2, \ldots \tag{34}$$

2 If a direction is imposed on the rotating system, as by the presence of a neighboring particle or application of an electric or magnetic field, the angular-momentum component in the direction of the field can have the values

$$m\frac{h}{2\pi} \qquad m = -l, -l+1, \ldots, 0, 1, \ldots, l \tag{35}$$

Thus, in this approach, each allowed state of the system is initially characterized by its total angular momentum and the value this total would project along an imposed direction. It is customary to depict these quantities by vectors and to draw diagrams like those of Fig. 3-13 to show for each value of l the possible states, each requiring an additional designation by means of a value of m, that can occur.

Thus, instead of describing states by wave functions, we do so by angular-momentum vectors.

These angular-momentum designations can, for simple rigid rotating systems such as linear molecules (but not for flexible rotating systems such as a hydrogen atom), lead simply and directly to values

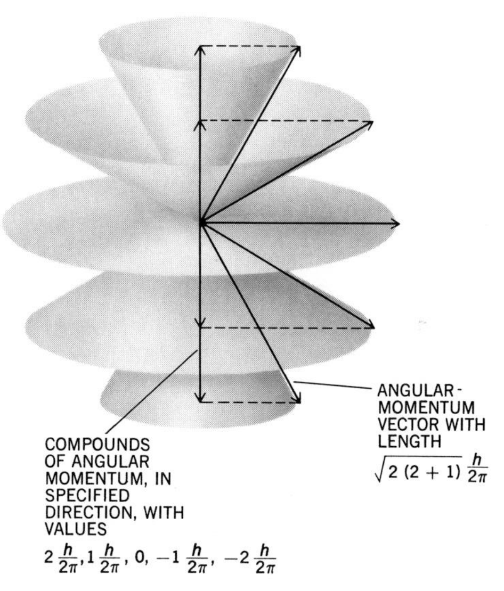

FIGURE 3-13
A vector representation of the five states that correspond to an angular-momentum quantum-number value of 2.

for the rotational energy of each state. One need only write

$$\text{Angular momentum} = I\omega = \sqrt{l(l+1)}\frac{h}{2} \qquad [36]$$

$$\begin{aligned}\text{Rotation energy} &= \tfrac{1}{2}I\omega^2 \\ &= \frac{(I\omega)^2}{2I} = l(l+1)\,\frac{h^2}{8\pi^2 I}\end{aligned} \qquad [37]$$

Thus a state designated by a value of l would have an energy $l(l+1)h^2/8\pi^2I$, which, if the moment of inertia were known, could be calculated.

In the following chapter this approach, as well as the wave-equation method, will be used to give us the information we need to describe the energies of molecules when motions, in addition to their overall translation, are considered.

Problems

1 Review the argument that water has the formula H_2O, and not HO, based on the experimental result that 2 volumes of water vapor would on decomposition yield 2 volumes of H_2 and 1 volume of O_2.

2 What are the wavelength and the frequency of radiation whose quanta have energies of 3×10^{-20} J? *Ans.* $\lambda = 66{,}200$ Å; $\nu = 4.53 \times 10^{13}\ \text{s}^{-1}$.

3 What is the energy in joules per Avogadro's number of quanta for the quanta of radiation in the ranges of electromagnetic radiation shown in Table 3-1? Compare the values with the room-temperature value of the average translational kinetic energy of a molecule.

4 The Lyman series of spectral lines from atomic hydrogen have frequencies given by the expression

$$\nu = 3.288 \times 10^{15}\left(\frac{1}{1^2} - \frac{1}{n_1{}^2}\right) \quad \text{Hz} \qquad n_1 = 2, 3, 4, \ldots$$

For the line with $n_1 = 2$:

a Calculate the frequency in hertz and the wavelength in angstroms of the radiation. *Ans.* $\nu = 2.47 \times 10^{15}$ Hz; $\lambda = 1210$ Å.

b Calculate the energy of the quanta of this radiation in joules per quantum and per Avogadro's number of quanta. *Ans.* 1.634×10^{-18} J per quantum, 982,000 J mol^{-1}.

c In what region of electromagnetic radiation does this spectral line fall?

5 Compare the energy necessary to excite the electron of the hydrogen atom from the $n = 1$ to the $n = 2$ state with the average translational energy of a molecule at 25°C. At what temperature is the average molecular translational energy equal to this hydrogen-atom excitation energy?

6 Calculate the energy and the wavelength of the radiation emitted when the electron of a hydrogen atom drops from the $n = 6$ orbit to the $n = 5$ orbit. *Ans.* $\Delta\epsilon = 2.66 \times 10^{-20}$ J; $\lambda = 74{,}600$ Å.

7 Plot, but on a scale that is linear in frequency, a spectrogram like that of Fig. 3-4 for the first four transitions of the Brackett series. Add an energy and a wavelength scale to the plot.

8 Calculate the wavelength of the de Broglie waves associated with:
a An electron accelerated by a voltage of 10,000 V. The velocity acquired by the electron is then about 6×10^7 m s^{-1}.
b A molecule of nitrogen moving with an average speed for a temperature of 25°C.
c A car weighing 1 ton traveling with a speed of 60 mph.

9 Use the de Broglie wavelength associated with a particle, and the idea that standing waves must form when a particle is confined to a region of space, to deduce the energies allowed for a particle confined to a square potential well like that of Fig. 3-7. Compare the results with those obtained by application of the Schrödinger equation.

10 Calculate the energies of the three lowest quantum states for an electron confined to an infinitely walled square potential well 10 Å wide. Show graphically the shape of ψ and ψ^2 for these three states.
Ans. 6.02×10^{-20}, 24.1×10^{-20}, 54.2×10^{-20} J.

11 Calculate an expression for the allowed energy levels of a baseball, of mass 5 oz, confined to a baseball park of length 350 ft. To what velocity jumps do these correspond?

12 The probability of finding a particle at a given position is given by ψ^2. If the wave function for a particle in a square well is $\psi = A \sin (n\pi x/a)$, find, by a suitable integration, what the total probability is that the particle will be found between $x = 0$ and $x = a$. Replace A by a suitable term so that $\int_{x=0}^{x=a} \psi^2 \, dx$ is always unity, i.e., so that one particle will be found somewhere in the well.

13 The average one-dimensional translational energy of an N_2 molecule was shown by the kinetic-molecular theory to have the value $\frac{1}{2}kT$. What would be the quantum number of an average N_2 molecule at 25°C in a container 0.1 m long? What would be the energy separation between successive allowed quantum states at this energy?
Ans. $n = 4.17 \times 10^9$; $\Delta\epsilon = 9.83 \times 10^{-31}$ J.

14 Draw an energy-level diagram to show the pattern formed by the four lowest energies of a rigid rotating linear molecule. Include an energy scale in terms of $h^2/8\pi^2 I$. Show the number of states corresponding to each energy level.

15 In Sec. 3-5, the energies of the allowed states of an electron confined to a 3-Å region were obtained, and the spacings of these energies were said to be indicative of that for electrons in atoms. Do a corresponding suggestive calculation by obtaining an expression that shows the energies that would be allowed to an electron if it were restricted to move in a circle with a circumference of 3 Å. Plot these results, and those from Eq. [18], for the first few energies, alongside one another.

References

SHERWIN, C. W.: "Basic Concepts of Physics," Holt, Rinehart and Winston, Inc., New York, 1961. Chapter 6 deals with some of the key experiments and ideas in the development of quantum mechanics.

HERZBERG, G.: "Atomic Spectra and Atomic Structure," Dover Publications, Inc., New York, 1944. A classic introduction to atomic spectra that provides the background in quantum mechanics necessary for an understanding of the principal features of atomic-spectral analysis.

RESNICK, R., and D. HALLIDAY: "Physics for Students of Science and Engineering," chaps. 47 and 48, John Wiley & Sons, Inc., New York, 1960.

SEMAT, H.: "Introduction to Atomic and Nuclear Physics," 4th ed., Chapman & Hall, Ltd., London. Includes a detailed treatment of the pre-Schrödinger equation material that is touched on in this chapter.

HEITLER, W.: "Elementary Wave Mechanics," Oxford University Press, Fair Lawn, N.J., 1945. A short, quite readable presentation that deals with some of the material of this chapter as well as that of Chap. 10.

SLATER, J. C.: "Quantum Theory of Atomic Structure," chaps. 1 and 2, McGraw-Hill Book Company, New York, 1960, and L. PAULING and E. B. WILSON, JR.: "Introduction to Quantum Mechanics," McGraw-Hill Book Company, New York, 1935. Introductory material in these two books deals with some of the topics of this chapter as a preliminary to further work on quantum mechanics.

HOFFMANN, B.: "The Strange Story of the Quantum," 2d ed., Dover Publications, Inc., New York, 1959. One of the best—accurate and amusing—of the popular stories of the development of quantum-mechanical concepts.

4

AN INTRODUCTION TO MOLECULAR ENERGIES

The kinetic-molecular model that was used in Chap. 2 as a basis for understanding the ideal PVT behavior of gases treated molecules as point particles characterized only by a mass and a velocity. The model was elaborated to give these particles size and mutual interaction to lead us to some appreciation of nonideal behavior as described, for example, by van der Waals' equation. Now we have enough understanding of the basis and nature of the results of quantum mechanics so that we can make use of the model of molecules in which molecules are recognized to be assemblies of atoms. We begin by describing the energies of such molecules. Our principal focus will be on the energies of molecules of gases, but much of what will be learned will also carry over to the liquid and solid states. In several chapters following this one, the relationship of the energies of individual molecules to the energy and some of the chemical properties of macroscopic samples of materials will be developed.

4-1 Categories of Thermal Energies of Molecules of Gases

An atom, as you know, consists of one or more electrons and a nucleus, and a molecule consists of some number of atoms held together in a somewhat flexible structural arrangement. The energy of the collection of the many electrons and the nuclei that constitute molecules could be treated in terms of the energies of the separate particles and their energies of interaction. But under many circumstances the electronic state of the molecule remains that of the lowest energy, or *ground,* state. Under these circumstances—which are those for most materials at not too high temperatures—we need not deal in detail with the electronic energy of the molecule, and we can treat molecules as if they were structures like that shown in Fig. 4-1.

Think of a molecule as a collection of atoms held, more or less firmly, in a particular spatial arrangement by chemical bonds. The model corresponds to a set of balls joined together by springs. We will seek to describe the energy of such a system, known as the *thermal* energy, that results from the jostling of any one such molecule by its neighbors—a picture that we began to develop in the kinetic-molecular-theory study.

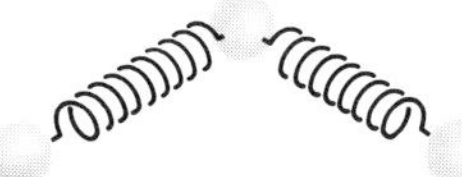

FIGURE 4-1
An example of the model of a molecule that is suitable for the treatment of the thermal energies of most molecules at not too high temperatures and under the condition where no chemical reactions are occurring.

(When we consider chemical reactions, in which molecules are torn apart and new ones are assembled, or temperatures so high that the excited electronic states are involved, we will have to consider also the energy associated with the electronic structure of the molecules. Such energy features will be considered in Chap. 7. Here the energies of molecules that are not exposed to conditions that disrupt their electronic structure will be dealt with.)

As would be immediately apparent if an actual ball-and-spring model were tossed around, the motions by which the system could have energy would be most conveniently described by terms such as *translational, rotational,* and *vibrational.* We need not repeat the detailed analysis of the dynamics of systems which leads to this separation of such types of motions and energies. It is enough to recognize that:

Translation consists of motion of the center of gravity of the molecule.

Rotation consists of a motion which provides a net angular momentum about the center of gravity.

Vibration involves the oscillation of the particles of the system, i.e., the atoms of the molecule, in such a way that there is no net contribution to the motion of the center of gravity nor to the angular momentum about this center.

Shortly, we will see how such motions are to be described for a system like a molecule, for which the quantum rules are significant.

But first we can ask, for an assembly of n particles, for example a molecule with n atoms, how many different motions are to be included in each of the above classes.

You have already seen in Chap. 2 that any translational motion of a particle can be treated in terms of components in three perpendicular directions, which are usually labeled as the x, y, and z directions. Thus, any velocity u can, as was shown in Fig. 2-1, be resolved into its three components, and it follows, as developed in Sec. 2-4, that the translational energy of the particle, $\frac{1}{2}mu^2$, can be interpreted according to the sum of contributions along the three orthogonal axes.

Thus we say that there are *three* translational *degrees of freedom* and that the translational energy of a particle is the sum of the translational energies along each of the coordinates corresponding to the three degrees of freedom.

In a similar way, as again would be obvious if an actual system were observed, the rotational motion of a general-shaped collection of particles, or general-shaped object, can be described in terms of rotational components about three perpendicular axes. Thus, for systems such as those of Fig. 4-1, there are *three* "molecular" rotational degrees of freedom.

For linear systems, which include all diatomic as well as a few polyatomic molecules, only the two axes, perpendicular to the molecular axis, as shown in Fig. 4-2, are to be included as a "molecular" rotation. Rotation about the third axis involves only motion of the electrons. The energy of rotation about this axis thus remains constant, unless the angular momentum of the electrons changes. This is indeed possible, but such changes, like other electronic excitations, involve very large energies and are to be treated separately from the remaining molecular motions now under study.

We must now see how many vibrational degrees of freedom a molecule with n atoms will have. If the n atoms were very loosely bound together to form the molecule, the total number of degrees of freedom of the molecule would be calculated as $3n$; that is, 3 translational degrees of freedom would be assigned to each atom. For actual molecules the bonds are sufficiently strong so that it is usually not convenient to think of the atoms moving independently. The coupling of the atoms does not, however, change the total number of degrees of freedom of the system. Only the way of counting these degrees of freedom is changed.

The number of vibrational degrees of freedom is now obtained by subtracting the number of translational and the number of rotational degrees from $3n$. It is concluded, therefore, that a molecule with n atoms will have $3n - 6$ (or $3n - 5$ if the molecule is linear) vibrational degrees of freedom.

Now the ways in which a molecule can acquire energy because of the thermal jostlings have been classified, and we must study, in turn, the details of the translational, rotational, and vibrational motions of molecules.

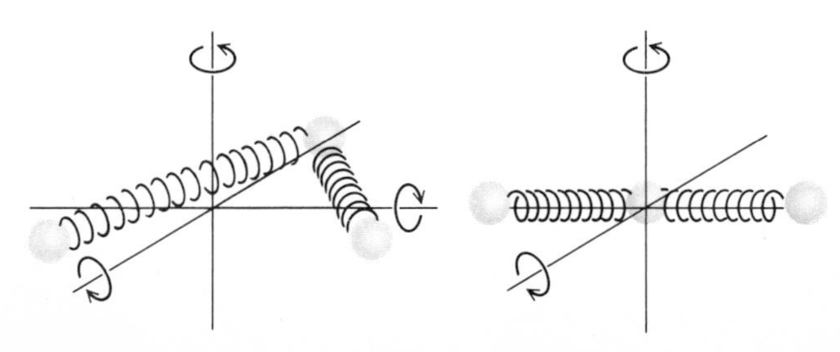

FIGURE 4-2
The 3 molecular rotational degrees of freedom of a generally shaped molecule compared with the 2 degrees of a linear molecule.

4-2 The Translational Energies of a Molecule of an Ideal Gas

Although the translational energies of gas molecules have been studied extensively in Chap. 2, that study was made without any recognition of the quantum restriction that can enter. These restrictions do not negate any of the results that were obtained, but the quantum-mechanical treatment of translational energies provides some additional insight.

The model of an ideal gas used in Chap. 2 pictures the molecules as moving freely within the confines of the impenetrable walls of the container. The corresponding quantum-mechanical problem treats a particle subject to a three-dimensional square-well potential with dimensions equal to that of the gas container. The basic problem has already been treated in Sec. 3-6, and now it is only necessary to apply it to the situation at hand.

Consider, to be specific, nitrogen gas, with N_2 molecules of mass $0.0280/6.02 \times 10^{23} = 4.65 \times 10^{-26}$ kg in a 1-liter container that we will take to be cubic with dimensions of 0.10 m. The energy-spacing factor $h^2/8ma^2$ is calculated to be

$$\frac{h^2}{8ma^2} = 1.2 \times 10^{-40} \text{ J}$$

and thus the energies of the allowed states are

$$\epsilon = (n_x^2 + n_y^2 + n_z^2)(1.2 \times 10^{-40}) \qquad \text{J per molecule} \qquad [1]$$

These allowed energies can also be expressed in units of joules per mole, and then we have

$$E = (n_x^2 + n_y^2 + n_z^2)(7.2 \times 10^{-17}) \qquad \text{J mol}^{-1} \qquad [2]$$

The allowed energies given by these equations can be compared with the average translational energy, at 25°C for example, of $\frac{3}{2}kT = 6.17 \times 10^{-21}$ J per molecule, or $\frac{3}{2}RT = 3700$ J mol^{-1}. The energy spacing is seen to be very small compared with this average energy.

Comparison of the energy spacing of the allowed states at an energy equal to the average thermal energy can also be made. First the value of $(n_x^2 + n_y^2 + n_z^2)$ to give a state with this energy is calculated from Eq. [1] as

$$6.17 \times 10^{-21} = (n_x^2 + n_y^2 + n_z^2)(1.2 \times 10^{-40})$$

or

$$(n_x^2 + n_y^2 + n_z^2) = 5.1 \times 10^{19} \qquad [3]$$

A variety of choices of n_x, n_y, and n_z would satisfy this equation. For the case of $n_x = n_y = n_z$ we have

$$3n_x^2 = 5.1 \times 10^{19}$$

and

$$n_x = 4.1 \times 10^{10} \qquad [4]$$

Now the energy gap between a state with n_x and one with $n_x + 1$ at this value of n_x can be calculated as

$$\begin{aligned}\Delta\epsilon &= \{[(n_x + 1)^2 + n_y^2 + n_z^2] - (n_x^2 + n_y^2 + n_z^2)\}\frac{h^2}{8ma^2} \\ &= (2n_x + 1)\,\frac{h^2}{8ma^2} \\ &\approx 2(4.1 \times 10^{10})(1.2 \times 10^{-40}) \\ &= 1 \times 10^{-29}\text{ J}\end{aligned} \qquad [5]$$

Again the energy spacing between the translational states is very small compared with the average translational energy, which is some 10^8 times larger. Such comparisons justify the assumption on which the kinetic-molecular treatment of Chap. 2 was based, namely, that *all* translational energies were possible for the molecules of a gas. (In studies of other properties of gases, in particular a property known as entropy, there is a consequence of the quantum restrictions on the translational energies even though no consequence is apparent when gas pressure and average translational energy are considered.)

4-3 The Allowed Rotational Energies of a Molecule of a Gas

Molecules have thermal energy not only by overall translational motion, but also by rotational motion. The rotation of a generally shaped molecule, as you saw in Sec. 4-1, occurs about three perpendicular axes, and the rotation of a linear molecule occurs about two such axes.

Now let us see what quantum restrictions enter to limit the rotational energies that can be adopted by a molecule in each of these rotational degrees of freedom. We will treat only the case of linear molecules which turns out to have a regular, easily described pattern of allowed rotational energies.

We could proceed from the Schrödinger equation and obtain wave functions and energies for the rotational states. Here we are primarily interested in the energies of the allowed rotational states, and as suggested in Sec. 3-7, these can be deduced most easily by working from the angular-momentum restrictions.

Each rotational state, i.e., each different way in which the molecule can rotate, is characterized by the angular momentum of the molecule when it is in that state and by the component of this angular momentum that would be projected along an imposed direction. The integers that index the total angular momentum of a rotating molecule are usually represented by J and the directional component by M or, to remind us of the total angular-momentum value, by M_J.

Thus each allowed rotational state of a molecule would produce an angular momentum along a direction of

$$M_J\frac{h}{2\pi} \qquad M_J = -J, \ldots, 0, \ldots, +J \qquad [6]$$

and would have a total angular momentum of

$$\sqrt{J(J+1)}\,\frac{h}{2\pi} \qquad J = 0, 1, 2, \ldots \qquad [7]$$

and a rotational energy (Sec. 3-7) of

$$\epsilon_{\text{rot}} = J(J+1)\,\frac{h^2}{8\pi^2 I} \qquad J = 0, 1, 2, \ldots \qquad [8]$$

where I is the moment of inertia of the linear molecule, a quantity to be dealt with in detail in Sec. 13-1.

Figure 4-3 shows the pattern of energy levels and the number of states, i.e., the degeneracy, at each level. This pattern applies to all linear molecules. Each molecule, however, because of the masses of its atoms and the distances between them, has a characteristic moment of inertia, and thus a characteristic energy-spacing factor $h^2/8\pi^2 I$.

The number of states g_J, an energy level specified by a value of J, is given, as can be seen from the values of Fig. 4-3, by the expression $g_J = 2J + 1$.

J	(NUMBER OF STATES)	ENERGY
4	(9)	$20(h^2/8\pi^2 I)$
3	(7)	$12(h^2/8\pi^2 I)$
2	(5)	$6(h^2/8\pi^2 I)$
1	(3)	$2(h^2/8\pi^2 I)$
0	(1)	0

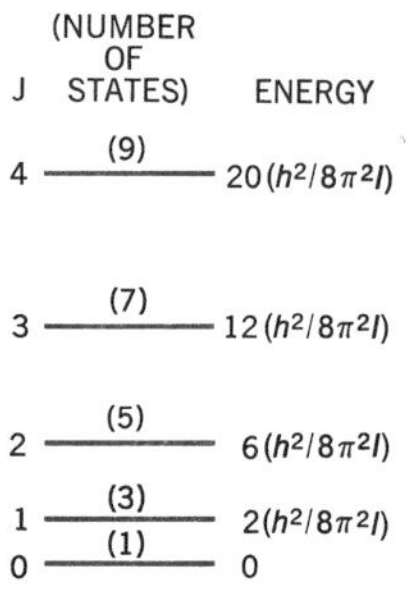

FIGURE 4-3
The first few rotational-energy levels allowed to a rotating linear molecule. Notice that the number of states g_J is related to J by the expression $g_J = 2J + 1$.

Molecular-spectra studies, to be described in Chap. 13, provide accurate values of the moments of inertia of molecules, and thus throw light on molecular dimensions. Here it will be enough to use a representative value to deduce the order of magnitude of energy separations between the allowed rotational states.

As you would see if you formed the product mr^2 with an m value of, say, $(0.060/\mathfrak{N})$ kg and an r value of several angstroms, values for the moment of inertia of relatively small molecules are of the order of 10^{-45} kg m^2. Thus a representative rotational-energy spacing factor is

$$\frac{h^2}{8\pi^2 I} = 5 \times 10^{-24} \text{ J per molecule}$$

$$= 3 \text{ J mol}^{-1} \qquad [9]$$

Again, comparison of these values with the room temperature value of kT or RT leads to the conclusion that rotational-energy levels are closely spaced and that for some purposes the classical (i.e., non-quantum-mechanical) assumption that all rotational energies are allowed is valid. For particularly small, light molecules with low moments of inertia, H_2 being the most noteworthy example, this generalization is, however, not valid.

Finally, since the rotational-energy spacings increase with rotational energy, as you can see in Fig. 4-3, this conclusion should also be tested by, for example, calculating the rotational-energy spacing at a rotational energy of kT. This test constitutes Prob. 7.

4-4 The Vibrational Energies of Gas-phase Molecules

The final major category of the thermal energies of molecules consists of the vibrational energy that can be described in terms of the oscillatory motion of the atoms of the molecule in relation to each other. The nature of this type of molecular motion is best introduced by considering first a classical ball-and-spring system.

The vibrational characteristics of such a ball, or particle, are determined by the mass of the particle and by the nature of the spring. For both ordinary-sized objects held by actual springs, and also for atoms held by chemical bonds, the simplest assumption, that the particle experiences a restoring force pulling or pushing it back to its equilibrium position that is proportional to the distance to which the particle has been displaced from its equilibrium position, turns out to be quite satisfactory. Such a force-displacement relation is known as *Hooke's law*. Since displacing the particle in one direction brings about a force in the opposite direction, Hooke's law is written

$$f = -kx \qquad [10]$$

where f is the restoring force, and x is the displacement from the equilibrium position. The proportionality constant k is known as the *force constant* and is a measure of the stiffness of the spring. The force constant is equal to the restoring force operating for a unit displacement from the equilibrium position. For molecular systems the force constant has usually been given in the units of dynes per centimeter, and these are related to the SI units of newtons per meter by 1 dyne $\text{cm}^{-1} = 10^{-3}$ N m^{-1}.

The classical motion of a particle, such as that of Fig. 4-4, can be deduced from Newton's law $f = ma$. If $f = -kx$ and $a = d^2x/dt^2$ are substituted, one obtains

$$m\frac{d^2x}{dt^2} = -kx \qquad \text{or} \qquad \frac{m}{k}\frac{d^2x}{dt^2} = -x \qquad [11]$$

A solution to this equation can be seen by inspection and verified by substitution to be

$$x = A \sin\sqrt{\frac{k}{m}}\,t \qquad [12]$$

A is a constant that is equal to the maximum value of x; that is, it is the vibrational amplitude. The position of the particle therefore varies sinusoidally with time, since every time t increases by $2\pi\sqrt{m/k}$, the quantity $\sqrt{k/m}\,t$ increases by 2π and the particle traces out one complete cycle. The time corresponding to one oscillation, or vibration, is therefore $2\pi\sqrt{m/k}$. More directly useful is the reciprocal of this quantity, which is the frequency of vibration, i.e., the number of cycles performed per second. If this quantity is denoted by ν_{vib}, we have

$$\nu_{vib} = \frac{1}{2\pi}\sqrt{\frac{k}{m}} \qquad [13]$$

For a system of ordinary dimension, there is therefore a natural frequency of oscillation that depends on the values of k and m. Any amount of energy can be imparted to the vibrating system, and this energy changes only the amplitude of the vibration.

The quantum-mechanical solution to this problem provided by the Schrödinger equation differs, of course, in that only certain amounts of vibrational energy are allowed. These can be deduced by entering the vibrational potential function U into the Schrödinger equation and solving for the wave functions and energies of the allowed vibrational states. Since $f = -dU/dx$ and $f = -kx$, we obtain, immediately upon integrating, the potential function as

$$U = \tfrac{1}{2}kx^2$$

The potential energy therefore rises parabolically on either side of the equilibrium position, as illustrated in Fig. 4-4.

Once a potential function for the motion to be studied has been arrived at, it is possible to substitute this function in the Schrödinger equation and to solve for the allowed energy-level pattern. The procedure is thus analogous to that used for the square well in Sec. 3-6. The parabolic potential function, however, makes determination of the solution wave functions rather more difficult, and these are shown, without derivation, in Fig. 4-5. Their similarity to the square-well functions is readily apparent.

The energies of the allowed vibrational states would also be given

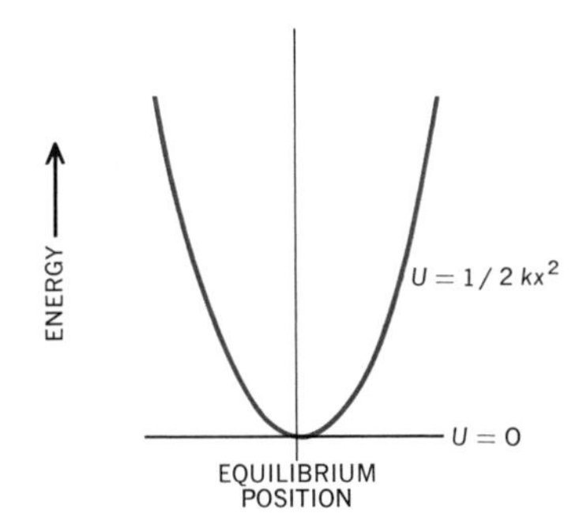

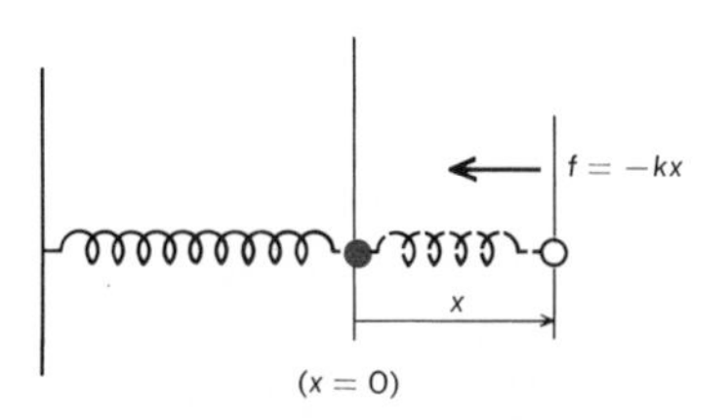

FIGURE 4-4
Hooke's law of force for a single particle.

by solution of the Schrödinger equation. Again, only the result is given, and with the symbol v to represent the integers that enter, the energies of the allowed states are

$$\epsilon_{\text{vib}} = (v + \tfrac{1}{2})\frac{h}{2\pi}\sqrt{\frac{k}{m}} \qquad v = 0, 1, 2, \ldots \tag{14}$$

This quantum-mechanical result therefore indicates a pattern of energy levels with a constant spacing $(h/2\pi)\sqrt{k/m}$, shown in Fig. 4-5.

It is interesting that the quantum-mechanical solution introduces the collection of terms $(1/2\pi)\sqrt{k/m}$ that correspond to the natural vibrational frequency of a classical oscillator. Equation [14] can therefore be written

$$\epsilon_{\text{vib}} = (v + \tfrac{1}{2})h\nu_{\text{vib}} \tag{15}$$

where the term ν_{vib} is interpreted according to Eq. [13].

The values of the vibrational-energy-spacing factor $h\nu_{\text{vib}}$ or $(h/2\pi)\sqrt{k/m}$ can be deduced from spectroscopic studies, as will be shown in Sec. 13-2. Again, here we need only see the importance of the quantum restrictions on the energy.

The vibrational-energy-spacing factors, which depend on the stiffness of the chemical bonds and on the masses of the vibrating atoms, are again characteristic of each molecule. A typical value, however, is 2×10^{-20} J per molecule, or about 12,000 J mol^{-1}. Here then, the quantum restrictions are very significant, and we must think of most molecules as having the least possible vibrational energy, or the next higher, and so forth. We cannot use the classical view that any vibrational energy is allowed.

Thus, although the quantum restrictions on translational motion

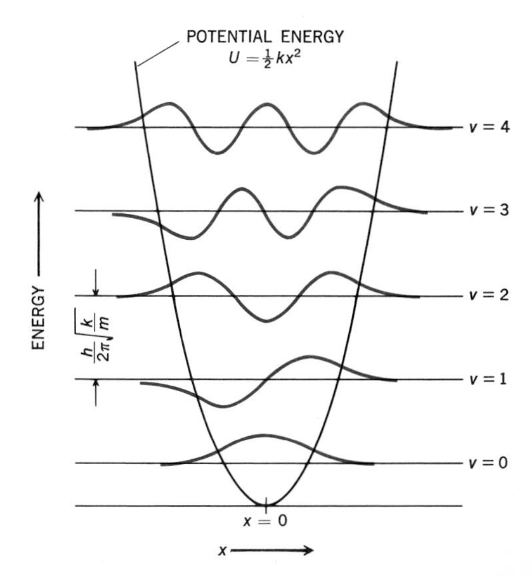

FIGURE 4-5
The potential-energy levels (black) and the wave functions (color) for the first few states of a vibrating particle. (Compare with the energies and states for a particle in a square potential well shown in Fig. 3-8.)

can be ignored in considerations of the energy of a collection of molecules, as they can for the rotational motion of most molecules, those on vibrational energies cannot.

4-5 Summary and a Comment on Electronic States of Molecules

The various molecular degrees of freedom—translation, rotation, and vibration—have been treated separately. Any gas-phase molecule will of course have energy of each type; i.e., as it moves about with some amount of translational energy, it will also be twirling in a rotational manner and the component atoms will simultaneously be vibrating against one another. The entire energy possibilities can be suggested by a schematic energy-level diagram like that of Fig. 4-6. Thus the thermal energy of a molecule is measured by the position it occupies

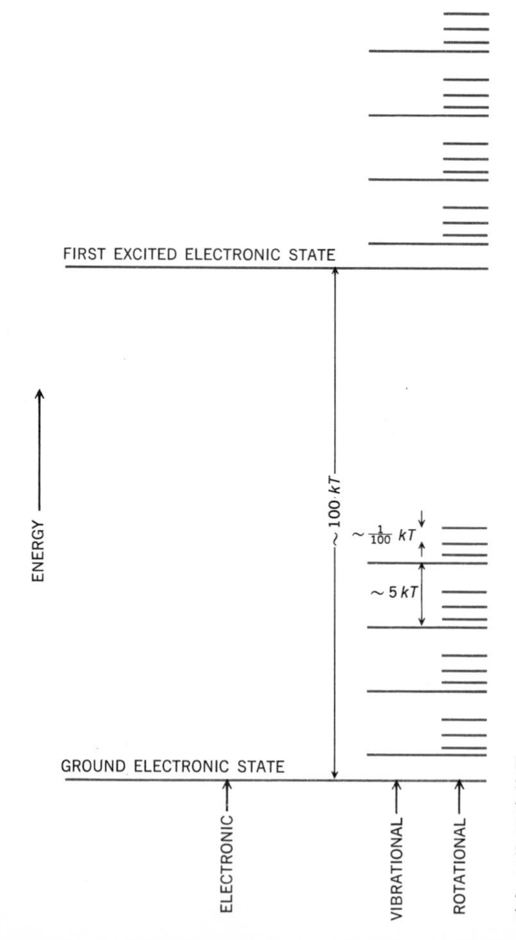

FIGURE 4-6
Schematic representation of typical molecular-energy levels. Translational levels, spaced by the order of $10^{-14}kT$, are not shown. A value of kT at room temperature is assumed.

in such a diagram measured from the lowest possible zero-point energy level.

It should be remembered that there are additional electronic states available to molecules that, at higher temperatures, must be considered if the thermal energy is to be deduced. These states correspond to different, higher-energy electronic arrangements than occur in the "ground state." For a few molecules these electronic states have energies not much in excess of the reference energy, the room temperature value of kT, but generally the energy spacing between different electronic states of molecules is relatively large. Furthermore, these states, in contrast to those for the hydrogen atom and those for other molecular degrees of freedom, have energies that present no regular, easily described pattern. They must be dealt with, as a result, molecule by molecule, and no generally applicable expressions such as those for the translational, rotational, and vibrational degrees of freedom are available.

Problems

1 How many translational, rotational, and vibrational degrees of freedom are there for the molecules He, N_2, CO_2 (linear), H_2O (bent), CH_4 (tetrahedral), and benzene, C_6H_6, which has a planar hexagonal shape?

2 Attempt to represent graphically and qualitatively the probability of a particle being at various positions in a cubic potential box for the quantum states $n_x = n_y = n_z = 1$ and $n_x = 2$, $n_y = n_z = 1$. What are the energies of these states if the box is cubic with dimension 5 Å and the particle is an electron?

3 Deduce an expression for the energies of the translational states of a molecule in a rectangular box with sides of length a, a, and b.

Draw energy-level diagrams at the left and at the right sides of a page to show the first dozen or so energy levels for a cubic box of dimension a and a rectangular box with dimension a, a, and $b = 2a$.

Add to the diagram "correlation lines" showing the qualitative way in which the energies of the states of a particle in a cubic box would change if some of the degeneracy were removed by elongating the box.

4 Calculate the energy-spacing factor $h^2/8ma^2$ and the translational-energy spacing at a translational energy of $\frac{3}{2}kT$ for an H_2 molecule in a cubic 0.10-m box at 25°C.

5 A general guide is that quantum restrictions become important when particles are confined to regions in space that are of the order of angstrom dimensions. Calculate the energy-spacing factors for an electron, a hydrogen molecule, and a nitrogen molecule confined to a 10-Å cubic box.

6 If radiation were absorbed or emitted by molecules when they changed from a state with one translational energy to a state with the next-higher or next-lower translational energy, what region of the electromagnetic spectrum would be involved? (In fact, translational energies are not directly affected by radiation.)

7 Calculate the energy gap between adjacent rotational energies at a rotational energy of about $\frac{1}{2}kT$ for a molecule with a moment of inertia of 10^{-45} kg m^2.

5

THE ENERGIES OF COLLECTIONS OF MOLECULES: THE MOLECULAR APPROACH

The introductory material of the preceding chapter has shown that molecules exist in a world of quantum restrictions. This implies that a molecule, or an atom, or an electron, must exist in one of the allowed states and must have the energy corresponding to that state. Collections of large numbers of molecules, such as one deals with in ordinary-sized, or *macroscopic,* systems, must therefore consist of these molecules distributed throughout the allowed states. Many of the properties of chemical materials can be deduced if the energies of the quantum states and the distribution throughout these states are known. Although in practice the deduction of properties of macroscopic samples from such detailed calculations is often impossibly difficult, the way in which this is done for such samples as ideal gases is revealing and allows one to understand the molecular basis of many properties. Here, therefore, we shall investigate the way in which some properties related to the energy of a gas sample are deduced.

5-1 The Boltzmann Distribution

The distribution expression that will be used to deduce the relative population of the energies available to the molecules of a gas has already been introduced in the study of Sec. 2-5 of the distribution of speeds and translational energies of gas molecules. Then energies were treated classically.

The recognition of quantized energies, as in Fig. 5-1, leads us to write in place, for example, of Eq. [24] of Chap. 2, the expression

$$\frac{N_i}{N} = (\text{const})\ e^{-(\epsilon_i - \epsilon_0)/kT}\, g_i$$

or

$$\frac{N_i}{N} = (\text{const})\ g_i\ e^{-(\epsilon_i - \epsilon_0)/kT} \qquad [1]$$

where N_i, the counterpart of dN, is the number of molecules with energy ϵ_i, the term $\epsilon_i - \epsilon_0$ is the energy difference between the energy ϵ_i and the lowest allowed energy ϵ_0, and g_i, the counterpart of du_x, is the number of states at the energy ϵ_i.

It is often more convenient to deal at first with the ratio of the populations of states at different energies, for this avoids the constant term of Eq. [1]. First recognize that if N_i is the number of molecules with energy ϵ_i, and g_i is the number of states at this energy, then N_i/g_i is the number of molecules *per state* at this energy. Then the ratio of the number of molecules per state at energy ϵ_i to the number per state at some other energy ϵ_j can be written from Eq. [1] as

$$\frac{N_i/g_i}{N_j/g_j} = e^{-(\epsilon_i - \epsilon_j)/kT} \qquad [2]$$

Often it is convenient to refer the populations of states to that of the lowest energy state. If this state is denoted by $j = 0$, we can write

$$\frac{N_i/g_i}{N_0/g_0} = e^{-(\epsilon_i - \epsilon_0)/kT}$$

$$= e^{-(\Delta\epsilon_i)/kT} \qquad [3]$$

where $\Delta\epsilon_i = \epsilon_i - \epsilon_0$.

Now you can notice that if $\Delta\epsilon_i$ is large compared with kT, the ratio $(N_i/g_i)/(N_0/g_0)$ will be very small. Such is the case, generally, for all states belonging to electronically excited states like that shown in Fig. 4-6. Thus the energies of the molecules of a gas can often be adequately dealt with in terms of the translational, rotational, and vibrational energies of the ground electronic state.

In the following two sections the origin of the Boltzmann distribution expression will be investigated. Then the expression will be applied to the various molecular motions of gas-phase molecules. You can proceed to those applications without working through Secs. 5-2 and 5-3 if you accept Eq. [23] as the form that Eq. [1] takes on when the constant term is evaluated so that the summation of N_i over all energy levels is N, the total of molecules in the sample being dealt with.

*5-2 Derivation of the Boltzmann Distribution Expression

A *distribution* is a statement of the number of molecules that occupy each of the energy levels that the quantum restrictions allow.

For the general energy-level pattern, including multiplicities, as shown in Fig. 5-1, we must now see if we can deduce (without depending on the Boltzmann distribution expression that was revealed in advance in the preceding section, or even on the hint given there that the temperature is involved) how an Avogadro's number of molecules distribute themselves throughout this pattern of energy levels. We will proceed by looking for the most probable distribution, and we begin by seeing how the probability of a distribution is expressed.

A problem which in this regard is identical with that of the distribution of molecules throughout the pattern of Fig. 5-1 is that in which one asks about the probabilities of various distributions of marbles thrown randomly into a box with various-sized compartments. Figure 5-2 suggests a specific marble-compartment analog to the molecule-energy-level problem. If one could construct such a compartmentalized box and have a suitably random throwing device, one could verify that the expression for the probabilities of the different arrangements (which is worked out in introductory chapters of texts dealing with probability) contains two factors.

The first factor involves the relative sizes of the compartments. The probability of each marble landing in a given compartment is proportional to the size of the compartment. This consideration implies that in the expression for the probability of a particular distribution of the four marbles of Fig. 5-2, there is a term that is the product of the sizes of the compartments occupied by each marble, or more conveniently, the product of the sizes of the compartments each raised to the power corresponding to the number of marbles occupying that compartment in that particular distribution. (The corresponding factor for the molecules distributed throughout energy levels will be written $g_1^{N_1}$, $g_2^{N_2}$, $g_3^{N_3}$, ...) This first factor, it will be noticed, implies that the most probable distribution will be the one in which all the marbles are in the largest compartment—and this is clearly not the case.

The second factor expresses the tendency of the marbles to distribute themselves. It can be formed by seeing the total number of ways the individual marbles can be rearranged without altering the total numbers in

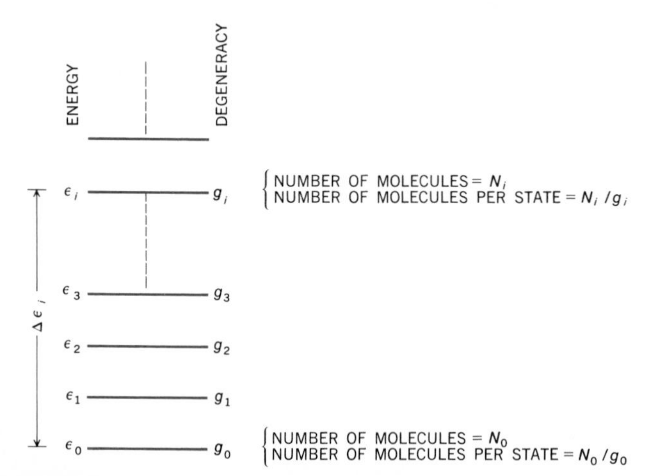

FIGURE 5-1
A generalized set of energies available to the molecules of a sample as a result of a particular type of motion.

DISTRIBUTION	PROBABILITY FACTORS		NET PROBABILITY
	First	Second	
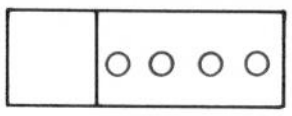 [∣ O O O O]	$(\frac{3}{4})^4$	$\frac{4!}{4!}$	0.32
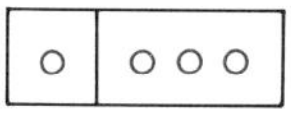 [O ∣ O O O]	$(\frac{1}{4})(\frac{3}{4})^3$	$\frac{4!}{1!3!}$	0.42
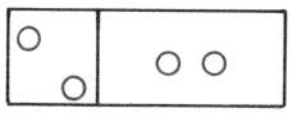 [O O ∣ O O]	$(\frac{1}{4})^2(\frac{3}{4})^2$	$\frac{4!}{2!2!}$	0.21
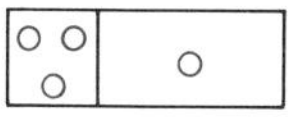 [O O O ∣ O]	$(\frac{1}{4})^3(\frac{3}{4})$	$\frac{4!}{3!1!}$	0.05
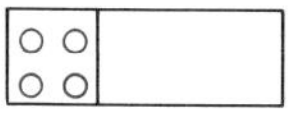 [O O O O ∣]	$(\frac{1}{4})^4$	$\frac{4!}{4!}$	0.004

MOLECULES IN ENERGY LEVELS WITH MULTIPLICITIES $g_1, g_2, \ldots$

FIRST FACTOR $= g_1^{N_1} g_2^{N_2} \ldots$

SECOND FACTOR $= \dfrac{N!}{N_1!\,N_2!\ldots}$

NET PROBABILITY $= g_1^{N_1} g_2^{N_2} \ldots \dfrac{N!}{N_1!\,N_2!\ldots}$

FIGURE 5-2
A marble-and-compartment illustration of the probability factors that enter into the expression for the probability of a distribution of molecules throughout an energy-level pattern.

each box. If you investigate the rearrangement of numbers in compartments, as [1|2|3|4| . . . , you will convince yourself that if you choose N numbers, you can make $N!$ different arrangements. In the marble example, some of the total of 4! rearrangements, however, correspond to rearrangements within a compartment, and these are not significant. The factor which expresses the significant ways in which the marbles can be rearranged, for a given distribution, is obtained in the case of the marbles of Fig. 5-2 by dividing 4! by the meaningless interchanges: 4! in the first distribution, 3! in the second, and so forth (that is, for the molecules in energy levels this factor has the form $N!/N_1!N_2!N_3! \ldots$).

The total probabilities for the marble distributions are shown in Fig. 5-2, and if you are not familiar with such probability expressions, you should investigate further examples to convince yourself that the combination of the two factors dealt with here leads to probabilities for various distributions that seem reasonable.

In a similar way, we should express the probability, which will now be designated by W, of a given distribution of a total of N molecules throughout energy levels as

$$W = (g_1^{N_1} g_2^{N_2} g_3^{N_3} \ldots) \frac{N!}{N_1!N_2!N_3!} \qquad [4]$$

The goal, it will be recalled, was to find the values of N_1, N_2, N_3, . . . that maximize W. It is mathematically more convenient to look for a maximum in ln W, which occurs also when W is a maximum. (The logarithmic expression allows Stirling's approximation $\ln x! = x \ln x - x$ for large numbers, which is derived in Appendix 2, to be applied.) When Stirling's approximation is used for the N_i terms, we obtain

$$\begin{aligned}\ln W &= N_1 \ln g_1 + N_2 \ln g_2 + \cdots + \ln N! \\ &\qquad - N_1 \ln N_1 - N_2 \ln N_2 - \cdots + N_1 + N_2 + \cdots \\ &= N_1\left(1 + \ln\frac{g_1}{N_1}\right) + N_2\left(1 + \ln\frac{g_2}{N_2}\right) + \cdots + \ln N! \end{aligned} \qquad [5]$$

To find the maximum in $\ln W$, we first take the derivative and, with the recognition that $N!$ is a constant, we obtain for any N_i

$$\frac{\partial \ln W}{\partial N_i} = \frac{\partial}{\partial N_i}\left[N_i\left(1 + \ln\frac{g_i}{N_i}\right)\right] = \ln\frac{g_i}{N_i} \qquad [6]$$

We cannot now proceed directly to the desired maximum by setting this derivative equal to zero. We must recognize that there are some limitations on the values the N_i's may take. (They cannot, for example, all go to zero, or infinity, as Eq. [6] might suggest.) The total number of molecules N is fixed, and thus, for any distribution,

$$N_1 + N_2 + N_3 + \cdots = \Sigma N_i = N \qquad [7]$$

At a given temperature, moreover, the total energy is also some fixed quantity, and this requirement can be expressed by stipulating that the energy in excess of that which the system would have for all molecules in the lowest energy level is some constant amount E_{thermal}. We write

$$\begin{aligned}&N_0(0) + N_1(\epsilon_1 - \epsilon_0) + N_2(\epsilon_2 - \epsilon_0) + \cdots \\ &\qquad = \Sigma(\epsilon_i - \epsilon_0)N_i = E_{\text{thermal}}\end{aligned} \qquad [8]$$

The method for imposing such conditions is *Lagrange's method of undetermined multipliers,* discussed in Appendix 3. In this procedure we introduce two parameters, the Lagrange multipliers, which we shall designate as α and β, and look for a maximum in

$$\ln W - \alpha\Sigma N_i - \beta\Sigma(\epsilon_i - \epsilon_0)N_i \qquad [9]$$

We find the maximum, with respect to each N_i, by forming

$$\frac{\partial}{\partial N_i}[\ln W - \alpha\Sigma N_i - \beta\Sigma(\epsilon_i - \epsilon_0)N_i] = 0$$

or

$$\frac{\partial \ln W}{\partial N_i} - \alpha - \beta(\epsilon_i - \epsilon_0) = 0 \qquad [10]$$

for each energy level i.

Now the values of N_i that produce the desired maximum in $\ln W$ can be obtained by substituting Eq. [6] in Eq. [10] to yield

$$\ln\frac{g_i}{N_i} - \alpha - \beta(\epsilon_i - \epsilon_0) = 0$$

or

$$N_i = g_i e^{-\alpha} e^{-\beta(\epsilon_i - \epsilon_0)} \qquad [11]$$

This result for the population of the states in the most probable distribution remains mysterious until we eliminate, or investigate, the parameters α and β.

The term involving α can be eliminated by considering a sample of given size; one containing an Avogadro's number $\mathfrak{N}$ of molecules is convenient. For the summation of all N_i to give $\mathfrak{N}$,

$$\mathfrak{N} = \sum_i N_i = e^{-\alpha} \sum_i g_i e^{-\beta(\epsilon_i - \epsilon_0)}$$

Rearrangement gives

$$e^{-\alpha} = \frac{\mathfrak{N}}{\Sigma g_i e^{-\beta(\epsilon_i - \epsilon_0)}} \qquad [12]$$

With this expression, the $e^{-\alpha}$ term can be eliminated from our expression Eq. [11] for the population of the states in a sample containing a total of $\mathfrak{N}$ molecules to give

$$N_i = \frac{\mathfrak{N}}{\sum_i g_i e^{-\beta(\epsilon_i - \epsilon_0)}} g_i e^{-\beta(\epsilon_i - \epsilon_0)} \qquad [13]$$

You should note that the term in parentheses is characteristic of the system and is not identified with any one of the states. As a result, we can form the easier-to-interpret ratio of the populations of two states, and for this quantity our derivation gives us

$$\frac{N_i}{N_j} = \frac{g_i e^{-\beta(\epsilon_i - \epsilon_0)}}{g_j e^{-\beta(\epsilon_j - \epsilon_0)}} = \frac{g_i}{g_j} e^{-\beta(\epsilon_i - \epsilon_j)}$$

or

$$\frac{N_i/g_i}{N_j/g_j} = e^{-\beta(\epsilon_i - \epsilon_j)} \qquad [14]$$

The result clearly parallels the Boltzmann distribution expression of Eq. [2], but to complete the derivation of the Boltzmann expression we must proceed, as in the following section, to show that β can be identified with $1/kT$.

*5-3 The Thermal Energy due to the Translational Motion of the Molecules of a Gas and the Deduction That $\beta = 1/kT$

Let us now apply Eq. [13], the expression for the population of each state when molecules are distributed throughout available states, to the molecules of a one-dimensional gas, to deduce the translational energy of an Avogadro's number of gas molecules. The answer will be in terms of the remaining unknown parameter β, and we will be able to use the fact that this energy must be equal to $\frac{1}{2}RT$ to evaluate the parameter. (In later sections we will follow a similar procedure for rotational and vibrational energies, but then, having already interpreted β, we will be able to deduce the thermal energies for these motions.)

The energy that we will calculate is $\mathrm{E}_{\text{thermal}}$, the energy of the system in excess of that which it would have if all molecules were in the lowest energy level. This energy can also be denoted by introducing E for the total energy of the system and E_0 for the energy of the system with all molecules in the lowest energy level. Then we can write

$$\mathrm{E}_{\text{thermal}} = \mathrm{E} - \mathrm{E}_0 \qquad [15]$$

The kinetic energy of translation of $\frac{3}{2}RT$, for example, is such thermal energy, since this is the average energy over and above that which the molecules would have in the lowest translational-energy level.

The thermal energy resulting from the distribution of molecules throughout the available energies can be written

$$\mathrm{E} - \mathrm{E}_0 = N_0(0) + N_1(\epsilon_1 - \epsilon_0) + N_2(\epsilon_2 - \epsilon_0) + \cdots$$
$$= \sum_i N_i(\epsilon_i - \epsilon_0) \qquad [16]$$

To carry out the calculation we must use the information that Eq. [13] gives for the N_i's. Substitution of Eq. [13] into Eq. [16] yields, with Avogadro's number $\mathfrak{N}$ as the total number of molecules,

$$\mathrm{E} - \mathrm{E}_0 = \frac{\mathfrak{N}}{\sum_i g_i e^{-\beta(\epsilon_i - \epsilon_0)}} \sum_i g_i(\epsilon_i - \epsilon_0) e^{-\beta(\epsilon_i - \epsilon_0)} \qquad [17]$$

This general expression can be applied to the one-dimensional translational motion for which the energy levels, as you saw in Sec. 4-2, are all single, that is, $g_i = 1$ for all i, and form a pattern according to the equation

$$\epsilon_i = \frac{i^2h^2}{8ma^2} \qquad i = 1, 2, 3, \ldots$$

Such large values of i will occur for most of the states that contribute to the summations of Eq. [17] that we can replace the lowest energy, $ih^2/8ma^2$, by zero, and thus obtain

$$\mathrm{E} - \mathrm{E}_0 = \frac{\mathfrak{N} \sum (i^2h^2/8ma^2) e^{-\beta i^2h^2/8ma^2}}{\sum e^{-\beta i^2h^2/8ma^2}} \qquad [18]$$

The steps corresponding to successive i values are small compared with the range over which the summation extends, and so the summations can be replaced by integrations, giving

$$\mathrm{E} - \mathrm{E}_0 = \frac{\mathfrak{N} \int_0^\infty (i^2h^2/8ma^2) e^{-\beta i^2h^2/8ma^2}\, di}{\int_0^\infty e^{-\beta i^2h^2/8ma^2}\, di} \qquad [19]$$

Both integrals are listed in Appendix 1, and these lead us to the result

$$\mathrm{E} - \mathrm{E}_0 = \tfrac{1}{2}\mathfrak{N}(\tfrac{1}{\beta}) \qquad [20]$$

Thus $1/\beta$ must be interpreted as a quantity which is proportional to the thermal energy. One can proceed from this result in a number of ways. One way is to realize that $1/\beta$ could be used in all the roles that we attribute to temperature and could thus be the basis for the establishment of a temperature function. More straightforward is the recognition that the temperature scale we have been using is such that the one-dimensional translational energy of a mole of gas is

$$\mathrm{E} - \mathrm{E}_0 = \tfrac{1}{2}RT = \tfrac{1}{2}\mathfrak{N}(kT) \qquad [21]$$

For the Boltzmann distribution result that we obtained in Eq. [20] to be consistent with this value, we must make the identification

$$\beta = \frac{1}{kT} \qquad [22]$$

Then, returning to Eq. [13], we see that we now have produced the result

$$N_i = \frac{\mathfrak{N}}{\Sigma g_i e^{-(\epsilon_i-\epsilon_0)/kT}} g_i e^{-(\epsilon_i-\epsilon_0)/kT} \qquad [23]$$

or from Eq. [14],

$$\frac{N_i/g_i}{N_j/g_j} = e^{-(\epsilon_i-\epsilon_j)/kT} \qquad [24]$$

5-4 The Partition Function

The distribution expression, Eq. [23], for a collection of particles is rather clumsy. In overcoming this fault, we are led to the introduction of a quantity that has considerable significance.

The summation that appears in the denominator of Eq. [23], and for the specific case of one-dimensional translational energies in Eq. [18], turns out to be important enough to warrant a symbol, q, and a name, the *partition function.* Thus we define q by

$$q = \sum_i g_i e^{-(\epsilon_i-\epsilon_0)/kT} \qquad [25]$$

You should notice that the calculation of a value, or an expression, for q for a particular type of motion requires information only about the pattern of allowed energies and the degeneracies of these levels.

Some of the implications of the name partition function can be seen by returning to Eq. [23] and writing it

$$N_i = \frac{\mathfrak{N}}{q} g_i e^{-(\epsilon_i-\epsilon_0)/kT} \qquad [26]$$

For the ground state, $i = 0$, this becomes

$$\frac{N_0}{g_0} = \frac{\mathfrak{N}}{q} \qquad [27]$$

The N_0 particles could be said to be distributed or partitioned throughout the g_0 states that are available to them at energy ϵ_0. Likewise, the total number of particles $\mathfrak{N}$ can be said, as a result of the Boltzmann distribution, to be distributed, or partitioned, throughout q states. Thus q is a measure of the *availability of states,* a sort of net degeneracy, that results from the energy pattern throughout which the molecules are distributed and from the temperature which controls the distribution.

The convenience, even elegance, that enters when q is used to denote the summation of Eq. [25] can be seen by developing the general expression for the thermal energy.

The thermal energy $\mathrm{E}_{\text{thermal}} = \mathrm{E} - \mathrm{E}_0$, the energy in excess of that which the system would have if all molecules were in their lowest energy states, is obtained, as in Eq. [16], as

$$\mathrm{E} - \mathrm{E}_0 = N_0(0) + N_1(\epsilon_1 - \epsilon_0) + N_2(\epsilon_2 - \epsilon_0) + \cdots$$

$$= \sum_i N_i(\epsilon_i - \epsilon_0)$$

Now, with Eq. [26] this can be developed as

$$\mathrm{E} - \mathrm{E}_0 = \sum_i \frac{\mathfrak{N}}{q} g_i(\epsilon_i - \epsilon_0)e^{-(\epsilon_i - \epsilon_0)/kT}$$

$$= \frac{\mathfrak{N}}{q} \sum_i g_i(\epsilon_i - \epsilon_0)e^{-(\epsilon_i - \epsilon_0)/kT} \qquad [28]$$

The first derivative of q, as well as q itself, turns out to be involved. The derivative is

$$\frac{dq}{dT} = \sum_i g_i e^{-(\epsilon_i - \epsilon_0)/kT} \frac{\epsilon_i - \epsilon_0}{kT^2}$$

and multiplication by kT^2 leads to the useful relation

$$kT^2 \frac{dq}{dT} = \sum_i g_i(\epsilon_i - \epsilon_0)e^{-(\epsilon_i - \epsilon_0)/kT} \qquad [29]$$

Comparison with Eq. [28] shows that the thermal energy can be expressed by

$$\mathrm{E} - \mathrm{E}_0 = \frac{\mathfrak{N}}{q} kT^2 \frac{dq}{dT}$$

$$= \frac{RT^2}{q} \frac{dq}{dT} \qquad [30]$$

Thus, for any type of motion, the thermal energy can be calculated if an expression which can be differentiated with respect to temperature is obtained for $q = \Sigma g_i e^{-(\epsilon_i - \epsilon_0)/kT}$.

In a similar way it will be found that other energy-related properties that result from the distribution of molecules throughout an energy pattern can be expressed in terms of the partition function.

In most of the molecular problems that we deal with, the state of a molecule can conveniently be described as the result of the translational, the rotational, the vibrational, and the electronic states, which make up the overall state. It is important to see that this separation of energies at the molecular level leads with Eq. [30] to a corresponding separation of the energy terms for the system of $\mathfrak{N}$ molecules.

Consider molecules whose energy is described in terms of two types of motion, say a and b, that they can undergo. Then, if these motions can be treated separately, we can write for some allowed energy, compared with the lowest energy,

$$\epsilon_i = \epsilon_j{}^a + \epsilon_k{}^b \qquad [31]$$

Furthermore, if the degeneracy of energy level $\epsilon_j{}^a$ is $g_j{}^a$ and that of level $\epsilon_k{}^b$ is $g_k{}^b$, the number of states with energy $\epsilon_j{}^a + \epsilon_k{}^b$ will be

$$g_i = g_j{}^a g_k{}^b \qquad [32]$$

It follows that the partition function q for the system can be interpreted in terms of the separate motions as

$$\begin{aligned} q &= \sum_i g_i e^{-\epsilon_i/kT} \\ &= \sum_{j,k} g_j{}^a g_k{}^b e^{-(\epsilon_j{}^a + \epsilon_k{}^b)/kT} \\ &= \sum_j g_j{}^a e^{-\epsilon_j{}^a/kT} \sum_k g_k{}^b e^{-\epsilon_k{}^b/kT} \\ &= q_a q_b \end{aligned} \qquad [33]$$

The energy consequences are seen by rewriting Eq. [30] as

$$\mathrm{E} - \mathrm{E}_0 = RT^2 \frac{d \ln q}{dT}$$

which, with $\ln q = \ln q_a q_b = \ln q_a + \ln q_b$, becomes

$$\begin{aligned} \mathrm{E} - \mathrm{E}_0 &= RT^2 \frac{d \ln q_a}{dT} + RT^2 \frac{d \ln q_b}{dT} \\ &= (\mathrm{E} - \mathrm{E}_0)_a + (\mathrm{E} - \mathrm{E}_0)_b \end{aligned} \qquad [34]$$

Thus, when separate energy patterns for molecular motions are recognized, the total partition function is the product of the partition functions for the separate motions, and the total energy of the system is calculated as the sum of energy contributions from the separate motions.

5-5 The Three-dimensional Translational Energy of a Mole of Gas

That the translational energy of an Avogadro's number of gas molecules is $\frac{3}{2}RT$ will come as no surprise. But the calculation of this result will illustrate the procedure that is based on partition-function expressions.

The translational states have energies that are expressed by

$$\begin{aligned}\epsilon &= \epsilon_x + \epsilon_y + \epsilon_z & i &= 1, 2, 3, \ldots \\ &= (i^2 + j^2 + k^2)\frac{h^2}{8ma^2} & j &= 1, 2, 3, \ldots \\ & & k &= 1, 2, 3, \ldots\end{aligned} \qquad [35]$$

The partition function, again with the assumption that the ground state has an energy so close to zero that a zero value can be assigned to it, is

$$\begin{aligned}q_{\text{trans}} &= \sum_{i,j,k} e^{-(i^2+j^2+k^2)h^2/8ma^2kT} \\ &= \sum_i e^{-i^2h^2/8ma^2kT} \sum_j e^{-j^2h^2/8ma^2kT} \sum_k e^{-k^2h^2/8ma^2kT} \\ &= q_x q_y q_z\end{aligned} \qquad [36]$$

(Notice that the summation is over all states, not over all energy levels. Thus the degeneracy that would enter for an energy-level summation does not appear.)

Replacing the summations by integrations, and using an integral of Appendix 1, gives

$$\begin{aligned}q_x = q_y = q_z &= \int_0^\infty e^{-i^2h^2/8ma^2kT}\,di \\ &= \frac{\sqrt{2\pi ma^2kT}}{h}\end{aligned} \qquad [37]$$

Thus

$$q_{\text{trans}} = \frac{(2\pi ma^2kT)^{3/2}}{h^3} \qquad [38]$$

It is convenient also to replace a by $\mathrm{v}^{1/3}$ to avoid the requirement of a cubic container, and this replacement gives

$$q_{\text{trans}} = \frac{(2\pi mkT)^{3/2}}{h^3}\mathrm{v} \qquad [39]$$

From this result and

$$\frac{dq}{dT} = \frac{(2\pi mk)^{3/2}}{h^3}\mathrm{v}\left(\frac{3}{2}T^{1/2}\right) \qquad [40]$$

we immediately obtain

$$\begin{aligned}(\mathrm{E} - \mathrm{E}_0)_{\text{trans}} &= \frac{RT^2}{q}\frac{dq}{dT} \\ &= \frac{RT^2}{[(2\pi mkT)^{3/2}\mathrm{v}/h^3]}\frac{(2\pi mk)^{3/2}\mathrm{v}}{h^3}\frac{3}{2}T^{1/2} \\ &= \tfrac{3}{2}RT\end{aligned} \qquad [41]$$

Now let us proceed to other molecular motions where the result is not so well known.

5-6 The Thermal Energy due to the Rotational Motion of the Molecules of a Gas

As in the preceding section, let us calculate the rotational energy of an Avogadro's number of molecules that, as for the molecules of a gas at not too high a pressure, are freely rotating. The allowed energies and degeneracies of the energy levels for a rotating linear molecule, with 2 rotational degrees of freedom, were shown in Sec. 4-3 to be given by

$$\epsilon_{\text{rot}} = J(J+1)\frac{h^2}{8\pi^2 I}$$

and the degeneracies by

$$g_J = 2J + 1 \qquad J = 0, 1, 2, \ldots$$

Now we must calculate

$$q_{\text{rot}} = \sum_{J=0}^{\infty} (2J+1)e^{-J(J+1)h^2/8\pi^2 IkT} \tag{42}$$

It will be seen that for most molecules at not too low a temperature, the values of J that will lead to most of the contributing terms in this summation are very large compared with unity, and we can write

$$q_{\text{rot}} \approx \sum_{J=0}^{\infty} 2Je^{-J^2h^2/8\pi^2 IkT} \tag{43}$$

Furthermore, the large number of summation terms that contribute allow us to treat J as a continuous variable so that the summation can be replaced by an integral. Thus we have

$$q_{\text{rot}} \approx 2\int_0^{\infty} Je^{-J^2(h^2/8\pi^2 IkT)}\, dJ \tag{44}$$

Use of the appropriate integral from Appendix 1 allows this to be evaluated to give

$$q_{\text{rot}} = \frac{8\pi^2 IkT}{h^2} \tag{45}$$

The derivative is

$$\frac{dq_{\text{rot}}}{dT} = \frac{8\pi_2 Ik}{h^2} \tag{46}$$

With these results we obtain

$$\begin{aligned}(\mathrm{E} - \mathrm{E}_0)_{\text{rot}} &= \frac{RT^2(8\pi^2 Ik/h^2)}{8\pi^2 IkT/h^2} \\ &= RT \end{aligned} \tag{47}$$

Thus we again achieve for the linear molecule considered here, with the 2 rotational degrees of freedom, the value expected on a classical basis of $\frac{1}{2}RT$ thermal energy per degree of freedom.

For a generally shaped molecule, rotation can be described in terms of component rotations about three perpendicular axes. The moments of inertia about these axes are known as the *principal moments of inertia* and are represented by I_A, I_B, and I_C. The pattern of allowed energies is now not so easily expressed, and here, therefore, the partition function is cited without derivation. The result is, for the 3 rotational degrees of freedom,

$$q_{\text{rot}} = \sqrt{\pi}\left(\frac{8\pi^2 kT}{h^2}\right)^{3/2}(I_A I_B I_C)^{1/2} \qquad [48]$$

The relation to the 2-degrees-of-freedom equation [45] is straightforward.

For molecules with very low moments of inertia, H_2 being the most conspicuous example, or for all molecules at rather low temperatures, the assumptions $J \gg 1$ for most of the terms in the summation, and the replacement of the summation by an integration, is not justified. In such a case, other approaches to the evaluation of the partition-function sum must be taken. A similar problem which forces us to a term-by-term summation occurs when the net vibrational energy of a collection of molecules is considered.

5-7 The Thermal Energy due to the Vibrational Motion of Molecules

The one vibrational mode of a diatomic molecule, and the $3n$-6 or $3n$-5 modes of a polyatomic molecule, each has associated with it a set of vibrational states with energies, as shown in Sec. 4-4, given by

$$\begin{aligned}\epsilon_{\text{vib mode}} &= \left(v + \frac{1}{2}\right)\frac{h}{2\pi}\sqrt{\frac{k}{m}} \\ &= \left(v + \frac{1}{2}\right)h\nu_{\text{vib}} \qquad v = 0, 1, 2, \ldots \end{aligned} \qquad [49]$$

For each vibrational mode we can thus write

$$\begin{aligned}q_{\text{vib mode}} &= \sum_{v=0}^{\infty} e^{-[(v+1/2)h\nu_{\text{vib}} - 1/2h\nu_{\text{vib}}]/kT} \\ &= \sum_{v=0}^{\infty} e^{-vh\nu_{\text{vib}}/kT}\end{aligned} \qquad [50]$$

Now, in contrast to the situation in the two preceding sections, the vibrational spacings are appreciable compared with kT, and therefore only a few of the terms in the series contribute. The sum cannot be replaced by an integral. We can, however, obtain an expression by developing the summation

$$q_{\text{vib}} = 1 + e^{-h\nu_{\text{vib}}/kT} + e^{-2h\nu_{\text{vib}}/kT} + \cdots \qquad [51]$$

With the introduction of the convenient symbol

$$x = \frac{h\nu_{\text{vib}}}{kT}$$

this becomes

$$\begin{aligned} q_{\text{vib mode}} &= 1 + e^{-x} + e^{-2x} + \cdots \\ &= 1 + (e^{-x})^1 + (e^{-x})^2 + \cdots \end{aligned} \qquad [52]$$

This series can be recognized as the binomial expansion of $(1 - e^{-x})^{-1}$, and thus we have

$$q_{\text{vib mode}} = \frac{1}{1 - e^{-x}} = \frac{1}{1 - e^{h\nu_{\text{vib}}/kT}} \qquad [53]$$

The derivative is

$$\begin{aligned} \frac{dq_{\text{vib mode}}}{dT} &= \frac{(h\nu_{\text{vib}}/kT^2)e^{-h\nu_{\text{vib}}/kT}}{(1 - e^{-h\nu_{\text{vib}}/kT})^2} \\ &= \frac{(x/T)e^{-x}}{(1 - e^{-x})^2} \end{aligned} \qquad [54]$$

The thermal energy contributed by a vibrational mode is then obtained, from Eq. [30], as

$$\begin{aligned} (\mathrm{E} - \mathrm{E}_0)_{\text{vib mode}} &= \frac{RT^2(x/T)e^{-x}}{1 - e^{-x}} \\ &= RT\frac{x}{e^x - 1} \end{aligned} \qquad [55]$$

Alternatively, with x replaced by $h\nu_{\text{vib}}/kT$ and R by $\mathfrak{N}k$, this can be written

$$(\mathrm{E} - \mathrm{E}_0)_{\text{vib mode}} = \frac{\mathfrak{N}(h\nu_{\text{vib}})}{(e^{h\nu_{\text{vib}}/kT} - 1)} \qquad [56]$$

Plots of $(\mathrm{E} - \mathrm{E}_0)_{\text{vib mode}}$, for various values of T as a function of the vibrational-energy-level spacing, are shown in Fig. 5-3.

We can note immediately that only as $x \to 0$, that is, as $h\nu_{\text{vib}}$ becomes very much less than kT, does this expression reduce to our classical expectations. For then we can write

$$e^x = 1 + x + \cdots$$

and

$$\begin{aligned} RT\frac{x}{e^x - 1} &= RT\frac{x}{(1 + x + \cdots) - 1} \\ &\approx RT \end{aligned}$$

Thus, in this limit, the average kinetic energy is $\frac{1}{2}RT$, and this, together with the equal potential energy that characterizes a harmonic oscillator, gives the deduced result RT.

A typical vibrational-state-energy separation (Sec. 4-4) is 2×10^{20} J per molecule, and at 25°C

$$x = \frac{h\nu_{\text{vib}}}{kT} = \frac{2 \times 10^{-20}}{(1.38 \times 10^{-23})(298)} = 5 \qquad [57]$$

The thermal vibrational energy for $\mathfrak{N}$ oscillators distributed throughout this energy pattern according to Boltzmann's distribution is, by Eq. [55],

$$(\text{E} - \text{E}_0)_{\text{vib}} = RT\frac{x}{e^x - 1}$$

$$= (8.315)(298)\frac{5}{e^5 - 1}$$

$$= 84 \text{ J mole}^{-1} \qquad [58]$$

Classically, where no quantum restrictions apply, an average potential and kinetic energy of RT, or about 2500 J mol^{-1}, would be expected at 25°C. The difference between 2500 and the value obtained in Eq. [58] is attributable to the effect of the rather widely spaced vibrational quantum levels.

Each of the vibrational degrees of freedom of a polyatomic molecule has a pattern of vibrational energies associated with it. The total vibrational energy of a polyatomic molecule is obtained by adding together the thermal energy contributed by each of these degrees of freedom, using Eq. [55] or [56] or Fig. 5-3 to calculate these contributions.

The thermal vibrational energy of a mole of SO_2 can be calculated as an illustration. Spectroscopic studies, which will be discussed in Sec.

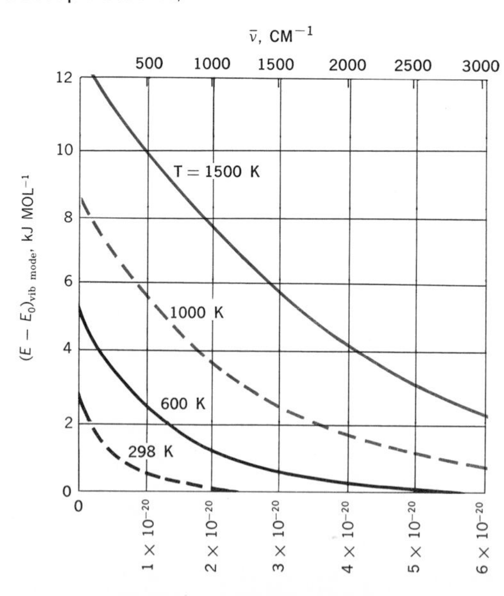

FIGURE 5-3
The thermal vibrational energy of a vibrational mode as a function of the vibrational-energy-level spacing.

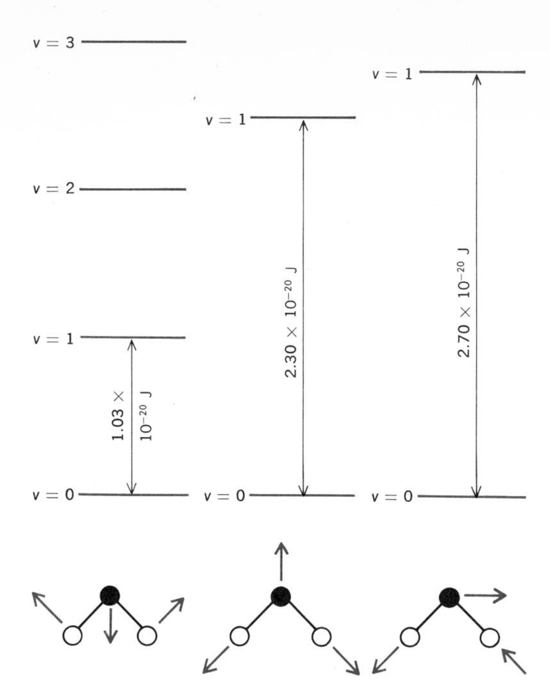

FIGURE 5-4
The three ($3n - 6$) ways in which SO_2 can vibrate and the lower energy levels allowed for each of these motions. (The arrows showing the directions of vibration can be thought of as the initial displacements that would occur.)

13-2, show that the 3 vibrational degrees of freedom ($3n - 6 = 3$) have the spacings indicated in Fig. 5-4. (The ways in which the atoms move in the three vibrations are not important here, but are included in Fig. 5-4.) The thermal vibrational energy can now be calculated from Eq. [55] or [56] or, for certain temperatures, read, approximately, off Fig. 5-3. At room temperature, for example, the 3 vibrational degrees of freedom contribute 550 + 51 + 22 = 623 J mol^{-1}. At 1000 K, where more molecules are spread throughout higher energy levels, the total thermal vibrational energy is 5580 + 3220 + 2670 = 11,470 J mol^{-1}.

5-8 Thermal Energies

The contributions of translational, rotational, and vibrational energies to the total molecular energy can now be summarized. The thermal energy for 1 mol of a typical gaseous compound can be written

$$\begin{aligned} \mathrm{E} - \mathrm{E}_0 &= \mathrm{E}_{\text{trans}} + \mathrm{E}_{\text{rot}} + \mathrm{E}_{\text{vib}} + \mathrm{E}_{\text{elec}} \\ &= \tfrac{3}{2}RT + \tfrac{3}{2}RT \text{ (or } RT) + E_{\text{vib}} + 0 \\ &= \begin{cases} 3RT + \mathrm{E}_{\text{vib}} & \text{nonlinear} \\ \tfrac{5}{2}RT + \mathrm{E}_{\text{vib}} & \text{linear} \end{cases} \end{aligned} \qquad [59]$$

These equations can be rearranged to show the energy content of $\mathfrak{N}$ molecules as

$$\begin{aligned} \mathrm{E} &= \mathrm{E}_0 + (3RT + \mathrm{E}_{\text{vib}}) & &\text{nonlinear molecules} \\ \mathrm{E} &= \mathrm{E}_0 + (\tfrac{5}{2}RT + \mathrm{E}_{\text{vib}}) & &\text{linear molecules} \end{aligned} \qquad [60]$$

TABLE 5-1 Typical molecular-energy spacing and thermal energies

	Representative energy-level spacing		Thermal energy, $E - E_0$
	J per molecule*	J mol^{-1}†	
Electronic	10^{-18}	600,000	0
Vibrational	10^{-20}	6,000	See Eq. [55] or [56] or Fig. 5-3
Rotational	10^{-23}	6	$\frac{3}{2}RT$, nonlinear; RT, linear
Translational	10^{-35}	6×10^{-12}	$\frac{3}{2}RT$

*Compare with the room-temperature value of kT of 4×10^{-21} J per molecule.
†Compare with the room-temperature value of RT of 2500 J mol^{-1}.

It is for such statements as these that the preliminary investigation of molecular structure of the preceding chapter has been introduced. Molecular-quantum restrictions and the related net thermal energies are summarized in Table 5-1.

The understanding of the allowed molecular energy levels and such derived quantities as the average energy of molecules distributed throughout such energy schemes will soon be seen to be a necessary preliminary to the full understanding and use of the powerful methods of thermodynamics which are now to be studied. Although the methods of thermodynamics are very different from the approaches used so far, and will, in fact, seem at first to be quite unrelated to our search for details of the molecular world, they will finally become invaluable as windows into this world.

Problems

1 Calculate the number of hydrogen atoms that have $n = 2$ compared with the number that have $n = 1$ at 25°C. At what temperature would 1 percent of the atoms have $n = 2$? *Ans.* $T = 25{,}700$ K.

2 Indicate by vertical lines erected on an energy abscissa the relative populations of the first five quantum states for an electron confined to a one-dimensional square potential well 100 Å wide. Make such a chart for $T = 298$ K and $T = 1000$ K.

3 For a large number of particles confined to a cubic container for which the size of the container, the mass of the particles, and the temperature are such that $h^2/8ma^2 = 0.1\,kT$, calculate the ratio of the number of particles:
a In a state with $n_x = 1$, $n_y = 2$, $n_z = 3$ compared with the number in a state with $n_x = n_y = n_z = 1$.
b At the energy $(1^2 + 2^2 + 3^2)h^2/8ma^2$ compared with the number at an energy $(1^2 + 1^2 + 1^2)h^2/8ma^2$.

4 For the states for the one-dimensional translational motion of N_2 molecules confined to a region of length 0.10 m at 25°C, calculate the population ratios N_i/N_0 for the states that have energies of 1×10^{-21}, 2×10^{-21}, 5×10^{-21}, and 10×10^{-21} J per molecule. Make an N_i/N_0 versus energy plot. Add a speed coordinate alongside the energy coordinate, and read off values of N_i/N_0 for speed values so that an N_i/N_0 versus speed graph can be prepared. Compare this curve with that given in Fig. 2-5*a*. *Optional:* Repeat for 1500 K.

5 Calculate the Boltzmann distribution spacing factor $e^{-\Delta\epsilon_i/kT}$ at 25°C for the representative electronic-, vibrational-, rotational- and translational-energy spacings given in Fig. 4-6.

6 Imagine a system consisting of three singly degenerate energy levels separated by spacings of 3×10^{-21} J per molecule. What would be the thermal energy of an Avogadro's number of molecules in such a system at 25°C? (The system might have states with other energies, but these energies are assumed to be so high that none of these other states are populated.)

7 Describe the molecular features that would give rise to large (*a*) translational, (*b*) rotational, and (*c*) vibrational partition functions.

8 To test that the summation we call the partition function is a measure of the number of available states, calculate numerical values for the translational, rotational, and vibrational partition functions for a representative small gas-phase molecule. If only a singly degenerate ground electronic state is needed to describe a molecule, what would be the electronic partition function? Do all these results support the idea that q measures the availability of states?

9 What is the thermal energy of an Avogadro's number of rigid, i.e., nonvibrating, diatomic molecules at 298 K? At 1000 K?
Ans. $(\mathrm{E} = \mathrm{E}_0)_{298} = 6190$; $(\mathrm{E} - \mathrm{E}_0)_{1000} = 20{,}800$ J mol^{-1}.

10 Consider a reaction in which two monatomic molecules combine to give a rigid, i.e., nonvibrating, diatomic molecule. Express the energy content of each species in terms of the zero-state energy and the translational- and rotational-energy contribution. Write an expression for the difference in energy between the product and the sum of the reactants. Will more or less energy be given out by the reaction as the temperature is raised?

11 Verify from the rotational partition function for generally shaped molecules given in Sec. 5-6 that the corresponding thermal energy is $\frac{3}{2}RT$.

12 The vibrational energy levels of HCl consist of an evenly spaced set with a separation of 5.94×10^{-20} J. Calculate the ratio of the number of molecules in one energy level to the number of molecules in the next-lower level at 25°C. Do the same for I_2, for which the vibrational energies are spaced by 0.43×10^{-20} J. *Ans.* $(N_1/N_0)_{\mathrm{HCl}} = 5.4 \times 10^{-7}$; $(N_1/N_0)_{I_2} = 0.35$.

13 Compare the limits, at $h\nu_{\mathrm{vib}}$ equal to zero or $T = \infty$, of the energy curves of Fig. 5-3 with the values that would be expected if a vibrational mode behaved classically.

14 Calculate the total thermal energy of a mole of gaseous SO_2 at 1500 K.

References

GUGGENHEIM, E. A.: "Boltzmann's Distribution Law," Interscience Publishers, Inc., New York, 1955. A short readable treatment of the Boltzmann distribution and of some developments of thermodynamic functions.

NASH, L. K.: "Elements of Statistical Thermodynamics," Addison Wesley Publishing Company, Inc., Reading, Mass., 1968. An excellent elementary introduction to the basis of the statistical expressions and the way in which they are related to thermodynamic quantities.

DAVIDSON, N.: "Statistical Mechanics," McGraw-Hill Book Company, New York, 1962. A detailed and mathematical development of mathematical procedures in Chap. 5, distributions in Chaps. 6 and 7, and application to the thermodynamic properties of diatomic molecules in Chap. 8 and of polyatomic molecules in Chap. 11.

GURNEY, R. W.: "Introduction to Statistical Mechanics," chap. 1, Probability in Molecular Systems, and chap. 2, The Method of Undetermined Multipliers and the Boltzmann Distribution Expression, McGraw-Hill Book Company, New York, 1949.

6

THE FIRST LAW OF THERMODYNAMICS

6-1 The Nature of Thermodynamics

A powerful method for studying chemical phenomena, which can be developed quite independently of the atomic and molecular theory of the preceding chapters and can be applied to systems of any complexity, is that of *thermodynamics*. The name implies a study of the flow of heat, but it will be seen that the subject treats the more general quantity energy. The energy changes associated with chemical reactions are themselves of considerable importance and will be dealt with in this and the following chapter. Even greater chemical interest, however, stems from the fact that the equilibrium position of a reacting system can be related to these energy changes. Much of the succeeding thermodynamic development will be directed toward associating thermal properties with the equilibrium state of a chemical system.

Thermodynamics is a logical subject of great elegance. Three concise statements, the three laws of thermodynamics, are made that sum up our experiences with energy and natural processes. From these statements logical deductions are then drawn that bear on almost every aspect of chemistry.

One should note that the great contribution of thermodynamics is that it *systematizes* the information we obtain from various experiments or measurements and allows us to draw conclusions about other aspects of the behavior of the system than that measured. The initial statements, for example, of the first law of thermodynamics may seem now to be so obvious as to be unnecessary and not very useful. As one proceeds through this and the following chapter, however, it will be noticed that the logical developments that follow from this first law allow important quantitative conclusions to be drawn that do not follow immediately or obviously from the statement of the first law.

The validity of thermodynamics depends only on the three generalizations and on the logic of the succeeding deductions. Thermodynamics is therefore independent of any model or theory, such as the molecular theory, for the nature of matter. Any alteration in our present ideas and theories about the nature of molecules would therefore in no way affect the validity of any thermodynamic result. Thermodynamics has a permanence which might, for example, be compared with that of Euclid's geometric theorems in plane geometry, but which is not shared by our ever-changing views on the nature of atoms and molecules.

Modern physical chemistry, however, attempts to understand the nature of the chemical world and does so primarily in terms of the atomic and molecular theory. Although thermodynamics can be kept aloof from these molecular ideas, it need not be. In practice, one deals with thermodynamics most frequently when detailed molecular ideas are used to try to explain or calculate some thermodynamic result. It will be found that the ability to understand thermodynamic quantities on a molecular basis, although not at all necessary for the study of thermodynamics, is a very valuable aid to the study of chemistry and of the molecular world.

The introduction from time to time of molecular explanations will provide a concrete model on which thermodynamic quantities can be understood. Most students find this quite helpful. It cannot be emphasized too strongly, however, that the molecular model need not be introduced in our study of thermodynamics.

6-2 Measurement and Relation of Thermal and Mechanical Energy

The forms of energy that are most easily and directly dealt with are *thermal* energy and *mechanical* energy.

The simplest, if rather special, device for measuring a change in thermal energy is an ice calorimeter, shown schematically in Fig. 6-1. If heat flows into the calorimeter from some outside source, some of the ice will melt, the total ice-water volume will decrease, and the indicator level will fall. We then can use the height of the indicator level as a measure of the thermal energy that the calorimeter has gained. Such a device, which could be imagined to be made with other materials or operated at other pressures, provides a means of measuring thermal energies. The type of calorimeter used in practice that is the counterpart of the ice calorimeter is one in which, when heat flows into it or out of it from a process under study, an amount of thermal energy

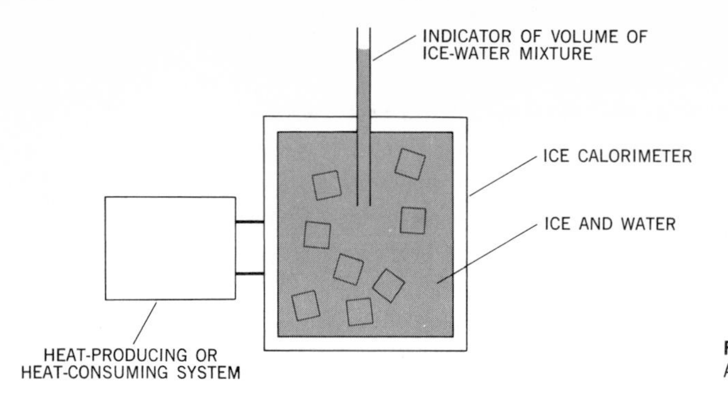

FIGURE 6-1
A schematic ice calorimeter.

is added to or taken from the calorimeter to maintain constant temperature. Then this energy is determined from the electrical devices used for the addition or removal of energy. In any case, you see that a measure of thermal energy is obtained, and this is done without any confusing or circuitous argument that depends on the use of a change in temperature.

We will frequently use schematic representations of *thermal reservoirs* that can exchange heat with some system in which we are interested. You can picture these as one or more isothermal calorimeters, or a series of calorimeters, like the ice calorimeter that can be set at any desired temperature.

Mechanical energy is best pictured, and satisfactorily defined, in terms of the potential energy that is stored when work is used to raise a weight against the force of gravity. The work that is done is defined as the force times the distance through which the force acts. Thus, if we accept, for simplicity, the constant gravitational pull at some point on the earth's surface, the storage of mechanical energy in a system like that of Fig. 6-2 is measured by the height to which a given mass is raised. This schematic device will be called a *mechanical reservoir* and will be used to measure the gain or loss of mechanical energy that

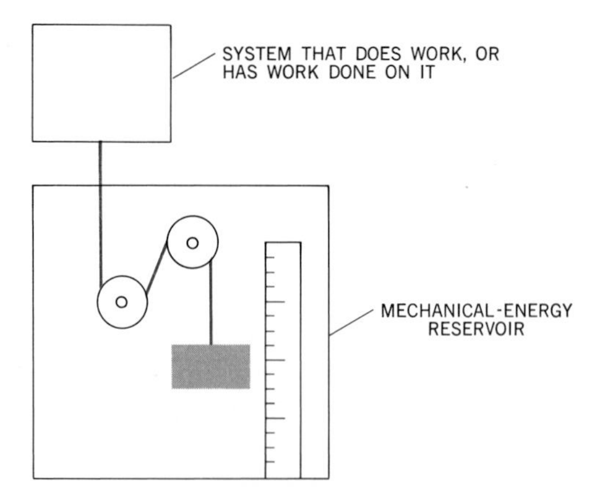

FIGURE 6-2
A schematic illustration of a mechanical-energy reservoir.

results when a system of interest does work or requires work to be done on it.

With thermal and mechanical reservoirs, measurements could be made that provide the first step toward the familiar conservation-of-energy principle and the first law of thermodynamics. Suppose various amounts of mechanical energy, as measured, say, by a change in position of the weights of Fig. 6-2, were converted to heat which flowed into the thermal reservoir. It would be found that the gain in thermal energy was proportional to the expenditure of mechanical energy. (This, you recall, was the result of James Joules' study of the mechanical equivalent of heat.) It is this proportionality that entitles us to use the common term *energy* for the changes that are measured by the two reservoirs. Furthermore, we can introduce a single energy unit that can be used for both types of energy. This unit, the joule, has already been introduced, and it happens to be defined in terms of mechanical energy. (If one proceeds by also introducing a thermal-energy unit, the calorie for example, the above studies allow the two energy units to be related, the relation being 1 cal = 4.184 J.)

Now we can introduce the symbols $\Delta E_{\text{thermal res}}$ and $\Delta E_{\text{mech res}}$ to denote changes in the energies of these reservoirs, and you see that these energies, expressed in the common unit of joules, can be interrelated.

6-3 The First Law of Thermodynamics

In practice it is not the energy changes that occur in the thermal and mechanical reservoirs that are of interest. Rather, it is what these energy changes can tell us about what is happening in a *system*—a collection of materials that are undergoing some physical, chemical, or biological change—that interests us. The first law of thermodynamics helps us to deduce the energy changes in the system from the measurable energy changes in the reservoirs. (When you reach the second law you will see the merits of keeping separate track of the thermal and mechanical energies.)

The system and the thermal-energy reservoir and the mechanical-energy reservoir that can be related to it are shown in the block diagram of Fig. 6-3. The three blocks provide the *universe* for the system and any process occurring in the system. Nothing else is affected by happenings in the system, and it is in this sense that the word universe is used.

Consider some change—a chemical reaction for instance—that occurs in the system and changes it from some state a, which implies certain chemical constituents at some pressure or temperature, to some other state b, implying different chemical constituents, or pressure, or temperature. Without any detailed information on the process that is

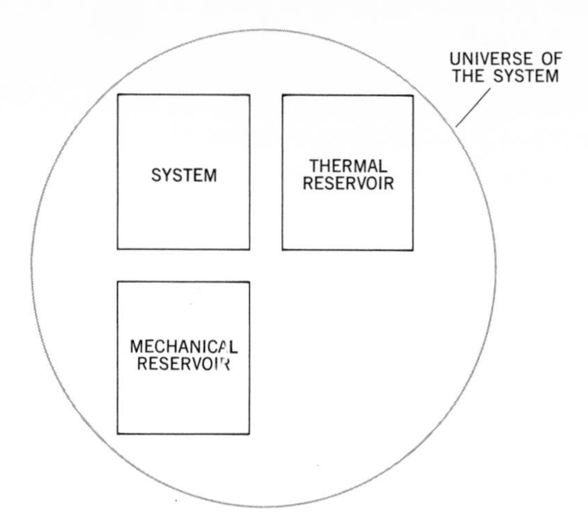

FIGURE 6-3
The relation between the system, the two energy reservoirs, and the universe of the system.

occurring in the system, we can measure $\Delta E_{\text{thermal res}}$ and $\Delta E_{\text{mech res}}$, and then we can use the conservation-of-energy principle to deduce the energy change $\Delta E_{a\to b}$ that has occurred in the system. The conservation principle applied to the universe of the process is expressed by

$$\Delta E_{\text{univ}} = 0$$

or

$$\Delta E_{\text{thermal res}} + \Delta E_{\text{mech res}} + \Delta E_{a\to b} = 0$$

The desired information about the system is then given by

$$\Delta E_{a\to b} = -\Delta E_{\text{thermal res}} - \Delta E_{\text{mech res}} \qquad [1]$$

It should be clear that for any change in the system, the energy change that must be attributed to the system to maintain the conservation-of-energy principle can be deduced. But in general, there are many ways of carrying a system from one state a to another state b, and at this stage, different values might be expected to be obtained for $\Delta E_{a\to b}$ for each of these different processes.

We would find, however, that if many different routes were taken from some initial state a to some final state b, that although $\Delta E_{\text{thermal res}}$ and $\Delta E_{\text{mech res}}$ might each be different for the various routes, the sum $-(\Delta E_{\text{thermal res}} + \Delta E_{\text{mech res}})$ would always be the same. In this way we could conclude that $\Delta E_{a\to b}$ depends only on the states a and b, and not on the way in which the system is brought from one state to the other. As a result, we can introduce E as the energy of the system, and since it is a function of the state of the system, we can write, for changes in E, the relation

$$\Delta E = E_b - E_a \qquad [2]$$

Note that no information is given, or needed, on how the system is changed from state a to state b.

An alternative, and more elegant, way of showing that $\Delta E_{a \to b}$ is dependent only on state a and b depends on imagining the system to be returned from b to a by a different path from that taken for the a to b process. If, then, it is imagined, to be specific, that $\Delta E_{a \to b}$ is positive and that for the return path $\Delta E_{b \to a} < -\Delta E_{a \to b}$, the reservoirs together would have had to gain some energy as a result of this cyclic process. The cycle could then be repeated to produce more reservoir energy, each cycle returning the system to its initial state. A perpetual-motion machine would result, and it is one of our generalizations of nature that this could not occur. Thus we must conclude that no matter what forward and reverse paths are taken, $\Delta E_{a \to b} = -\Delta E_{b \to a}$, or $\Delta E_{a \to b} = E_b - E_a$ and $\Delta E_{b \to a} = E_a - E_b$.

We now can state the conservation-of-energy principle in the useful form that constitutes the *first law of thermodynamics*.

The energy change of a system for any process that takes the system from state a to state b can be deduced from

$$\Delta E = -\Delta E_{\text{thermal res}} - \Delta E_{\text{mech res}} \qquad [3]$$

and ΔE is dependent only on the states a and b and can be treated in terms of the *state function* E according to

$$\Delta E = E_b - E_a \qquad [4]$$

The quantity E, which you now see implies a property of the system, just as does the volume V, is known as the *internal energy*.

In writing, and using, the first-law expression

$$\Delta E = -\Delta E_{\text{thermal res}} - \Delta E_{\text{mech res}}$$

or for infinitesimal changes $dE = -dE_{\text{thermal res}} - dE_{\text{mech res}}$, you must recognize that E is a function of the state of the system and that $E_{\text{thermal res}}$ and $E_{\text{mech res}}$ are functions of the states of the thermal and mechanical reservoirs. Thus, when processes carry the system from state a to state b, definite values for $\Delta E_{\text{thermal res}}$ and $\Delta E_{\text{mech res}}$ are not implied, even though the first law relates the sum of these quantities to ΔE. Thus the variables that determine changes in the state of the system cannot be used to treat changes in the reservoirs. The reservoir energies are determined by the "states" of the reservoirs, and these are simply dependent on, in one case, an indicator such as the volume index of an ice calorimeter and, in the other case, by the height of a weight.

Often, in thermodynamic treatments, we take a "system-centered" view, rather than one that treats the system and the thermal and mechanical surroundings on an equal basis. Then it is customary to limit state functions to those of the system. With this point of view and the new variables

$q =$ heat gained by system

$w =$ work gained by, or "done on," the system

we can write, instead of

$$\Delta E = -\Delta E_{\text{thermal res}} - \Delta E_{\text{mech res}}$$

the expression

$$\Delta E = q + w \qquad [5]$$

The absence of subscripts on the new terms q and w leads us to write them with lowercase letters to indicate that they are not functions of the state of the system. In a similar way, if an infinitesimal change in the state of the system is considered, we write

$$dE = \delta q + \delta w \qquad [6]$$

where the symbols remind us that δq and δw are not derivatives of system state functions.

As we proceed we will often use the relation

$$\Delta E = -\Delta E_{\text{thermal res}} - \Delta E_{\text{mech res}}$$

and have in mind the components of the universe of the system as shown in Fig. 6-3. At other times the system-centered viewpoint, and the more convenient words "heat" and "work" and symbols q and w, will be adequate.

6-4 Determination of ΔE: Reversible and Irreversible Processes

The determination of ΔE from measurements on the thermal and mechanical reservoirs and the fact that it is only the net effect of these reservoirs that is fixed for a given change in the system is illustrated by the three examples of Fig. 6-4. Of the infinite variety of intermediate processes that could be imagined, two extreme cases described as *irreversible* and *reversible* are shown. The former word is applied to any process that occurs at other than a state of balance; the latter to a process in which the driving forces that make the reaction proceed are offset only infinitesimally from a state of balance. The term reversible for this latter case is used because, as can be seen in the examples of Fig. 6-4, an infinitesimal change from the state of balance could reverse the process, i.e., raise the weight, compress the gas, or electrolyze the HCl solution. The importance of the reversible and irreversible classifications of processes will be apparent in second-law studies. Here they serve to categorize the many different ways in which processes can occur.

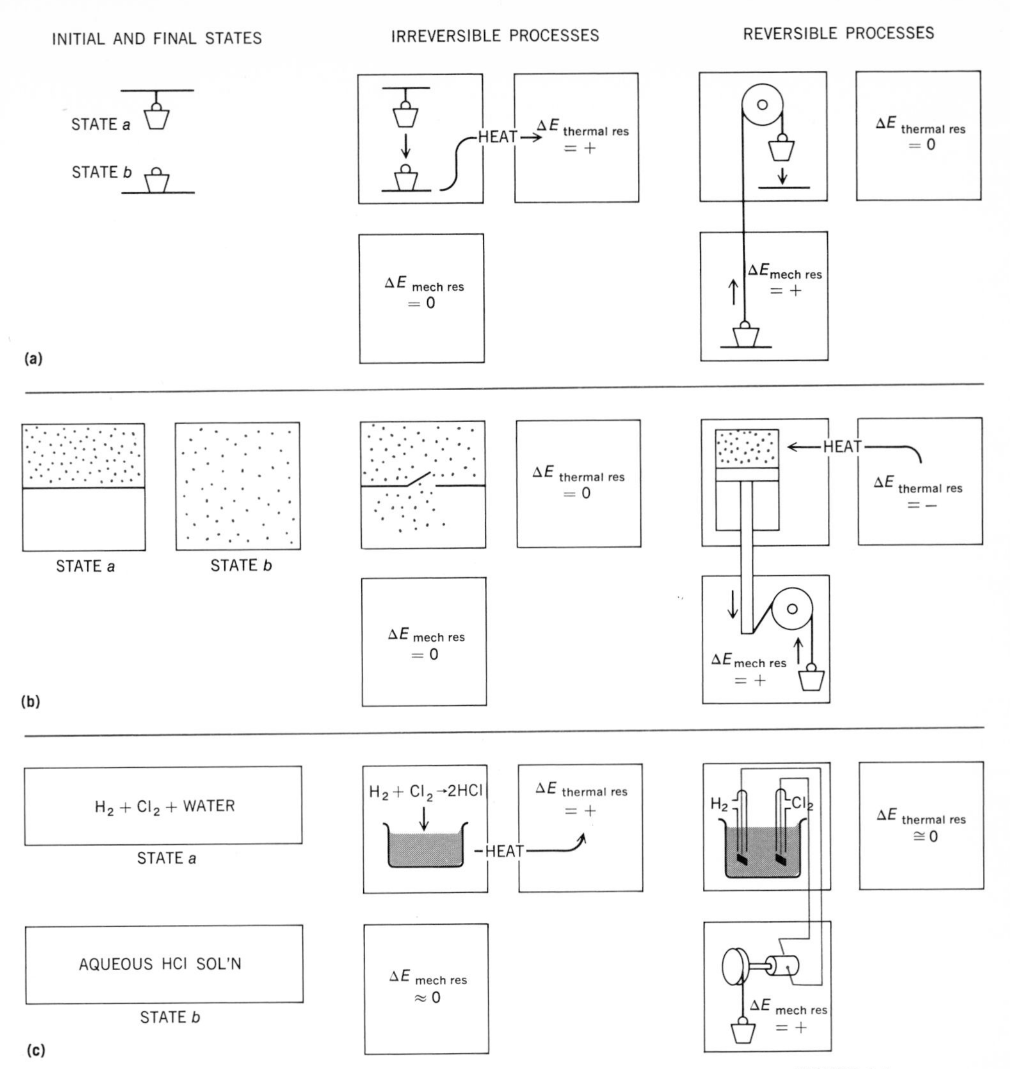

FIGURE 6-4
Some examples of the different processes that can be used to change the state of a system for which it can be seen, qualitatively, that $\Delta E = -\Delta E_{\text{thermal res}} - \Delta E_{\text{mech res}}$ is the same regardless of path.

6-5 $\Delta E_{\text{mech res}}$ for Expansions and Contractions of the System

In almost all the chemical examples to be considered in the first half of this book, the mechanical surroundings will gain or lose energy only as a result of the work done as the system expands or contracts against the confining pressure. (In later chapters electrochemical cells will be treated, and then the electrical output provides an additional route for the transfer of energy to the mechanical reservoir.) Here we will develop an expression for the energy changes corresponding to expansions and contractions.

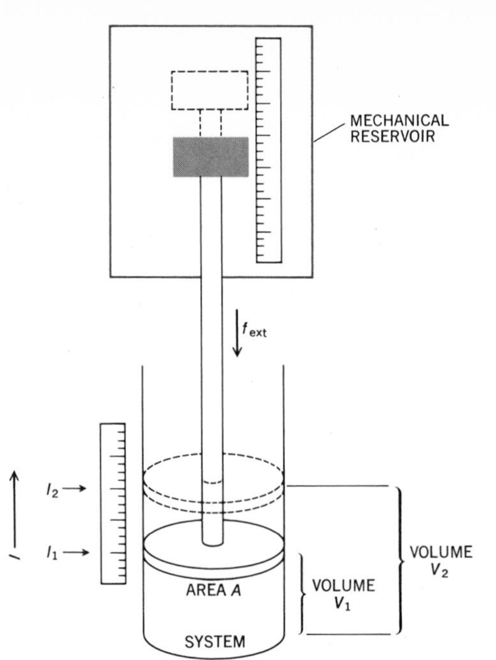

FIGURE 6-5
Illustration for the deduction that, for an expansion,

$$\Delta E_{\text{mech res}} = \int_{V_1}^{V_2} P_{\text{ext}}\, dV$$

Consider the system to occupy the cylinder of a piston arrangement, as in Fig. 6-5. The energy that can be transferred to the mechanical reservoir depends on the force that the connection to this reservoir exerts, called f_{ext} because it is external to the system, and the distance through which this force acts. Work is done by the mechanical reservoir if this force drives the piston down and compresses the system. Work is done by the system and stored in the mechanical reservoir if the piston shaft exerts a force that overbalances f_{ext} and drives it up. Thus we can write

$$\Delta E_{\text{mech res}} = \int_{l_1}^{l_2} f_{\text{ext}}\, dl$$

Insertion of the piston cross-section area A into the denominator and numerator converts this to

$$\Delta E_{\text{mech res}} = \int_{l_1}^{l_2} \frac{f_{\text{ext}}}{A} A\, dl$$

Since pressure is force per unit area, the quantity f_{ext}/A can be written P_{ext}, the pressure that acts on the piston from the outside. Furthermore, $A\, dl$ can be recognized to be the change in volume of the system and can be replaced by the symbol dV. Thus we can write

$$\Delta E_{\text{mech res}} = \int_{V_1}^{V_2} P_{\text{ext}}\, dV \qquad [7]$$

or

$$w = -\int_{V_1}^{V_2} P_{\text{ext}}\, dV$$

Calculations with this expression can be made only if the nature of the expansion or compression, i.e., the relation between P_{ext} and V, is specified.

For example, expansion against a constant external pressure—as is the case for systems which are open to the atmosphere—leads to

$$\Delta E_{\mathrm{mech\ res}} = -w = P_{\mathrm{ext}} \int_{V_1}^{V_2} dV = P_{\mathrm{ext}}(V_2 - V_1)$$
$$= P_{\mathrm{ext}}\, \Delta V \qquad [8]$$

Other special and important expansions are ones in which the expansion is balanced, or reversible, implying that the external pressure is only infinitesimally less than the external pressure throughout the expansion. The integration of Eq. [7] then requires a relation between P_{ext}, or P, where the absence of a subscript implies a system property, and the volume. If, for example, the system consists of n mol of an ideal gas, this relation is $PV = nRT$, and if the expansion is at constant temperature, the integration gives

$$\Delta E_{\mathrm{mech\ res}} = \int_{V_1}^{V_2} \frac{nRT}{V}\, dV$$
$$= nRT \int_{V_1}^{V_2} d(\ln V)$$
$$= nRT \ln \frac{V_2}{V_1} \qquad [9]$$

A numerical example will illustrate other features of the use of the basic relation given in Eq. [7].

Example Calculate ΔE for the conversion, at 100°C and 1 atm pressure, of 1 mol of water to steam. The latent heat of vaporization of water is 40,670 J mol^{-1}, the density of liquid water can be taken as 1 g ml^{-1}, and water vapor can be treated as an ideal gas.

The thermal reservoir must supply the latent heat of vaporization, and thus we have immediately

$$\Delta E_{\mathrm{thermal\ res}} = -40{,}670\ \mathrm{J}$$

The expansion is against a constant pressure of 1 atm, and so Eq. [7] gives

$$\Delta E_{\mathrm{mech\ res}} = P_{\mathrm{ext}}(V_2 - V_1)$$

The required quantities are

$$P_{\mathrm{ext}} = 1\ \mathrm{atm} = 101{,}325\ \mathrm{N\ m^{-2}}$$

$$V_1\ (= \text{volume of 1 mol of liquid water}) = 18\ \mathrm{ml} = 18 \times 10^{-6}\ \mathrm{m^3}$$

$$V_2\ (= \text{volume of 1 mol of water vapor}) = 0.0224 \times 373/273 = 0.0306\ \mathrm{m^3}$$

Thus

$$\Delta E_{\text{mech res}} = 101{,}325\,(0.0306 - 18 \times 10^{-6}) \\ = 101{,}325\,(0.0306) \\ = 3100 \text{ J}$$

Finally,

$$\Delta E = -\Delta E_{\text{thermal res}} - \Delta E_{\text{mech res}} \\ = -(-40{,}670) - (3100) \\ = +37{,}570 \text{ J}$$

You can notice that most of the energy received by the water from the thermal reservoir goes into raising the internal energy of the water as it goes from liquid to vapor. A small amount, shown by the increase in energy of the mechanical surroundings, is used in pushing back the atmosphere.

6-6 The Enthalpy Function

In studies of the energy changes associated with reactions of chemical interest, a great variety of chemical materials and transformations will be dealt with. The ways in which the transformations are carried out will also be of many types, but it is convenient to consider two conditions that are special and often encountered. One is that in which the *volume* of the system is kept constant, and the other that in which the *pressure* on the system is held constant. The latter situation is, for example, that existing for reactions, or other processes, carried out in containers open to the atmosphere.

For a constant-volume process the schematic thermal and mechanical surroundings, as in Fig. 6-3, can be used without any complication. We write

$$\Delta E = -\Delta E_{\text{thermal res}} - \Delta E_{\text{mech res}} \qquad [10]$$

For a constant-volume process, no PV work of the type treated in Sec. 6-5 is done, and therefore only some other form of work, as could result from an electrical output, for example, could lead to a $\Delta E_{\text{mech res}}$ term. If, as is often the case, no such work is done,

$$\Delta E_{\text{mech res}} = 0 \qquad [11]$$

and for such constant-volume processes we have the result, which is specially convenient for calorimetric studies,

$$\Delta E = -\Delta E_{\text{thermal res}} = q \qquad \text{[const volume]}^1 \qquad [12]$$

where q is the heat absorbed by the system.

[1] To emphasize the special conditions under which equations apply, reminders such as this will be added to the equations where these restrictions are easily overlooked.

Constant-pressure processes are different in that, generally, the volume of the system changes and work is done on or by the surroundings, and this leads to a $\Delta E_{\text{mech res}}$ term. In particular, if the system expands by an amount ΔV against a constant external pressure P, as shown in Fig. 6-5, an amount of energy $P\,\Delta V$ is gained by the mechanical reservoir. If we let $\Delta E'_{\text{mech res}}$ denote all other ways in which mechanical energy is gained by the reservoir, we can write

$$\Delta E = -\Delta E_{\text{thermal res}} - \Delta E'_{\text{mech res}} - P\,\Delta V \qquad [13]$$

But again, in many processes carried out at constant pressure, no mechanical work, other than through $P\,\Delta V$, is done. For such constant-pressure processes, we have

$$\Delta E = -\Delta E_{\text{thermal res}} - P\,\Delta V \qquad [14]$$

If a constant-pressure process involves an expansion, $P\,\Delta V$ will be positive, and in addition to any energy exchanged with the thermal surroundings, this much mechanical energy will be transferred to the surroundings. For a volume contraction, $P\,\Delta V$ is negative, and the system gains this amount of energy from the surroundings. We could proceed with these ideas and Eq. [14], but it is, in fact, a nuisance for all reactions carried out open to the atmosphere to have to include this $P\,\Delta V$ term along with the calorimeter term $\Delta E_{\text{thermal res}}$. It is more convenient to write

$$\Delta E + P\,\Delta V = -\Delta E_{\text{thermal res}} \qquad [15]$$

This lumping together of ΔE and $P\,\Delta V$ can be made part of our thermodynamic treatment by introducing a new energy term, labeled H, called the *enthalpy*, or *heat content*, defined by

$$H = E + PV \qquad [16]$$

For any process the change in H will then be

$$\begin{aligned}\Delta H &= \Delta E + \Delta(PV) \\ &= \Delta E + P\,\Delta V + V\,\Delta P \qquad [17]\end{aligned}$$

For any *constant-pressure* process, the final term is zero and we have, with Eq. [15],

$$\begin{aligned}\Delta H &= \Delta E + P\,\Delta V \\ &= -\Delta E_{\text{thermal res}} = q \qquad \text{[const pressure]} \qquad [18]\end{aligned}$$

Thus, for constant-pressure processes, calorimetric measurements which measure the energy change of the thermal reservoir lead directly to ΔH, that is, to a value of the change in $H = E + PV$ of the system.

Very generally, it is convenient to deal with E in connection with constant-volume processes and with $H = E + PV$ in connection with constant-pressure processes. Of course, for either type of process, one can calculate and deal with either or both the quantities ΔE and ΔH.

The enthalpy, like the energy, is a state function. That is, for a particular state of a system, the enthalpy has some particular value. This follows directly from the definition of H as the sum of two other state functions E and PV.

6-7 Some Properties of State Functions

To describe the properties of chemical substances and of the reactions they undergo, we will need to know how some of these properties are related to each other and how they are related to other variables. Some quite general mathematical relations are helpful.

The systems we will deal with, which so far have been left pretty much unspecified, will contain one or more chemical species at some temperature and pressure, and these species may change, or be considered to change, to other species or to other conditions of temperature and pressure. The nature and properties of such systems can be specified by specifying the type and amount of each chemical species and the physical conditions. The state of the system, i.e., all the properties of the material, are found to be fixed if, in addition to the amount of each of the materials, two variables, such as the temperature and pressure, are fixed. If the amounts of the materials are represented by the number of moles, $n_1, n_2, n_3, \ldots$, any property of the system could be expressed as a function of these molar amounts and two other variables. These variables are usually taken to be the temperature and pressure, and thus we can write, with f denoting some mathematical function,

$$\text{Property} = f(T,P,n_1,n_2,\ldots) \qquad [19]$$

For a system consisting of a single component, a term that will be carefully defined in Chap. 20, this would simplify to

$$\text{Property} = f(T,P,n) \qquad [20]$$

Properties that require the inclusion of amounts, as by $n_1, n_2, \ldots$, or n, in this function, are known as *extensive* properties, and examples are E, H, V, mass, and so forth.

Properties also belong to the class known as *intensive*, and these do not depend on the amount of material. Examples are the temperature, the density, and the quantities such as E, H, and V for a fixed amount of material. Thus the molar quantities E, H, and V would be examples of intensive properties. For a single-component material an intensive property is found to be adequately expressed by a function of two variables. Then, for a fixed quantity of a substance,

$$E = f(T,P) \qquad [21]$$

For multicomponent systems, which we will not treat in detail until Chap. 21, the composition dependence of an intensive property can be shown by including the mole fractions x_1, x_2, ... Thus, for a fixed amount of material

$$E = f(T,P,x_1,x_2, \ldots)$$

(Although the above comments are adequate for very many chemical systems, others need the specification of more than two variables to determine the state of the system. Thus, for example, a magnetic material would have properties that are appreciably affected by the magnetic field that is imposed, and to fix the state of such a system, three variables, such as T, P, and the magnetic field, would have to be assigned.)

It is not necessary to choose T and P as the two *independent* variables. For example, we could specify T and V, or temperature and density, and then, in general, the state, i.e., all the other properties including the pressure, would be fixed.

Let us now see what follows from the fact that we can write, for a variety of properties of the system, relations like Eq. [21] for the internal energy.

This analytical expression could, alternatively, be shown graphically, as is suggested in Fig. 6-6. From either the analytical or geometric approach—if the actual dependence were known—one could determine the separate dependence of E on P and T. In this way one would obtain the partial derivatives

$$\left(\frac{\partial E}{\partial T}\right)_P \quad \text{and} \quad \left(\frac{\partial E}{\partial P}\right)_T$$

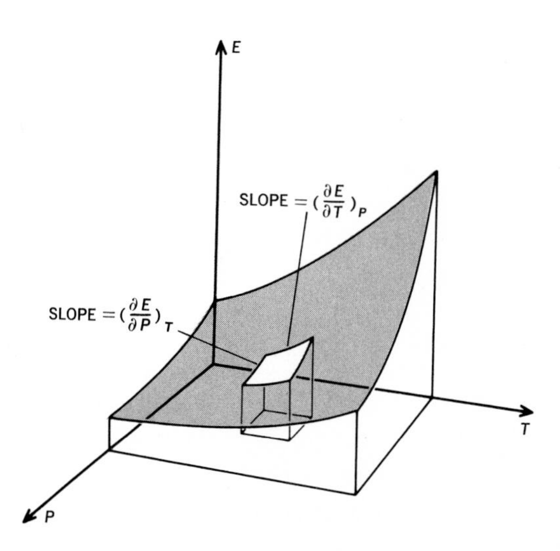

FIGURE 6-6
Two partial derivatives of the state function E.

where, it will be recalled, the subscript signifies the variable that is held constant. A general infinitesimal change in E can be written in terms of these slopes, or rates of change, and of the changes dT and dP that are involved in the general process. This leads to the *total differential*

$$dE = \left(\frac{\partial E}{\partial T}\right)_P dT + \left(\frac{\partial E}{\partial P}\right)_T dP$$

An often important relationship between the two partial derivative coefficients of such a total differential of a state function is

$$\frac{\partial}{\partial P}\left(\frac{\partial E}{\partial T}\right)_P = \frac{\partial}{\partial T}\left(\frac{\partial E}{\partial P}\right)_T$$

As new state functions are encountered, you will see that the total differential can be of value in the deduction of needed relations.

Since energies are usually more conveniently dealt with as E if T and V are the variables of interest, and as H if T and P are of interest, the most useful starting points for the development of many relationships are

$$dE = \left(\frac{\partial E}{\partial V}\right)_T dV + \left(\frac{\partial E}{\partial T}\right)_V dT \qquad [22]$$

and

$$dH = \left(\frac{\partial H}{\partial P}\right)_T dP + \left(\frac{\partial H}{\partial T}\right)_P dT \qquad [23]$$

For processes in which the total differential is known to be zero, we have a relation between the partial differential coefficients and the independent variables. For example, for $dE = 0$,

$$\left(\frac{\partial E}{\partial V}\right)_T dV = -\left(\frac{\partial E}{\partial T}\right)_V dT$$

or

$$\frac{dV}{dT} = -\frac{(\partial E/\partial T)_V}{(\partial E/\partial V)_T} \quad \text{or} \quad \frac{dT}{dV} = -\frac{(\partial E/\partial V)_T}{(\partial E/\partial T)_V} \qquad [24]$$

The total differential expression can also be used to show the way in which the dependent variable depends on one of the independent variables. For example, one can write, from Eq. [23],

$$\frac{dH}{dT} = \left(\frac{\partial H}{\partial T}\right)_P + \left(\frac{\partial H}{\partial P}\right)_T \frac{dP}{dT} \qquad [25]$$

Applications of expressions such as these that stem from total differential expressions will be made in the next section, in which ideal-gas behavior is explored, and in Sec. 6-10, where nonideal behavior is investigated.

6-8 The Expansion of an Ideal Gas: The Dependence of Internal Energy and Enthalpy of an Ideal Gas on Pressure

An important experiment by Joule in 1843 was directed at the study of the internal energy of a gas. These early insensitive experiments led to a conclusion that we now recognize as characterizing ideal-gas behavior. The experiment performed in an apparatus such as that depicted in Fig. 6-7 consisted in filling one of the two bulbs with air at a pressure of about 20 atm, the other bulb being evacuated. When the stopcock between the bulbs was opened, no change in the temperature of the water bath was observed.

It follows, since neither heat nor work was passed to the surroundings from the system (the system being the gas in the two-bulb assembly), that the first-law expression leads to a value of ΔE of zero for the expansion process. We thus have the situation that led us to Eq. [24], and again we can write, now with the explicit statement of a constant-E process,

$$\left(\frac{\partial T}{\partial V}\right)_E = -\frac{(\partial E/\partial V)_T}{(\partial E/\partial T)_V} \qquad [26]$$

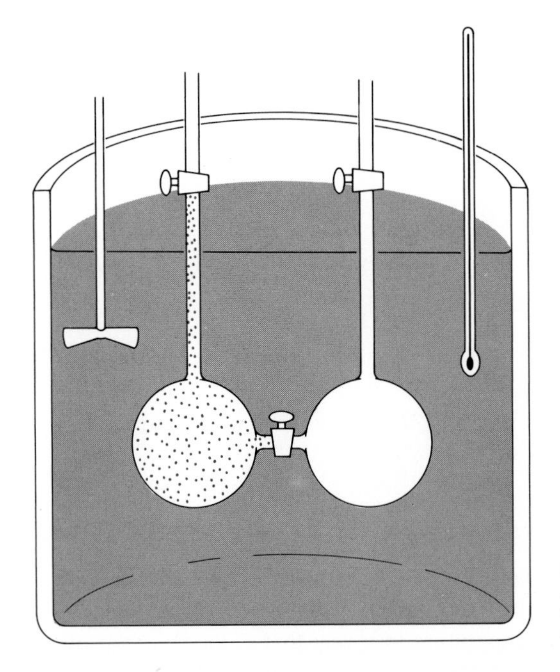

FIGURE 6-7
Apparatus for the Joule experiment on the heat effect of the expansion of a gas.

Joule's experimental result that a constant-E expansion produced no change in temperature means that the partial derivative $(\partial T/\partial V)_E$ is zero, which implies with Eq. [26] that $(\partial E/\partial V)_T$ is zero. This deduced result, that at a fixed temperature the internal energy of a gas is independent of its volume, is now taken as an additional characteristic of ideal behavior. The mathematical formulation of the consequences of this is seen by returning to the total differential for E, shown in Eq. [22], and inserting this ideal-gas characteristic $(\partial E/\partial V)_T = 0$. Then one has

$$dE = \left(\frac{\partial E}{\partial T}\right)_P dT \quad \text{or} \quad \frac{dE}{dT} = \left(\frac{\partial E}{\partial T}\right)_P \quad \text{[ideal gas]} \tag{27}$$

Thus, for an ideal gas, E changes with T for *any* process in the same way that it does for the pressure held constant.

This result is immediately understandable on the basis of our kinetic-molecular model and the discussion of thermal energy in the preceding chapter. The absence of attractions and repulsions between the molecules of a gas was seen to be the basis for obedience to the $PV = nRT$ law. The absence of such interactions, furthermore, means that the energy of the gas will be unaffected by the average distance between molecules and will depend only on their distribution throughout their allowed energy levels. For nonideal gases and for liquids and solids, on the other hand, these interactions do occur, and the molecules have a different average potential energy when the average intermolecular distance is changed. In general, therefore, E does depend on P and V as well as on T.

For an ideal gas, H, as well as E, is a function only of the temperature. From the defining equation $H = E + PV$, the derivative of H with respect to pressure at constant temperature can be formed as

$$\left(\frac{\partial H}{\partial P}\right)_T = \left(\frac{\partial E}{\partial P}\right)_T + \left[\frac{\partial (PV)}{\partial P}\right]_T \tag{28}$$

Recognition that, for an ideal gas, $(\partial E/\partial P)_T = 0$ and, from $PV = nRT$, that $[\partial(PV)/\partial P]_T = 0$ leads us to

$$\left(\frac{\partial H}{\partial P}\right)_T = 0 \tag{29}$$

A similar result would be found for $(\partial H/\partial V)_T$. Thus we can write, for the enthalpy of an ideal gas,

$$\left(\frac{\partial H}{\partial T}\right)_P = \left(\frac{\partial H}{\partial T}\right)_V = \frac{dH}{dT} \tag{30}$$

Again, ideal-gas behavior has converted the specific temperature dependence of H to the simpler result of Eq. [30], which shows that H, as well as E, varies with temperature in a way that is independent of any restriction, or lack of restriction, on the pressure and volume.

6-9 The Dependence of E and H on Temperature: The Heat Capacities C_P and C_V of Ideal Gases

The energies of chemical systems are influenced mostly by chemical reactions and by temperature changes. Reactions will be dealt with in the following chapter. Temperature effects can be treated here, and are done so in terms of the *heat capacities*.

As for the energy content, it is convenient to deal separately with heat capacities measured at constant volume, indicated by C_V, and those measured at constant pressure, C_P. The definitions of these heat capacities, which can be appreciated by imagining the type of experiment that provides heat-capacity data, are

$$C_V = \lim_{\Delta T \to 0} \left(\frac{-\Delta E_{\text{thermal res}}}{\Delta T} \right)_{V \text{ of system const}} \tag{31}$$

$$C_P = \lim_{\Delta T \to 0} \left(\frac{-\Delta E_{\text{thermal res}}}{\Delta T} \right)_{P \text{ of system const}} \tag{32}$$

That is, the heat capacities are the thermal energy that must be added to the system per unit temperature rise under the specified conditions.

Furthermore, for all constant-volume processes, we can write $\Delta E = -\Delta E_{\text{thermal res}}$, and for all constant-pressure processes, for which only PV work is involved, we can write $\Delta H = -\Delta E_{\text{thermal res}}$. These relations allow the definitions of C_V and C_P to be converted to

$$C_V = \left(\frac{\partial E}{\partial T} \right)_V \tag{33}$$

$$C_P = \left(\frac{\partial H}{\partial T} \right)_P \tag{34}$$

Notice that these are quite generally applicable, no assumption of ideal-gas behavior having been made.

Now let us see how C_V and C_P are related. Let us form the difference and then reduce the result to one involving only E, instead of E and H, by using $H = E + PV$. We have

$$\begin{aligned} C_P - C_V &= \left(\frac{\partial H}{\partial T} \right)_P - \left(\frac{\partial E}{\partial T} \right)_V \\ &= \left(\frac{\partial E}{\partial T} \right)_P - P\left(\frac{\partial V}{\partial T} \right)_P - \left(\frac{\partial E}{\partial T} \right)_V \end{aligned} \tag{35}$$

You could, at this stage, see the simplification that results from the assumption of ideal-gas behavior, but let us proceed to a more generally useful equation before this is done. The first term of Eq. [35] can be converted to an equation involving a $(\partial E/\partial T)_V$ term by using the total differential result, like Eq. [25], namely,

$$\frac{dE}{dT} = \left(\frac{\partial E}{\partial V} \right)_T \frac{dV}{dT} + \left(\frac{\partial E}{\partial T} \right)_V \tag{36}$$

or stipulating constant pressure,

$$\left(\frac{\partial E}{\partial T}\right)_P = \left(\frac{\partial E}{\partial V}\right)_T \left(\frac{\partial V}{\partial T}\right)_P + \left(\frac{\partial E}{\partial T}\right)_V \qquad [37]$$

Substitution of this result for the first term of Eq. [35] and rearranging gives

$$C_P - C_V = \left[P + \left(\frac{\partial E}{\partial V}\right)_T\right]\left(\frac{\partial V}{\partial T}\right)_P \qquad [38]$$

This general relation can be used now to investigate ideal-gas behavior. The term $(\partial E/\partial V)_T$ is then zero, and the remaining term $P(\partial V/\partial T)_P$ is found, with the relation $P\text{v} = RT$ for 1 mol of an ideal gas, to have the value R. Thus

$$\text{c}_P - \text{c}_V = R \qquad \text{[1 mol of an ideal gas]} \qquad [39]$$

For real gases, as well as liquids and solids, the internal energy does depend on the volume, and thus $(\partial E/\partial V)_T$ cannot be set equal to zero, and furthermore, $(\partial V/\partial T)_P$ cannot be assigned a value independent of the particular material. In Sec. 19-3 you will see how the difference in heat capacities is related to measurable properties of materials.

6-10 The Expansion of Real Gases: The Joule-Thomson Coefficient

Following the experiments of Joule that failed to detect a heat effect when a gas was allowed to expand freely, more refined and sensitive experiments of a different type, but directed toward the same end, were performed by Joule and Thomson. Since these results on the temperature change experienced by gases, as these gases expand freely, are of great practical importance in the liquefaction of gases, as we shall see later, it is now necessary to report the type of experiment and the results of Joule and Thomson.

In their experiments a flow system was used which allowed the gas to pass from a high pressure to a low pressure through a throttling valve. The system was thermally insulated, so that, as the gas passed through the valve, no heat could be absorbed or given off; that is, q, or $\Delta E_{\text{thermal res}}$, equals zero. Such an absence of heat transfer is frequently encountered, as in reactions that occur very rapidly or in reactions that occur in insulated vessels, and such processes for which $\Delta E_{\text{thermal res}}$, or q, equals zero, are said to be *adiabatic*.

In a Joule-Thomson experiment, measurements of the pressure and temperature on either side of the throttling valve are made, and the results reported are for the temperature change per unit pressure change, as shown in Table 6-1. In contrast to the expansions dealt

TABLE 6-1 Results obtained from Joule-Thomson experiments at 1 atm pressure
These experiments can be seen to give constant enthalpy values, and thus the data are for the partial derivative $(\partial T/\partial P)_H$, also denoted by the Joule-Thomson coefficient μ_{JT}.

t°C	$\frac{dT}{dP}$ for He (deg atm^{-1})	$\frac{dT}{dP}$ for N_2 (deg atm^{-1})
−100	−0.058	0.649
0	−0.062	0.266
100	−0.064	0.129
200	−0.064	0.056

with in Sec. 6-8, measurable, but generally small, temperature changes can be achieved.

Experimental results show that a gas can cool down or heat up in a Joule-Thomson expansion. At room temperature all gases except hydrogen and helium are found to be cooled by such an expansion; i.e., the entries for Table 6-1 would be positive. Studies of the temperature dependence of the expansion effect show that all gases change over as the temperature is lowered from a heating to a cooling effect, the changeover temperature being known as the *inversion temperature*. The inversion temperature for hydrogen is found to be 193 K, and that for helium is 53 K. Below these temperatures these gases also are cooled by a Joule-Thomson expansion. The practical importance of this will be pointed out in Sec. 8-6.

For this adiabatic-expansion process, for example that experienced by 1 mol of the gas as it moves through the throttling valve, $\Delta E_{\text{thermal res}}$ is zero; but what is $\Delta E_{\text{mech res}}$? On the left of the valve in Fig. 6-8, the mechanical reservoir expends energy as the gas is pushed at a constant pressure P_1 from a volume of v_1 to zero volume. Thus

$$(\Delta E_{\text{mech res}})_1 = \int_{\text{v}_1}^{0} P_1\, dV = P_1 \int_{\text{v}_1}^{0} dV = -P_1\text{v}_1$$

On the right, work is done by the 1 mol of emerging gas as it pushes

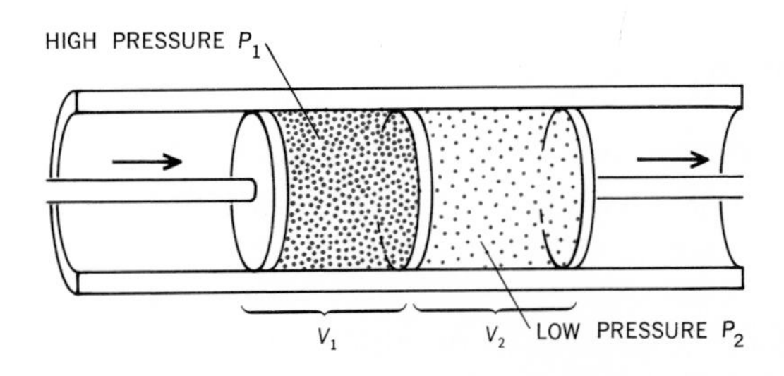

FIGURE 6-8
Schematic representation of the Joule-Thomson throttling process.

the piston, which exerts a pressure P_2 from zero volume to a volume v_2. For this

$$(\Delta E_{\text{mech res}})_2 = \int_0^{\mathrm{V}_2} P_2\, dV = P_2 \int_0^{\mathrm{V}_2} dV = P_2\mathrm{v}_2$$

Thus, for the passage of 1 mol of gas,

$$\Delta E_{\text{mech res}} = P_2\mathrm{v}_2 - P_1\mathrm{v}_1 \qquad [40]$$

Further, since $\Delta E_{\text{thermal res}}$ is zero, the first law gives us

$$\Delta E = -\Delta E_{\text{mech res}} = -(P_2\mathrm{v}_2 - P_1\mathrm{v}_1) \qquad [41]$$

or

$$\mathrm{E}_2 - \mathrm{E}_1 = -(P_2\mathrm{v}_2 - P_1\mathrm{v}_1)$$

This can also be written

$$\mathrm{E}_2 + P_2\mathrm{v}_2 = \mathrm{E}_1 + P_1\mathrm{v}_1$$

or finally,

$$\mathrm{H}_2 = \mathrm{H}_1 \qquad \text{or} \qquad \Delta H = 0 \qquad [42]$$

With this understanding we can use the expression that we obtain by setting the total differential for H equal to zero to obtain the equation, comparable with Eq. [24],

$$\left(\frac{\partial T}{\partial P}\right)_H = -\frac{(\partial H/\partial P)_T}{(\partial H/\partial T)_P} \qquad [43]$$

Now you can recall that $(\partial H/\partial T)_P = \mathrm{c}_p$, and if μ_{JT} is introduced as the *Joule-Thomson coefficient,* the quantity that is evaluated from the throttling experiment, we obtain

$$\left(\frac{\partial H}{\partial P}\right)_T = -\mu_{\mathrm{JT}}\mathrm{c}_p \qquad [44]$$

We have with this result, and for the other partial derivative $(\partial \mathrm{H}/\partial T)_P = \mathrm{c}_p$, a complete description of the TP surface for H, a fact that is shown analytically by the total differential expression, which now can be written

$$d\mathrm{H} = (-\mu_{\mathrm{JT}}\mathrm{c}_p)\, dP + (\mathrm{c}_p)\, dT \qquad [45]$$

6-11 Adiabatic Expansions of Ideal Gases

Adiabatic processes are of sufficient importance in the development of thermodynamic concepts so that it is worthwhile here investigating, for the simple case of an ideal gas, the PVT behavior of a gas expanding in this way. These results will be most immediately appreciated by comparing them with the PVT behavior for an isothermal expansion, namely, T constant and $PV =$ constant.

For n mol of an ideal gas, we can write, according to Sec. 6-8,

$$\frac{dE}{dT} = n\text{c}_V \qquad \text{or} \qquad dE = n\text{c}_V\, dT \tag{46}$$

Likewise, for a reversible expansion process the energy transferred to the mechanical reservoir is

$$dE_{\text{mech res}} = P\, dV = \frac{nRT}{V}\, dV \tag{47}$$

For an adiabatic process $dE_{\text{thermal res}} = 0$, and the first law requires $dE = -dE_{\text{mech res}}$, or

$$n\text{c}_V\, dT = -nRT\frac{dV}{V}$$

or

$$\frac{\text{c}_V}{R}\frac{dT}{T} = -\frac{dV}{V} \tag{48}$$

For a process that takes the gas from a volume V_1 at a temperature T_1 to a new volume V_2 at a temperature T_2, one has

$$\frac{\text{c}_V}{R}\int_{T_1}^{T_2}\frac{dT}{T} = -\int_{V_1}^{V_2}\frac{dV}{V}$$

or

$$\frac{\text{c}_V}{R}\ln\frac{T_2}{T_1} = -\ln\frac{V_2}{V_1} \tag{49}$$

On rearrangement, and after antilogarithms have been taken, this result can be written

$$V_1 T_1^{\text{c}_V/R} = V_2 T_2^{\text{c}_V/R} \tag{50}$$

A frequently more useful arrangement is obtained by substituting for T_1 and T_2 the ideal-gas-law expression $T = PV/nR$ and $\text{c}_P = \text{c}_V + R$ to get

$$P_1 V_1^{\text{c}_P/\text{c}_V} = P_2 V_2^{\text{c}_P/\text{c}_V} \tag{51}$$

The ratio of the two heat capacities frequently occurs, and is given the designation

$$\gamma = \frac{\text{c}_P}{\text{c}_V} \tag{52}$$

With this notation the variation of pressure and volume of a reversible adiabatic process involving an ideal gas is given by

$$P_1 V_1^{\gamma} = P_2 V_2^{\gamma} \tag{53}$$

or

$$PV^{\gamma} = \text{const} \tag{54}$$

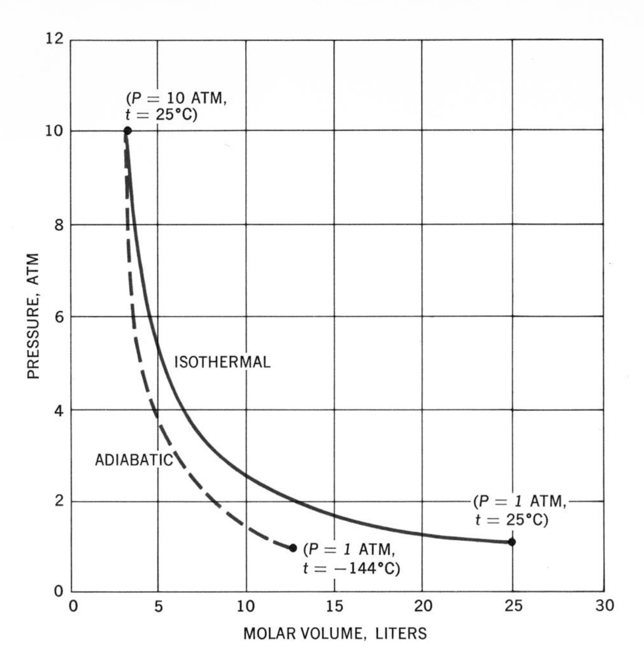

FIGURE 6-9
Reversible isothermal and adiabatic expansions of 1 mol of N_2 from 10 atm and 25°C to 1 atm. (For N_2 at 25°C, $\gamma = 1.40$.)

On plots of P versus V, curves for adiabatic processes are steeper than are those for isothermal processes, as is indicated in Fig. 6-9. When a gas expands isothermally, heat is absorbed to make up for the work done by the gas. For an adiabatic expansion the work of expansion uses up the thermal energy of the gas. As a result the temperature falls, and the pressure change, for a given expansion, is greater than in the corresponding isothermal expansion.

6-12 A Molecular Interpretation of E and H of Ideal Gases

In the preceding chapter it was pointed out that in the molecular interpretation of energies it is helpful to divide the energy of a molecule into two terms. The first of these terms represents the energy of the system for all molecules in their lowest allowed energy level. In anticipation of the introduction of the internal-energy property, this was denoted by E_0. The second term gives the thermal energy, denoted by $\text{E} - \text{E}_0$, which represents the thermal, or excitation, energy. The use of the symbols E_0 and $\text{E} - \text{E}_0$, the sum of which must be E, is in keeping with the use of this symbol for the thermodynamic internal energy.

The enthalpy function can also be written for molecular interpretation. The defining equation $\text{H} = \text{E} + P\text{V}$ for 1 mol of gas leads, with $\text{E} = \text{E}_0 + (\text{E} - \text{E}_0)$, to

$$\text{H} = \text{E}_0 + (\text{E} - \text{E}_0) + P\text{V} \tag{55}$$

or for an ideal gas,

$$\begin{aligned} H &= E_0 + (E - E_0) + RT \\ &= E_0 + [(E - E_0) + RT] \end{aligned} \qquad [56]$$

where the square brackets indicate the temperature-dependent, thermal contribution to the enthalpy.

A calculation of the thermal-energy functions $E - E_0$ and $H - E_0$ for NO_2 will illustrate the molecular approach to thermodynamic quantities. For this molecule, which will later be shown to be nonlinear, there will be 3 translational degrees of freedom, 3 rotational degrees, and $3n - 6 = 3$ vibrational degrees. A spectroscopic study shows that the three vibrational modes have energy-level schemes with spacings of 1.49, 2.63, and 3.21×10^{-20} J per molecule.

With these data we can calculate the energy that 1 mol of NO_2 gas has as a result of the molecules occupying energy levels other than the lowest available ones. The calculation is summarized in Table 6-2. As will be recalled from Chap. 5, only the vibrational contribution requires a detailed calculation. The thermal-enthalpy function $H - E_0$ is obtained, by reference to Table 6-2 and Eq. [56], as

$$\begin{aligned} (E - E_0) + RT &= 7718 + 2479 \\ &= 10{,}197 \text{ J} \end{aligned} \qquad [57]$$

At first it is perhaps more informative to write

$$\begin{aligned} E &= E_0 + 7718 \text{ J} \\ H &= E_0 + 10{,}197 \text{ J} \end{aligned} \qquad [58]$$

In this form the breakdown of the internal energy and the enthalpy into a lowest-level energy and a thermal-energy component is more apparent.

The calculation of such quantities should further emphasize that, although in thermodynamic calculations changes in E and H are often calculated from measured changes in the thermal and mechanical

TABLE 6-2 The calculation of the thermal energy of NO_2 at 25°C

Contributing term			*Energy (J mol⁻¹)*	
Translation				3718
Rotation				3718
Vibration:				
$h\nu_{vib} = 1.49 \times 10^{-20}$		$x = 3.62$	$\dfrac{RTx}{e^x - 1} = 247$	
$= 2.63 \times 10^{-20}$		$= 6.39$	27	
$= 3.21 \times 10^{-20}$		$= 7.80$	8	282
			$E - E_0 =$	7718

TABLE 6-3 The enthalpy function $H - E_0 = E - E_0 + RT$, in J mol^{-1}, at various temperatures*

	298.15 K	600 K	1000 K	1500 K
H_2	8467	17,274	29,145	44,744
O_2	8660	17,904	31,367	49,272
CO	8672	17,612	30,361	47,525
CO_2	9364	22,269	42,769	71,145
H_2O	9906	20,427	36,016	57,940
CH_4	10,029	23,217	48,367	88,408
Ethane	11,950	33,539	76,484	144,348
Ethylene	10,565	28,167	61,756	113,386
Acetylene	10,008	25,635	50,585	85,969
Benzene	14,230	51,400	126,202	239,952

*F. A. Rossini et al. (eds.), Tables of Selected Values of Chemical Thermodynamic Properties, *Natl. Bur. Std. (U.S.) Circ.* 500, 1952.

reservoirs, the functions E and H are state functions and can be calculated for particular conditions without regard to changes or paths. The calculations leading to Eqs. [58], furthermore, again show that, *for ideal gases,* E and H are functions only of the temperature. The thermal contributions to E and H for an ideal gas can be calculated if, in addition to the properties of the molecules of the gas, only the value of the temperature is given.

Although the lowest-level energy term E_0 has yet to be dealt with, one should see that the molecular approach to thermodynamic properties provides an additional valuable way in which to look at and understand the internal energy and the enthalpy. It is also true that such calculations provide very useful information on thermodynamic properties that are difficult to obtain by the classical thermodynamic methods. The thermal-enthalpy terms of Table 6-3, for example, are taken from an extensive tabulation that has been prepared for practical use from calculations such as that illustrated above. One should notice, after having followed through the calculation for NO_2, that the calculations of such functions can be done at any temperature, such as 1000 and 1500 K, and that in this way thermodynamic data for relatively inaccessible conditions can easily be obtained.

Use of these functions for problems of chemical interest will be postponed until the following chapter. There the procedures for dealing with the E_0 term, which our molecular approach has introduced but failed to treat, will be developed.

6-13 Molecular Interpretation of C_V and C_P

Since the heat capacities C_V and C_P are given by the derivatives with respect to temperature of E and H, the explicit expressions for the

TABLE 6-4 The factors that enter into the calculation of c_V for 1 mole of an ideal gas

Translation	$\frac{d}{dT}(\frac{3}{2}RT) = \frac{3}{2}R$
Rotation:	
Nonlinear	$\frac{d}{dT}(\frac{3}{2}RT) = \frac{3}{2}R$
Linear	$\frac{d}{dT}(RT) = R$
Vibration $\left(x = \frac{h\nu_{\text{vib}}}{kT}\right)$	$\frac{d}{dT}\left(\frac{\mathfrak{N}\, h\nu_{\text{vib}}}{h\nu_{\text{vib}}/kT - 1}\right) = \frac{Rx^2e^x}{(e^x - 1)^2}$
Electronic	$= 0$

thermal contributions to these latter quantities can be used to obtain the corresponding contributions to c_P and c_V. The derivatives that are needed are shown in Table 6-4. With these expressions and information on the linear or nonlinear structure of the molecule and on the energy spacings in the vibrational modes, the heat capacities can be calculated according to

$$c_V = c_{\text{trans}} + c_{\text{rot}} + c_{\text{vib}} + c_{\text{elec}} \qquad [59]$$

and

$$c_P = c_V + R \qquad [60]$$

Only the c_{vib} term appears to require appreciable calculations to be made, it being assumed, as is the case for most molecules at moderate temperatures, that $c_{\text{elec}} = 0$. For some purposes the graphical treatment of the c_{vib} function, as is given in Fig. 6-10, provides adequate

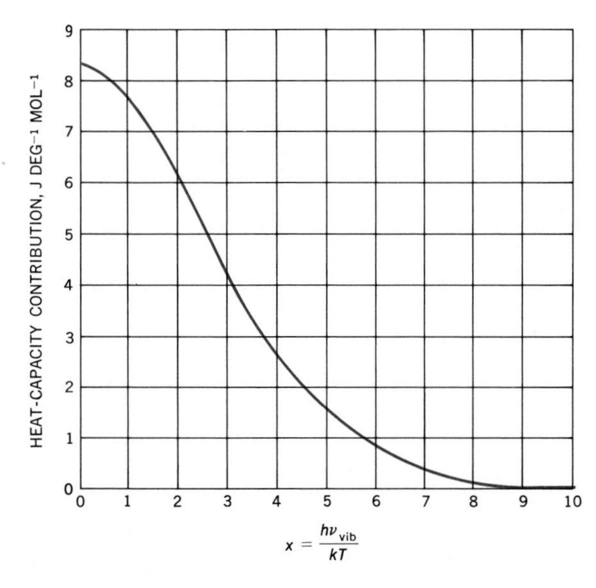

FIGURE 6-10 The heat-capacity contribution of a vibrational mode as a function of $x = h\nu_{\text{vib}}/kT$.

values. Such graphs, or the extensive tables of values of the function for c_{vib} that are available, often allow the numerical calculation of this quantity to be avoided.

More informative than Fig. 6-10 is Fig. 6-11, in which the heat-capacity contribution of a vibrational mode is shown as a function of temperature for various vibrational-energy spacings. Notice that for any spacing the heat-capacity contribution approaches the classical value of R at high temperatures, but that this limit is reached more easily by the vibrational modes with small steps between the allowed energies. Notice also that at low temperatures the heat-capacity contributions approach zero, a result that is understood by the inability of the molecules, at these temperatures, to acquire enough energy to move to any level higher than that of the lowest $v = 0$ state.

Again, NO_2 vapor can be used as an example to illustrate the calculation of thermodynamic properties from molecular data. The contributions to c_V are evaluated in Table 6-5 and give the result $c_V = 28.79$ J deg^{-1} mol^{-1}. With this result, furthermore,

$$\begin{aligned} c_P &= c_V + R \\ &= 28.79 + 8.31 = 37.10 \text{ J deg}^{-1} \text{ mol}^{-1} \end{aligned}$$

Table 6-6 shows results for some other simple molecules.

These heat-capacity calculations are the first of a number of valuable thermodynamic results that can be completely deduced by a molecular approach. The calculation of a property, such as the heat capacity, from the properties of the molecules which make up the gas is a considerable feat. Such calculations, furthermore, are of very great practical value. Heat capacities at high temperatures, for example, are frequently needed and are not easily measured. The molecular calculation, however, often provides a fairly easy way of obtaining otherwise inaccessible values of c_V and c_P.

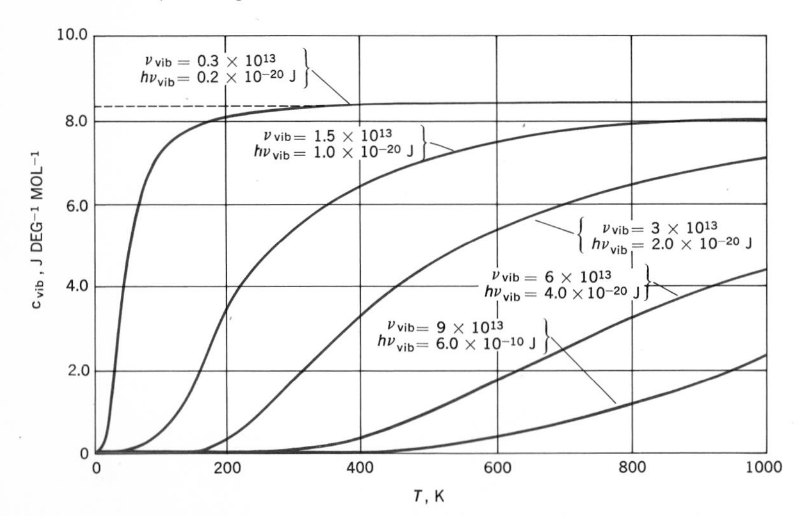

FIGURE 6-11
The heat-capacity contributions of vibrational modes with various energy spacings as a function of temperature.

TABLE 6-5 The calculation of the heat capacity of NO_2 at 25°C

Contributing term		Heat-capacity contribution ($J\ deg^{-1}\ mol^{-1}$)
Translation		$\frac{3}{2}R = 12.47$
Rotation		$\frac{3}{2}R = 12.47$
Vibration:		
$h\nu_{vib} = 1.49 \times 10^{-10}$	$x = 3.62$	$\frac{Rx^2e^x}{(e^x - 1)^2} = 3.07$
$h\nu_{vib} = 2.63$	$x = 6.39$	$\frac{Rx^2e^x}{(e^x - 1)^2} = 0.57$
$h\nu_{vib} = 3.21$	$x = 7.80$	$\frac{Rx^2e^x}{(e^x - 1)^2} = 0.21$
Electronic		$= 0$
		$c_V = 28.79$

In some cases, on the other hand, it is easier to make direct thermal measurements of c_P for gases than it is to do the calculation like that illustrated for NO_2. For example, spectroscopic data might not be available to show the energy-level spacing in one of the vibrational degrees of freedom, as is the case for the "torsional" motion in ethane in which one CH_3 group executes a "hindered" rotation relative to the other CH_3 group. In such cases, measured values for the thermodynamic quantity c_P can be used to deduce something about the molecular motion.

Finally, you should notice that the heat capacity is completely determined by the distribution of the molecules of a sample throughout the available energy levels. This contrasts with the internal-energy and enthalpy properties for which the thermal, temperature-dependent energies $E - E_0$ and $H - E_0$, with numerical values as given for NO_2 in Eq. [58], generally play minor roles when changes in the total internal energy $E = E_0 + (E - E_0)$ and enthalpy $H = E_0 + (H - E_0)$ are studied. Yet to be introduced is another thermodynamic property, the *entropy,* that stands with internal energy or enthalpy in importance but, like the heat capacity, is completely determined by the distribution of molecules throughout the available states.

TABLE 6-6 Constant-pressure heat capacities calculated from molecular properties*

Gas	$c_P(J\ deg^{-1}\ mol^{-1})$ *at 25° C*
H_2	28.8
N_2	29.1
O_2	29.4
HCl	29.1
CO	29.2
H_2O	33.6
CO_2	37.1
SO_2	39.8
NH_3	35.6
CH_4	35.7

* From F. A. Rossini et al. (eds.), Tables of Selected Values of Chemical Properties, *Natl. Bur. Std. (U.S.) Circ.* 500, 1952.

Problems

1 Calculate the work that can be done by a mass of 400 g, that is, 0.400 kg, falling a distance of 2.75 m. How much heat would be produced if this mass were allowed to fall freely this distance? What would be the value of $\Delta E_{\text{mech res}}$ in the first case and $\Delta E_{\text{thermal res}}$ in the second?

2 What is the heat of vaporization of benzene if 1.34 A of current passing through an electric heater of 50 Ω resistance for 5.62 min vaporizes 78.1 g of benzene? (Electrical heat is calculated from I^2R, where I is the current and R the resistance.) *Ans.* 30,300 J mol^{-1}.

3 One mole of an ideal gas is allowed to expand against a piston that supports 0.4 atm pressure. The initial pressure is 10 atm, and the final pressure is 0.4 atm, the temperature being kept constant at 0°C.

- **a** How much energy is transferred to the mechanical surroundings from the gas during the expansion? *Ans.* 2180 J.
- **b** What is the change in internal energy and in the enthalpy of the gas? *Ans.* $\Delta E = \Delta H = 0$.
- **c** How much heat is absorbed from the thermal surroundings? *Ans.* 2180 J.

4 One mole of an ideal gas is allowed to expand reversibly, i.e., against a confining pressure that is at all times infinitesimally less than the gas pressure, from an initial pressure of 10 atm to a final pressure of 0.4 atm, the temperature being kept constant at 0°C.

- **a** How much work is done by the gas? *Ans.* 7310 J.
- **b** What is the change in E and in H? *Ans.* $\Delta E = \Delta H = 0$.
- **c** How much heat is absorbed by the gas? *Ans.* 7310 J.

5 A chemical reaction in a gas mixture at 500°C decreases the number of moles of gas which can be assumed to behave ideally by 0.347. If the internal-energy change is +23.8 kJ, what is the enthalpy change?

6 The densities of ice and water at 0°C are 0.9168 and 0.9998 g ml^{-1}, respectively. Calculate the difference between ΔH and ΔE of fusion for 1 mol under atmospheric pressure. The density of liquid water at 100°C is 0.9584 and of water vapor at 100°C and 1 atm is 0.000596 g ml^{-1}. Calculate the difference between ΔH and ΔE for the vaporization of water at atmospheric pressure.

7 A cylinder containing 1 mol of liquid water at 100°C is heated until the liquid is converted to vapor. The cylinder is fitted with a piston which just resists a pressure of 1 atm. How much work is done by the expanding gas? If the heat of vaporization of water is 40,670 J, what is the change in internal energy? *Ans.* 3100 J; $\Delta E = 37{,}570$ J.

8 Three-tenths mole of CO is heated at a constant pressure of 10 atm from 0 to 250°C. The molar heat capacity of CO at constant pressure is $c_P = 26.86 + 6.97 \times 10^{-3}T - 8.20 \times 10^{-7}T^2$ J. If CO can be assumed to behave ideally, calculate $\Delta E_{\text{thermal res}}$, $\Delta E_{\text{mech res}}$, ΔE, and ΔH for the process.

9 Calculate $\Delta E_{\text{thermal res}}$, $\Delta E_{\text{mech res}}$, ΔE, and ΔH when 3.45 g of liquid CCl_4 is heated from 0 to 25°C at a pressure of 1 atm. The coefficient of thermal expansion of CCl_4 is 0.00118 per degree Celsius, the density at 20°C is 1.595 g ml^{-1}, and the molar heat capacity at constant pressure of liquid CCl_4 is 129 J deg^{-1} mol^{-1}.

10 Hydrogen gas is expanded reversibly and adiabatically from a volume of 1.43 liters, at a pressure of 3 atm and temperature of 25°C, until the volume is 2.86 liters. The heat capacity c_P of hydrogen can be taken to be 28.8 J deg^{-1} mol^{-1}.

- **a** Calculate the pressure and temperature of the gas, assumed to be ideal, after the expansion. *Ans.* 1.137 atm, 226 K.
- **b** Calculate $\Delta E_{\text{thermal res}}$, $\Delta E_{\text{mech res}}$, ΔE, and ΔH for the gas.
Ans. $\Delta E_{\text{thermal res}} = 0$; $\Delta E_{\text{mech res}} = +62.0$; $\Delta E = -259$ J; $\Delta H = -364$ J.

11 The value of γ for CH_4 is 1.31, and near room temperature and at pressures less than about 1 atm the gas behaves ideally. An adiabatic and reversible expansion is performed which reduces the pressure of a CH_4 sample, initially at 100°C and with volume 3 liters, from 1 to 0.1 atm.

a What are the final temperature and volume of the gas?
b How much work is done by the gas?
c What is the difference between ΔH and ΔE for the process?

12 Calculate the thermal contribution $E - E_0$ to the internal energy and $H - E_0$ to the enthalpy of Cl_2 at 25 and at 1000°C. Assume ideal-gas behavior, and use the vibrational-energy-level spacing of 1.11×10^{-20} J per molecule.
Ans. $(E - E_0)_{0°C} = 6535$ J; $(H - E_0)_{0°C} = 9012$ J.

13 Calculate the molar heat capacity of Cl_2 at 25 and at 1000°C using the data of Prob. 12, and compare with the values obtained from the empirical equation $C_P = 37.0 + 0.67 \times 10^{-3}T - 2.84 \times 10^5 T^{-2}$ J.

14 Using the thermal-energy data of Prob. 12 and assuming ideal-gas behavior, calculate ΔE, ΔH, $\Delta E_{\text{thermal res}}$, and $\Delta E_{\text{mech res}}$ when 1 mol of Cl_2 is heated at a constant pressure of 1 atm from 25 to 1000°C. Repeat the calculation using the empirical heat-capacity expression of Prob. 13.

15 Using the data of Table 6-3, determine the *difference in* ΔH of the reaction at 1500 and at 298 K, that is, $\Delta H_{1500} - \Delta H_{298}$, for the reactions:
a $2CO + O_2 \rightarrow 2CO_2$
b $2H_2 + O_2 \rightarrow 2H_2O$

16 Calculate the number of Br_2 molecules that are in the $v = 0$, $v = 1$, $v = 2$, and $v = 3$ energy levels when 1 mol of Br_2 is held at 100°C. The vibrational-energy-level spacing is 0.64×10^{-20} J.

Calculate at this temperature the vibrational contribution to $E - E_0$ of the molecules in the $v = 1$, in the $v = 2$, and in the $v = 3$ levels. Compare the sum of these contributions with the total thermal vibrational energy obtained using Eq. [55] of Chap. 5. *Ans.* $(E - E_0)_{\text{vib}} = 1452$ J.

17 Obtain the expression for the heat-capacity contribution of a vibrational degree of freedom given in Table 6-4 by differentiating the thermal-vibrational-energy expression with respect to temperature.

Verify that the limit of C_{vib} as x goes to zero, i.e., for small $h\nu_{\text{vib}}$ or high T, is the value that would be expected classically.

18 Calculate the heat capacity of SO_2 at 25 and at 1000°C using the data of Fig. 5-4.

References

Some of the large number of textbooks on chemical thermodynamics that treat the subject in greater detail and, in some cases, from different points of view than that taken in this text are listed below. These books will provide additional material related to topics of this chapter and to the material on thermodynamics in the following four chapters.

NASH, L. K.: "Elements of Chemical Thermodynamics," Addison-Wesley Publishing Company, Inc., Reading, Mass., 1962. A short elementary, but precise, introduction to chemical thermodynamics.

MAHAN, B. H.: "Elementary Chemical Thermodynamics," W. A. Benjamin, Inc., New York, 1963. A treatment of chemical thermodynamics at a level similar to that of Nash.

KLOTZ, I.: "Chemical Thermodynamics," W. A. Benjamin, Inc., New York, 1964. A very careful, readable account of chemical thermodynamics.

PITZER, K. S., and L. BREWER: "Thermodynamics," McGraw-Hill Book Company, New York, 1961. Brings up to date the classic treatment of thermodynamics by G. N. Lewis and M. Randall. Includes very valuable summaries of available data, references to original literature, and examples of the application of thermodynamics to chemical problems.

DENBIGH, K.: "The Principles of Chemical Equilibrium," Cambridge University Press, New York, 1961. A very clear and careful development, with emphasis, as the title indicates, on equilibria in chemical systems. The last third of the book deals with the calculation of thermodynamic properties on the basis of molecular properties.

ZEMANSKY, M. W.: "Heat and Thermodynamics," 4th ed., McGraw-Hill Book Company, New York, 1957. A treatment with rather more emphasis on systems of interest in physics and engineering than chemistry, but the presentation enables the reader to follow this extension readily.

A rather complete bibliography, with informative comments, of articles that are pertinent to the study of chemical thermodynamics is given by L. K. Nash, *J. Chem. Educ.*, **42**:64–75 (1965). Although this collection was prepared primarily as an aid to teachers of the subject, it also provides the student with the source of many interesting comments on the subject and its presentation.

7

THERMOCHEMISTRY

The specific application of the first law of thermodynamics to the study of chemical reactions is referred to as *thermochemistry*. Thermochemistry deals with the measurement or calculation of the heat absorbed or given out in chemical reactions. The subject has therefore great immediate practical importance. Thermochemistry also provides the data from which the relative energy or enthalpy contents of chemical compounds can be deduced. This aspect implies that thermochemistry is basic to the study of chemical bonding, and as we shall see, it also provides data necessary for the thermodynamic study of chemical equilibria.

7-1 Measurements of Heats of Reaction

It will make approaches used in thermochemistry more understandable if some of the methods and the scope of the more common experimental techniques are first mentioned.

Only a very few of the many possible chemical reactions are such that their heat of reaction can be accurately determined directly. To be suitable for a precise calorimetric study, a reaction must be fast, complete, and clean. A fast reaction is required so that the heat of the reaction will be given out or absorbed in a short period of time. It is then easier to prevent heat from flowing away from the reaction system or into it from the surroundings while the measurement of the temperature change of the system is being made. The completeness of the reaction is required so that difficult corrections for unreacted material need not be made. Finally, a clean reaction implies one that goes completely to a given set of products with no complicating side reactions. These stipulations are rather severe and rule out all but a few types of reactions.

In the field of organic chemistry the combustion reactions are of great utility. The burning of a compound containing only carbon, hydrogen, and oxygen usually leads, in the presence of excess oxygen, to the nice formation of the sole products CO_2 and H_2O. For organic compounds containing other elements, the products are not always so well defined, but combustion reactions are frequently practical.

The heat of combustion is usually determined in a "bomb calorimeter." Figure 7-1 shows some of the features of the bomb in which the combustion occurs. A weighed sample is placed in the cup within the reaction chamber, or bomb, and the bomb is then filled with oxygen under a pressure of about 20 atm. A fine wire dipping into the sample is heated by an electric current to start the reaction. Once started, the reaction proceeds rapidly with the evolution of a large amount of heat. This heat is determined by the temperature rise of the water around the calorimeter. It is customary to calibrate the apparatus by the measurement of the temperature rise resulting from the combustion of a sample with a known heat of combustion. The bomb and water chamber are carefully insulated from the surroundings so that the system is insulated and no heat can flow into or out of it. Such an arrangement is called an *adiabatic* calorimeter.

Heats of combustion are usually very large, of the order of hundreds of kilocalories per mole, and with careful work these can be measured with an accuracy of better than 0.01 percent. This accuracy is necessary because it often is the difference in energy contents of two com-

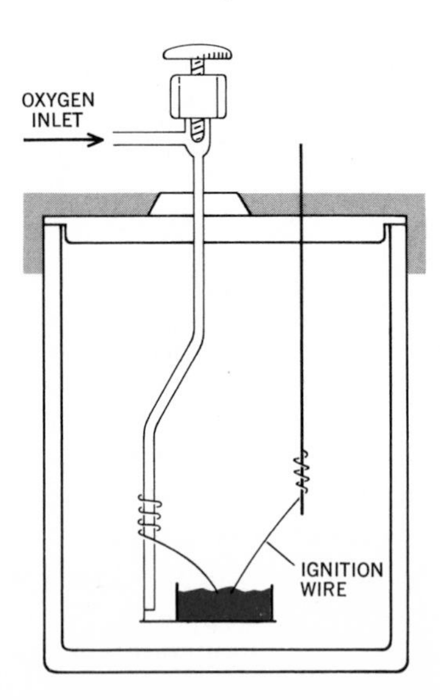

FIGURE 7-1
A combustion bomb.

pounds that is of interest. This is often a small difference between two large numbers.

Table 7-1 shows some heats of combustion of a number of organic compounds.

A second important type of reaction that is occasionally suitable for organic compounds is hydrogenation. Unsaturated materials can sometimes be made to add hydrogen in a reaction that is suitable for calorimetric study. Such reactions give out less heat than oxidation reactions and are therefore more useful in giving information on small energy differences between compounds.

The heats of inorganic reactions, in aqueous solution for instance, can frequently be measured in a calorimeter that is essentially an insulated container open to the atmosphere. Heats of neutralization, solution, and complex formation can be so studied. The same stringent requirements as previously mentioned still apply, of course, if accurate results are to be obtained.

TABLE 7-1
Enthalpies of combustion, i.e., the heat *evolved* in the constant-pressure combustion reaction, at 25°C, in kJ mol^{-1}
Products are $CO_2(g)$ and $H_2O(l)$.

H_2	285.84
C(graphite)	393.51
$CO(g)$	282.99
$CH_4(g)$	890.35
$C_2H_6(g)$	1559.88
$C_3H_8(g)$	2220.07
n-Butane(g)	2878.51
i-Butane(g)	2871.65
n-Heptane(g)	4811.2
Ethylene(g)	1410.97
Acetylene(g)	1299.63
Benzene(g)	3301.51
Ethanol(l)	1366.95
Acetic acid(l)	872.4
Glucose(s)	2815.8
Sucrose(s)	5646.7

7-2 Internal-energy and Enthalpy Changes in Chemical Reactions

Thermochemistry deals with the particular process of a chemical reaction. If such reactions are represented as

$$\text{Reactants} \rightarrow \text{products} \qquad [1]$$

the internal-energy and enthalpy changes for the process are related to the energy and enthalpy contents of the reactants and products by

$$\Delta E = E_{\text{products}} - E_{\text{reactants}} \qquad [2]$$

and

$$\Delta H = H_{\text{products}} - H_{\text{reactants}} \qquad [3]$$

For example, the enthalpy change ΔH for the reaction

$$C + \tfrac{1}{2}O_2 \rightarrow CO \qquad [4]$$

gives the enthalpy of 1 mol of carbon monoxide compared with the enthalpy of 1 mol of carbon plus one-half the enthalpy of 1 mol of oxygen; that is,

$$\Delta H = \text{H}_{CO} - \text{H}_{C} - \tfrac{1}{2}\text{H}_{O_2} \qquad [5]$$

When chemical reactions are dealt with, the Δ notation, which previously signified any change, means the difference of some property for the products and the reactants.

If heat is absorbed in the reaction and the products contain more energy than the reactants, then ΔE, if the process is at constant volume, and ΔH, if the process is at constant pressure, are positive; that is, H and E increase as a result of the reaction. Such reactions are called *endothermic*. Reactions for which ΔE and ΔH are negative proceed with a decrease in E and H. Heat is therefore given out, and the reaction is said to be *exothermic*.

To summarize:

ΔH or ΔE	positive	heat absorbed	endothermic
ΔH or ΔE	negative	heat given out	exothermic

The combustion experiments previously mentioned are all exothermic, and therefore for these reactions ΔH and ΔE are negative.

7-3 Relation between ΔE and ΔH

A measurement of a heat of reaction usually gives directly either the internal-energy change or the enthalpy change. Either datum can, however, be used to calculate the other. If the reaction is performed in a constant-volume apparatus, such as the bomb calorimeter, no work of expansion is performed and the heat of the reaction is equal to the internal-energy change. If a constant-pressure system is used, the heat of the reaction, as was pointed out in Sec. 6-6, is equal to the enthalpy change.

The difference between ΔH and ΔE for a reaction depends on the volume change that occurs when the reaction is performed at constant pressure and on this pressure. If all reactants and products are solids or liquids, the change in volume will be quite small. Since 1 mol of a typical simple compound is likely to have a volume of no more than 0.1 liter, it is unlikely that a reaction would cause a volume change of more than about 0.01 liter. At ordinary pressures the PV term corresponding to this volume change is only

$$\Delta(PV) = P\,\Delta V = (101{,}325)(0.01) = 1 \text{ J}$$

The enthalpy change would be greater or less than the internal-energy change by this amount. However, a joule is often negligible compared with the experimental error of a measured heat of reaction.

If gases are involved in the reaction, an appreciable value of $\Delta(PV)$ can occur and ΔH and ΔE will be significantly different. Suppose that the reaction produces a net change Δn in the number of moles of gas. To the accuracy of the ideal-gas laws and neglecting the volumes of liquids and solids, the PV term will be greater for the products than for the reactants by an amount

$$\Delta(PV) = \Delta(nRT) = RT\,\Delta n \qquad [6]$$

The enthalpy change for the reaction will differ from the internal-energy change by $RT\,\Delta n$, or for the reaction

$$\Delta H = \Delta E + RT\,\Delta n \qquad [7]$$

Consider, as an example, the reaction

$$2CO(g) + O_2(g) \rightarrow 2CO_2(g) \quad [8]$$

where g stands for gas. A bomb-calorimetric study would give the heat of this reaction, i.e., the combustion of 2 mol of carbon monoxide, as 563,500 J given out. Since this is a constant-volume experiment, we have immediately

$$\Delta E = -563{,}500 \text{ J} \quad [9]$$

The product contains 2 mol of gas, and the reactants contain 3 mol. The value of Δn is -1, and at 25°C the enthalpy change is calculated as

$$\begin{aligned}\Delta H &= \Delta E + RT\,\Delta n \\ &= -563{,}500 + (8.3143)(298.15)(-1) \\ &= -563{,}500 - 2480 = 565{,}980 \text{ J} = 565.98 \text{ kJ}\end{aligned} \quad [10]$$

It has now been pointed out that the heats of some reactions can be measured and that ΔE and ΔH for these reactions can be calculated.

7-4 Thermochemical Equations

An elaboration of the form in which chemical-reaction equations are usually written is sometimes advisable in thermochemical work.

Since the heat of a reaction depends on whether a reagent is solid, liquid, or gas, it is necessary to specify the state of the reagents. This is usually done by adding s, l, or g after the compound. Occasionally a more careful description is necessary. When carbon is involved in a reaction, for instance, it is necessary to state whether it is graphite or diamond. One must be especially careful with water, which can quite reasonably be involved as a gas or a liquid.

Another characteristic of thermochemical equations is the appearance of fractional numbers of moles of some of the reactants or products. If one were interested in the heat of a reaction which produces 1 mol of water, for instance, one would write the reaction and the enthalpy change as

$$H_2(g) + \tfrac{1}{2}O_2(g) \rightarrow H_2O(l) \qquad \Delta H = -285.84 \text{ kJ} \quad [11]$$

The reaction written with integers corresponds to the formation of 2 mol of water, and twice as much heat evolved. Such a reaction would be written

$$2H_2(g) + O_2(g) \rightarrow 2H_2O(l) \qquad \Delta H = -571.68 \text{ kJ} \quad [12]$$

Combustion reactions are usually written for the combustion of 1 mol of material. For benzene, for example, one would usually write

$$C_6H_6(g) + \tfrac{15}{2}O_2(g) \rightarrow 6CO_2(g) + 3H_2O(l) \qquad \Delta H = -3301.51 \text{ kJ} \quad [13]$$

One additional comment is necessary. Unless the information is otherwise given, it is convenient to attach the temperature and pressure at which the reported enthalpy applies. A superscript degree sign indicates that the pressure is 1 atm, a usual reference, or standard, state. (This superscript, meaning standard state, must be distinguished from the previously used subscript zero, meaning the lowest energy-level quantity.) The temperature is indicated by a subscript giving the absolute temperature. An equation for the formation of water at 25°C and 1 atm would then be

$$H_2(g) + \tfrac{1}{2}O_2(g) \rightarrow H_2O(l) \qquad \Delta H^\circ_{298} = -285.84 \text{ kJ} \tag{14}$$

7-5 Indirect Determination of Heats of Reaction

For reactions which are not suitable for direct calorimetric study, the internal-energy or enthalpy changes can often be obtained by an indirect method. This was originally suggested by Hess in 1840, and is often known as *Hess's law of heat summation.* We now recognize that it is merely an application of the first law of thermodynamics.

The determination of the enthalpy change when carbon is converted from graphite to diamond, i.e.,

$$\text{C(graphite)} \rightarrow \text{C(diamond)} \qquad \Delta H = ? \tag{15}$$

will illustrate the method. Although this reaction can be made to occur, it is certainly very unsuitable for any direct calorimetric study.

The combustion of both graphite and diamond can be conveniently studied, and these reactions and the heats of combustion are

$$\text{C(graphite)} + O_2(g) \rightarrow CO_2(g) \qquad \Delta H^\circ_{298} = -393.51 \text{ kJ} \tag{16}$$

$$\text{C(diamond)} + O_2(g) \rightarrow CO_2(g) \qquad \Delta H^\circ_{298} = -395.40 \text{ kJ} \tag{17}$$

The enthalpy changes of these reactions clearly differ as a result of the different enthalpies of graphite and diamond, as is shown by writing

$$-393.51 \text{ kJ} = \text{H}_{CO_2} - \text{H}_{\text{C(graphite)}} - \text{H}_{O_2} \tag{18}$$

$$-395.40 \text{ kJ} = \text{H}_{CO_2} - \text{H}_{\text{C(diamond)}} - \text{H}_{O_2} \tag{19}$$

Subtraction of these algebraic equations with cancellation of the enthalpies of CO_2 and O_2 gives, on rearrangement,

$$\begin{aligned} \text{H}_{\text{C(diamond)}} - \text{H}_{\text{C(graphite)}} &= -393.51 + 395.40 \\ &= +1.89 \text{ kJ} \end{aligned} \tag{20}$$

The result of 1890 J is the enthalpy change of the original graphite-to-diamond reaction.

It is important to notice that this same result could have been obtained by subtracting the two combustion-reaction equations (Eqs.

[16] and [17]) as though they were algebraic equations. Canceling the O_2 and CO_2 terms and moving the C(diamond) term to the right side with a change of sign gives

$$\text{C(graphite)} \rightarrow \text{C(diamond)} \quad \Delta H = -393.51 - (-395.40) = +1.89 \text{ kJ} \qquad [21]$$

This treatment of the reaction equation is always possible and saves writing out the heat contents of each of the species before the subtraction is made.

Combining reactions by either of these methods is equivalent to calculating ΔH for the desired reaction by an indirect but experimentally more feasible path. The preceding example can be illustrated as

$$\begin{array}{ccc} & CO_2 & \\ \Delta H = -393.51 \text{ kJ} \nearrow & & \searrow \Delta H = +395.40 \text{ kJ} \\ \text{C(graphite)} + O_2 & \rightarrow & \text{C(diamond)} + O_2 \end{array}$$

The fact that the enthalpy is a thermodynamic property requires ΔH to be the same by the indirect path, that is, $-393.51 + 395.40 = +1.89$ kJ, as by the direct path. It is this fact which allows the previous reaction subtractions to be performed.

One additional example can be given. The heat of the reaction by which a compound is formed from its elements is, as we shall see, of special interest. In few cases can such reactions be directly studied. Consider, for example, the formation of methane according to the equation

$$\text{C(graphite)} + 2H_2(g) \rightarrow CH_4(g) \qquad [22]$$

Again combustion reactions can be used and give

$$\begin{array}{lll} (a) & CH_4(g) + 2O_2(g) \rightarrow CO_2(g) + 2H_2O(l) & \Delta H^\circ_{298} = -890.35 \text{ kJ} \\ (b) & H_2(g) + \tfrac{1}{2}O_2(g) \rightarrow H_2O(l) & \Delta H^\circ_{298} = -285.84 \text{ kJ} \\ (c) & \text{C(graphite)} + O_2(g) \rightarrow CO_2(g) & \Delta H^\circ_{298} = -393.51 \text{ kJ} \\ \hline -(a) + 2(b) + (c) & & \\ & \text{C(graphite)} + 2H_2(g) \rightarrow CH_4(g) & \Delta H^\circ_{298} = -74.84 \text{ kJ} \end{array}$$

7-6 Standard Heats of Formation

Methods for obtaining, either directly or indirectly, the enthalpies of many reactions have been given. Although it is not at all practical to compile a table of all the reactions for which such data are available, it would be feasible to tabulate information for each of the compounds for which thermal data have been obtained. The *standard heats of formation* provide a means of making such a table. Furthermore, from these data it is possible to work out the heats of any reactions involving the listed compounds.

If, for the moment, we consider only the elements, we are at liberty to choose some *standard* state for each element and to specify enthalpies in other states relative to that state. It is customary to take this *standard state as that of 1 atm pressure and 25°C with the element in the physical state and stable form under these conditions.* The enthalpy of the elements in the standard state is arbitrarily given the value zero. Thus the standard enthalpy of H_2 at 1 atm and 25°C is zero. At higher temperatures the enthalpy, referred to the zero value at 1 atm and 25°C, will be some positive quantity; at lower temperatures it will be some negative quantity. Enthalpies based on this standard state are called *standard enthalpies.* The assignment of the arbitrary value of zero for the standard enthalpy at 1 atm and 25°C to all elements is allowed because no chemical reaction converts one element into another.

Once standard enthalpies are assigned to the elements, it is possible to determine standard enthalpies for compounds. These enthalpies are usually called *standard heats of formation.* Consider, for example, the reaction

$$\text{C(graphite)} + O_2(g) \rightarrow CO_2(g) \qquad \Delta H^\circ_{298} = -393.51 \text{ kJ} \qquad [23]$$

The enthalpy change for the reaction is equal to the standard enthalpy of CO_2 less the standard enthalpies of C and O_2. Since the latter are elements in their standard state, their standard enthalpies are zero. The standard enthalpy, or standard heat of formation, of CO_2 must therefore be -393.51 kJ. The enthalpy of the reaction by which the compound is formed from its elements, all in their standard states, is seen to be equal to the standard heat of formation of the compound. As a result of this, one uses the symbol ΔH_f° for this standard heat of formation.

By combining reactions one can frequently deduce the enthalpy of formation of a compound from its elements, as was illustrated for CH_4 in Sec. 7-5. Furthermore, when the standard enthalpies of some compounds have been determined, they can be used in working out the standard enthalpies of other compounds.

Appendix 5 includes standard enthalpies of some common compounds. Such a table of enthalpies can be used to determine the enthalpy for any reaction, at 1 atm and 25°C, involving the elements and any of the compounds appearing in the table.

The heat of hydrogenation of ethylene can be calculated as an illustration. One writes

$$\begin{array}{lccccc} & CH_2CH_2(g) & + & H_2(g) & \rightarrow & CH_3CH_3(g) \\ \text{At } 25°\text{C} & \Delta H_f^\circ = +52.30 & & \Delta H_f^\circ = 0 & & \Delta H_f^\circ = -84.68 \end{array} \qquad [24]$$

The enthalpy of the reaction is the difference in enthalpy of the products and the reactants. Thus

$$\begin{aligned}\Delta H^\circ_{298} &= \Delta H^\circ_f(CH_3CH_3) - \Delta H^\circ_f(H_2) - \Delta H^\circ_f(CH_2CH_2) \\ &= -84.68 - 0 - 52.30 \\ &= -136.98 \text{ kJ} \qquad [25]\end{aligned}$$

The calculated enthalpy for the reaction has, of course, no necessary connection with the arbitrary assignment of zero enthalpy to the elements in their standard states. It is the enthalpy change occurring when 1 mol of ethylene is hydrogenated.

7-7 Standard Heats of Formation of Ions in Aqueous Solutions

It might be noticed that so far the reactions that have been dealt with have presumed that the reagents were in the form of pure solids, liquids, or gases or that, if a solution was involved, the solution process involved a negligible heat effect. Many reactions of chemical interest do not fall in this category, and the reactions involving ions in aqueous solution constitute the most important type of reactions not so far dealt with. Here it will be shown that as long as we deal with very dilute solutions—or to be strictly correct, infinitely dilute solutions—a procedure similar to that developed in Sec. 7-6 can be used to provide a table of standard heats of formation of ions. These standard heats can then be used to calculate the heats of reactions involving ions in dilute aqueous solutions.

The direct approach to the heat of formation of ions would be that of considering reactions in which such ions are formed. We might, for example, consider the solution of 1 mol of HCl gas in a large amount of water. In solution the dissociation is complete and hydrated H^+ (or if you like, H_3O^+) and Cl^- ions are formed. The reaction can be represented as

$$HCl(g) \xrightarrow{\text{water}} H^+(aq) + Cl^-(aq) \qquad [26]$$

where the symbol (aq) implies that the ion is present in a large amount of water. The heat evolved in the reaction, at 25°C, is 75.14 kJ, and thus for the reaction one writes

$$\Delta H = -75.14 = \Delta H^\circ_f[H^+(aq) + Cl^-(aq)] - \Delta H^\circ_f[HCl(g)]$$

With the value of ΔH°_f for $HCl(g)$ given in Appendix 5, this can be arranged to

$$\begin{aligned}\Delta H^\circ_f[H^+(aq) + Cl^-(aq)] &= -75.14 + (-92.30) \\ &= -167.44 \text{ kJ} \qquad [27]\end{aligned}$$

In this way we have obtained the standard heat of formation of the *pair* of ions H^+ and Cl^- in aqueous solution, and furthermore, we see a method for obtaining the standard heats of formation of the groups of ions that result from the solution of any acid, base, or salt.

The fact that all reactions such as these lead to an electrically neutral solution means that no procedure, involving ordinary chemical reactions, will give only a single type of ion as a product. Therefore only the heats of formation of collections of ions can be measured.

We wish, of course, to be able to tabulate ΔH_f° values for individual ions. First, we must be sure that the enthalpy of a dilute solution can, in fact, be treated in terms of contributions attributed to the separate types of ions in the solution. Experiments in which dilute solutions of nonreacting electrolytes are mixed, for example the mixing of dilute solutions of HCl and KBr, show no heats of reaction. It follows that the heat of a sufficiently dilute solution can be attributed to separate contributions from the ions that are present, and each contribution is independent of the other ions that are present.

The absence of reactions involving only a single type of ion does not frustrate our attempt to tabulate the standard heats of formation of ions. Rather, it permits the assignment of an arbitrary value to the standard heat of formation of any one ion. Values for the heats of formation of all other ions will then follow.

It is generally agreed to assign the value of zero to the heat of formation of H^+ ions in dilute aqueous solution at 25°C; that is,

$$\Delta H_f^\circ[H^+(aq)] = 0$$

Once this step has been taken, one can use, for example, Eq. [27] to obtain the heat of formation of the chloride ion as

$$\begin{aligned}\Delta H_f^\circ[Cl^-(aq)] &= -167.44 - 0 \\ &= -167.44 \text{ kJ}\end{aligned} \qquad [28]$$

Likewise, the heat of solution of potassium chloride is 17.18 kJ at 25°C, and the standard heat of formation of KCl(s) is −435.87 kJ. From these data, one obtains

$$\begin{aligned}\Delta H_f^\circ[K^+(aq)] + \Delta H_f^\circ[Cl^-(aq)] &= 17.18 - 435.87 \\ &= -418.65 \text{ kJ}\end{aligned}$$

The value for the chloride ion can now be used to obtain

$$\begin{aligned}\Delta H_f^\circ[K^+(aq)] &= -418.65 - (-167.44) \\ &= -251.21 \text{ kJ}\end{aligned}$$

This procedure, which depends on the arbitrary assignment of zero for the value of ΔH_f° for the H^+ ion, provides data such as that of Appendix 6. With such data the heats of reactions at 25°C of any reaction involving dilute solutions of these ions can be calculated.

As an example of the use of standard enthalpy data we can calculate the heat of reaction in which calcium carbonate precipitates from a solution as a result of the addition of CO_2 to an aqueous solution containing Ca^{++} ions in low concentration. The reaction can be written

$$Ca^{++}(aq) + CO_2(g) + H_2O(l) \rightarrow CaCO_3(s) + 2H^+(aq) \qquad [29]$$

We first write

$$\Delta H = 2\Delta H_f^\circ[H^+(aq)] + \Delta H_f^\circ[CaCO_3(s)] - \Delta H_f^\circ[Ca^{++}(aq)] - \Delta H_f^\circ[CO_2(g)] - \Delta H_f^\circ[H_2O(l)]$$

and then use the data of Appendixes 5 and 6 to obtain

$$\Delta H = 2(0) + (-1206.87) - (-542.96) - (-393.51) - (-285.84) = +15.44 \text{ kJ}$$

Perhaps it should be pointed out that only one water is shown explicitly in the equation. No doubt each Ca^{++} and each H^+ ion is strongly hydrated, and such hydrations involve an appreciable heat effect. All this, however, is taken into account by the procedure used to set up the standard heats of formation of the ions. These heats include the contribution of the ion and all the water molecules involved in the hydration of the ion.

7-8 Temperature Dependence of Heats of Reaction

The enthalpies of reactions calculated from a table of standard heats of formation apply to a temperature of 25°C. For these data to be of wider value, a means for determining the heats of reactions at other temperatures must be available. This can be done by writing the enthalpy of the reaction as

$$\Delta H = H_{\text{products}} - H_{\text{reactants}} \qquad [30]$$

and differentiating with respect to temperature to get

$$\left[\frac{\partial(\Gamma H)}{\partial T}\right]_P = \left(\frac{\partial H_{\text{products}}}{\partial T}\right)_P - \left(\frac{\partial H_{\text{reactants}}}{\partial T}\right)_P \qquad [31]$$

The change in enthalpy with respect to temperature at constant pressure was shown in Sec. 6-9 to be the heat capacity at constant pressure, so that we can write

$$\left[\frac{\partial(\Delta H)}{\partial T}\right]_P = (C_P)_{\text{products}} - (C_P)_{\text{reactants}} = \Delta C_P \qquad [32]$$

and in a similar manner,

$$\left[\frac{\partial(\Delta E)}{\partial T}\right]_V = \Delta C_V \qquad [33]$$

It is necessary, therefore, to know only the difference in the heat capacities of the products and the reactants in order to determine how the enthalpy of a reaction changes with temperature for temperature ranges that do not encompass any phase changes. For small ranges of temperature, ΔC_p can be taken as constant, and integration gives

$$\Delta H_2 - \Delta H_1 = \Delta C_P(T_2 - T_1) \quad [34]$$

A diagrammatic derivation of this result is perhaps more revealing. Consider a constant-pressure reaction that proceeds with an enthalpy change ΔH_1 at temperature T_1 and an enthalpy change ΔH_2 at temperature T_2, as in Fig. 7-2. If the reactants are heated from T_1 to T_2, the heat absorbed, or enthalpy change, will be $(C_P)_{\text{reactants}}(T_2 - T_1)$. If the products are similarly heated, the enthalpy change will be $(C_P)_{\text{products}}$ $(T_2 - T_1)$. Now one can equate the enthalpy involved in going from state a to state b of Fig. 7-2 by the two different paths to get

$$\Delta H_1 + (C_P)_{\text{products}}(T_2 - T_1) = \Delta H_2 + (C_P)_{\text{reactants}}(T_2 - T_1)$$

or

$$\Delta H_2 - \Delta H_1 = \Delta C_P (T_2 - T_1) \quad [35]$$

as previously obtained.

If heats of reaction over a wide range of temperatures are needed, it is necessary to take into account the dependence of the heat capacities themselves on temperature. Experimental heat capacities could be expressed by an empirical relation such as

$$c_P = a' + b'T + c'T^2 + \cdots \quad [36]$$

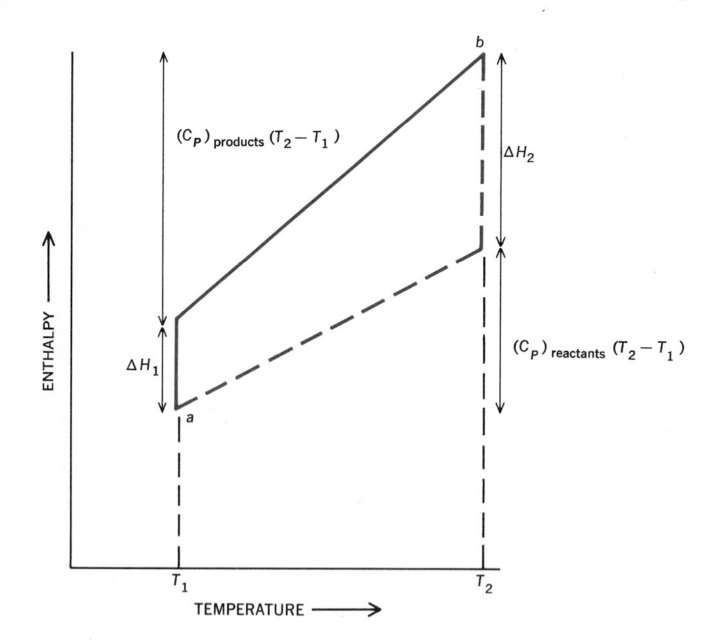

FIGURE 7-2
The temperature dependence of the heat of reaction (assuming that C_P of products and reactants are independent of T in the temperature interval T_1 to T_2).

In practice, however, it is somewhat more satisfactory to use an equation of the form

$$c_P = a + bT + cT^{-2} \tag{37}$$

Representative values of the coefficients are given in Table 7-2.

If each of the c_P's of the products and the reactants is so written, one can see that ΔC_P will have the same form as that shown for c_P. The previous differential equation, Eq. [32], which corresponds to the integral form

$$\Delta H_{T_2} - \Delta H_{T_1} = \int_{T_1}^{T_2} \Delta C_P \, dT \tag{38}$$

can then be integrated. One obtains a relation of the form

$$\Delta H_{T_2} - \Delta H_{T_1} = \Delta a(T_2 - T_1) + \tfrac{1}{2}\Delta b(T_2^2 - T_1^2) - \Delta c\left(\frac{1}{T_2} - \frac{1}{T_1}\right) \tag{39}$$

Thus, from equations based on experimental data for the heat capacities of all the reagents involved in the reaction, the difference in heats of a reaction at two different temperatures can be calculated from the appropriate integrated form of Eq. [38].

Likewise, an expression for the general dependence of the heat of reaction on the temperature can be obtained from the indefinite integral corresponding to Eq. [38]. One has, then,

$$\Delta H_T = \int \Delta C_P \, dT + \text{const} \tag{40}$$

and if an empirical equation of the form of Eq. [37] is used,

$$\Delta H_T = \Delta a T + \tfrac{1}{2}\Delta b T^2 - \Delta c \frac{1}{T} + \text{const} \tag{41}$$

It is then most convenient to refer to 25°C, so that the standard enthalpies can be used to determine the integration constant.

This procedure can be illustrated by obtaining an equation for the heat of reaction

$$\text{C(graphite)} + 2\text{H}_2(g) \rightarrow \text{CH}_4(g)$$

as a function of the temperature. The heat-capacity data of Table 7-2 are given in the form of Eq. [37] and can be used to obtain the coefficients in the heat-of-reaction expression of Eq. [41]. One obtains

$$\Delta H_T(\text{joules}) = -47.78T + 18.28 \times 10^{-3}T^2 - 5.61 \times 10^5 \frac{1}{T} + \text{const}$$

To evaluate the constant, a value of ΔH_T at any temperature in the range within which the equation is applicable must be available.

TABLE 7-2 Heat capacities, in J deg^{-1} mol^{-1}, at constant pressure*
Parameters for the equation $c_P^\circ = a + bT + cT^{-2}$

	a (J deg^{-1} mol^{-1})	b (J deg^{-2} mol^{-1})	c (J deg mol^{-1})
GASES (IN TEMPERATURE RANGE 298 TO 2000 K)			
He, Ne, Ar, Xr, Xe	+20.79	0	0
S	+22.01	-0.42×10^{-3}	$+1.51 \times 10^5$
H_2	+27.28	+3.26	+0.50
O_2	+29.96	+4.18	−1.67
N_2	+28.58	+3.76	−0.50
S_2	+36.48	+0.67	−3.76
CO	+28.41	+4.10	−0.46
F_2	+34.56	+2.51	−3.51
Cl_2	+37.03	+0.67	−2.84
Br_2	+37.32	+0.50	−1.25
I_2	+37.40	+0.59	−0.71
CO_2	+44.22	+8.79	−8.62
H_2O	+30.54	+10.29	0
H_2S	+32.68	+12.38	−1.92
NH_3	+29.75	+25.10	−1.55
CH_4	+23.64	+47.86	−1.92
TeF_6	+148.66	+6.78	−29.29
LIQUIDS (FROM MELTING POINT TO BOILING POINT)			
I_2	+80.33	0	0
H_2O	+75.48	0	0
NaCl	+66.9	0	0
$C_{10}H_8$	+79.5	$+407.5 \times 10^{-3}$	0
SOLIDS (FROM 298 K TO MELTING POINT, OR 2000 K)			
C(graphite)	+16.86	$+4.77 \times 10^{-3}$	-8.54×10^5
Al	+20.67	+12.38	0
Cu	+22.63	+6.28	0
Pb	+22.13	+11.72	+0.96
I_2	+40.12	+49.79	0
NaCl	+45.94	+16.32	0
$C_{10}H_8$	−115.9	+937	0

*Calculated from data of G. N. Lewis and M. Randall, "Thermodynamics," 2d ed. (rev. by K. S. Pitzer and L. Brewer), McGraw-Hill Book Company, New York, 1961.

Here, as is often the case, this datum is provided by the tabulated standard heats of formation at 298.15 K. For CH_4 the value of ΔH_f° is given in Appendix 5 as -74.85 kJ, and with this value one rearranges the preceding equation to give

$$\text{const} = -74{,}850 + 47.78(298.15) - (18.28 \times 10^{-3})(298.15)^2 + \frac{5.61 \times 10^5}{298.15}$$

Substitution of the value of the constant term given by this expression now gives the heat of formation of CH_4 from its elements as a function of temperature as

$$\Delta H_T = -47.78T + 18.28 \times 10^{-3}T^2 - 5.61 \times 10^5 \frac{1}{T} - 60{,}347 \text{ J}$$

In the above example empirical heat-capacity data have been used to express the heat of a reaction as a function of temperature. We shall next see how it is often possible to express energy changes occurring during a reaction in terms of the energies of the molecules involved in the reaction.

7-9 Heats of Reaction and the Molecular Model

It has been shown in the preceding chapter that the enthalpy of a compound can be composed into a temperature-independent part called E_0 and a temperature-dependent part $H - E_0$. The former depends on the lowest allowed energy and reflects the strength of the chemical binding in the compound, and the latter measures the thermal excitation of the compound. Our detailed understanding of molecular energies can now be carried over to chemical reactions.

If the molecules involved are not too complex, it is possible to calculate $H - E_0$ for each molecule in the reaction. The difference $\Delta(H - E_0)$ can then be formed. The enthalpy difference for the reaction depends on this term through the relation

$$\Delta H = \Delta E_0 + \Delta H - \Delta E_0 = \Delta E_0 + \Delta(H - E_0) \qquad [42]$$

If the $H - E_0$ terms can be calculated and if a heat of reaction ΔH is known at some temperature, the lowest-level energy difference ΔE_0 can be calculated. This term gives the energy change for the reaction when all reactants and products are in the lowest energy levels.

If we are interested in the reagents in their standard state of 1 atm pressure, the enthalpy of the reaction can be written

$$\Delta H^\circ = \Delta E_0^\circ + \Delta(H^\circ - E_0^\circ) \qquad [43]$$

For ideal gases the pressure has no effect on the enthalpy or energy, and this designation of the standard state does not affect any of the terms. The table of standard enthalpies can be used directly for ideal

TABLE 7-3 The thermal contributions $H - E_0$ to the enthalpy of N_2, O_2, and NO_2, treated as ideal gases
For N_2 and O_2, $h\nu_{vib}$ is 4.69×10^{-20} and 3.14×10^{-20} J, respectively. Values for NO_2 are given in Table 6-2.

	N_2	O_2	NO_2
298 K:			
Translation	$\frac{3}{2}RT = 3718$	$\frac{3}{2}RT = 3718$	$\frac{3}{2}RT = 3718$
Rotation	$\frac{2}{2}RT = 2479$	$\frac{2}{2}RT = 2479$	$\frac{3}{2}RT = 3718$
Vibration	0	9	282
Electronic	0	0	0
PV term	$RT = 2479$	$RT = 2479$	$RT = 2479$
	8676 J	8685 J	10,197 J
1500 K:			
Translation	$\frac{3}{2}RT = 18{,}708$	$\frac{3}{2}RT = 18{,}708$	$\frac{3}{2}RT = 18{,}708$
Rotation	$\frac{2}{2}RT = 12{,}472$	$\frac{2}{2}RT = 12{,}472$	$\frac{3}{2}RT = 18{,}708$
Vibration	3,280	5,320	19,920
Electronic	0	0	0
PV term	$RT = 12{,}472$	$RT = 12{,}472$	$RT = 12{,}472$
	46,930 J	48,970 J	69,810 J

gases to determine the ΔH term of Eq. [42] which is identical with the ΔH° term of Eq. [43].

A molecular interpretation of the heat of the reaction

$$N_2(g) + 2O_2(g) \rightarrow 2NO_2(g) \qquad [44]$$

can be given as an illustration. The thermal contributions to the enthalpy of the three species, assumed to behave as ideal gases, can be calculated as illustrated previously for NO_2. The terms of these calculations are shown in Table 7-3. From these data one obtains

$$\begin{aligned}\Delta(H - E_0)_{298} &= 2(H - E_0)_{NO_2} - 2(H - E_0)_{O_2} - (H - E_0)_{N_2} \\ &= 2(10{,}192) - 2(8678) - 8670 \\ &= -5642 \text{ J}\end{aligned} \qquad [45]$$

The heat of reaction at 298 K can be calculated from the standard heats of formation. These give

$$\begin{aligned}\Delta H_{298} &= 2H_f^\circ(NO_2) - H_f^\circ(N_2) - 2H_f^\circ(O_2) \\ &= 2(33{,}853) - 0 - (2)(0) \\ &= 67{,}706 \text{ J}\end{aligned} \qquad [46]$$

The results of Eqs. [45] and [46] can be inserted into a rearranged form of Eq. [42] to give

$$\begin{aligned}\Delta E_0 &= \Delta H_{298} - \Delta(H - E_0)_{298} \\ &= 67{,}706 - (-5642) \\ &= 73{,}348 \text{ J}\end{aligned} \qquad [47]$$

In this way, the energy change that would occur when the reagents in their lowest allowed energy state react to form the product molecules, also in their lowest energy state, is obtained. One can now write

$$\Delta H = 73{,}348 \text{ J} + \Delta(H - E_0) \qquad [48]$$

where ΔH is the enthalpy change of the reaction at some temperature, and $\Delta(H - E_0)$ is the calculable thermal-enthalpy term at that temperature.

With Eq. [48] it is a relatively simple matter to calculate the heat of reaction for the formation of NO_2 at some new and even experimentally difficult temperature. The terms for the oxidation of N_2 at 1500 K, for example, are also given in Table 7-3 and yield

$$\begin{aligned}\Delta(H - E_0)_{1500} &= (2)(69{,}830) - (2)(48{,}950) - 46{,}900 \\ &= -5140 \text{ J}\end{aligned} \qquad [49]$$

with which we obtain

$$\begin{aligned}\Delta H_{1500} &= \Delta E_0 + \Delta(H - E_0)_{1500} \\ &= 73{,}348 - 5140 \\ &= 68{,}208 \text{ J}\end{aligned} \qquad [50]$$

It is clear that the heat of reaction can be calculated by this method at any temperature for which the thermal-energy terms $\mathrm{H} - \mathrm{E}_0$ can be calculated. At high temperatures, usually above 1500 K, it is necessary also to take into account higher electronic states which become appreciably populated.

Calculations such as this can give the heats of reactions at any temperature if a single heat of reaction is known. Great use of this approach has been made, particularly in the field of hydrocarbon reactions. Predictions of the heats of rearrangements, of combustion, and so forth, of petroleum-fraction molecules can be made at temperatures at which actual measurements would be very difficult.

This type of calculation, based on our knowledge of the energies of molecules, gives a particularly clear insight into the basis for the dependence of ΔH for a reaction on the temperature. We shall see later, in Chap. 10, that a similar understanding of the quantities that determine the equilibrium position reached by reacting molecules is obtained by separating the thermal and lowest-level components of thermodynamic quantities.

7-10 Bond Enthalpies and Bond Energies

A type of reaction that provides further insight into the molecular basis of the energetics of chemical reactions is that in which molecules are completely disrupted to give free atoms. The enthalpies of such reactions lead to *bond enthalpies,* quantities that interpret the enthalpy of the molecule in terms of the enthalpies of the chemical bonds that are drawn to represent the bonding. This molecular-enthalpy component has until now been left as a part of E_0 and ΔE_0, and only the enthalpy of the motions of the intact molecule has been treated.

The enthalpy of atomization of a molecule can be calculated from the values of ΔH_f° for the parent molecule and the free gaseous atoms, as given in Appendix 5. The necessary data for these species, for those elements whose standard state consists of gaseous diatomic molecules, come from spectroscopic studies of the dissociation of the parent diatomic molecule.

The value of ΔH_f° for the atomic gaseous state for elements whose standard state is not that of gaseous diatomic molecules must be obtained by other means. Most important is the element carbon, with graphite as its standard state. Measurements of the heat of sublimation of carbon, and analysis of the equilibrium gas-phase species, which include molecules like C_2 as well as atomic carbon, lead to the value 718.38 kJ mol^{-1} for ΔH_f° for free carbon atoms.

For diatomic species, the *bond enthalpy* is obtained directly from the listed enthalpies. Thus, for O_2,

$$O_2(g) \rightarrow 2O(g) \qquad \Delta H^\circ = 2(247.52) = 495.04 \text{ kJ} \qquad [51]$$

The energy of the reaction is the bond enthalpy for the O_2 molecule.

In a similar way the bond enthalpies of polyatomic molecules containing a single type of chemical bonds can be calculated. For CH_4, for example, we can write, with all data applying to 25°C,

$$\begin{array}{lccc} & CH_4(g) \rightarrow & C(g) \;+ & 4H(g) \\ \Delta H_f^\circ & -74.85 & 718.38 & 4(217.94) \end{array}$$

$$\Delta H^\circ = +1590.14 - (-74.85) = +1664.99 \text{ kJ} \qquad [52]$$

This is the enthalpy that, with the molecular view of CH_4 that we usually adopt, is required to break the four C—H bonds. The C—H bond energy in methane is thus obtained as

$$\frac{1664.99}{4} = 416.25 \text{ kJ} \qquad [53]$$

In a similar way the bond enthalpy for other bonds can be obtained by considering molecules with a single type of bond.

Difficulties arise, however, when we attempt to determine bond enthalpies for bonds that occur in molecules that have, in addition, other bond types. For example, consider the C—O bond of methyl alcohol, CH_3OH. We have, using the data of Appendix 5,

$$CH_3OH(g) \rightarrow C(g) + 4H(g) + O(g)$$

$$\Delta H^\circ = 1837.66 - (-201.25) = +2038.91 \text{ kJ} \qquad [54]$$

This result, the enthalpy required to overcome the bonding in CH_3OH, must now be interpreted as the enthalpy required to break three C—H, one C—O, and one O—H bonds. One way of assigning the total to the different bond types is to assume that the C—H and O—H bonds have the same bond enthalpies as the CH_4 and H_2O. On this basis, and with the value 463 kJ for the O—H bond enthalpy, the C—O bond enthalpy in CH_3OH is

$$2039 - 3(416) - 463 = 328 \text{ kJ} \qquad [55]$$

Carryover of values, such as the C—O bond energy deduced here, to other molecules, CH_3—O—CH_3 for example, depends on the useful, but inexact assumption that a particular bond has an energy that is independent of its molecular environment. The difficulties that are introduced often lead to discrepancies, of several kilojoules per bond, between the actual enthalpy of atomization of a molecule and that calculated from bond energies. As a result, bond enthalpies that are tabulated are often "best" values that apply reasonably well to a variety of molecular types. Examples are given in Table 7-4.

Bond energies can be obtained in a similar way if the thermal-enthalpy contributions are removed so that values for ΔE_0° for atomization reactions are obtained. Such energies correspond, for diatomic molecules, to the dissociation energy shown in Fig. 7-3. Thus, for the $O_2 \rightarrow 2O$ reaction, we calculate, using the simplifying assumption that the vibrational contribution to the thermal energy will be small and can be neglected,

$$\begin{aligned}\Delta(H^\circ - E_0^\circ)_{298} &= 2(\tfrac{3}{2}RT + RT) - (\tfrac{3}{2}RT + RT + RT) \\ &= \tfrac{3}{2}RT = 3.72 \text{ kJ} \qquad [56]\end{aligned}$$

TABLE 7-4 Bond enthalpies for chemical bonds, in kJ mol^{-1}*

SINGLE BONDS

H—H	436	C—H	415	N—H	391	O—H	463	F—F	158
H—F	563	C—C	344	N—N	159	O—O	143	Cl—Cl	243
H—Cl	432	C—Cl	328	N—O	175	O—F	212	Br—Br	193
H—Br	366	C—Br	276	N—F	270	S—H	368	I—I	151
H—I	299	C—O	350	N—Cl	200	S—S	266	Cl—F	251
		C—N	292					Br—Cl	218
								I—Cl	210
								I—Br	178

MULTIPLE BONDS

C═C	615	N—N	418
C≡C	812	N≡N	946
C═O	724		

*From a more extensive table by L. Pauling, "General Chemistry," W. H. Freeman and Company, San Francisco, 1970.

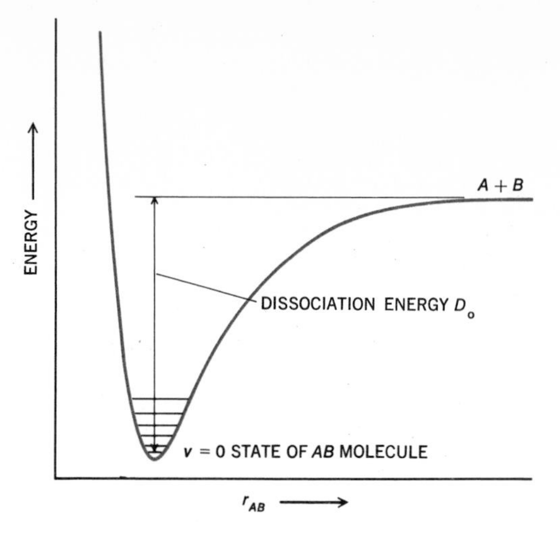

FIGURE 7-3
The relation of the potential-energy curve for a diatomic molecule to its dissociation energy D_0.

It follows that for the dissociation of O_2

$$\Delta E_0^\circ = \Delta H_{298}^\circ - \Delta(H^\circ - E_0^\circ)_{298}$$
$$= 495.04 - 3.72 = 491.32 \text{ kJ} \qquad [57]$$

In a similar way the thermal-enthalpy contributions can be removed from the atomization reactions of polyatomic molecules, and values for the bond energies for all species in their lowest allowed energy levels can be obtained.

Problems

1 A sample of liquid acetone weighing 0.5680 g is burned in a bomb calorimeter for which the heat capacity, including the sample, is 5640 J deg^{-1}. A temperature rise from 22.87 to 24.56°C is observed.

- **a** How much heat is given out per gram of acetone?
- **b** What are the values of $\Delta E_{\text{thermal res}}$, $\Delta E_{\text{mech res}}$, and ΔE per mole of acetone?
- **c** What is the change in the number of moles of gaseous reagents per mole of acetone?
- **d** What is the value of ΔH for the reaction?

2 When 0.532 g of benzene is burned at 25°C in a constant-volume system with an excess of oxygen, 22.3 kJ of heat is given out and the products are $CO_2(g)$ and $H_2O(l)$. What is the heat of combustion of benzene? For this combustion process, what are $\Delta E_{\text{thermal res}}$, $\Delta E_{\text{mech res}}$, ΔH, and ΔE per mole of benzene?

3 Deduce values for ΔE for the combustion reactions for which enthalpies are listed in Table 7-1. Assume that all gases behave ideally.

4 Deduce the enthalpy change for the reaction

$2CH_4(g) \rightarrow CH_3CH_3(g) + H_2(g)$

by combining the equations and the heats for the combustion reactions of the three reagents. *Ans.* $\Delta H = 65.02$ kJ.

5 The standard heat of formation of carbon monoxide cannot be determined conveniently from the reaction of carbon and a limited supply of oxygen. Show how the heat of formation of carbon monoxide from the elements can be deduced from conveniently measurable heats of combustion.

6 A flow process converts 0.5 mol of acetylene per minute into benzene by passing acetylene gas over a catalyst bed. At what rate must heat be added or removed in order to keep the catalyst bed and the exit benzene vapor at the same temperature, 25°C, as the incoming acetylene?

Ans. 99.75 kJ removed per minute.

7 Calculate the heat of combustion of benzene from the standard heats of formation given in Appendix 5.

8 Calculate the standard heat of formation of *n*-butane from the heat-of-combustion data of Table 7-1. *Ans.* -124.72 kJ.

9 The heat of combustion of cyclopropane has been reported as 2091.2 kJ mol^{-1}. Calculate the standard heat of formation at 25°C.

10 Combustion at a constant pressure of diborane, B_2H_6, proceeds according to the equation

$$B_2H_6(g) + 3O_2(g) \rightarrow B_2O_3(s) + 3H_2O(g)$$

and 2020 kJ is liberated per mole of B_2H_6. Combustion of elemental boron also proceeds to the product B_2O_3 and gives out 1264 kJ mol^{-1}. What is the standard heat of formation of diborane?

11 Deduce the heat of the reaction whereby ethyl alcohol is formed from ethylene and water. Use (*a*) the heats of combustion of Table 7-1 and (*b*) the heats of formation of Appendix 5. Specify states of reagents.

12 Calculate, from the heats of combustion in Table 7-1 and the results of Prob. 10, the heat per pound that can be released by the reagents of the following reactions (ΔH_{vap} of O_2 is 6.820 kJ mol^{-1}):

$$C_7H_{16}(n\text{-heptane})(g) + 11O_2(l) \rightarrow 7CO_2(g) + 8H_2O(g)$$

$$B_2H_6(g) + 3O_2(l) \rightarrow B_2O_3(s) + 3H_2O(g)$$

13 Calculate the value of ΔH for the solution of 1 mol of solid silver-chloride in a large amount of dilute ammonium hydroxide solution to produce the soluble salt of the silver ammonia complex ion. Make use of the data of Appendixes 5 and 6.

14 Calculate the maximum flame temperature, i.e., the adiabatic value that corresponds to all the heat of reaction going into heating the product molecules, when H_2 is burned in just sufficient oxygen to lead to the product H_2O. Make use of the heat-capacity data of Table 7-2, but note that the data are not really applicable to the temperature range of the problem.

Ans. 4600 K.

15 Using the data of Appendix 5 and Table 7-2, calculate the heat of reaction

$$C(\text{graphite}) + O_2(g) \rightarrow CO_2(g) \text{ at } 1500 \text{ K}$$

16 The heat of sublimation of NaCl can be estimated to be 755 kJ mol^{-1} at 25°C. Using a vibrational-level spacing for NaCl vapor molecules of 0.755×10^{-20} J, calculate the enthalpy of NaCl(*g*) at 1000°C compared with the crystal at 25°C, assuming ideal-gas behavior.

17 The enthalpy change of the reaction $Cl_2(g) \rightarrow 2Cl(g)$ at 25°C is 242.7 kJ. Calculate what the internal-energy change and the enthalpy change would be for all reagents in their lowest allowed quantum states. The vibrational-energy-level spacing for Cl_2 is found spectroscopically to be 1.13×10^{-20} J. What would the internal-energy change and the enthalpy change for the dissociation of Cl_2 be at 1500 K?

Ans. $\Delta E_0 = 239.5$ kJ; $\Delta H_{1500\,K} = 248.9$ kJ.

18 At high temperatures all molecules tend to break up into smaller fragments. Do the energy considerations of the Cl_2 example in Prob. 17 explain this?

19 Draw to scale a diagram like that of Fig. 7-2 for the reaction $CO(g) + \frac{1}{2}O_2(g) \rightarrow CO_2(g)$, showing the enthalpies of reaction at 25 and at 1500 K. Make use of the data of Appendix 5 and Table 7-2. Read off other values of ΔH°, and plot ΔH° as a function of temperature over this temperature range.

20 Plot curves showing $H^\circ - E_0^\circ$ for CO, $\frac{1}{2}O_2$, and CO_2 over the temperature range 25°C to 1500 K. The vibrational spacings for CO and O_2 are 4.31×10^{-20} and 3.14×10^{-20} J, respectively. CO_2 is a linear molecule and has four modes of vibration with the spacings 1.32, 1.32, 2.75, and 4.67, all times 10^{-20}. Plot $\Delta(H^\circ - E_0^\circ)$ for the reaction $CO(g) + \frac{1}{2}O_2(g) \rightarrow CO_2(g)$ versus temperature. Using the standard heats of formation, calculate ΔE_0° for the reaction. Using values of $\Delta(H^\circ - E_0^\circ)$ and the value of ΔE_0°, obtain values of ΔH°, and plot against temperature. Compare with the curve obtained in Prob. 19.

21 Using data from Prob. 20, calculate the heat capacity of CO_2 at various temperatures between 298 and 1500 K. Compare these values with those from the empirical expression $C_p = a + bT + cT^{-2}$ and the data of Table 7-2.

22 Using the data of Prob. 9, calculate a value for the $-\overset{|}{\underset{|}{C}}-\overset{|}{\underset{|}{C}}-$ bond energy for cyclopropane. *Ans.* 305 kJ.

23 Verify the value given for the $-\overset{|}{N}-H$ bond enthalpy in Table 7-4.

24 Verify the $-\overset{|}{C}=\overset{|}{C}-$ bond enthalpy of Table 7-4. Instead of assuming that the $-\overset{|}{\underset{|}{C}}-H$ bond of methane had the same enthalpy as that in ethylene, if it were assumed that $-\overset{|}{C}=\overset{|}{C}-$ had exactly twice the enthalpy of a $-\overset{|}{\underset{|}{C}}-\overset{|}{\underset{|}{C}}-$ single bond, what would be the deduced value for the enthalpy of the $=\overset{|}{C}-H$ bond in ethylene?

25 If the $-\overset{|}{\underset{|}{C}}-H$ bond enthalpy of Table 7-4 is assumed to apply to both n-butane and i-butane, what values of the $-\overset{|}{\underset{|}{C}}-\overset{|}{\underset{|}{C}}-$ bond enthalpy are deduced from the data of Table 7-1?

26 From the bond-enthalpy data of Table 7-4, deduce ΔH_f° for ethanol, and compare with the value in Appendix 5. *Ans.* 240 kJ.

27 Explain why the $-\overset{|}{\underset{|}{C}}-\overset{|}{\underset{|}{C}}-$ bond enthalpy in diamond is half the value of the heat of sublimation of diamond.

References

The general treatments of chemical thermodynamics listed at the end of Chap. 6 contain material on thermochemistry. References to compilations of thermochemical data can be found particularly in the books by Pitzer and Brewer and by Klotz. The following books are also pertinent to the subject of thermochemistry.

COTTRELL, T. L.: "The Strengths of Chemical Bonds," Butterworth & Co. (Publishers), Ltd., London, 1958. Presents the experimental methods for obtaining bond-dissociation energies in diatomic and polyatomic molecules and summarizes the available data.

JANAF Thermochemical Tables, 2d ed., Natl. Std. Ref. Data Ser., *Natl. Bur. Std. (U.S.)*, no. 37, 1971. Thermodynamic data for many inorganic compounds.

ROSSINI, F. W., et al. (eds.): Tables of Selected Values of Chemical Thermodynamic Properties, *Natl. Bur. Std. (U.S.) Circ.* 500, 1952. An excellent collection of thermodynamic properties of chemical substances, with references to original sources.

8

ENTROPY AND THE SECOND AND THIRD LAWS OF THERMODYNAMICS

A very large part of chemistry is concerned, in one way or another, with the state of equilibrium and the tendency of systems to move in the direction of the equilibrium state. Thermodynamics is the basic approach to the study of equilibria. Enthalpy and internal-energy changes are, themselves, not reliable indications of the tendency of a reaction to proceed; i.e., they do not indicate where the equilibrium lies. The thermodynamic and molecular treatments that we now take up are concerned with the following questions:

1 Can the equilibrium state of a chemical system be determined by the use of some new thermodynamic function?

2 If so, can this function, and therefore the equilibrium state, be understood in terms of the properties of the molecules involved?

After a preliminary discussion of the general statements of the second law of thermodynamics, a new thermodynamic function will, rather arbitrarily, be introduced. This function will allow the second law to be applied to chemical systems, and it will be seen that the second law and the new function are concerned with the equilibrium state and the tendency of processes, or reactions, to occur spontaneously. A molecular interpretation of the new thermodynamic function will then be suggested.

Not until the following chapter will the thermodynamic study of equilibria be completed by the introduction of another, more convenient function.

8-1 General Statements of the Second Law of Thermodynamics

Although the second law can be stated in a number of different ways, all statements can ultimately be shown to generalize our knowledge that natural processes tend to go to a state of equilibrium. The second law sums up our experiences with equilibria, just as the first law summed up our experience with energy. The general statements of the second law, like the conservation-of-energy statement of the first law, are not immediately applicable to chemical problems. After the law is introduced in general terms, it will be shown that it can be expressed in a chemically useful form.

Two important statements of the second law have been given. One due to Lord Kelvin is that "it is impossible by a cyclic process to take heat from a reservoir and convert it into work without at the same time transferring heat from a hot to a cold reservoir." Another statement, given by Clausius, is that "it is impossible to transfer heat from a cold to a hot reservoir without at the same time converting a certain amount of work into heat."

The first of these two statements is illustrated by the fact that a ship cannot derive work from the energy in the sea on which it moves. A moment's thought about all types of engines will show that there is always a hot and a cold source. A steam engine, for example, could not be made to produce work if it were not for the high pressure and high temperature of the steam *compared* with the surroundings.

This statement of the second law is related to equilibria when it is realized that work can be obtained from a system only when the system is not already at equilibrium. If a system is at equilibrium, no process tends to occur spontaneously and there is nothing to harness to produce work. A nonchemical example is the production of hydroelectric power. Here work is obtained when the spontaneous tendency of water to flow from a high to a low level is used. Lord Kelvin's statement recognizes that the spontaneous process is the flow of heat from a higher to a lower temperature and that only from such a spontaneous process can work be obtained.

The second statement is readily illustrated by the operation of a refrigerator. Again we have the recognition that the spontaneous flow of heat is from a high to a low temperature and that the reverse is possible only when work is expended.

A rather more sophisticated approach than will be used here concerns itself with how the high and low temperatures referred to in these statements are defined. The statements of Kelvin and Clausius, in fact, provide fundamental definitions for temperature. Temperature so defined, however, can be shown to be identical with the temperature scale that makes the relation $PV = nRT$ hold for ideal gases.

The chemist's interest in the second law of thermodynamics is aroused by the possibility of this law saying something about the position of equilibrium in a chemical process.

8-2 Entropy and Another Statement of the Second Law of Thermodynamics

In considering the thermodynamics of changes, chemical or otherwise, we must focus on the initial and the final states that are connected by a real or imagined process. We now would like to know whether there is a property (and if there is, how it can be evaluated) that lets us deduce if a natural, spontaneous process might change the system from the one state to the other.

Chemical terminology would have it that spontaneous processes are those in which there is a *driving force* that tends to make the reaction or process occur. The study of any property that might be associated with the tendency of processes, or chemical reactions, to proceed, and thus might give a measure to the driving force of chemical reactions, is of obvious chemical interest.

(A process or reaction for which there is no tendency to proceed in either direction has been described as balanced, or *reversible*. By contrast, a spontaneous process is unbalanced, or *irreversible*—no small change in the conditions could overcome the natural driving force.)

It turns out to be profitable, in the search for such a property, to deal with $\Delta E_{\text{thermal res}}/T$ rather than with q or $\Delta E_{\text{thermal res}}$ itself. Such a quantity turns out to be sufficiently important to merit a symbol $\Delta S_{\text{thermal res}}$ and a name, the change in *entropy* of the thermal reservoir. More generally, there will be thermal reservoirs at a number of different temperatures, or the reservoir must operate over a range of temperatures. In the latter case the entropy change of the thermal reservoir would be calculated by a suitable integration of

$$dS_{\text{thermal res}} = \frac{dE_{\text{thermal res}}}{T} \qquad [1]$$

We are at liberty, you should recognize, to ascribe any features to this new entropy function that we like, the requirement being that we construct a function that is self-consistent and allows us to form a useful expression for the second law. In this vein we further specify that, for all processes,

$$dS_{\text{mech res}} = 0 \qquad [2]$$

Now we are at a stage comparable with that in the development of the first law, where we knew how to determine the energy changes in the thermal and mechanical surroundings. The next step is to use this information to learn about the change in the property, now the entropy, of the system and to see how it can be used.

We proceed by making two statements that together express a law of nature, which we call the *second law of thermodynamics*. (A law of nature is, you recall, a generalization, and you accept it if you find that it agrees with observations.)

The two statements that constitute the second law are:

1 *When a process is carried out reversibly, the entropy change in the universe of the process is zero.* (Notice that, since we know how to calculate $\Delta S_{\text{thermal res}}$, and we have required $\Delta S_{\text{mech res}}$ to be zero, this statement provides a means for the deduction of the entropy change ΔS of the system.)

2 *For processes that proceed irreversibly, i.e., out of balance and therefore spontaneously, the entropy of the universe of the process increases.* (Thus, if by some means ΔS and $\Delta S_{\text{thermal res}}$ are known, the possibility of this process occurring spontaneously can be deduced by inspecting the sign of $\Delta S + \Delta S_{\text{thermal res}}$.)

That the first statement can lead to a value for the entropy change of the system can first be illustrated.

Example 1 What is the entropy change suffered by 1 mol of liquid water at 1 atm and 100°C when it is converted to vapor at the same temperature and pressure?

The problem can be visualized by Fig. 8-1, and the numerical entries result from the use of the heat of vaporization, 40,670 J, to obtain, first,

$$\Delta S_{\text{thermal res}} = \frac{-40{,}670}{373} = -109 \text{ J deg}^{-1} \qquad [3]$$

Then, since the vaporization can be imagined to occur reversibly with only an infinitesimal temperature difference between the reservoir and the system, the requirement that $\Delta S_{\text{univ}} = 0$ leads to

$$\Delta S_{\text{thermal res}} + \Delta S = 0$$

or

$$\Delta S = -\Delta S_{\text{thermal res}} = +109 \text{ J deg}^{-1} \qquad [4]$$

Now let us turn to a process which we know to be spontaneous and see if, as the second statement claims, the entropy of all concerned increases.

Example 2 Consider a vessel containing liquid water and steam, at 1 atm, and therefore at 100°C, to be brought in contact with a vessel

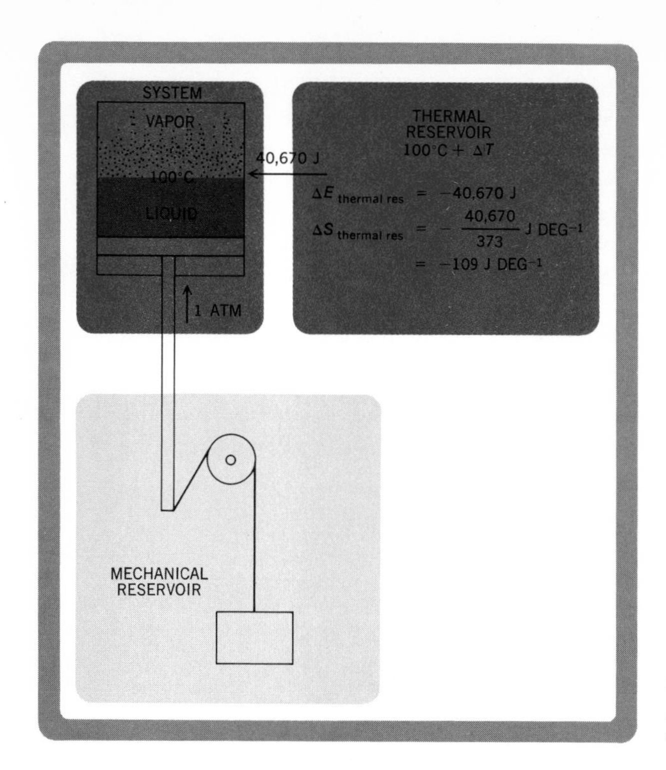

FIGURE 8-1
The formation of 1 mol of steam from 1 mol of liquid water at 1 atm and 100°C.

containing liquid water and ice, at 1 atm, and therefore at 0°C. What is the entropy change for the flow of some amount of heat Q from the hot to the cold vessel? (The water-and-steam and water-and-ice containers are used so that their temperatures will remain constant even though heat, if not too much, is transferred. If temperatures were allowed to change, the calculations would be similar in principle, but a little more involved.)

First a reversible way of performing this energy transfer must be

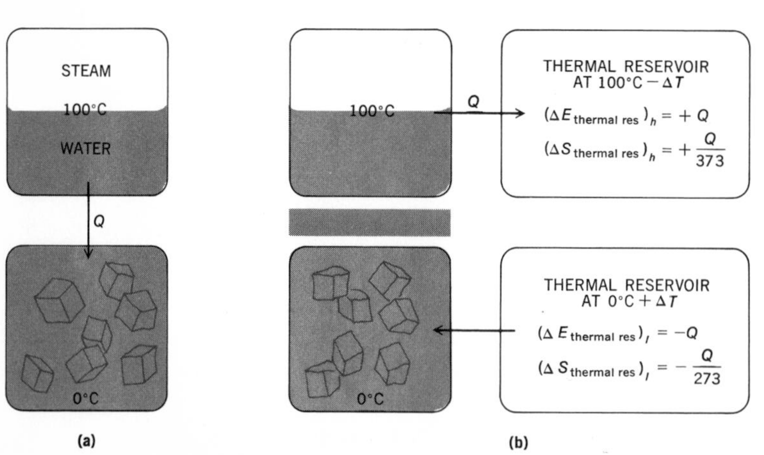

FIGURE 8-2
The (*a*) irreversible and (*b*) reversible transfer of heat Q from a water-steam container at 100°C to a water-ice container at 0°C.

devised so that the constancy of the entropy of all concerned in a reversible process can be used to calculate the entropy change of the system, which consists of the two containers. (Heat flowing directly from a high to a low temperature corresponds to an irreversible process, and for such processes we have no direct way of calculating this entropy change.) A reversible process results if we connect each of the two parts of the system to separate heat reservoirs as shown in Fig. 8-2*b*, one at a temperature infinitesimally lower than 100°C, the other infinitesimally higher than 0°C. The heat flows indicated in the figure give the net result, as far as the system is concerned, of transferring energy from the hot to the cold body. The addition of the thermal reservoirs, however, makes the process reversible.

The entropy change suffered by the thermal reservoirs is now calculated as

$$\Delta S_{\text{thermal res}} = \Delta S_{\text{hot res}} + \Delta S_{\text{cold res}} = \frac{Q}{373} - \frac{Q}{273}$$

$$= -0.00098Q \qquad [5]$$

Since the process of Fig. 8-2*b* is reversible, $\Delta S_{\text{univ}} = 0$, and thus

$$\Delta S = -\Delta S_{\text{thermal res}} = +0.00098Q \qquad [6]$$

The entropy change of the system has been calculated by means of the arrangement of Fig. 8-2*b*, but as will be shown, the entropy change of the *system* is independent of the way in which the process is performed and depends only on the initial and final states. The result gives, therefore, the entropy change ΔS of the system when heat Q is transferred from the hot end to the cold end by *any* process, including that of Fig. 8-2*a*.

Now we can consider the entropy change of the universe for the direct heat transfer, as in the arrangement of Fig. 8-2*a*. No heat reservoirs are involved, and an entropy change results, therefore, only from that which occurs in the system. This we have deduced to be the positive quantity $+0.00098Q$. Thus, for the direct heat transfer,

$$\Delta S_{\text{univ}} = \Delta S = +0.00098Q = \text{a positive quantity} \qquad [7]$$

If we did not know that heat would flow spontaneously from hot to cold, this result, and the statements of the second law, would tell us that this would happen!

Let us consider one final example.

Example 3 Determine the entropy change for the isothermal expansion of n mol of an ideal gas at temperature T from a volume V_1 to a volume V_2.

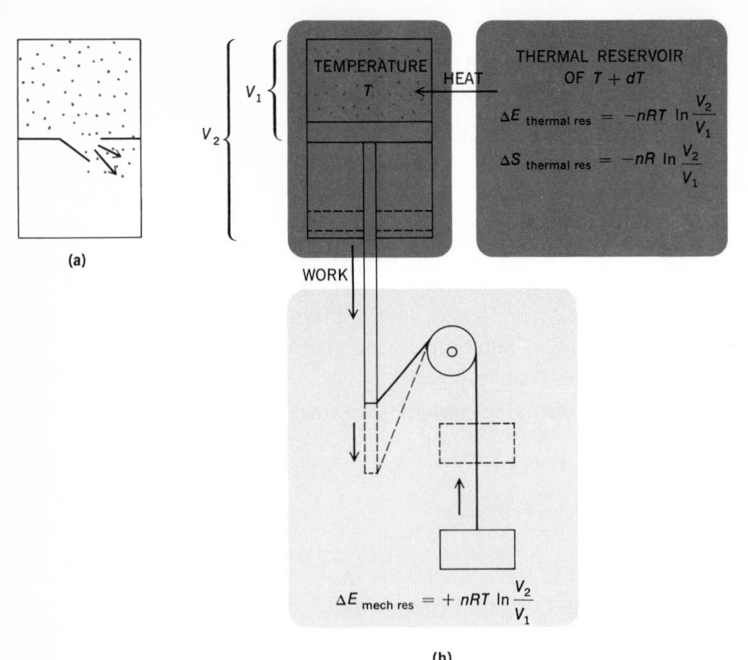

FIGURE 8-3
The (*a*) irreversible and (*b*) reversible isothermal expansion of n mol of an ideal gas.

The expansion can be performed irreversibly by simply opening a stopcock and allowing the gas to rush into the previously evacuated compartment, as represented by Fig. 8-3*a*. We have no method for analyzing this process directly to find the entropy difference between the expanded gas and the initial gas. It is necessary to devise an alternative process that can be carried out reversibly and that takes the system from the same initial to the same final state. Figure 8-3*b* shows one possibility.

For this isothermal ideal-gas expansion, the internal energy does not change and the thermal surroundings give up an amount of energy equal to that gained by the mechanical surroundings. Thus we calculate, since for the reversible expansion $P_{\text{ext}} = P$,

$$\Delta E_{\text{mech res}} = \int P_{\text{ext}}\, dV = \int P\, dV = \int_{V_1}^{V_2} \frac{nRT}{V}\, dV$$

$$= nRT \ln \frac{V_2}{V_1} \qquad [8]$$

and

$$\Delta E_{\text{thermal res}} = -nRT \ln \frac{V_2}{V_1} \qquad [9]$$

This allows us to obtain

$$\Delta S_{\text{thermal res}} = \frac{\Delta E_{\text{thermal res}}}{T} = -nR \ln \frac{V_2}{V_1} \qquad [10]$$

Now, for this reversible process to give the required $\Delta S_{\text{univ}} = 0$, we must have

$$\Delta S = -\Delta S_{\text{thermal res}} = +nR \ln \frac{V_2}{V_1}$$

With this result, we can return to the free expansion of the gas as in Fig. 8-3a. The system changes from the same initial to the same final states as in the alternative reversible expansion, but no thermal surroundings are involved. Thus

$$\Delta S_{\text{univ}} = \Delta S = +nR \ln \frac{V_2}{V_1}$$

so that

$$\Delta S_{\text{univ}} = +nR \ln \frac{V_2}{V_1}$$
$$= \text{a positive quantity} \qquad [11]$$

Again, if we did not recognize the spontaneity of the escape of a gas into an evacuated chamber, this result, an increase in the entropy of all that is affected by the process, would lead us to this expectation.

Some of the importance attached to entropy, as a result of such deductions, can be seen by its position alongside energy in the famous maxim of Clausius: "The energy of the universe is constant; the entropy of the universe always tends toward a maximum." Since all natural processes are spontaneous, they must occur with an increase of entropy, and therefore the sum total of the entropy in the universe is continually increasing. Recognition of this trend leads to some interesting philosophical discussion, as, for example, Sir Arthur Eddington's idea that "entropy is time's arrow."

We are now in a position to summarize the results of this and the preceding section and thereby to indicate the use that can be made of the function called entropy. Suppose one wishes to investigate the possibility of a reaction, either chemical or physical, proceeding from one state a to another state b. If the entropy difference ΔS_{univ} for the process can be calculated, use can be made of the statements:

If ΔS_{univ} *is positive, the reaction will tend to proceed spontaneously from state* a *to state* b.

If ΔS_{univ} *is zero, the system is at equilibrium and no spontaneous process will occur.*

If ΔS_{univ} *is negative, the reaction will tend to go spontaneously in the reverse direction, i.e., from* b *to* a.

That these properties of entropy sum up our experience with naturally occurring phenomena has been illustrated by the simple examples given above.

For the type of process used in these examples, it is certainly cumbersome and unnecessary to introduce the entropy function. On the other hand, when dealing with a chemical reaction

Reactants $\rightarrow$ products

one would be greatly aided by a thermodynamic property that could be determined for the reactants and the products, and any surroundings that are involved, and that would tell whether or not the reaction would tend to proceed spontaneously.

The above considerations all depend on the use of results obtained from measurements on the thermal surroundings to deduce, for a reversible process, the entropy change that is occurring in the system. (The procedure is parallel to that used in applications of the first law, where measurements on the energy changes in the thermal and mechanical surroundings allowed the change in energy in the system to be deduced.) But again, there generally are a number of reversible ways of going from some initial state of the system to some final state. Have we any assurance that the values calculated for ΔS for all these will be the same; i.e., are we entitled to interpret a value for ΔS deduced for some process that takes the system from state a to state b as $S_b - S_a$? (There is, you should note, some good reason to be uncertain of the answer, because in our first-law deduction that E was a state function, we used the fact that the sum of $\Delta E_{\text{thermal res}}$ and $\Delta E_{\text{mech res}}$ was independent of path, even though the separate terms were quite path-dependent.)

That entropy is a state function can be deduced by considering a specific cyclic process from which the results can be extended to a large class of processes.

8-3 The Carnot Cycle

Although chemists are not necessarily interested in the conversion of heat into work, consideration of a particularly simple engine for doing this leads us to the recognition that a system in a given state has a certain amount of entropy; i.e., entropy, like internal energy or enthalpy, is a state function.

The engine that we consider was analyzed originally by the French engineer Sadi Carnot, in 1824. This engine is very convenient for analysis, but although the pattern of operation is not entirely different from that of a steam or internal-combustion engine, it is not a practical device. One cycle of this engine has the net effect of using heat from a hot thermal reservoir, producing some mechanical energy and giving

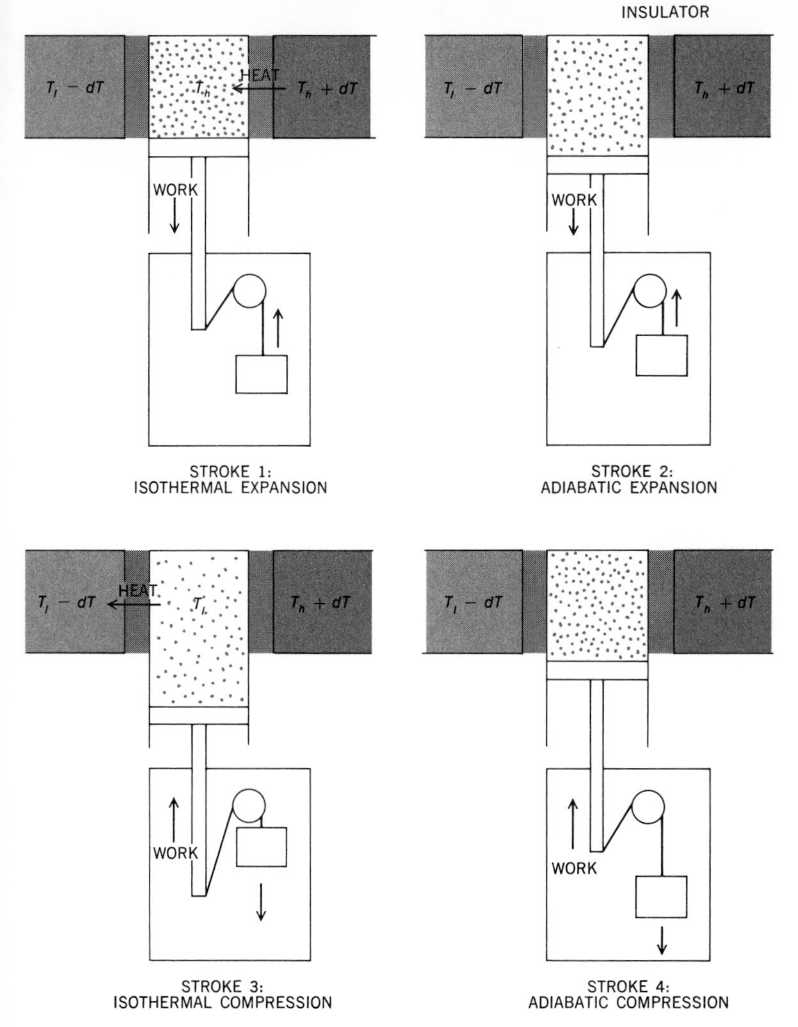

FIGURE 8-4
The four strokes of the Carnot cycle.

some heat to a cold thermal reservoir. The engine would operate by continually repeating this cycle.

A diagram of a Carnot engine system and the thermal and mechanical reservoir is shown in Fig. 8-4. Any material could be assumed to be present in the cylinder, but the analysis is greatly simplified if the working material in the cylinder is an ideal gas. We shall assume further that there is 1 mol of this gas.

The Carnot cycle consists of four steps, shown in Fig. 8-5. Each step is performed reversibly; i.e., the pressure of the gas is only infinitesimally different from that of the piston, and the heat flows across an infinitesimal temperature gradient. First the energy exchanged between the system and the thermal and mechanical reservoirs will be deduced for each step.

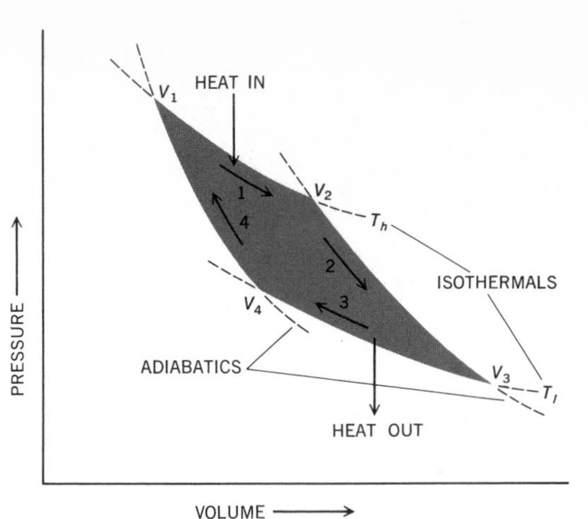

FIGURE 8-5
The Carnot cycle on a PV diagram. (The four strokes are numbered as in Fig. 8-4.)

Step 1 The gas is expanded isothermally at a high temperature T_h from an initial volume V_1 to a volume V_2. Work is done by the gas, and heat flows from the hot reservoir to the gas.

Since the gas is ideal and the temperature is constant, $\Delta E = 0$ and the energy changes of the surroundings during this first step can be expressed as

$$\Delta(E_{\text{mech res}})_1 = -(\Delta E_{\text{thermal res}})_h$$
$$= \int P\,dV = RT_h \int_{V_1}^{V_2} \frac{dV}{V} = RT_h \ln \frac{V_2}{V_1} \qquad [12]$$

Step 2 The gas is expanded adiabatically from a volume V_2 to a volume V_3, and the temperature drops to a low temperature T_l. Work is done by the gas, and since the insulators are in place for this adiabatic step, no heat is transferred.

The fact that the working substance is an ideal gas lets us write

$$\Delta E_2 = c_V(T_l - T_h)$$

and the first law then gives

$$\Delta(E_{\text{mech res}})_2 = -\Delta E_2 = -c_V(T_l - T_h) = c_V(T_h - T_l) \qquad [13]$$

Step 3 The gas is compressed isothermally at the lower temperature T_l from the volume V_3 to a volume V_4. Work is now done *on* the gas by the piston, and heat flows from the gas to the cold reservoir.

Since the gas is ideal and the temperature is constant, $\Delta E = 0$ and

$$\Delta(E_{\text{mech res}})_3 = -(\Delta E_{\text{thermal res}})_l = \int P\,dV = RT_l \int_{V_3}^{V_4} \frac{dV}{V}$$
$$= RT_l \ln \frac{V_4}{V_3} \qquad [14]$$

We note that, since v_4 is less than v_3, this corresponds to a negative $\Delta E_{\text{mech res}}$, and thus to work being done on the gas and heat being given off from the gas.

Step 4 The gas is compressed adiabatically from the volume v_4 to the original volume v_1, and the temperature rises to the higher temperature T_h at which the cycle was started. Work is again done *on* the gas during this compression, and no heat is absorbed.

The first law gives

$$\Delta(E_{\text{mech res}})_4 = -\Delta E_4 = c_V(T_h - T_l) \quad [15]$$

Here also it can be noted that, corresponding to work being done on the gas, $\Delta E_{\text{mech res}}$ is negative.

That the entire cycle obeys the first law of thermodynamics can first be checked. The net mechanical energy delivered to the surroundings, since the effects in steps 2 and 4 nullify each other, is given by the sum from steps 1 and 3, i.e.,

$$\Delta E_{\text{mech res}} = RT_h \ln \frac{v_2}{v_1} + RT_l \ln \frac{v_4}{v_3} \quad [16]$$

This expression can be simplified by the recognition that the adiabatics of steps 2 and 4 relate the four volumes according to equations

$$T_h^{c_V/R} v_2 = T^{c_V/R} v_3 \qquad \text{and} \qquad T_h^{c_V/R} v_1 = T^{c_V/R} v_4 \quad [17]$$

Division of the first of these by the second gives

$$\frac{v_2}{v_1} = \frac{v_3}{v_4} \quad [18]$$

Thus Eq. [16] reduces to

$$\Delta E_{\text{mech res}} = R(T_h - T_l) \ln \frac{v_2}{v_1} \quad [19]$$

The energy change of the thermal reservoirs is calculated as the sum of the values for the two isothermal steps as

$$\Delta E_{\text{thermal res}} = -RT_h \ln \frac{v_2}{v_1} - RT_l \ln \frac{v_4}{v_3} \quad [20]$$

or with Eq. [18],

$$\Delta E_{\text{thermal res}} = -R(T_h - T_l) \ln \frac{v_2}{v_1} \quad [21]$$

The first law can now be seen to be obeyed because, for a cycle in which the working substance is returned to its initial state, the energy given up by the thermal reservoirs, Eq. [21], is equal to that gained by the mechanical reservoirs, Eq. [19].

8-4 The Efficiency of the Transformation of Heat into Work

The performance of an engine is usually computed on the basis of the net mechanical energy produced for a given consumption of thermal energy from the hot reservoir. In an internal-combustion engine, for example, one measures the mechanical energy produced from a given amount of fuel, which is the counterpart of heat from the hot reservoir. The efficiency of the Carnot engine is defined, therefore, as

$$\text{eff} = \frac{\Delta E_{\text{mech res}}}{-(\Delta E_{\text{thermal res}})_h} \qquad [22]$$

Use of Eqs. [19] and [12] allows this efficiency to be shown as

$$\text{eff} = \frac{R(T_h - T_l)\ln(\text{v}_2/\text{v}_1)}{RT_h \ln(\text{v}_2/\text{v}_1)} = \frac{T_h - T_l}{T_h} \qquad [23]$$

This formula represents the results of the analysis of the operation of the Carnot engine. The relation of this particular result to the previously given general statements of the second law can now be investigated.

Kelvin's statement that thermal energy cannot be completely converted to mechanical energy leads us to investigate the possibility of an engine with 100 percent thermodynamic efficiency. According to Eq. [23], this can happen only if T_h is infinite or T_l is zero. The first is obviously an impracticality—although engines are run at as high a temperature as possible to increase their efficiencies. The second, as we will see when the third law of thermodynamics is dealt with, is another impossibility.

The Clausius statement requires us to think of a Carnot cycle that is run in the reverse direction to that of the preceding derivation. Since all the steps in the cycle are reversible, this could occur if the temperatures of the heat reservoirs were changed by infinitesimal amounts. We now have a refrigerator in which an amount of mechanical energy is expended, some thermal energy is taken from the cold reservoir, and energy equal to the sum of these is delivered to the hot reservoir. Then both the numerator and denominator in Eq. [22] have signs opposite from that for operation as an engine. The efficiency is therefore still given by Eq. [23]. Now it is interpreted as the ratio of the mechanical energy spent to the thermal energy delivered to the hot reservoir. This can only be zero, implying unassisted flow of heat from the cold to the hot reservoir for $T_h = T_l$, a result that is in accordance with the Clausius statement of the second law.

It can be shown (see Prob. 11) that no other engine working between the same two temperatures can convert thermal to mechanical energy with a greater efficiency than does the Carnot engine. Other reversible engines, in fact, will have the same efficiency as the Carnot engine; the Carnot assumption of 1 mol of an ideal gas as a working fluid merely allows us to use available expressions for the expansions and compressions in each of the engine's strokes. The Carnot cycle efficiency can therefore be used to make an estimate of the maximum conversion of heat to work that can be expected for a real engine.

A steam engine, for instance, can be taken as operating between some high temperature around 120°C and a condenser temperature of about 20°C. For such an engine the efficiency can be estimated as

$$\text{eff} = \frac{393 - 293}{393}(100) = 25\% \qquad [24]$$

Thus, for every four units of heat supplied in the steam, the equivalent of one unit of work can be obtained, and three units of heat is given off at the low temperature. This efficiency is the maximum that could be expected if there were no other inefficiencies in the operation. In addition, of course, there is a mechanical efficiency that limits the available work to some fraction of this theoretically possible amount.

Such a calculation indicates the desirability of operating an engine with as high a value of T_h and as low a value of T_l as possible.

8-5 Entropy Is a State Function

Now the results of the Carnot cycle derivation can be used to prove the previous assumption that the entropy of the system is a function of the state of the system, so that the entropy difference between two states of a system is the same for *any* process connecting these two states.

A relation that leads to this conclusion is obtained by equating the Carnot cycle efficiency expressions of Eqs. [22] and [23] and, in addition, recognizing that

$$\Delta E_{\text{mech res}} = -(\Delta E_{\text{thermal res}}) = -[(\Delta E_{\text{thermal res}})_h + (\Delta E_{\text{thermal res}})]$$

This leads to

$$\frac{(\Delta E_{\text{thermal res}})_h + (\Delta E_{\text{thermal res}})_l}{(\Delta E_{\text{thermal res}})_h} = \frac{T_h - T_l}{T_h} \qquad [25]$$

Rearrangement then gives

$$\frac{(\Delta E_{\text{thermal res}})_h}{T_h} + \frac{(\Delta E_{\text{thermal res}})_l}{T_l} = 0 \qquad [26]$$

or more suggestively,

$$\sum_{\text{cycle}} \frac{\Delta E_{\text{thermal res}}}{T} = 0 \qquad [27]$$

The special significance of this result follows from the fact that any reversible cycle involving an ideal gas can be thought of as being

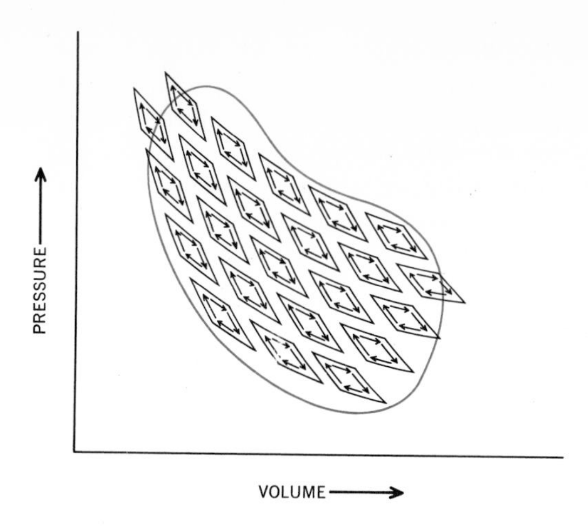

FIGURE 8-6
The cyclic process approximated by a set of schematic Carnot cycles.

made up of a large number of Carnot cycles, as illustrated by Fig. 8-6. A grid of isotherms and adiabatics is drawn through the heavy curve, which represents any reversible cyclic process. The grid can be used to construct a set of Carnot cycles so that the outer parts of the set trace out a curve that approximates that of the general process. If one were to perform all the Carnot cycles in the set, the new result would be almost the same as performing the general process. This results from the fact that all the Carnot cycle steps that are inside the boundary are canceled out because each is traced in both a forward and a reverse direction. In the limit of an infinitely closely spaced grid, an outline with infinitesimal steps that corresponds to the general cyclic process is obtained, and this, furthermore, encloses the same area on the PV diagram as does the general process. The net effect of performing all the Carnot cycles, it can be shown, will be the same as performing the general process.

Since any cyclic reversible process can be replaced by a set of Carnot cycles, one can write, in general,

$$(\Delta S_{\text{thermal res}})_{\text{cycle}} = \sum_{\text{cycle}} \frac{\Delta E_{\text{thermal res}}}{T} = 0$$

This result can be written in the more generally applicable form

$$(\Delta S_{\text{thermal res}})_{\text{cycle}} = \oint_{\text{rev}} \frac{dE_{\text{thermal res}}}{T} = 0 \qquad [28]$$

where the sign $\oint$ signifies integration around a complete cycle and the subscript reminds us that the result is limited to reversible processes.

Thus, for any *cyclic reversible* process, the change in entropy of not only the universe, but also the thermal surroundings, is zero. It follows that over the cycle the entropy change of the *system* must also be zero. If the cycle includes the points a and b, the return from b to a nullifies the entropy change that occurred in the a to b leg of the cycle. It follows that *all* paths from a to b must produce the same entropy change. Thus proof of the zero value for the cyclic integral is equivalent to proof that the entropy difference between two states is a function only of those states and is not dependent on the path used to go from one to the other.

Thus, even if this entropy change is calculated for a reversible process, as by the first statement of Sec. 8-2, it can be a valid measure of the change in entropy of the system no matter how the process is carried out—reversibly or irreversibly. The entropy, like the internal energy, enthalpy, volume, and so forth, is a property of the system, a state function.

8-6 The Unattainability of Absolute Zero

The third and final of the great summations of our experiences with nature on which all the deductions of thermodynamics are based can now be introduced. Full use of thermodynamics can then be made in our subsequent studies of chemical systems.

The second law of thermodynamics has introduced entropy, and this function has been shown to be important when the directions of spontaneous changes are investigated. The second law, moreover, shows how differences in entropy of two states of a system can be determined. The third law gives a method for obtaining the absolute value of the entropy of a system. As with the first two laws, the third is an expression of our experiences with nature. The basic experiences for this law come about from attempts to achieve very low temperatures. Such activity is not as readily undertaken as are the experiments that form the basis of the first and second laws, and for this reason the third law will appear to be a less general principle. All attempts that have been made to obtain lower and lower temperatures, however, lead to the general statement that *the absolute zero of temperature is unattainable*. It is this statement that can be used as a basis for the expression in the next section for the third law of thermodynamics. First, however, it may be of interest to mention briefly some of the steps that have been taken in the direction toward absolute zero.

It is not the mere attainment of low temperatures that stimulates attempts to achieve such temperatures. It will be shown that the measurement of absolute values for entropies requires measurements at temperatures approaching absolute zero. Furthermore, as absolute zero is approached, the reduction of thermal energy leads to the appearance of a number of very interesting phenomena that are obscured or nonexistent at higher temperatures. Interest in such effects has led to the production of lower and lower temperatures.

The principal refrigeration technique is based on the cooling resulting from a Joule-Thomson expansion, as discussed in Sec. 6-10. It will be recalled that, when a gas is below its inversion temperature, an expansion produces a cooling effect as a result of work being done by the gas to overcome the mutual attraction of the molecules of the gas.

The countercurrent arrangement, introduced by Siemens in 1860, whereby the expanded, and therefore cooled, gas is passed back over the compressed gas, leads to cooling which can be continued until condensation occurs. By such a process liquid air can be formed. The liquid air so produced can be distilled to give oxygen, boiling point 90 K, and nitrogen, boiling point 77 K. Liquid nitrogen can now be readily obtained commercially, and the attainment of a temperature of about 77 K presents no problem to the research worker.

Still lower temperatures are obtained by performing such a Joule-Thomson expansion on hydrogen. This gas must first, however, be cooled below its inversion temperature of 193 K, by means, for example, of liquid nitrogen. Expansion then allows liquid hydrogen, boiling point 22 K, to be formed. Finally, liquid hydrogen can be used to cool helium below its inversion temperature, and subsequent expansion produces liquid helium, which boils at 4 K. Temperatures of somewhat less than 1 K can be produced by reducing the pressure over the helium, but this technique is limited by the large amounts of vapor that must be pumped off. For many low-temperature research problems the temperatures reached by liquid hydrogen or helium are satisfactory and can be reached with commercially available liquid helium or a helium liquefier.

There are now several techniques for reaching temperatures below the 1 deg obtainable with liquid helium. Some depend on the special properties of the helium isotopes ^{3}He and ^{4}He, and others make use of the magnetic properties of materials that result from atomic magnetic moments, due to unpaired electrons, or from nuclear magnetic moments. The use of low-temperature magnetization can serve to illustrate the techniques and their limitations.

The temperature-entropy curves for *paramagnetic* salts, like gadolinium sulfate octahydrate, which contain unpaired electrons, and materials like copper, whose nuclei have a magnetic moment, are illustrated in Fig. 8-7. (In Chap. 10 you will see the basis on which the lower entropy for the magnetized, and thus more *ordered,* forms is to be expected.)

The *adiabatic-demagnetization* procedure for producing low temperatures can now be seen by considering the two steps of the process, shown in Fig. 8-8, and the consequences of these steps, shown in Fig. 8-9. The second reversible and isentropic step produces the desired reduction in the temperature of the sample. If, furthermore, another heat-removal reservoir is available at this lower temperature, the magnetization and demagnetization procedure can be repeated to attain yet a lower temperature.

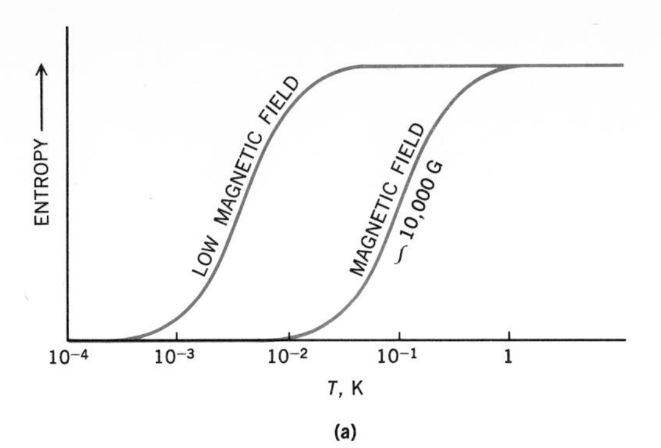

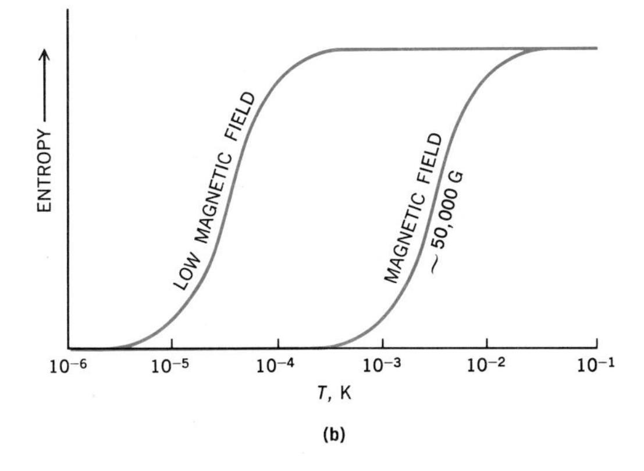

FIGURE 8-7
Representative ST diagrams for a material that (a) is paramagnetic because of unpaired electrons and (b) depends on magnetic effects of nuclei.

If a paramagnetic salt is used, it need only be precooled to about 1 K, which can be done with liquid helium, to reach a state in which the entropy difference between the magnetized and demagnetized states is appreciable. (At higher temperatures the thermal agitation in the crystal is so great that the applied magnetic field is ineffective in producing the necessary lining up of the magnetic moments of the unpaired electrons.) Adiabatic demagnetization then leads typically to temperatures of about 0.005 K.

When the procedure depends on nuclear magnetic properties, the

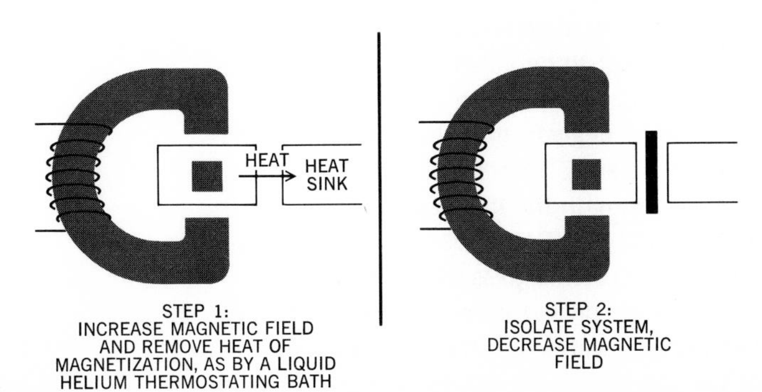

FIGURE 8-8
The steps of the adiabatic demagnetization procedure shown schematically.

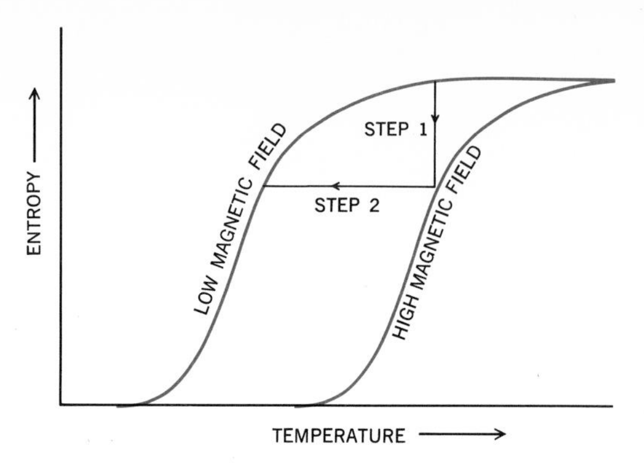

FIGURE 8-9
The effects on an ST diagram of the steps of Fig. 8-8.

substance must be precooled to about 0.01 K, nuclear magnets being about 2000 times smaller than an electron magnet, before satisfactory nuclear alignment can be achieved. However, the difficulty is repaid by the lower temperature, as low as about 10^{-6} K, to which the adiabatic demagnetization step could then lead.

Low-temperature, or *cryogenic,* work shows that temperatures very near absolute zero can be obtained. It appears, however, that in spite of the remarkably close approaches to absolute zero that have been achieved, absolute zero cannot be attained. In addition to direct attempts in this direction, one draws on other evidence, such as that provided by the molecular interpretation of matter, to support this generalization.

The chemist's interest in the unattainability of absolute zero stems primarily from the implication of this generalization on the entropy of crystals as the temperature approaches absolute zero.

8-7 Entropy and the Third Law of Thermodynamics

In the magnetic-cooling method a difference in entropy is used to reduce the temperature. More generally, any process that operates between states that at a given temperature have different entropies can be used to perform such a cooling. Furthermore, if any such entropy differences existed down to absolute zero, and a reversible process existed that could connect them, one might expect to be able to use these entropy changes to reduce the temperature to absolute zero. The generalization that absolute zero is unattainable suggests that the entropies of all materials at absolute zero are the same. This conclusion must, however, be restricted to materials that are in the thermodynamically most stable state for this temperature range. (One finds, for example, that many materials are frozen into a metastable glassy state as the temperature is reduced, and this state may persist even as absolute zero is approached because of the slowness with which the crystalline form is produced. The entropy of the glassy state could be different, in fact higher, than that of the crystal at absolute zero. Since the metastable state cannot be converted directly to the stable state by a reversible process, this entropy difference could not be used in attempts to reach absolute zero.) We can conclude, therefore, that *the entropy of all perfect crystalline substances must be the same at absolute zero.* Furthermore, to be consistent with the molecular interpretation of entropy, we take the entropy at absolute zero to be zero. In this way we come to the chemically useful statement of the third law of thermodynamics, quoted from the classic thermodynamics text by G. N. Lewis and M. Randall: "If the entropy of each element in some crystalline state be taken as zero at the absolute zero of temperature, every substance has a finite positive entropy; but at the absolute zero of temperature the entropy may become zero, and does so become in the case of perfect crystalline substances."

The third law makes it possible to obtain absolute entropies of chemical compounds from calorimetric measurements. The difference in entropy between 0 K and a temperature T can be deduced from the defining equation for entropy, by considering nearly reversible additions of heat from a variable-temperature heat reservoir. Then, for the sample under study, we have

$$S_T - S_0 = -\int_0^T \frac{dE_{\text{thermal res}}}{T} = \int_0^T \frac{\delta q}{T} \qquad [29]$$

The third law states that the S_0 can be assigned the value of zero for materials that form perfect crystals. The integral of Eq. [29] can be evaluated from measured heat capacities and heats of transition. If a given phase is heated from T_1 and T_2 at constant pressure, it gains entropy according to the expression

$$S_T - S_0 = \int_{T_1}^{T_2} \frac{dH}{T} = \int_{T_1}^{T_2} \frac{C_P\,dT}{T} = \int_{T_1}^{T_2} C_P\,d(\ln T) \qquad [30]$$

The integration can be performed if the necessary values for C_P are measured. The integration is usually performed graphically from a plot of either C_P/T versus T or C_P versus ln T. In either treatment the area under the curve between two temperatures gives, according to Eq. [30], the entropy increment for that temperature range. The method is illustrated in Fig. 8-10. Since heat-capacity measurements are usually not taken down below about 15 K, attainable with liquid hydrogen, an extrapolation to absolute zero is necessary. (The basis for the extrapolation, which introduces a comparatively small term in the total entropy, will be given when the nature of solids is studied.)

In taking a compound from near absolute zero to some temperature such as 25°C, a number of phase transitions are generally encountered. At each of these transitions the heat capacity will, for the most part, change abruptly, and heat will be absorbed. The entropy change corre-

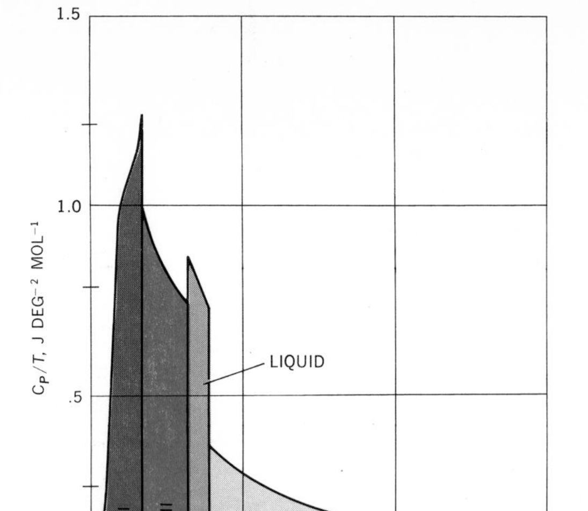

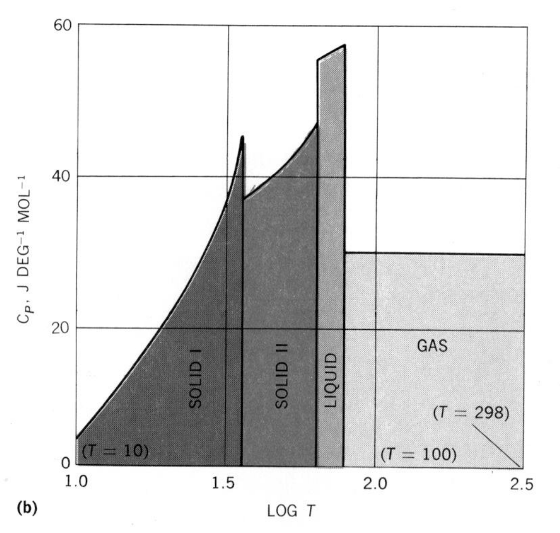

FIGURE 8-10
Graphs of C_P/T versus T and C_P versus ln T for N_2 (at $T = 298$, log $T = 2.475$). [*Data of W. F. Giauque and J. O. Clayton, J. Am. Chem. Soc.,* **55:***4875 (1933).*]

sponding to such transitions can be calculated, as illustrated in Example 1 of Sec. 8-2, from the measured enthalpy change for the transition by the expression

$$\Delta S_{\text{trans}} = \frac{\Delta H_{\text{trans}}}{T_{\text{trans}}}$$

The entropy obtained by adding up all the contributions from absolute zero is usually reported at 25°C. Table 8-1 shows the terms that go into such a determination for nitrogen, and Appendix 5 gives the results that have been obtained for a number of compounds.

Such third-law entropy values are of interest to the chemist in two

TABLE 8-1 The entropy of nitrogen from heat-capacity data*

	J deg⁻¹ mol⁻¹
0–10 K extrapolation	1.92
10–35.61 K (graphical integration)	25.25
Transition 228.9/35.61	6.43
35.61–63.14 K (graphical integration)	23.38
Fusion 720.9/63.14	11.42
63.14-bp (graphical integration)	11.41
Vaporization 5535.0/77.32	72.13
Correction for gas imperfection	0.92
Ideal gas 77.32–298.2 K (calculated, see Secs. 10-2 and 10-3)	39.20
Entropy of ideal gas at 298.2 K and 1 atm	192.06

* From W. F. Giauque and J. O. Clayton, *J. Am. Chem. Soc.*, **55**:4875 (1933).

important areas. First, they provide the data with which molecular calculations of entropy can be compared. Second, along with values for the enthalpies of reactions, they constitute the thermodynamic basis for treating chemical equilibria.

The following chapter will introduce one further, very important thermodynamic function defined in terms of entropy and enthalpy. This new function will allow chemical equilibria to be more conveniently studied and molecular explanations of thermodynamic quantities to be more quantitatively given.

Problems

1 Show the system and the surroundings that can be used to perform the reversible conversion of 1 mol of benzene from solid to liquid, the temperature staying constant at the freezing point of benzene, 5.4°C. What is the entropy change of the benzene and of the surroundings? (The heat of fusion of benzene is 126 J g^{-1}.)

2 What is the entropy of vaporization of:

	Boiling point (°C)	ΔH_{vap} (kJ mol^{-1})
Argon	−185.7	7.86
Mercury	356.6	64.85
Carbon tetrachloride	76.7	30.00
Benzene	80.1	30.75

3 Five moles of an ideal gas is expanded from an initial pressure of 2 atm to a final pressure of 1 atm, the temperature being maintained at 25°C.

a Devise an arrangement so that the process can be carried out reversibly. What are the entropy changes of the thermal reservoir and of the gas itself? *Ans.* $\Delta S_{\text{thermal res}} = -28.83$; $\Delta S = +28.83$ J deg^{-1}.

b Devise an arrangement such that the expansion occurs in an unbalanced, irreversible way so that no work is done on the mechanical surroundings. What are the entropy changes of the thermal surroundings, the system, and the universe?

c Suppose the expansion occurs against a piston that maintains a pressure of 1 atm throughout the expansion. What would be the entropy changes of the thermal surroundings, the system, and the universe?
d Make a statement summarizing the relation of the entropy changes in the thermal surroundings, the system, and the universe to the degree of unbalance, or irreversibility, of the process.

4 Calculate the entropy change that occurs when 500 J of heat flows from a hot body at 100°C to a cold one at 0°C, the temperature of both bodies not being appreciably changed by the heat flow.

5 Ten grams of ice at 0°C is added to 20 g of water at 90°C in a Dewar flask. The heat of fusion of water is 5980 J mol^{-1}; the specific heat of water can be taken as independent of temperature; and the heat capacity of the Dewar can be ignored.
a What is the final temperature of the water? *Ans.* 33.5°C.
b How could the process be performed reversibly, and what would the entropy changes of the surroundings and of the system then be?
Ans. $\Delta S_{\text{thermal res}} = -2.93$ J deg^{-1}; $\Delta S = +2.93$ J deg^{-1}.
c What is the entropy change of the system for the direct addition?
Ans. -2.93 J deg^{-1}.
d What is the entropy change of the surroundings for the direct addition?

6 A 0.276-liter gas bulb at 25°C holding 0.046 mol of hydrogen is connected by a tube with a stopcock to an evacuated bulb with a volume of 0.5 liter. Calculate the entropy change when the stopcock is opened. Assume that hydrogen behaves ideally.

7 Compare the theoretical efficiency of a steam engine operating at 5 atm pressure, at which pressure water boils at 152°C, with one operating at 100 atm, at which pressure water boils at 312°C. The condenser in each case is at 30°C. *Ans.* 29 percent at 5 atm, 48 percent at 100 atm.

8 A Carnot cycle uses 1 mol of an ideal gas, for which $c_v = 25$ J deg^{-1} mol^{-1}, as the working substance and operates from a most compressed stage of 10 atm pressure and 600 K. It expands isothermally to a pressure of 1 atm, and then adiabatically reaches a most expanded stage at a temperature of 300 K.
a Obtain numerical values for $\Delta E_{\text{thermal res}}$ and $\Delta E_{\text{mech res}}$ for each stroke.
Ans. Stroke 1: $(\Delta E_{\text{thermal res}})_h = -11{,}490$ J; $(\Delta E_{\text{mech res}})_1 = +11{,}490$ J.
Stroke 2: $(\Delta E_{\text{mech res}})_2 = +7500$ J.
Stroke 3: $(\Delta E_{\text{thermal res}})_l = +5740$ J; $(\Delta E_{\text{mech res}})_3 = -5740$ J.
Stroke 4: $(\Delta E_{\text{mech res}})_4 = -7500$ J.
b Repeat the calculation for a maximum compression of 100 atm at 600 K and an expansion to 1 atm and then to 300 K. Compare the efficiency with that of the first calculation.

9 Plot the Carnot cycles of Prob. 8 on a graph of P versus V.

10 The net work delivered to the mechanical surroundings of a Carnot cycle is the summation of $P\,dV$ for all strokes. Obtain this by graphical integration on the figure of Prob. 9, and compare with the value calculated in Prob. 8.

11 Assume that an engine of a type different from the Carnot engine exists which can operate between heat reservoirs at temperatures T_h and T_l with greater efficiency than that calculated for the Carnot engine. Imagine this more efficient engine to be coupled to the Carnot engine so that one engine drives the other to make the second engine operate as a refrigerator. Show that the supposition of an engine with an efficiency greater than that calculated for a reversible engine operating between the temperatures T_h and T_l leads to a violation of the initial statements of the second law of thermodynamics.

12 On graphs of T versus $S_{\text{thermal res}}$ and T versus S:
   a Sketch a typical Carnot cycle.
   b Plot the data from the Carnot cycles of Prob. 8.

13 Since the relation $\Delta S = -\Delta S_{\text{thermal res}} = -\Delta E_{\text{thermal res}}/T$, or $\Delta E_{\text{thermal res}} = T\,\Delta S_{\text{thermal res}} = -T\,\Delta S$ holds for reversible processes, the total of $\Delta E_{\text{thermal res}}$ for a Carnot cycle is the cyclic integral $\oint T\,dS_{\text{thermal res}}$, or $-\oint T\,dS$. Obtain this value by graphical integration on the figure of Prob. 12, and compare with the value calculated in Prob. 8.

14 If the entropy of 1 mol of N_2 at 1 atm and 25°C is denoted by s°_{298}, what would be the entropy of 1 mol of N_2 at 1 atm and 150°C if the heat capacity per mole could be taken to have the value 29 J deg^{-1} mol^{-1}?

15 Calculate the increase in entropy of 3 mol of methane, CH_4, when the temperature is raised from 300 to 1000 K, the pressure remaining constant at 1 atm. The heat capacity is given in Table 7-2. *Ans.* 185.1 J deg^{-1}.

16 By a suitable graphical integration, determine the entropy of metallic silver at 25°C from the data at the right for the heat capacity per mole which are calculated from those reported by Meads, Forsythe, and Giauque, *J. Am. Chem. Soc.*, **63**:1902 (1941). Assume that the heat capacity approaches absolute zero according to a T^3 relation; that is, $c_P = \text{const}\ T^3$, and $c_P = 0$ at $T = 0$. *Ans.* $S^{\circ}_{298.1} = 42.7$ J deg^{-1} mol^{-1}.

$T\,(K)$	c_P ($J\ deg^{-1}$)
15	0.67
30	4.77
50	11.65
70	16.33
90	19.13
110	20.96
130	22.13
150	22.97
170	23.61
190	24.09
210	24.42
230	24.73
250	24.73
270	25.31
290	25.44
300	25.50

17 At 25°C and 1 atm the entropy of 12.01 g of diamond is 2.45 J deg^{-1} and that of 12.01 g of graphite is 5.71 J deg^{-1}. If equilibrium could be established in an isolated system, which implies that there be no heat of reaction, which form of carbon would result?

The heat evolved in the combustion of diamond is 395.40 kJ mol^{-1} and of graphite is 393.51 kJ mol^{-1}. If equilibrium could be established and only the entropy effect produced by this energy difference were important, which form of carbon would result?

18 Calculate the heat capacity of N_2 over the temperature range 77 to 298 K. Only the translational and rotational contributions are significant at such temperatures. Using these heat-capacity data, calculate the entropy gained by 1 mol of N_2 as the temperature is increased from 77 to 298 K, and compare with the value shown in Table 8-1. *Ans.* 39.4 J deg^{-1} mol^{-1}.

References

BENT, H. A.: "The Second Law: An Introduction to Classical and Statistical Thermodynamics," Oxford University Press, Fair Lawn, N.J., 1965. A very stimulating treatment of the formulation and application of the second law, with many quotations from, and references to, original work.

FAST, J. D.: "Entropy," McGraw-Hill Book Company, New York, 1963. A treatment of thermodynamics from the viewpoint that entropy is an important property and should not be considered to play a role secondary to energy.

WILKE, J.: "The Third Law of Thermodynamics," Oxford University Press, Fair Lawn, N.J., 1961. A monograph dealing entirely with the third law. The inclusion of a considerable amount of statistical material, however, makes this treatment more easily followed if read after Chaps. 9 and 10 have been studied.

HOARE, F. E., L. C. JACKSON, and N. KURTI (eds.): "Experimental Cryophysics," Butterworth & Co. (Publishers), Ltd., London, 1961. A collection of review articles on the methods of attaining and measuring very low temperatures.

MENDELSSOHN, K.: "The Quest for Absolute Zero," McGraw-Hill Book Company, New York, 1966. A nonmathematical account of both the early and the more recent aspects of science that relate to the achievement of low temperatures.

LOUNASMAA, O. V.: New Methods for Approaching Absolute Zero, *Sci. Am.*, **221**(6): 26–35 (1969).

9

FREE ENERGY AND CHEMICAL EQUILIBRIA

In the preceding chapters the thermodynamic functions that have a bearing on the direction in which a process can proceed spontaneously, i.e., without being driven by some externally applied force, have been introduced. Now a single and more convenient function that indicates this direction will be introduced. This function, *the free energy,* moreover, lends itself to the treatment of the equilibrium state toward which the process moves. The interrelation of thermodynamic properties and the equilibrium states of chemical reactions is, for chemistry, the most important accomplishment of thermodynamics.

9-1 A Convenient Measure of the Driving Force of a Reaction: The Free Energy

The entropy change that must be considered if the direction of chemical change is to be deduced is that of the universe of the change, and this is the sum of that occurring in the system and that occurring in the thermal surroundings. But all the information that is used to calculate these two quantities comes from tables of data, like those of Appendixes 5 and 6, of properties of the reagents. These allow the entropy change of the system to be calculated, and the heat of reaction, and thus the entropy change, of the thermal surroundings to be deduced. One can therefore make calculations using these properties of the system without thinking explicitly about the surroundings.

Consider first a reaction, occurring at constant temperature, for which a balanced equation can be written and the necessary heat of formation and entropy data are available. We can calculate

$$\Delta S_{\text{univ}} = \Delta S_{\text{system}} + \Delta S_{\text{thermal res}}$$
$$= \Delta S - \frac{\Delta H}{T} \qquad [1]$$

It is customary to express this quantity in a somewhat different way. First, to convert the expression to one with the more familiar units of energy, we multiply it by T, the constant temperature of the reaction, to give

$$T\,\Delta S - \Delta H$$

Then we introduce a symbol and a name for this collection of terms by writing

$$-\Delta G = T\,\Delta S - \Delta H$$

or

$$\Delta G = \Delta H - T\,\Delta S \qquad [T \text{ const}] \qquad [2]$$

where ΔG is called the change in *free energy,* or more completely, the *Gibbs free energy.* The function ΔG offers certain conveniences. We notice, first, that the sign change has reversed the entropy-of-the-universe implications, so that now we have

$$\begin{array}{ll} \Delta G = - & \textit{the reaction can proceed as written} \\ \Delta G = + & \textit{the reverse reaction can proceed} \end{array} \qquad [3]$$

The quantity ΔG leads also to an additional way of looking at the driving force of a reaction. Consider, to be specific, a reaction in which the entropy of the system decreases. The final term in the above ΔG equation, $-T\,\Delta S$, is then positive. The reaction can proceed spontaneously only if ΔH is negative and large enough to overcome $-T\,\Delta S$, thus giving ΔG a negative sign. Or if we are thinking of converting the energy from the reaction for mechanical or other purposes, we see that at least enough energy to compensate for the $T\,\Delta S$ term must be left as ΔH and delivered to the thermal surroundings. Only energy in excess of this amount can be used as a special driving force. Thus the free energy is a measure of the work that can be drawn from a system after the entropy demands have been met by payment of heat to the thermal reservoir. This is the basis for the adjective "free."

The free-energy function itself, from which these constant-temperature changes in free-energy stem, is defined as

$$G = H - TS \qquad [4]$$

Since H and TS are properties of the system, so also is G.

Let us work from this defining equation to come again on a descriptive character of changes in free energy. First the defining equation of H is substituted in Eq. [4] to give

$$G = E + PV - TS \qquad [5]$$

For an infinitesimal change in G, one now has

$$dG = dE + P\,dV + V\,dP - T\,dS - S\,dT \qquad [6]$$

These many terms are reduced when we limit our considerations to reversible processes, so that $dS = -dE_{\text{thermal res}}/T$, or $T\,dS = -dE_{\text{thermal res}} = \delta q$, and furthermore, to those occurring at constant temperature, so that $dT = 0$, and constant pressure, so that $dP = 0$. Then Eq. [6] becomes

$$dG = dE + P\,dV - \delta q \qquad [7]$$

The first law, in the form $dE = -dE_{\text{thermal res}} - dE_{\text{mech res}} = \delta q + \delta w$, can then be used to replace $dE - \delta q$ by δw to give

$$dG = \delta w + P\,dV$$

or

$$\begin{aligned} -dG &= -\delta w - P\,dV \\ &= dE_{\text{mech res}} - P\,dV \qquad \text{[reversible; } T, P \text{ const]} \end{aligned} \qquad [8]$$

Since a reversible process has been stipulated, and for such processes a maximum amount of work, $-\delta w$, can be drawn from the system, we see that the decrease in free energy is a measure of the maximum useful work that the process can produce, the adjective "useful" implying work over and above that involved in expansions or compressions.

Since the free-energy change measures the useful work that might be obtained from a process at constant temperature and pressure, it is a measure of the spontaneity of the process. We have the very important result that *the decrease in free energy of a constant-temperature, constant-pressure process is the measure of the tendency of the process to proceed spontaneously.* If a change in a system is considered and a *decrease* in free energy is calculated for that change, spontaneous processes that carry out that change can be expected, and moreover, any such process could be harnessed to deposit energy, over and above that of expansion against the confining pressure, in the mechanical-energy reservoir. If a different change in the system is considered and an *increase* in free energy is calculated for that change, no spontaneous process can occur to carry out that change. In fact, processes carrying the system in the reverse direction will be those that can occur spontaneously. Finally, for a considered change, if the free energy is calculated to be unaffected, the system will show no tendency to proceed from one state to the other.

The qualitative relations, for constant-pressure processes, of free energy to spontaneity and equilibrium can be summarized as follows:

For $\Delta G = -$ *the process tends to proceed spontaneously*
For $\Delta G = 0$ *the system is at equilibrium*
For $\Delta G = +$ *the process would proceed spontaneously only in the opposite direction*

One should notice that in these statements the free-energy change is that for the system, and the role of the surroundings is implicitly included.

Stating that a system is at equilibrium means only that it is in equilibrium with respect to the reactions or processes that are being considered. For chemical systems there will frequently be reactions which are thermodynamically possible but which will not occur for the particular conditions of the systems and therefore need not be considered. A careful statement will always say that a system is necessarily in equilibrium only with respect to certain specified processes.

For the chemist the free energy is probably the most important of the thermodynamic quantities.

It will only be mentioned that, although the free-energy function G, the Gibbs free energy, is suitable for direct application to constant-pressure processes, another free-energy function is more convenient for constant-volume processes. This function, known as the Helmholtz free energy, is represented by A and is defined as

$$A = E - TS \qquad [9]$$

A development such as is performed on G in this section would show that in a constant-volume process the decrease in A corresponds to the driving force of the reaction. Thus H and G are functions that are convenient for constant-pressure processes, whereas E and A are more convenient for constant-volume processes. In the introduction to thermodynamics that is presented here, it will be sufficient to develop the applications of H and G.

9-2 Some Free-energy Calculations

The use of free energy can be illustrated by some simple examples.

Example 1 Calculate the free-energy change for the process of converting 1 mol of water at 100°C and 1 atm to steam at the same temperature and pressure.

We return to the defining equation for the free energy, $G = H - TS$, and calculate the free-energy change from the enthalpy and entropy contributions. For a constant-temperature process the change in free energy is given by

$$\Delta G = \Delta H - T\,\Delta S \qquad [10]$$

Since the process is carried out at constant pressure, the enthalpy change is equal to the heat absorbed. This heat is the heat of vaporization ΔH_{vap}, and thus

$$\Delta H = \Delta H_{vap}$$

The entropy change can also be calculated. Since T is constant and a heat reservoir at a temperature just above 100°C can be imagined, the result for the reversible process is simply

$$\Delta S = -\Delta S_{\text{thermal res}} = -\frac{\Delta H_{vap}}{T}$$

and thus

$$T\,\Delta S = \Delta H_{vap} \qquad [11]$$

The free energy for the process is therefore

$$\Delta G = \Delta H_{vap} - \Delta H_{vap} = 0$$

This result, in view of the previous discussion, could have been written down immediately. Since water and steam are in equilibrium at 100°C and 1 atm, the free-energy change for the conversion of one to the other must be zero.

In the formation of steam from water at these conditions, the energy factor ΔH works, by demanding heat and lowering the entropy of the thermal surroundings, to oppose the process, whereas the entropy factor $T\,\Delta S$ of the system itself increases and favors the process. At 100°C and 1 atm these effects just balance, and water and steam are in equilibrium. The free-energy equation shows how the two factors reach a compromise.

Example 2 Calculate the free-energy change when 1 mol of water is formed from its elements at 25°C and 1 atm pressure.

The reaction to be considered is

$$H_2 + \tfrac{1}{2}O_2 \rightarrow H_2O(l) \qquad [12]$$

Again we make use of the defining equation $G = H - TS$. In this reaction, as in that of the preceding example, the constant-temperature feature eliminates any term like $S\,dT$ or $S\,\Delta T$ from the expression for a change in G, and one has, for the process of the reaction,

$$\Delta G = \Delta H - T\,\Delta S$$

The heat of this reaction, ΔH, is the standard heat of formation of water, which is given in Appendix 5 as -285.84 kJ. The entropies of the three reagents, given by the third law of thermodynamics, are also given in Appendix 5. These values lead to

$$\begin{aligned}\Delta S &= s_{H_2O(l)} - \tfrac{1}{2}s_{O_2} - s_{H_2} \\ &= 69.94 - (\tfrac{1}{2})(205.03) - 130.59 \\ &= -163.16 \text{ J deg}^{-1}\end{aligned} \qquad [13]$$

Substitution of these values gives

$$\Delta G = -285{,}840 - (298.15)(-163.16) = -237{,}190 \text{ J} \qquad [14]$$

The calculation shows that the reaction proceeds with a large decrease in free energy, and this agrees with the known fact that the reaction tends to proceed spontaneously. (The fact that mixtures of oxygen and hydrogen do not react appreciably until a spark or flame starts the reaction is a question of the rate of the reaction and is of no concern in thermodynamics.)

Example 3 Calculate the difference in free energy of n mol of an ideal gas at pressure P_1 and n mol of the gas at a lower pressure P_2. The temperature remains constant.

The necessary enthalpy and entropy terms are calculated with reference to a cylinder-and-piston arrangement that can take the gas reversibly from P_1 to P_2. Since the temperature is constant, ΔE and $\Delta(PV)$ are zero, and therefore so also is ΔH. Furthermore, as in the third example of Sec. 8-2, the entropy change of the system is

$$\Delta S = nR \ln \frac{P_1}{P_2} \qquad [15]$$

The free-energy change is now calculated as

$$\begin{aligned}\Delta G &= \Delta H - \Delta(TS) = \Delta H - T\,\Delta S = 0 - nRT \ln \frac{P_1}{P_2} \\ &= nRT \ln \frac{P_2}{P_1}\end{aligned} \qquad [16]$$

Since P_1 is greater than P_2, ΔG has a negative value.

In Sec. 9-1 it was shown that for a constant-pressure, constant-temperature process the value of $-\Delta G$ was a measure of the tendency of the process to occur.

To apply this criterion to the pressure-changing isothermal expansion of a gas, we must return to Eq. [6] and retain the $V\,dP$ term as we proceed. Then, in place of Eq. [8], we reach

$$-dG = -\delta w - d(PV) \qquad [17]$$

For the constant-temperature expansion of an ideal gas being dealt with here, $d(PV) = d(nRT) = 0$, and the decrease in free energy is equal to the work that could be obtained from the system. It follows, therefore, that since ΔG for the isothermal expansion of an ideal gas is negative, work could be drawn from the system, and the expansion will tend to proceed spontaneously. Again, a rather powerful method has been used to obtain an obvious answer. The value of such a method when applied to chemical systems should, however, be apparent.

It is convenient to recognize that this result, Eq. [16], gives a quantitative expression for the fact that a gas at high pressure is capable of doing more useful work than one at low pressure if both lead to products at the same pressure. A gas at a high pressure has a high free energy, and spontaneous processes, with their accompanying decreasing free energy, can occur. This result will be mentioned again later in the chapter.

9-3 Standard Free Energies

The preceding examples of free-energy calculations show that changes in free energies can be calculated and that these changes correlate with the tendency of the system to proceed to a state of equilibrium. In view of this fact, it would be very useful to have a tabulation of free energies of chemical compounds so that the free-energy change of a possible reaction could be easily calculated.

Free energies, like any other energies, must have some reference point. The same procedure is followed as for enthalpies. A zero value is assigned to the free energies of the stable form of the elements at 25°C and 1 atm pressure. These, and the free energies of compounds based on these references, are known as *standard free energies of formation.* The data can be determined from free energies of reactions in exactly the same way as were standard heats of formation. Example 2 of Sec. 9-2 gives the standard free energy of formation of liquid water as -237.19 kJ. Appendix 5 shows the standard free energies of a number of compounds.

A use of the data of Appendix 5 can be indicated by applying them to the question of whether or not it would be worthwhile to look for a catalyst to promote the hydrogenation of ethylene at 25°C and 1 atm. One writes

$$\begin{array}{lccc} & H_2C{=}CH_2(g) & + \; H_2(g) \rightleftharpoons & CH_3CH_3(g) \\ \Delta G_f^\circ(25°C) & +68.12 & 0 & -32.89 \end{array} \qquad [18]$$

and calculates

$$\Delta G^\circ_{298} = -32.89 - (+68.12) = -101.01 \text{ kJ} \qquad [19]$$

The large negative free-energy change shows that the reaction tends to proceed spontaneously, and if a way is found to make it proceed fast enough, i.e., if a catalyst is available, the reaction will occur.

9-4 The Dependence of Free Energy on Pressure and Temperature

The standard free energies, such as appear in Appendix 5, allow predictions to be made of the possibility of a reaction for the single conditions of 25°C and 1 atm. For these free-energy data to be of real use, a means must be available for calculating free energies at other pressures and temperatures.

Since G is a state function, we can show how it varies with P and T by writing the total differential

$$dG = \left(\frac{\partial G}{\partial P}\right)_T dP + \left(\frac{\partial G}{\partial T}\right)_P dT \qquad [20]$$

In Eq. [6], however, we already have found

$$dG = dE + P\,dV + V\,dP - T\,dS - S\,dT \qquad [21]$$

Again, as in Sec. 9-1, let us suppose that we are dealing with states of the systems that can be connected by a reversible process so that we can write $T\,dS = \delta q$. Now, however, T and P will not be held constant, and the only remaining restriction will be to reversible processes in which the only work is PV work, i.e., $P\,dV = -\delta w$. The first law $dE = \delta q + \delta w$ is, under these circumstances, $dE = T\,dS - P\,dV$, or $dE - T\,dS + P\,dV = 0$. Three terms of Eq. [23] thus cancel, and this leaves

$$dG = V\,dP - S\,dT \qquad [22]$$

Comparison with Eq. [20] now gives the desired relations involving the coefficients as

$$\left(\frac{\partial G}{\partial T}\right)_P = -S \qquad [23]$$

$$\left(\frac{\partial G}{\partial P}\right)_T = V \qquad [24]$$

[Reversible process, only PV work]

These two results show how the free energy of a chemical compound depends on the pressure and the temperature. For the present, only the pressure dependence is considered.

Since liquids and solids are quite incompressible, the free-energy change corresponding to an isothermal increase in pressure, ΔP, for some not too large pressure change can be written, according to Eq. [24] and on the assumption of a constant volume, as $V\,\Delta P$. Since the molar volumes of solids and liquids are relatively small, this change in free energy resulting from the application of ordinary pressures to liquids and solids is also relatively small, and for many purposes the pressure dependence of the free energy of liquids and solids can be neglected.

For gases the dependence of free energy on pressure is appreciable and important. For an ideal gas, P and V are related by the ideal-gas law, and the integration of Eq. [24] can be performed to give the free-energy change when the pressure is changed, at constant temperature, from P_1 to P_2. Thus

$$G_2 - G_1 = \int V\,dP = nRT\int_{P_1}^{P_2} \frac{dP}{P}$$

$$= nRT \ln \frac{P_2}{P_1} \qquad [25]$$

This result is that obtained in Example 3 of Sec. 9-2, in which the separate enthalpy and entropy changes were calculated.

Of particular interest is the extent to which the free-energy changes from its standard-state value when the pressure changes from 1 atm. If state 1 is the standard state, then

$$P_1 = 1 \text{ atm} \quad \text{and} \quad G_1 = G^\circ$$
$$P_2 = P \quad \text{and} \quad G_2 = G$$

With this notation for states 1 and 2, Eq. [25] can be rewritten for 1 mol as

$$\text{G} - \text{G}^\circ = RT \ln \frac{P}{1}$$

$$\text{G} = \text{G}^\circ + RT \ln P_{\text{atm}} \qquad [T \text{ const}, P \text{ atm}] \qquad [26]$$

Note that here pressures are conveniently expressed in atmospheres, rather than in the SI units of newtons per square meter, because the standard state has been specified to be that with 1 atm pressure. The unit-identifying subscript will be used, where necessary, to indicate this deviation from SI units.

Equation [26] is strictly applicable to ideal gases, since $PV = nRT$ was assumed for the P versus V relation in the integration of Eq. [25], but if the details of nonideal behavior are ignored, it can be used, and assumed to apply approximately, for all gases.

Equation [26] shows that the free energy of a gas at pressure P is made up of the free energy that it has at 1 atm plus an additional term that is positive for P larger than 1 atm and negative for P less than 1 atm.

9-5 The Pressure Dependence of the Free Energy of Nonideal Gases: The Fugacity

A straightforward treatment of nonideal gases would use a suitable equation of state, such as van der Waals' equation, to allow the integration of Eq. [26] to be performed. Such a procedure, however, results in an expression for ΔG that is a complicated and unwieldy function of P. Since the simplicity of the ideal-gas results, Eqs. [25] and [26], will be found to lead to other important equations with a similar simple form, we avoid, for nonideal gases, the integration of Eq. [24] with an empirical equation of state and, instead, use what at first will seem a rather devious approach.

A satisfactory procedure is the introduction of a new function called the *fugacity,* denoted by f. If G_1 and G_2 are the molar free energies of a gas at two pressures P_1 and P_2, the fugacities f_1 and f_2 of the gas at these pressures are *defined* so that

$$G_2 - G_1 = RT \ln \frac{f_2}{f_1} \tag{27}$$

This procedure insists on the free-energy equation having the convenient form of Eq. [27]. The nonideal complications are hidden in the fugacity terms. Comparison of this defining equation for fugacity with the ideal-gas equation [25] shows that for ideal behavior the fugacity is proportional and can be set equal to the pressure. For nonideal behavior we must expect the fugacity of a gas to deviate from its pressure. It is now necessary to show how the fugacity of a gas at a particular pressure and temperature can be deduced.

A number of manipulations are necessary. In the thermodynamic equation for 1 mol of gas at constant temperature,

$$G_2 - G_1 = \int_{P_1}^{P_2} V\,dP \tag{28}$$

the quantity RT/P can be added to and subtracted from the integrand to give

$$\begin{aligned} G_2 - G_1 &= \int_{P_1}^{P_2} \frac{RT}{P} + \left(V - \frac{RT}{P}\right) dP \\ &= \int_{P_1}^{P_2} \frac{RT}{P}\,dP + \int_{P_1}^{P_2} \left(V - \frac{RT}{P}\right) dP \\ &= RT \ln \frac{P_2}{P_1} + \int_{P_1}^{P_2} \left(V - \frac{RT}{P}\right) dP \end{aligned} \tag{29}$$

Comparison of this expression with Eq. [27] gives

$$RT \ln \frac{f_2}{f_1} = RT \ln \frac{P_2}{P_1} + \int_{P_1}^{P_2} \left(V - \frac{RT}{P}\right) dP$$

or

$$RT \ln \frac{f_2/P_2}{f_1/P_1} = \int_{P_1}^{P_2} \left(V - \frac{RT}{P}\right) dP \tag{30}$$

Since all gases tend to become ideal as the pressure approaches zero, that is,

$$\frac{f}{P} \to 1 \quad \text{as} \quad P \to 0$$

the ratio f_1/P_1 becomes unity when P_1 approaches zero. Furthermore, if P and f are written instead of P_2 and f_2 for the arbitrary pressure and fugacity in Eq. [30], we have

$$RT \ln \frac{f}{P} = \int_{P=0}^{P=P} \left(\mathrm{v} - \frac{RT}{P} \right) dP \qquad [31]$$

If sufficient data are available for v as a function of P, at some temperature, values for f/P, and therefore of f, at some pressure P, can be obtained by integration of Eq. [31]. If the data are expressed in a convenient analytical form, the integration can be carried out analytically. On the other hand, if tabular data are available, graphical integration must be used.

As an illustration, PV data for methane at $-50°C$ are given in Table 9-1. These data allow the plot of $\mathrm{v} - RT/P$ in Fig. 9-1 to be made, and graphical integration then gives the integration results and the values of f/P and f included in Table 9-1. (Special care must be taken with units when Eq. [31] is used. The ratio f/P must be dimensionless; if P is in atmospheres, so is f. The integrand terms v and RT/P must have the same units, and moreover, the units of the right side of the equation must be the same as those on the left side, i.e.,

TABLE 9-1 Calculations of the fugacity of methane at $-50°C$

P (atm)	v (liters mol^{-1})	RT/P (liters mol^{-1})	$\mathrm{v} - RT/P$ (liters mol^{-1})	$\int_0^P \left(\mathrm{v} - \frac{RT}{P}\right) dP$* (liter atm mol^{-1})	f/P	f (atm)
0			−0.080†	0.00	1.00	0
10	1.747	1.831	−0.084	−0.82	0.96	10
20	0.830	0.915	−0.085	−1.66	0.91	18
40	0.366	0.458	−0.092	−3.48	0.83	33
60	0.208	0.305	−0.097	−5.34	0.75	45
80	0.129	0.229	−0.100	−7.36	0.67	54
100	0.092	0.183	−0.091	−9.20	0.60	60
120	0.076	0.153	−0.077	−10.84	0.55	66
160	0.064	0.114	−0.050	−13.28	0.48	77
200	0.0591	0.0915	−0.0324	−14.88	0.44	88
300	0.0525	0.0610	−0.0085	−16.83	0.40	120
400	0.0491	0.0458	+0.0033	−16.99	0.40	160
600	0.0451	0.0305	+0.0146	−15.08	0.44	260
800	0.0427	0.0229	+0.0198	−11.61	0.53	420
1000	0.0410	0.0183	+0.0227	−7.31	0.67	870

* By graphical integration on Fig. 9-1.
† Calculated from van der Waals' equation.
SOURCE: Data from R. H. Perry, C. H. Chilton, and S. D. Kirkpatrick (eds.), "Chemical Engineers' Handbook," 3d ed., McGraw-Hill Book Company, New York, 1950.

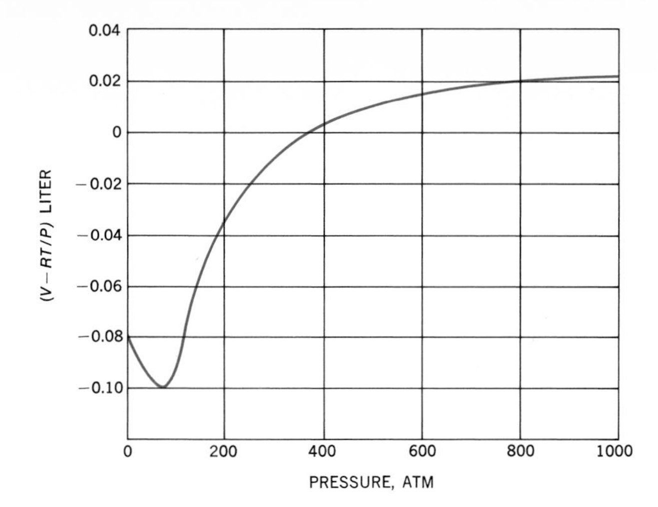

FIGURE 9-1
Graph for the integration of v − RT/P for methane at −50°C.

of RT. All this can be accomplished by using v in liters, P in atmospheres, and R with the value 0.08205 liter atm mol^{-1} deg^{-1}.)

With PVT data such as those used in the methane example, one can deduce, for any pressure, the fugacity of the gas. Then the dependence of the free energy on the pressure can be calculated, even if the gas shows nonideal behavior, by means of Eq. [27].

For gases for which molar-volume measurements have not been made and an equation of state is not available, the law of corresponding states can be used, if data on the critical point are available, to estimate the fugacities at various temperatures and pressures. It will be recalled that this law states that, in terms of the reduced variables P_R, v_R, and T_R, all gases follow the same equation of state. This means that at the same value of P_R and T_R all gases have the same imperfection and therefore the same nonideality. Furthermore, the variation of the compressibility factor Z with the reduced pressure has been represented for various values of T_R in Fig. 1-11. These data are all that are necessary for the integration of Eq. [31]. Figure 1-11 gives values of

$$Z = \frac{Pv}{RT}$$

from which one obtains

$$v = \frac{RT}{P} Z \qquad [32]$$

and Eq. [31] can be written

$$RT \ln \frac{f}{P} = \int_0^P \left(\frac{RT}{P} Z - \frac{RT}{P} \right) dP$$

$$= RT \int_0^P (Z - 1) \frac{dP}{P}$$

or

$$\ln \frac{f}{P} = \int_0^P (Z - 1)\frac{dP}{P}$$

$$= \int_0^P (Z - 1)\frac{dP_R}{P_R} \qquad [33]$$

The data of Z as a function of P_R for a given value of T_R then allow graphical integrations to be performed to give curves such as those of Fig. 9-2.

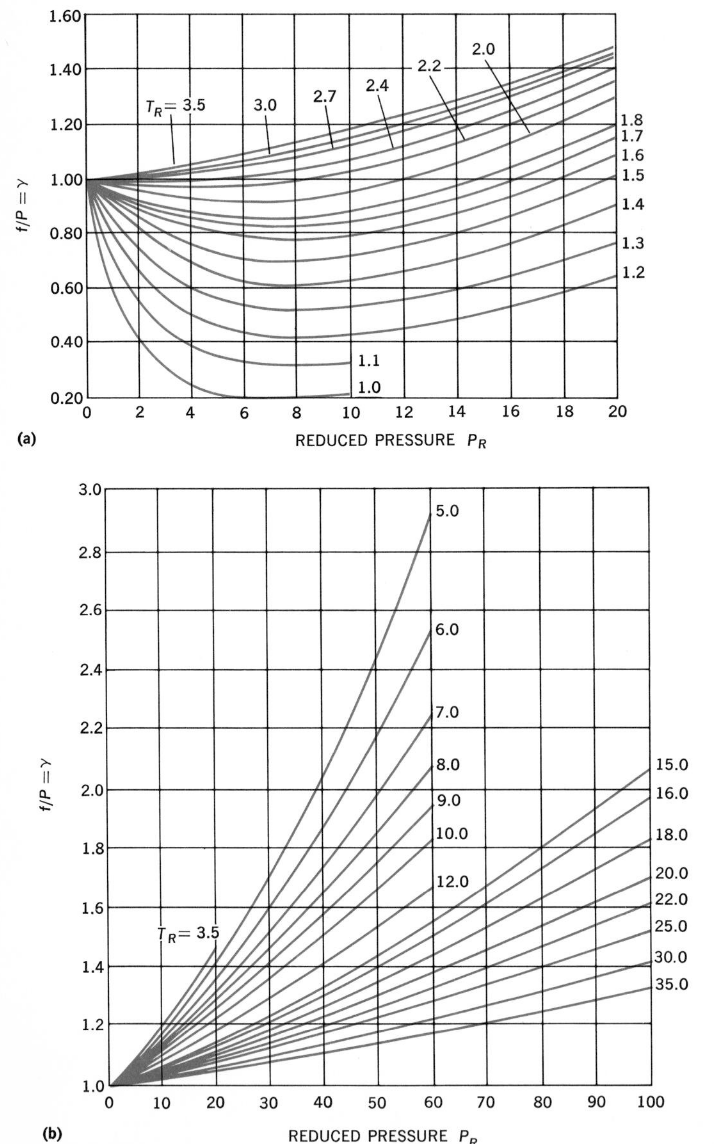

FIGURE 9-2
The ratio f/P for gases as a function of the reduced pressure $P_R = P/P_c$ and the reduced temperature $T_R = T/T_c$ (*a*) at pressures and temperatures near the critical point; (*b*) at high temperatures and pressures. [*From R. H. Newton, Ind. Eng. Chem.*, **27**:*302 (1935), and R. H. Perry (ed.), "Chemical Engineers' Handbook," 3d ed., McGraw-Hill Book Company, New York, 1950.*]

9-6 The Standard State for Nonideal Gases: The Activity and Activity Coefficient

The convenience of choosing standard states for which the properties of compounds can be tabulated has been mentioned, with respect to free energies, in Sec. 9-3, and the variation of free energy from the value for the standard state has been expressed for ideal gases, in Sec. 9-4, by the formula

$$G = G^\circ + RT \ln P_{\text{atm}}$$

For ideal gases, it will be recalled, the standard-state molar free energy, denoted by G°, is chosen as that of the gas at 1 atm pressure. We now need a procedure for choosing a standard state for real gases and for expressing the free energy of such gases as the conditions vary from this standard state.

Since the fugacities of all gases become equal to the pressures in the limit of very low pressure, it might seem suitable to choose the standard state as that of some low pressure, say 10^{-5} atm. We could then assume that f°, the fugacity at this pressure, was equal to the pressure and could calculate the changes from the free energy at this pressure by means of Eq. [27] and fugacity data such as that of Table 9-1. Such a procedure, however, would not approach that used for ideal gases when the gas under consideration showed negligible nonideality.

A more satisfactory procedure is to choose the standard state as that at which the fugacity would be equal to 1 atm if the gas followed ideal behavior from zero pressure up to this fugacity. This selection of a hypothetical state for the standard state of nonideal gases can be appreciated with reference to Fig. 9-3.

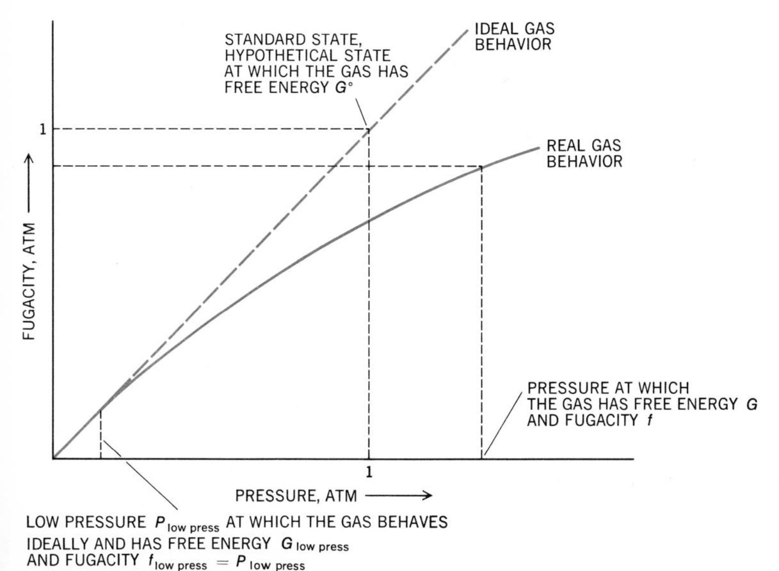

FIGURE 9-3
A schematic representation of the fugacity-pressure relation for a real gas and the relation that would exist if the gas behaved ideally up to a pressure of 1 atm.

Although the standard state is not one that is ever reached by the real gas, it is a state for which the free energy is perfectly well defined.

Suppose, for example, that the free energy of a gas were determined at some pressure P, as shown in Fig. 9-3, at which the gas behavior is not ideal. The free energy G at this pressure can be related, if a fugacity value at pressure P is available, to the free energy $G_{\text{low pressure}}$ at some low pressure $P_{\text{low pressure}}$ by

$$G - G_{\text{low pressure}} = RT \ln \frac{f}{f_{\text{low pressure}}}$$

$$= RT \ln \frac{f}{P_{\text{low pressure}}} \qquad [34]$$

The relation between the free energy $G_{\text{low pressure}}$ and the standard-state free energy follows from ideal behavior and is

$$G^\circ - G_{\text{low pressure}} = RT \ln \frac{1 \text{ atm}}{P_{\text{low pressure}}} \qquad [35]$$

Subtraction of Eqs. [34] and [35] eliminates the common "low-pressure" terms and gives

$$G - G^\circ = RT \ln \frac{f_p}{1 \text{ atm}} \qquad [36]$$

or with the assumption that the fugacity is expressed in atmospheres,

$$G - G^\circ = RT \ln f \qquad [37]$$

Thus, although G° is based on a hypothetical state, free energies of the gas at any pressure can be related to this standard free energy if the fugacity of the gas at that pressure is known.

Comparison of Eq. [37] with the ideal gas relation

$$G - G^\circ = RT \ln P$$

suggests the introduction of a term to exhibit the nonideality of a gas by comparing f with P. The *activity coefficient,* denoted by γ and defined as

$$\gamma = \frac{f}{P} \qquad [38]$$

is introduced.

With this factor the free-energy expression for real gases can be written

$$G - G^\circ = RT \ln \gamma P \qquad [39]$$

The special convenience of γ is that it shows explicitly the importance of nonideality.

The activity coefficient for gases has already been tabulated as f/P for the example of methane in Table 9-1. Furthermore, the same ratio, which we here label as the activity coefficient, is given graphically on the basis of the law of corresponding states in Fig. 9-2.

* 9-7 The Dependence of Entropy, Enthalpy, and Heat Capacity on Pressure

The direct relation between the free-energy changes in a chemical reaction and the position of the chemical equilibrium justifies the extended treatment of the pressure dependence of free energy given in Secs. 9-5 and 9-6. Here, more briefly, the dependence of other thermodynamic properties on the pressure will be considered.

We can begin with the relation between the coefficients of the total differential for G, as given for any state function in Sec. 6-7,

$$\frac{\partial}{\partial P}\left(\frac{\partial G}{\partial T}\right)_P = \frac{\partial}{\partial T}\left(\frac{\partial G}{\partial P}\right)_T$$

which, with Eqs. [23] and [24], can immediately be developed to the first of our desired relations,

$$\left(\frac{\partial S}{\partial P}\right)_T = -\left(\frac{\partial V}{\partial T}\right)_P \quad [40]$$

This is one of the frequently useful equations known as *Maxwell's relations*. Data for the temperature dependence of the volume at constant pressure are generally available for gases, liquids, and solids, although sometimes such data are not available in a convenient analytical form. Such data can, however, be used to deduce the pressure dependence of the entropy.

This result, Eq. [40], can also be used to obtain a relation between the pressure dependence of the enthalpy and PVT data. We return to the defining equation

$$H = E + PV \quad [41]$$

or

$$dH = dE + P\,dV + V\,dP \quad [42]$$

As in the development of Sec. 9-4, we again specify a reversible process $dS = -dS_{\text{thermal res}}$ and $T\,dS = \delta q$, and only PV work; so $\delta w = -P\,dV$. The first law, $dE = \delta q + \delta w$, then becomes $dE = T\,dS - P\,dV$, and substitution in Eq. [41] yields

$$dH = T\,dS + V\,dP \quad [43]$$

Division of Eq. [43] by dP and stipulation of constant temperature yield

$$\left(\frac{\partial H}{\partial P}\right)_T = T\left(\frac{\partial S}{\partial P}\right)_T + V \quad [44]$$

or with Eq. [40],

$$\left(\frac{\partial H}{\partial P}\right)_T = -T\left(\frac{\partial V}{\partial T}\right)_P + V \quad [45]$$

Again the pressure dependence can be deduced if PVT data are available.

As an illustration of the use of Eq. [45], we can return to the Joule-Thomson coefficient, which, in Sec. 6-9, was related to other thermodynamic quantities by

$$\mu_{JT} = -\frac{1}{c_p}\left(\frac{\partial H}{\partial P}\right)_T$$

Now we see that the partial derivative is given by Eq. [45], and we can write

$$\mu_{JT} = -\frac{1}{c_p}\left[v - T\left(\frac{\partial v}{\partial T}\right)_P\right] \qquad [46]$$

It is this result that makes understandable the positive and negative values that μ_{JT} can take on and allows the inversion temperature to be interpreted as that for which v equals $T(\partial v/\partial T)_P$. For ideal gases you can note that this equality always holds, and thus $\mu_{JT} = 0$ for all conditions of pressure and temperature.

Finally, a thermodynamic property of gases that can be measured at various pressures is the heat capacity c_p. Its dependence on pressure is obtained by recognizing that $c_p = (\partial H/\partial T)_P$ and writing

$$\left(\frac{\partial c_p}{\partial P}\right)_T = \frac{\partial}{\partial P}\left(\frac{\partial H}{\partial T}\right)_P$$

$$= \frac{\partial}{\partial T}\left(\frac{\partial H}{\partial P}\right)_T \qquad [47]$$

Then use of Eq. [45] gives

$$\left(\frac{\partial c_p}{\partial P}\right)_T = -T\left(\frac{\partial^2 v}{\partial T^2}\right)_P \qquad [48]$$

The expressions for the pressure dependence of S, H, and c_p that now have been obtained can be used if the necessary PVT data are available to perform the integrations that allow the values of these properties at different pressures to be deduced. Furthermore, by proceeding as was done for free energies in Sec. 9-6, values for the standard states can be obtained.

9-8 Quantitative Relation of ΔG and the Equilibrium Constant

Methods are now available for the determination of the pressure dependence of the free energies of the reagents in a chemical reaction, and qualitative arguments have been given for deciding, on the basis of the free-energy change, whether a reaction will proceed in one direction or the other. Chemical experience tells us, however, that a reaction will proceed in a given direction only until the system reaches a state of equilibrium. It will now be seen that free energies can be used to show not only the direction in which a reaction tends to proceed, but also the equilibrium state to which this reaction carries the system.

Consider a reaction involving four gases A, B, C, and D, all at temperature T, which now will be assumed to behave ideally.

$$aA + bB \rightarrow cC + dD \qquad [49]$$

where a, b, c, and d are the number of moles of each reagent involved.

The free energies of a mol of A at a pressure P_A, b mol of B at P_B, etc., can be written, in view of Eqs. [26], as

$$\begin{aligned}
&\text{Free energy of } a \text{ mol of } A = a\text{G}_A = a\text{G}^\circ_A + aRT \ln P_A \\
&\text{Free energy of } b \text{ mol of } B = b\text{G}_B = b\text{G}^\circ_B + bRT \ln P_B \\
&\text{Free energy of } c \text{ mol of } C = c\text{G}_C = c\text{G}^\circ_C + cRT \ln P_C \\
&\text{Free energy of } d \text{ mol of } D = d\text{G}_D = d\text{G}^\circ_D + dRT \ln P_D
\end{aligned} \qquad [50]$$

Here G_A, G_B, etc., and G°_A, G°_B, etc., are the free energies of 1 mol of the reagent A, B, etc.

The free-energy change for the reaction, when the four species have the arbitrary pressures P_A, P_B, P_C, and P_D, can now be calculated as

$$\begin{aligned}
\Delta G &= G_{\text{products}} - G_{\text{reactants}} \\
&= c\text{G}_C + d\text{G}_D - a\text{G}_A - b\text{G}_B \\
&= c\text{G}^\circ_C + d\text{G}^\circ_D - a\text{G}^\circ_A - b\text{G}^\circ_B + RT \ln \frac{(P_C)^c(P_D)^d}{(P_A)^a(P_B)^b}
\end{aligned}$$

or

$$\Delta G = \Delta G^\circ + RT \ln \frac{(P_C)^c(P_D)^d}{(P_A)^a(P_B)^b} \qquad \left[\begin{matrix} T \text{ const,} \\ \text{pressures arbitrary} \end{matrix}\right] \qquad [51]$$

This equation relates the free-energy change of the reaction for the reagents at the pressures P_A, P_B, P_C, P_D to a term ΔG° involving the free energies of all reagents at 1 atm pressure and a term for the free-energy pressure dependences.

If, instead of arbitrary pressures, each of the pressures corresponds to a condition of equilibrium among the reagents A, B, C, and D, the free-energy change ΔG under these conditions will be zero, and we can write

$$0 = \Delta G^\circ + RT \ln \left[\frac{(P_C)^c(P_D)^d}{(P_A)^a(P_B)^b}\right]_{\text{equilibrium}} \qquad [52]$$

or

$$\Delta G^\circ = -RT \ln \left[\frac{(P_C)^c(P_D)^d}{(P_A)^a(P_B)^b}\right]_{\text{equilibrium}} \qquad [53]$$

where the subscript is added to indicate that the equality holds only when the partial pressures are a set existing when the system is at equilibrium with respect to this reaction. Notice that the left side of Eq. [53] gives the free-energy change for all the reagents at their standard states; the right side is calculated for some set of partial pressures that produce the equilibrium condition. You can look on the right side, or the second term on the right of Eq. [52], as being the partial pressures necessary to undo the driving force that exists when the reagents are in their standard states.

Equation [53] constitutes a thermodynamic derivation of the familiar equilibrium-constant expression. Since $\Delta G°$ for a particular reaction at a given temperature is a fixed quantity, the argument of the logarithmic factor must have some constant value that is independent of the individual pressures. It is customary to call this constant the *equilibrium constant* and to denote it by the symbol K or, since pressures are involved, by K_p. Thus

$$K_p = \left[\frac{(P_C)^c(P_D)^d}{(P_A)^a(P_B)^b}\right]_{\text{equilibrium}} \qquad [54]$$

and Eq. [53] becomes, with this notation,

$$\Delta G° = -RT \ln K_p \qquad [55]$$

This equation represents one of the most important results of thermodynamics. By it, the equilibrium constant of a reaction is related to a thermochemical property. The immediate and obvious value of the equation is that it allows the calculation not only of the direction in which a reaction will proceed but also of the equilibrium state which the reacting system will finally attain. For many systems it is much easier to measure or calculate the thermodynamic property of free energy than it would be to measure the equilibrium constant. On the other hand, it is sometimes easier to determine an equilibrium constant and from it to calculate the free-energy change.

This quantitative relationship can readily be seen to be consistent with the previous qualitative statements about the significance of the free-energy change. If $\Delta G°$ is very negative, for instance, the argument of Sec. 9-1 leads to the expectation of a spontaneous reaction. Equation [55] confirms this by showing that the equilibrium constant would be a large positive quantity. The reaction therefore would proceed until a large partial pressure of products relative to reactants was built up.

The above derivation applies strictly only to ideal gases. For nonideal gases the free energy, as shown in Secs. 9-6 and 9-7, does not differ from its standard value exactly according to the equation $G = G° + RT \ln P$. In Sec. 9-9 the effect of nonideal behavior on the equilibrium relationship will be dealt with. Here no great error will be introduced by applying the present equations to most gaseous reagents at relatively low pressures.

Furthermore, since many equilibria are studied in solution, it is very desirable to have a result that can be applied to the concentration of reagents in addition to one that treats gaseous reagents. Only a minor extension of the present treatment is needed to obtain the corresponding equation with the equilibrium constant expressed in terms of concentrations. This will, however, be postponed to a later chapter.

As an example of the use of the free-energy–equilibrium relation, the industrially important process of the formation of ammonia from its elements can be considered. The reaction is

$$N_2 + 3H_2 \rightleftharpoons 2NH_3 \qquad [56]$$

and the standard free-energy terms are

$$\begin{aligned} N_2&: \quad \Delta G^\circ_{298} = 0 \\ 3H_2&: \quad 3\Delta G^\circ_{298} = 0 \\ 2NH_3&: \quad 2\Delta G^\circ_{298} = (2)(-16{,}635) = -33{,}270 \text{ J} \end{aligned} \qquad [57]$$

Thus, for the reaction,

$$\Delta G^\circ_{298} = -33{,}270 \text{ J} \qquad [58]$$

and Eq. [55] gives

$$\log K_p = -\frac{\Delta G^\circ_{298}}{2.303RT} = 5.83$$

and

$$K_p = \frac{(P_{NH_3})^2}{(P_{N_2})(P_{H_2})^3} = 6.8 \times 10^5 \text{ atm}^{-2} \text{ at } 25°\text{C} \qquad [59]$$

It should be pointed out that if the reaction had been written

$$\tfrac{1}{2}N_2 + \tfrac{3}{2}H_2 \rightleftharpoons NH_3 \qquad [60]$$

the value of ΔG°_{298} would have been $-16{,}635$ J, and the equilibrium constant calculated from Eq. [53] would have been

$$K_p \text{ (Eq. [60])} = \frac{P_{NH_3}}{(P_{N_2})^{1/2}(P_{H_2})^{3/2}} = 8.2 \times 10^2 \text{ atm}^{-1} \qquad [61]$$

This result is the square root of Eq. [59].

The synthesis of ammonia would seem to be indicated as certainly feasible by the calculation. It turns out, however, that it is very difficult to get the reagents to react fast enough, i.e., to come to the calculated equilibrium position fast enough, to make the process feasible. To increase the speed of the reaction, it is usually run at higher temperatures, and it is necessary, therefore, that we be able to calculate the equilibrium constant at temperatures other than 25°C.

Such an interest in reactions at various temperatures is sufficiently general so that it will be necessary to develop a method for finding the temperature dependence of the equilibrium constant.

First, however, the ammonia-synthesis example can be used to indicate the need for calculations of the equilibrium pressures at high total pressures. Inspection of Eqs. [59] or [61], or even the qualitative arguments based on Le Chatelier's principle, indicate that the equilibrium percentage of ammonia in the reacting mixture, at any temperature, will be favored by high pressures. Under such conditions, however, when quantitative results are required, assumption of ideal behavior cannot be made. The way in which equilibria are treated in nonideal-gas systems, where the free-energy–equilibrium-constant relation cannot be developed from the ideal-gas relations (Eqs. [50]), will first be presented. The following section, like Secs. 9-5 and 9-6 on nonideal systems, could be omitted, however, and the temperature dependence of the equilibrium constant could now be studied.

9-9 Equilibrium Constants for Systems of Real Gases

The development of expressions comparable with those of Eqs. [54] and [55] in a way that allows for nonideal-gas behavior begins with relations of the type

$$G - G^\circ = nRT \ln f$$

for each reagent. One then obtains, by treatment like that followed through for ideal gases, the result

$$\Delta G^\circ = -RT \ln \left[\frac{(f_C)^c(f_D)^d}{(f_A)^a(f_B)^b}\right]_{\text{equilibrium}} \qquad [62]$$

Now the factor that must remain constant involves fugacities rather than pressures. Since the treatment here is thermodynamically exact (it will be recalled that f is defined so that $\text{G} - \text{G}^\circ = RT \ln f$), the constant term will now be labeled K_{therm}; that is,

$$K_{\text{therm}} = \left[\frac{(f_C)^c(f_D)^d}{(f_A)^a(f_B)^b}\right]_{\text{equilibrium}} \qquad [63]$$

This thermodynamically exact expression is more frequently used with the substitution, for each of the components, of the relation $f = \gamma P$, which gives

$$K_{\text{therm}} = \frac{(\gamma_C P_C)^c(\gamma_D P_D)^d}{(\gamma_A P_A)^a(\gamma_B P_B)^b}$$
$$= \frac{(\gamma_C)^c(\gamma_D)^d}{(\gamma_A)^a(\gamma_B)^b} K_p \qquad [64]$$

Only for ideal gases will all the activity coefficients be unity, and only then will $K_{\text{therm}} = K_p$. Furthermore, the activity-coefficient term may well change as the pressure of the system or as the individual pressures change. Then only K_{therm} can be expected to be a constant for any arrangement of pressures, and the pressure term K_p will be a nonconstant "equilibrium constant."

An illustration of the effects of nonideality on the equilibrium constant is provided again by the reaction

$$\tfrac{1}{2}N_2 + \tfrac{3}{2}H_3 \rightleftharpoons NH_3$$

TABLE 9-2 Equilibrium constants for the reaction $\frac{1}{2}N_2 + \frac{3}{2}H_2 \rightleftharpoons NH_3$ at 450°C*

Total pressure (atm)	Equilibrium pressures (atm)			K_p	$\frac{\gamma_{NH_3}}{\gamma_{N_2}^{1/2}\gamma_{H_2}^{3/2}}$	K_{therm}
	NH_3	N_2	H_2			
10	0.204	2.44	7.35	0.0066	0.99	0.0065
50	4.58	11.3	34.1	0.0068	0.94	0.0064
100	16.35	20.9	62.7	0.0072	0.88	0.0063
300	106.5	48.4	145	0.0088	0.69	0.0061
600	322	69.5	208	0.0129	0.50	0.0064
1000	694	76.5	229	0.0231	0.43	0.0099

*Data from A. T. Larson, *J. Am. Chem. Soc.*, **46**:367 (1924).

The reaction is generally carried out at high pressures and at a temperature of about 450°C. One is interested in knowing the equilibrium partial pressure of NH_3 for various partial pressures of N_2 and H_2.

The data of Table 9-2 for the pressure expression K_p have been calculated from the measured partial pressures of the components at equilibrium. It is clear that at these pressures nonideal effects are important and that use of the ideal expression K_p is not satisfactory.

For the individual reagents the activity coefficients can be calculated, as indicated in Sec. 9-6. In this way one can obtain the activity-coefficient expression listed in Table 9-2 and the results for K_{therm}. A considerable improvement in constancy is seen to result from the use of activities rather than pressures.

The result at the highest pressure indicates that our treatment, as entered in Table 9-2, is still somewhat approximate. This does not imply any approximation in the formation of K_{therm}, but stems from the evaluation of the activity coefficients for the individual gases as if they were pure gases at the total pressure of the reaction system. A correct treatment would make use of PVT data on nonreacting gas mixtures so that the activity coefficients of the components in the mixture could be evaluated. The treatment illustrated in Table 9-2 is satisfactory, however, at all but very high pressures.

9-10 Temperature Dependence of the Free Energy of a Reaction and the Equilibrium Constant

The free energy of each compound involved in a reaction is dependent on the temperature, according to Eq. [23], by the relation

$$\left(\frac{\partial G}{\partial T}\right)_P = -S$$

For a chemical reaction it is the free energy of the products less that of the reactants that is of interest. Application of Eq. [23] to each reagent allows the expression

$$\left[\frac{\partial(\Delta G)}{\partial T}\right]_P = -\Delta S \qquad [65]$$

to be written, where

$$\Delta G = G_{\text{products}} - G_{\text{reactants}} \qquad [66]$$

and

$$\Delta S = S_{\text{products}} - S_{\text{reactants}} \qquad [67]$$

An expression for the temperature dependence of ΔG that is easier to use results if ΔS is eliminated from Eq. [65].

At any constant temperature the changes of free-energy, enthalpy and entropy for any reaction are related by

$$\Delta G = \Delta H - T\,\Delta S \quad \text{or} \quad \Delta S = \frac{\Delta H - \Delta G}{T} \qquad [T \text{ const}] \qquad [68]$$

The latter expression can be used to eliminate ΔS from Eq. [65] to give

$$\left[\frac{\partial(\Delta G)}{\partial T}\right]_P = \frac{-\Delta H + \Delta G}{T} = -\frac{\Delta H}{T} + \frac{\Delta G}{T}$$

or

$$\left[\frac{\partial(\Delta G)}{\partial T}\right]_P - \frac{\Delta G}{T} = -\frac{\Delta H}{T} \qquad [69]$$

The two terms on the left side of Eq. [69] can be shown to be equivalent to

$$T\left[\frac{\partial(\Delta G/T)}{\partial T}\right]_P = T\frac{T[\partial(\Delta G/\partial T)]_P - \Delta G}{T^2}$$

$$= \left[\frac{\partial(\Delta G)}{\partial T}\right]_P - \frac{\Delta G}{T} \qquad [70]$$

Now, comparison of Eqs. [69] and [70] lets us write

$$T\left[\frac{\partial(\Delta G/T)}{\partial T}\right]_P = -\frac{\Delta H}{T} \qquad [71]$$

When this relation is applied to the reagents of a reaction, each at the constant pressure corresponding to the standard states, it becomes

$$T\frac{d(\Delta G^\circ/T)}{dT} = -\frac{\Delta H^\circ}{T} \qquad \left[\begin{matrix}\text{Standard-state pressure,} \\ T \text{ const}\end{matrix}\right] \qquad [72]$$

Finally, the relation between ΔG° and the equilibrium constant, Eq. [55] for ideal-gas behavior or Eq. [64] if nonideal behavior is to be allowed for, can be inserted to give, on rearrangement,

$$\frac{d(\ln K)}{dT} = \frac{\Delta H^\circ}{RT^2} \qquad [73]$$

This important formula is the goal of the derivation. The rate of change of the equilibrium constant with temperature is seen to depend on the standard heat of the reaction.

The change of ln K, or K, over not too large a temperature range can be obtained from this derivative result by an integration of this expression either with the assumption of a constant value of ΔH° or with the temperature dependence of this quantity expressed by the empirical expressions developed in Sec. 7-8. Integrations can be carried out by first rearranging Eq. [73] to

$$\frac{d(\ln K)}{d(1/T)} = -\frac{\Delta H^\circ}{R} \qquad \text{or} \qquad \frac{d(\log K)}{d(1/T)} = -\frac{\Delta H^\circ}{2.303R} \qquad [74]$$

The integrated form of these equations, on the assumption that ΔH° is temperature-independent, is

$$\log K = -\frac{\Delta H^\circ}{2.303R}\frac{1}{T} + \text{const} \qquad [75]$$

Both the integrated and differential forms show that a plot of log K versus $1/T$ should give a straight line with a slope equal to $-\Delta H^\circ/2.303R$. The linearity shown by good measurements can be judged by the example of Fig. 9-4. The straight line, furthermore, has been drawn with the slope $\Delta H^\circ/R$, with a value of ΔH° from the data of Appendix 5.

Thus a measured value of ΔH° can be used to calculate the equilib-

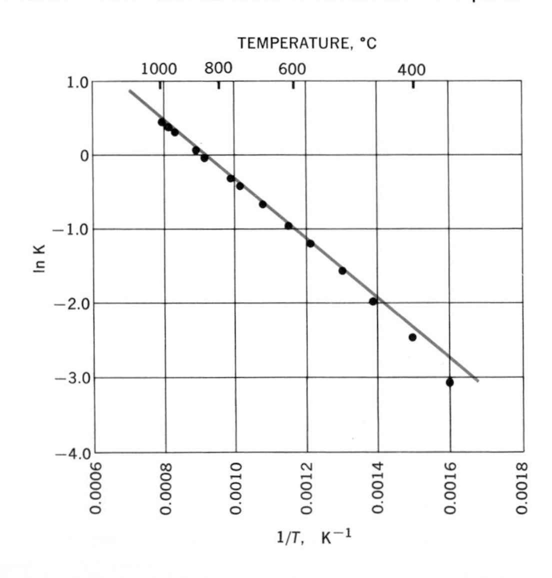

FIGURE 9-4 The temperature dependence of the equilibrium constant for the reaction $CO_2 + H_2 \rightleftharpoons CO + H_2O$. (*From L. P. Hammett, "Introduction to the Study of Physical Chemistry," McGraw-Hill Book Company, New York, 1952.*)

rium constant at temperatures other than that for which it is given. Conversely, it is possible to use measurements of the equilibrium constant at a number of temperatures to evaluate the standard enthalpy change for the reaction.

When much larger temperature ranges are considered, the basis of the dependence of the equilibrium constant on temperature can more clearly be seen by returning to the expressions

$$\Delta G^\circ = \Delta H^\circ - T\,\Delta S^\circ$$

and

$$\Delta G^\circ = -RT \ln K$$

or

$$RT \ln K = -\Delta H^\circ + T\,\Delta S^\circ$$

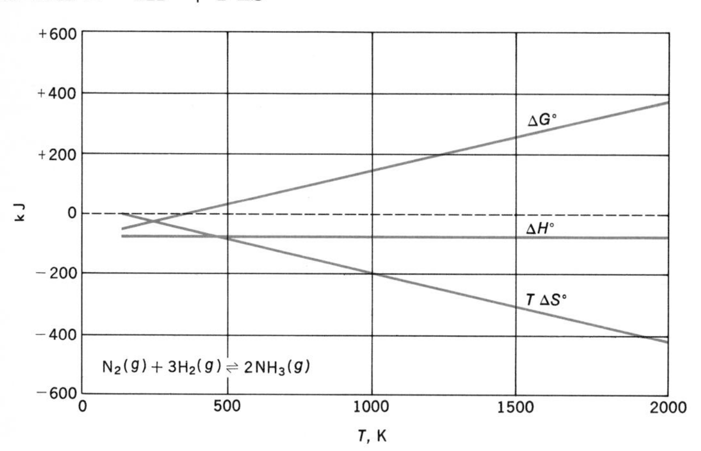

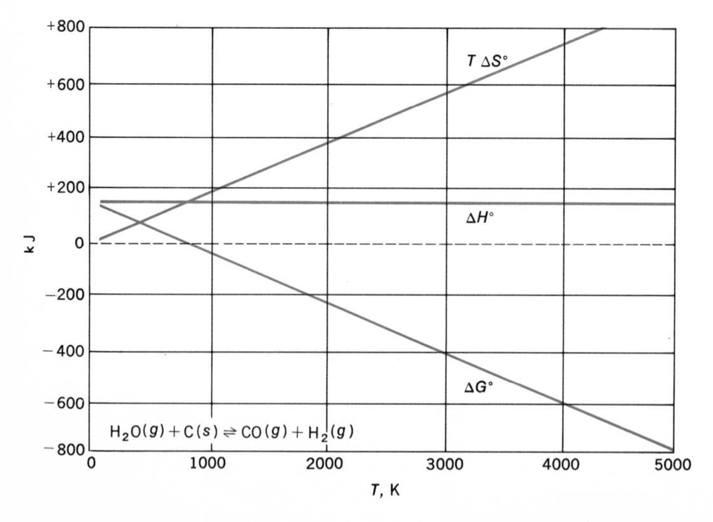

FIGURE 9-5
The temperature dependence of ΔG°, ΔH°, and $T\,\Delta S^\circ$ over large temperature ranges.

This equation is valid at any temperature when values of ΔH° and $T\,\Delta S^\circ$ appropriate to that temperature are used. Generally, the $T\,\Delta S^\circ$ term, as might be expected from the presence of the explicit T factor, is the more temperature-dependent. At high temperatures this term dominates the ΔH° term to give an $RT \ln K$ value that is increasingly positive or negative, depending on whether ΔS for the reaction is positive or negative. Thus at high temperatures the equilibrium constant generally becomes increasingly greater if ΔS is positive, or smaller if ΔS is negative. Examples of these behaviors are shown by the reactions used as illustrations in Fig. 9-5.

In general, the more gas-phase molecular or atomic particles there are, the higher the entropy. This fact and the overwhelming importance of the entropy of the system at high temperatures lead to the general breakup or dissociation of species at these temperatures. One can generalize that, at high temperatures, the side of the equation with more gas-phase species will be dominant—a generalization that is not valid unless the $T\,\Delta S^\circ$ term dominates the ΔH° term in contributing to ΔG°.

Problems

1 The values of ΔH and ΔS for a chemical reaction are -94.5 kJ and -189.1 J deg^{-1}, respectively, and these values are not changed much by temperature changes.
a What is ΔG for the reaction at 300 K? *Ans.* -37.8 kJ.
b What is ΔG for the reaction at 1000 K? *Ans.* $+94.5$ kJ.

2 Calculate ΔH, ΔS, and ΔG when 1 mol of water is converted from liquid at 100°C and 1 atm to vapor at the same temperature and pressure. Discuss the influence of the entropy and energy factors on the spontaneity of the reaction.

3 What is the free-energy change when 4.50 g of water is converted from liquid at 25°C to vapor at 25°C and a pressure of 2×10^{-5} atm? (The equilibrium vapor pressure of water at 25°C is 0.0313 atm, and the vapor can be assumed to behave ideally.) Will the process from liquid to this vapor tend to proceed spontaneously? *Ans.* 4560 J.

4 Using enthalpy and entropy data of Appendix 5, confirm the value of the standard free energy given there for CO_2 at 25°C.

5 Calculate the free energies at 25°C, based on the standard free energies of Appendix 5, of:
a Ethylene at 10^{-5} atm pressure.
b Hydrogen at 10^{-5} atm pressure.
c Ethane at 1 atm pressure.
What would be the free-energy change for the formation of ethane from hydrogen and ethylene if the reagents were at these partial pressures?
Compare with the value obtained in Eq. [19], and comment on the spontaneity of the reaction in each situation.

6 Calculate the free energies at 25°C, based on the standard heats of formation and the entropies of C_2H_4, H_2, and C_2H_6 at 1 atm and at the pressures indicated in Prob. 5. Calculate ΔH, ΔS, and ΔG for the reaction $H_2 + C_2H_4 \rightarrow C_2H_6$ for the reagents at 1 atm pressure and for the pressure conditions of Prob. 5. Comment on the factors that affect the spontaneity of the reaction in the two situations.

7 The standard heat of formation and the entropy of n-pentane, $CH_3CH_2CH_2CH_2CH_3$, at 25°C and 1 atm are -146.44 kJ mol^{-1} and 349.0 J deg^{-1} mol^{-1}, respectively. The values for neopentane, $(CH_3)_4C$, are 165.98 kJ mol^{-1} and 306.4 J deg^{-1} mol^{-1}.

a What is the free-energy difference for these compounds at 25°C and 1 atm pressure?

b What pressure would neopentane be taken to in order to have the same entropy as n-pentane has at 1 atm and the same free energy as n-pentane has at 1 atm? *Ans.* 6.0×10^{-3} atm, 16 atm.

c An amount of n-pentane to produce 1 atm pressure is put in a reaction chamber with a catalyst that brings about equilibrium with neopentane. What are the pressures of the two isomers at equilibrium, the temperature being held at 25°C?

Ans. $P_{n\text{-pentane}} = 0.059$; $P_{\text{neopentane}} = 0.941$ atm.

8 What is the equilibrium constant at 25°C for the formation of benzene from acetylene by the reaction $3(C_2H_2) \rightleftharpoons C_6H_6$? What would be the equilibrium pressure in a reaction vessel which initially held acetylene at 1 atm pressure at 25°C?

9 Assuming that ΔH for the reaction of Prob. 8 is essentially constant over the temperature range 25 to 500°C, obtain an expression for the equilibrium constant as a function of temperature in this range. What is the equilibrium constant at 500°C? *Ans.* $K_p = 1.3 \times 10^{23}$.

10 Calculate the logarithm of the equilibrium constant at 25°C for the reaction $CO_2(g) + H_2(g) \rightleftharpoons CO(g) + H_2O(g)$, using the free-energy data of Appendix 5.

Obtain an expression for the heat of reaction as a function of temperature from the standard heats of formation and the heat-capacity data of Table 7-2.

Deduce an expression for $\log K$ as a function of temperature in the temperature range 300 to 1000 K. Plot and compare with the experimental results shown in Fig. 9-4 (note that Fig. 9-4 is in terms of natural logarithms).

11 What is the equilibrium constant at 25°C for the reaction of oxygen and hydrogen to form water vapor? Equal volumes of oxygen and hydrogen are mixed and put in a constant-volume container. If the initial total pressure is 0.01 atm, what will be the pressures of all reagents if a spark is passed through the mixture and equilibrium is established?

12 At 3000 K the equilibrium pressures of CO_2, CO, and O_2 are 0.6, 0.4, and 0.2 atm, respectively.

a Calculate the equilibrium constant for the reaction $2CO_2 \rightleftharpoons 2CO + O_2$.

b What is the value of ΔG°_{3000} for this reaction? *Ans.* 60.7 kJ.

13 The equilibrium constant at 400°C for the reaction $\frac{3}{2}H_2 + \frac{1}{2}N_2 \rightleftharpoons NH_3$ is 0.0129.

a Calculate ΔG°_{673} for this reaction.

b Using the standard heats of formation and the heat capacities of Table 7-2, obtain an expression for the heat of reaction as a function of temperature.

c Deduce the equilibrium constant and ΔG° at 298 K.

d With the result of part c and the standard heats of formation of the reagents, calculate ΔS° for the reaction at 298 K.

e Compare this value of ΔS° from the measured equilibrium constant and thermal data with the value given in Appendix 5.

14 At not too high pressures the PV behavior of gases conforms to the equation $Pv = RT + BP$. For 1 mol of oxygen at 25°C and for pressures up to about 1 atm, the expression becomes $Pv = R(298) - 0.0211P$ liter atm.

a Calculate the fugacity of oxygen at 1 atm.
b At what pressure does the fugacity have the value 1 atm?

15 Verify, by graphical integrations of Fig. 1-11, some points on the curves of Fig. 9-2.

16 Explain, in terms of the source of nonideal behavior introduced by van der Waals, what might cause an activity coefficient of a nonideal gas to be greater than unity and what might cause it to be less than unity.

17 The critical data for H_2, N_2, and NH_3 are given in Table 1-3. Estimate, with the aid of Fig. 9-2, the activity coefficients of these gases at 1000 atm and at (*a*) 250°C and (*b*) 500°C. Calculate the activity-coefficient factor of the equilibrium-constant expression for nonideal gases, and compare with the value reported for 450°C in Table 9-2.

18 At 800 K the values at the right are reported for the density of steam [Tables of Thermal Properties of Gases, *Natl. Bur. Std. (U.S.) Circ.* 564]. What are the fugacity and the activity coefficient of steam at 800 K and 300 atm? Compare this value with those obtained from the critical data and Fig. 9-2.

Ans. $\gamma = 0.799$.

Pressure (atm)	*Density* ($g\ ml^{-1}$)
1	0.00027464
10	0.027648
20	0.0055709
40	0.011312
80	0.023344
120	0.036184
160	0.049937
200	0.064724
240	0.08070
280	0.09803
300	0.1073

19 The equilibrium constant for the reaction $H_2 + CO_2 \rightarrow H_2O + CO$ at 986°C is 1.60 at rather low pressures, where all the gases behave essentially ideally. Estimate the value that the expression for K_p, expressed in pressures, would have at a total gas pressure of 500 atm and the same temperature.

***20** For most systems of gases, liquids, or solids, the volume increases with temperature. What qualitative feature does this imply for the pressure dependence of the entropy? What familiar system behaves contrary to this generalization?

***21** Use the expressions of Sec. 9-7 to deduce the pressure dependence of G, S, H, and C_P of 1 mol of an ideal gas.

***22** An equation of state for gases known as Berthelot's equation can be written, with parameters expressed in terms of the critical-point quantities, as

$$P\mathrm{v} = RT + \frac{9RPT_c}{128P_c}\left(1 - 6\frac{T_c^2}{T^2}\right)$$

Although this equation does not conform very closely to the actual PVT relations of real gases, it is convenient for the estimation of the pressure dependence of thermodynamic quantities.

Obtain from this equation of state an expression for the value of the inversion temperature.

What values are calculated for helium and for hydrogen? Compare with the observed values.

***23** Van der Waals' equation, for 1 mol of a gas, can be written in the form $P\mathrm{v} = RT - a/\mathrm{v} + bP + ab/\mathrm{v}^2$. Furthermore, the volume terms in the denominators are present in relatively small terms and can be replaced by using the ideal-gas relation $P\mathrm{v} = RT$. Use this guide to obtain an expression for the temperature dependence of the Joule-Thomson coefficient.

Discuss, on the basis of this relation, the molecular-level interpretation for the occurrence of positive and negative values for the Joule-Thomson coefficient.

***24** Because of hydrogen bonding, the vapor of alcohols shows considerable nonideal behavior. The PVT data for methyl alcohol can be represented by a virial-type equation $V = RT/P + B + DP^2$ with $B = -0.100 - 0.002148e^{1986/T}$ liter and $D = -0.835 \times 10^{-14}e^{10{,}750/T}$ liter atm^{-2}.

Calculate and plot the dependence of the molar enthalpy of the vapor on pressure at the normal boiling point, 64.6°C, from zero pressure up to the vapor pressure of 1 atm.

***25** Using the data of Prob. 24, deduce the heat capacity of methyl alcohol in the ideal-gas state at 341 K from the experimental result that at this temperature and a pressure of 0.987 atm the heat capacity is 113.0 J deg^{-1} mol^{-1}. (Calculations using molecular data suggest a value for C_P° at this temperature of 61.5 deg^{-1} mol^{-1}.)

***26** Use the data of Prob. 24 to perform an analytical integration of Eq. [31] and thus obtain data on the ratio of fugacity to pressure for methyl alcohol vapor up to 1 atm pressure at a temperature of 64.6°C.

***27** The difference in entropy of a real gas at some pressure P and the entropy the gas would have at 1 atm pressure if it behaved ideally up to this pressure can be obtained by integrating Eq. [40] for the real gas from zero pressure to pressure P and subtracting the integral, for ideal-gas behavior from zero to 1 atm pressure. In this way, set up a general expression for the entropy of a real gas compared with that of the ideal gas at 1 atm based on the virial equation of state.

***28** Using the expression of Prob. 27 and the virial data for methyl alcohol given in Prob. 24, calculate the entropy difference between methyl alcohol vapor at its vapor pressure of 0.161 atm at 25°C and that of the standard state and 25°C.

10

ENTROPY, FREE ENERGY, AND CHEMICAL EQUILIBRIA: THE MOLECULAR APPROACH

Chemical reactions tend to proceed to equilibrium states, and these, as you have seen, are characterized by maximum entropies of the universe of the reaction. This total entropy driving force is the result of the entropy change of the system itself and the entropy change in the thermal surroundings, the latter stemming from the enthalpy change in the system. You have already seen that the enthalpy change in a system can, to some extent, be interpreted on the basis of the molecular model. Now we must see how the entropy change of the system, and thus the free-energy change and the equilibrium position, can be so interpreted.

10-1 The Molecular Interpretation of Entropy

Let us begin with a postulate that will give us a way of calculating the entropy of a system, 1 mol of an ideal gas for example, from the detailed molecular-level properties of the system. The validity of the postulate will be confirmed when such calculated results are found to be in agreement with those deduced by thermodynamic methods.

According to the approach of Boltzmann, the entropy of a system is given by the elegantly simple expression

$$S = k \ln W \qquad [1]$$

where k is Boltzmann's constant, with value 1.3806×10^{-23} J deg^{-1}, and W is the number of different molecular-level arrangements, or the *probability* of the macroscopic state of the system.

Now let us investigate W so that we obtain an equation that allows, for some actual system, the calculation of S and thus gives us the opportunity of testing Eq. [1].

The number of arrangements of marbles in boxes, as in Fig. 5-2, or of molecules in quantum states, leads to an expression for the number of arrangements, or probability, that is to be associated with that distribution. This probability is found, if you work through to Eq. [5] of Sec. 5-2, to be given, for an Avogadro's number of particles, by

$$\ln W = \ln \mathfrak{N}! + \Sigma N_i \left(1 + \ln \frac{g_i}{N_i}\right) \qquad [2]$$

where N_i is the number of molecules with energy ϵ_i, at which energy there are g_i states. We must now look a little more carefully at this expression before we use it to deduce the properties of gases, as we will in the next few sections.

Equation [2] is obtained, as is obvious from Fig. 5-2, when we label the marbles, or particles, that we distribute throughout the boxes, or quantum states. We assumed that the marbles or molecules were *distinguishable* and thus could be kept track of. If they were not distinguishable, the number of arrangements, or probability, that would be associated with each distribution would be greatly reduced. As can be seen by comparing the number of arrangements $N!$ of the integers 1, 2, 3, . . . , N with the single arrangement that would be recognized for N 1's, 1, 1, . . . or N 2's, 2, 2, 2, . . . , and so forth, the probability of N items must be reduced by a factor of $N!$ if these items are indistinguishable instead of distinguishable. (Since all distributions are equally affected by this recognition of indistinguishability, this reduction was of no consequence in Sec. 5-2, where we sought the most probable distribution.) To calculate a value for $\ln W$ that will apply to the distribution of gas molecules over the allowed energy levels, we must recognize that the molecules cannot be labeled, and therefore that the previous value of W must be divided by $\mathfrak{N}!$, or that $\ln \mathfrak{N}!$ must be subtracted from Eq. [2] to give

$$\ln W = \sum_i N_i \left(1 + \ln \frac{g_i}{N_i}\right) \qquad [3]$$

Now the value of $\ln W$ must be deduced from the most probable arrangement of molecules throughout the energy-level pattern. This distribution is given by the Boltzmann equation, Eq. [26] of Chap. 5, as

$$\frac{N_i}{g_i} = \frac{\mathfrak{N}}{q} e^{-\Delta\epsilon_i/kT} \qquad [4]$$

where q is the partition summation $\Sigma_i g_i e^{-\Delta\epsilon_i/kT}$ and $\Delta\epsilon_i$ is the difference in energy between the ith energy level and the lowest energy level.

The logarithmic form of the reciprocal of Eq. [4], which is needed for Eq. [3], is

$$\ln \frac{g_i}{N_i} = \frac{\Delta\epsilon_i}{kT} + \ln \frac{q}{\mathfrak{N}} \qquad [5]$$

Substitution in Eq. [3] leads then to the molar entropy

$$\begin{aligned} \mathrm{s} &= k \ln W \\ &= k\left[\sum_i N_i\left(1 + \frac{\Delta\epsilon_i}{kT}\right) + \ln \frac{q}{\mathfrak{N}}\right] \\ &= \frac{1}{T}\sum_i (N_i\,\Delta\epsilon_i) + k\mathfrak{N}\left(1 + \ln \frac{q}{\mathfrak{N}}\right) \end{aligned} \qquad [6]$$

The remaining summation will be recognized as giving the familiar thermal energy which we have been representing by $\mathrm{E} - \mathrm{E}_0$. The last term can be rearranged and presented, with R written for $k\mathfrak{N}$, so that we reach the desired result

$$\mathrm{s} = \frac{\mathrm{E} - \mathrm{E}_0}{T} + R \ln \frac{q}{\mathfrak{N}} + R \qquad [7]$$

This expression will be used in the following sections to calculate the entropies of several gases and, by the comparison of the results with the third-law entropy values of these compounds, to confirm the basic equation $S = k \ln W$ used in the molecular calculations.

Although the thermal-energy term $\mathrm{E} - \mathrm{E}_0$ is not dependent, for ideal gases, on the pressure, we will see that q and s are pressure-dependent. When the standard state of 1 atm is specified, Eq. [7] will be written

$$\mathrm{s}^\circ = \frac{\mathrm{E}^\circ - \mathrm{E}_0^\circ}{T} + R \ln \frac{q^\circ}{\mathfrak{N}} + R \qquad [8]$$

Here you should notice that all the terms in Eq. [7] or [8] are calculable if the energy pattern of the allowed states is known. The entropy is entirely dependent on the distribution of the molecules of the sample throughout these states. This distribution does not, as it did in our calculations of energies and enthalpies, lead to only a minor component of the total quantity.

The applicability of Eq. [8], it should be pointed out, is limited, and these limitations stem from two sources.

First, we have again the question of distinguishability of the particles. The displayed result, Eq. [8], is applicable to gases for which the molecules can in no way be distinguished one from the other. Perfect crystalline materials, on the other hand, are different in that the order and rigidity of the crystal lattice allow, at least in principle, the atoms or molecules to be labeled, as by their position relative to a point in the crystal. Liquids present some intermediate difficult-to-analyze cases. In the applications to follow, only gases will be dealt with. The qualitative conclusions regarding the factors that influence the entropy can, however, be carried over to liquids and solids, even if quantitative results will not be obtained for these states.

The second limitation sets in when, as for example happens at low temperatures, the particles are crowded into the states of the lower energy levels. Then the distribution does not conform to the Boltzmann distribution. This distribution is generally followed when, as for gases at not too low a temperature, there are many more available states than there are molecules to be accommodated. When the available states are heavily populated, the distribution that results depends on the fundamental nature of the particles of the system. Electrons, and other particles which, like electrons, have half-integral spin quantum numbers, appear to be limited to two particles per quantum state. (This restriction is, in fact, familiar in the context of the electron configurations of atoms.) Particles that behave in this way are said to obey *Fermi-Dirac* statistics, and in Chap. 18, the importance of this behavior in the treatment of the free electrons in metals will be seen. Other fundamental particles, those with integral spin quantum numbers, have no such restriction on the number per state, and are said to obey *Bose-Einstein* statistics, and even these, when crowded into available states, will, because of their indistinguishability, be then found not to conform to the Boltzmann distribution. The systems that we deal with in the calculation of thermodynamic properties here will, however, have very many more available quantum states than particles. In such situations most of the states will be unoccupied; the special features of "fermions" or "bosons" will be of no consequence. The Boltzmann distribution will then be effectively followed, and when the probability expression is corrected for indistinguishable particles, the resulting $k \ln W$ and entropy result (Eq. [8]) are, as you will see, valid.

10-2 The Translational Entropy of an Ideal Gas

It is particularly important to obtain a qualitative understanding of the molecular basis for entropy and energy and, furthermore, an appreciation of how these factors determine the free-energy difference and the equilibrium constant for reacting species. It is satisfying also to see, however, that the molecular expressions that have been obtained here in a rather simple manner do lead to quantitative values for thermodynamic functions. In particular, it can be shown that a value can be calculated for the entropy of an ideal gas and that such calculated values agree with the thermodynamic third-law values. It should be pointed out that in this and the following several sections the treatment of a number of complications that are present is avoided. The more detailed treatment of the molecular interpretation of thermodynamic functions depends on a more thorough discussion of probability. Advanced texts on statistical mechanics or statistical thermodynamics can be consulted for this material.

The calculation of entropy from molecular properties makes use of Eq. [7]:

$$\mathrm{S} = \frac{\mathrm{E} - \mathrm{E}_0}{T} + R \ln \frac{q}{\mathfrak{N}} + R$$

The translational energy of an Avogadro's number of molecules has been shown on both classical and quantum-mechanical grounds to be $\frac{3}{2}RT$. This value can be substituted for the first term of Eq. [7] to give

$$\mathrm{S}_{\text{trans}} = \frac{\frac{3}{2}RT}{T} + R \ln \frac{q_{\text{trans}}}{\mathfrak{N}} + R$$

$$= \tfrac{5}{2}R + R \ln \frac{q_{\text{trans}}}{\mathfrak{N}} \qquad [9]$$

The translational partition function has been obtained, in Sec. 5-5, as

$$q_{\text{trans}} = \left(\frac{2\pi mkT}{h^2}\right)^{3/2} \mathrm{V} \qquad [10]$$

Now this expression for the translational partition function can be inserted into Eq. [9] to give

$$\mathrm{S}_{\text{trans}} = R\left[\frac{5}{2} + \ln \frac{(2\pi mkT)^{3/2}}{h^3}\mathrm{V}\right] \qquad [11]$$

This equation was obtained by Sackur and Tetrode by an early and rather unsatisfactory derivation. It has now been frequently checked against third-law entropies and can be relied on to give the translational contribution to the entropy of an ideal gas.

The experimental third-law entropy of argon gas at 1 atm pressure and 87.3 K, its normal boiling point, has been reported as 129.1 J deg^{-1} mol^{-1}. The values, in SI units, necessary for the calculation of $\mathrm{S}_{\text{trans}}$ from Eq. [11] are

$m = 6.63 \times 10^{-26}$ kg $\qquad h = 6.62 \times 10^{-34}$ J s

$k = 1.38 \times 10^{-23}$ J deg^{-1} $\qquad \mathrm{V} = \frac{87.3}{273}(0.0224) = 0.0716\ \text{m}^3$

$T = 87.3$ K

The entropy calculated for these conditions is therefore

$$\begin{aligned} S &= R[\tfrac{5}{2} + 2.303 \log (4.61 \times 10^5)] \\ &= 8.315\,(2.500 + 13.04) \\ &= 129.2\ \text{J deg}^{-1}\ \text{mol}^{-1} \end{aligned} \qquad [12]$$

The calculated value agrees nicely, therefore, with that of 129.1 based on the third-law calorimetric method.

Such agreement of calculated and third-law entropies can be taken as support either for the molecular postulates of Schrödinger and Boltzmann or for the thermodynamic choice of a zero entropy at absolute zero.

The general expression of Eq. [11] can be put in a form more convenient for numerical calculations by stipulating standard conditions of 1 atm pressure and 25°C. Then, when numerical values are inserted, we obtain the standard molar entropy as

$$\mathrm{s}^\circ_{298}\ (\mathrm{J\ deg^{-1}\ mol^{-1}}) = 108.65 + 28.72 \log M \qquad [1\ \mathrm{atm},\ 25^\circ\mathrm{C}] \tag{13}$$

where M is the mass of a mole, expressed in grams.

10-3 The Rotational Entropy of the Molecules of an Ideal Gas

As was pointed out in Sec. 4-3, a rotating molecule has a set of allowed rotational-energy levels. For a diatomic, or any linear, molecule the allowed rotational energies of a molecule of moment of inertia I are given approximately by

$$\epsilon_{\mathrm{rot}} = \frac{J(J+1)h^2}{8\pi^2 I} \qquad J = 0, 1, 2, \ldots \tag{14}$$

Furthermore, the rotational energy corresponding to a given value of J will be found to have a degeneracy of $2J + 1$. These features of the rotational-energy patterns allowed, in Sec. 5-6, the rotational partition function to be deduced, and this result can now be used to obtain the rotational-entropy contribution.

When the motions of the molecule can be treated in terms of separate degrees of freedom, the energy levels are expressed as a sum, such as $\epsilon = \epsilon_{\mathrm{trans}} + \epsilon_{\mathrm{rot}}$, and the degeneracies of the levels as the products, such as $g = g_{\mathrm{trans}} g_{\mathrm{rot}}$. The partition function, as shown in Sec. 5-4, is then the product of the partition function for the separate motions, and thus here

$$q = q_{\mathrm{trans}} q_{\mathrm{rot}}$$

When this product is inserted in Eq. [7], and the $\mathfrak{N}$ in the denominator of the second term, and the additional R term is included with the translational terms, we have

$$\mathrm{s} = \frac{\mathrm{E} - \mathrm{E}_0}{T} + R \ln \frac{q}{\mathfrak{N}} + R$$

$$= \left[\frac{(\mathrm{E} - \mathrm{E}_0)_{\mathrm{trans}}}{T} + \frac{(\mathrm{E} - \mathrm{E}_0)_{\mathrm{rot}}}{T}\right] + \left(R \ln \frac{q_{\mathrm{trans}} q_{\mathrm{rot}}}{\mathfrak{N}}\right) + R$$

$$= \left[\frac{(E - E_0)_{trans}}{T} + R \ln \frac{q_{trans}}{\mathfrak{N}} + R\right] + \left[\frac{(E - E_0)_{rot}}{T} + R \ln q_{rot}\right]$$

$$= S_{trans} + S_{rot} \qquad [15]$$

We see that the rotational contribution to the entropy, which must be added to the translational contribution calculated in Sec. 10-2, is given by

$$S_{rot} = \frac{(E - E_0)_{rot}}{T} + R \ln q_{rot} \qquad [16]$$

The partition function for rotation of a linear molecule was obtained in Sec. 5-6 as

$$q_{rot} = \frac{8\pi^2 IkT}{h^2} \qquad [17]$$

For a linear molecule, which has just 2 rotational degrees of freedom, the value of $E - E_0$ for rotation was found, with this expression, to be RT. With these results and Eq. [16], the rotational entropy of a heteronuclear diatomic molecule or unsymmetric linear molecule can be written

$$S_{rot} = \frac{(E - E_0)_{rot}}{T} + R \ln q_{rot}$$

$$= R\left(1 + \ln \frac{8\pi^2 IkT}{h^2}\right) \qquad [18]$$

For CO, as determined by the method to be treated in Sec. 13-1, the moment of inertia is 1.45×10^{-46} kg m^2. The rotational-entropy contribution at 25°C can therefore be calculated as

$$\begin{aligned} S_{rot} &= R(1 + 2.303 \log 107.2) \\ &= R(1 + 4.676) \\ &= 47.2 \text{ J deg}^{-1} \text{ mol}^{-1} \end{aligned} \qquad [19]$$

For comparison, the translational entropy of CO at 1 atm and 298 K can be calculated from Eq. [11] or [13] as

$$S^\circ_{trans} = 149.6 \text{ J deg}^{-1} \text{ mol}^{-1} \qquad [20]$$

The much greater translational-entropy contribution as compared with the rotational-entropy contribution can be understood in terms of the much closer spacing of the translational-energy levels and, therefore, the much larger number of translational states throughout which the molecules are distributed.

Equation [18] is the special result derived from Eq. [16] that is applicable to all heteronuclear diatomic molecules and all linear molecules that have nonidentical ends. More generally shaped molecules, with 3 rather than 2 rotational degrees of freedom, are encompassed by the use of $\frac{3}{2}RT$ for the rotational energy and a rotational partition function, as given in Sec. 5-6,

$$q_{\text{rot}} = \sqrt{\pi}\left(\frac{8\pi^2 kT}{h^2}\right)^{3/2}(I_A I_B I_C)^{1/2} \qquad [21]$$

One additional feature needs to be recognized to make the partition functions, Eq. [17] for diatomic and linear molecules and Eq. [21] for generally shaped polyatomic molecules, applicable to molecules containing identical atoms.

When a molecule, diatomic or linear, has like atoms, as do homonuclear diatomic molecules and linear molecules, like acetylene H—C≡C—H, a further restriction on the allowed rotational states enters. The restriction has the same origin as the *Pauli exclusion principle* mentioned in a different connection in Sec. 10-1, in the statistics followed by different types of fundamental particles. An alternative expression of this exclusion principle is that the wave function for an allowed state of a many-particle atom or molecule must have a particular symmetry; it must change sign or not change sign, depending on the nature of the identical particles, when the roles of equivalent particles are interchanged. This restriction, in a way we cannot deal with here, enters to eliminate half the rotational states when, as for homonuclear diatomic molecules N_2, O_2, H_2, etc., and linear molecules like H—C≡C—H, a rotation can turn the molecule so that it appears unchanged. (The restriction, in fact, enters only when the like nuclei are truly identical, i.e., are isotopically the same.) For such molecules the availability of rotational states as expressed by the partition function is reduced to half that given above.

This reduction in rotational states is often shown by introducing the *symmetry number* σ, for the molecule, and including this in the denominator of the previously obtained rotational-partition-function expression. Thus, for diatomic and linear molecules, we write

$$q_{\text{rot}} = \frac{8\pi^2 IkT}{\sigma h^2} \qquad [22]$$

The symmetry number represents the number of ways that the molecule can be rotated to give back the original molecule, identical except for identifications attached to indistinguishable nuclei. Thus, for molecules such as HCl and HCN, σ is unity, but for N_2, CO_2, and H—C≡C—H, σ is 2.

Again, for calculations it is convenient to introduce numerical values into the general expression. When this is done, the result of substituting Eq. [22] in Eq. [16], for temperature of 25°C, is

$$s^\circ_{298}\ (\text{J deg}^{-1}\ \text{mol}^{-1}) = 924.75 + 19.145 \log I - 19.15 \log \sigma \qquad [23]$$

where I has the SI units of kg m^2, and the symmetry number has the value 1 or 2.

For a generally shaped molecule that has identical atoms that can be interchanged by a rotation of the molecule, a similar insertion of a symmetry number in the denominator of Eq. [21] is necessary. The symmetry number is again the number of different ways the molecule can be turned to give back an original position. The symmetry number of H_2O is 2, of NH_3 is 3, and of CH_4 is 12.

10-4 The Vibrational Entropy of the Molecules of an Ideal Gas

The vibrational contribution to the entropy separates from the translational and from the rotational contributions in the same way that was shown in Sec. 10-3 for rotation. Thus, for each vibrational mode, one adds to the translational and rotational contributions the term

$$\mathrm{S}_{\mathrm{vib}} = \frac{(\mathrm{E} - \mathrm{E}_0)_{\mathrm{vib}}}{T} + R \ln q_{\mathrm{vib}} \tag{24}$$

The first term, the vibrational thermal energy, has been shown in Eq. [55] of Chap. 5 to be given by

$$\mathrm{E} - \mathrm{E}_0 = RT\frac{x}{e^x - 1} \tag{25}$$

where

$$x = \frac{h\nu_{\mathrm{vib}}}{kT}$$

The vibrational partition function for each vibrational degree of freedom was obtained in Sec. 5-7 as

$$q_{\mathrm{vib}} = \frac{1}{1 - e^{-h\nu_{\mathrm{vib}}/kT}} \tag{26}$$

or with $x = h\nu_{\mathrm{vib}}/kT$ as

$$q_{\mathrm{vib}} = \frac{1}{1 - e^{-x}} \tag{27}$$

Thus, for each vibrational degree of freedom of the molecules of a mole of gas, the entropy contribution is

$$\mathrm{S}_{\mathrm{vib}} = R\frac{x}{e^x - 1} + R \ln \frac{1}{1 - e^{-x}} \tag{28}$$

or with numerical values,

$$\mathrm{S}_{\mathrm{vib}}\ (\mathrm{J\ deg^{-1}\ mol^{-1}}) = 8.3143\left(\frac{x}{e^x - 1} + 2.303 \log \frac{1}{1 - e^{-x}}\right) \tag{29}$$

Application of Eq. [29] to the CO example leads, with the vibrational-energy-spacing value of 4.31×10^{-20} J, to an entropy contribution that is negligible compared with the translational and rotational contributions.

Qualitatively, it can be recognized that such a small fraction of the molecules are in vibrational states other than the lowest available vibrational level that the entropy contribution from the vibrational states is, in this example, effectively zero. In general, the vibrational-entropy contribution is small but, except for wide vibrational spacings as in CO, not negligible.

It follows, therefore, that the total entropy of CO at 1 atm and 298 K is calculated, according to the expressions given here, as

$$\begin{aligned} S^\circ &= S^\circ_{trans} + S^\circ_{rot} + S^\circ_{vib} + S^\circ_{elec} \\ &= 149.5 + 47.2 + 0.00 + 0.00 \\ &= 196.8 \text{ J deg}^{-1} \text{ mol}^{-1} \end{aligned} \qquad [30]$$

Calculations that allow for the fact that the rotational-energy term $E - E_0$ is not exactly the classical value give the calculated result 197.9 J deg^{-1} mol^{-1}.

The calculation of the entropy of CO is an example of the results that can be deduced for thermodynamic functions from a knowledge of molecular properties. For larger gas-phase molecules, the procedure is usually limited by the difficulty in deducing the energy-level spacings for the $3n - 6$ vibrational modes. For liquids and solids, so little is known about the allowed energy-level patterns that it is not generally possible to perform the summations over energy levels and obtain values for thermodynamic properties.

10-5 The Molecular Interpretation of Free Energy and Equilibria

Now that the molecular basis for the thermal energy and for the entropy of ideal gases can be given, the useful thermodynamic-function free energy can be given a molecular-level description.

The defining equation for the free energy of a mole of material is

$$G = H - TS$$

or for the standard state,

$$G^\circ = H^\circ - TS^\circ$$

Now, with the molecular interpretations, the molar properties are given by

$$H^\circ = E^\circ_0 + (E^\circ - E^\circ_0) + R$$

and

$$\mathrm{S}^\circ = \frac{\mathrm{E}^\circ - \mathrm{E}_0^\circ}{T} + R \ln \frac{q^\circ}{\mathfrak{N}} + R$$

and we obtain

$$\mathrm{G}^\circ = \mathrm{E}_0^\circ - RT \ln \frac{q^\circ}{\mathfrak{N}} \qquad [31]$$

This very important expression shows how the free energy is to be understood in terms of molecular energies. For example, a compound can have a high free energy and tend to be reactive as a result of having a large value of E_0° or a low value of q°. The reactivity can therefore stem from a high-energy base or a low probability. The energy and probability can, of course, work against each other. The free-energy function gives the net effect of these terms. Again, even if calculations cannot be made, these factors will operate, and recognition of the energy and probability terms provides an understanding of free energy.

Let us first confirm the free-energy expression [31]. Consider a simple reaction

$$A \rightleftharpoons B$$

which involves species for which complete information on the patterns of allowed energies is available.

We now can write, for species A and species B, each in their standard states,

$$\begin{aligned} \mathrm{G}_A^\circ &= (\mathrm{E}_0^\circ)_A - RT \ln \frac{q_A^\circ}{\mathfrak{N}} \\ \mathrm{G}_B^\circ &= (\mathrm{E}_0^\circ)_B - RT \ln \frac{q_B^\circ}{\mathfrak{N}} \end{aligned} \qquad [32]$$

Then the standard free-energy change for the $A \rightleftharpoons B$ reaction is

$$\begin{aligned} \Delta G^\circ &= \mathrm{G}_B^\circ - \mathrm{G}_A^\circ \\ &= [(\mathrm{E}_0^\circ)_B - (\mathrm{E}_0^\circ)_A] - RT \ln \frac{q_B^\circ}{q_A^\circ} \\ &= \Delta E_0^\circ - RT \ln \frac{q_B^\circ}{q_A^\circ} \end{aligned} \qquad [33]$$

The thermodynamic relation $\Delta G^\circ = -RT \ln K$ can now be used to obtain

$$-RT \ln K = \Delta E_0^\circ - RT \ln \frac{q_B^\circ}{q_A^\circ}$$

or

$$K = e^{-\Delta E_0^\circ/RT} \frac{q_B^\circ}{q_A^\circ} \qquad [34]$$

This important result shows that the equilibrium constant can be interpreted in terms of two types of molecular-energy terms. The first, ΔE_0°, is the difference in the lowest-level energies of the products and the reactants, and the second, the q_B/q_A term, is the ratio of the availability of states provided by the products and reactants.

The simple but important molecular-level interpretation of the equilibrium constant for a chemical reaction, given by Eq. [34], has been obtained from a rather extended development that has led through thermal energy, partition functions, entropy, and finally free energy. The starting point for the calculation of all these quantities has been the Boltzmann distribution equation. Now, to emphasize the role of this equation in all calculations of macroscopic properties from molecular affairs, it will be used to show that Eq. [34] can also be obtained in a very direct way.

Consider the generalized patterns of the energies of the allowed states of the chemical species A and B in their standard states as shown in Fig. 10-1. These general descriptions for chemical compounds have already been used in Fig. 5-1, and here all that is added is the relative energy $\Delta\epsilon_0^\circ$ of the two energy patterns. This also is familiar as the molar quantity ΔE_0°, the difference in energy between a mole of A and a mole of B if all the molecules of both species were in their lowest possible energy states.

On a molecular basis the question of the position of the equilibrium between A and B is phrased in this way: If a large number of molecules

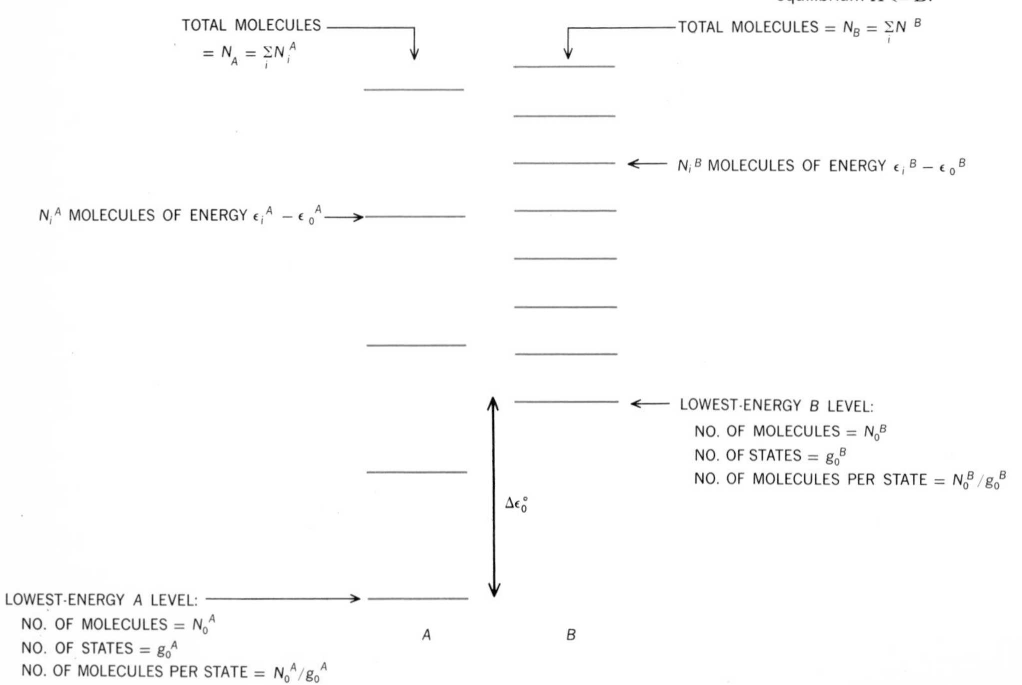

FIGURE 10-1
Schematic energy diagrams for the molecules of reagents A and B connected by the equilibrium $A \rightleftharpoons B$.

are allowed to equilibrate and distribute themselves throughout the energy-level pattern of Fig. 10-1, how many will end up as A molecules, i.e., occupy the A levels, and how many as B molecules, i.e., occupy the B levels? The question is answered by application of the Boltzmann distribution expression.

Let $N_0{}^A$ be the number of molecules which, at equilibrium, occupy the lowest energy level. This happens, in the example of Fig. 10-1, to be an A level.

The total number of molecules N_A in the A levels is, with Eq. [3] of Sec. 5-1,

$$N_A = \sum_i N_i{}^A = \sum_i g_i{}^A \frac{N_0{}^A}{g_0{}^A} e^{-\Delta\epsilon_i{}^A/kT}$$

$$= \frac{N_0{}^A}{g_0{}^A} \sum_i g_i{}^A e^{-\Delta\epsilon_i{}^A/kT} = \frac{N_0{}^A}{g_0{}^A} q_A^\circ \qquad [35]$$

In a similar way the number of molecules N_B distributed throughout the B levels is related to the number distributed throughout the lowest-energy B states as given by

$$N_B = \frac{N_0{}^B}{g_0{}^B} q_B^\circ \qquad [36]$$

Since equilibrium is established between the distribution throughout the A and the B states, the population of the lowest B state is related to the population of the lowest A state by the Boltzmann expression

$$\frac{N_0{}^B/g_0{}^B}{N_0{}^A/g_0{}^A} = e^{-\Delta\epsilon_0^\circ/kT}$$

or

$$\frac{N_0{}^B}{g_0{}^B} = \frac{N_0{}^A}{g_0{}^A} e^{-\Delta\epsilon_0^\circ/kT} \qquad [37]$$

The population of the B levels, as given by Eq. [36], can now be rewritten

$$N_B = \frac{N_0{}^A}{g_0{}^A} e^{-\Delta\epsilon_0^\circ/kT} q_B^\circ \qquad [38]$$

The equilibrium constant for the reaction of A to B might be expressed as the ratio of the pressure, or the concentration, of B to A. Both these terms will be dependent on, and proportional to, the number of moles, or molecules, of the two reagents. We can write, therefore,

$$K = \frac{N_B}{N_A} \qquad [39]$$

The expressions for N_B of Eq. [38] and N_A of Eq. [35] can now be substituted to give

$$K = e^{-\Delta\epsilon_0^\circ/kT} \frac{q_B^\circ}{q_A^\circ} \quad [40]$$

If the difference in lowest-level energies is written per mole instead of per molecule, $\Delta\epsilon_0^\circ$ is replaced by ΔE_0° and k by R. This yields

$$K = e^{-\Delta E_0^\circ/RT} \frac{q_B^\circ}{q_A^\circ}$$

Thus we arrive again by a direct route at Eq. [34].

A very simple example, involving highly artificial energy-level schemes, can be used to illustrate the calculations that can be made with the expression for the free energy. Consider now the molecules A and B each to have only one allowed energy level. That for A consists of a level of multiplicity 2; that for B, of multiplicity 3, as shown in Fig. 10-2. Furthermore, the multiple energy level for B is taken as 1200 J mol^{-1} higher than that for A; that is, ΔE_0° for the reaction A to B is 1200 J mol^{-1}. The free-energy difference between A and B and the equilibrium constant for the system can be calculated at two temperatures, say 25 and 1000°C.

The partition functions are very simply calculated as

$$\begin{aligned} q_A^\circ &= \sum_i (g_i e^{-\Delta\epsilon_i/kT})_A = g_0{}^A e^{-0/kT} = (2)(1) = 2 \\ q_B^\circ &= \sum_i (g_i e^{-\Delta\epsilon_i/kT})_B = g_0{}^B e^{-0/kT} = (3)(1) = 3 \end{aligned} \quad [41]$$

Now Eq. [33] can be used to give

$$\begin{aligned} \Delta G_{298}^\circ &= 1200 - (8.314)(298)(2.303)(\log \tfrac{3}{2}) \\ &= 1200 - 1000 \\ &= 200 \text{ J} \end{aligned} \quad [42]$$

and

$$\begin{aligned} \Delta G_{1273}^\circ &= 1200 - (8.314)(1273)(2.303)(\log \tfrac{3}{2}) \\ &= 1200 - 4300 \\ &= -3100 \text{ J} \end{aligned} \quad [43]$$

These values of ΔG° can be used to calculate the equilibrium constants at the two temperatures. Alternatively, we can start over again and calculate the equilibrium constant directly from Eq. [34]; i.e.,

$$K = e^{-\Delta E_0^\circ/RT} \frac{q_B^\circ}{q_A^\circ}$$

Thus

$$\begin{aligned} K_{298} &= e^{-1200/(8.314)(298)}(1.5) = 0.92 \\ K_{1273} &= e^{-1200/(8.314)(1273)}(1.5) = 1.33 \end{aligned} \quad [44]$$

It is well worthwhile, even for this artificial example, to notice how the energy and entropy, or probability, factors combine to determine

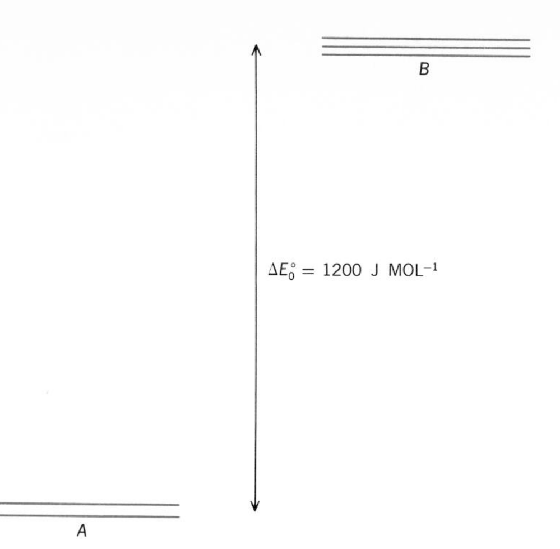

FIGURE 10-2
A simple energy pattern for the molecules A and B to illustrate the effects of q_B/q_A and ΔE_0° on the equilibrium position for the reaction $A \rightleftharpoons B$.

the equilibrium constant. At the lower temperature, the energy term ΔE_0° dominates and leads to the favoring of A over B. At higher temperatures, this factor becomes less important, and the larger number of states that constitute B swings the equilibrium over to the side of B.

Additional care is necessary when the number of moles of reagents connected by the equilibrium are not equal. Consider the gas-phase reaction

$$A + B \rightarrow AB \qquad [45]$$

The entropy of each reagent in its standard state can be expressed by Eq. [7], and thus, with $\Delta(H^\circ - E_0^\circ) = \Delta(E_0 - E_0^\circ) + R$,

$$\Delta S^\circ = \Delta\left(\frac{H^\circ - E_0^\circ}{T}\right) + R \ln \frac{q_{AB}^\circ}{q_A^\circ q_B^\circ} + R \ln \mathfrak{N} \qquad [46]$$

The standard free energy at the reaction temperature T is $\Delta G^\circ = \Delta H^\circ - T\,\Delta S^\circ$, or with Eq. [46],

$$\Delta G^\circ = \Delta E_0^\circ - RT \ln \frac{q_{AB}^\circ}{q_A^\circ q_B^\circ} - RT \ln \mathfrak{N} \qquad [47]$$

Comparison of this expression with the thermodynamic relation $\Delta G^\circ = -RT \ln K$ can now be made only if it is recognized that K denotes, as Eq. [26] of Chap. 9 shows, the ratio of the concentrations of the reagents at equilibrium to the values at the standard state. When this is made explicit, we have, for the $A + B \rightleftharpoons AB$ reaction,

$$\begin{aligned}\Delta G^\circ &= -RT \ln \frac{[AB]/[AB]^\circ}{[A]/[A]^\circ[B]/[B]^\circ} \\ &= -RT \ln \left(\frac{[AB]}{[A][B]} \frac{[A^\circ][B^\circ]}{[AB^\circ]}\right) \\ &= -RT \ln \left(K \frac{[A]^\circ[B]^\circ}{[AB]^\circ}\right) \qquad [48]\end{aligned}$$

Equating the free-energy changes of Eqs. [47] and [48] and rearranging gives

$$K = \frac{[AB]^\circ}{[A]^\circ [B]^\circ} \frac{q_{AB}^\circ}{q_A^\circ q_B^\circ} e^{-\Delta E_0^\circ / RT} \qquad [49]$$

The awkward standard-state concentrations cancel the $\mathfrak{N}/v$ terms that occur, as Eqs. [9] and [10] show, in the translational component of the partition functions. (This follows because the standard-state concentrations are the same for all reagents, and are expressed here in molecules per unit volume.) Thus the equilibrium constant becomes

$$K = \frac{(2\pi m_{AB} kT/h^2)^{3/2}\, q_{AB}{}^r q_{AB}{}^v q_{AB,\mathrm{elec}}}{(2\pi m_A kT/h^2)^{3/2}\, q_A{}^r q_A{}^v q_{A,\mathrm{elec}} (2\pi m_B kT/h^2)^{3/2}\, q_B{}^r q_B{}^v q_{B,\mathrm{elec}}} e^{-\Delta E_0^\circ / RT} \qquad [50]$$

Such equations are comparable with the equation obtained for the $A \rightarrow B$, except for the absence of the volume term in the translational-partition-function terms.

When equilibrium constants are to be calculated from tabulated quantities for the reagents, one returns to the free energies of the reagents. Rearrangement of Eq. [31] gives the "free-energy function"

$$\mathrm{G}^\circ - \mathrm{E}_0^\circ = -RT \ln \frac{q^\circ}{\mathfrak{N}} \qquad \text{or} \qquad \frac{\mathrm{G}^\circ - \mathrm{E}_0^\circ}{T} = -R \ln \frac{q^\circ}{\mathfrak{N}} \qquad [51]$$

If such functions can be calculated for all the reagents in the reaction, one can calculate $\Delta(G^\circ - E_0^\circ)$ for the reaction and write

$$\Delta G^\circ = \Delta E_0^\circ + \Delta(G^\circ - E_0^\circ)$$

The temperature-independent term ΔE_0° can be obtained from thermal measurements, as was discussed in Sec. 7-9. The calculated values of $\Delta(G^\circ - E_0^\circ)$ at various temperatures can be used to obtain ΔG°, and then the equilibrium constant of the reaction at these temperatures.

The use of calculated free-energy functions $\mathrm{G}^\circ - \mathrm{E}_0^\circ$ can be illustrated by deducing the equilibrium constant for the reaction

$$2CO(g) + O_2(g) \rightleftharpoons 2CO_2(g) \qquad [52]$$

at two different temperatures. The free-energy functions that would be obtained from Eq. [51] are given for 298 and 1500 K in Table 10-1. We shall also need the fact that would be derived, as shown in Sec. 7-9, from the calculations of $H - E_0$ for each reagent and from the heat of the reaction that for this reaction $\Delta E_0^\circ = -558.69$ kJ. From these data, the free-energy change for the combustion of carbon monoxide is obtained as

$$\begin{aligned} \Delta G_{298}^\circ &= \Delta E_0^\circ + \Delta(G^\circ - E_0^\circ)_{298} \\ &= -558.69 + [2(58.07) - 2(50.17) - 61.09] \\ &= -603.98 \text{ kJ} \end{aligned} \qquad [53]$$

TABLE 10-1 Values of $G^\circ - E_0^\circ$, in kJ mol^{-1}, as would be calculated from partition functions and Eq. [51]

	298 K	1500 K
CO	50.17	324.97
O_2	61.09	346.35
CO_2	58.07	363.67

Now Eq. [55] of Chap. 9, which relates this thermodynamic value to the equilibrium constant for the reaction, can be used to obtain

$$\log K_{298} = -\frac{\Delta G^\circ_{298}}{2.303\,RT} = 105$$

and [54]

$$K_{298} = 10^{105}$$

Similarly, for 1500 K, from the value of ΔE_0° and the data of Table 10-1,

$$\Delta G^\circ_{1500} = -827.64 \text{ kJ}$$

and [55]

$$K_{1500} = 10^{28}$$

These results illustrate that the values of $G^\circ - E_0$, which can be calculated from the pattern of allowed energy levels for the molecules involved in a reaction, can be used to obtain the equilibrium constant for the reaction. Thermal data must be available only to the extent that ΔE_0° can be deduced. Instead of the thermal data which allow ΔE_0° to be deduced, a value of K for a temperature that allows the measurement of K to be made conveniently might be available. This K value could be used along with the calculated $G^\circ - E_0^\circ$'s to give ΔE_0°. Then, as in the example here, one can readily obtain values for the equilibrium constant at other temperatures, perhaps very high ones, at which measurements cannot easily be made.

The calculation of values for ΔG° and K for various temperatures either from one thermal result or from one measured equilibrium constant is a step of considerable practical importance.

10-6 Molecular Interpretation of the Third Law

The molecular deductions of the preceding sections have led to the same conclusion as that stated in the third law of thermodynamics, namely, that an absolute value can be assigned to the entropy of a chemical compound. When the entropy values calculated from the details of the molecular energies are compared with those obtained from calorimetric third-law measurements, agreement within experimental error is usually found. But there are some exceptions. Thus the third-law entropies for CO and N_2O are too low by about 4.6 J deg^{-1} mol^{-1}; that for H_2O is too low by about 3.3 J deg^{-1} mol^{-1}; and so forth. These discrepancies can now be attributed to the failure of these materials to form the perfect crystalline state that is required at absolute zero for the third law to be applied.

It is the perfectly ordered state of the crystal, with all the molecules in the same lowest energy level, that is the molecular basis for the third-law result that the entropy is zero at absolute zero.

(The positive values for the entropies of all compounds at temperatures above absolute zero result from the fact that, as the temperature is raised, more and more energy levels become accessible to the molecules. The entropies at such temperatures are, of course, very characteristic of the individual molecule, since each molecule has its own particular energy-level pattern.)

The discrepancies between calculated and third-law entropies can now be attributed to a nonzero value of the entropy, because of imperfect ordering, as absolute zero is approached. Thus we must explain an absolute-zero entropy of, for example, about 4.6 J deg^{-1} mol^{-1} for CO.

A disorder to be expected for such a material is that in which the molecular alignment in the crystal is not (CO CO CO CO ···) but rather a disordered pattern such as (CO CO OC CO ···). A crystal formed initially in this way could have the disorder "frozen" in as the temperature is lowered, there being too little thermal energy for the molecules to rearrange to the ordered structure. Thus, instead of each molecule having a single state to occupy, the randomness makes two states available to each molecule. The entropy of such a crystal can then be expected to be greater by $k \ln 2^{\mathfrak{N}} = R \ln 2 = 5.8$ J deg^{-1} mol^{-1} than it would be for a perfect crystal. This is, in fact, the approximate discrepancy found for CO.

Other types of disorder can now be expected to persist at absolute zero and to lead to apparent discrepancies in the third law. A glassy material at absolute zero, for example, will not have the necessary molecular order to guarantee an entropy of zero at absolute zero. In view of difficulties such as these, the third-law statement must include the restriction that only perfectly ordered crystalline materials have zero entropy at absolute zero.

Problems

1 Calculate the entropy of helium at 25°C and 1 atm pressure. Compare this result with the third-law entropy value of 124.7 J deg^{-1} mol^{-1}.

2 Plot the entropy contribution to a molecule from a rotational degree of freedom as a function of the moment of inertia of the molecule. Prepare plots at 298 and 1500 K. (Moments of inertia of relatively simple molecules to which the equations developed here apply are in the range 0.03×10^{-45} to about 10.00×10^{-45} kg m^2.)

3 Plot the entropy contribution to a molecule from a vibrational degree of freedom as a function of the vibrational-level spacing. Prepare plots for 298 and 1500 K. (Vibrational levels are found spectroscopically to be spaced by energies up to about 8×10^{-20} J.)

Compare these curves with the qualitative statement that the entropy increases as the number of available states increases.

4 Calculate the entropy of 1 mol of chlorine at the standard conditions of 1 atm pressure and 25°C. The moment of inertia of the Cl_2 molecule is 1.15×10^{-45} kg m^2, and the vibrational-energy-level spacing is 1.11×10^{-20} J. Compare the answer with the value listed in Appendix 5.

5 Calculate the entropy of N_2O at 25°C and 1 atm pressure. The molecule is linear and has a moment of inertia of 6.69×10^{-46} kg m^2. The four vibrational modes have spacings 1.17, 1.17, 2.56, 4.45, all $\times 10^{-20}$ J. Compare the calculated value with the third-law result, corrected for R ln 2 residual entropy at absolute zero, of 220.1 J deg^{-1} mol^{-1}.

6 Calculate the entropy of 1 mol of carbon disulfide at 1 atm and 25°C, and compare with the value listed in Appendix 5. The CS_2 molecule is linear and has a moment of inertia of 2.56×10^{-45} kg m^2, and the energy spacings in the four vibrational modes are 0.8, 0.8, 1.3, and 3.0, all $\times 10^{-20}$ J.

7 Calculate the equilibrium constants at 25 and 1000°C for hypothetical molecules A and B such as in Fig. 10-1 but with A having a single level and B having a doubly degenerate one 2000 J mol^{-1} higher than that of A.

8 Calculate ΔH, ΔS, ΔG, and K at 298 and 1500 K for the reaction forming 1 mol of B from 1 mol of A. A molecule of A is characterized by a series of levels spaced by the constant amount 4000 J mol^{-1}, and a molecule of B by a similar series with a constant spacing of 8000 J mol^{-1}. The lowest B level is 800 J mol^{-1} above the lowest A level. (The series that must be summed can be handled by a term-by-term numerical summation.)

References

HINSHELWOOD, C. N.: "The Structure of Physical Chemistry," parts I and II, Oxford University Press, Fair Lawn, N.J., 1958. An essentially nonmathematical presentation of the molecular basis of entropy and an interpretation, in terms of molecular behavior, of physical and chemical phenomena.

SHERWIN, C. W.: "Basic Concepts of Physics," Holt, Rinehart and Winston, Inc., New York, 1961. Chapter 7 offers a very clear presentation of some of the basic features of statistical mechanics. No mathematical complexities obscure the treatment, and many simple systems are used to illustrate the concepts.

ANDREWS, F. C.: "Equilibrium Statistical Mechanics," John Wiley & Sons, Inc., New York, 1963. Treatment of probability and distributions followed by applications, thermodynamic functions, and ideal and real gases.

KITTEL, C.: "Elementary Statistical Physics," John Wiley & Sons, Inc., New York, 1958. Treatment of many topics of this chapter dealt with in a particularly straightforward manner.

MAYER, J. E., and M. G. MAYER: "Statistical Mechanics," chaps. 5–9, John Wiley & Sons, Inc., New York, 1940. A more detailed development of the relation between thermodynamic functions and molecular properties.

DAVIDSON, N.: "Statistical Mechanics," McGraw-Hill Book Company, New York, 1962. A rather complete presentation of the statistical basis of chemical thermodynamics.

MOELWYN-HUGHES, E. A.: "Physical Chemistry," chaps. 7–9, Cambridge University Press, New York, 1951. A treatment of the calculation of thermodynamic properties from molecular properties that provides a very suitable extension to the material of this chapter.

11

INTRODUCTION TO THE THEORY OF CHEMICAL BONDING

The subject of atomic and molecular structure has been introduced in Chap. 3 only in sufficient detail to allow a molecular interpretation to be given to the thermodynamic functions. A further direct look into the molecular world will now be taken.

One of the most exciting endeavors in man's investigation of the world in which he lives has been his attempts to understand the basic units of matter that make up the material world. For the chemist the basic units are the molecules and the atoms of which they are composed. Some of the long-sought-for answers to the questions of why and how atoms are held together into molecules can now be given, and in some respects the description of the nature of chemical bonding that is achieved represents the culmination of one aspect of man's efforts to unravel the secrets of matter.

Only atoms and relatively simple molecules will be dealt with in this chapter. This will be adequate, however, to show the basis of chemical bonding and to provide a guide for the extension of bonding ideas to more complex molecules, an extension that will be made in the following chapter. The formidable obstacles that have stood in the way of applications of the basic approaches to molecules of interest are now, with the use of high-speed computers, being overcome. A remarkable increase in our understanding of molecules and molecular reactions, and therefore of macroscopic materials and processes, is resulting from such calculations.

We begin by extending the study of atoms that was begun in Chap. 3. Although detailed studies of atoms, which are based primarily on results from atomic spectroscopy, are the concern of specialists, you will see that our understanding of molecules is based to a large extent on our knowledge of their simpler components. It is for this reason that the following study of atoms is undertaken here.

ATOMIC STRUCTURE

11-1 The Solution of the Schrödinger Equation for the Hydrogen Atom

The problem that we begin with, the nature of the hydrogen atom, has far-reaching implications in studies of atoms and molecules because the solutions that are obtained can be used in studies of the electrons of all atoms. The essential feature that leads to this generality is the action on the electron of "central forces." This implies that the forces acting are independent of angular factors and depend only on the distance of the electron from a central point. In the hydrogen atom, and with some approximation in all atoms, orbital electrons experience such forces.

The Schrödinger equation, introduced in Sec. 3-4, allows us to impose the restrictions on the behavior of a confined particle that stems from the wave nature of the particle. In atomic and molecular problems these restrictions are imposed on particles—electrons and nuclei—that interact with each other through electrostatic forces that are described by Coulomb's law. Thus the hydrogen atom is treated in terms of the familiar attractive forces between the orbital electron and the proton, which constitutes the nucleus. The special atomic features, such as the failure of the electron to fall into the nucleus, arise from the wave-nature and Schrödinger-equation requirements.

The Schrödinger equation can be set up for the electron moving in the electric field of the nucleus. If the nuclear charge is $+Ze$, where Z is the atomic number, and the electronic charge is $-e$, the potential energy of the electron at a distance r from the nucleus is $-Ze^2/r$. Substitution of this potential-energy function in the three-dimensional Schrödinger equation (Sec. 3-6) gives

$$-\frac{h^2}{8\pi^2 m}\left(\frac{\partial^2\psi}{\partial x^2}+\frac{\partial^2\psi}{\partial y^2}+\frac{\partial^2\psi}{\partial z^2}\right)-\frac{Ze^2}{r}\psi=\epsilon\psi \quad [1]$$

The spherical symmetry of the potential-energy function about the nucleus suggests that solutions will be more readily found if the equation is written in the polar coordinates r, θ, and ϕ shown in Fig. 11-1. These are related to the cartesian coordinates, as can be seen by inspection of that figure.

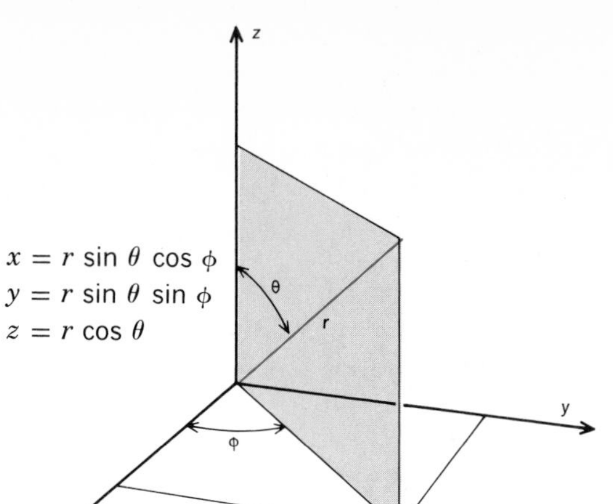

FIGURE 11-1
Location of a point in space by polar coordinates.

The Schrödinger equation in polar coordinates, which is displayed but not worked with, is

$$-\frac{h^2}{8\pi^2 m}\frac{1}{r^2 \sin\theta}\left[\sin\theta\frac{\partial}{\partial r}\left(r^2\frac{\partial\psi}{\partial r}\right)+\frac{\partial}{\partial\theta}\left(\sin\theta\frac{\partial\psi}{\partial\theta}\right)\right.$$
$$\left.+\frac{1}{\sin\theta}\frac{\partial^2\psi}{\partial\varphi^2}\right]+U(r)\psi=\epsilon\psi \quad [2]$$

where $U(r)$ is the potential-energy function which, in preparation for the central field problems to be dealt with here, is shown to be a function of r and to be independent of θ and φ.

An important simplification results by investigating a solution function ψ of the form

$$\psi = R(r)Y(\theta,\varphi) \quad [3]$$

that is, ψ is the product of some function of r and some other function of the angles θ and φ.

Then

$$\frac{\partial}{\partial r}\left(r^2\frac{\partial\psi}{\partial r}\right) = Y(\theta,\varphi)\left[\frac{d}{dr}r^2\frac{dR(r)}{dr}\right]$$
$$= Y(\theta,\varphi)\left[2r\frac{dR(r)}{dr}+r^2\frac{d^2R(r)}{dr}\right]$$
$$\frac{\partial}{\partial\theta}\left(\sin\theta\frac{\partial\psi}{\partial\theta}\right) = R(r)\left[\frac{\partial}{\partial\theta}\sin\theta\frac{\partial Y(\theta,\varphi)}{\partial\theta}\right]$$
$$\frac{\partial^2\psi}{\partial\varphi^2} = R(r)\ \frac{\partial^2 Y(\theta,\varphi)}{\partial\varphi^2}$$

Substitution of these terms in the Schrödinger equation and rearrangement gives

$$\frac{1}{R(r)}\frac{d}{dr}\left(r^2\frac{dR}{dr}\right) + \frac{8\pi^2 m}{h^2}[\epsilon - U(r)]r^2$$
$$= -\frac{1}{Y(\theta,\varphi)\sin\theta}\frac{\partial}{\partial\theta}\left[\sin\theta\frac{\partial Y(\theta,\varphi)}{\partial\theta}\right] - \frac{1}{Y(\theta,\varphi)\sin^2\theta}\frac{\partial^2 Y(\theta,\varphi)}{\partial\varphi^2} \quad [4]$$

Since the left side of this equation is a function only of r and the right only of θ and φ, the two sides of the equation can remain equal to each other for all values of r, θ, and φ only if both sides are equal to some constant.

Thus the angular features of the wave function can be deduced from the equation formed from the right-hand side of Eq. [4], and this can be done without specifying the potential energy any further than to say that it is a function only of r.

Solution of the radial equation formed from the left side of Eq. [4] can be carried out only with the insertion of a specific function for $U(r)$. For hydrogenlike atoms, $U(r) = -Ze^2/4\pi\epsilon_0 r$. Solutions of this radial equation are then found, for a bound electron, only for values of ϵ given by

$$\epsilon = -\frac{2\pi^2 me^4 Z^2}{(4\pi\epsilon_0)^2 h^2 n^2} \quad \text{where } n = 1, 2, 3, \ldots \quad [5]$$

Thus one quantum number enters when the radial equation is solved. It is represented by n, and only this quantum number shows up in the energy expression for the free atom (Eq. [5]) as well as in the wave function. It plays a role very similar to that of the rather arbitrarily introduced Bohr atom quantum number. It determines the electronic energy and, as will be shown, is primarily responsible for determining the average distance of the electron from the nucleus.

Two more quantum numbers, completing the set for this three-dimensional system, appear when the angular equation is solved. The symbols, names, possible values, and principal properties of all three quantum numbers that the Schrödinger equation introduces are listed in Table 11-1. An additional quantum number, the spin quantum number, which will be discussed in Sec. 11-3, is included to complete the table.

The azimuthal quantum number l and the magnetic number m

TABLE 11-1 Quantum numbers for electrons in atoms

Symbol	Name	Allowed values	Property principally determined
n	Principal	$1, 2, 3, \ldots$	Size and energy of orbital
l	Azimuthal	$0, 1, 2, \ldots, n-1$	Shape of orbital
m	Magnetic	$-l, -l+1, \ldots, 0, \ldots, l-1, l$	Orientation of orbital
m_s	Spin	$\frac{1}{2}, -\frac{1}{2}$	Spin of electron

determine the angular factors of the wave function. These two quantum numbers, like the principal quantum number n, introduce themselves in just the same way as did the quantum numbers in the solutions of Secs. 3-5 and 3-6 for the particle-in-a-box. The quantum numbers l and m are different from the number n, however, in that, for the hydrogen atom, they specify the wave function that describes the position of the electron but they do not occur in the expression for the energy of the electron.

The significance of the quantum numbers is better appreciated by considering the wave functions that the Schrödinger equation produces as descriptions of the positions that an electron can assume when it is held to a nucleus of charge Ze.

11-2 The Hydrogen-atom Wave Functions

The mathematical forms of some of the wave functions that are solutions to the hydrogen-atom wave equation are shown in Table 11-2. The inclusion of variable nuclear charge Z makes these wave functions appropriate to any one electron system, and they are therefore "hydrogenlike" wave functions. Again, one should remember that these expressions represent the same kind of result as did the trigonometric functions of the particle-in-a-box problem. In particular, the square of these functions gives the probability of the electron being in a volume element at some position designated by r, θ, and ϕ. The values of ψ and ψ^2 implied by the functions of Table 11-2 are best shown by dia-

TABLE 11-2 The hydrogenlike-atom wave functions $\psi = R(r)Y(\theta,\varphi)$ for the $n = 1$ and $n = 2$ shells
$a_0 = h^2/4\pi^2 me^2 = 0.529$ Å. Z is the effective nuclear charge, which, for the hydrogen atom, has the value 1.

n	l	m	$R(r)$	$Y(\theta,\varphi)$	Wave-function symbol
1	0	0	$2\left(\frac{Z}{a_0}\right)^{3/2} e^{-Zr/a_0}$	$\left(\frac{1}{4\pi}\right)^{1/2}$	1s
2	0	0	$\left(\frac{Z}{2a_0}\right)^{3/2}\left(2 - \frac{Zr}{a_0}\right) e^{-Zr/2a_0}$	$\left(\frac{1}{4\pi}\right)^{1/2}$	2s
2	1	0	$\frac{1}{\sqrt{3}}\left(\frac{Z}{2a_0}\right)^{3/2}\left(\frac{Zr}{a_0}\right) e^{-Zr/2a_0}$	$\left(\frac{3}{4\pi}\right)^{1/2} \cos\theta$	$2p_z$
2	1	+1	$\frac{1}{\sqrt{3}}\left(\frac{Z}{2a_0}\right)^{3/2}\left(\frac{Zr}{a_0}\right) e^{-Zr/2a_0}$	$\left(\frac{3}{8\pi}\right)^{1/2} \sin\theta\, e^{-i\varphi}$	$2p_x$, $2p_y$
2	1	−1	$\frac{1}{\sqrt{3}}\left(\frac{Z}{2a_0}\right)^{3/2}\left(\frac{Zr}{a_0}\right) e^{-Zr/2a_0}$	$\left(\frac{3}{8\pi}\right)^{1/2} \sin\theta\, e^{+i\varphi}$	

grams that give separately the radial and the angular parts of the wave functions.

The radial distribution of the wave functions is controlled primarily by the quantum number n. The principal factor affecting the radial extent of a wave function, or *orbital,* is the exponential factor, which has the form e^{-Zr/na_0}, where Z is the atomic number of the nucleus and a_0 is the collection of constants $h^2/4\pi^2me^2$ that equals 0.529 Å and is called the *Bohr radius.* This exponential term is such that for larger values of the principal quantum number n, the wave function falls off less rapidly with the distance from the nucleus. With the larger nuclear charges, which will be encountered when atoms other than hydrogen are considered, the fall-off is more rapid and the electron is held more closely to the nucleus. The radial part of the wave function and the square of this function are shown for the three lowest energy orbitals $n = 1$, $n = 2$, and $n = 3$ in Fig. 11-2a. The detailed form is seen to depend on l, but the overall extension on n.

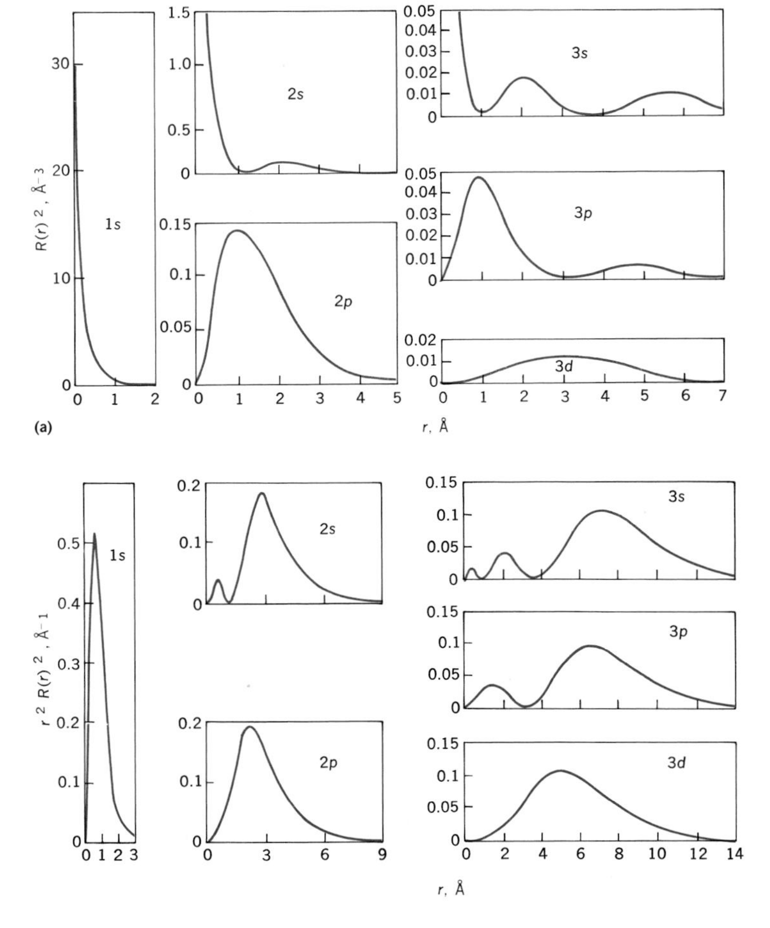

FIGURE 11-2
(a) The square of the radial part of the wave function for the hydrogen atom. (b) The radial distribution function $r^2R^2(r)$ for the hydrogen atom.

A better picture, for some purposes, of the radial distribution of the electron orbits is obtained by showing the relative probabilities of the electron being at various distances from the nucleus rather than, as in Fig. 11-2a, being in unit volumes at various distances. A distinction exists because the volume of an annular element, a distance r from the nucleus, is proportional to the area $4\pi r^2$ of a sphere of radius r. The probability of the electron being at a distance between r and $r + dr$ from the nucleus is therefore given by $4\pi r^2\,dr$ times the probability of its being in a unit volume at a distance r from the nucleus. (A similar situation, it will be recalled, was encountered in studies of molecular velocities in Sec. 2-6, and there also we were led to consider spherical-shell elements.)

Figure 11-2b shows the *radial-distribution functions* that include the r^2 factor. It is interesting to note that the distance of the electron from the nucleus at the maximum in the radial-distribution curves for $1s$, $2p$, $3d$, etc., where there are single maxima, is exactly equal to the radius of the corresponding orbit calculated by the Bohr theory.

The actual wave function for the electron of the hydrogen atom is given by the product of the radial part and the angular part. The wave function with $l = 0$, known as an *s orbital,* and that with $l = 1$, known as a *p orbital,* are to be considered here.

As the wave functions of Table 11-2 show, the s wave functions, i.e., those with $l = 0$, $m = 0$, have no dependence on either of the angles. The wave function varies in the same way for all directions from the nucleus, and the radial-distribution curves of Fig. 11-2 are therefore complete descriptions. The constancy of the angular factor or the square of this factor can, however, be represented as the sphere of Fig. 11-3, so that the total wave function is the radial function times a factor that is independent of the angles.

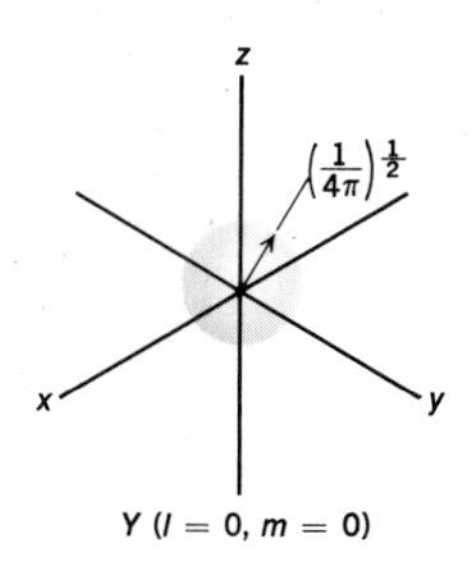

Y ($l = 0$, $m = 0$)

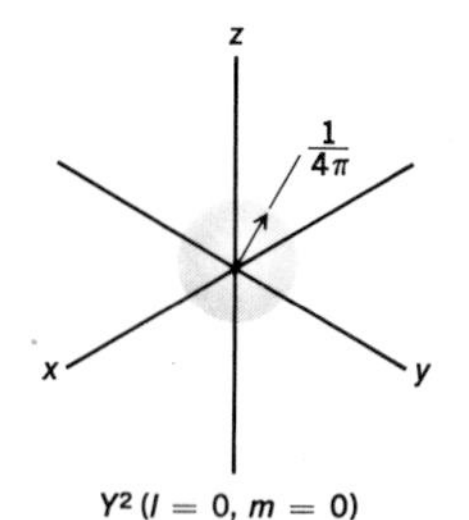

Y^2 ($l = 0$, $m = 0$)

FIGURE 11-3
Angular factors, and probability factors for s orbitals.

As the name angular factor implies, generally the shape of the orbitals is determined by this factor, and thus by the values of the quantum numbers l, and in a way by m. For many chemical purposes it is the orbital-shape consequences of l that are important, but in other cases it is the angular-momentum implication, which this quantum number also carries, that is more significant. In this regard it must be recognized that l implies an orbital angular momentum of $\sqrt{l(l+1)}(h/2\pi)$. Thus an s electron has zero orbital angular momentum, a p electron has an orbital angular momentum of $\sqrt{1(2)}(h/2\pi)$, and so forth.

The magnetic quantum number m also carries angular-momentum implications, a value of m indicating an angular-momentum component in any direction that is imposed in the atom of $m(h/2\pi)$. Thus a p electron of an atom in an applied magnetic field could have its total angular-momentum vector of $\sqrt{1(2)}(h/2\pi) = 1.414(h/2\pi)$ oriented so that the value along the field direction would be one of the three values of $m(h/2\pi)$, namely, $(1)h/2\pi$, 0, $(-1)(h/2\pi)$. A diagram that might be drawn to represent these orientations is shown in Fig. 11-4.

Now let us return to the p wave functions themselves to see what spatial features they imply.

The set of p wave functions introduce the fact that wave functions can be real or imaginary. The corresponding angular probability functions must, however, be real. Although for real wave functions we have been calculating the probability function as the square of the wave function, now we must make use of the more general postulate that for a wave function ψ, the probability function is $\psi\psi^*$, where ψ^* is the complex conjugate of ψ. (ψ^* is obtained from ψ by replacing i, which denotes $\sqrt{-1}$, by $-i$, wherever it appears.) For real numbers, $\psi\psi^*$ becomes simply ψ^2, and for imaginary or complex numbers, the function obtained will be real.

The angular probability factors corresponding to the p wave functions of Table 11-2 can be calculated and plotted, as is done in Fig. 11-5. Notice that the $m = +1$ and $m = -1$ probability functions are identical. The distinction between states described by the same values of n and l but values of m of opposite sign would be seen by the effect of a magnetic field on an atom.

In such a field, the energy of the $m = 0$ state would be unaffected, whereas those for $m = +1$ and $m = -1$ would be changed, and changed in opposite directions. The picture that this result suggests is that the $m = +1$ and $m = -1$ states correspond to the electron

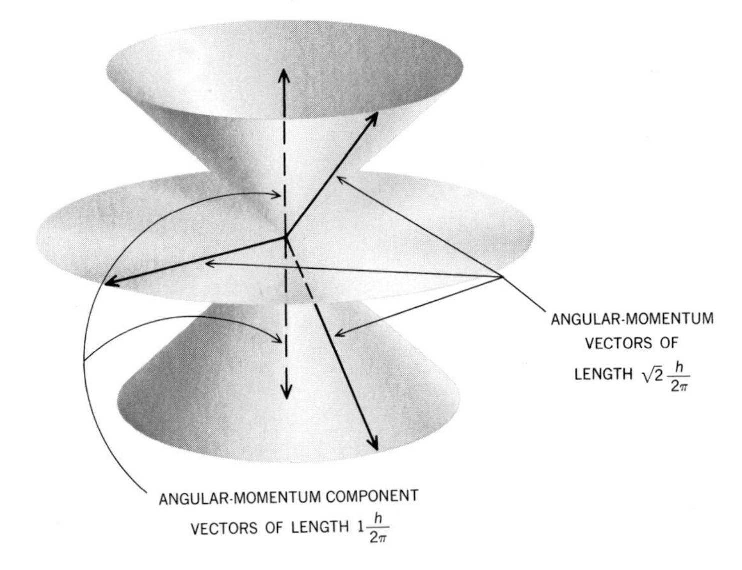

FIGURE 11-4
The vectors associated with the orbital angular momentum of an electron in a p orbital.

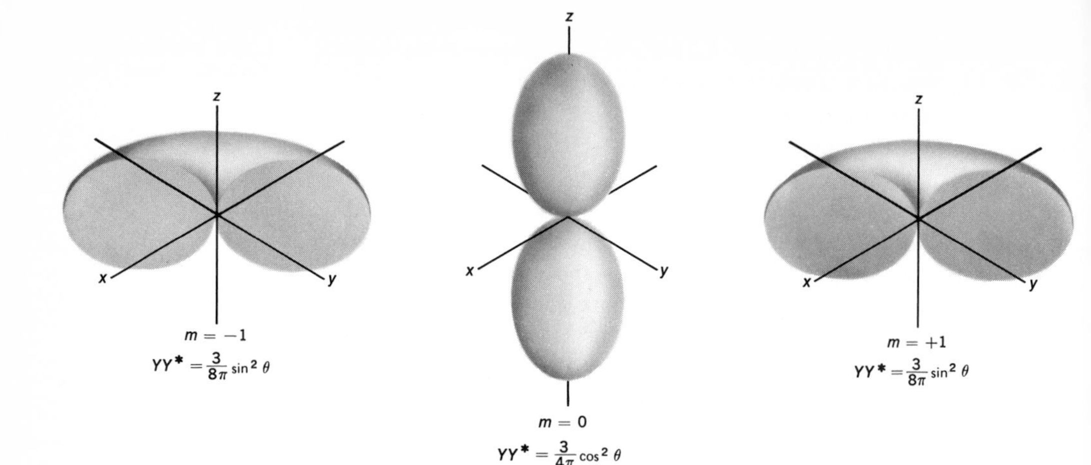

FIGURE 11-5
Plots of the angular factors YY^* for the p-orbital functions of Table 11-2.

moving about the nucleus in a region corresponding to the probability function of Fig. 11-5, in opposite directions. In this way you can see that the spatial and angular-momentum implications of the quantum numbers l and m can be brought together.

For most chemical uses, it is convenient to have, instead of the wave functions of Fig. 11-5, three identical probability functions for the $l = 1$, or p, orbitals. These can be constructed, since the l functions are degenerate, by taking linear combinations of those listed in Table 11-2. With the identities

$$\sin\varphi = \frac{e^{i\varphi} - e^{-i\varphi}}{2i} \qquad \text{and} \qquad \cos\varphi = \frac{e^{i\varphi} + e^{-i\varphi}}{2}$$

and appropriate normalizing factors, we can construct the new angular-wave-function factors

$$p_x = \frac{1}{\sqrt{2}i}[Y(l = 1, m = +1) - Y(l = 1, m = -1)]$$

$$= \left(\frac{3}{4\pi}\right)^{1/2} \sin\theta \sin\varphi \qquad [6]$$

and

$$p_y = \frac{1}{\sqrt{2}}[Y(l = 1, m = +1) + Y(l = 1, m = -1)]$$

$$= \left(\frac{3}{4\pi}\right)^{1/2} \sin\theta \sin\varphi \qquad [7]$$

Consideration of a few values of θ and φ will show that the resulting wave functions lead to the three equivalent p wave and probability functions, as shown in Fig. 11-6.

Finally, to complete the introduction of the quantum numbers of Table 11-1 that specify the state of an electron of a hydrogenlike atom, the electron-spin quantum number m_s must be described. From spectral studies, such as those to be treated in Sec. 11-3, it is found that the orbital descriptions implied by the quantum numbers n, l, and m are not adequate to account for all the observed spectral features. These demand the inclusion of a contribution of the electron itself, in addition to that resulting from its orbital role. In this way the idea of electron spin enters, and it is found that it is adequate to associate a spin quantum number of $\frac{1}{2}$ with this electron spin so that, in the presence of any directional field, the electron can occupy either of the two spin states that have angular momentum of $+\frac{1}{2}(h/2\pi)$ or $-\frac{1}{2}(h/2\pi)$ in the field direction. These states can be treated in a parallel manner to the orbital states by introducing the spin quantum number m_s and allowing it to take on only the two values $+\frac{1}{2}$ and $-\frac{1}{2}$.

FIGURE 11-6
The equivalent p-orbital functions that are obtained from linear combinations of those given in Table 11-2.

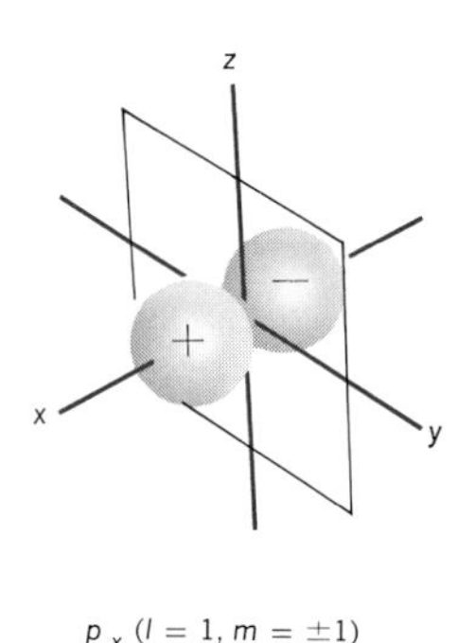

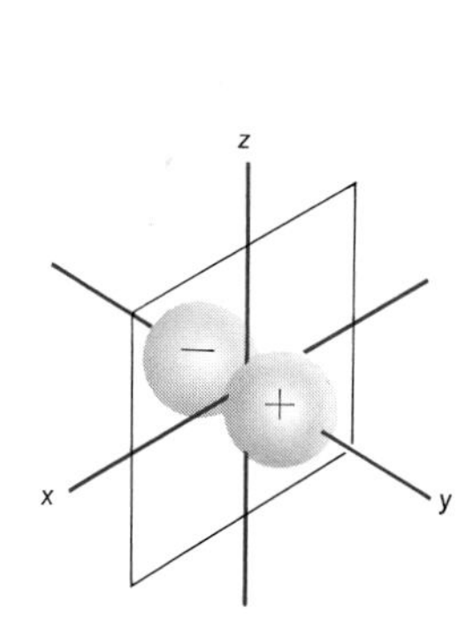

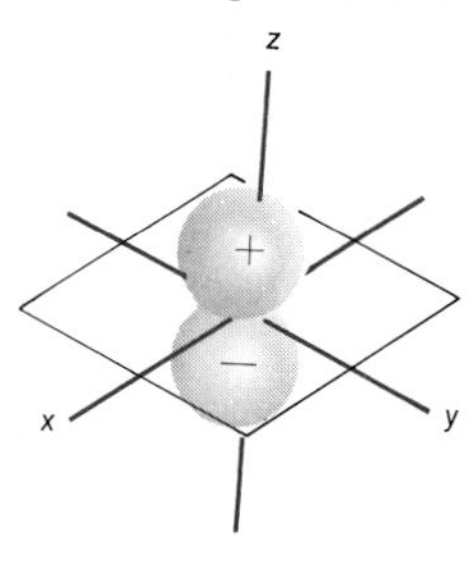

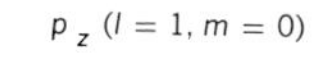

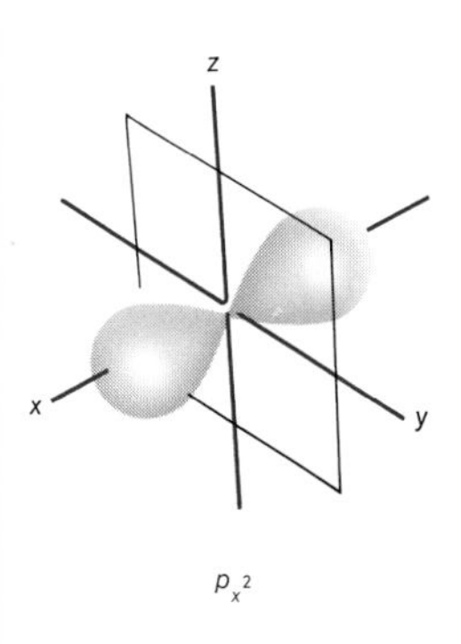

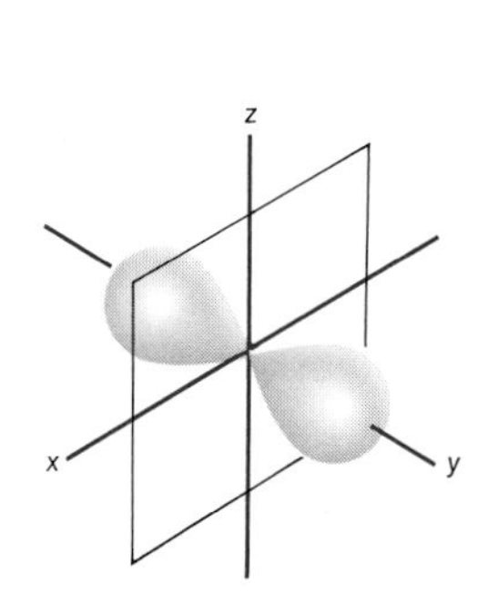

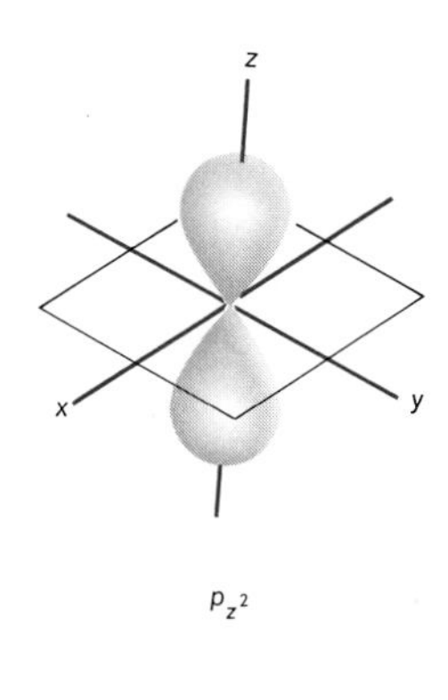

11-3 Experimental Characterization of Atomic States

Let us now turn to many-electron atoms and see how the results for the one-electron, hydrogenlike atom can be extended to these more complex systems. First we list the features of many-electron atoms that we would hope to be able to deduce or measure.

Energy. The relative energies of different states of the atom and the energy change involved in the removal of an electron from the atom, i.e., the *ionization energy,* or the addition of an electron to the atom, i.e., the *electron affinity.*

Size. Some measure of the significant extension of the electrons from the nucleus, particularly for the lowest energy, or *ground state,* of the atom.

Angular momentum. The results for hydrogenlike atoms suggest that angular momentum will be important, and will be quantized, for many-electron atoms.

Let us begin by considering the experimental results that can be obtained that bear on these properties of many-electron atoms.

Atomic absorption or emission spectra and the idea that spectral lines result from the transition of an atom from a state with one energy to a state with a different energy provide information on atomic energies that is the basis for the construction of an empirical atomic-energy-level diagram. Some of the features of these diagrams for many-electron atoms are illustrated by the spectrum and energy diagram for the sodium atom in Fig. 11-7a.

Such energy diagrams are of limited value without descriptions of the atomic states that correspond to the allowed energies. These descriptions are given most simply and directly in terms of the angular momenta of the atomic states. The idea of quantized angular momenta and the coupling of angular-momentum contributions of the components of the atom, i.e., the "vector" model of the atom, leads, with observation of the details, or fine structure, of each spectral line to the assignment of an angular momentum of the atom to the states with the indicated energies. This angular-momentum characterization of each state of the atom can be broken down into two components, one due to the orbital motion of the electrons about the nucleus and the other due to the net spin angular momentum of the electrons.

It is customary to indicate the first of these angular-momentum characteristics of atomic states by using the capital letters S, P, D, F, . . . to represent L values of 0, 1, 2, 3, . . . , respectively, in the total orbital-angular-momentum expression $\sqrt{L(L+1)}(h/2\pi)$. The spin angular momentum $\sqrt{S(S+1)}(h/2\pi)$, where S is the net electron-spin angular-momentum quantum number, is usually not reported

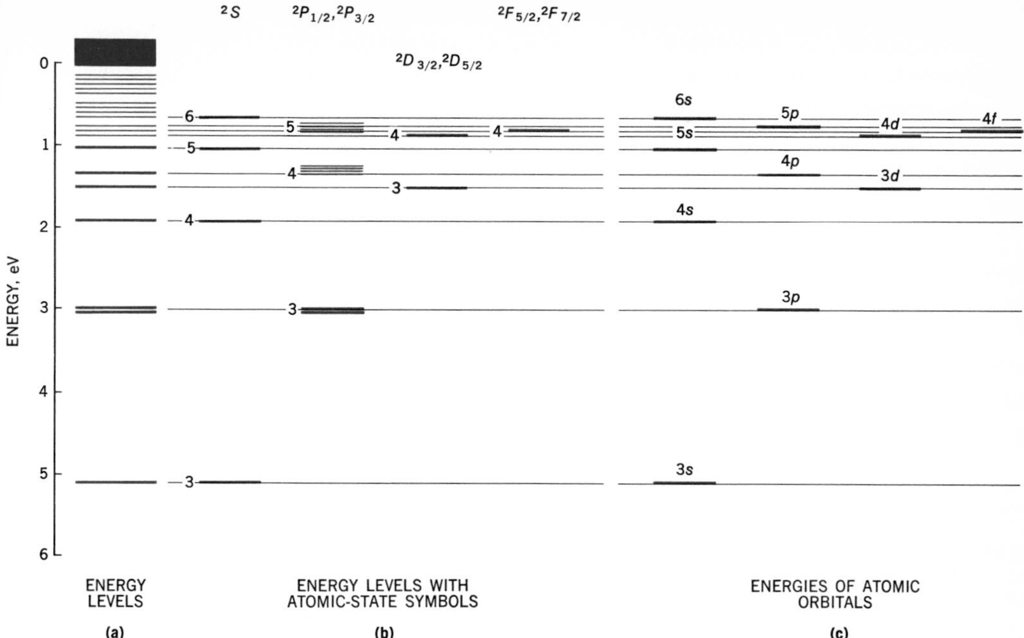

FIGURE 11-7
The energy-level diagram for the sodium atom deduced from the observed atomic spectral lines. All the spectral lines can be accounted for by transitions between the energy levels of (*a*). When the angular-momentum characteristics of the state of the atom are recognized, the character of the transitions allow the angular-momentum symbols in (*b*) to be added. If the states of the sodium atom are interpreted in terms of the single outer electron, the orbital energy pattern for that electron, shown in (*c*), is obtained.

explicitly, but rather is indicated through two of its consequences. First, the *total* angular momentum, given the quantum-number symbol J, which is the quantized vector sum of the orbital and spin angular momenta, is reported as a subscript. Thus Fig. 11-8 illustrates the example of $P_{1/2}$ and $P_{3/2}$ states, which imply $L = 1$ and $S = \frac{1}{2}$ and $J = \frac{1}{2}$ and $\frac{3}{2}$. Second, the states, such as $P_{1/2}$ and $P_{3/2}$, which are the same except for the relative orientation of the orbital- and spin-angular-momentum vectors, are often little different in energy and appear close

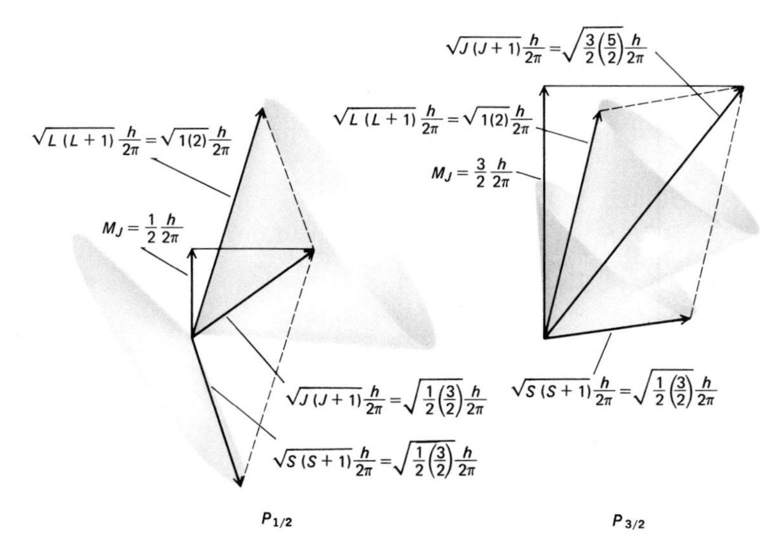

FIGURE 11-8
The vector diagrams that show the two states $P_{1/2}$ and $P_{3/2}$ that can be formed by combining the orbital and spin-angular-momentum vectors of a single p electron so that the angular momenta subtract or add.

to one another on an energy diagram. In this example a "doublet" is formed, and this is indicated as a superscript on the symbol, and we write $^2P_{1/2}$ and $^2P_{3/2}$ for the two states. The superscript, as this example illustrates, is in fact the value of $2S + 1$.

It is such symbols as are illustrated in Fig. 11-7*b* that show the net orbital angular momenta and the net electron-spin angular momenta that summarize the characterization of the atomic states that can be placed on an energy diagram as a result of an analysis of atomic spectra. Most important here will be the description of the ground states of the atoms.

Although the deduction of the angular-momentum features of the states of atoms from the very complex atomic spectra that are produced by most elements is quite difficult, characterization of at least the ground state and determination of the energy difference between this state and the ionized atom have been achieved for pretty well all the elements.

This *ionization energy,* an atomic quantity of great importance, is the energy that is absorbed in reactions such as

$$\text{Na}(g) \rightarrow \text{Na}^+(g) + e^-(g)$$

This energy is usually reported in units of electron volts ($1 \text{ eV} = 1.6021 \times 10^{-19}$ J per molecule $= 96.49 \text{ kJ mol}^{-1}$). Values for this energy, and the symbols describing the ground electronic state, are given for some of the elements in Table 11-3.

A second chemically important atomic quantity is the *electron affinity*. This is the energy, also usually expressed in electron volts,

TABLE 11-3 Ionization energies of some atoms,* in eV

Element	Ionization energy		Element	Ionization energy	
	First	Second		First	Second
H $^2S_{1/2}$	13.595		Mg $^1S_0$	7.644	15.03
He $^1S_0$	24.580	54.50	Al $^2P_{1/2}$	5.984	18.82
Li $^2S_{1/2}$	5.390	75.62	Si $^3P_0$	8.149	16.34
Be $^1S_0$	9.320	18.21	P $^4S_{3/2}$	11.0	19.65
B $^2P_{1/2}$	8.296	25.15	S $^3P_2$	10.357	23.4
C $^3P_0$	11.264	24.38	Cl $^2P_{3/2}$	13.01	23.80
N $^4S_{3/2}$	14.54	29.60	Ar $^1S_0$	15.755	27.62
O $^3P_2$	13.614	35.15	K $^2S_{1/2}$	4.339	31.81
F $^2P_{3/2}$	17.42	34.98	Ca $^1S_0$	6.111	11.87
Ne $^1S_0$	21.599	41.07	Sc $^2D_{3/2}$	6.56	12.80
Na $^2S_{1/2}$	5.138	47.29	Ti $^3F_2$	6.85	13.63

*Mostly from C. E. Moore, Atomic Energy Levels, *Natl. Bur. Std. (U.S.) Circ.* 467, 1949, 1952, and 1958.

with which a neutral atom binds another electron. It is therefore the energy *given out* in reactions such as

$$F(g) + e^-(g) \rightarrow F^-(g)$$

or to make the quantity more comparable with the ionization potential, it is the energy *absorbed* in reactions such as

$$F^-(g) \rightarrow F(g) + e^-(g)$$

There are only a few atoms, in the gas phase, that bind an extra electron tightly enough so that the electron affinity has an appreciable, positive value. It is, moreover, a matter of considerable difficulty to deduce values for the electron affinity. Mass-spectrometer studies of the equilibrium between atoms and their negatively charged ions have, however, yielded some values, as shown in Table 11-4.

TABLE 11-4
Some atomic electron affinities*
(Values of the energy, in electron volts, released in the process atom + electron → negative ion)

Element	*Electron affinity*
H	0.747
F	3.45
Cl	3.61
Br	3.36
I	3.06
O	1.47
S	2.07
C	1.25

*For some other elements rather unreliable values are available. Most such values are small, less than 1 eV, and some are negative.
SOURCE: Compilation by H. B. Gray, "Electrons and Chemical Bonding," W. A. Benjamin, Inc., New York, 1964.

11-4 The Basis for Descriptions of Many-electron Atoms: Central Forces, Electron Orbitals, and the Pauli Exclusion Principle

On what basis can the observed atomic states, ionization energies, and electron affinities of many-electron atoms be understood?

Only for the simplest extension from the hydrogen atom, i.e., to the helium atom, has it been feasible to treat the interactions of the nucleus and all the electrons exactly. But it has been found very useful to describe the electrons of a many-electron atom one at a time, each electron being subject to the effect of the nucleus and the other electrons *averaged* over the positions they occupy. This allows us to use one-electron descriptions to build up a total description of the atom. Such one-electron descriptions are known as *orbitals,* and although the description of an electron orbital in many-electron atoms depends on an averaging of the effect of all other electrons, the concept makes manageable the otherwise intractable problems raised by atoms and molecules.

Since each electron of a many-electron atom is treated as being subject to central forces, as is the electron of the hydrogen atom, the angular features of atomic orbitals, i.e., the angular factors in the electron distribution and their specification by values of l and m, and the angular-momentum implications of these quantum numbers, are exactly as in the one-electron atom. The central forces themselves, however, are affected by the other electrons, and thus the energy and the extension of the electron from the nucleus are no longer the same as in the hydrogen-atom case.

The form of the hydrogen-atom energy expression of Eq. [5] can be maintained for the orbitals of a many-electron atom if either the principal quantum number or the effective charge is given an "effective" value for each orbital. If the effective-charge route is taken, the orbital energies are expressed as

$$\epsilon = -R\frac{(Z_{\text{eff}})^2}{n^2} \qquad [8]$$

Then the spectral data such as those for sodium given in Fig. 11-7c can be used to deduce the nuclear charge that the valence electron would see as it looked back to the nucleus and the $n = 1$ and $n = 2$ closed shells from its position in various outer orbitals. Calculated values for Z_{eff} for this example are given in Table 11-5.

Now one sees, first, that for a given l value the orbitals with the lower values of n have the greater effective nuclear charge. Only for large n values is the unit charge value of the nucleus-and-closed-shell assembly approached. The interpretation is that the smaller orbitals, those with lower values of n, are to some extent successful in penetrating the inner shells, and thus are affected by more than the net $+1$ charge of the nucleus and inner shells. These shells are said to *screen* the nucleus, and this screening is not completely effective in the case of the smaller orbitals.

Second, one notices that, for a given value of n, the orbitals with higher l values are more effectively screened from the nucleus. We thus say that the lower the l value, the more *penetrating* is the orbital and, as a result, the lower is its energy. This effect makes us recognize the orbital-energy dependence on l as well as on n. The general dependence is that s orbitals penetrate more and have lower energy than p orbitals, p penetrate more and have lower energy than d, and so forth. The dependence of orbital energies on the quantum numbers n and l that can be deduced from spectroscopic data and quantum-mechanical calculations is illustrated in Fig. 11-9.

TABLE 11-5 The effective nuclear charges Z_{eff} that would be assigned for the energies of sodium shown in Fig. 11-7 when these energies are interpreted in terms of the hydrogen-atom energy expression $\epsilon = -R\dfrac{(Z_{\text{eff}})^2}{n^2}$

n ↑	0 (s orbitals)	1 (p orbitals)	2 (d orbitals)	3 (f orbitals)
8	1.20	1.12	1.002	1.000
7	1.24	1.14	1.002	1.000
6	1.29	1.17	1.002	1.000
5	1.37	1.21	1.003	1.000
4	1.51	1.28	1.003	1.000
3	1.84	1.42	1.003	

—l⟶

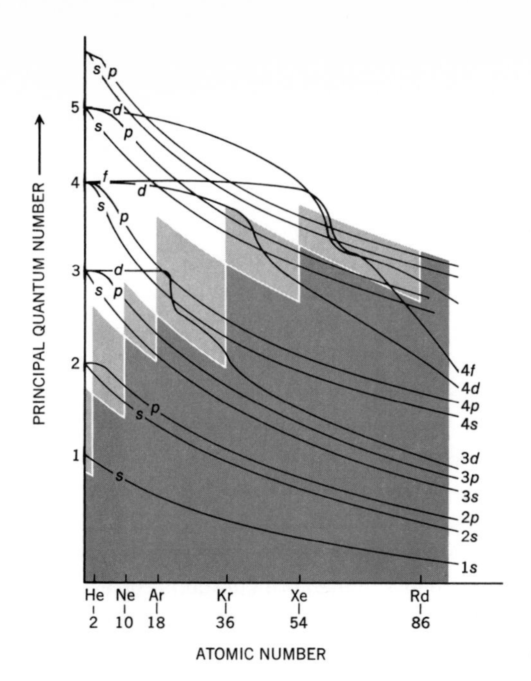

FIGURE 11-9
Orbital energies for various amounts of filling of inner orbitals. (Dark color areas indicate filled orbitals; light color indicate partially filled orbitals.)

To proceed with the electron orbital approach to many-electron atoms we must have a guide to the assignment of electrons to these orbitals. The Boltzmann distribution can be called on, and because the energy differences for electrons in different orbits are often large, it is the arrangement in which electrons occupy the lowest-energy available orbitals, i.e., the *ground state of the atom,* that is often of principal concern.

This ground state is not, however, that in which all electrons occupy the $1s$ orbital. A restriction that prevents this, and leads to atomic structures that explain the periodic-table arrangement of the elements, was stated by W. Pauli and is known as the *Pauli exclusion principle.* In the present application it may be stated that, in a single atom, no orbital may accommodate more than two electrons, or only two electrons can be assigned the same values for n, l, and m, or, finally, only one electron can have a given set of values of n, l, m, and m_s. We have already mentioned a consequence of this principle in Sec. 10-1, where particles such as electrons were said to obey Fermi-Dirac statistics and to be restricted to one per quantum state, or two, if the electron spin is not used in characterizing the states.

On the basis of this exclusion principle, two electrons can be assigned to an s orbital, each can be described by the same value of n, by the l value of zero, by the m value of zero, and by m_s values of $+\frac{1}{2}$ and $-\frac{1}{2}$. Six electrons can be assigned to a set of p orbitals since, with $l = 1$, two can have $m = +1$ and opposite spins, two can have

$m = 0$ and opposite spins, and two can have $m = -1$ and opposite spins. In a similar way the number of electrons in a state with given n and l values can be seen to be $2(2l + 1)$.

With this exclusion principle and the use of electron orbitals, the electronic states of many-electron atoms can be described.

11-5 Electron Configurations and the Periodic Table

Now let us use the approach of Sec. 11-4 and the experimental results of Sec. 11-3, as summarized in Fig. 11-9, to describe the atoms of the elements. The procedure is so familiar that its basis and limitations are easily overlooked. Each of the hydrogen-atom orbitals $1s$, $2s$, $2p_x$, $2p_y$, $2p_z$, and so forth, can accommodate two electrons, and thus s orbitals can accommodate 2, p orbitals 6, d orbitals 10, and so forth. Orbital descriptions of the electrons of atoms can be indicated by the *electron configurations* of the elements shown, for the ground states, in Table 11-6. For the lighter elements, these configurations correspond to the successive filling of orbitals with higher n values and, within each n shell, the successive filling of the s, then the p, orbitals.

Farther along in the list of elements the inner-shell effects become greater and, in addition, the energy spacing between highest occupied states with different values of n becomes less. The result is an irregularity in the electron configurations that first shows up in the placement of an electron in the $4s$ rather than $3d$ orbital in the ground state of potassium, atomic number 19. Only for somewhat higher nuclear charges, starting at scandium, atomic number 21, does the ground-state electron configuration contain a $3d$ electron. Such electron configurations are based, you should recall, on the results of spectroscopic studies of the atom, for such studies allow the deduction of the angular momentum of the ground state. On this basis the electronic configuration of this state can be deduced.

The basis for such observed results can be seen from Fig. 11-11, which shows that, at potassium and calcium, atomic numbers 19 and 20, the $4s$ and $4p$ orbitals are of lower energy than the $3d$, but when the end of the first transition-metal series is reached, say at zinc, atomic number 30, the reverse is true. Similar reversals occur with increasing nuclear charge in later series.

The dependence of orbital energies on the quantum number l, as well as n, as shown in Fig. 11-9, also indicates the basis for the important concept of closed shells of electrons and the observed relative inertness of the electron configurations adopted by the noble gases. As Fig. 11-9 shows, so-called closed shells occur at electron configurations in which an l shell is complete, and moreover, there is an appreciable energy gap between this configuration and that with next-higher energy.

11-6 Size and Energy of Atomic Orbitals and Atoms: Approximate Methods, Self-consistent Field Results

The relative energies of atomic orbitals and electron configurations are of importance in the considerations of Sec. 11-5. Also important are the size and energy of each electron orbital in a given electron configuration of an atom and the net size and energy of the entire atom.

These quantities can be estimated by a calculation procedure that depends on the use of the approximate electron orbitals.

First, recall that in the simple particle-in-a-box problem, the Schrödinger equation was set up with the appropriate potential-energy term, and solutions to the resulting differential equation were sought. It was not difficult to solve the equation, i.e., to see that a trigonometric function would be a solution. In many-electron atom problems the potential-energy term can be formed from the appropriate coulombic potential-energy terms, and we can again arrive at a differential equation. This equation, however, has no readily apparent solution. It is, in fact, very unlikely that any reasonably simple expression exists that solves such differential equations. To proceed it is necessary to have some way of finding an approximate solution to the equation. In almost all molecular applications of quantum mechanics a similar situation arises, and the technique of using approximate descriptions for chemical systems has become a basic part of modern chemical theory. Most often these approximate solutions or descriptions are based on the hydrogen-atom wave functions, which were given in Table 11-2.

The Schrödinger equation can be written in a compact form that allows approximate solutions to be more easily used. For one electron in a three-dimensional system the Schrödinger equation in cartesian coordinates is

$$-\frac{h^2}{8\pi^2 m}\left(\frac{\partial^2\psi}{\partial x^2}+\frac{\partial^2\psi}{\partial y^2}+\frac{\partial^2\psi}{\partial z^2}\right)+U(x,y,z)\psi=\epsilon\psi \qquad [9]$$

It is now necessary to put this equation in a form so that a value of the energy ϵ can be obtained for a given potential function $U(x,y,z)$ when the equation is such that an exact solution function cannot be found.

The set of differentials is conveniently abbreviated as

$$\nabla^2=\frac{\partial^2}{\partial x^2}+\frac{\partial^2}{\partial y^2}+\frac{\partial^2}{\partial z^2} \qquad [10]$$

and this abbreviation allows the Schrödinger equation to be written

$$-\frac{h^2}{8\pi^2 m}\nabla^2\psi+U\psi=\epsilon\psi \qquad [11]$$

where U is now understood to be a function of the three coordinates. The symbol ∇^2 is known as an *operator* and is said to operate on ψ.

TABLE 11-6 Electronic configurations of the elements as free, gaseous atoms*

Z	Element	1	2		3			4				5				6				7
		s	s	p	s	p	d	s	p	d	f	s	p	d	f	s	p	d	f	s
1	H	1																		
2	He	2																		
3	Li	2	1																	
4	Be	2	2																	
5	B	2	2	1																
6	C	2	2	2																
7	N	2	2	3																
8	O	2	2	4																
9	F	2	2	5																
10	Ne	2	2	6																
11	Na	2	2	6	1															
12	Mg	2	2	6	2															
13	Al	2	2	6	2	1														
14	Si	2	2	6	2	2														
15	P	2	2	6	2	3														
16	S	2	2	6	2	4														
17	Cl	2	2	6	2	5														
18	Ar	2	2	6	2	6														
19	K	2	2	6	2	6		1												
20	Ca	2	2	6	2	6		2												
21	Sc	2	2	6	2	6	1	2												
22	Ti	2	2	6	2	6	2	2												
23	V	2	2	6	2	6	3	2												
24	Cr	2	2	6	2	6	5	1												
25	Mn	2	2	6	2	6	5	2												
26	Fe	2	2	6	2	6	6	2												
27	Co	2	2	6	2	6	7	2												
28	Ni	2	2	6	2	6	8	2												
29	Cu	2	2	6	2	6	10	1												
30	Zn	2	2	6	2	6	10	2												
31	Ga	2	2	6	2	6	10	2	1											
32	Ge	2	2	6	2	6	10	2	2											
33	As	2	2	6	2	6	10	2	3											
34	Se	2	2	6	2	6	10	2	4											
35	Br	2	2	6	2	6	10	2	5											
36	Kr	2	2	6	2	6	10	2	6											
37	Rb	2	2	6	2	6	10	2	6			1								
38	Sr	2	2	6	2	6	10	2	6			2								
39	Y	2	2	6	2	6	10	2	6	1		2								
40	Zr	2	2	6	2	6	10	2	6	2		2								
41	Nb	2	2	6	2	6	10	2	6	4		1								
42	Mo	2	2	6	2	6	10	2	6	5		1								
43	Tc	2	2	6	2	6	10	2	6	6		1	...	?						
44	Ru	2	2	6	2	6	10	2	6	7		1								
45	Rh	2	2	6	2	6	10	2	6	8		1								
46	Pd	2	2	6	2	6	10	2	6	10										
47	Ag	2	2	6	2	6	10	2	6	10		1								
48	Cd	2	2	6	2	6	10	2	6	10		2								
49	In	2	2	6	2	6	10	2	6	10		2	1							
50	Sn	2	2	6	2	6	10	2	6	10		2	2							

*From M. J. Sienko and R. A. Plane, "Chemistry," McGraw-Hill Book Company, New York, 1957.

TABLE 11-6 *(continued)*

Z	Element	1	2		3			4				5				6				7
		s	s	p	s	p	d	s	p	d	f	s	p	d	f	s	p	d	f	s
51	Sb	2	2	6	2	6	10	2	6	10		2	3							
52	Te	2	2	6	2	6	10	2	6	10		2	4							
53	I	2	2	6	2	6	10	2	6	10		2	5							
54	Xe	2	2	6	2	6	10	2	6	10		2	6							
55	Cs	2	2	6	2	6	10	2	6	10		2	6			1				
56	Ba	2	2	6	2	6	10	2	6	10		2	6			2				
57	La	2	2	6	2	6	10	2	6	10		2	6	1		2				
58	Ce	2	2	6	2	6	10	2	6	10	2	2	6			2 . . . ?				
59	Pr	2	2	6	2	6	10	2	6	10	3	2	6			2 . . . ?				
60	Nd	2	2	6	2	6	10	2	6	10	4	2	6			2				
61	Pm	2	2	6	2	6	10	2	6	10	5	2	6			2 . . . ?				
62	Sm	2	2	6	2	6	10	2	6	10	6	2	6			2				
63	Eu	2	2	6	2	6	10	2	6	10	7	2	6			2				
64	Gd	2	2	6	2	6	10	2	6	10	7	2	6	1		2				
65	Tb	2	2	6	2	6	10	2	6	10	9	2	6			2 . . . ?				
66	Dy	2	2	6	2	6	10	2	6	10	10	2	6			2 . . . ?				
67	Ho	2	2	6	2	6	10	2	6	10	11	2	6			2 . . . ?				
68	Er	2	2	6	2	6	10	2	6	10	12	2	6			2 . . . ?				
69	Tm	2	2	6	2	6	10	2	6	10	13	2	6			2				
70	Yb	2	2	6	2	6	10	2	6	10	14	2	6			2				
71	Lu	2	2	6	2	6	10	2	6	10	14	2	6	1		2				
72	Hf	2	2	6	2	6	10	2	6	10	14	2	6	2		2				
73	Ta	2	2	6	2	6	10	2	6	10	14	2	6	3		2				
74	W	2	2	6	2	6	10	2	6	10	14	2	6	4		2				
75	Re	2	2	6	2	6	10	2	6	10	14	2	6	5		2				
76	Os	2	2	6	2	6	10	2	6	10	14	2	6	6		2				
77	Ir	2	2	6	2	6	10	2	6	10	14	2	6	7		2				
78	Pt	2	2	6	2	6	10	2	6	10	14	2	6	9		1				
79	Au	2	2	6	2	6	10	2	6	10	14	2	6	10		1				
80	Hg	2	2	6	2	6	10	2	6	10	14	2	6	10		2				
81	Tl	2	2	6	2	6	10	2	6	10	14	2	6	10		2	1			
82	Pb	2	2	6	2	6	10	2	6	10	14	2	6	10		2	2			
83	Bi	2	2	6	2	6	10	2	6	10	14	2	6	10		2	3			
84	Po	2	2	6	2	6	10	2	6	10	14	2	6	10		2	4 . . . ?			
85	At	2	2	6	2	6	10	2	6	10	14	2	6	10		2	5 . . . ?			
86	Rn	2	2	6	2	6	10	2	6	10	14	2	6	10		2	6			
87	Fr	2	2	6	2	6	10	2	6	10	14	2	6	10		2	6			1 . . . ?
88	Ra	2	2	6	2	6	10	2	6	10	14	2	6	10		2	6			2
89	Ac	2	2	6	2	6	10	2	6	10	14	2	6	10		2	6	1		2 . . . ?
90	Th	2	2	6	2	6	10	2	6	10	14	2	6	10		2	6	2		2
91	Pa	2	2	6	2	6	10	2	6	10	14	2	6	10	2	2	6	1		2 . . . ?
92	U	2	2	6	2	6	10	2	6	10	14	2	6	10	3	2	6	1		2
93	Np	2	2	6	2	6	10	2	6	10	14	2	6	10	4	2	6	1		2 . . . ?
94	Pu	2	2	6	2	6	10	2	6	10	14	2	6	10	5	2	6	1		2 . . . ?
95	Am	2	2	6	2	6	10	2	6	10	14	2	6	10	7	2	6			2 . . . ?
96	Cm	2	2	6	2	6	10	2	6	10	14	2	6	10	7	2	6	1		2 . . . ?
97	Bk	2	2	6	2	6	10	2	6	10	14	2	6	10	8	2	6	1		2 . . . ?
98	Cf	2	2	6	2	6	10	2	6	10	14	2	6	10	9	2	6	1		2 . . . ?
99	E	2	2	6	2	6	10	2	6	10	14	2	6	10	10	2	6	1		2 . . . ?
100	Fm	2	2	6	2	6	10	2	6	10	14	2	6	10	11	2	6	1		2 . . . ?
101	Mv	2	2	6	2	6	10	2	6	10	14	2	6	10	12	2	6	1		2 . . . ?

The convenient operator notation can be extended if the symbol $\mathcal{H}$, known as the *Hamiltonian,* is introduced for the terms:

$$\mathcal{H} = -\frac{h^2}{8\pi^2 m}\nabla^2 + U \qquad [12]$$

With this notation the Schrödinger equation is very compactly written

$$\mathcal{H}\psi = \epsilon\psi \qquad [13]$$

(Although $\mathcal{H}$ is here treated simply as a symbol for the two terms of the Schrödinger equation, it can be recognized as the quantum-mechanical *operator* which, when applied to the wave function for a particle, yields the energy of the particle. In fact, the first term of Eq. [12] arises from the kinetic energy of the system and, more obviously, the second term arises from the potential energy of the system. This interpretation of $\mathcal{H}$ is particularly valuable in constructing $\mathcal{H}$ for a system which is a collection of particles.)

Now the compact expression of Eq. [13] can be multiplied on the left by the function ψ to give

$$\psi\mathcal{H}\psi = \psi\epsilon\psi \qquad [14]$$

Since ϵ is nothing more than a number giving the calculated energy of the system, it commutes, and Eq. [14] can be written

$$\psi\mathcal{H}\psi = \epsilon\psi^2 \qquad [15]$$

(Care is being taken here because $\mathcal{H}$, by comparison, does not commute and $\mathcal{H}\psi \neq \psi\mathcal{H}$.)

If a solution were inserted in Eq. [15] for ψ, one would obtain an equation that showed that, for any value of the spatial coordinates, a fixed value of ϵ would satisfy the equation. Averaging such values over all values of the coordinates would then give the value of ϵ that corresponds to the approximate wave function. This result can also be obtained from an integration over all these coordinates to give

$$\int\psi\mathcal{H}\psi\, d\tau = \epsilon\int\psi^2\, d\tau$$

or

$$\epsilon = \frac{\int\psi\mathcal{H}\psi\, d\tau}{\int\psi^2\, d\tau} \qquad [16]$$

It is this form of the Schrödinger equation that can be most conveniently used when the energy corresponding to a function that approximates the solution function is sought.

For a given system the potential-energy function can usually be written down, and therefore so can the Hamiltonian operator $\mathcal{H}$, defined according to Eq. [12]. If a well-behaved expression (such as the appropriate product of the electron orbitals, each based on a hydrogen-atom wave function) that is expected to approximate the true wave function for the system can be constructed, it can be inserted in the right side of Eq. [16]. The necessary mathematical operations of differentiation and integration can be performed, and a value of ϵ will be obtained. In this way an approximate energy for the system can be calculated when a function that approximates the true wave function is specified.

This method of calculating the energy of a system would, however, be unduly hazardous if no way existed for checking the reliability of such energy calculations. An important theorem, known as the *variation theorem,* provides the desired criterion.

The theorem is stated, here without proof, as showing that the value of ϵ obtained from an approximate wave function is less negative, i.e., the system *seems* less stable, than the value obtained from the true wave function. The best value of ϵ that can be obtained from various approximate wave functions is therefore the lowest value yielded by Eq. [16]. The approach, then, is to devise an approximate wave function and to calculate a value of ϵ from it. Variations on this wave function that improve it will produce, when inserted in Eq. [16], a lower and better value of ϵ. The lowest value of ϵ that can be obtained is the best theoretical value for the energy of the system.

Now a procedure that yields spatial and energy information for the electron orbitals can be described.

Consider one of the electrons of a many-electron atom. Its potential energy results from interaction with the nucleus and with the other electrons of the atom. We proceed by expressing the potential-energy term for the latter in terms of an averaged interaction of the one electron with all the other electrons distributed in space according to approximate orbitals assumed for them. With this approximate one-electron $\mathcal{H}$, Eq. [16] can be solved to obtain a best approximate orbital and energy for this electron. Another electron is then singled out and its orbital description is refined. Repetition of this procedure leads ultimately to a *self-consistent* set of descriptions for the electrons of the atom that provides a good approximation to the structures of many-electron atoms. Results that have been obtained by computer calculations are shown in Fig. 11-10.

A number of features of such diagrams should be noticed. Examples are the smallness of the inner shells like the $1s$ in atoms of high atomic number and the similar, but greatly reduced, contraction of orbitals such as the $2s$, as a result of the increased nuclear charge partially screened by other electrons of the same or smaller principal quantum number.

MOLECULAR THEORY

The principal subject of this chapter can now be dealt with. An understanding of atomic structure and of the methods of wave mechanics

FIGURE 11-10
Contour diagrams for the electron orbitals of atoms. For explanatory details see Arnold C. Wahl, Molecular Orbital Densities: Pictorial Studies, *Science*, **151**:961 (1966), or Chemistry by Computer, *Sci. Am.*, **222**:54 (1970). (*Chart by Arnold C. Wahl. Copyright © 1970 by McGraw-Hill, Inc. Used with permission of McGraw-Hill Book Company.*)

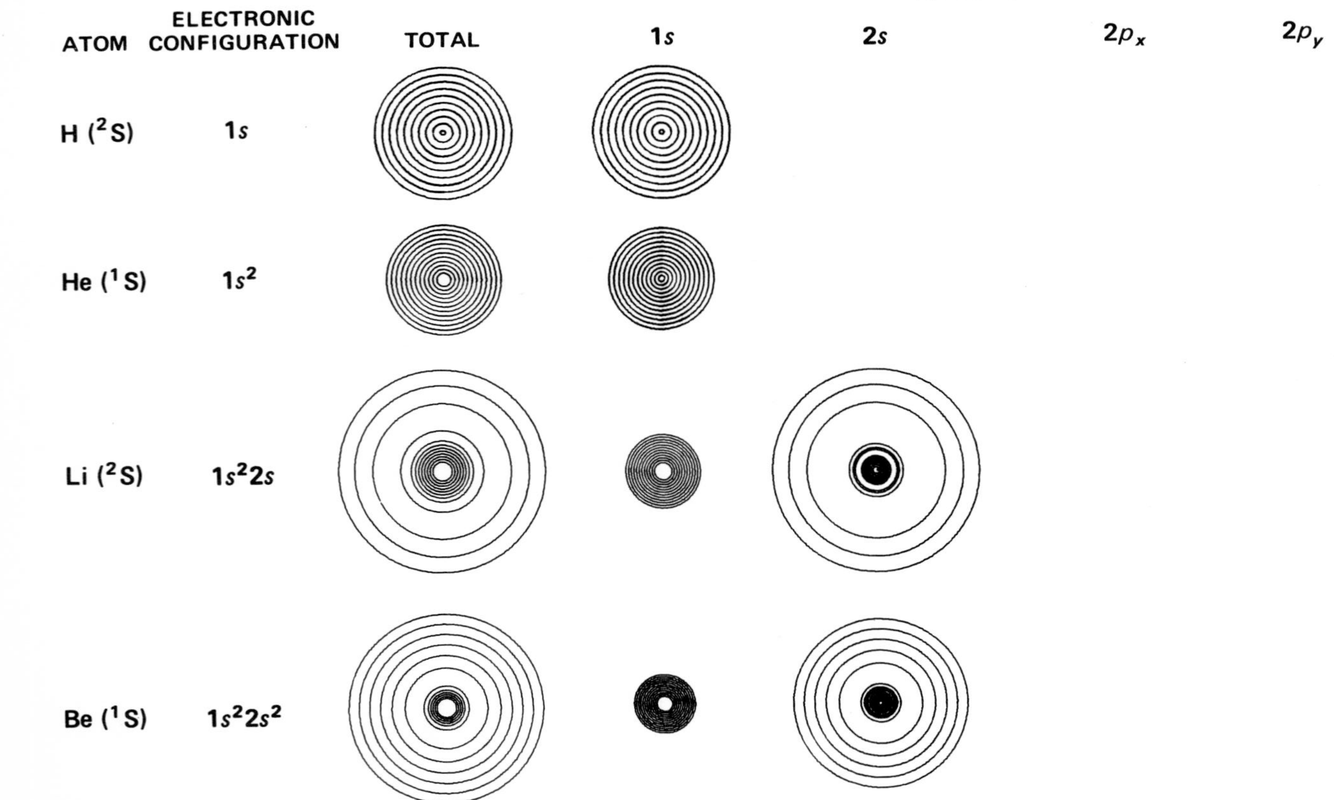

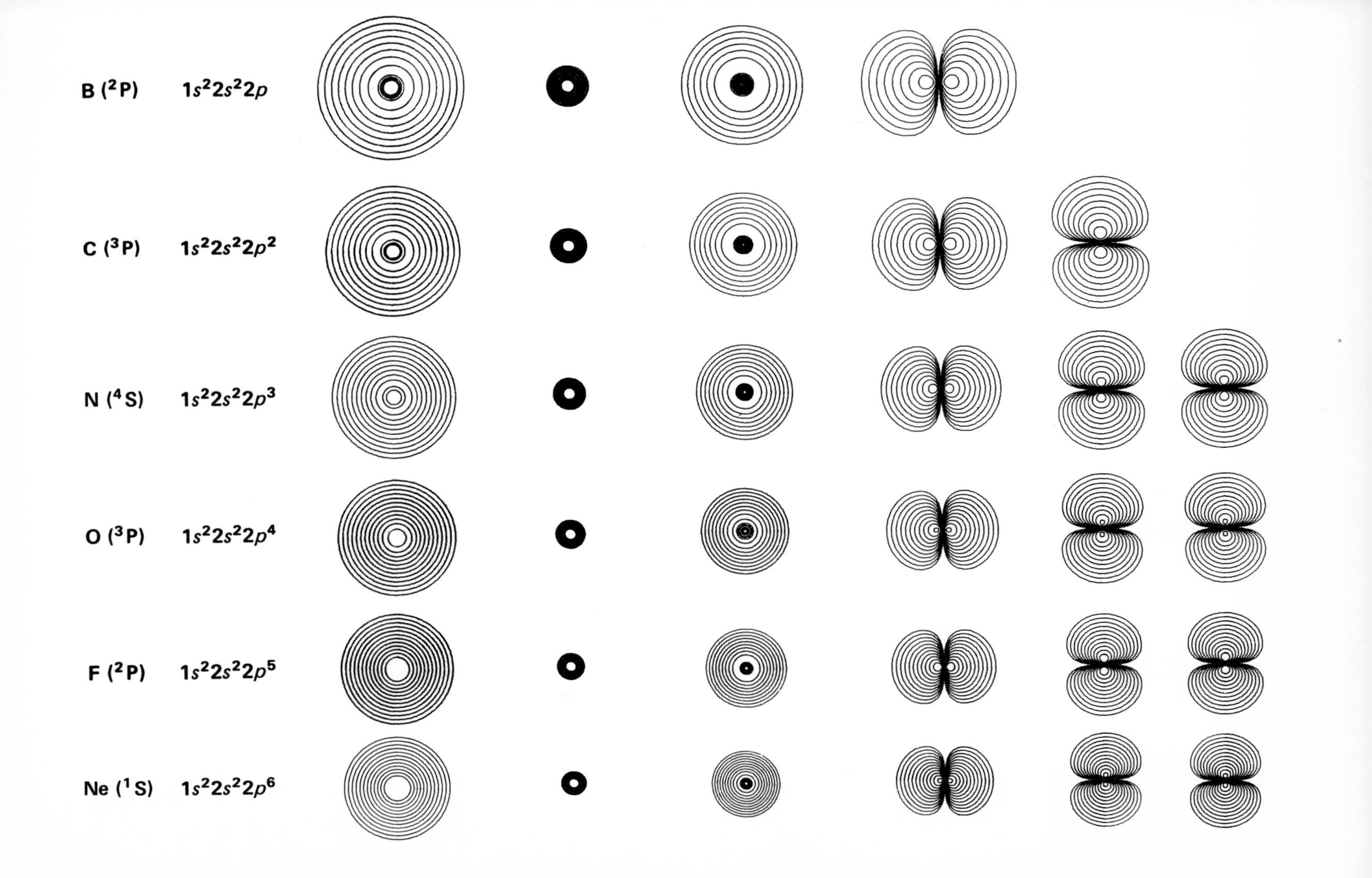

B (²P) 1s²2s²2p
C (³P) 1s²2s²2p²
N (⁴S) 1s²2s²2p³
O (³P) 1s²2s²2p⁴
F (²P) 1s²2s²2p⁵
Ne (¹S) 1s²2s²2p⁶

provides the means for tackling one of the fundamental questions of chemistry: what binds atoms together into molecules? This question has existed since the beginnings of chemistry, and a clear answer would be a culmination of much of the theoretical work of chemistry. The current solution will be illustrated by the simplest covalent molecule, H_2. It will be seen that molecular bonding is at present treated only with considerable awkwardness and that our knowledge of wave mechanics, rather than terminating the study of chemical theory, leads to many new and exciting problems.

In this section of the chapter, the quantum-mechanical approach to covalent bonding will be shown by a consideration, except for the evaluation of some integrals, of the bonding in the hydrogen-molecule ion and the hydrogen molecule. From these very simple systems it will be possible to extend the theory of chemical bonding, in a semiquantitative and semiempirical way, as is done in the following chapter, to molecular structures of more chemical interest. It is this extension which has become a basic and necessary part of the approaches and language in all branches of chemistry.

11-7 The Beginnings of Bonding Theory

The introduction of the modern view of matter as made up of atoms and molecules by Dalton in 1808 led immediately to questions as to the nature of the forces that hold the atoms together when they are combined into a molecule. At that time the contemporary studies of the effect of an electric current on chemical compounds by Davy probably contributed to the dominating theory of Berzelius that chemical union stems from an electrical attraction between particles of opposite charge. This electrovalent theory persisted, in part at least, because of the lack of any alternative theory. The recognition of molecules such as H_2 and N_2 and similar difficulties with many organic compounds led to the use of a schematic representation of chemical bonding and a tendency, particularly in the rapidly developing subject of organic chemistry, to a bypassing of the question of the source of chemical affinity.

It will be recalled that in the latter half of the nineteenth century, although no adequate theory of chemical bonding was available, the synthesis and structural studies of organic compounds rapidly proceeded. The theory and representation of aromatic systems by Kekule and the recognition of the tetrahedral carbon atom by Le Bel and by van't Hoff show the achievements that were made. Not only could compounds of very considerable complexity be synthesized and analyzed, but a representation of the geometric arrangement of all the atoms of the molecule could often be made. The development of this structural theory of organic compounds and the organization of the immense body of facts of organic chemistry must be ranked as one of the greatest accomplishments of science.

A similar development in inorganic chemistry occurred toward the end of the century in the work of Alfred Werner. His introduction of the idea of coordination number brought to inorganic chemistry a system that allowed for the same progress as structural theory had made possible for organic chemistry.

All these developments took place with little understanding of the chemical forces that, for instance, caused the four hydrogen atoms to arrange themselves tetrahedrally about a carbon atom in methane. These chemical developments occurred before what logically would be prerequisite information had been unraveled. The chemists' curiosity about the nature of the chemical world could not wait for the work, around 1900, of J. J. Thomson and Rutherford to provide the logical starting point: an understanding of simple atomic structure.

The investigation at the turn of the century of the basic chemical units of matter, the electron and the nucleus, and their arrangement in atoms led to a renewed interest in the nature and source of the "affinities" exerted by atoms that were so successfully represented in structural theory. Knowledge that atoms consist of a nucleus with some outer arrangement of electrons was quickly coupled with the implications of the periodic table and led to some primitive but still valuable ideas on atomic and molecular structure. At this stage it is convenient to treat separately the two different types of chemical bonding that came to be recognized.

11-8 The Ionic Bond

The detailed interpretation of the electronic structure of the periodic arrangement of the atoms was not available in the early 1900s. The special stability of the inert gases, however, led to the idea that an outer shell of eight electrons was particularly stable. That electrons of atoms might be *transferred* from one atom to another so that they both could achieve inert-gas configurations was suggested at an early stage by J. J. Thomson and was developed by W. Kossel in 1916 into a theory of the ionic bond. Once the tendency for electrons to be transferred was postulated, the original theory of Berzelius could be resurrected to explain the binding of the ions that results.

The ionic bond can be used to introduce the factors that enter into the energy–versus–internuclear-distance curve that is used to display the energetic consequences of bonding between atoms.

The ionic bond is most easily understood with reference to a gas-phase molecule, such as NaCl. At internuclear distances not far from the equilibrium distance it appears to be satisfactory to treat the mole-

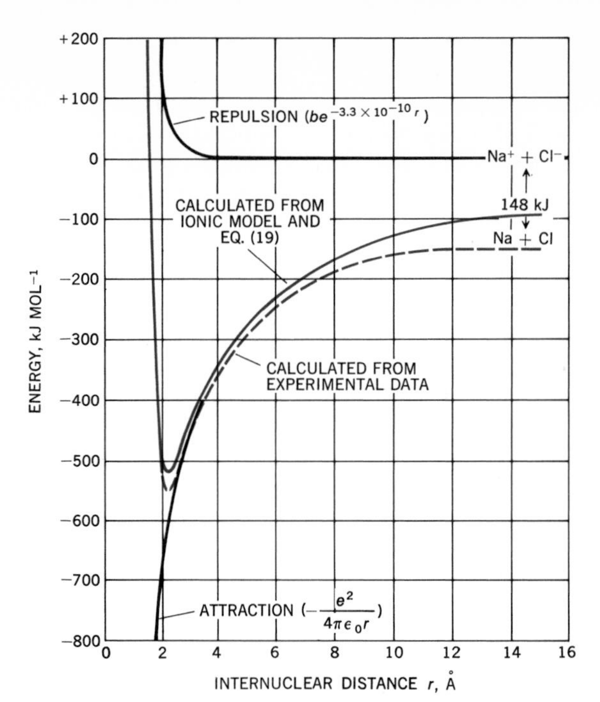

FIGURE 11-11 Calculated and experimental energy curves for the gas-phase NaCl molecule.

cule as an Na^+ ion bound to a Cl^- ion by the coulombic attraction of the opposite charges.

If the charges on the ions are q_1 and q_2 (for NaCl the value of q_1 is $+e$ and that of q_2 is $-e$), the electrostatic attraction leads to the attractive-energy term, shown in Fig. 11-11, $q_1q_2/4\pi\epsilon_0 r$, where r is the internuclear separation. The effect of this energy term is to draw the ions together.

An opposing effect exists in the form of a repulsion between the nuclei and their closed shells of electrons.

This repulsion term cannot easily be deduced, and it is satisfactory here to use an empirical expression to represent the repulsion that sets in at small internuclear distances. The form of the variation of this repulsive-energy contribution with internuclear distance, also shown in Fig. 11-11, has been shown by Born and Mayer to be satisfactorily represented by an empirical equation of the form

$$U_{\text{rep}} = be^{-r/p} \qquad [17]$$

where p and b are empirical constants. Furthermore, to a quite good approximation, the constant p can be taken to be the same for all ionic molecules and is equal to 0.30×10^{-10} m. The total energy function for an ionic bond can then be written

$$U = \frac{q_1q_2}{4\pi\epsilon_0 r} + be^{-(3.3\times10^{10})r} \qquad [18]$$

For molecules such as NaCl, the expression is

$$U = -\frac{e^2}{4\pi\epsilon_0 r} + be^{-(3.3\times10^{10})r} \qquad [19]$$

The value of the remaining empirical constant b can be assigned, for a given molecule, so that the total potential-energy curve constructed from Eq. [19] will have a minimum of an internuclear separation equal to the experimentally determined equilibrium internuclear distance.

The pure ionic or electrostatic bond, as reference to the previous tables of ionization potentials and electron affinities shows, cannot provide a complete description of the bonding in gas-phase molecules such as NaCl. At very long internuclear distances, i.e., for separate free sodium and chlorine particles, the energetically favored species are the atoms Na and Cl, rather than the ions Na^+ and Cl^-. Since, for gas-phase reactions,

$$Na \rightarrow Na^+ + e^- \qquad \Delta H = +496 \text{ kJ mol}^{-1}$$
$$Cl + e^- \rightarrow Cl^- \qquad \Delta H = -348 \text{ kJ mol}^{-1}$$

and therefore

$$Cl + Na \rightarrow Na^+ + Cl^- \qquad \Delta H = +148 \text{ kJ mol}^{-1} \qquad [20]$$

An experimental potential-energy curve for the NaCl molecule can be deduced by methods to be discussed in Chap. 13. This curve is shown in Fig. 11-11 for comparison with that deduced on the basis of a completely ionic system. This comparison shows that, although the ionic description is adequate at short distances, it must go over into some atomic description at very long distances if it is to represent the energy of the fragments of dissociation.

The nature of the forces that hold two atoms like Na and Cl at their equilibrium separation is, however, essentially understandable in terms of an ionic bond.

It is now necessary to investigate the source of attraction that is responsible for the stability of molecules such as H_2, where this electron transfer and an ionic description are obviously not appropriate.

11-9 Introduction to the Covalent Bond

Even with the initial ideas on atomic structure and the inert-electron configurations, it is not immediately clear how atoms could be held together in molecules such as H_2 and N_2 and most organic molecules. The Bohr theory that had such a remarkable success in explaining the behavior of the hydrogen atom seemed of no help in explaining the hydrogen molecule. Simultaneously with the more physical developments that led to the Bohr theory, G. N. Lewis, at the University of California, was developing a theory of the covalent bond from a more chemical point of view.

Lewis was led to explain covalent bonds as resulting from the *sharing of pairs of electrons* in such a way that the participating atoms all achieved the stable noble-gas configuration. It is interesting to look back at the now-primitive-looking model of the atom on which his theory was based. The special stability of an outer shell of eight electrons suggested to Lewis that the electrons might occupy corners of a cube about the nucleus. Although this model is no longer accepted, the ideas of Lewis and the diagrams he developed to represent the arrangement of the outer shell of electrons are still invaluable in presenting chemical theory.

Lewis diagrams illustrate the role of the electrons of the outer shell of an atom. Both the inner shells and the nucleus are represented by the symbol of the element. The sharing of the outer electrons to form chemical bonds can then easily be depicted by writing a pair of electrons between the bonded atoms. This pairing, according to Lewis, should lead to a noble-gas configuration about each atom. Thus each hydrogen atom should have a share in two electrons, and most other atoms a share in eight electrons. In some cases the sharing of two or three pairs of electrons in a bond is necessary to give the desired configurations. A few examples will illustrate these useful diagrams; for covalent compounds

$$\cdot\overset{\cdot}{\underset{\cdot}{C}}\cdot + 4H\cdot \rightarrow H:\overset{H}{\underset{H}{C}}:H$$

$$:\overset{\cdot\cdot}{\underset{\cdot}{O}}\cdot + 2H\cdot \rightarrow :\overset{\cdot\cdot}{\underset{\underset{H}{:}}{O}}:H$$

$$\cdot\overset{\cdot}{\underset{\cdot}{C}}\cdot + 4:\overset{\cdot\cdot}{\underset{\cdot\cdot}{Cl}}\cdot \rightarrow :\overset{\cdot\cdot}{\underset{\cdot\cdot}{Cl}}:\overset{:\overset{\cdot\cdot}{Cl}:}{\underset{:\underset{\cdot\cdot}{Cl}:}{C}}:\overset{\cdot\cdot}{\underset{\cdot\cdot}{Cl}}:$$

$$2\cdot\overset{\cdot}{\underset{\cdot}{C}}\cdot + 2H\cdot \rightarrow H:C:::C:H$$

and for ionic compounds, where electron transfer leads to the desired inert-gas configurations,

$$Na\cdot + \cdot\overset{\cdot\cdot}{\underset{\cdot\cdot}{Cl}}: \rightarrow [Na]^{+} + [:\overset{\cdot\cdot}{\underset{\cdot\cdot}{Cl}}:]^{-}$$

Lewis diagrams and the bonding arrangements represented by these diagrams represent a considerable advance over the earlier representation of molecular formulas, in which a line was drawn to represent the bond between bonded atoms. The Lewis theory, for example, allows the number of covalent bonds, or shared pairs of electrons, associated with an atom to be understood on the basis of its atomic number or its position in the periodic table.

Lewis diagrams are still used as the most convenient, and frequently but not always as an adequate, representation of the electron configuration in a molecule. The sharing of electrons in covalent bonds as proposed by Lewis, however, is only a step in the direction of explaining the covalent bond. After Lewis' theory, published in 1916, it remained to show how the electrons were arranged in space about the atom or molecule, and particularly to do this in a way consistent with Bohr's theory of the hydrogen atom. Also left unanswered was the original question of why covalent bonds form, or, following Lewis, why they form and why the sharing of pairs of electrons is involved in most molecules.

Many attempts by Lewis and others to answer such fundamental questions were made in the years following 1916. With the appearance of the Schrödinger wave equation in 1926 and its immediate success in dealing with atomic problems, all attempts to understand more clearly the covalent bond were turned to this new approach.

The complexity which develops when the Schrödinger equation is applied to molecular systems means that only simple systems can be adequately treated. An understanding of the important example of the H_2 molecule and the even simpler H_2^+ ion, however, allows the chemist to extend quantum-mechanical approaches in a qualitative manner to many chemical systems. It is therefore very worthwhile to appreciate the quantum-mechanical theory of the covalent bond as illustrated by these simple species.

11-10 The Hydrogen-molecule Ion

The simplest, though certainly not the most familiar, example of a covalent chemical bond is provided by the H_2^+ ion. Some properties of this ion, consisting of two protons held together by a single electron, can be deduced from spectroscopic observations of highly excited H_2 molecules. The equilibrium bond length of this molecule ion and the variation of its potential energy with internuclear distance derived from such spectroscopic measurements and from refined theoretical calculations are illustrated by the dashed curve in the potential-energy diagram of Fig. 11-12. Now we must see if we can reach some understanding of the source of the binding energy of this system. A number of approaches are available, and the one to be used here is chosen because it leads most naturally into the methods used for other molecules.

We can begin by making use of our knowledge that for an electron bound to a proton the lowest-energy wave function is designated as a $1s$ orbital. If A designates one of the two nuclei with which we now deal, this function, as shown in Table 11-2, will have the form

$$1s_A = \frac{1}{\sqrt{\pi}}\left(\frac{1}{a_0}\right)^{3/2} e^{-r_A/a_0} \qquad [21]$$

The corresponding wave function centered on nucleus B will be

$$1s_B = \frac{1}{\sqrt{\pi}}\left(\frac{1}{a_0}\right)^{3/2} e^{-r_B/a_0} \qquad [22]$$

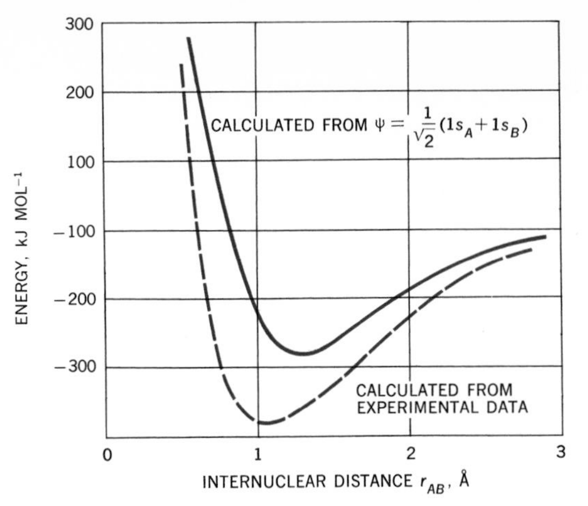

FIGURE 11-12
Calculated and experimental potential-energy curves for the hydrogen-molecule ion.

A suitable trial function for the behavior of an electron associated with two nuclei, as represented in Fig. 11-13, can be based on these two mathematical functions. We can combine the descriptions for the electron in the two separate nuclei to give an expression that may be a satisfactory approximation for the system involving both nuclei and the one electron. A linear combination, as will be checked, turns out to give an expression of the right form. We can write the trial function

$$\psi = \frac{1}{\sqrt{2}}(1s_A + 1s_B) \qquad [23]$$

where the factor $1/\sqrt{2}$ normalizes the function to the extent that, for large separations between nucleus A and nucleus B, the total probability of the system containing one electron is unity. (One sees this from

$$\int\left[\frac{1}{\sqrt{2}}(1s_A + 1s_B)\right]^2 d\tau = \tfrac{1}{2}\int(1s_A{}^2 + 1s_B{}^2 + 2s_A 1s_B)\, d\tau$$

Since in the limit of infinite separation between A and B the functions $1s_A$ and $1s_B$ will not have appreciable values in the same region, the term $1s_A 1s_B$ will vanish. Furthermore, $1s_A$ and $1s_B$ are normalized $1s$ wave functions and the integrals of each of the first two terms over all space give unity for each term. It follows that as $r_{AB} \rightarrow \infty$, the wave function of Eq. [23] is normalized.)

To proceed to the use of Eq. [16] to calculate an energy corresponding to the trial function, we must first write down the Hamiltonian

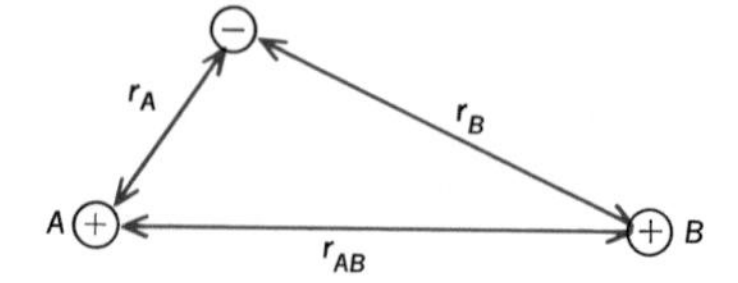

FIGURE 11-13
Coordinates of the two electrons of the hydrogen-molecule ion.

$\mathcal{H}$. In view of the potential-energy interactions, which can be recognized in Fig. 11-13, and leaving the internuclear repulsion to be included later, $\mathcal{H}$ can be written

$$\mathcal{H} = -\frac{h^2}{8\pi^2 m}\nabla^2 - \frac{e^2}{4\pi\epsilon_0 r_A} - \frac{e^2}{4\pi\epsilon_0 r_B} \qquad [24]$$

Now the variation-method energy can be calculated from

$$\epsilon = \frac{\int \psi \mathcal{H} \psi \, d\tau}{\int \psi^2 \, d\tau}$$

$$= \frac{\frac{1}{2}\int (1s_A + 1s_B)\left(-\frac{h^2}{8\pi^2 m}\nabla^2 - \frac{e^2}{4\pi\epsilon_0 r_A} - \frac{e^2}{4\pi\epsilon_0 r_B}\right)(1s_A + 1s_B)\, d\tau}{\frac{1}{2}\int (1s_A + 1s_B)^2 \, d\tau}$$

Making use of the equivalence of $1s_A$ and $1s_B$ to pair up and reduce the total number of terms gives

$$\epsilon = \frac{\begin{array}{l}\int 1s_A\left(-\frac{h^2}{8\pi^2 m}\nabla^2\right)1s_A\,d\tau + \int 1s_A\left(-\frac{e^2}{4\pi\epsilon_0 r_A}\right)1s_A\,d\tau + \int 1s_A\left(-\frac{e^2}{4\pi\epsilon_0 r_B}\right)1s_A\,d\tau \\ \qquad + \int 1s_A\left(-\frac{h^2}{8\pi^2 m}\nabla^2\right)1s_B\,d\tau + \int 1s_A\left(-\frac{e^2}{4\pi\epsilon_0 r_A}\right)1s_B\,d\tau\end{array}}{1 + \int 1s_A 1s_B\,d\tau}$$

$$= \frac{\epsilon_{\mathrm{H}} + \int 1s_A\left(-\frac{e^2}{4\pi\epsilon_0 r_B}\right)1s_A\,d\tau + \int 1s_A\left(-\frac{h^2}{8\pi^2 m}\nabla^2 - \frac{e^2}{4\pi\epsilon_0 r_A}\right)1s_B\,d\tau}{1 + \int 1s_A 1s_B\,d\tau} \qquad [25]$$

The evaluation of the remaining integrals presents some difficulty, and such difficulties, which are most severe in the case of the "two-center" type of integrals that involve functions measured from nucleus A and nucleus B, are typical of the situation encountered in studies of larger molecules. Here we shall not follow through this purely mathematical step, but will simply state the results. When the internuclear repulsion $e^2/4\pi\epsilon_0 r_{AB}$ is subtracted from the energy given by Eq. [25], a potential-energy curve for the H_2^+-molecule ion is found that does have a minimum. This minimum occurs at an internuclear distance of 1.32 Å and has a depth of 1.77 eV. Comparison with the experimental curve of Fig. 11-12 shows that although the approach has some success, it is not able to provide energies in good agreement with the experimental potential-energy curve.

The variation approach can now be followed to see if the simple approximate function of Eq. [23] can be varied so as to approximate more closely the behavior of the electron of the H_2^+ system and therefore lead to better agreement with the observed equilibrium bond length and bond energy. The simplest procedure that can be followed consists in treating the orbital exponent corresponding to the nuclear charge as a variable parameter which must be adjusted at each internuclear distance to give the lowest possible energy. When this is done one finds that the greatest binding energy occurs at an internuclear distance of 1.06 Å, at which distance the effective nuclear charge is 1.23, and that the dissociation energy is 2.25 eV. Substantial improvement has occurred.

A further type of modification that is important in many modern calculations depends on the recognition that the atomic orbitals used in the trial function need not be restricted to those of lowest energy for the atoms; i.e., the trial function need not be based on 1*s* functions. One can, in principle, add in contributions from the complete set of atomic orbitals, relying on the variation theorem to decide how much of a contribution leads to an improved wave function. The best wave function and the observed dissociation energy could then be approached as closely as computational facilities permit.

We have seen now that the simplest covalently bound system can be treated by an approximate method, making use of trial functions that are "linear combinations of atomic orbitals" (LCAO). Although in its simplest form this LCAO method leads to results that are in rather poor agreement with the observed binding energy, methods are available for improving the wave function. Thus satisfactory values for the properties of molecules are often obtained, even if the binding energy is in relatively poor agreement with the experimentally obtained value.

11-11 The Hydrogen Molecule: The Electron-pair Bond

How can the procedure of Sec. 11-10 be extended to describe the way in which two electrons are accommodated between the bonded atoms, as in the simplest molecular example, H_2?

The most direct extension is the recognition that the Pauli exclusion principle allows two electrons, if they have opposite spins, to be described by the same orbital. Thus the electrons of two hydrogen atoms can be "placed," in the bond orbital shown in Fig. 11-14. The corresponding mathematical description can be given if we first label one electron number 1 and the other number 2 and then form the product

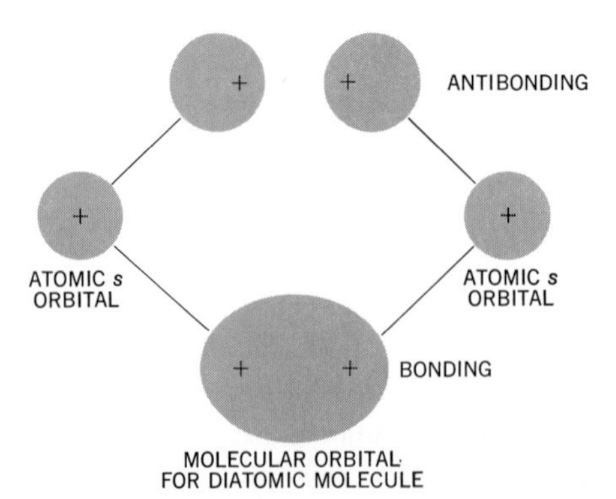

FIGURE 11-14
Molecular orbitals constructed from atomic *s* orbitals.

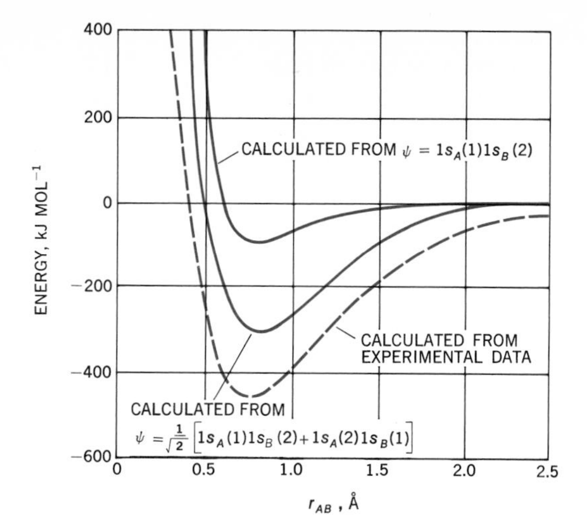

FIGURE 11-15
The binding energy of the H_2 molecule based on the approximate wave functions.

of the wave functions for the two electrons. Thus, for H_2, we form the approximate *molecular orbital* wave function

$$\begin{aligned}\psi &= \tfrac{1}{2}\varphi(1)\varphi(2)\\ &= \tfrac{1}{2}[1s_A(1) + 1s_B(1)][1s_A(2) + 1s_B(2)]\\ &= \tfrac{1}{2}[1s_A(1)\,1s_B(2) + 1s_A(2)\,1s_B(1)\\ &\qquad + 1s_A(1)\,1s_A(2) + 1s_B(1)\,1s_B(2)]\end{aligned} \qquad [26]$$

Use of this trial function and the Hamiltonian $\mathcal{H}$ appropriate to the H_2 molecule leads to the prediction of bonding between the two hydrogen atoms.

Notice that the wave function contains terms that correspond to the electrons being distributed over both nuclei and terms that correspond to the electrons both being on the same nuclei. In this approach the total wave function is the composite of these terms.

A second type of bond orbital can be constructed by subtracting rather than adding the two atomic wave functions. The molecular orbital $1s_A - 1s_B$ can be used to describe one or two electrons. Occupation of such an orbital is found to lead to a molecule with a higher energy than that of the atoms from which it is constructed. The orbital is said to be an *antibonding orbital.* The nature of such orbitals is suggested in Fig. 11-14. Their importance arises from the need to describe molecules in excited states and, as will be seen in the following chapter, molecules with many electrons.

An alternative, even simpler trial wave function that might be considered, at least for short internuclear distances, is

$$\psi = 1s_A(1)\,1s_B(2) \qquad [27]$$

Calculations with this function, as Fig. 11-15 shows, lead to little binding energy. Thus $1s_A(1)\,1s_B(2)$ is a poor approximation to the correct description for the system. The principal fault, as W. Heitler and F. London recognized, is that, since electrons are indistinguishable particles, different roles cannot be ascribed to the two electrons.

It is not in accord with our knowledge of the system, therefore, to suppose, as is implied by Eq. [27], that one electron can be recognized as electron 1 and that it is known to remain about a nucleus called nucleus A. Heitler and London pointed out that the approximate wave function for the molecule should recognize this indistinguishability of the electrons, and they suggested an approximate function which allows both arrangements of the electrons. Their approximation to the wave function for the electrons of the hydrogen molecule is

$$\psi = \frac{1}{\sqrt{2}}[1s_A(1)1s_B(2) + 1s_A(2)1s_B(1)] \qquad [28]$$

where $1/\sqrt{2}$ is a factor that ensures that the wave function is normalized at infinite r_{AB}.

You should note that it is acceptable to label electrons in the construction of a trial wave function. The resulting function must, however, attribute identical roles to each of the electrons of the system.

When this trial function is used in Eq. [16] and the repulsion of the nuclei is allowed for, the potential minimum, as shown in Fig. 11-15, is found to be 300 kJ mol^{-1} below that of the independent hydrogen atoms. Furthermore, this minimum occurs at an internuclear distance of 0.80 Å. Comparison of these results with the corresponding experimental values of 458 kJ mol^{-1} and 0.740 Å shows that the relatively simple Heitler-London wave function has been successful in accounting for much of the observed bonding.

In this approach the chemical bond can be largely attributed to the fact that the electrons are not confined to one or the other atoms of the bond, as they are in the less satisfactory function $1s_A(1)1s_B(2)$, but rather, as is said, can *exchange*.

11-12 The Nature of Chemical Bonds: The Virial Theorem

You have now seen the mechanics by which the bond energy and the distribution of the electrons in a covalent bond can be calculated. Such calculations lead to the introduction of terms such as "overlap" and "exchange," but these describe the approximate procedure that is used rather than the nature or basis of the bond itself.

The terms in the Schrödinger equation, or in the Hamiltonian $\mathcal{H}$, suggest, however, that no matter how the wave functions are described, there will be kinetic- and potential-energy contributions to the total energy. The kinetic energy is a consequence of the wave nature of the particle, and as the one-dimensional square-well example showed, the kinetic energy corresponding to a given wave function is greater the smaller the region of available space. The potential energy arises from all coulombic interactions among charged particles. The nature of these two energy terms is not dependent on the approximations used in describing the bond, and as a result, analysis of these terms provides insight into the nature of the bond itself.

One basis for unscrambling the contributions to the total energy is the *virial theorem*. The theorem says that for any collection of particles for which the forces follow an inverse-square force law, as do electrostatic forces, the relation

$$2(\text{KE}) = -U \qquad [29]$$

holds; i.e., twice the kinetic energy is equal but of opposite sign to the potential energy. The total energy of the collection of particles, which, for example, could be the two protons and the one or two electrons of the H_2^+ or H_2 species, is

$$\epsilon = \text{KE} + U \qquad [30]$$

which, with $2(\text{KE}) = -U$, becomes

$$\epsilon = \tfrac{1}{2}U$$

or

$$\epsilon = -\text{KE} \qquad [31]$$

These remarkable and simple relations can be applied immediately to the equilibrium condition of an atom or of a molecule. For H_2, for example, we see, from the net value of $\epsilon = -458$ kJ mol^{-1} relative to two hydrogen atoms, that $(\text{KE})_{H_2} = +458$ kJ mol^{-1} and $U_{H_2} = -916$ kJ mol^{-1}, both relative to two hydrogen atoms. We see that the bond energy in H_2 results from the very large decrease in potential energy. This effect more than compensates for the increase in kinetic energy that results from the confinement of the electrons to these relatively small regions. Similar insights can be gained by applying the virial theorem to other systems in their equilibrium configuration.

Now let us see how we can interpret the energy of a chemical bond at internuclear separations other than that for the minimum in the binding-energy curve. Again H_2 will serve as an example. Consider a coordinate system centered on one atom of an H_2 molecule. The second atom will be at some distance r, the internuclear separation. For this distance to be other than the equilibrium one, some external force must act, and in acting, it contributes an additional potential-energy term to the molecule. The force that must be exerted is just the derivative of the total energy with respect to internuclear distance; i.e., it is the slope of the experimentally based curve such as that of Fig. 11-15. The potential energy corresponding to this force is just $r\,(d\epsilon/dr)$, and thus the virial theorem becomes

$$2(\text{KE}) = -U - r\frac{d\epsilon}{dr} \qquad [32]$$

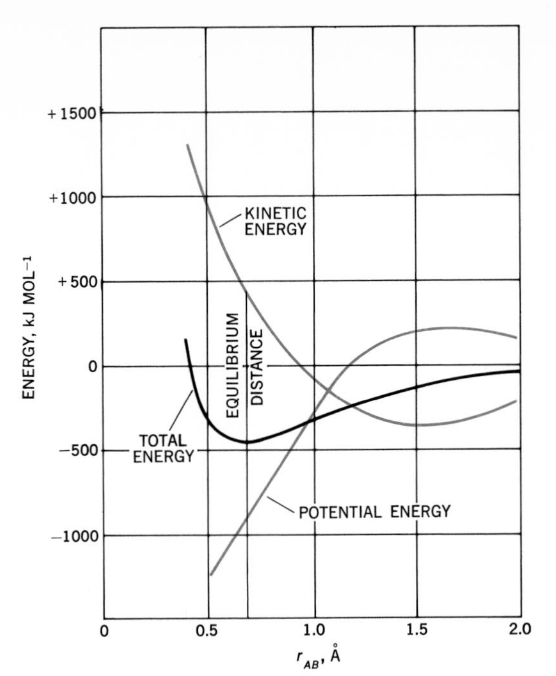

FIGURE 11-16 Average kinetic, potential, and total energies for H_2 as a function of internuclear distance. The difference between the value for the molecule and that for the separated atoms is given for each energy. (*From K. S. Pitzer, "Quantum Chemistry," Prentice-Hall, Inc., Englewood Cliffs, N.J., 1953.*)

The kinetic- and potential-energy contributions to the total energy ϵ are now deduced from $\epsilon = \text{KE} + U$ and Eq. [32] as

$$\text{KE} = -\epsilon - r\frac{d\epsilon}{dr}$$

and

$$U = 2\epsilon + r\frac{d\epsilon}{dr} \qquad [33]$$

Since experimental ϵ versus r curves can be deduced, as will be shown in Chap. 13, the derivative terms can be evaluated. Thus KE versus r and U versus r curves can be constructed. Examples are shown for H_2 in Fig. 11-16. Now you see that as the bond begins to form at relatively large internuclear distances, it is the kinetic energy that contributes to the binding, a result of the somewhat greater region in which the electrons can move. At this stage additional electron-electron and proton-proton repulsions work to oppose the bonding. Only at shorter internuclear distances does the reverse occur.

Problems

1 Plot the relative probabilities of the electron being at various distances from the nucleus for the $1s$ wave function for the hydrogen atom, and see graphically that the most probable distance is the same as that obtained by the Bohr theory.

Confirm by differentiating the probability function that this maximum is at a_0.

2 The differential-volume element at a distance r from the nucleus is $4\pi r^2\,dr$. By a suitable integration find the average distance of a $1s$ electron of the H atom from the nucleus.

$\int_0^\infty x^3 e^{-x}\,dx = 6.$ *Ans.* $\frac{3}{2}a_0$.

3 Show by differentiation that the maximum in the radial-distribution curve for the $2p$ orbital occurs at the same radius as given by the simple Bohr theory for an $n = 2$ orbit.

4 Plot, on the same graph, the electron density along the x axis and along a line at 45° to the x axis for the p_x orbit for $n = 2$ of a hydrogenlike atom.

5 From the expression for $Y(\theta,\varphi)$ of a p_z orbital given in Table 11-2, prepare graphs showing the values of $Y(\theta,\varphi)$ and $Y(\theta,\varphi)^2$ in the xz plane and in the yz plane.

6 Show by a suitable integration that the $1s$ wave function of Table 11-2 is normalized.

7 Calculate the effective nuclear charge for the outer electrons of lithium, sodium, and potassium from the given values of the ionization potentials of these elements.

8 Calculate the effective nuclear charges for one of the outer electrons of fluorine, neon, and sodium and rationalize the relative values of the results.

9 Use Eq. [16] to calculate the energy of the lowest allowed state of an electron in a 10-Å one-dimensional square-well potential function. Guess that a good wave function is the solution function of $\psi = A \sin(\pi x/a)$.

10 A well-behaved function that goes to zero at $x = 0$ and at $x = a$ and that might be used as an approximate solution to the square-well problem (if the exact solution could not be found) is $Bx(a - x)$, where B is some constant. Calculate the energy of the electron in the 10-Å potential well on the basis of this approximate function.

11 Assuming that the potential-energy function of Eq. [19] is satisfactory:

a Evaluate the constant b for KCl so that the minimum in the potential-energy curve occurs at the observed equilibrium internuclear distance of 2.8 Å. *Ans.* 0.915×10^{-15} J.

b Prepare a plot of this potential-versus-internuclear separation.

c What energy does this potential function predict for the separation of KCl from its equilibrium internuclear distance into the ions K^+ and Cl^-? *Ans.* 444 kJ mol^{-1}.

d According to the ionization-potential and electron-affinity values of Tables 11-3 and 11-4, how much energy is required for the reaction $K + Cl \rightarrow K^+ + Cl^-$ for infinitely separated particles? *Ans.* 70 kJ mol^{-1}.

e What are the energetically favored products of the dissociation of KCl?

f Compare the value calculated for the dissociation to stable products from the results of part c or d with the observed value of 4.42 eV = 426 kJ mol^{-1}. *Ans.* 374 kJ mol^{-1}.

12 Consider the nature of the terms that arise when Eqs. [23] and [24] are inserted in Eq. [16], and see that as r_{AB} approaches infinity, the energy becomes ϵ_H, and therefore that the trial function of Eq. [23] is the correct function for separated nuclei, and thus an acceptable trial function for finite separations.

13 Calculate, using the approximate function of Eq. [23], the probability of the electron of the H_2^+ molecule ion being in unit volumes at various positions along the internuclear axis for $r_{AB} = 1.06$ Å. Show these results by a plot of these probabilities as ordinates, with the nuclei located along the abscissa. Also plot the probability $\frac{1}{2}(\psi_A^2 + \psi_B^2)$ that would correspond to two nearby hydrogen atoms unperturbed by one another.

14 Verify that at infinite nuclear separation the Heitler-London trial function written in Eq. [28] is normalized.

15 Plot the probability function for the electrons of the H_2 molecule along the internuclear axis for the functions of Eqs. [27] and [28] for the internuclear separation of 0.740 Å.

16 A somewhat simpler model of the ionic bond than that used in Sec. 11-7 treats the oppositely charged ions as structureless particles that attract each other by coulombic electrostatic forces. They further are pictured as "hard spheres" that exert no repulsive forces until the spheres touch and then they are "hard" or incompressible. Recognizing that the virial theorem can be used only where inverse-square forces exist, indicate what you can about the kinetic- and potential-energy features of this model, and compare the qualitative feature with those of a covalent bond as represented by that of H_2.

References

BORDASS, W. T., and J. W. LINNETT: A New Way of Presenting Atomic Orbitals, *J. Chem. Educ.*, **47**:672 (1970). Very effective three-dimensional-appearing, computer-generated contour diagrams for atomic orbitals.

CARTMELL, E., and G. W. A. FOWLES: "Valency and Molecular Structure," Butterworth & Co. (Publishers), Ltd., London, 1961. An elementary and largely pictorial treatment of quantum theory, atomic structure, and the bonding and structure of organic and inorganic compounds.

GRAY, HARRY B.: "Electrons and Chemical Bonding," W. A. Benjamin, Inc., New York, 1964. A very suitable supplement to this and the following chapter. Atoms and diatomic and simple polyatomic molecules are first treated. The final chapters contain material on more complex systems, including those in which *d* orbitals are involved in the bonding.

LINNETT, J. W.: "Wave Mechanics and Valency," Methuen & Co., Ltd., London, 1960. A short account of some of the principles of quantum mechanics, followed by clear, careful treatments of the hydrogen atom, the helium atom, and the hydrogen-molecule ion and the hydrogen molecule. Some discussion of more complex systems is also included.

COULSON, C. W.: "Valence," 2d ed., Oxford University Press, Fair Lawn, N.J., 1961. A treatment of chemical bonding that avoids much of the mathematical developments and emphasizes the qualitative aspects of chemical-bond theory that are of value for chemical considerations.

KARPLUS, M., and R. N. PORTER: "Atoms and Molecules: An Introduction for Students of Physical Chemistry," W. A. Benjamin, Inc., New York, 1970. A somewhat terse but good extension of the material presented here.

Further study of the methods used to investigate the nature of covalent bonds will require a more complete and mathematical knowledge of quantum-mechanical methods. Several of the many texts that provide the background necessary are:

PILAR, F. L.: "Elementary Quantum Chemistry," McGraw-Hill Book Company, New York, 1968.

HANNA, M. W.: "Quantum Mechanics in Chemistry," W. A. Benjamin, Inc., New York, 1965.

PITZER, K. S.: "Quantum Chemistry," Prentice-Hall, Inc., Englewood Cliffs, N.J., 1953.

PAULING, L., and E. B. WILSON, JR.: "Introduction to Quantum Mechanics," McGraw-Hill Book Company, New York, 1935.

EYRING, J., J. WALTER, and G. E. KIMBALL: "Quantum Chemistry," John Wiley & Sons, Inc., New York, 1944.

SLATER, J. C.: "Quantum Theory of Atomic Structure" and "Quantum Theory of Molecules and Solids," vol. 1, "Electronics of Molecules," McGraw-Hill Book Company, New York, 1960 and 1963.

12

THE NATURE OF THE BONDING IN CHEMICAL COMPOUNDS

12-1 Bonding in Homonuclear Diatomic Molecules

The general ideas developed in the preceding chapter with regard to the bonding in the H_2^+ molecule ion and the H_2 molecule can be extended to other homonuclear diatomic molecules. To do this, we must first expand our investigation of linear combinations of the orbitals of the two atoms that will give orbitals for the molecule. In the jargon of quantum mechanics, we want to use the MO-LCAO method, i.e., molecular orbitals approximated by linear combinations of atomic orbitals. Once these orbitals that provide approximate descriptions of the behavior of the electrons in the molecule are constructed, electrons can be assigned to the orbitals in a way that is consistent with the Pauli exclusion principle.

The two combinations that can be made from $1s$ atomic orbitals were suggested by Fig. 11-14. These molecular orbitals can accommodate a total of four electrons and thus can be used to describe any bonding in species such as H_2^+, H_2, He_2^+, He_2. From the assignment of electrons to the orbital diagrams, as shown in Fig. 12-1, the net number of bonding electrons, 1, 2, 1, and 0 in these four species, can be deduced. We thus come on a simple qualitative guide to the bonding in such species.

The description of bonding of atoms that have electrons in the $2s$ atomic orbitals, Li and Be, requires molecular orbitals to be constructed from the $2s$ as well as the $1s$ atomic orbitals. This is shown pictorially in Fig. 12-2. The presence of pairs of $1s$ electrons on each atom leads, as you see, to equal numbers of bonding and antibonding electrons,

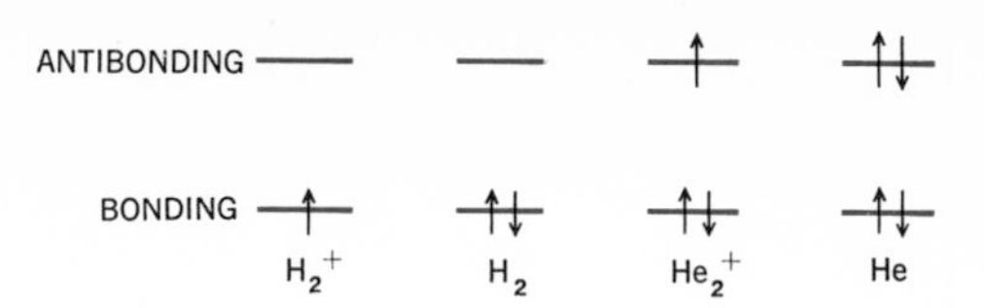

FIGURE 12-1
The assignment of electrons to the molecular orbitals for the species H_2^+, H_2, He_2^+, and He.

and thus such inner shells produce, in this approximation, no net bonding effect.

For diatomic molecules of other second-row elements molecular orbitals must be constructed from atomic p orbitals to develop enough molecular orbitals to accommodate the electrons. With p orbitals, two different situations occur.

Atomic p orbitals that lie along the molecular axis and, as mentioned in Sec. 11-2, give to an electron no angular momentum along this axis can be combined, i.e., added or subtracted, to form molecular orbitals, as shown diagrammatically in Fig. 12-3. These are comparable with the orbitals obtained from atomic s orbitals in that they have no additional nodes perpendicular to the molecular axis and give an electron no angular momentum about the axis. Such orbitals are designated as σ molecular orbitals to emphasize these similarities with atomic s orbitals. These characteristics of molecular orbitals depend only on the cylindrical symmetry of the molecule, and thus, although their energy and extension implications are not easily deduced, their node and angular-momentum features are immediately determined.

Atomic p orbitals that project perpendicular to the molecular axis lead to bonding and antibonding molecular orbitals, as Fig. 12-4 shows, that have an additional node perpendicular to the molecular axis. An electron in such an orbital, which is related to the $l = 1$, $m = \pm 1$ atomic p orbitals of Table 11-2, is designated a π orbital.

The molecular orbitals of homonuclear diatomic molecules, such as those shown in Figs. 12-3 and 12-4, can also be characterized by the effect of inverting the function through the center of symmetry of the molecule, i.e., through the midpoint of the bond. This operation either leaves the function unchanged, in which case the orbital is said

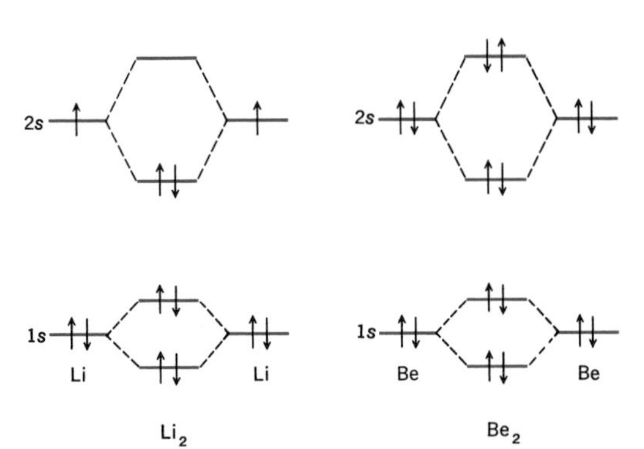

FIGURE 12-2
Molecular-orbital descriptions for Li_2 and Be_2 showing a net pair of bonding electrons for Li_2 but no net bonding electrons for Be_2.

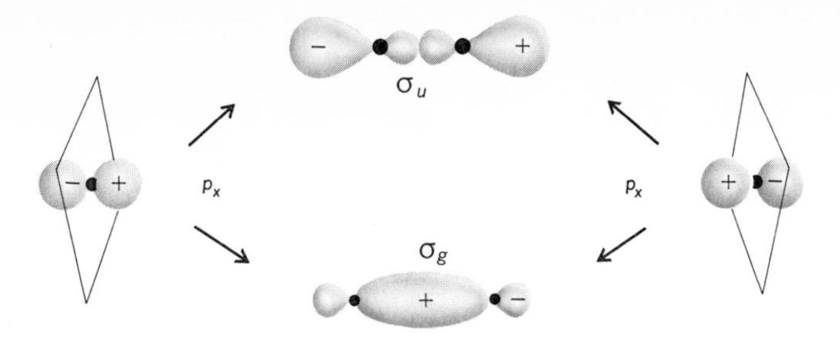

FIGURE 12-3
The formation of σ molecular orbitals from atomic p orbitals.

to be of the g (*gerade,* even) type, or it changes the sign of the function, in which case the orbital is said to be of the u (*ungerade,* odd) type. The symbols g and u are often included in the orbital symbol, as in Figs. 12-3 and 12-4.

Just as the orbital energy diagram for atoms, shown in Fig. 11-9, became complex because of the subtleties that determine the energies of the orbitals, so also is the energy pattern of the molecular orbitals of diatomic molecules somewhat uncertain. For the second-row diatomic molecules, however, the patterns of Fig. 12-5a and b are adequate to suggest many of the bonding properties of the molecules. A correlation of some properties with bonding predictions from Fig. 12-5

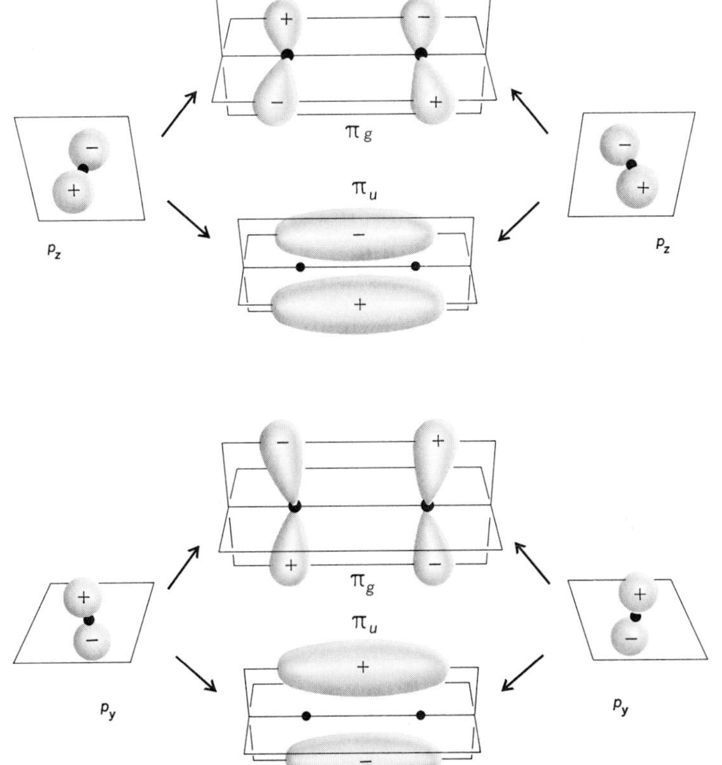

FIGURE 12-4
The formation of π molecular orbitals from atomic p orbitals.

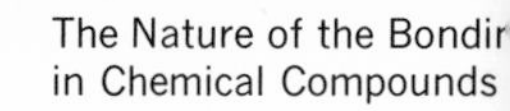

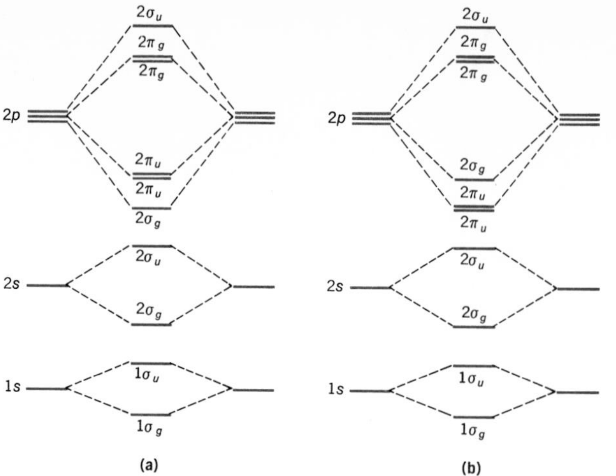

FIGURE 12-5 Representative molecular-orbital diagrams from homonuclear diatomic molecules. Diagram (*a*) can be used to describe molecules or ions formed from O, F, or Ne; (*b*) can be used for molecules or ions from Li, Be, B, C, or N.

for such molecules is given in Table 12-1. You see there that considerable organization of the bonding features of such molecules is obtained.

Energy patterns for the molecular orbitals of diatomic molecules can be shown by a diagram like that of Fig. 12-6 in which the orbital energies from the extreme of the completely separated atoms of the molecule are shown related to the orbital energies of the completely merged, or "united," atoms. The energy lines are drawn so that the pattern of orbital energies of a particular diatomic molecule can be read off at some intermediate position along the abscissa.

TABLE 12-1 Bond lengths and bond energies for some homonuclear diatomic molecules and ions
These are arranged according to the net number of bonding electrons, as deduced from a diagram like that of Fig. 12-5.

No. of bonding electrons					
6	5	4	3	2	1
				H_2 0.741 Å 432 kJ	H_2^+ 1.06 Å 255 kJ
	O_2^+ 1.123 Å	O_2 1.207 Å 493 kJ	O_2^{3-} 1.26 Å	O_2^{--} 1.49 Å	
N_2 1.100 Å 941 kJ	N_2^+ 1.116 Å				
				Cl_2^+ 1.891 Å	Cl_2 1.988 Å 239 kJ

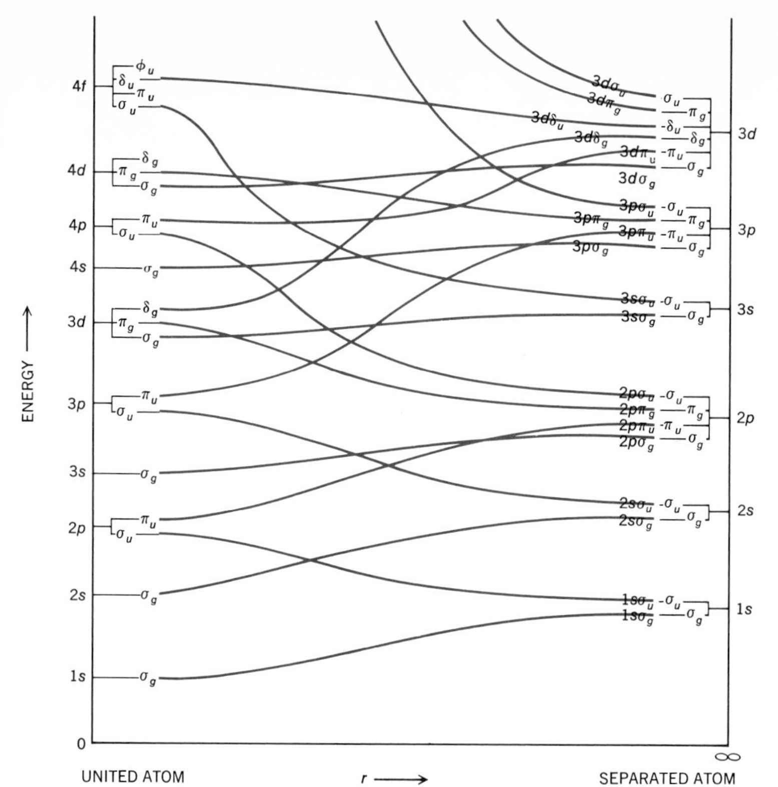

FIGURE 12-6 Correlation diagram showing the relation of molecular orbitals to the atomic orbitals that would be formed at zero internuclear distance (the *united atom*) and at infinite internuclear distance (the separated atoms). (*From M. Kasha.*)

To proceed to more quantitative results, one must use these molecular orbitals, or other similar functions, in a repetitive variation treatment to yield self-consistent results like those achieved for atoms. When calculations are done, the orbitals and atomic states that are deduced are illustrated by the diagrams of Fig. 12-7.

12-2 Heteronuclear Bonds and the Ionic Character of Bonds

The ideas derived from our study of H_2 regarding the overlap of atomic orbitals and the molecular-orbital treatment of Sec. 12-1 that classifies the electrons of a molecule can be carried over to heteronuclear diatomic molecules. Let us now see what special treatments such molecules, with the HCl molecule serving as an example, will require.

Any description of a heteronuclear bond should recognize the possibility of the bonding electrons being more closely associated with one nucleus than the other. In the HCl example, for instance, it is customary to think of the bonding pair of electrons as being more closely associated with the Cl atom than with the H atom.

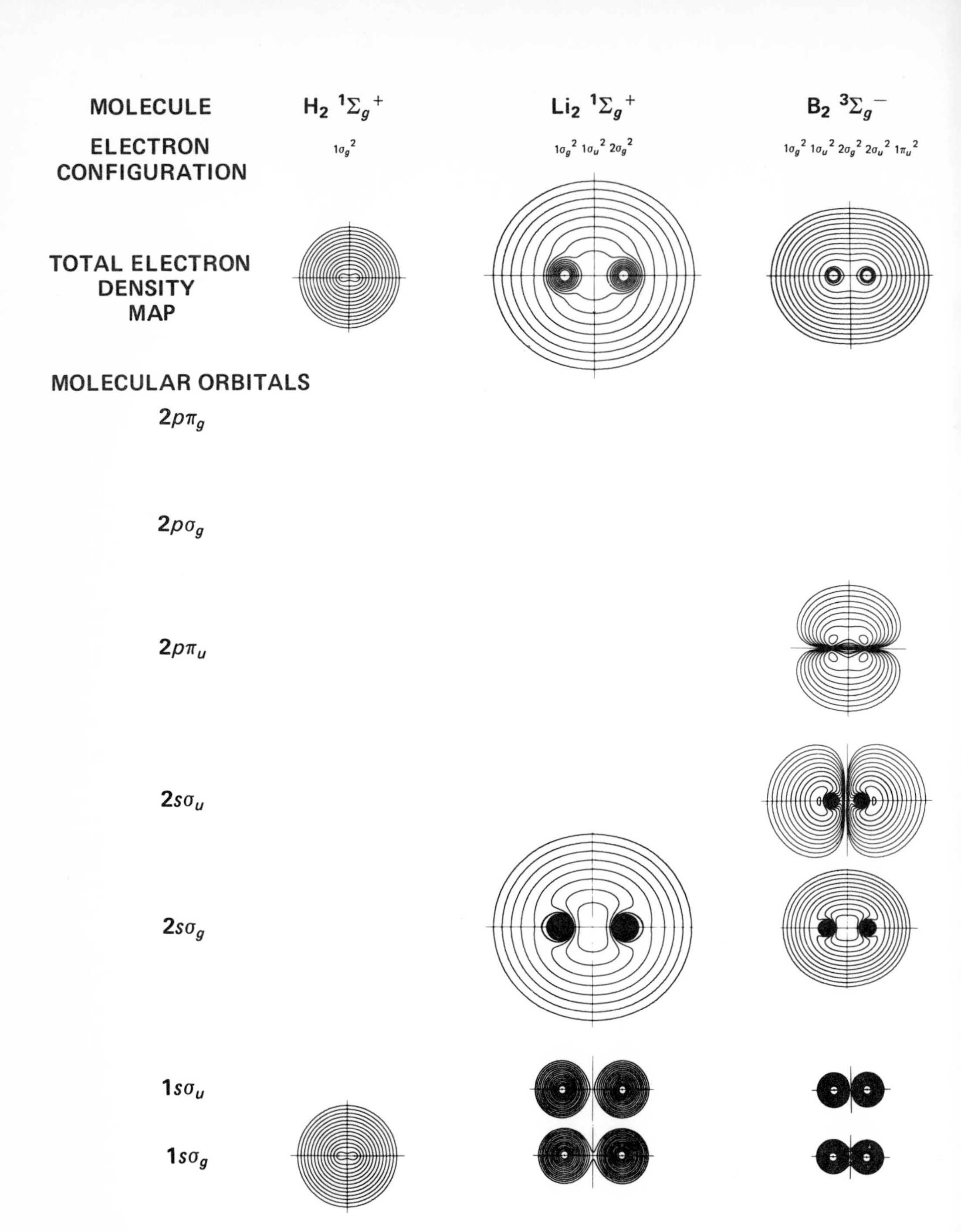

FIGURE 12-7
Contour diagrams for the electron orbitals of diatomic molecules. For explanatory details see Arnold C. Wahl, Molecular Orbital Densities: Pictorial Studies, *Science*, **151**:961 (1966), or Chemistry by Computer, *Sci. Am.*, **272**:54 (1970). (*Chart by Arnold C. Wahl. Copyright © 1970 by McGraw-Hill, Inc. Used with permission of McGraw-Hill Book Company.*)

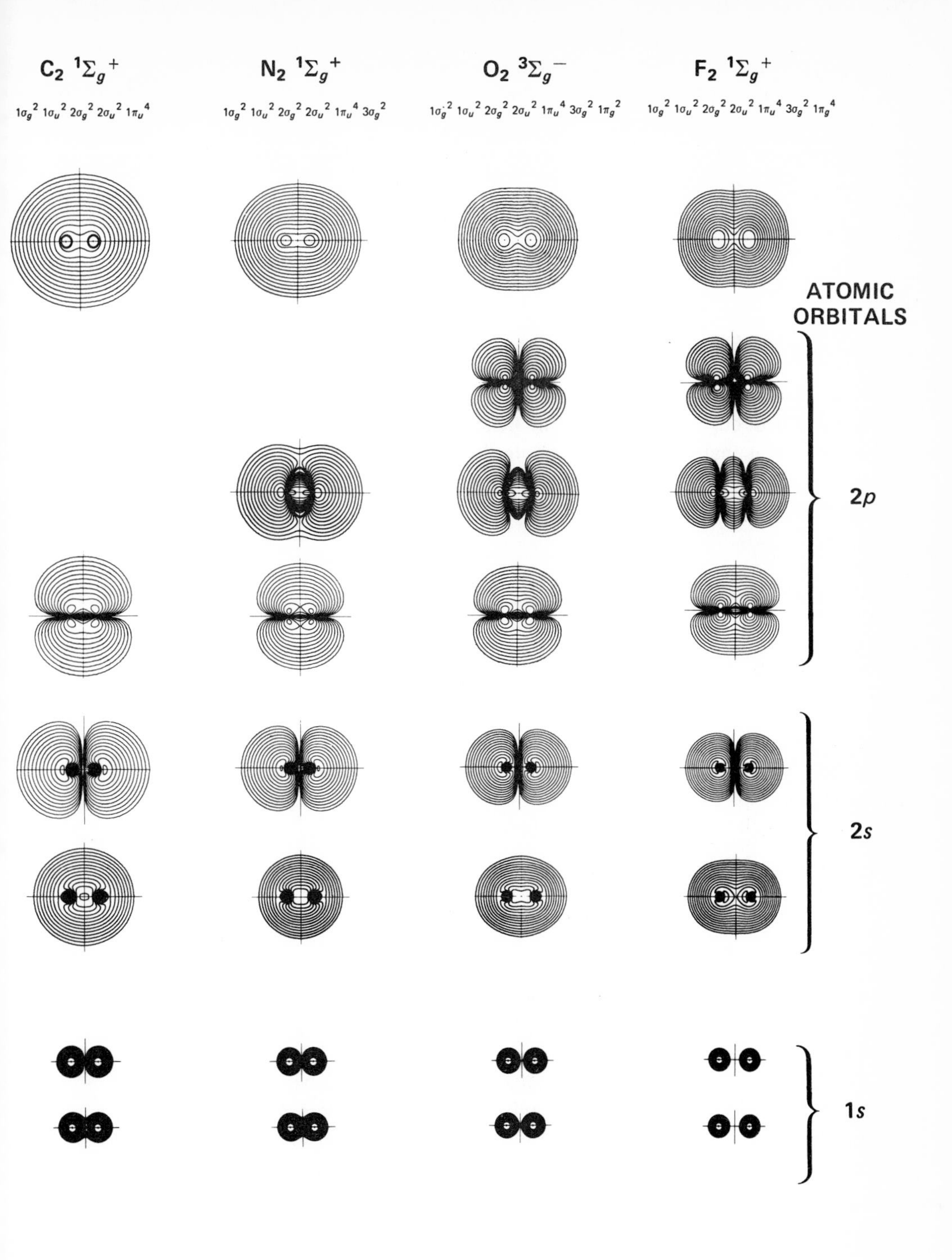

C2 1Σg+
1σg2 1σu2 2σg2 2σu2 1πu4
N2 1Σg+
1σg2 1σu2 2σg2 2σu2 1πu4 3σg2
O2 3Σg−
1σg2 1σu2 2σg2 2σu2 1πu4 3σg2 1πg2
F2 1Σg+
1σg2 1σu2 2σg2 2σu2 1πu4 3σg2 1πg4
ATOMIC ORBITALS
2p
2s
1s

The bond in HCl can be interpreted in terms of the sharing of the electron originally in a $1s$ orbital in the hydrogen atom and the unpaired electron originally in one of the $3p$ orbitals of chlorine. A suitable approximation to the distribution of each of these electrons, numbered 1 and 2, in the HCl bond can be written as the molecular orbitals

$$\varphi(1) = \frac{1}{\lambda}[1s_{\mathrm{H}}(1)] + \lambda[3p_{\mathrm{Cl}}(1)]$$

and [1]

$$\varphi(2) = \frac{1}{\lambda}[1s_{\mathrm{H}}(2)] + \lambda[3p_{\mathrm{Cl}}(2)]$$

The form of these equations is important, but since no calculations will be carried through, a normalization factor need not be treated. Our expectation is that the factor λ is greater than unity, and therefore the $3p_{\mathrm{Cl}}$ term is more important in the description than is the $1s_{\mathrm{H}}$ term. The two-electron bond is then described by the wave function

$$\begin{aligned}\psi &= \varphi(1)\varphi(2)\\ &= \left\{\frac{1}{\lambda}[1s_{\mathrm{H}}(1)] + \lambda[3p_{\mathrm{Cl}}(1)]\right\}\left\{\frac{1}{\lambda}[1s_{\mathrm{H}}(2)] + \lambda[3p_{\mathrm{Cl}}(2)]\right\}\\ &= \frac{1}{\lambda^2}[1s_{\mathrm{H}}(1)1s_{\mathrm{H}}(2)] + [1s_{\mathrm{H}}(1)3p_{\mathrm{Cl}}(2) + 1s_{\mathrm{H}}(2)3p_{\mathrm{Cl}}(1)]\\ &\qquad + \lambda^2[3p_{\mathrm{Cl}}(1)3p_{\mathrm{Cl}}(2)] \end{aligned} \quad [2]$$

If the electrons are, in fact, held more closely by the Cl atom than by the H atom, the first term, which corresponds to both electrons being on the H atom, will be small, on account of the $1/\lambda^2$ factor, and can be ignored. The bond description can then be written

$$\psi = [1s_{\mathrm{H}}(1)3p_{\mathrm{Cl}}(2) + 1s_{\mathrm{H}}(2)3p_{\mathrm{Cl}}(1)] + \lambda^2[3p_{\mathrm{Cl}}(1)3p_{\mathrm{Cl}}(2)] \quad [3]$$

This is usually written with the notation

$$\psi = a\psi_{\mathrm{cov}} + b\psi_{\mathrm{ionic}} \quad [4]$$

where ψ_{cov} implies the equally shared description of the first term in Eq. [3], and ψ_{ionic}, the description of the bond that puts electrons on the more electron-attracting atom.

Thus the bonding electrons of a heteronuclear bond can be described in terms of a covalent factor and an ionic factor. More pictorially, the bond in HCl is described in terms of wave-function terms corresponding to

H—Cl (cov) and H^+ :Cl^- (ionic)

Homonuclear bonds represent one extreme in that the electrons are shared equally and the covalent term alone is satisfactory. For heteronuclear bonds, however, all degrees of importance of the ionic term are encountered, even up to the extreme of NaCl-type molecules,

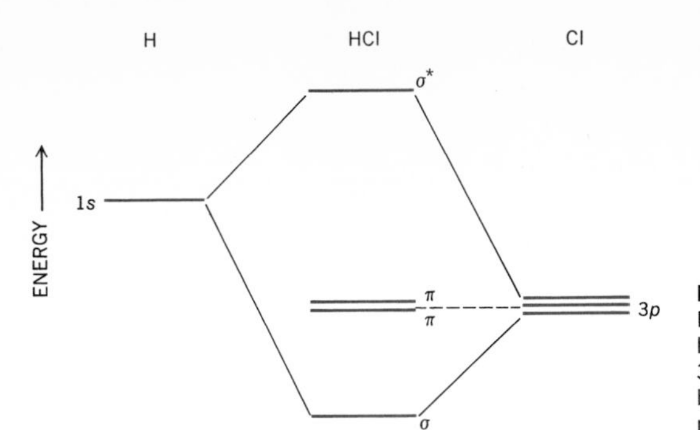

FIGURE 12-8
Effect on the energies of the hydrogen $1s$ and the chlorine $3p$ orbitals as a result of bond formation in the HCl molecule.

where it is the covalent part of the description that has become negligible.

Molecular-orbital diagrams can be constructed for heteronuclear diatomic molecules, and these are understandable in terms of the diagrams for homonuclear molecules and the enhanced importance of the orbitals of the more electron-attracting atom, as illustrated in the HCl example. Figure 12-8 provides an example of such diagrams.

12-3 Electronegativities

In view of the unequal sharing of bond electrons revealed in Sec. 12-2, it seems desirable to try to assign to each atom a number which measures the tendency of the bonding electrons to be drawn toward that atom. *Electronegativity* is the name given to the index that attempts to represent this electron-attracting tendency. A large number of methods have been suggested to arrive at electronegativity values, but only two of these need be mentioned.

A rather direct, but somewhat limited, method due to Mulliken makes use of the ionization potential and the electron-affinity data. The attraction that an atom, or really the ion, exerts on a pair of electrons that are in a bond between that atom and another atom can be expected to be some average of the attraction of the free ion for an electron, i.e., the ionization potential, and the attraction of the neutral atom for an electron, i.e., the electron affinity. Numerical values are obtained that coincide with values from other methods if electronegativities, designated as x, are calculated from

$$x = \frac{I + A}{5.6} \qquad [5]$$

where I and A are the ionization potential and electron affinity, in electron volts. The factor of 5.6 is an arbitrary-scale factor. In this way, for example, one calculates

$$x_{\text{Cl}} = \frac{13.01 + 3.61}{5.6} = 3.0 \qquad [6]$$

Similarly, values can be obtained for other elements for which ionization-potential and electron-affinity data are available.

An alternative method, due to Pauling, makes use of bond energies and a treatment of bond energies that is influenced by the *variation theorem.*

A description of the bonding in a heteronuclear bond, such as is implied by Eq. [4], which allows for the ionic character of the bond, is expected, in view of the variation theorem, to lead to a better, i.e., greater, calculated bond energy than an incomplete description, as implied, for example, by ψ_{cov}.

An important step was made by Pauling in suggesting that the extra energy corresponding to the better description of the bonding electrons could be deduced from bond energies such as those obtained from thermal data in Sec. 7-10. In this method, the energy that a bond would have if the electrons were equally distributed between the nuclei is calculated from the average of the covalent-bond energies of the atoms of the bond. Thus this hypothetical bond energy is calculated for HCl, with D representing bond energies, as

$$\begin{aligned} D_{Cl_2} &= 243 \text{ kJ mol}^{-1} \\ D_{H_2} &= 436 \\ & 2\overline{)679} \\ (D_{HCl})_{cov} &= 340 \text{ kJ mol}^{-1} \end{aligned} \qquad [7]$$

It turns out generally to be better, for no satisfactory reason, to take the geometric mean instead of the arithmetical average. In this way one calculates

$$(D_{HCl})_{cov} = \sqrt{(243)(436)} = 325 \text{ kJ mol}^{-1} \qquad [8]$$

The actual bond energy is identified with the more complete description, and for HCl, Table 7-4 gives

$$D_{HCl} = 432 \text{ kJ mol}^{-1} \qquad [9]$$

Comparison of the values 432 and 325 confirms our ideas that a covalent description for the bond of a heteronuclear molecule is inadequate and that a good description of a heteronuclear chemical bond must include an ionic term. For almost all heteronuclear bonds a similar result is found, namely, that the actual energy is greater than the calculated covalent value.

The energy difference Δ between the actual and covalent-bond energies can be taken as a measure of the ionic character of the bond. For HCl one calculates

$$\Delta = 432 - 325 = 107 \text{ kJ mol}^{-1} = 1.11 \text{ eV} \qquad [10]$$

Bonds in which the ionic term is important are expected to have large values of Δ, and vice versa. The value of Δ for a given bond can be taken as a measure of the electronegativity difference of the two atoms of the bond. Pauling found that a self-consistent set of electronegativities could best be deduced if one used the relation

$$x_B - x_A = \sqrt{\Delta} \qquad [11]$$

where Δ is expressed in electron volts. The use of the square root is quite arbitrary, but leads, for example, to essentially the same value for $x_{Cl} - x_I$ from the data for ICl as from the data for HCl and HI.

Pauling's method allows differences in electronegativities to be deduced from bond-energy data, as shown in Table 12-2. To obtain values for the individual atoms, it is necessary to pick an arbitrary reference point. If x_H is taken as 2.2, electronegativity values range from about 1.0 to 4.0, and this is considered a convenient range. Results obtained in this way are compared in Table 12-2 with values from Mulliken's method, and general agreement is noticed.

The electronegativity values for some elements arranged in the periodic-table form are shown in Fig. 12-9, and are of considerable interest. One should recognize that the electronegativity is at least a semiquantitative parameter that reflects much of the chemical behavior of the elements. Electronegativities have widespread and important

TABLE 12-2 Electronegativity differences and electronegativities according to the methods of Pauling and of Mulliken

	Bond energies (kJ mol^{-1})		Δ		
Bond	Hypothetical cov (geometric mean of covalent-bond energies)	Observed	kJ	eV	$\sqrt{\Delta} = x_B - x_A$
H—F	262	563	301	3.1	1.7
H—Cl	326	432	105	1.1	1.0
H—Br	290	366	76	0.8	0.9
H—I	256	299	43	0.4	0.6
O—H	250	463	213	2.3	1.5
N—H	264	391	127	1.3	1.1
C—H	387	415	28	0.3	0.5
C—O	222	350	128	1.4	1.2
C—F	234	441	207	2.1	1.4

Atom	x_{Pauling}	x_{Mulliken}
H	(2.2)	2.5
F	3.9	3.8
Cl	3.2	3.0
Br	3.1	2.7
I	2.8	2.4

			H 2.2				
Li 1.0	Be 1.5	B 2.0		C 2.5	N 3.0	O 3.5	F 4.0
Na 0.9	Mg 1.2	Al 1.5		Si 1.8	P 2.1	S 2.5	Cl 3.0
K 0.8	Ca 1.0	Sc 1.3	Ti-Ga 1.7 ± 0.2	Ge 1.8	As 2.0	Se 2.4	Br 2.8
Rb 0.8	Sr 1.0	Y 1.2	Zr-In 1.9 ± 0.3	Sn 1.8	Sb 1.9	Te 2.1	I 2.5
Cs 0.7	Ba 0.9	La-Lu 1.1	Hf-Tl 1.9 ± 0.4	Pb 1.8	Bi 1.9	Po 2.0	At 2.2
Fr 0.7	Ra 0.9	Ac 1.1	Th⟶ 1.3⟶				

FIGURE 12-9
Electronegativities of the elements arranged in periodic order. *(After L. Pauling, "The Nature of the Chemical Bond," 3d ed., Cornell University Press, Ithaca, N.Y., 1960.)*

uses in chemistry. It is important to appreciate what it is that they are intended to represent and how they are deduced.

12-4 The Valence Bond: Hybridization

Although molecular orbitals can be constructed by making appropriate linear combinations of atomic orbitals for molecules of any shape and complexity, and although such orbitals are often used as the basis for computer calculations of molecular properties, chemists have found it convenient to treat many molecules in terms of the individual bonds that are represented by the line between adjacent atoms. The basis for such diagrams can be seen by using atomic orbitals to construct valence, or bond, orbitals.

An attempt to describe the bonding in CH_4 illustrates the procedure. Chemical evidence, in the hands of Le Bel and van't Hoff, showed long ago that the four hydrogens are equivalent and, furthermore, that they are placed in the most symmetric, tetrahedral arrangement about the central carbon atom, as shown in Fig. 12-10. An electronic picture of the bonding must at least lead to a tetrahedral arrangement of the bond orbitals.

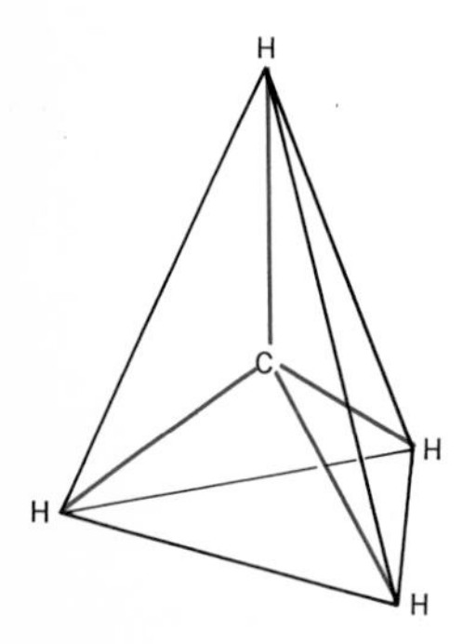

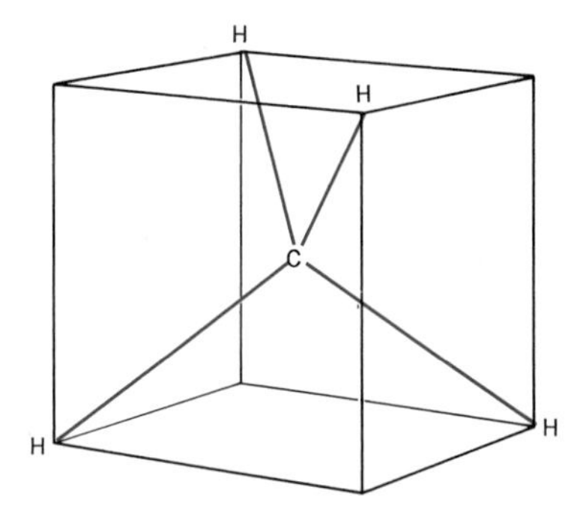

FIGURE 12-10
Two representations of the tetrahedral structure of methane. The angle between any pair of bonds is $109\frac{1}{2}°$.

Each of the four hydrogen atoms presents a $1s$ orbital containing one electron. Bonding to the carbon atom will occur, as it did in the H_2 example, when these orbits overlap with carbon-atom orbitals containing one electron.

The lowest energy state of a gaseous carbon atom has the electronic structure that can be represented as

Only two bonds, instead of the expected four bonds, can be understood immediately on the basis of this atomic configuration. It is observed spectroscopically, however, that an energy of 272 kJ mol^{-1} will excite one of the $2s$ electrons to the empty $2p$ orbit to give the structure

The expenditure of this *promotional energy* is allowed since four electrons are available for bond formation and the two extra bonds that can be formed will more than make up for the energy required to promote the electron. The chemical fact is that carbon tends normally to share four pairs of electrons as in CH_4 but that the methylene radical $:CH_2$, which does not need to expend any promotional energy, is also a species of considerable consequence.

A difficulty still exists in that the atomic wave functions of carbon would seem to indicate one bond of one type, based on the $2s$ orbital, and three bonds of another type, based on the $2p$ orbitals. The solution to this dilemma was given by Pauling, who pointed out that, in the approximation that tries to explain the bonded system in terms of the free-atom orbitals, the most suitable orbitals of the atom should be used as a basis for the description. Furthermore, linear combinations of the $2s$ and $2p$ wave functions can be made to give four new wave functions that are satisfactory atomic functions for the description of the bonded system. This procedure of combining orbitals to form new ones is called *hybridization,* and the new sets are called *hybrid orbitals.* The most suitable set, according to Pauling, consists of those wave functions which *project out farthest from the central atom* and can therefore concentrate in the region between two nuclei. When the four orbitals that project farthest are constructed from the $2s$ and $2p$ orbitals, one finds, in fact, that these are concentrated along tetrahedral directions. Their angular functions are indicated in Fig. 12-11. Again the complete description requires a radial factor that shows a gradual fall-off of the function away from the nucleus. The overlap of these sp^3 hybrid orbitals with the $1s$ orbitals of the hydrogen atoms will lead to the observed tetrahedral geometry.

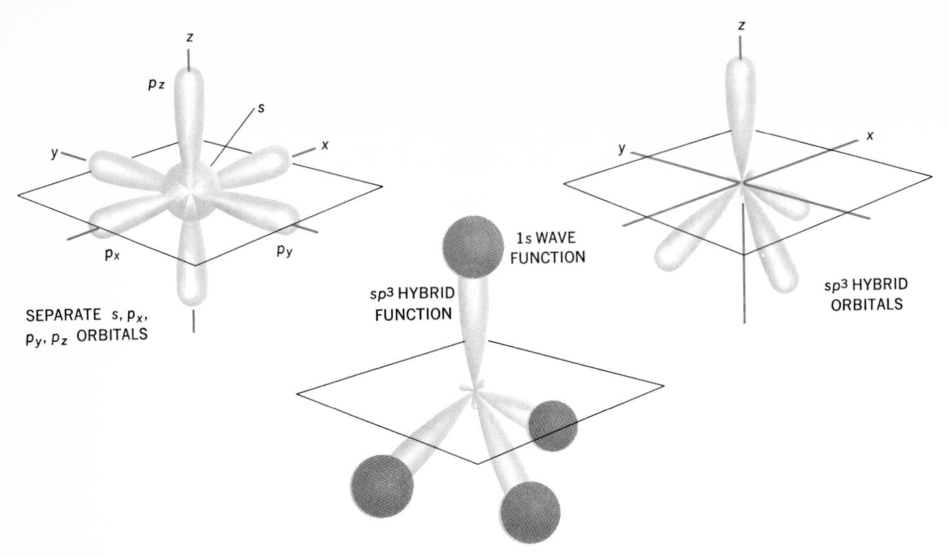

FIGURE 12-11
The sp^3 hybrid orbitals that can be formed from linear combinations of the s and p orbitals and the overlapping orbitals that lead to bonding in methane. (The angular part of the wave functions is shown, distorted for clarity.)

No further description of the bonding in molecules such as CH_4 will be given here. The bonds can be understood, in principle, from the overlap of the sp^3 hybrid orbitals with the $1s$ orbitals of the four hydrogen atoms. The detailed, exact, or nearly exact, calculation of the electron distribution and the bond energies in a molecule of even this complexity—the molecule that seems small to an organic chemist all but overwhelms the theoretician, with its array of five nuclei and ten electrons—is possible, but only with great difficulty. For many chemical purposes, however, even diagrams of the type of Fig. 12-11 are a major step forward from the Lewis diagram representation and the stick model of chemical bonds.

A similar but rather more subtle problem arises when one attempts to describe the bonding in a molecule such as H_2O. The ground-state oxygen-atom configuration is represented as

and as Fig. 12-12a shows, this atomic basis leads to a bonding description with a bond angle of 90° between the two O—H bonds. In view of the CH_4 example, however, the possibility of hybridizing the orbitals in order to form better bonds must be considered as an alternative description. If tetrahedral hybrid orbitals are formed, the bonded system is represented in Fig. 12-12b. The better bonds are formed, however, at the expense of promoting a pair of electrons from the low-lying $2s$ orbit to the somewhat higher-energy sp^3 hybrid orbital. The merits of the two descriptions of Fig. 12-12 cannot easily be decided by quantum-mechanical attacks, again because of the complexity of the sys-

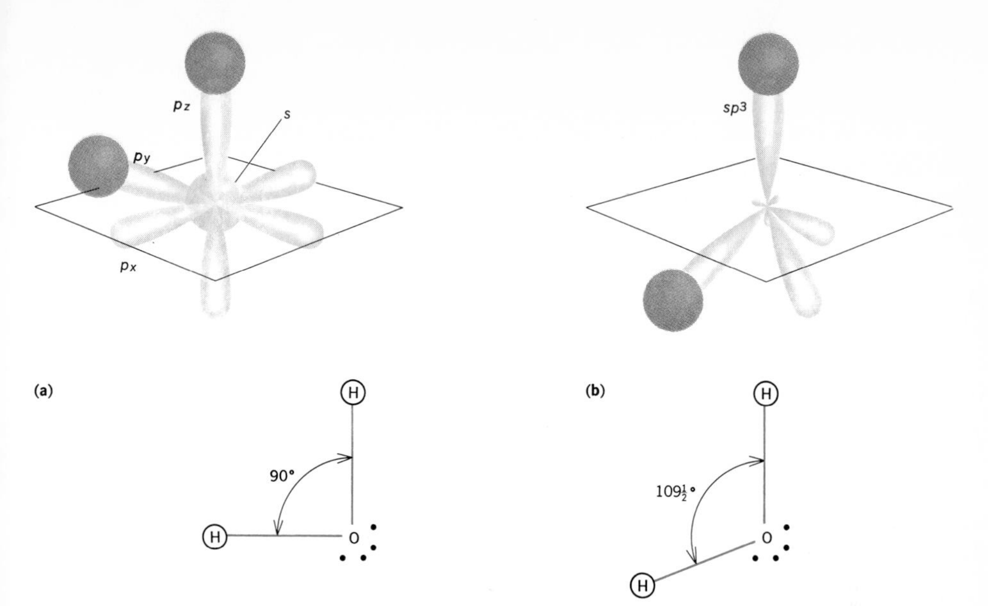

FIGURE 12-12
Two descriptions of the bonding in H_2O. The observed angle between the two O—H bonds is about 105°. (*a*) H_2O based on s, p_x, p_y, and p_z orbitals of oxygen. (*b*) H_2O based on sp^3 hybrid orbitals of oxygen.

tem. The experimental result that the bond angle in water is 105°, as we shall see in the following chapter, suggests that some intermediate description is preferable.

Finally, it can be mentioned that in a molecule such as HF a similar difficulty arises. Two possible descriptions of the bonding are given in Fig. 12-13. No different nuclear arrangements are suggested by these different pictures, but different electronic distributions are implied. The electron positions, particularly those of the nonbonding electrons, create considerable havoc when attempts are made to interpret the dipole moment of a molecule simply in terms of its ionic character. This problem will be encountered in the following chapter.

You should recognize that hybridization is a procedure used to generate orbitals that are suitable to describe the bonding in molecules, from orbitals that were constructed from studies of hydrogenlike atoms. An alternative, more direct route to the description of the electrons of molecules depends on the *postulate* that orbitals, more or less spherical, can be constructed around an atom and that these

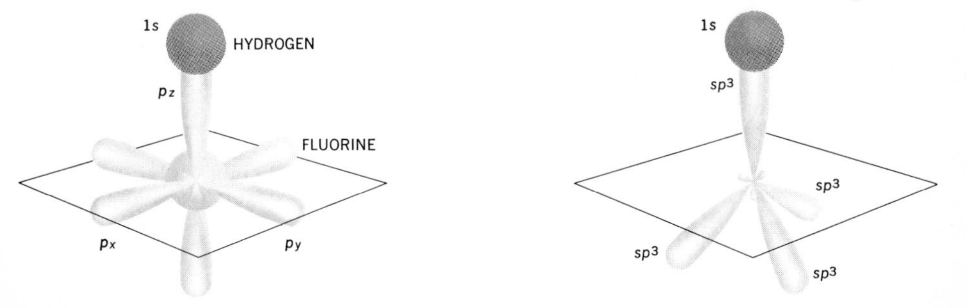

FIGURE 12-13
Two descriptions of the bonding in HF.

can accommodate pairs of electrons in bonding or nonbonding situations. The atoms of most second-row elements are postulated as being able to accommodate four such electron-pair orbitals. Larger atoms are able to accommodate five or six such pairs. The approach is similar to the valence-bond-hybridization treatment, but it has some special values in that it is clearly a model whose success in predicting structures is its justification. This also is the justification of the hybridization approach, but its use of more mathematically described functions can obscure this.

12-5 π Bonding

The chemist's tendency to describe bonds in terms of atomic orbitals leads him to distinguish two ways in which a directed orbital, such as a p orbital, can overlap with a similar orbital of an adjacent atom. As a result, as was shown in Sec. 12-1, bonds stemming from atomic p orbitals can be σ or π bonds. These possibilities exist, as we will now see, in polyatomic as well as diatomic molecules.

The bonding in a compound such as ethylene can be most neatly described—there are, of course, any number of other ways of attempting a description—in terms of carbon-atom hybrids that combine the $2s$ and two of the $2p$ orbitals into *trigonal hybrids*. These hybrids are planar arrangements, as shown in Fig. 12-14. Such trigonal hybrids on each of the carbon atoms can overlap, as indicated, to form the molecular skeleton from σ bonds. The remaining p orbitals on each

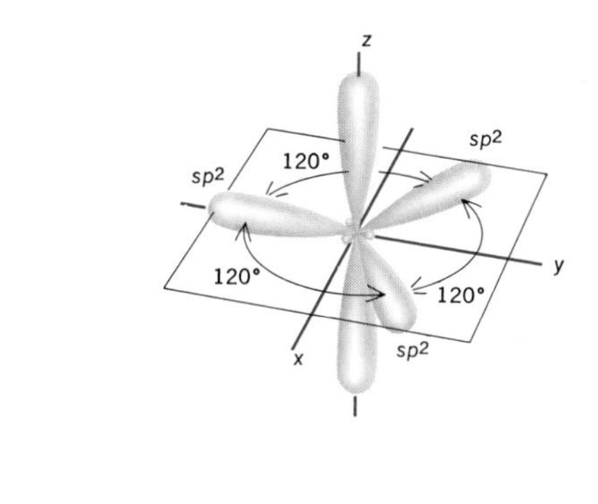

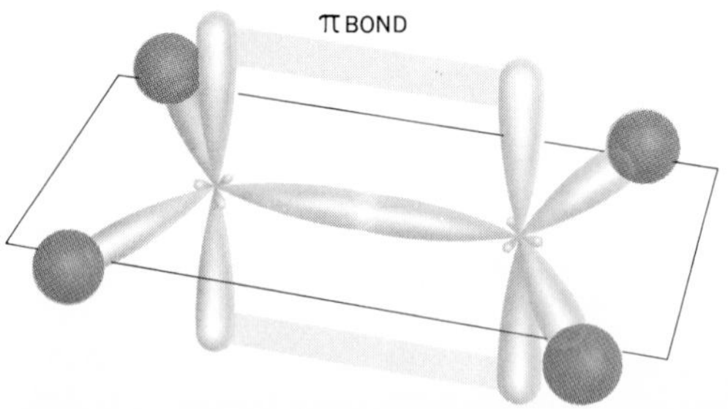

FIGURE 12-14
sp^2 trigonal hybrids and their use in a description of the bonding in ethylene $H_2C{=}CH_2$. (The distorted orbitals fail to show that the two p_z orbitals do overlap and form a bond.)

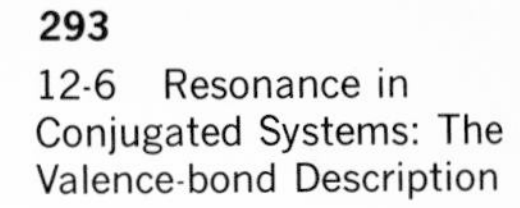

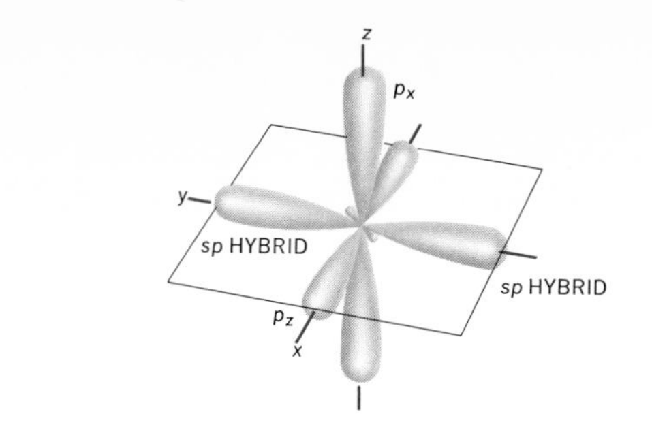

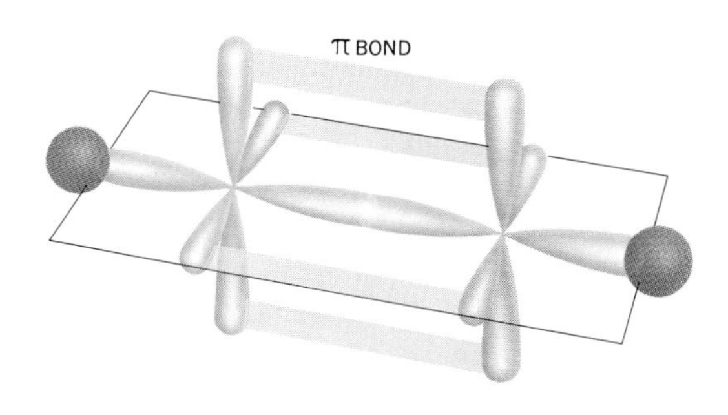

FIGURE 12-15
sp hybrid orbitals and their use in a description of the bonding in acetylene H—C≡C—H.

carbon atom are perpendicular to the plane of the trigonal hybrid and are in a position to form a π bond. In this way, the observed planar configuration of ethylene and the resistance of the double bond to rotation are nicely accounted for.

In a similar manner the triple bond of acetylene can be described. Now the carbon atoms form σ bonds from the linear *sp* hybrid of the 2*s* and one of the 2*p* orbitals. The remaining two 2*p* orbitals of each carbon atom are at right angles to this linear arrangement. The observed linear structure of acetylene is accounted for by the bonding of σ bonds and two π bonds in perpendicular planes, as indicated in Fig. 12-15.

The π-bond type of description of multiple bonds has the advantage for chemical interpretations and theoretical treatments of separating the electrons of the system into a set that forms the fairly ordinary σ bonds and a set that lends the characteristic unsaturation features to the system. The advantages of this division will be seen in the following section and in later discussions on the mechanisms of reactions.

12-6 Resonance in Conjugated Systems: The Valence-bond Description

Aromatic compounds present an area that has long been of great interest to organic chemists and of special interest also to physical chemists. The preliminary task of the theoretician has been, since 1865, that of providing a quantum-mechanical basis for the Kekule description of aromatic systems as *resonance hybrids* of the various bonded arrangements that can be drawn. The representation of benzene by Kekule as

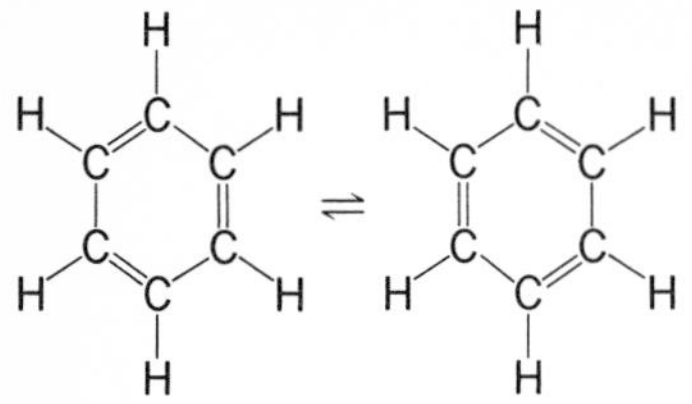

greatly advanced the chemistry of aromatics, and it is necessary now to analyze, in quantum-mechanical terms, the significance of this representation.

The π-bond description is here convenient because, as Fig. 12-16*a* shows, the benzene skeleton can be constructed from carbon sp^2 hybrids that form σ bonds to each other and to the hydrogen atoms. It remains to investigate the role of the six electrons in the six p orbitals perpendicular to the plane of the molecule. If we desire to describe the bonding in terms of the now familiar overlap of adjacent p orbitals to form π bonds, it is found that there are two different ways of overlapping the p orbitals, as indicated in Fig. 12-16*b*. An approximate description of the bonded system might make use of the overlap arrangement of ψ_{I}, which would tend to build up the electron density between atoms 1 and 2, 3 and 4, and 5 and 6. The alternative description, ψ_{II}, which builds up electrons between 2 and 3, 4 and 5, and 6 and 1, is equally good, however. It would be better, instead of using either ψ_{I} or ψ_{II}, to form an approximate wave function for the system

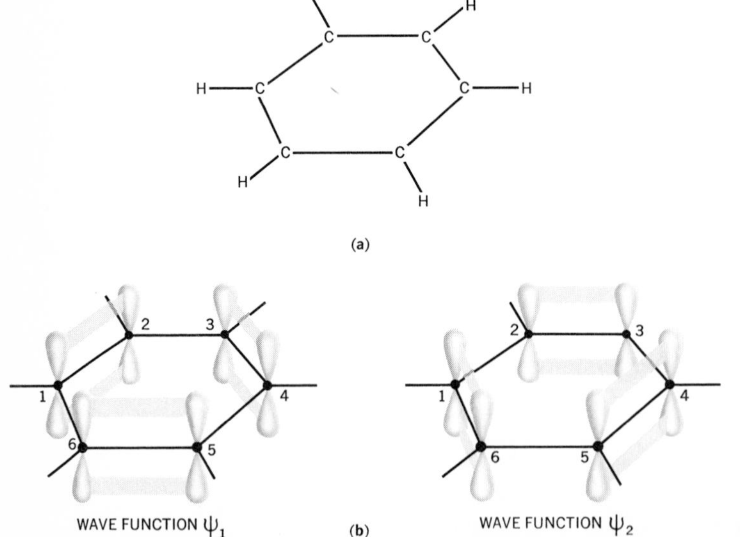

FIGURE 12-16
A description of the bonding in benzene, using sp^2p_z orbitals on the carbon atoms and π-bond formation between adjacent p_z orbitals. (*a*) The bonded skeleton of benzene based on sp^2 trigonal hybrids of the carbon atoms. (*b*) Two ways of overlapping adjacent p orbitals to form π bonds in the benzene molecule.

that embodied the features of both I and II. One might try, as an approximate description,

$$\psi = \psi_{\mathrm{I}} + \psi_{\mathrm{II}} \qquad [12]$$

An even better description would include wave-function terms corresponding to the overlap implied by the Dewar structures

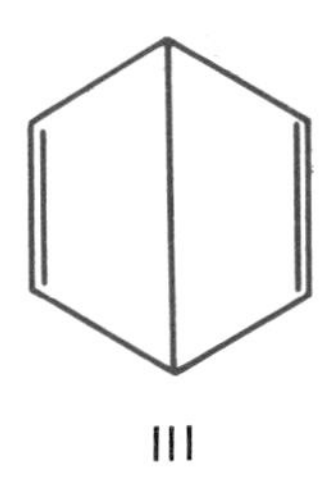

III

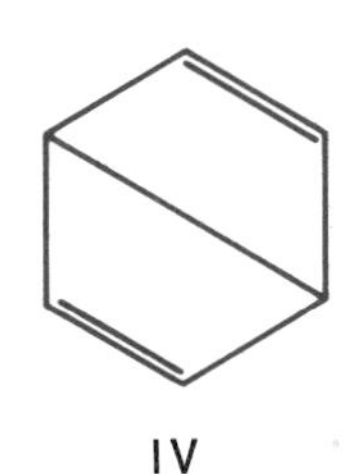

IV

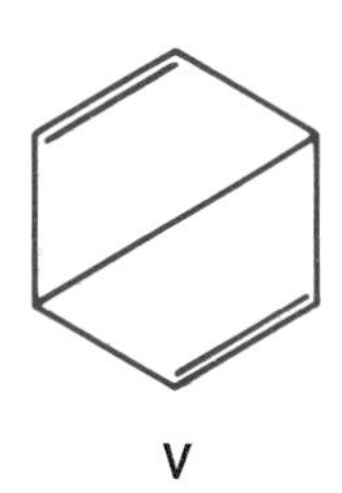

V

A better approximation to the π electron distribution in benzene would then be

$$\psi = a\psi_{\mathrm{I}} + a\psi_{\mathrm{II}} + b\psi_{\mathrm{III}} + b\psi_{\mathrm{IV}} + b\psi_{\mathrm{V}} \qquad [13]$$

where b is presumably less than a.

The chemical description of this approximate procedure, which attempts to describe the system in terms of familiar, but not too satisfactory, bonding arrangements, is that the structures ψ_{I} to ψ_{V} are *resonance structures* and that the molecule is a *resonance hybrid* of these structures. The quantum-mechanical approach shows that resonance structures are merely partial descriptions, based on the valence-bond approach, that can be used to build up a more complete description.

Our previous experience with improvements of electronic descriptions, in the Heitler-London treatment of the H_2 molecule and the introduction of ionic terms in heteronuclear-bond descriptions, has illustrated the variation-theorem result that better descriptions correspond to lower energies. This principle can again be checked, although a number of interesting problems are present which cannot be dealt with here, by comparing the actual heat of formation of a molecule such as benzene with that which would be expected on the basis of the bonds drawn in one of the Kekule structures. From Appendix 5 the heat-of-formation reaction of benzene is given as

$$6\mathrm{C(graphite)} + 3\mathrm{H_2}(g) \rightarrow \mathrm{C_6H_6}(g) \qquad \Delta H = 83\ \mathrm{kJ} \qquad [14]$$

Furthermore,

$$6\mathrm{C(graphite)} \rightarrow 6\mathrm{C}(g) \qquad \Delta H = (6)(718) = 4308\ \mathrm{kJ} \qquad [15]$$

and

$$3\mathrm{H_2}(g) \rightarrow 6\mathrm{H}(g) \qquad \Delta H = (3)(440) = 1320\ \mathrm{kJ} \qquad [16]$$

From these data the heat of formation of benzene from the gaseous atoms is obtained as

$$6C(g) + 6H(g) \rightarrow C_6H_6(g)$$
$$\Delta H = -4308 - 1320 + 83 = -5545 \text{ kJ} \quad [17]$$

The energy of the hypothetical structure I or II is estimated from the bond energies of Table 7-4 as

$$\begin{aligned} 6D_{\text{C-H}} &= (6)(415) = 2490 \\ 3D_{\text{C-C}} &= (3)(344) = 1032 \\ 3D_{\text{C=C}} &= (3)(615) = \underline{1845} \\ &\qquad\qquad\quad 5367 \text{ kJ} \end{aligned}$$

The difference between these two results, 178 kJ, known as the *resonance energy,* is in the direction expected from the variation theorem; i.e., more heat is evolved in the formation of benzene than would be expected on the basis of one of the Kekule structures.

Although benzene offers the nicest and the classical example of resonance, the same phenomena can be recognized in other aromatic and unsaturated systems. It is necessary only that there be more than one reasonable way in which a bond diagram for the molecule can be drawn. This is always the case when the molecule can be said to have an alternating arrangement of multiple and single bonds, known as *conjugation.*

12-7 The Molecular Orbital Approach to Aromatic and Conjugated Molecules

Descriptions of the π-electron structure of aromatic and conjugated molecules can also be based on molecular orbitals. This approach, in fact, is much more amenable to extension to quantitative calculations and to studies of excited electronic states.

We begin again with the benzene example at the stage where the π-orbital system of the molecule is to be treated separately from the σ-bonded framework, as in Fig. 12-16. Now molecular orbitals are to be constructed from the six carbon-atom p orbitals so that six electrons can be accommodated. Here we will proceed pictorially, counting on the number of nodes in the resulting molecular orbitals to be a guide to their energies. The lowest- and highest-energy combinations, shown at the bottom and top of Fig. 12-17, can be constructed on the basis used in the diatomic examples of Sec. 12-1. The wave functions corresponding to these diagrams are

$$\psi_1 + \psi_2 + \psi_3 + \psi_4 + \psi_5 + \psi_6$$

and

$$\psi_1 - \psi_2 + \psi_3 - \psi_4 + \psi_5 - \psi_6$$

ENERGY ⟶

$$\frac{1}{\sqrt{6}}(\psi_1 - \psi_2 + \psi_3 - \psi_4 + \psi_5 - \psi_6)$$

$$\begin{cases} \dfrac{1}{\sqrt{12}}(2\psi_1 - \psi_2 - \psi_3 + 2\psi_4 - \psi_5 - \psi_6) \\ \dfrac{1}{2}(\psi_2 - \psi_3 + \psi_5 - \psi_6) \end{cases}$$

$$\begin{cases} \dfrac{1}{\sqrt{12}}(2\psi_1 + \psi_2 - \psi_3 - 2\psi_4 - \psi_5 + \psi_6) \\ \dfrac{1}{2}(\psi_2 + \psi_3 - \psi_5 - \psi_6) \end{cases}$$

$$\frac{1}{\sqrt{6}}(\psi_1 + \psi_2 + \psi_3 + \psi_4 + \psi_5 + \psi_6)$$

FIGURE 12-17
Combinations of the atomic π orbitals that provide suitable molecular-orbital functions for benzene. Factors are included so that, if the atomic orbitals are normalized and orthogonal, the molecular orbitals will be normalized.

Use of these functions, with each ψ assigned an atomic p-orbital function located on the appropriate carbon atom, would lead, with a variation-type calculation of Eq. [16] of Chap. 11, to the relative energies implied in Fig. 12-17.

Two remaining types of orbitals of intermediate energy can also be constructed. To maintain the symmetry of the hexagon, pairs of these must be constructed so that linear combinations of the members of the pair will avoid the implication of any distinguishing character for one or more of the atoms or bonds. Sets of such orbitals, placed on the energy scale according to their number of nodes, are included in Fig. 12-17. At each of the two intermediate energy levels there are therefore two molecular orbitals, and these energies thus have a degeneracy of 2.

Assignment of the six π electrons can now be made to suggest the ground state of benzene, as in Fig. 12-18. Again the energy of the molecular system compared with that with no delocalization of π electrons can be treated. Furthermore, the spectral results for the energies of electronic transitions to higher states can be interpreted.

It is instructive to consider also an approach to molecular orbitals that does not depend on an LCAO procedure. Here "free-electron" molecular orbitals will be illustrated. The example of benzene could again be used, but for variety a nonaromatic conjugated system will be treated. Consider a fairly long conjugated system such as β-carotene:

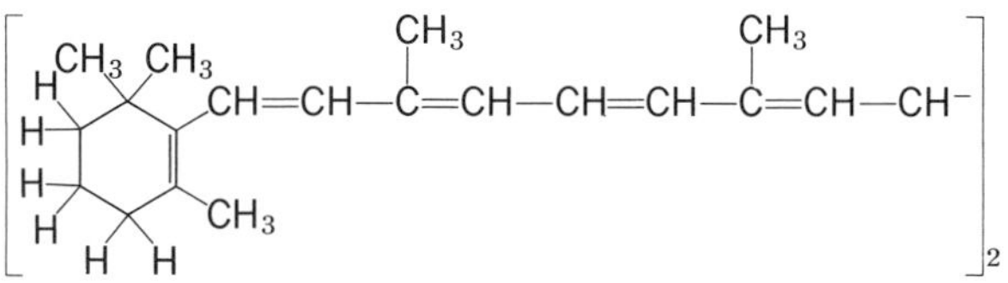

$$\left[\text{(ring)}\!-\!CH{=}CH{-}\overset{CH_3}{\overset{|}{C}}{=}CH{-}CH{=}CH{-}\overset{CH_3}{\overset{|}{C}}{=}CH{-}CH{-} \right]_2$$

If resonance structures are drawn, it will be noticed that each carbon-carbon bond along the chain has appreciable double-bond character. The π electrons are therefore not localized in one bond, but are

FIGURE 12-18
Occupation of the molecular orbitals of Fig. 12-17 in the ground state of benzene.

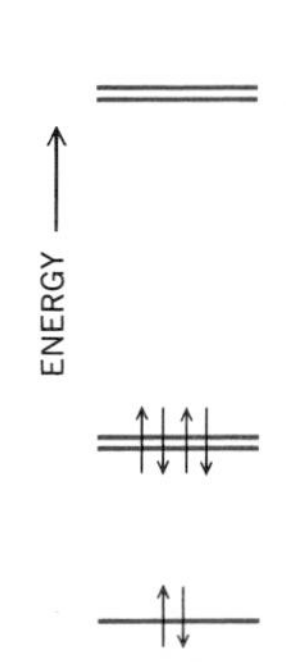

relatively free to move throughout the whole carbon skeleton. This suggests that the skeleton be considered as a roughly uniform region of low potential bounded, at the ends of the molecule, by regions of infinitely high potential. The resulting square potential well is to be the receptacle for the 22π electrons. An expression for the allowed energies of these electrons, electron-electron repulsions being ignored, has been obtained in Sec. 3-5. The energies are given by the expression

$$\epsilon = \frac{n^2h^2}{8ma^2} \qquad n = 1, 2, 3, \ldots \qquad [18]$$

where a is the effective length of the molecule. Two electrons, one with each spin direction, can be placed in each quantum level. The electron description so obtained is shown in Fig. 12-19.

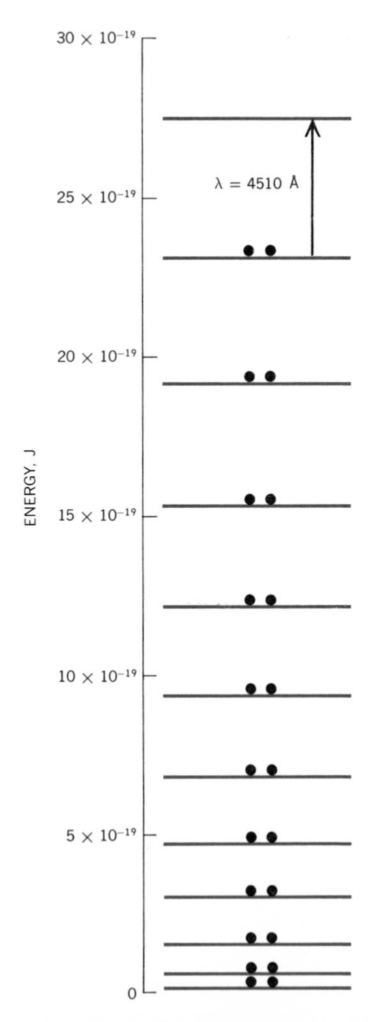

FIGURE 12-19
The energy levels and the observed spectroscopic transition for the 22π electrons of β-carotene according to the free-electron or square-potential-well model.

The chief merit of this treatment of conjugated systems is that it offers an easy approach to relating the calculation of the wavelength of light that is absorbed to the length of the conjugated system. In Fig. 12-19, for example, the transition indicated gives rise to an absorption band that is found to be centered at 4510 Å. Absorption at this wavelength implies the absorption of quanta of energy

$$\Delta\epsilon = \frac{(6.62 \times 10^{-34})(3 \times 10^{8})}{4510 \times 10^{-10}}$$

$$= 4.41 \times 10^{-19}\ \text{J} \qquad [19]$$

According to the free-electron model for the π-electron system, this result can be related to the energy difference

$$\epsilon_{12} - \epsilon_{11} = (12^2 - 11^2)\frac{h^2}{8ma^2} \qquad [20]$$

Substitution of numerical values leads to the deduction of a, the length of the region in which the electrons are free to move, as

$$a = 17.7\ \text{Å} \qquad [21]$$

This value, as we shall see when experimental methods for determining molecular dimensions are developed in the following chapters, is very close to that which would be expected for the extended conjugated carbon chain of the β-carotene molecule.

This molecular-orbital approach, generally making use of linear combinations of atomic orbitals, is the one most often used when quantitative quantum-mechanical treatments of conjugated systems are made.

12-8 Bonding with d Orbitals

In the preceding sections some of the approaches used to describe bonding between atoms that have partially filled s and p orbits have been described. These approaches are appropriate to organic compounds, except perhaps those involving elements such as chlorine, sulfur, and phosphorus. Inorganic compounds, on the other hand, frequently contain an atom, or atoms, that have partially filled d orbits. For the study of such compounds, an understanding of the geometry of d orbits is necessary.

Algebraic expressions can be given for the Schrödinger equation atomic solutions with $l = 2$ in the same way as they were given in Table 11-2 for $l = 0$ and $l = 1$. For qualitative bond descriptions, however, the angular-function diagrams, like those for $l = 0$ and $l = 1$ in Fig. 11-3, are adequate.

The five functions that arise for $l = 2$ can be combined in a variety of ways to give a resulting function suitable for various purposes. For

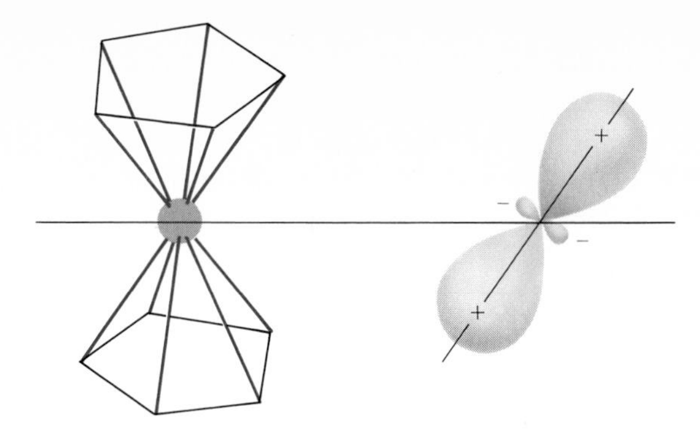

FIGURE 12-20
The directions and shapes, as given by the angular factor, for the five equivalent d orbitals that can be constructed. [*See R. E. Powell, J. Chem. Educ.,* **45**:*45* (*1968*).]

angular-momentum considerations, when an atom is subjected to an imposed direction, those that correspond to angular-momentum components of ± 2, ± 1, and 0 along this direction would be used. On the other hand, to illustrate the fivefold degeneracy of the set of d orbitals, in the absence of any disturbance, the five equivalent functions of Fig. 12-20 would be shown.

Most useful, however, are d-orbital functions that have a simple, direct relation to the coordinate axes, for it is these that can be used, pictorially, to treat octahedral and tetrahedral situations. Such functions are shown in Fig. 12-21. Three of the orbitals project between the cartesian axes and lie, as shown by the subscript identifications, primarily in planes defined by pairs of axes. The remaining two orbitals are concentrated along the axes.

One might proceed in a straightforward manner to describe bonding involving transition metals, where the partially filled d orbitals play a dominant role, in terms of the d-orbital atomic wave functions. There

FIGURE 12-21
A representation of the angular dependence of the five d orbitals that is convenient for consideration of octahedral systems.

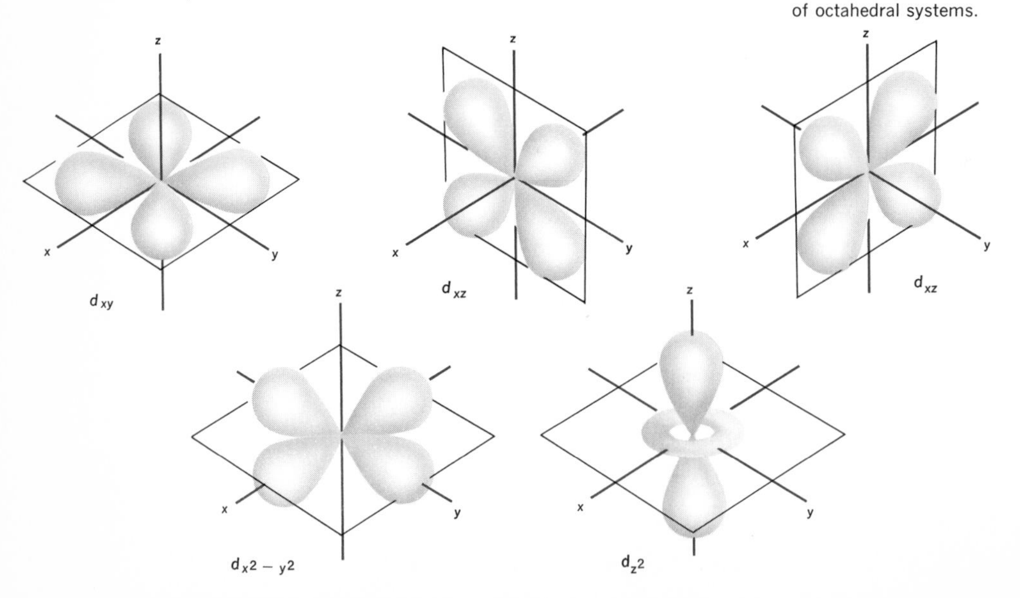

is, however, the possibility that hybrid orbitals provide a suitable basis for describing bonded systems. If it is assumed that, as Fig. 11-9 suggests, the $3d$ orbital has approximately the same energy as the $4s$ and $4p$ orbitals, then some of these can be combined to form hybrid orbitals, just as s and p orbitals were previously combined. A combination that is suitable for the description of the important octahedral class of compounds is that of d^2sp^3, for which the six hybrid orbitals are equivalent and project toward the corners of an octahedron.

12-9 Bonding in Coordination Compounds

The largest class of compounds in which d orbitals are involved in bonding is that of coordination compounds of transition-metal ions. Compounds such as these consist of a central transition-metal ion with attached electron-rich groups called *ligands*. Transition elements are characterized by incompletely filled $3d$, $4d$, or $5d$ orbitals, and it is therefore the role of these d orbitals that must be understood if the bonding in coordination compounds is to be described.

Three stages in the development of a description of such compounds can be recognized. The first has us consider, specifically, the pairs of electrons on each ligand but has us characterize the metal ion to which they are attached as a simple, positively charged sphere that in some undefined way accommodates the pairs of ligand electrons. This model suggests the Lewis acid-base character of coordination compounds. It also leads us to expect nothing other than the symmetric tetrahedral or octahedral packing of ligands around the metal ion, and these arrangements are, in fact, the ones most commonly adopted. But this simple model is in many ways inadequate. It fails to account for the most obvious transition—metal coordination-compound feature—the variety in color. Likewise, it gives us no basis for expecting, for example, four-coordinated planar and distorted octahedral structures that occur in some cases.

A second approach takes, in a way, an opposite viewpoint. Having accepted the overall attraction of the charged, or at least polar, ligands for the positive metal ion, this approach treats the electrons of the metal ion in some detail. In doing so, it assumes that we need be concerned only with the shape or symmetry of the electric field imposed by the surrounding ligands and need not deal in detail with the electron pairs of the ligands. This approach is known as the *crystal-field theory*. (The name stems from the initial use of the approach to describe the effect on a metal ion in a crystal of the electric field exerted as a result of the presence and arrangement of neighboring ions in the crystal.) The theory is most easily introduced by considering the effect of the octahedral field of six ligands on a metal ion with a single outer d electron.

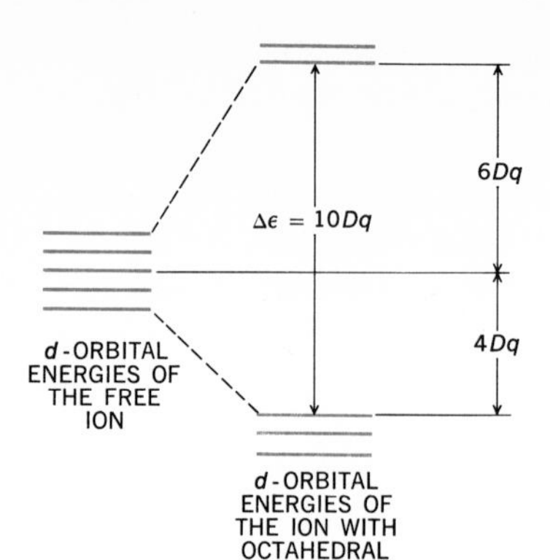

FIGURE 12-22
Crystal-field splitting of d orbitals by an octahedral field.

The Ti^{3+} ion has such an electron configuration. This free gaseous Ti^{3+} ion is spherical because the single d electron can occupy each of the five $3d$ orbitals with equal probability, the orbitals having equal energies.

Now, how are orbital energies affected by the octahedral field that ligands would impose, as in the $Ti(OH_2)_6{}^{3+}$ ion, for example? The effect is to split the five d orbitals into two sets, one of three orbitals and the other of two orbitals. For negatively charged, electron-repelling ligands, the set of three orbitals corresponds to a lower energy, and the set of two orbitals to a higher energy. This splitting of the d-orbital set is illustrated in Fig. 12-22. (The splitting and the energy effect can be described in terms of two higher-energy orbitals that point at the electron-repelling ligands, and three lower-energy orbitals that point between and avoid these ligands, as can be seen from the diagrams of Fig. 12-21.)

This simple idea of a crystal-field splitting of the d orbitals—$3d$ orbitals in the case of the $Ti(OH_2)_6{}^{3+}$ ion—provides a basis for describing the electronic transition that produces the violet color of this species. The comparison, furthermore, allows the value of the crystal-field splitting to be deduced. The procedure consists in assigning the metal-ion electron to the lower-energy orbital set, as in Fig. 12-23, and attributing the color-producing electronic transition to the process shown there. In this way, the model is further developed by the assignment of a value to the crystal-field d-orbital splitting, which is represented as $10Dq$, a collection of terms that here will be treated as a single parameter.

FIGURE 12-23
The assignment of the d electron of Ti^{3+} in the $Ti(OH_2)_6{}^{+}$ ion to the split d orbitals and the transition responsible for the color of the ion.

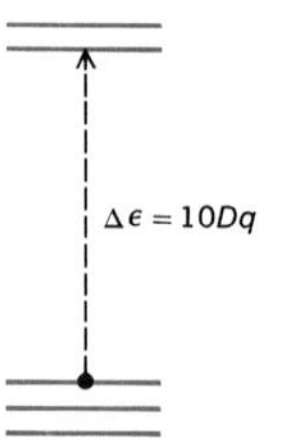

When other one-electron metal ions are considered, different amounts of splitting of the $3d$ orbital degeneracy must be expected. A diagram that then can be used is shown in Fig. 12-24. Each coordination compound with one d electron and octahedrally placed ligands can be located on the abscissa by using spectral results for the extent of splitting. Similar but more complicated diagrams can be drawn for metal ions with more than one d electron, the complexity arising from the variety of atomic states that then can exist and the degeneracy of some of these states that is split by the field of the ligands.

The third step in the refinement of descriptions for the bonding in coordination compounds takes account of the detailed involvement of both the metal-ion and ligand orbitals and electrons. Two quite different approaches have been used. The earlier one, due primarily to the work of Pauling, has its origins in the valence-bond approach, the later one, known as the ligand-field method, in the molecular-orbital approach.

The first approach attempts to describe the bonding in much the same way, using covalent and ionic terms, as has already been done for bonds using s and p orbitals. If, according to Pauling, the covalent contribution is important, the metal atom will present empty projecting hybrid orbitals in the direction of the ligands so that overlap can occur and a good bond can be formed. The electron pair that enters the bonding orbital is that of the ligand, as, for example, is suggested by the diagram

$$Co^{3+}(:NH_3)_6$$

The bonds so formed produce the structure

$$\left[\begin{array}{ccc} & NH_3 \quad NH_3 & \\ H_3N & — Co — & NH_3 \\ H_3N & NH_3 & \end{array}\right]^{3+}$$

Such bonds are examples of coordinate or coordinate covalent bonds in contrast to the bonds previously studied in which one electron was donated by each atom forming the bond. According to Pauling, such ligand attachment can occur, and when covalency is sufficiently important, the metal atom will present the suitable projecting hybrid orbitals to the ligands. Thus the example $Co(NH_3)_6^{3+}$ is described in terms of a Co^{3+} ion with outer electronic structure

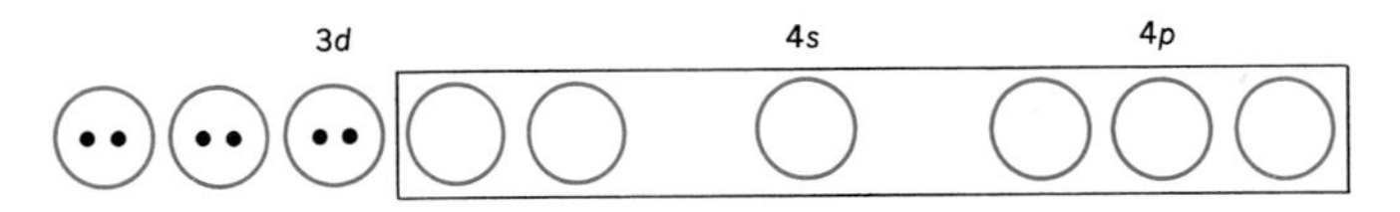

where the six outer electrons are pushed back into three of the d orbitals to allow the formation of empty d^2sp^3 octahedral hybrid orbitals for covalent-bond formation with the ligands.

Alternatively, according to Pauling, each ligand may have less of a tendency to donate a pair of electrons to a covalent bond, and the ligands are then attracted to the central metal atom by essentially electrostatic, or ionic, forces. Such a situation appears to correspond to the ion CoF_6^{3-}, and in this case the most stable arrangement for

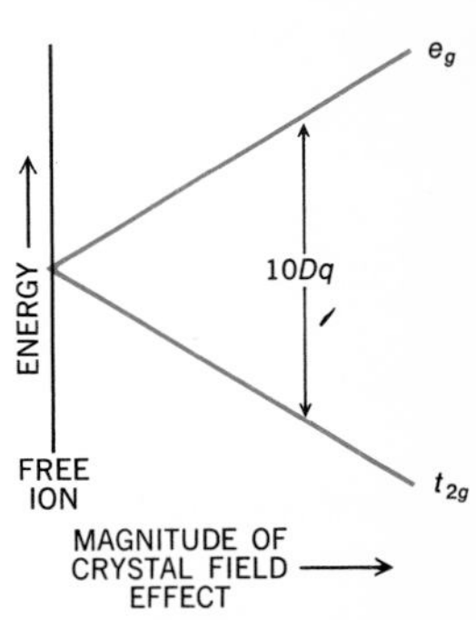

FIGURE 12-24
Splitting of the d-orbital energies as a function of the field produced by octahedrally placed ligands. The symmetry designations e_g and t_{2g} and the splitting parameter $10Dq$ are included. In symmetry notation e implies a doubly degenerate set, t a triply degenerate set, and the subscript g denotes a function that is symmetric about a center of symmetry.

the metal atom has the electrons unpaired and represented by

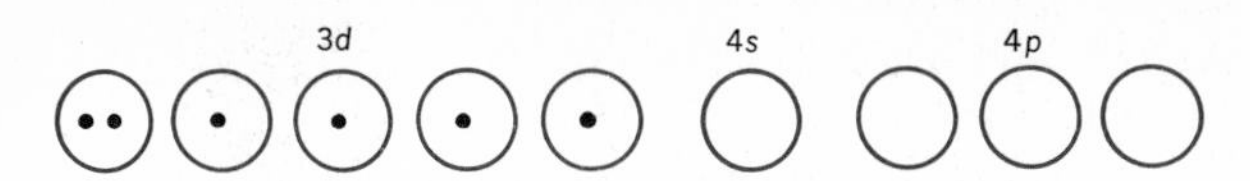

In Chap. 15 an experimental method will be presented that allows the deduction of the number of unpaired electrons in a compound. Such experiments indicate no unpaired electrons in $Co(NH_3)_6^{3+}$ and four unpaired electrons in CoF_6^{3-}. It is with such data in mind that Pauling set up his theory of coordination compounds, and the explanation of the possible different numbers of unpaired electrons in different compounds of an element is seen to be neatly given.

The molecular-orbital approach, which since its development has steadily gained favor over the Pauling valence-bond treatment, requires the development of a set of orbitals that will accommodate the involved electrons of the metal ion and of the ligands. This can be done most easily, at least diagrammatically, by assuming that in the transition-metal ions only the outer d-orbital electrons and the pair of electrons in the orbital of each ligand projecting toward the metal ion need to be accommodated.

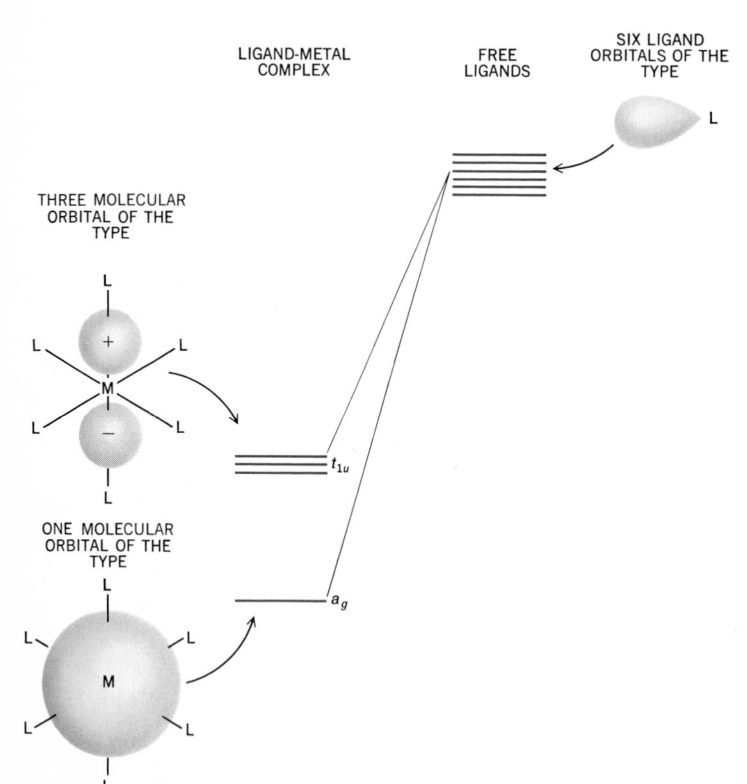

FIGURE 12-25
The two types of molecular orbitals formed in an octahedral complex by ligand orbitals for which no combination with metal d orbitals can occur. (Metal s and p orbitals can interact, but this is not considered here.) The symmetry designation a implies a symmetric, nondegenerate function, and the subscript u as in t_{1u}, a function that is antisymmetric with respect to a center of symmetry.

Consider the specific case of an octahedral complex. What orbitals will describe how the $3d$ or $4d$ electrons of the metal ion and the 12 electrons of the ligands behave? Two relatively low-energy orbital sets can be made by combining the ligand orbitals to give no-node and one-node functions, as shown in Fig. 12-25, that are the counterparts of the lowest-energy atomic or particle-in-a-cubic-box orbitals. The next-higher functions that might be made in this way from ligand orbitals are the four-node cloverleaf-type functions. These, however, are compatible with the metal d orbitals of Fig. 12-21 that project along the coordinate axes. It follows that we must make bonding and antibonding combinations to show the complete set of molecular orbitals. Remaining unaffected, in this simple treatment, by the ligand orbitals are the three d orbitals that project between the ligand directions and thus do not lead to bonding and antibonding combinations.

The result of this entire orbital-construction procedure is the set of molecular-orbital energies for octahedral complexes shown in Fig. 12-26. This diagram, you should recognize, is the counterpart of the hydrogen-atom orbital diagram that can be used to describe the electronic structures of the elements of the periodic table. Now we have a diagram that can be used to describe the electronic structure of any octahedral coordination compound.

The uses of the diagram can be illustrated by considering again the two complexes of Co^{3+}, namely, CoF_6^{3-} and $Co(NH_3)_6^{3+}$. The four unpaired electrons of the former, in contrast to no unpaired electrons of the latter, can be accounted for by attributing different energy-spacing effects to the different ligands. The energy separation between the bonding and antibonding orbitals of Fig. 12-26 will thus be different. The result, as in the Co^{3+} example of Fig. 12-27, accounts for the various numbers of unpaired electrons in different complexes of a given transition-metal ion.

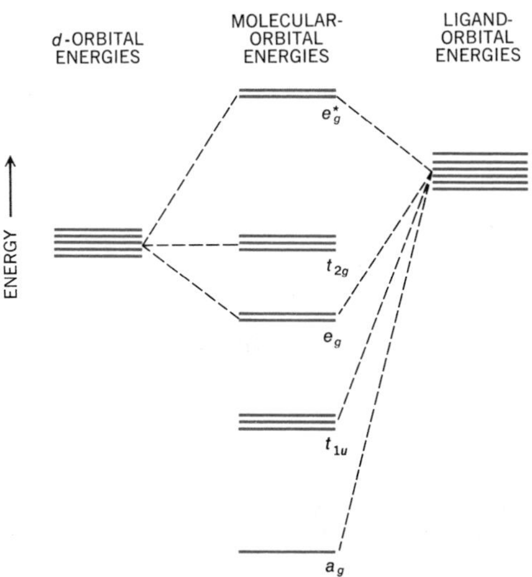

FIGURE 12-26
The pattern of molecular-orbital energies of an octahedral complex deduced from the effect of symmetry and the interactions between the ligand orbitals and the d orbitals of the metal.

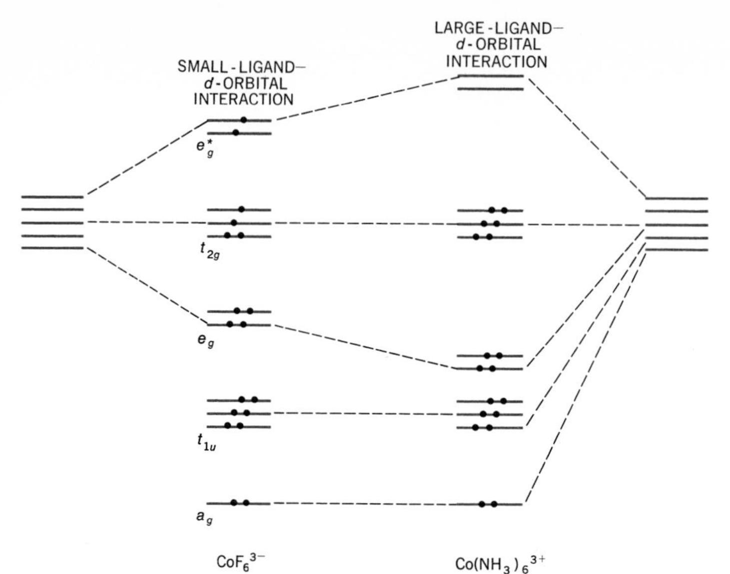

FIGURE 12-27 The use of the orbital diagram to explain the presence of four unpaired electrons in CoF_6^{3-} and none in $Co(NH_3)_6^{3+}$, both being complexes of the Co^{3+} ion which has six d electrons. (In fact, other important factors affecting the orbital-energy pattern must be taken account of.)

As for the crystal-field model, but in contrast to the valence-bond treatment, orbital diagrams like that of Fig. 12-26 can be used to interpret the colors of coordination compounds and the spectral result for the energy separation between states of the complex. It is the relation of the ligand-field theory to the extensive results from spectral studies, as well as to the data on structures and magnetic properties, that in part is responsible for its general adoption.

Problems

1 Explain, on the basis of the molecular-orbital diagram of Fig. 12-5, why the O_2^- ion, as found in the superoxide salts like KO_2, which has one more electron than does O_2 and has a longer bond than O_2, whereas for N_2, which has one more electron than the ion N_2^+, the N_2 bond is shorter than that of N_2^+.

2 Show the equivalence between the molecular-orbital and the valence-bond descriptions of the bonding in N_2.

3 Verify the differences in electronegativity shown in Table 12-2 for the atoms of the bonds H—F and —C—H (with C showing two further vertical bonds), using the bond energies of Table 7-4.

Assigning a value of 2.2 to the electronegativity of H, calculate values for C and F. Compare the difference in electronegativity obtained for C and F with the value that results from considerations of the —C—F bond (with C showing two further vertical bonds).

4 Using the ionization potentials and the electron affinities given in Tables 11-3 and 11-4, verify the electronegativities given in Table 12-2 for H, Cl, and F.

5 In view of the electronegativity values of Fig. 12-9, characterize metals by a statement about their electronegativities.

6 The hybridization of atomic orbitals to provide new functions that are a more suitable base for the description of bonding can be illustrated by a similar combination of wave functions of the one-dimensional particle-in-a-box model. Form functions that are the sum and the difference of the two lowest-energy solution functions. Plot these new hybrid functions, and plot the probability functions to which they correspond. Use each hybrid function in a variation-method treatment, and obtain the energy to which it corresponds. Plot these energies along with those of the original solution functions, and compare the two sets of values.

7 Verify that the angle between any pair of bonds in a tetrahedral structure is $109\frac{1}{2}°$.

8 Represent the bonding in the molecules NH_3, BF_3, H—C(=O)—H, and H—C≡N by:

a Lewis diagrams.

b Diagrams, such as those of Figs. 12-12 and 12-15, which indicate the angular factor in the wave function and therefore something of the position in space occupied by the electrons.

9 The bonding in ethylene and in acetylene can be described in many ways other than that which uses sp^2 or sp orbitals forming σ bonds and p orbitals forming π bonds. A quite satisfactory description can be based on sp^3 hybrid orbitals. Show bonding pictures based on these hybrid orbitals for ethylene and for acetylene.

10 The heat liberated when 1 mol of hydrogen is added to 1 mol of cyclohexene is 120 kJ mol^{-1}. The heat liberated when 3 mol of hydrogen is added to 1 mol of benzene is 208 kJ. What value do these data suggest for the resonance energy of benzene? *Ans.* 153 kJ mol^{-1}.

11 Use a stick-model diagram to show the σ bonds in butadiene. Draw in the additional p orbitals on each carbon atom, and show how resonance requires the planarity of the molecule. Indicate the way in which the orbitals are occupied in each of the three most important resonance descriptions.

12 Using a stick-model diagram of the σ bonds, draw the most important resonance structures, and indicate the expected geometry for CO_2, CH_3COOH, $CH_3CO_2^-$, phenol, and acetamide, ($CH_3C(=O)NH_2$).

13 The effective length of the conjugated molecule

CH_3—(CH=CH—)$_4$$CH_3$

is about 9.8 Å. Calculate and plot the energies of the first five allowed states, using a one-dimensional square-well potential to represent the molecular skeleton. Place the eight π electrons in the energy levels that would normally be occupied. Indicate the transition that would correspond to the absorption of radiation, and calculate the energy of quanta necessary to cause this excitation. Compare the wavelength of the radiation that has quanta of this energy with the wavelength of 3000 Å, which is the wavelength of the observed absorption band. *Ans.* $\Delta\epsilon = 56.3 \times 10^{-20}$ J; $\lambda = 3510$ Å.

14 Magnetic measurements indicate that in the ion $Fe(CN)_6^{4-}$ there are no unpaired electrons, whereas in the ion FeF_6^{4-} there are four unpaired electrons. Describe the bonding of each of these ions so as to explain the observed numbers of unpaired electrons. Use both the Pauling method of description and that of the ligand-field theory. Both ions presumably have their ligands arranged octahedrally about the Fe^{++} ion.

15 Using the d-orbital diagrams of Fig. 12-21, show the d-orbital energy pattern that would be expected in the crystal-field approach for the tetrahedral arrangement of four ligands about a central metal ion. (*Hint:* Bring the ligands in along directions that avoid the d orbitals that lie along the cartesian axes.)

References

Many of the references given at the end of Chap. 11 contain material on the systems of more general chemical interest dealt with in this chapter. To these references can be added the following, which deal primarily with such systems:

LEWIS, G. N.: "Valence and the Structure of Atoms and Molecules," Dover Publications, Inc., New York, 1966. A paperback reprint of a classic from the days when modern descriptions of atoms and molecules were beginning to emerge.

LINNETT, J. W.: "The Electronic Structure of Molecules: A New Approach," Methuen & Co., Ltd., London, 1964.

PRITCHARD, J. O., and H. A. SKINNER: The Concept of Electronegativity, *Chem. Rev.*, **55**:745 (1955).

PAULING, L.: "The Nature of the Chemical Bond," 3d ed., Cornell University Press, Ithaca, N.Y., 1960.

STREITWIESER, A., JR.: "Molecular Orbital Theory for Organic Chemists," John Wiley & Sons, Inc., New York, 1961.

ORGEL, L. E.: "An Introduction to Transition Metal Chemistry," John Wiley & Sons, Inc., New York, 1960.

BALLHAUSEN, C. J.: "Introduction to Ligand Field Theory," McGraw-Hill Book Company, New York, 1962.

13

EXPERIMENTAL STUDY OF MOLECULAR STRUCTURE: SPECTROSCOPIC METHODS

Grouped together here and in the following two chapters are treatments of various experimental methods that give information on the geometry and electronic structures of molecules. The difficulties that were encountered in the application of a completely theoretical approach to molecular bonding and structure lead one to refer frequently to experimentally determined properties in order to understand molecular phenomena. Knowledge of the size, shape, rigidity, and electronic structure of molecules deduced from the experimental methods that are treated here goes hand in hand with the theoretical approaches of the preceding two chapters. Spectroscopy is the measurement and interpretation of electromagnetic radiation absorbed or emitted when the molecules, or atoms, or ions, of a sample move from one allowed energy to another. These allowed energies have been used throughout Chaps. 4 to 10 in our interpretation of the thermodynamic properties of materials. Now the origin of the values used there for the spacing of some of the energy levels will be seen. Our principal concern will be with the areas of molecular spectroscopy that stem from changes in the rotational, vibrational, and electronic energies. In addition, energies that were not considered in our thermodynamic studies, resulting from energy differences that arise when a sample is placed in a magnetic or electric field, are susceptible to spectroscopic studies. Nuclear-magnetic-resonance spectroscopy and electron-spin spectroscopy will illustrate such studies.

13-1 Rotational Spectra

The rotational energies of a rotating linear molecule were shown in Sec. 4-3 to be given by the expression

$$\epsilon_{\text{rot}} = \frac{h^2}{8\pi^2 I} J(J+1) \qquad J = 0, 1, 2, \ldots \tag{1}$$

and the degeneracies of the rotational levels by $g_J = 2J + 1$. These features, at this stage lacking an associated energy scale, are shown in Fig. 13-1.

Before transitions between these allowed rotational energies are considered, the moment of inertia I must be treated in more detail than was necessary earlier.

The simplest rotary system is a particle moving in a circular path, with radius r about a fixed point. Its kinetic energy is $\frac{1}{2}mv^2$, and this can be arranged to

$$\begin{aligned}\epsilon_{\text{rot}} &= \tfrac{1}{2}mv^2 \\ &= \tfrac{1}{2}(mr^2)\left(\frac{v}{r}\right)^2\end{aligned} \tag{2}$$

The two terms in parentheses are convenient for the treatment of rotary systems. Thus we introduce the moment of inertia $I = mr^2$ and the angular velocity $\omega = v/r$.

With these symbols, the rotational energy is given by

$$\epsilon_{\text{rot}} = \tfrac{1}{2}I\omega^2 \tag{3}$$

When I is suitably defined for an assembly of particles like the atoms of a molecule, this expression can be extended beyond its initial application to a single particle rotating about a fixed point. The necessary definition of I is

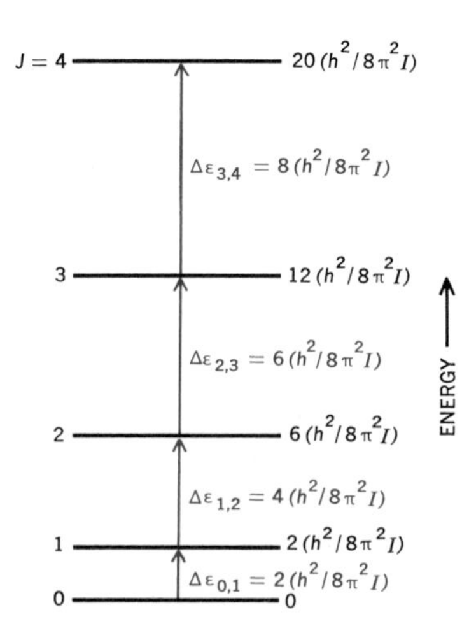

FIGURE 13-1
The energies of the allowed rotational states of a linear molecule according to the expression $\epsilon_{\text{rot}} = (h^2/8\pi^2 I)J(J+1)$.

$$I = \sum_i m_i r_i^2 \qquad [4]$$

where i numbers off the atoms of the molecule.

That this definition of I leads to a rotational-energy expression of the form of Eq. [3] can be illustrated for the case of an assembly like the diatomic molecules with which we will be primarily concerned.

For a diatomic molecule the center of gravity, as shown in Fig. 13-2, is located so that $m_1 r_1 = m_2 r_2$. Combination with the relation $r_1 + r_2 = r$ gives

$$r_1 = \frac{m_2}{m_1 + m_2} r$$

and [5]

$$r_2 = \frac{m_1}{m_1 + m_2} r$$

As the molecule rotates about its center of gravity, the atoms will revolve with the same angular velocity ω. The velocities of each will be, according to $\omega = v/r$,

$$v_1 = r_1\omega \qquad v_2 = r_2\omega \qquad [6]$$

The rotational energy then can be written

$$\begin{aligned}\epsilon_{\text{rot}} &= \tfrac{1}{2}m_1 v_1^2 + \tfrac{1}{2}m_2 v_2^2 \\ &= \frac{1}{2}\left[\frac{m_1 m_2^2}{(m_1 + m_2)^2} + \frac{m_1^2 m_2}{(m_1 + m_2)^2}\right] r^2\omega^2 \\ &= \frac{1}{2}\left[\frac{m_1 m_2}{m_1 + m_2} r^2\right]\omega^2 \qquad [7]\end{aligned}$$

The term in brackets can be seen to be the moment of inertia for this diatomic example by returning to $I = \Sigma_i m_i r_i^2$, which with Eqs. [5] gives

$$\begin{aligned}I &= \left[\frac{m_1 m_2^2}{(m_1 + m_2)^2} + \frac{m_1^2 m_2}{(m_1 + m_2)^2}\right] r^2 \\ &= \frac{m_1 m_2}{m_1 + m_2} r^2 \qquad [8]\end{aligned}$$

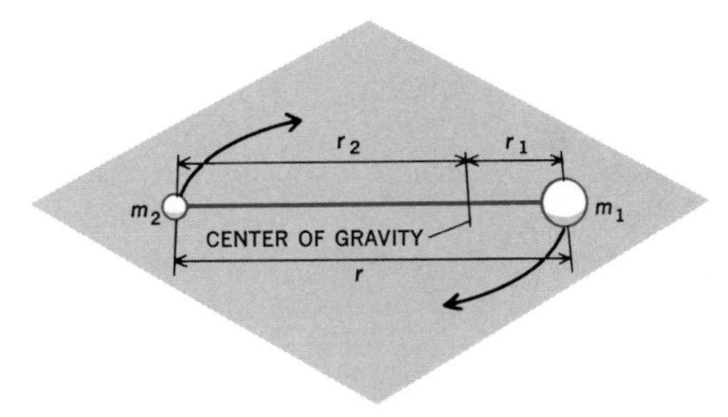

FIGURE 13-2
The rotation of a diatomic molecule about its center of gravity.

The collection of mass terms of Eqs. [7] and [8] enters into treatments of diatomic molecules so frequently that it is called the *reduced mass,* with symbol μ; that is,

$$\mu = \frac{m_1 m_2}{m_1 + m_2} \qquad [9]$$

With this notation the convenient expression $I = \mu r^2$ can be written for the diatomic case. For polyatomic molecules, I is related to the atomic masses and molecular geometry by Eq. [4].

Thus the structural implications of the moment of inertia are apparent. Now let us see how studies of electromagnetic radiation yield information on the energy spacing of rotational states, and thus on the energy factor $h/8\pi^2 I$ and on the moment of inertia.

A rotating molecule can withdraw energy from electromagnetic radiation or give up energy to the radiation if it can interact with the oscillating electric field associated with the radiation. The molecule can do this if it has a dipole moment. The rotating dipole provides a coupling with the oscillating electric field of the radiation and allows energy to be transferred from the radiation to the molecule, or vice versa. It is concluded, therefore, that a molecule must have a dipole moment in order to give rise to a rotational spectrum. One observes, in fact, that electrically symmetric molecules like H_2, N_2, and CO_2 give rise to no absorptions that can be attributed to only changes in the rotational energy of the molecules.

Even when a molecule has a dipole moment and can interact with the radiation, there is a restriction on the rotational transitions that can be induced. This restriction, an example of a *selection rule,* is that a molecule can increase or decrease its rotational energy only to the next higher or lower energy level when it absorbs or emits a quantum of electromagnetic radiation. This selection rule is written

$$\Delta J = \pm 1 \qquad [10]$$

The restriction can be understood in terms of the spin of radiation photons and the principle of the conservation of angular momentum, or its existence can be easily recognized from the nature of the observed rotation spectra. Rotational spectra are almost always studied by observing the radiation that is absorbed by the sample. For such *absorption spectra,* the only part of the selection rule that is of interest is $\Delta J = +1$.

Since the rotational-energy levels are closely spaced compared with kT, the molecules will be distributed throughout many of the lower allowed levels, such as those depicted in Fig. 13-1. The transitions which can occur, therefore, are between adjacent levels indicated by the $\Delta\epsilon$ terms of Fig. 13-3. These energy differences correspond, then, to the energies of quanta of radiation that can be absorbed to bring about $\Delta J = +1$ transitions.

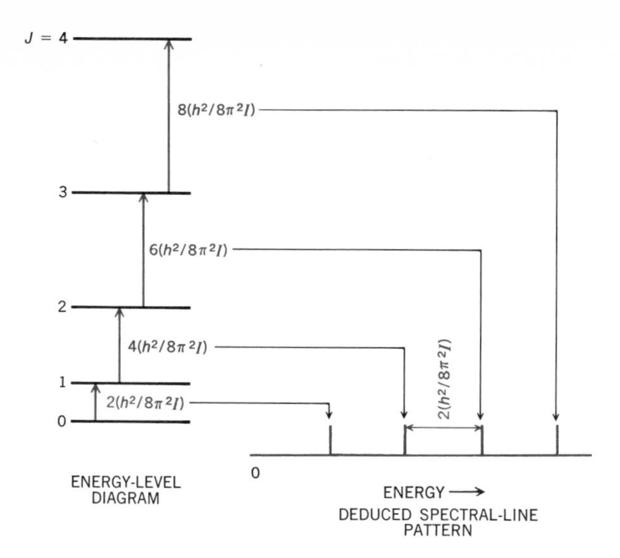

FIGURE 13-3
Rotational spectral transitions for a linear molecule.

It is more customary to deal with the frequency or wavelength of the absorbed radiation rather than with the energy of the quanta. Recalling Planck's relation $\Delta\epsilon = h\nu$, we can convert the quantum energies to frequencies by dividing by h. Thus the radiation which would be absorbed could be expressed in terms of multiples of the energy term $h^2/8\pi^2I$, as is done in Fig. 13-3, or in terms of multiples of radiation frequencies of $h/8\pi^2I$ Hz.

A feature of the frequencies predicted for absorbed radiation (not the rotational energies, but the transitions between them) that is apparent from Fig. 13-3 is that the values of $\Delta\epsilon$ are spaced by a constant factor of $2(h^2/8\pi^2I)$. Thus we expect to find in the microwave region a series of absorptions of radiation spaced by an equal energy amount, which can be identified with $2(h^2/8\pi^2I)$, or an equal frequency amount $2(h/8\pi^2I)$ for the molecule under investigation.

The rotational spectra of most molecules occur in the microwave spectral region. Diatomic molecules have, generally, such low moments of inertia that their rotational transitions, which have energies proportional to $h^2/8\pi^2I$, tend to occur on the high-energy, infrared end of the microwave spectrum. In fact, for rather low moment-of-inertia molecules like HCl, infrared spectral techniques can be used to obtain spectra like that of Fig. 13-4.

Microwave measurements generally are of much greater resolution and frequency precision, as the data for CO in Table 13-1 illustrate. Note that the different isotopic species give rise to rotational transitions that are detected as separate lines, a feature not true for the lower-resolution results of Fig. 13-4.

To a good first approximation Fig. 13-4 and the data of Table 13-1 for $C^{12}O^{16}$ conform to the rigid rotor interpretations indicated in Fig. 13-3. With this interpretation of the spectral lines one can proceed to

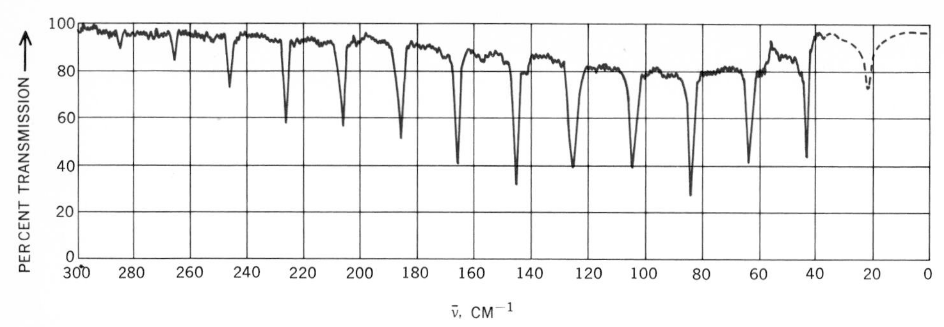

FIGURE 13-4
The observed absorption spectrum of HCl (g) showing that the rotational spectrum of a linear molecule gives a set of absorptions with approximately a constant spacing. This spacing can be identified with $2(h/8\pi^2 cI)$ cm^{-1} spectrum obtained with a Beckman IR-11 spectrometer.

molecular properties. For $C^{12}O^{16}$, for example, an averaging of the data of Table 13-1 yields approximately

$$2\frac{h}{8\pi^2 I} = 115{,}260 \text{ MHz}$$
$$= 1.1526 \times 10^{11} \text{ Hz} \qquad [11]$$

The calculated moment-of-inertia value is thus

$$I = 1.4560 \times 10^{-46} \text{ kg m}^2 \qquad [12]$$

Then with

$$\mu = \frac{(12.0000)(15.9941)}{12.0000 + 15.9941} \times \frac{10^{-3}}{6.0222 \times 10^{23}}$$
$$= 1.1384 \times 10^{-26} \text{ kg} \qquad [13]$$

the internuclear distance r is obtained from $I = \mu r^2$ as

$$r = 1.131 \times 10^{-10} \text{ m} = 1.131 \text{ Å} \qquad [14]$$

Full use of the precise data such as that of Table 13-1 can be made if it is recognized that the molecule cannot, in fact, be characterized as a rigid rotor and that it elongates as a result of centrifugal distortion that occurs as it rotates with higher frequencies. Thus the moment of inertia increases somewhat with increasing rotational energy.

With this recognition, an extrapolation procedure can be used to

TABLE 13-1 The absorptions of carbon monoxide in the microwave region and their assignment to rotational transitions*

CO species	J''	J'	Megahertz
$C^{12}O^{16}$	0	1	115,271.20
	1	2	230,537.97
	2	3	345,795.90
$C^{12}O^{18}$	0	1	109,782.18

* Tabulated in Microwave Spectral Tables, *Natl. Bur. Std. Monograph* 70, 1964–1969.

deduce moment of inertia and bond length for the nonrotating molecule of the $J = 0$ state.

For a general polyatomic molecule, a rather complicated energy-level pattern and rotational spectrum results. The pattern, and spectrum, depends upon all three principal moments of inertia, and when the spectrum is interpreted in terms of the energy pattern, values can be obtained for the three principal moments of inertia. (These have already appeared in the expression of Sec. 10-3 for the rotational contribution to the entropy of a generally shaped molecule.) From these results bond angles and bond distances can again be determined.

Some results obtained from studies of rotational spectra are shown in Table 13-2. As these data indicate, the method, when applicable, leads to extremely precise internuclear distances. One should be im-

TABLE 13-2 Some molecular dimensions from analyses of rotational spectra*

	Molecule	Bond	Bond distance (Å)
Diatomic	HF	H—F	0.917
	HCl	H—Cl	1.275
	HBr	H—Br	1.414
	HI	H—I	1.604
	CO	C—O	1.128
	FCl	F—Cl	1.628
	NaCl	Na—Cl	2.361
	CsCl	Cs—Cl	2.904
Polyatomic linear	HCN	C—H	1.064
		C≡N	1.156
	OCS	C=O	1.164
		C=S	1.558
	OCSe	C=O	1.159
		C=Se	1.709
	HCCCl	C—H	1.052
		C≡C	1.211
		C—Cl	1.632
Polyatomic nonlinear	CH_3Cl	H—C—H (angle)	$110°20' \pm 1°$
		C—H	1.103 ± 0.010
		C—Cl	1.782 ± 0.003
	CH_2Cl_2	H—C—H (angle)	$112°0' \pm 20'$
		Cl—C—Cl (angle)	$111°47' \pm 1'$
		C—H	1.068 ± 0.005
		C—Cl	1.7724 ± 0.0005

* From Gordy, Smith, and Trambarulo, "Microwave Spectroscopy," John Wiley & Sons, Inc., New York, 1953.

pressed by the fact that rotational spectra allow the measurement of distances within molecules relatively more accurately than one can measure the length of a desk top with a meterstick.

There are, of course, limitations to this method of structure determination. Only molecules with dipole moments can be studied. The moment-of-inertia data for polyatomic molecules can be analyzed into bond-angle and bond-length values only if additional isotopically substituted molecules are studied to provide as many moment-of-inertia "knowns" as there are structural-feature "unknowns." Finally, all measurements must be made on gases where the rotation is free and behaves according to Eq. [1]. Molecules of liquids or solids are interfered with by their neighbors to such an extent that no well-defined, discrete rotational energy levels exist, and no rotational spectra are therefore observed.

13-2 Vibrational Spectra

The molecular motion that has the next-larger energy-level spacing after the rotation of molecules is the vibration of the atoms of the molecule with respect to one another.

The allowed energies for a single particle of mass m vibrating against a spring with force constant k, that is, experiencing a potential energy $U = \frac{1}{2}kx^2$, where x is the displacement from equilibrium, were shown in Sec. 4-4 to be given by the expression

$$\epsilon_{\text{vib}} = (v + \tfrac{1}{2})\frac{h}{2\pi}\sqrt{\frac{k}{m}} \qquad v = 0, 1, 2, \ldots$$

$$= (v + \tfrac{1}{2})h\nu_{\text{vib}} \tag{15}$$

where ν_{vib}, the frequency of the classical oscillator, represents the term $(h/2\pi)\sqrt{k/m}$. This quantum-mechanical result indicates a pattern of energy levels with a constant spacing $(h/2\pi)\sqrt{k/m}$. It is this result that was used in Sec. 5-7 for the calculation of the average vibrational energy per degree of freedom.

Now let us investigate the details of the vibrational motion of the atoms of a molecule, and again the simplest case of a diatomic molecule will be our chief concern.

As the single-particle vibrational problem showed, it is necessary first to determine the form of the forces that restore the particles to their equilibrium position, or what is equivalent, to specify the potential energy of the system as a function of the internuclear distance. The difficulties encountered in the determination of the energy of molecular systems were seen in Chap. 11 to prevent potential-energy diagrams from being deduced for any but the simplest molecule. Figure 13-5 indicates the type of results that are obtained in such cases, and these can be taken as representative of the form of the potential-energy

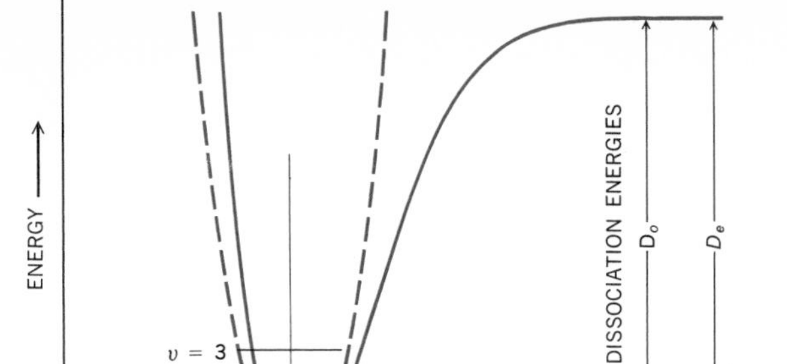

FIGURE 13-5
The solid curve is a typical potential-energy curve. The dashed curve is the Hooke's law approximation, which is satisfactory near the minimum, where the first few vibrational-energy levels occur.

versus internuclear-distance curves for molecular bonds. Vibrational energies, as we shall see, generally are sufficiently small so that the problem concerns itself with the portion of the potential-energy curve near the minimum. This portion, it turns out, can be satisfactorily approximated by a parabola, and the expression

$$U = \tfrac{1}{2}k(r - r_e)^2 \tag{16}$$

can be written. Here k is the force constant and measures the stiffness of the chemical bond, r is the variable internuclear distance, and r_e is the equilibrium internuclear distance. Use of Eq. [16] is said to constitute the *harmonic-oscillator* model because in classical systems this potential produces a harmonic oscillation.

The classical solution can again be obtained from Newton's $f = ma$ relation. If the bond is distorted from its equilibrium length r_e to a new length r, the restoring forces on each atom are $-k(r - r_e)$. These forces can be equated to the ma terms for each atom as

$$m_1\frac{d^2r_1}{dt^2} = -k(r - r_e) \qquad \text{and} \qquad m_2\frac{d^2r_2}{dt^2} = -k(r - r_e) \tag{17}$$

where r_1 and r_2 are the positions of atoms 1 and 2 relative to the center of gravity of the molecule. The relations between r_1, r_2, and r that keep the center of gravity fixed have been given by Eq. [5], and substitution in the $f = ma$ equation for particle 1, for example, gives

$$\frac{m_1 m_2}{m_1 + m_2}\frac{d^2r}{dt^2} = -k(r - r_e) \tag{18}$$

which, since r_e is a constant, can also be written

$$\frac{m_1 m_2}{m_1 + m_2}\frac{d^2(r - r_e)}{dt^2} = -k(r - r_e) \tag{19}$$

The term $r - r_e$ is the displacement of the bond length from its equilibrium position, and if the symbol x is introduced as $x = r - r_e$ and the reduced mass of μ of Eq. [9] is inserted for the mass term, Eq. [19] becomes

$$\mu \frac{d^2x}{dt^2} = -kx \qquad [20]$$

This expression is identical with the corresponding equation for a single particle, except for the replacement of the mass m by the reduced mass. It follows that the classical vibrational frequency for a two-particle system is given by

$$\nu_{\text{classical}} = \frac{1}{2\pi}\sqrt{\frac{k}{\mu}} \qquad [21]$$

and that the quantum-mechanical vibrational-energy-level result is

$$\epsilon_{\text{vib}} = (v + \tfrac{1}{2})\frac{h}{2\pi}\sqrt{\frac{k}{\mu}} \qquad v = 0, 1, 2, \ldots \qquad [22]$$

or

$$\epsilon_{\text{vib}} = (v + \tfrac{1}{2})h\nu_{\text{classical}} \qquad v = 0, 1, 2, \ldots \qquad [23]$$

The vibrational energies of a diatomic molecule consist of a set of levels as shown with the potential-energy functions in Fig. 13-5.

The spacing of these levels, according to the harmonic-oscillator, or parabolic-potential, approximation is a constant energy amount

$$\Delta\epsilon_{\text{vib}} = \frac{h}{2\pi}\sqrt{\frac{k}{\mu}} \qquad [24]$$

At room temperature the value of kT is sufficiently small compared with typical values of $\Delta\epsilon_{\text{vib}}$ so that most of the molecules are in the lowest allowed vibrational state. In a spectroscopic study, therefore, one investigates the absorption of radiation by these $v = 0$ state molecules.

Coupling with electromagnetic radiation can occur if the vibrating molecule produces an oscillating dipole moment that can interact with the electric field of the radiation. It follows that homonuclear diatomic molecules like H_2, N_2, and O_2, which necessarily have a zero dipole moment for any bond length, will fail to interact. The dipole moment of molecules such as HCl, on the other hand, can be expected to be some function, usually unknown, of the internuclear distance. The vibration of such molecules leads to an oscillating dipole moment, and a vibrational spectrum can be expected.

Even when interaction between vibrating molecule and the radiation occurs, a further selection rule applies. This rule restricts transitions resulting from the absorption or emission of a quantum of radiation by the relation

$$\Delta v = \pm 1 \qquad [25]$$

Vibrational spectra are usually determined by absorption spectroscopy, and then $\Delta v = +1$ is the only part of this selection rule which is pertinent.

One application of vibrational spectra to obtain molecular properties can now be illustrated by again using the example of HCl. One observes in the infrared region an absorption band centered at a frequency of about 8.67×10^{13} Hz. In practice, instead of reporting such frequencies, it is customary to report the related quantity, known as the *wave number,* represented by $\bar{\nu}$, which is the frequency divided by the speed of light, or equivalently, the reciprocal of the wavelength. Thus

$$\bar{\nu} = \frac{\nu}{c} = \frac{c/\lambda}{c} = \frac{1}{\lambda} \qquad [26]$$

In this use c or λ is given the cgs unit of centimeters, and thus, for HCl, the vibrational band would be reported at

$$\bar{\nu} = \frac{8.67 \times 10^{13}}{3 \times 10^{10}} = 2890 \text{ cm}^{-1} \qquad [27]$$

The relation of the wave number of the vibrational bond to the parameters that describe the vibrating molecule is obtained from Eq. [22] and $\Delta\epsilon = h\nu$ or $\Delta\epsilon = hc\bar{\nu}$ as

$$\bar{\nu}_{\text{vib}} = \frac{1}{2\pi c}\sqrt{\frac{k}{\mu}} \qquad [28]$$

This observed HCl absorption can be assigned to the vibrational transition from $v = 0$ to $v = 1$ and can be equated to $(1/2\pi c)\sqrt{k/\mu}$ cm^{-1}. With $\mu = 1.628 \times 10^{-27}$ kg, this yields

$$k = 483 \text{ N m}^{-1} \qquad [29]$$

The theory of vibrational spectra, together with the observed absorption, has now led to a value for the force constant of a chemical bond. The force constant, it will be recalled, measures the force required to sketch a bond by a given distance. The qualitative feature to be appreciated from results such as that worked out for HCl is that molecules are flexible. Although it is at first difficult to appreciate the significance of the numerical values obtained for bond-force constants, one can make interesting comparisons of the stiffness of different bonds. Some results for diatomic molecules and for bonds of polyatomic molecules are shown in Table 13-3. The increased stiffness of multiple bonds compared with single bonds is apparent and is in line with the greater strength of multiple bonds. It should be observed from the data of the table that the observed frequency, being determined by both the reduced mass and the force constant, according to Eq. [28], is not itself a simple measure of the bond stiffness.

Polyatomic molecules exhibit vibrational spectra which can be inter-

TABLE 13-3 Some bond-stretching force constants
Except for those marked with an asterisk, values have been obtained by deducing the curvature, and thus the force constant, at the minimum of the potential-energy curve for the bond. Asterisks indicate values calculated directly from the observed $v = 0$ to $v = 1$ spectral data.

Molecule	Bond	Force constant (N m^{-1})	Bond	Force constant (N m^{-1})
HF	F—H	970		
H_2O	O—H	840		
NH_3	N—H	710		
CH_4	C—H	580,520*		
CH_3CH_3	—C—H	530,490*	—C—C—	460,450*
CH_2CH_2	=C—H	620,510*	—C=C—	1090,930*
HCCH	≡C—H	630,600*	—C≡C—	1630,1590*
C_6H_6	C—H	590*	—C⩵C—	770*
H_2CO	=C—H	520,440*	—C=O	1300,1280*
CO_2			=C=O	1730
HCN	≡C—H	620	—C≡N	1880

SOURCE: M. D. Newton, W. A. Lathan, W. J. Henre, and J. A. Pople, *J. Chem. Phys.*, **52**:4070 (1970).

preted as arising from transitions within each of a number of energy-level patterns, each like that of Fig. 13-5. Each energy-level pattern corresponds to one of the $3n - 6$ (or $3n - 5$ for linear molecules) characteristic, or *normal,* vibrations of the molecule. One finds, for example, for H_2O vapor, absorptions centered at 1595, 3652, and 3756 cm^{-1}. For molecules with many atoms, $3n - 6$ becomes large, and one expects very many vibrational transitions and a very complicated spectral pattern.

The presence of any amount of symmetry in a molecule greatly simplifies the vibrational spectrum and the study of its modes of vibration. There is a general theorem, which cannot be dealt with here, that each of the $3n - 6$ (or $3n - 5$) vibrations must be either symmetric or antisymmetric with respect to any symmetry element (such as a plane of symmetry or a center of symmetry) of the molecule. For H_2O, for example, it can be shown that the two lower frequencies correspond to symmetric vibrations, while the highest frequency corresponds to an antisymmetric mode. The diagrams of Fig. 13-6 can therefore be drawn to show how the atoms might move in each of these vibrations. These diagrams are intended to suggest pure vibrational motions, and the arrows have therefore been drawn so that there is no overall translation or rotation. Such diagrams can easily be drawn to represent the symmetry of the actual vibrational modes. The exact motion of the atoms in a vibrational mode depends, however, on the masses of the atoms and the force constants of the molecule.

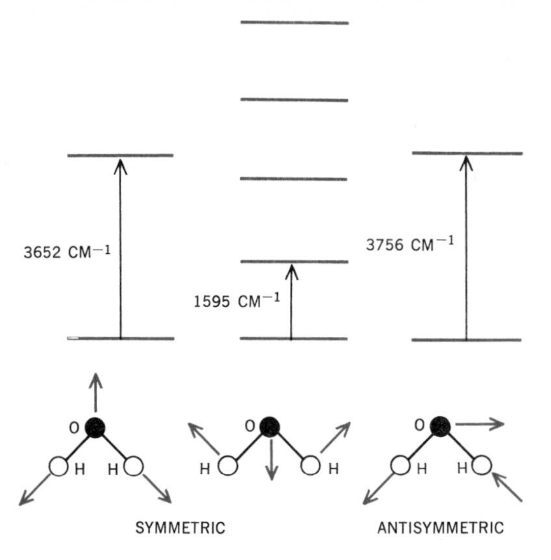

FIGURE 13-6
The symmetry of the three modes of vibrations of the water molecule and the associated vibrational-energy-level patterns.

A practical use, and one of great value, particularly in the field of organic chemistry, is that in which the infrared absorption spectrum of a large molecule is used to identify the compound or to indicate the presence of certain groups in the molecule. Bonds or groups within a molecule sometimes vibrate with a frequency, i.e., have an energy-level pattern with a spacing, that is little affected by the rest of the molecule. Absorption at a frequency that is characteristic of a particular group can then be taken as an indication of the presence of that group in the compound being studied. Table 13-4 shows a few of the groups that have useful *characteristic frequencies.*

An even simpler use of vibrational spectra consists in verifying the identity of a compound by matching its infrared spectrum to that of a known sample. Large molecules have such complicated spectra, as shown in Fig. 13-7, for example, that identical spectra can be taken as a sure indication of identical compounds. Thus, although for large molecules the complete vibrational spectrum can seldom be understood in terms of the nature of the $3n - 6$ vibrations, there are many uses to which such spectra can be put.

TABLE 13-4
Some characteristic bond-stretching frequencies

Group	$\bar{\nu}$ *(cm^{-1})*
—O—H	3500–3700
>N—H (—N—H with bonds above and below)	3300–3500
≡C—H	3340
=C—H (with bond above)	3000–3120
—C—H (with bonds above and below)	2880–3030
—C≡C—	2200–2260
—C≡N	2250
>C=O	1660–1870
>C=C	1600–1680

13-3 Rotation-Vibration Spectra

Gas molecules, as we know from the interpretations given to thermodynamic properties of gases, are simultaneously rotating and vibrating. It follows that an absorption spectrum of a gas might show the effects of changes in both rotational and vibrational energies. Such is indeed

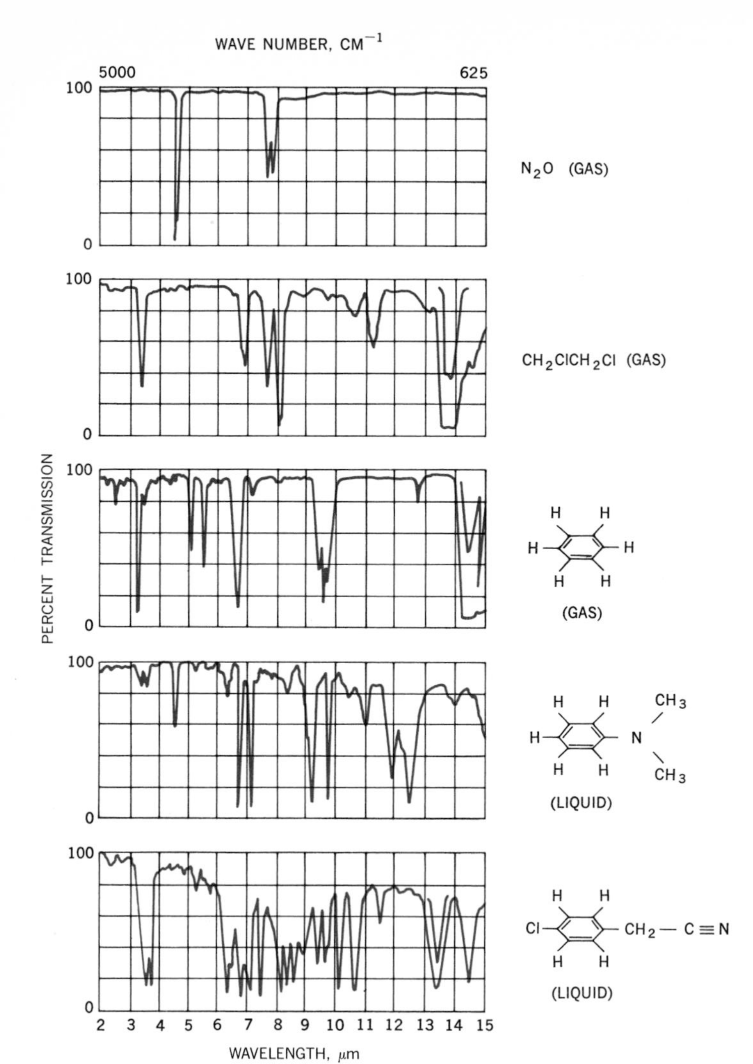

FIGURE 13-7 The infrared absorption spectra of several compounds.

the case, and the expanded view of the infrared absorption band of HCl in Fig. 13-8 shows the band structure that must be attributed to the rotational-energy changes that accompany this $v = 0$ to $v = 1$ vibrational transition. A diagrammatic interpretation of such a rotation-vibration spectrum is carried out in Fig. 13-9. The only feature that must be added to the interpretation of the spectra of the separate motions is the consequence of the $\Delta J = \pm 1$ selection rule. Now you can see that the $\Delta J = -1$ changes can occur, even though the spectrum is an absorption spectrum. The energy pattern, and the selection rules, of Fig. 13-9 lead to a predicted spectral pattern that is in general accord with the observed HCl spectrum of Fig. 13-8. (The additional

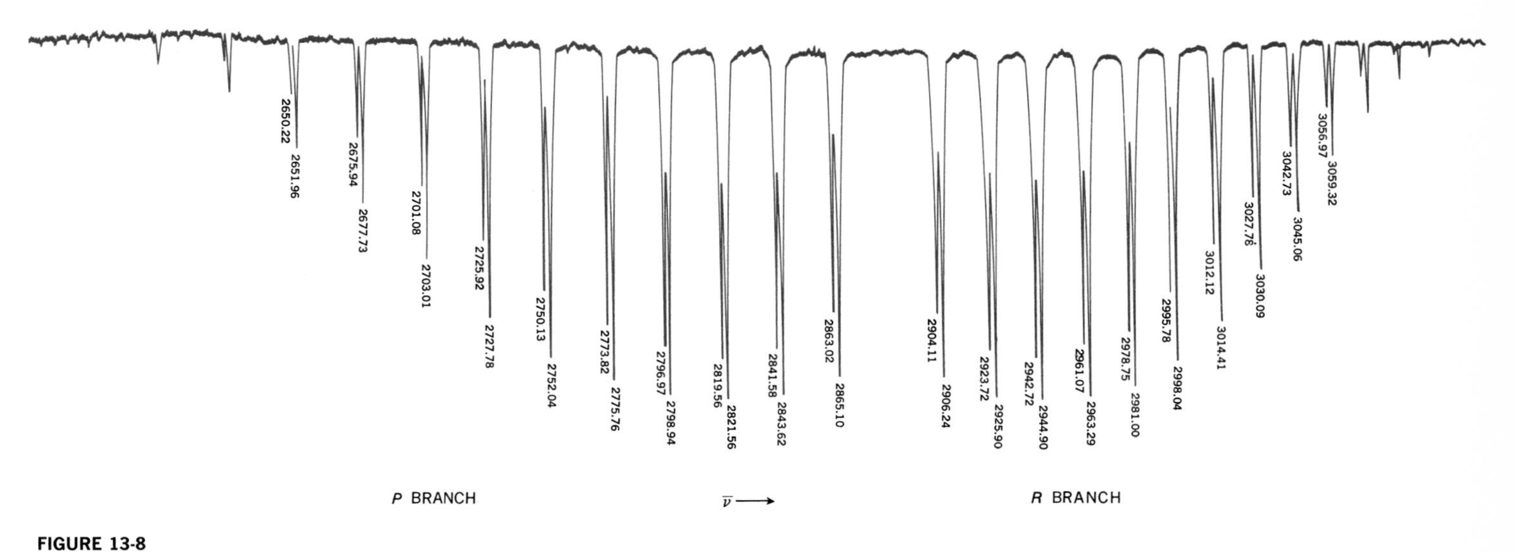

FIGURE 13-8
The fundamental absorption band for HCl under high resolution. (The lines are doubled due to the presence of the two isotopes Cl^{35} and Cl^{37} in the ratio 3:1.) The low-frequency side of the band is known as the P branch; the high-frequency side as the R branch.

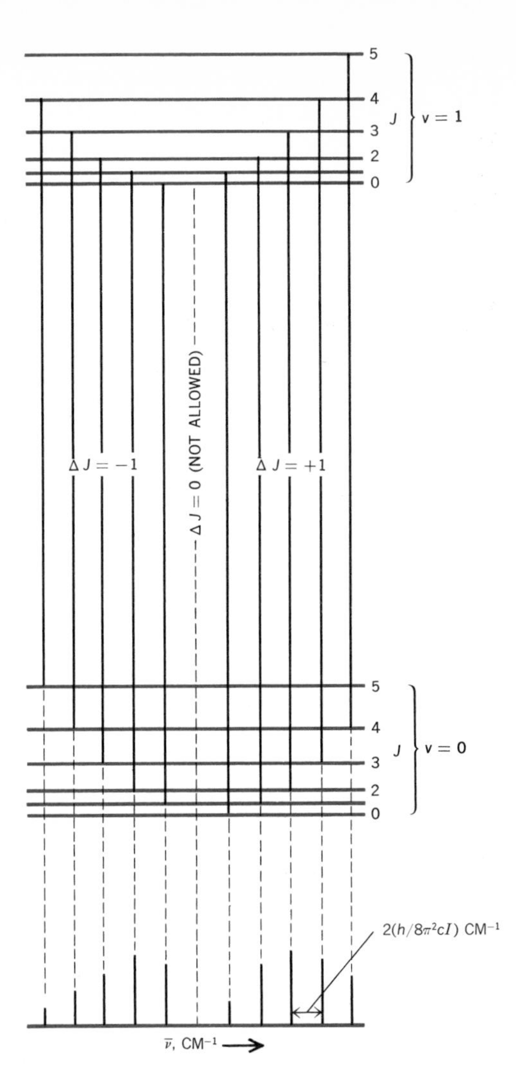

FIGURE 13-9
The rotational levels of the $v = 0$ and $v = 1$ vibrational levels of a diatomic molecule.

splitting in Fig. 13-8 is readily interpreted by recognizing that the gas sample consists of two important molecular species HCl^{35} and HCl^{37} and that, because of their somewhat different reduced masses, these species will have energy levels and spectral transitions that are displaced one from the other.)

From such rotation-vibration spectra, which have in the past been more easily obtained than microwave spectra, moment-of-inertia and molecular-structure data can be deduced. The treatment of Fig. 13-9 suggests, for example, that the spacing of the adjacent components of the spectral band can be identified with an energy of $2(h^2/8\pi^2 I)$ or a wave-number term $2(h/8\pi^2 cI)$ and the central gap with twice these values.

Closer inspection of Fig. 13-8 shows that the spacings of consecutive spectral lines is in fact not constant. The principal source of the general convergence at the high-energy end can be accounted for by attributing a somewhat longer effective bond length, greater moment of inertia, and smaller rotational constant, $(h/8\pi^2cI)$, to the molecules in the $v = 1$ state compared with the $v = 0$ state. The bond length in each vibrational state can, in fact, be deduced from a spectrum like that of Fig. 13-8, and moreover, an extrapolation to the bond length that the molecule would have if it were in the "equilibrium state" at the minimum of the potential curve like that of Fig. 13-5 can be calculated. It is in fact this *equilibrium* bond length r_e that is reported in Table 13-2.

Similar analyses of the rotation-vibration spectra can be carried out to yield rotational and vibrational information for polyatomic molecules. Detailed analysis is, however, generally restricted to small and symmetric cases.

13-4 Raman Spectroscopy

Rotation and vibration spectra can be obtained as a result of an alternative to the interaction of the electric field of radiation with the oscillating dipole, the basis of absorption spectroscopy. In *Raman spectroscopy* the interaction stems from the electric field and an oscillating *polarizability*. The nature of such effects will be dealt with in more detail in Chap. 15, but here it will be enough to recognize that an electric field can act on a molecule to distort the electron distribution and thus to induce a dipole. If the ease of distortion, or polarizability, oscillates, so also will the induced dipole. In the Raman effect this mechanism is made use of by radiating the sample with a monochromatic beam of visible or ultraviolet radiation. The wavelength, or the samples, are chosen so that this exciting radiation is not absorbed. It does, however, through the induced oscillating dipoles that it stimulates, lead to the transfer of energy with the rotation and vibration modes of the sample molecules.

The experimental arrangement now often uses a laser exciting beam, as from a helium-neon laser, which has the necessary high power and monochromaticity. The scattered photons, which are increased or decreased in energy from that of the exciting beam by the energy jumps produced in the rotational or vibrational modes of the sample, are generally viewed at right angles to the exciting beam.

Raman rotational spectra of gases contrast to absorption rotational spectra principally in that the molecules of the sample need not have a permanent dipole. All molecules are polarizable, and their rotation allows a rotating-induced dipole to occur. An example is shown in Fig. 13-10. An additional difference is that in the rotational Raman effect the selection rule for linear molecules is $\Delta J = 0, \pm 2$. It follows that

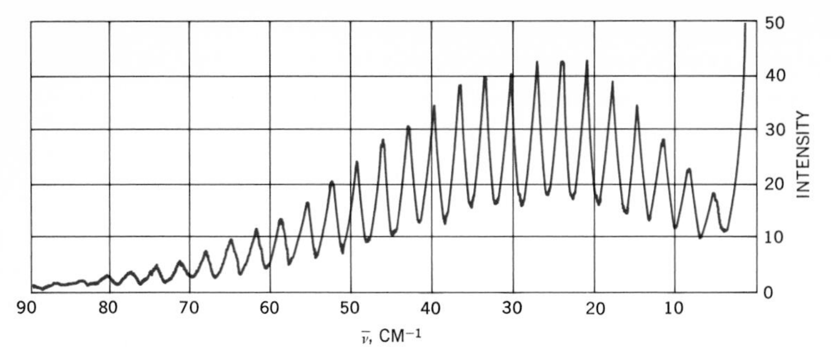

FIGURE 13-10
Raman spectrum showing the rotational-band structure of CO_2. Spectrum taken with a 25-300 laser Raman spectrometer.

the spacing of adjacent Raman rotational spectral lines is to be identified with an energy $4(h^2/8\pi^2 I)$ rather than the $2(h^2/8\pi^2 I)$ factor that applies to absorption spectra.

Vibrational Raman spectra have, for complex molecules, the general appearance of the corresponding absorption spectra, and often the same vibrational energy-level separation shows up as a spectral line in both procedures. A dramatic difference between the two spectral techniques, however, shows up when symmetric molecules are considered. The vibration of a homonuclear diatomic molecule, N_2 for example, produces an oscillating polarizability—the molecule generally will be more, or less, polarizable when it is lengthened than when it is shortened. As a result, there will be an oscillating-induced dipole, and thus a Raman vibrational spectrum. For such a molecule, as we saw in Sec. 13-2, there is no oscillating dipole and thus no absorption spectrum. For molecules with a center of symmetry, Raman spectra provide information on the symmetric vibrations of molecules, and absorption spectra on the antisymmetric. Generally, the two approaches can at times be brought together to reveal the symmetry, or shape, of a molecule of unknown structure. An illustration is provided by the data of Fig. 13-11, which are seen to be compatible with a linear symmetric structure for CO_2, and not with a bent or asymmetric structure.

FIGURE 13-11
The effect of symmetry on the activity of the normal vibrations of a molecule, CO_2, with a center of symmetry.

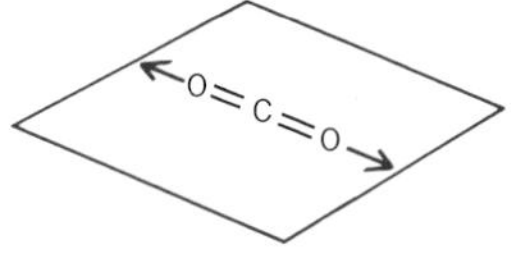

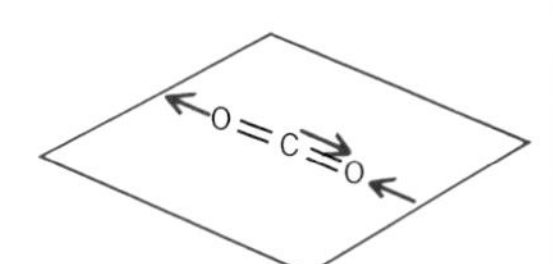

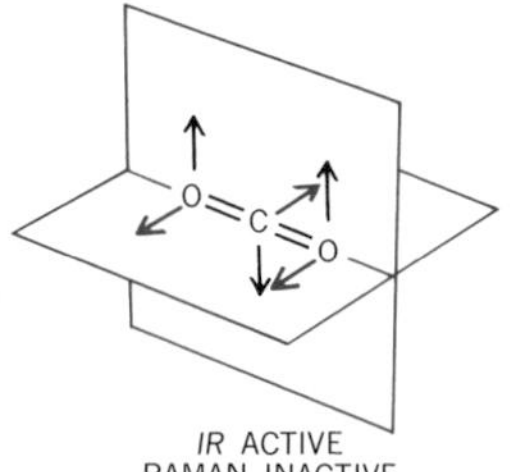

13-5 Electronic Spectra

The final type of energy levels of free molecules that lead to spectroscopically observable transitions is that in which the electron arrangement of the molecule is altered. The electronic spectra of atoms in which the electronic state, as described by the quantum numbers of the electrons of the atom, is changed have already been mentioned. In a similar manner, the electrons of a molecule can be excited to higher energy states, and the radiation that is absorbed in this process or the energy emitted in the return to the ground state can be studied. The energies involved are generally large, and electronic spectra are usually found in the ultraviolet region of the electromagnetic radiation.

Some of the information which would be necessary for a theoretical approach to the understanding of molecular electronic spectra has already been introduced. Various molecular-energy diagrams were drawn in Chap. 12, and these provide the qualitative descriptions of the electronic states of molecules that must be assigned to the initial and final states of an electronic transition. When this assignment is accomplished, the energy difference between the states involved is determined by the measured energy of the radiation quanta absorbed or emitted.

Some of the details of an electronic absorption or emission spectrum require the vibrational and rotational states of the molecule in its initial and final electronic states to be recognized. The vibrational aspects can be depicted by showing the energies of the electronic states as a function of internuclear distance.

In general, the bond strength in the excited state is less than that in the ground state, and often the equilibrium internuclear distance in the excited state is longer than in the ground state. The electronic spectra of many simple molecules can be explained in terms of a number of excited electronic states with potential-energy curves like those of Fig. 13-12.

The observed spectral transitions are related to such electronic-energy diagrams on the basis of the *Franck-Condon* principle. This principle stems from the idea that electrons move and rearrange themselves much faster than can the nuclei of molecules. For example, the time for an electron to circle a hydrogen nucleus can be calculated from Bohr's model to be about 10^{-16} s, whereas a typical period of vibration of a molecule is a thousand times longer, or about 10^{-13} s. Comparison of these times suggests that an electronic configuration will change in a time so short that the nuclei will not change their positions. The

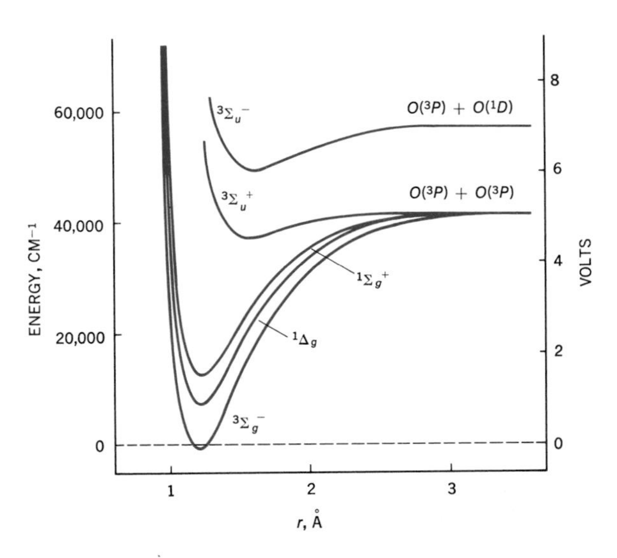

FIGURE 13-12
Potential-energy curves for some of the electronic states of the O_2 molecule. (*From G. Herzberg, "Molecular Spectra and Molecular Structure," vol. 1, "Spectra of Diatomic Molecules," D. Van Nostrand, Inc., Princeton, N.J., 1950.*)

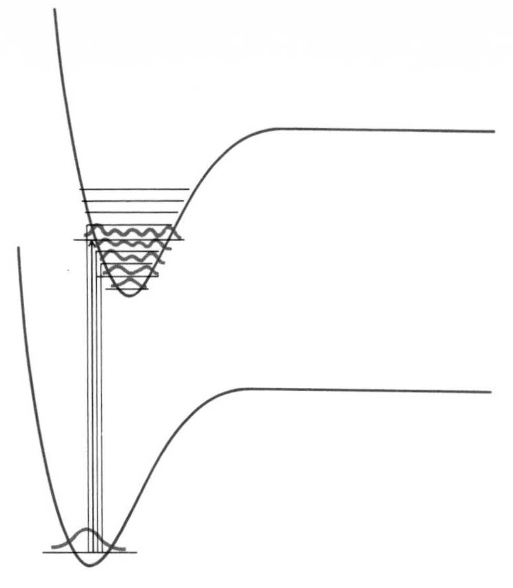

FIGURE 13-13
The most important transitions between the vibrational levels of electronic states and their relation to the potential-energy curves, vibrational wave functions, and energies for those states.

spectral transitions must be drawn vertically in Fig. 13-13, and not, as one might otherwise be tempted to do, from the potential minimum of the lower curve to that of the upper curve.

One further feature of the transitions between electronic-energy curves must be mentioned. Since a molecule vibrates, even when it is in the lowest vibrational-energy level, a range of internuclear distances must be considered. The quantum-mechanical solution for a vibrating molecule, as Fig. 13-13 indicates, shows that in the lowest energy state, contrary to classical ideas, the most probable internuclear distance is that corresponding to the equilibrium position. For the higher energy states, on the other hand, the quantum-mechanical result is more like the classical result that the most probable configuration is at the ends of the vibration, where the atoms must stop and reverse their direction. The transitions of Fig. 13-13 have been drawn with these ideas in mind. Transitions are expected to have greater probability of starting near the midpoint of the lowest vibrational level of the ground electronic state.

The most probable transitions will be those that reach vibrational states of the excited electronic state that have high probability functions at the internuclear separation equal to that of the ground state.

It follows, as Fig. 13-13 shows, that an electronic transition, in absorption, may show a series of closely spaced lines corresponding to different vibrational and rotational energies of the upper state.

Most molecules have a ground-state electronic configuration in which the spin of each electron is opposed to, or paired up with, the spin of another electron which otherwise has the same spatial quantum numbers. Such ground states are known as *singlet states* because, with no net spin angular momentum, the imposition of a reference direction by an applied electric or magnetic field can produce only the single component of zero angular momentum in the field direction. The ground state of the H_2 molecule is the simplest example of such a singlet state. Molecules will have, in addition to a ground state that is usually a singlet state, a number of excited states that also have all the electrons paired and that are therefore also singlet states.

Whether or not the ground state is a singlet state, there will be, for molecules with an even number of electrons, excited states in which a pair of electrons have their spins in the same direction, giving the molecule a net spin angular momentum of $\sqrt{1(2)}(h/2\pi)$. Angular components along a specified direction can now have the values $1(h/2\pi)$, 0, and $-1(h/2\pi)$, and such an electronic configuration is known as a *triplet state.*

Electronic spectra of compounds containing no heavy atoms, e.g., most organic compounds, indicate that the absorption of electromagnetic radiation does not unpair the electrons of the molecule. The important selection rule that *transitions occur between states of like multiplicity* is obtained. This rule is a powerful guide to the deduction of the nature of excited electronic states.

For molecules exposed to strong magnetic fields or containing a high-atomic-number atom whose nucleus exerts such a field, this selection rule is broken down, and the spin coupling is readily broken down in an electromagnetic transition.

The assignment of the multiplicity to the state reached by an absorption of radiation and to the states that the molecule goes into as it loses its high energy is a matter of great importance in the study of fluorescence, phosphorescence, and photochemistry. These subjects, however, will be postponed until Chap. 17.

The study of electronic spectra leads to a wealth of information about the electronic states and energies of molecules and to the bond distances and force constants of the molecule in excited electronic states. Such analyses are usually limited to diatomic or small linear molecules. For larger molecules, however, a more limited goal must be set. It is frequently sufficient to attempt to decide which of the electrons of the molecule are primarily responsible, i.e., which electron has its quantum number altered, for the observed transition.

13-6 Electronic Energies of Polyatomic Molecules

For larger and generally shaped molecules detailed descriptions such as those given in Fig. 13-12 cannot be reached. Spectroscopic techniques still provide a considerable insight into electronic properties.

The electronic absorptions of organic compounds, usually found in the ultraviolet region, can often be identified as a group within the molecule. First it is recognized that electrons in single covalent bonds

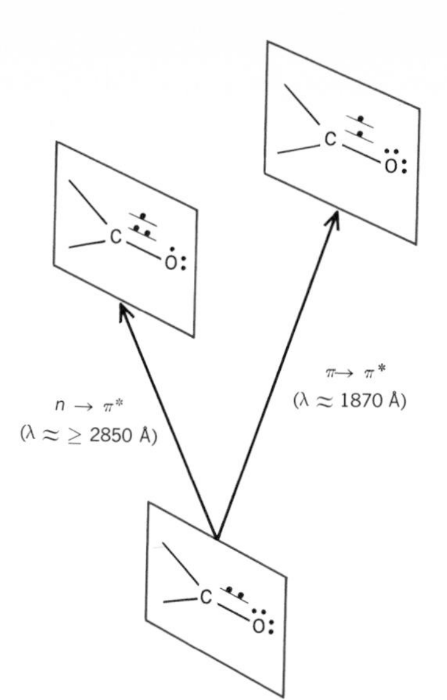

FIGURE 13-14
Schematic representation on the $n \rightarrow \pi^*$ and $\pi \rightarrow \pi^*$ transitions such as those exhibited by a carbonyl group.

such as C—C and C—H require very large energies to produce electronic excitation. Saturated hydrocarbons absorb only very high energy radiation, usually beyond 1600 Å, far in the ultraviolet. A simple olefin, however, has an absorption band at around 1700 Å, and this can be attributed to the excitation of the π electrons from the electron-paired bonding configuration to a high-energy, or antibonding, state. Such a transition is referred to as a $\pi \rightarrow \pi^*$ transition, the asterisk (*) implying an antibonding orbital.

Some molecules have electronic configurations which can be altered in different ways to lead to an excited, or high-energy, electronic state. This situation arises, for example, with compounds containing a carbonyl group C=Ö:. For such a group the possibility of exciting the π electrons to the excited π^* state exists, as with an olefin, to give a $\pi \rightarrow \pi^*$ transition. Alternatively, the nonbonding electrons of the oxygen might be excited to the higher-energy π^* electron state, and the absorption would then be characterized as an $n \rightarrow \pi^*$ transition, where the n signifies a nonbonding electron. The two types of transitions are represented in Fig. 13-14.

It should be pointed out also that not all electronic transitions of organic compounds occur in the ultraviolet region. The occurrence of colored compounds indicates absorption of radiation in the visible spectrum. Such absorption requires the electronic energy levels to be more closely spaced than in most molecules. The most frequently encountered type of organic molecule that absorbs in the visible region, i.e., is colored, consists of a conjugated system, frequently involving aromatic rings. The qualitative explanation for the closer spacing that results from the delocalization of the conjugated electrons is most easily given by regarding such electrons as being free particles within the potential box of the molecule, as discussed in Sec. 12-7. For sufficiently long "boxes" the spacing is small enough to bring the absorption of radiation into the visible part of the spectrum.

Still other colored compounds, those that contain transition-metal ions, owe their colors to transitions between states that involve the metal d orbitals, which are affected by the nature and geometry of the ligands.

All the above deals with the energy changes of the outermost, or highest-energy, electrons of the molecule. The energies of inner electrons of the atoms of a molecule can also be investigated. Much higher-energy quanta, in the far ultraviolet or in the x-ray region, are then needed. Measurement of the energies for ionization from inner shells is accomplished by a procedure known as *photoelectron spectroscopy* if ultraviolet radiation such as the 584-Å radiation from a helium discharge is used, or electron spectroscopy for chemical analysis (ESCA) if x-rays are used. The procedure in either case consists in using monochromatic ionizing radiation and measuring the kinetic energy that the dislodged electron carries away. The energy difference is that retained by the molecule as a result of this inner-shell ionization. A great deal of information on the energies of electrons in, for example, $n = 1$ orbitals, that is, K shells, can be obtained. These energies depend, as Fig. 13-15 shows, not only on the atom to which the K shell belongs, but also on the chemical environment of the atom.

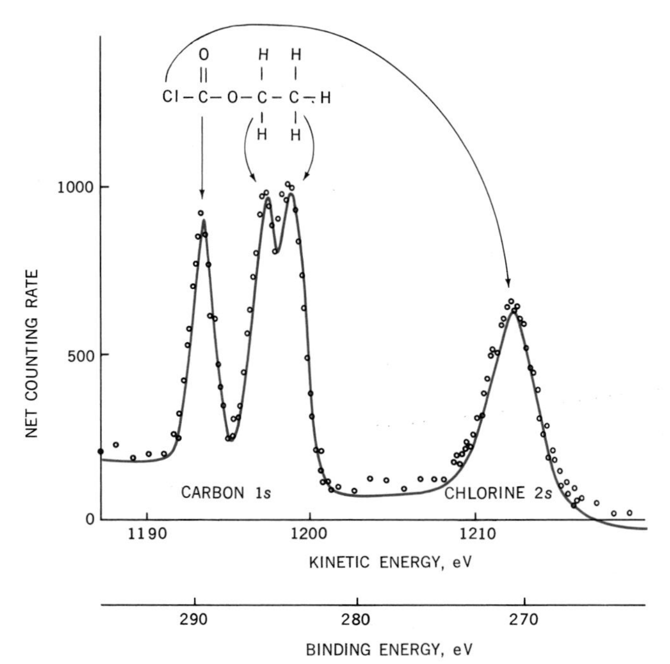

FIGURE 13-15 The photoelectron spectrum of ethyl chloroformate. [*From T. L. James, Photoelectron Spectroscopy, J. Chem. Educ.,* **48:**712 *(1971).*]

ELECTRON AND NUCLEAR MAGNETIC SPECTROSCOPY

The spectroscopic methods treated earlier in the chapter dealt with the study of transitions between energy levels of free, or nearly free, molecules. The spacing between such molecular-energy levels is a characteristic of the electronic structure and atomic makeup of the molecule. The spectroscopic methods now to be studied treat transitions between energy levels whose spacing is dependent on the magnetic field that is applied to the sample. Present-day studies make use of energy levels that arise in two different ways. The first type of energy level, and the transitions between such energy levels, to be discussed arises because the nuclei of some atoms have a magnetic moment, and different orientations of such nuclear magnets relative to the applied field have different energies. The second type of energy level to be dealt with arises from the magnetic moment of the electron. An electron that does not have a counterpart with opposite spin direction can also line up with a magnetic field in different directions, and because the electron has a magnetic moment associated with it, these different orientations correspond to different energies.

The transitions between both the energy levels due to nuclear orientation and those due to electron orientation are studied by means of a resonance method, which will be briefly described later, and one identifies these types of spectroscopic studies as nuclear-magnetic-resonance, or *nmr*, spectroscopy, and electron-spin-resonance, or *esr*, spectroscopy.

13-7 The Energy Levels of Nuclei in Magnetic Fields

Many nuclei have spin angular momentum, and this can be pictured as resulting from the spinning of the nucleus about an axis in much the same way as an electron has a spin angular momentum of $\sqrt{s(s+1)}(h/2\pi)$, where s for the electron must be $\frac{1}{2}$. The angular momentum of other atomic nuclei is also quantized and comes in units of $h/2\pi$. The nuclear spin quantum number I can therefore be introduced and allows the spin angular momentum to be written $\sqrt{I(I+1)}(h/2\pi)$. The spin quantum number is a characteristic of the nucleus and can be zero or can have various integral or half-integral values. These values can to some extent be correlated to the neutron and proton makeup of the nucleus. Most *nmr* studies have been concerned with the hydrogen nucleus, which has $I = \frac{1}{2}$, and the method of *nmr* spectroscopy can be satisfactorily illustrated by restricting our attention to this nucleus.

Along any defined direction in space, and now the applied magnetic field will specify this direction, the angular momentum of the spinning nucleus must present quantized components. A nucleus with an angular momentum $\sqrt{\frac{1}{2}(\frac{1}{2}+1)}(h/2\pi)$ must therefore line itself up, as indicated in Fig. 13-16, in such a way that the angular momentum in the direction of the magnetic field is $+\frac{1}{2}(h/2\pi)$ or $-\frac{1}{2}(h/2\pi)$. The allowed orientations of a nucleus with spin quantum number $\frac{3}{2}$ are also shown in Fig. 13-16 to illustrate the more general case.

The number of different allowed orientations of the nuclear spin direction is seen from Fig. 13-16 to be determined by the nuclear spin-quantum number. The difference in the energies of these different

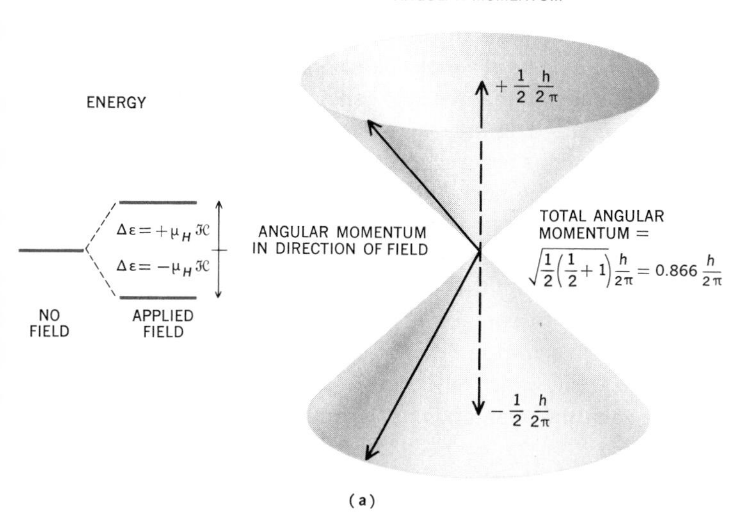

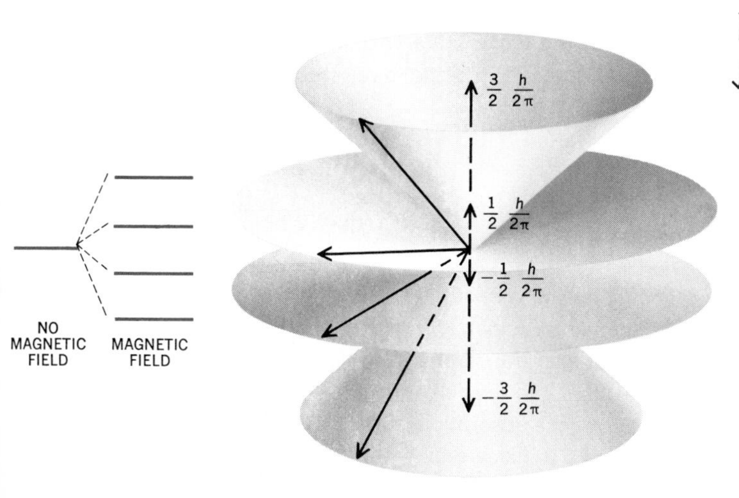

FIGURE 13-16
The allowed orientations of the angular-momentum vector in a magnetic field and the corresponding energy levels. (*a*) The hydrogen-atom nucleus with $I = \frac{1}{2}$ and magnetic moment μH. (*b*) A nucleus with $I = \frac{3}{2}$.

orientations is dependent on the interaction of the nuclear magnetic moment with the magnetic field. A nucleus, which is a charged particle, can be pictured as spinning on its axis and can be expected to have a magnetic moment in a manner analogous to that in which a coil of wire carrying a current has, according to Ampère's law, a magnetic moment. Our lack of understanding of the details of the charge distribution in a nucleus prevents us, however, from obtaining a theoretical value for the nuclear magnetic moment by this approach.

If, however, the magnetic moment of the nucleus of the hydrogen atom is denoted by μ_H and the magnetic field acting on the proton by $\mathcal{H}$, lining up the nuclear moments with and against the magnetic field will produce the energy levels indicated in Fig. 13-16. These values are calculated from the energy of a magnet lined up at various directions to the magnetic field and the quantum stipulation that the spin be lined up with or opposed to the field. Before proceeding to a more detailed treatment of the method used to study transitions between these energy levels and the complications that arise when the nuclei being studied are part of a molecule, a few general features will be reported.

Magnetic fields used in *nmr* spectroscopy usually have values of about 14,000 G. With this field strength it is found that radiation with a frequency of about 60 MHz has quanta with energies of the order of magnitude of the separation of the levels shown in Fig. 13-16. The nuclei may therefore absorb such radiation, which is in the radiofrequency range, and jump, by a change in the orientation of the spin, from the lower to the higher energy state.

It is of interest to notice that the energy-level separation is extremely small. The quanta of radiation of frequency 60 MHz have energies of 0.00004×10^{-21} J compared with a room-temperature value of kT of 4.1×10^{-21} J. The magnitude of this separation is spectroscopically important in that it implies a population in the lower of the two states that is little greater than that of the upper state. According to the Boltzmann distribution, the excess population in the lower state is calculated as

$$\frac{N(\text{lower})}{N(\text{upper})} = e^{\Delta\epsilon/kT}$$

$$= e^{\frac{0.00004 \times 10^{-21}}{4.1 \times 10^{-21}}} = 1.00001$$

It has not been necessary to point out in previous spectroscopic work that incident radiation induces not only transition to higher energy levels, for which radiation is absorbed, but also transitions from higher to lower levels, for which radiation is emitted. In *nmr* experiments the two levels are nearly equally populated, and the absorption of radiation is only slightly the more important effect. It follows that only weak absorption of radiation will occur and therefore that a very sensitive experimental arrangement will be necessary.

The frequency of the radiation that corresponds to the nuclear magnetic-energy-level spacings and the weakness of the radiation absorption that must be expected lead to a spectrometer of a radically different kind from those prism instruments which are used for electronic and vibrational spectral analyses. The arrangement that is most frequently used is shown in Fig. 13-17. The principal magnetic field acts on the nuclei of the sample to produce energy levels such as those indicated in Fig. 13-16. Transitions between these levels are stimulated by radiation from the radio-frequency transmitter, which sends out electromagnetic radiation from the transmitter coil. Radiation will be absorbed by the sample if the frequency of the radiation is such that the quanta of radiation have an energy matching the nuclear-energy-level spacing. When such radiation is absorbed, it can be thought of as producing nuclei in the excited state, which will then tend to reemit the radiation in order to approach the Boltzmann distribution ratio. It is this emitted radiation that is detected by the receiver coil, which, being oriented at right angles to the transmitter, receives no signal

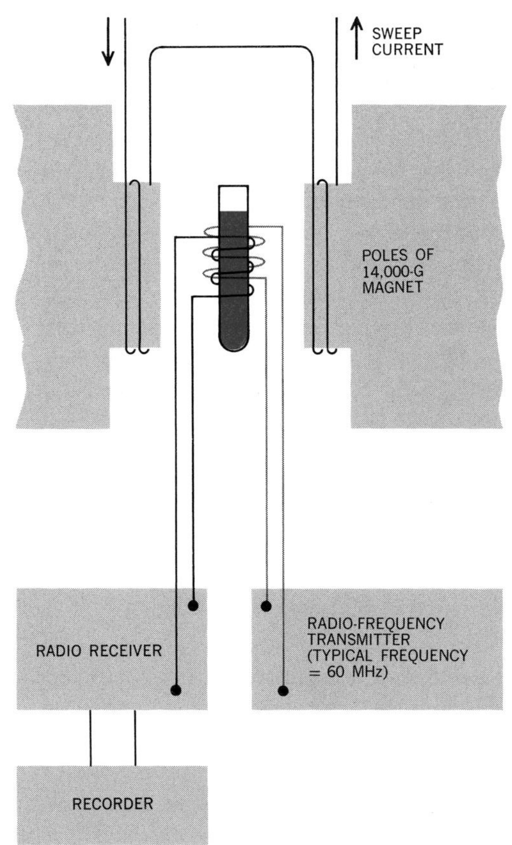

FIGURE 13-17
Schematic representation of an *nmr* spectrometer.

unless the sample provides this coupling with the transmitter. The signal from the receiver coil can be displayed on an oscilloscope or a recorder.

This indication of the operation of an *nmr* spectrometer implies that a fixed magnetic field is imposed on the sample and that the frequency of the radiation is varied. Thus, if the energy spacing is $2\mu_H\mathcal{H}$, as indicated in Fig. 13-16, and the radiation has frequency ν and quanta with energy $h\nu$, absorption of radiation can occur when

$$2\mu_H\mathcal{H} = h\nu \tag{30}$$

Since it is here possible to control the energy-level spacing by manipulating $\mathcal{H}$, the equality of Eq. [30] can be brought about either by adjusting ν after some fixed value of $\mathcal{H}$ is chosen or by adjusting $\mathcal{H}$ after some fixed value of ν has been selected. The latter procedure turns out to be experimentally more satisfactory. A fixed frequency, usually about 60 MHz, is supplied by the transmitter, and the magnetic field is varied through a small range until Eq. [30] is satisfied. At this point the sample absorbs radiation, the transmitter and receiver are coupled, the circuit can be said to be in resonance, and a signal is produced from the receiver circuit.

The signal that is obtained as a function of magnetic field for a fixed frequency of 60 MHz is shown in Fig. 13-18*a* and *b* for several simple compounds. The identification of the hydrogen atom, or groups of hydrogen atoms that produce a given signal, can be made by a simple comparison of these spectra or can be more definitely established by the use of deuterium-substituted derivatives.

If spectra are obtained at higher resolution, a considerable complexity appears, as is shown by the solid curves of (*a*) and the lower curve of (*b*).

The theory of Sec. 13-8 suggested that the nuclear-energy-level splitting is dependent on the nuclear magnetic moment and the magnetic field strength. The experimental results indicate that even if the absorption of only hydrogen atoms is studied, a number of closely spaced absorptions are observed. It is now necessary to see whether or not these finer details of *nmr* spectroscopy, which contain the information of principal interest to the chemist, can be understood.

13-9 Chemical Shifts and Nuclear Magnetic Interactions

The factors which lead to the different absorptions of Fig. 13-18 can often be treated separately from the factors that lead to the finer splittings indicated there.

The separation in the positions of the spectral lines associated with hydrogen atoms in different chemical environments is called the *chemical shift*. These shifts can be conveniently reported by means of the

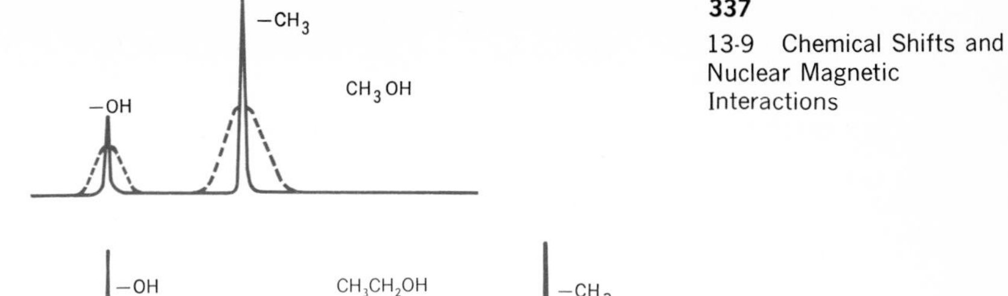

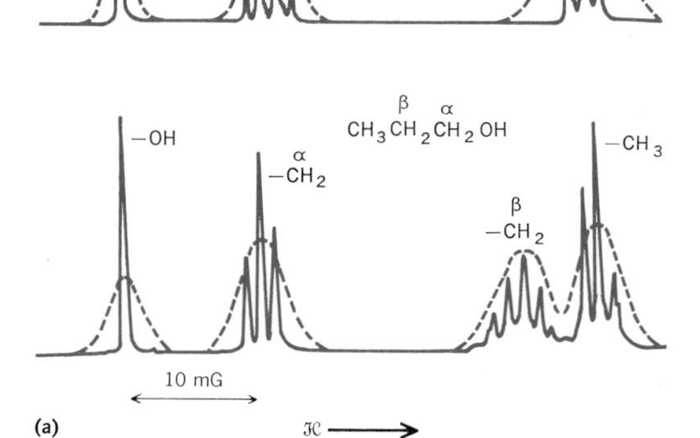

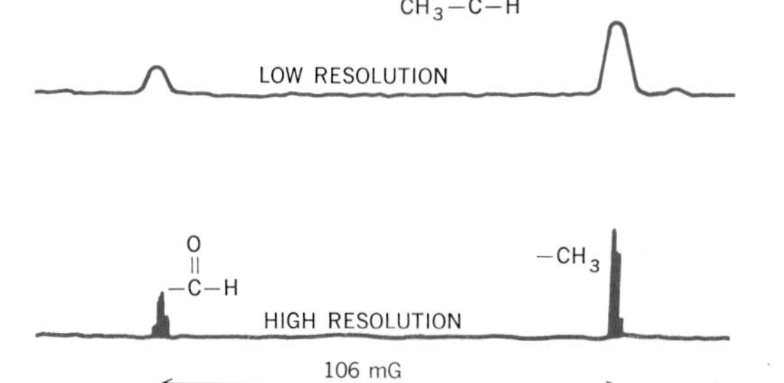

FIGURE 13-18
The *nmr* spectra of several simple compounds at a frequency of 60 MHz and a magnetic field of 14,000 G. In (*a*) the solid lines give the high-resolution spectra. The dashed lines show the appearance of the spectra at low resolution where the splitting arising from the interactions of the nuclei would not be observed.

difference in magnetic field necessary for absorption compared with that necessary for absorption by some reference. This difference is usually reported as the chemical shift δ, defined as

$$\delta = \frac{\mathcal{H}_{\text{ref}} - \mathcal{H}_{\text{sample}}}{\mathcal{H}_{\text{ref}}} \times 10^6 \qquad [31]$$

where the reference chosen is now usually tetramethyl silane, $(CH_3)_4Si$, because of the conveniently located, well-defined absorption that it produces. Since hydrogen atoms in different samples show absorption, or resonance, at fields that differ by the order of milligauss when the magnetic field is 14,000 G, the values of δ are made of convenient size by the inclusion of the factor 10^6 in Eq. [31].

The existence of the chemical shift can be attributed to the screening effect that the electrons about a nucleus exert. Thus, although the external magnetic field is the same for all hydrogen atoms of a sample of CH_3OH, for example, the electron distribution in the C—H and O—H bonds screens the nuclei from the applied field to different extents. Some correlations have succeeded in showing that the more the electrons of the bond to hydrogen are drawn to the bonding atom, the more exposed is the nucleus of the hydrogen atom. Such exposed nuclei generally absorb at lower magnetic fields than do well-shielded nuclei.

The *nmr* spectrum is, as a result of chemical shifts, a portrayal of the chemical environment of the various hydrogen atoms of the material. It follows that an analysis of a spectrum of an unknown material can lead to information on the types of bonding to hydrogen atoms, and often to the molecular structure of the sample. In this respect *nmr* complements infrared and ultraviolet spectroscopy in the elucidation of the structures of large molecules. Some of the characteristic chemical shifts that are used in such analyses are shown in Table 13-5.

The high-resolution detail, as shown schematically in Fig. 13-19, is also characteristic of the hydrogen-atom arrangement of the molecule and is therefore also helpful in structural determinations. Only some features of the source of these additional splittings can be given.

The magnetic field at a nucleus in a molecule is determined not only by the external magnetic field as modified by the shielding electrons, but also by the presence and orientation of the other nuclei in the molecule that behave as magnets, i.e., have magnetic moments. Since both O^{16} and C^{12} have zero spin and zero magnetic moment, the magnetic nuclei of many organic compounds consist only of the hydrogen atoms. It is the interaction between the nuclei of these atoms that leads to the additional splitting beyond that of the chemical shifts.

The nature and effect of these interactions can be illustrated by reference to the spectrum of acetaldehyde, $CH_3—\overset{\displaystyle O}{\overset{\|}{C}}—H$. The hydrogen atoms of the methyl group experience a magnetic field that depends on the applied field, on the chemical-shift effect of the shielding electrons, and on the influence of the magnetic field of the nucleus of the hydrogen atom adjacent to the carbonyl group. This nucleus, as Fig. 13-19 indicates, can line up with or against the principal magnetic field. The methyl hydrogen atoms will experience, therefore, a slightly greater or lesser magnetic field, depending on the orientation of the lone hydrogen atom. The methyl absorption will therefore be split into a doublet.

The single hydrogen atom also experiences a magnetic field that depends on the applied field, on the shielding provided by its bonding

TABLE 13-5 Characteristic values for the chemical shift δ for hydrogen in organic compounds*
Tetramethyl silane is the reference compound.

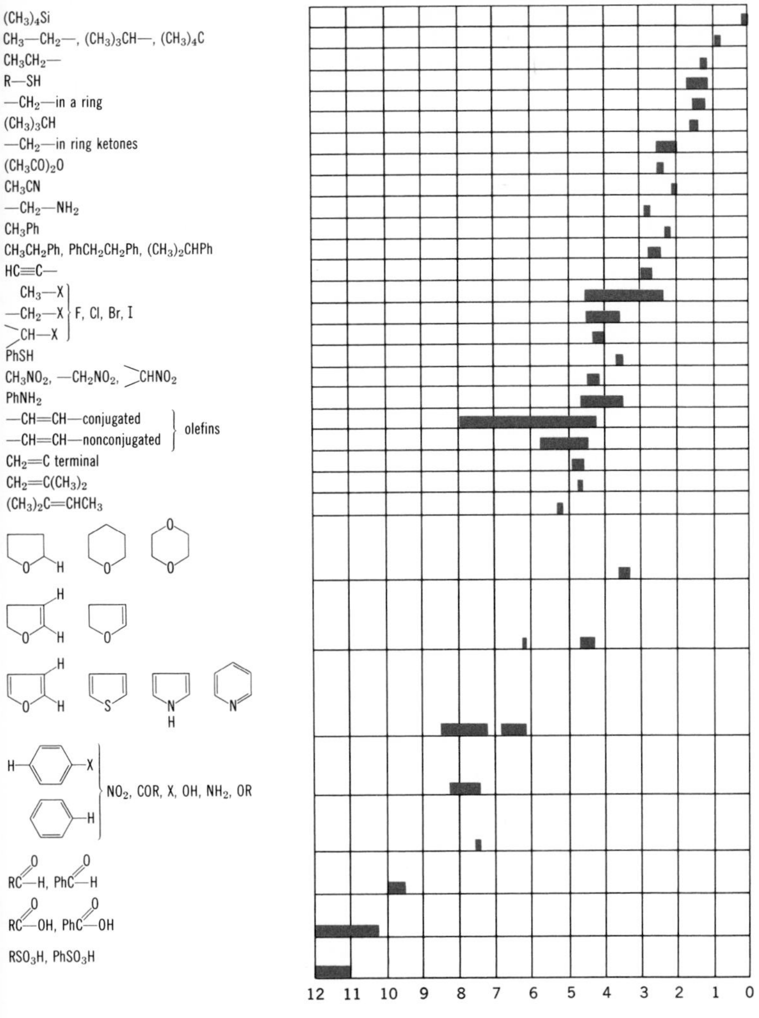

$(CH_3)_4Si$
$CH_3—CH_2—$, $(CH_3)_3CH—$, $(CH_3)_4C$
$CH_3CH_2—$
$R—SH$
$—CH_2—$in a ring
$(CH_3)_3CH$
$—CH_2—$in ring ketones
$(CH_3CO)_2O$
CH_3CN
$—CH_2—NH_2$
CH_3Ph
CH_3CH_2Ph, $PhCH_2CH_2Ph$, $(CH_3)_2CHPh$
$HC{\equiv}C—$
$CH_3—X$, $—CH_2—X$, $>CH—X$ } F, Cl, Br, I
PhSH
CH_3NO_2, $—CH_2NO_2$, $>CHNO_2$
$PhNH_2$
$—CH{=}CH—$conjugated, $—CH{=}CH—$nonconjugated } olefins
$CH_2{=}C$ terminal
$CH_2{=}C(CH_3)_2$
$(CH_3)_2C{=}CHCH_3$

$H—C_6H_4—X$, $C_6H_5—H$ } NO_2, COR, X, OH, NH_2, OR

$RC(=O)—H$, $PhC(=O)—H$
$RC(=O)—OH$, $PhC(=O)—OH$
RSO_3H, $PhSO_3H$

12 11 10 9 8 7 6 5 4 3 2 1 0

δ

*From E. Mokacsi, *J. Chem. Educ.*, **41**:38 (1964).

electrons, and on the influence of the three magnetic nuclei in the methyl group. There are four different ways in which the three magnets can arrange themselves relative to the applied field. These are shown in Fig. 13-19, where it is indicated that two of the ways are three times as probable as the other two. The lone hydrogen atom can experience, therefore, four slightly different magnetic fields, depending on the orientation of the spins of the methyl hydrogen nuclei. Four different resonance frequencies would be expected for the lone hydrogen nucleus, or in view of the experimental arrangement, four slightly different ap-

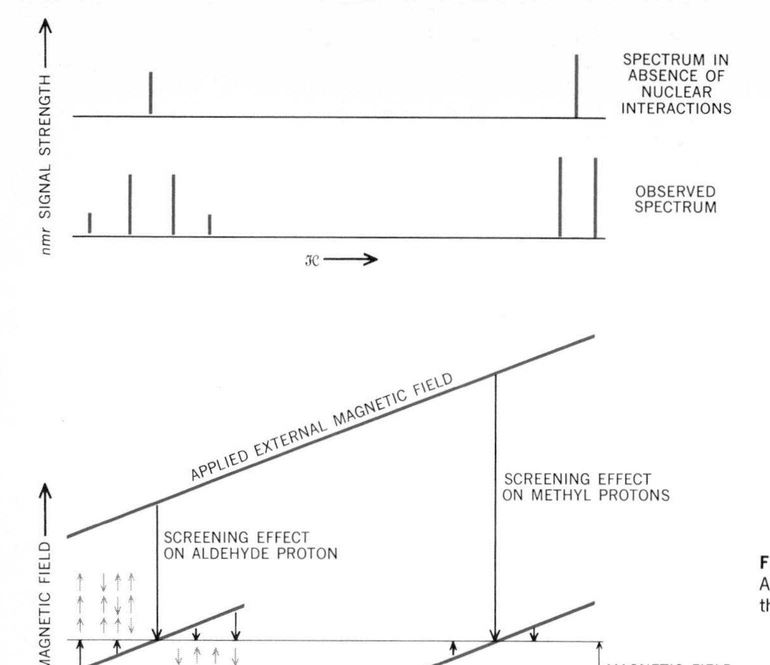

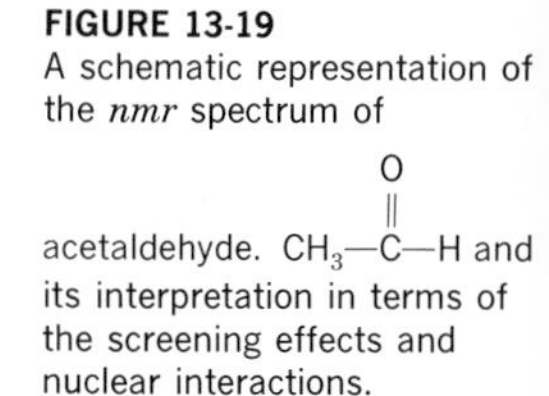
FIGURE 13-19 A schematic representation of the *nmr* spectrum of acetaldehyde. CH_3—C(=O)—H and its interpretation in terms of the screening effects and nuclear interactions.

plied fields at a fixed radiation frequency. The spectrum of Fig. 13-19 bears out these analyses.

This simple example should illustrate that the magnetic nuclear interactions give information on the type of neighbors of any hydrogen atom or group of atoms in the molecule. Such intimate details can be obtained even for quite large molecules, and it is this aspect which makes the fine splittings of *nmr* spectra of great value in molecular-structure studies.

A number of important features of *nmr* spectroscopy have not been dealt with in this brief introduction. It is frequently of interest, for example, to examine the mechanism by which the radiation is able to interact with the magnetic nuclei to turn them to a different orientation. This has not been treated here. Likewise, no mention has been made of the fact that if atoms, such as the hydrogen atoms of a water–sodium hydroxide solution, move their position from one molecule to another, so that they occupy a given position for less than about 10^{-2} s, the *nmr* spectrum shows a single absorption at a position characteristic of the average of the environments of the nuclei. If the nuclei change position less rapidly, the *nmr* spectrum will indicate two absorptions, one characteristic of the one environment and the other characteristic of the other. It follows that *nmr* techniques can be used to study the rate of very fast reactions, and this, in fact, is one of the most interesting aspects of *nmr* spectroscopy.

13-10 Electron-spin-resonance Spectroscopy

The presence of an unpaired electron in a molecule or ion allows energy levels to be produced from the interaction of the magnetic moment of the electron with an applied magnetic field. The electron, like the proton, has a half unit of spin angular momentum, and the spin angular momentum is quantized along the direction defined by the magnetic field, if the component in this direction is $+\frac{1}{2}(h/2\pi)$ or $-\frac{1}{2}(h/2\pi)$. These two states will have energies that are separated from the original state with no applied field by the amounts $+\mu_e\mathcal{H}$ and $-\mu_e\mathcal{H}$, where μ_e is the magnetic moment of the spinning electron and $\mathcal{H}$ is the magnetic field acting on the unpaired electron.

The electron-spin magnetic moment, however, is about a thousand times greater than a typical nuclear magnetic moment. The energy of interaction of the magnetic moment of the unpaired electron with the applied field will be greater than the corresponding interaction between the nuclear magnetic moment and the applied field. It is found that when a magnetic field of 3000 G is used, the energy spacing between the levels with different spin orientation relative to this field is such that transitions are caused by radiation of about 30-mm wavelength, a wavelength of the microwave region. The energy separation, even in the relatively low field of 3000 G, is therefore

$$\frac{3 \times 10^8}{0.03}(0.66 \times 10^{-33}) = 0.007 \times 10^{-21} \text{ J}$$

a value to be compared with nuclear-energy spacings of about 4×10^{-21} J in a field of 14,000 G.

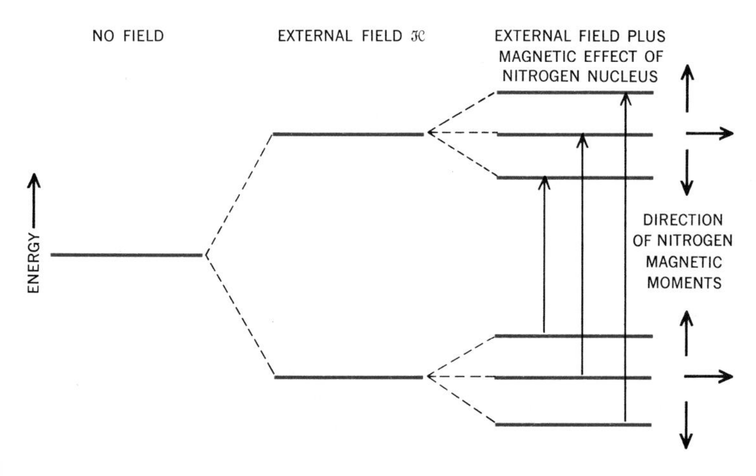

FIGURE 13-20 The energy-level diagram for the unpaired electron of the radical ion $(SO_3)_2NO^-$, showing the splitting of the two electron-spin states by the magnetic moment of the nitrogen nucleus.

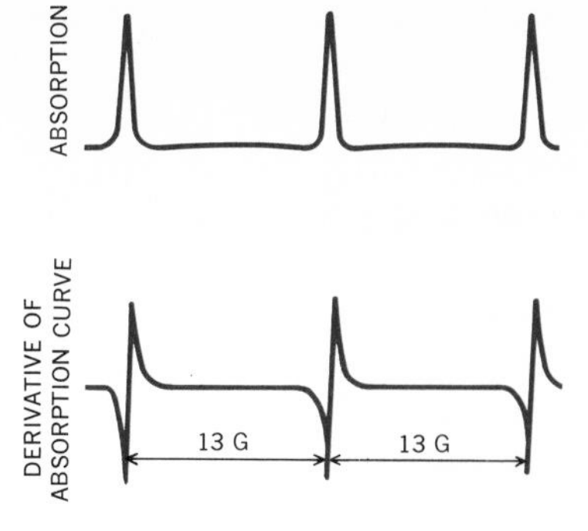

FIGURE 13-21
The *esr* spectrum of the radical ion $(SO_3)_2NO^-$ at a frequency of 9500 MHz and a magnetic field of about 3400 G.

The most prominent and revealing feature of electron-spin-resonance spectra is the splitting caused in the transition between the two electron-orientation states by the interaction of the magnetic moment of the spinning electron with the magnetic moments of those nuclei in the molecule which have magnetic moments. The electron-spin energies and the splitting of these energies due to the nitrogen nucleus, which has one unit of spin, are shown for the ion $(SO_3)_2NO^-$ in Fig. 13-20. Transitions occur which change the orientation of the electron spin relative to the applied magnetic field. The interaction between the electron and the magnetic nucleus is sufficiently small so that the transitions do not also change the magnetic-moment direction. With this selection rule the transitions of Fig. 13-20 can be drawn. The observed spectrum does in fact show three absorption bands. It should be mentioned that, because of the experimental arrangement used in *esr*, the derivative of the usual spectral absorption or emission curve

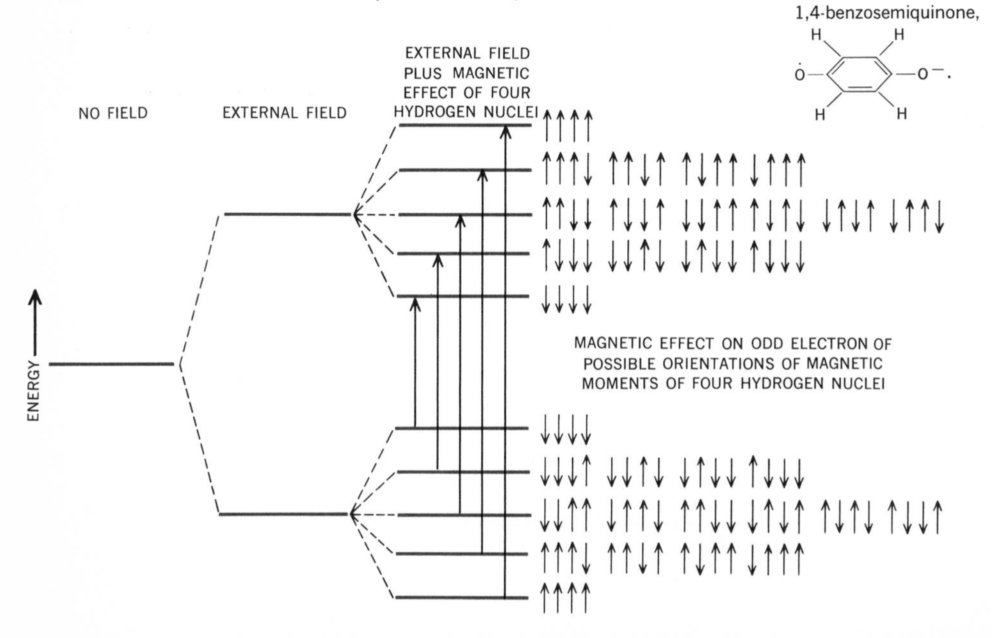

FIGURE 13-22
The energy-level diagram for the odd electron of the free-radical ion 1,4-benzosemiquinone,

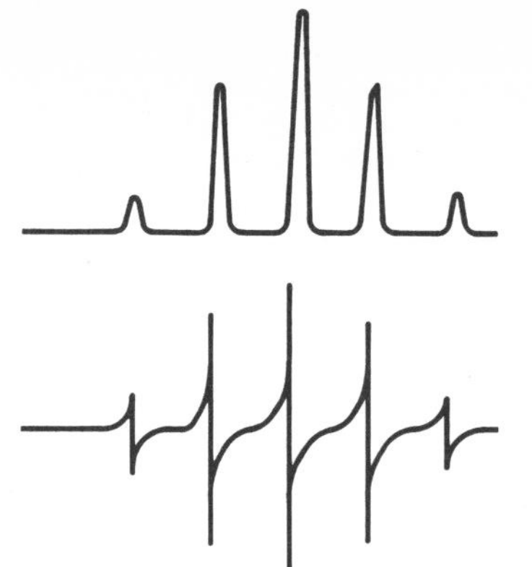

FIGURE 13-23
The *nmr* spectrum (*a*) integral, (*b*) differential of 1.4 benzosemiquinone,

H H
Ȯ— —O⁻,
H H

at a frequency of 9500 MHz and a magnetic field of about 3400 G.

is often shown. Figure 13-21 shows this presentation of the recorded spectrum.

The splittings due to interactions with the magnetic nuclei can be treated in much the same way as were the nuclear magnetic interactions in *nmr* spectroscopy. Thus, in the free-radical ion

H H
Ȯ— —O⁻
H H

the odd electron can move throughout the molecule, and it experiences the effect of the nuclear moments of the four equivalent hydrogens. The expected splittings and transitions and the observed spectrum are shown in Figs. 13-22 and 13-23.

Electron-spin-resonance spectroscopy provides a powerful tool for the study of chemical species with unpaired electrons. It gives information not only on the presence and number of such electrons, as measurements of paramagnetism often do, but also on the distribution of the electron in the molecule. The splitting due to interactions with a nuclear magnetic moment depends on the odd electron being distributed throughout the molecule so that it is to some extent near that nucleus. Such details of electronic configuration in free-radical-type molecules are one of the important features treated by *esr*.

Problems

1 The high-temperature microwave spectrum of KCl vapor shows an absorption at 7687.94 MHz that can be identified with the $J = 0$ to $J = 1$ transition of $K^{39}Cl^{35}$ molecules in the lowest, $v = 0$, vibrational state. What are the moment of inertia and the bond length?

2 Plot, side by side, the energies of some of the lower rotational states of

the CO molecule. which has a rather low moment of inertia, and the energies of some of the rotational states of CO_2, which has $I = 7.11 \times 10^{-46}$ kg m^2 and represents molecules with larger moments of inertia.

Calculate the populations, relative to the number n_0 in the lowest state, of some of the energy levels of both CO and CO_2, and represent them by a bar graph.

3 Confirm, by a Boltzmann distribution calculation, that any one of the first few rotational-energy levels of CO at 25°C has a population that is appreciable compared with that of the lowest energy level.

4 Derive an expression for the most populated rotational energy level for a linear molecule. (Assume that the rotational spacings are small compared with kT so that many levels are occupied and the maximum can be found by forming a derivative.)

5 The bond length of the gas-phase NaCl molecule is 2.36 Å. Show by vertical lines on a frequency abscissa the positions of the three lowest-frequency rotational transitions that will occur. Indicate the rotational quantum numbers involved in each transition.

6 Obtain a general expression for the energies of the $\Delta J = +1$ transitions which start from the J level and end in the $J + 1$ level. Base the derivation on a linear molecule for which the rotational-energy levels are given by Eq. [1].

7 For a chemical bond with a typical force constant k of 500 N m^{-1} and equilibrium bond length of 1.5 Å, plot the potential energy as a function of bond length for bond-length changes of up to 10 percent of the equilibrium bond length.

Draw a line indicating the value of the energy kT at 25°C. What percent distortion of the bond could be produced by this amount of energy if the bond behaved as a classical spring?

8 Calculate the energies of the three lowest-energy vibrational states for HF for which the force constant is 970 N m^{-1}. What is the energy spacing between these levels, and what would be the wave number of the radiation that would cause the transition from the $v = 0$ to the $v = 1$ level?

Ans. $\Delta\epsilon = 8.27 \times 10^{-20}$ J; $\bar{\nu} = 4160$ cm^{-1}.

9 The infrared spectrum, at low resolution, of CO shows an absorption band centered at 2170 cm^{-1}. What is the force constant of CO?

10 The vibrational-energy-level spacings of H_2, HD, and D_2 can be deduced from Raman spectra to be 4395, 3817, and 3118 cm^{-1}, respectively. Calculate the force constant of each isotopic species, and verify that isotopic substitution, which would not be expected to alter the electronic behavior in a molecule, does not change the bond force constant. (The small variation can be attributed to a derivation of the bond from a Hooke's law force relation.)

11 Since HCl consists of 75 percent HCl^{35} and 25 percent HCl^{37}, a spectrum of HCl shows absorption bands due to the two isotopic types.

- **a** Calculate the difference in frequency expected for the $v = 0$ to $v = 1$ vibrational transition of HCl^{35} and HCl^{37}, assuming that the force constants of the two molecules are identical and equal to 484 N m^{-1} dyne^{-1} cm^{-1}.
- **b** Compare this difference with the splitting of the vibration-rotation lines of Fig. 13-8.
- **c** Identify the components due to HCl^{35} and those due to HCl^{37}.

12 Calculate the relative populations of the $v = 1$ and $v = 0$ states for HCl at room temperature, and confirm that we are justified in treating room-temperature infrared spectra in terms of the $v = 0$ to $v = 1$ transition.

13 The microwave spectrum of KCl referred to in Prob. 1 is obtained on a vapor sample at a high enough temperature so that vibrational levels above the $v = 0$ level are also appreciably populated. As a result, additional absorptions at 7640.78, 7593.83, and 7547.07 MHz are observed, and these can be assigned to the $J = 0$ to $J = 1$ transitions of the $v = 1$, $v = 2$, and $v = 3$ states. Calculate moment-of-inertia and bond-length values for these vibrational states. What would the moment of inertia and bond length be for the hypothetical molecule that has no vibrational energy and is identified with the equilibrium position at the minimum of the potential-energy curve of Fig. 13-5?

14 Plot, showing the rotational-energy-level spacings to scale, some of the rotational-energy levels of the $v = 0$ and $v = 1$ states of the CO molecule. Indicate the transitions that are responsible for the rotational-vibrational spectrum.

15 From the rotational Raman spectrum of Fig. 13-10, calculate the moment of inertia of the CO_2 molecule. What is the CO bond distance in this molecule?

16 A frequently used empirical expression that generates a potential-energy curve like that expected for a diatomic molecule has been given by Morse as $U(r) = D_e(1 - e^{-\beta(r-r_e)})^2$, where D_e is the dissociation energy measured from the minimum of the potential curve, r_e is the equilibrium bond length, and β is a constant that is related to the molecular properties by $\beta = \sqrt{k/2D_e}$.

a Draw the potential curve for HCl according to the Morse function.

b Obtain the relation $\beta = \sqrt{k/2D_e}$ by forming $d^2U(r)/dr^2$, and compare this expression with the expression for the region around the potential minimum $[d^2U(r)/dr^2]_{r=r_e} = k$.

17 An electronic transition in CO is responsible for an absorption band around 1400 Å in the ultraviolet. A photograph of this band shows that it consists, in part, of a series of lines, expressed in wave numbers, at 64,703, 66,231, 67,675, 69,088, 70,470 cm^{-1}, and so forth. From the fact that these are absorption lines and are observed at fairly low temperatures, they can be assumed to arise from CO molecules in the lowest vibrational state. The abrupt beginning of the series at 64,703 cm^{-1} suggests that this transition leads to the $v = 0$ level of the excited electronic state.

a Draw a diagram like Fig. 13-13 to illustrate these transitions.

b The assumption of Hooke's law for chemical bonds leads to Eqs. [22] and [23], which imply that vibrational levels have a constant spacing. Recognize that electronic spectra allow this expression to be checked.

c Calculate a force constant from the $v = 0$, $v = 1$ spacing of the excited electronic state of the CO molecule. Compare this with the force constant for the normal, or ground, electronic state. *Ans.* 940 N m^{-1}.

References

WHIFFEN, D. H.: "Spectroscopy," John Wiley & Sons, Inc., New York, 1966. An excellent introductory treatment of the areas of spectroscopy that are of interest to chemists.

BARROW, G. M.: "Introduction to Molecular Spectroscopy," McGraw-Hill Book Company, New York, 1962. An extension of the material presented in this text on rotational, vibrational, and electronic spectra.

BAUMAN, R. P.: "Absorption Spectroscopy," John Wiley & Sons, Inc., New York, 1962. A treatment of infrared, visible, and ultraviolet absorption spectroscopy. Emphasis is on the experimental methods of absorption spectroscopy and the analysis and application of the spectral data.

SZYMANSKI, H. A. (ed.): "Raman Spectroscopy, Theory and Practice," vol. 2, Plenum Press, New York, 1970. Modern aspects of Raman spectroscopy including the use of laser sources.

ROBERTS, J. D.: "Nuclear Magnetic Resonance," McGraw-Hill Book Company, New York, 1959, and "An Introduction to Spin-Spin Splittings in High Resolution Nuclear Magnetic Resonance Spectra," W. A. Benjamin, Inc., New York, 1961. Two nonmathematical introductions to the use of *nmr* spectroscopy in the elucidation of the structures of organic molecules.

JAFFE, H. H., and M. ORCHIN: "Theory and Applications of Ultraviolet Spectroscopy," John Wiley & Sons, Inc., New York, 1962. An introduction to the theoretical expressions for the electronic energies of organic molecules and consideration of the correlation of experimental results with ideas on electronic structure.

HECHT, H. G.: "Magnetic Resonance Spectroscopy," John Wiley & Sons, Inc., New York, 1967. A short treatment of the fundamentals of magnetic-resonance techniques and discussions of broad-line and high-resolution *nmr* and of electron paramagnetic-resonance and double-resonance techniques.

More advanced reference-type books that deal with areas of spectroscopy in greater depth are:

HERZBERG, G.: "Spectra of Diatomic Molecules," "Infrared and Raman Spectra of Polyatomic Molecules" and "Electronic Spectra of Polyatomic Molecules," D. Van Nostrand Company, Inc., Princeton, N.J., 1950, 1955, and 1967.

WILSON, E. B., JR., J. C. DECIUS, and P. C. CROSS: "Molecular Vibrations," McGraw-Hill Book Company, New York, 1955.

KING, G. W.: "Spectroscopy and Molecular Structure," Holt, Rinehart and Winston, Inc., New York, 1964.

POPLE, J. A., W. G. SCHNEIDER, and H. J. BERNSTEIN: "High-resolution Nuclear Magnetic Resonance," McGraw-Hill Book Company, New York, 1959.

BOVEY, F. A.: "Nuclear Magnetic Resonance Spectroscopy," Academic Press, Inc., New York, 1969. A brief theoretical introduction followed by a clear and comprehensive treatment of high-resolution nuclear magnetic spectroscopy.

14

EXPERIMENTAL STUDIES OF MOLECULAR STRUCTURE: DIFFRACTION METHODS

Much of our information on the angles and distances within and between molecules, other than for the relatively simple molecules that can be well treated spectroscopically, comes from diffraction experiments. The way in which the interference effects in the scattered beams of various types, from both gaseous and crystalline samples, can be used to deduce the structure of the molecules or ions of the sample will now be studied. The experimental methods studied here depend on the constructive and destructive interference that occurs when an electron or an x-ray beam is scattered by the sample. We shall see that some features of the analyses are common to all diffraction experiments. The different types of beams and the different states of the sample will, however, lead to appreciable differences in the way the data are analyzed.

The diffraction effects produced by gas-phase molecules will be dealt with first. In these experiments a beam of electrons is used and the scattering pattern that is obtained yields information about the bond lengths and bond angles of the molecule. In the second half of the chapter the scattering by crystalline materials will be studied. Most such work makes use of a beam of x-rays, which are more penetrating than are electron beams. However, for some special studies a neutron beam has some advantages. Again bond lengths and angles for the molecules or ions of the crystal are obtained, but now also the way in which these units pack together is a product of the analysis.

To begin with, the basis for the effects to be used to obtain molecular data will be introduced by means of a simple example.

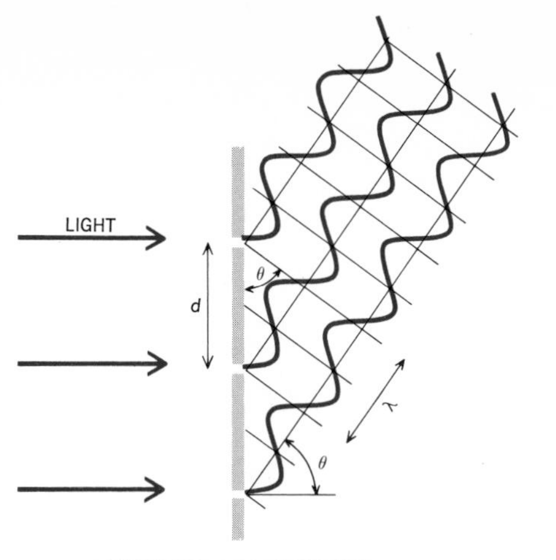

CONSTRUCTIVE INTERFERENCE

$$\sin\theta = \frac{\lambda}{d}$$

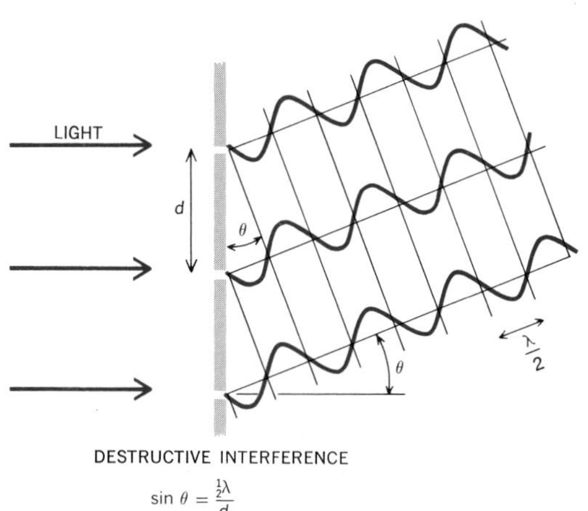

DESTRUCTIVE INTERFERENCE

$$\sin\theta = \frac{\frac{1}{2}\lambda}{d}$$

FIGURE 14-1
Interference effects from an illuminated set of slits that act as sources. Illustration of the result that, for constructive interference, $\sin\theta = n\lambda/d$, where $n = 0, 1, 2, \ldots$.

14-1 The Interference Phenomenon

The general principle of diffraction methods depends on the phenomenon of interference, which occurs when any wave motion is scattered from a number of centers. This phenomenon is, for example, exhibited by visible radiation when a beam of light passes through a series of closely spaced slits, as illustrated in Fig. 14-1. If the light is monochromatic, i.e., consists of radiation of only a single wavelength, the wave motions of the light emerging from the slits will add together in only certain directions. In these directions *constructive interference* is said to occur, and at these directions a beam of diffracted light will appear. At other directions the diffracted waves will be out of phase to various extents, *destructive interference* will occur, and less light will be seen. For the pattern of scattering units provided by the slits of Fig. 14-1 it is easy to see, as shown in the figure, that constructive interference occurs in directions defined by the angle θ, which are related to the spacing d between the slits and the wavelength λ of the light by the relation

$$\sin\theta = \frac{n\lambda}{d} \qquad [1]$$

where n is an integer. One sees from this, furthermore, that d and λ must be of the same order of magnitude if $n\lambda/d$ is to take on a number of values between 0 and 1 when n assumes various small-integral values. Under these conditions Eq. [1] yields several values of θ at which constructive interference will occur. When such is the case the angles for constructive interference can be measured, and if λ is known, Eq. [1] can be turned around to give

$$d = \frac{n\lambda}{\sin\theta}$$

The experiment could thus be used to deduce the value of the slit spacing d. We have come, therefore, with this very simple example, to see the principle on which the determination of structure by diffraction is based.

One further general procedure in the study of interference effects should be taken up before detailed studies of diffraction processes are begun. It is quite clear from the diagrams of waves, such as those of Figs. 14-1 and 14-2, that constructive interference, in which the amplitudes of the waves simply add together, occurs when two waves in phase combine and that destructive interference, in which the amplitudes of the waves must be simply subtracted, occurs when the two waves exactly out of phase combine. Expressions will be needed later for the wave that results when two waves, of perhaps different ampli-

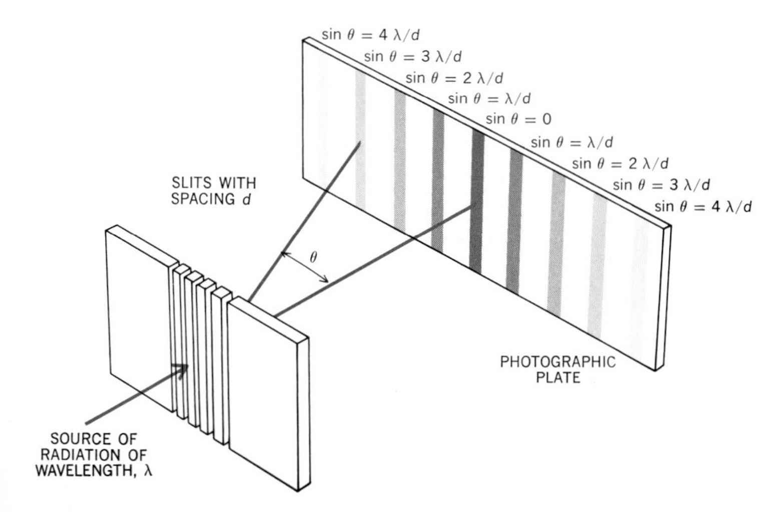

FIGURE 14-2
Illustrating that, by measurement of the angles for constructive interference, the spacing d of the slit system can be deduced. (If $\lambda \approx d$, then a number of diffraction lines will be observed on the photographic plate for reasonable values of $\sin\theta = n\lambda/d$.)

tudes, come together neither exactly out of phase nor exactly in phase, but rather with some intermediate phase relation.

The combination can be simply performed, considering first waves with equal amplitudes, by adding the two trigonometric expressions

$$A \cos (\omega t + \varphi_1)$$

and

$$A \cos (\omega t + \varphi_2)$$

for waves both of which have the same frequency ν, or angular velocity $\omega = 2\pi\nu$, the same amplitude A, but possibly different phases φ_1 and φ_2. The resultant wave produced by the combination of these two components can be described by

$$A_R = A[\cos (\omega t + \varphi_1) + \cos (\omega t + \varphi_2)]$$

Recollection of the trigonometric formula

$$\cos x + \cos y = 2 \cos \tfrac{1}{2}(x - y) \cos \tfrac{1}{2}(x + y)$$

allows this result to be converted to

$$A_R = 2A \cos \tfrac{1}{2}(\varphi_1 - \varphi_2) \cos [\omega t + \tfrac{1}{2}(\varphi_1 + \varphi_2)] \qquad [2]$$

The resultant wave is seen to have an amplitude of $2A \cos \frac{1}{2}(\varphi_1 - \varphi_2)$, to have the same frequency, or angular velocity, ω as the component waves, and to have a new phase that is the average of those of the components.

For most of our later purposes we shall be interested primarily in the amplitude factor. We shall be able to rely in subsequent studies of wave amplitudes on the result shown here, that when waves of the same frequency combine, they will produce a resultant wave that also has this frequency.

The amplitude factor shows clearly the dependence of the resultant on the phase difference of the components. One can notice, for example, that the largest amplitude, either positive or negative, and therefore constructive interference, occurs for $\frac{1}{2}(\varphi_1 - \varphi_2) = n\pi$ or $\varphi_1 - \varphi_2 = n(2\pi)$, where n is an integer. Likewise, the resultant wave has zero amplitude for $\frac{1}{2}(\varphi_1 - \varphi_2) = [(2n + 1)/2]\pi$, or $\varphi_1 - \varphi_2 = (2n + 1)\pi$, that is, an odd multiple of π.

We shall often be concerned with the square of the amplitude of the resultant beam, and if A_R represents the resultant amplitude, this quantity

$$A_R^2 = 4A^2 \cos^2 \tfrac{1}{2}(\varphi_1 - \varphi_2) \qquad [3]$$

can be rewritten by employing another half-angle formula, $2 \cos^2 x = 1 + \cos 2x$, as

$$A_R^2 = 2A^2[1 + \cos (\varphi_1 - \varphi_2)] \qquad [4]$$

ne can again investigate the special cases of constructive and destruc-ive interference, and one can see that intermediate situations are also andled by the substitution of the appropriate phase difference $\varphi_1 - \varphi_2$.

When we wish to combine interfering waves that have different mplitudes, say A_1 and A_2, the useful expressions are most conven-ently obtained by starting with the wave expressed by the exponential orms

$$A_1 e^{i(\omega t + \varphi_1)} \qquad \text{and} \qquad A_2 e^{i(\omega t + \varphi_2)}$$

hen these expressions are added, we obtain the description of the esulting wave as

$$\begin{aligned} A_R &= A_1 e^{i(\omega t + \varphi_1)} + A_2 e^{i(\omega t + \varphi_2)} \\ &= (A_1 e^{i\varphi_1} + A_2 e^{i\varphi_2}) e^{i\omega t} \end{aligned} \qquad [5]$$

he magnitude of the square of amplitude, $(A_i e^{i\varphi_1} + A_2 e^{i\varphi_2})$, of this esultant beam is now obtained from the product of the amplitude and ts complex conjugate. We thus obtain

$$\begin{aligned} |A_R|^2 &= (A_1 e^{i\varphi_1} + A_2 e^{i\varphi_2})(A_1 e^{-i\varphi_1} + A_2 e^{-i\varphi_2}) \\ &= A_1^2 + A_2^2 + A_1 A_2 (e^{i(\varphi_1 - \varphi_2)} + e^{-i(\varphi_1 - \varphi_2)}) \end{aligned}$$

he final term can be simplified by use of

$$\begin{aligned} e^{ix} + e^{-ix} &= (\cos x + i \sin x) + (\cos x - i \sin x) \\ &= 2 \cos x \end{aligned}$$

hich allows the square of the amplitude of the resultant wave to be ritten

$$A_R^2 = A_1^2 + A_2^2 + 2A_1 A_2 \cos(\varphi_1 - \varphi_2) \qquad [6]$$

his important result, which we shall make use of in our studies of iffraction, can be seen to reduce, when $A_1 = A_2$, to that obtained reviously and shown in Eq. [4].

For these results to be useful in the study of diffraction experi-ments, it only remains to point out that when the phase difference $\varphi_1 - \varphi_2$ results from the interfering components traveling different istances, the phase difference depends on the ratio of this distance to the wavelength of the waves according to the relation

$$\varphi_1 - \varphi_2 = \frac{\delta}{\lambda}(2\pi) \qquad [7]$$

gain the special cases of constructive and destructive interference, here δ is an integral or half-integral multiple of 2π, can be considered, nd it can also be recognized that intermediate situations are again andled.

The first use of interference effects to investigate structures fol-owed from the suggestion of Max von Laue in 1912 that the wave-engths of the then newly discovered x-rays were in the angstrom range and that interference effects might be produced when a beam of such radiation passed through a crystal in which the crystal planes were known to be spaced by amounts of the order of angstroms. That a similar interference effect might be observed with a beam of electrons followed from de Broglie's suggestion in 1923 that any particle moving with a momentum mv has associated with it a wavelength $\lambda = h/mv$. The statement that a beam of electrons has an associated wavelength implies that, if the wavelength is of the same order of magnitude as the spacing between some set of slits or some set of centers that scatter electrons, interference effects can be observed with an electron beam also.

ELECTRON DIFFRACTION: DIFFRACTION BY GASES

14-2 The Wave Nature and Scattering of a Beam of Electrons

An electron beam is produced by drawing electrons out of a cathode plate by means of an applied voltage and directing them to an anode. If such a beam is made to pass across a potential difference of $\mathcal{V}$ volts, each electron acquires kinetic energy as a result of the acceleration in this electric field. The potential difference, or voltage drop, is defined as the energy given to a unit charge when it falls through the potential difference. Thus, ignoring relativistic effects that begin to be important at the accelerating voltages that are used,

$$\tfrac{1}{2}mv^2 = e\mathcal{V}$$

This relation leads to the momentum expression

$$mv = \sqrt{2me\mathcal{V}}$$

which, for an electron, gives the de Broglie wavelength as

$$\lambda = \frac{h}{mv} = \frac{h}{\sqrt{2me\mathcal{V}}} = \frac{12.25}{\sqrt{\mathcal{V}}} \quad \text{Å} \qquad [8]$$

An accelerating potential of 40,000 V therefore corresponds to an electron-beam wavelength of 0.06 Å. Such a wavelength leads one to expect that interference effects will be observed when a beam so accelerated passes through a sample containing scattering centers separated by the distances between the atoms of a molecule.

When a beam of such high-energy electrons passes through a chamber containing gas molecules, the charges of the nuclei, and to some extent the charges of the electrons, of the molecules will interact with the incoming beam, and each of the atoms of the molecules of the gas will act as a radiation-scattering center in much the same way as each of the slits of Fig. 14-2 acts as a center of radiation. Since

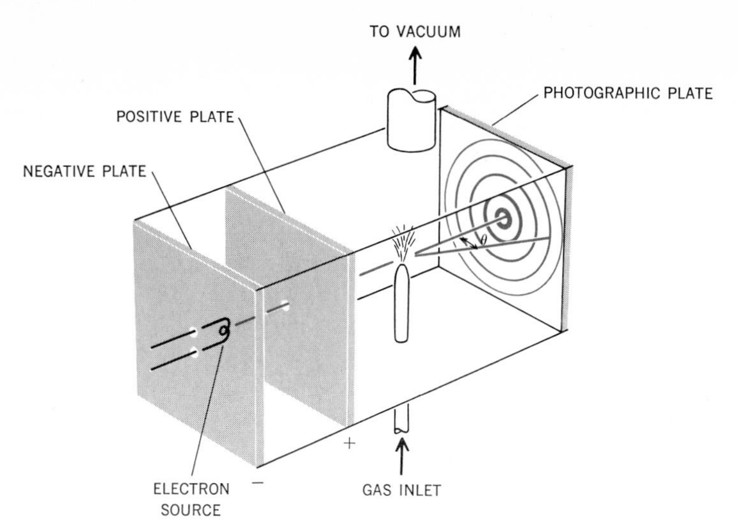

FIGURE 14-3
Arrangement for the study of gases by electron diffraction.

the particles of the electron beam carry a charge, the amount of scattering resulting from the interaction of the electron beam and the gas molecules is relatively large, and few molecules are needed to produce a detectable amount of scattered beam. (With an x-ray beam, on the other hand, the amount of scattering produced by each molecule would be much less, and crystals, with their higher concentration of scattering centers, are generally studied.)

The detailed mechanism by which the incoming electron beam, in an apparatus such as that shown schematically in Fig. 14-3, interacts with the atoms of the sample molecules cannot be dealt with here and need not be understood for an appreciation of many features of electron-diffraction studies. It can be mentioned, however, that the scattering of the beam can be attributed to coherent scattering, which implies no exchange of energy or phase between the beam and the scattering centers, and incoherent scattering, in which there is energy exchange and a resulting change in the wavelength and phase of the scattered electron beam. Both types of scattering lead to more forward than lateral scattering, and a detailed analysis shows that the intensity of such scattered beams falls off from the incident direction according to $1/\sin^4 \theta$, where θ is the scattering angle such as that in Fig. 14-2.

It is in the coherent, or elastic, scattering, in which the phase of the waves is unaffected, that the interference effects show up. These effects are therefore superimposed on a background of scattering that is not dependent on the structure of the molecule. The interference, or diffraction, effects can, however, be sorted out from the background on the photographic plate, and this part of the scattering, which is here of interest, can be dealt with.

To an electron beam passing through a gas, as in Fig. 14-3, each molecule appears to be made up of a number of scattering centers at some fixed distance from one another. It is now necessary to see how the interference effects between the scattered beams from the different atoms lead to a diffraction pattern from which the distances between the atoms can be deduced. The relation between the molecular geometry and the diffraction pattern, the counterpart of the equation (Eq. [1]) for the simple slit assembly of Figs. 14-1 and 14-2, is given by the Wierl equation, which will now be derived.

14-3 The Wierl Equation

The essential features of electron diffraction can be studied, and the basic relation, the Wierl equation, can be obtained, by considering the scattering pattern produced by a homonuclear diatomic molecule. The structure, i.e., the length of the bond between the atoms, will be deduced from the observed diffraction pattern.

A beam of electrons with wavelength λ, as Fig. 14-3 illustrates, is passed through a sample, actually a jet, of gas. Much of the beam will be unaffected by the gas, and will form a strong central spot on the photographic plate. Some of the beam, however, will be scattered by the molecules of the gas jet. Because of interference between the beams scattered from the different atoms of each molecule, a set of darkened rings will appear on the photographic plate.

For the scattering-interference effects produced by the simple slit assembly of Fig. 14-2 it was easy to deduce the relative amounts of the incident beam that would be diffracted and appear at various angles θ. It is now necessary to obtain this relation when the atoms of a molecule, which can have all orientations in space, take the place of the set of slits. Figure 14-4 shows a molecule, which in the derivation

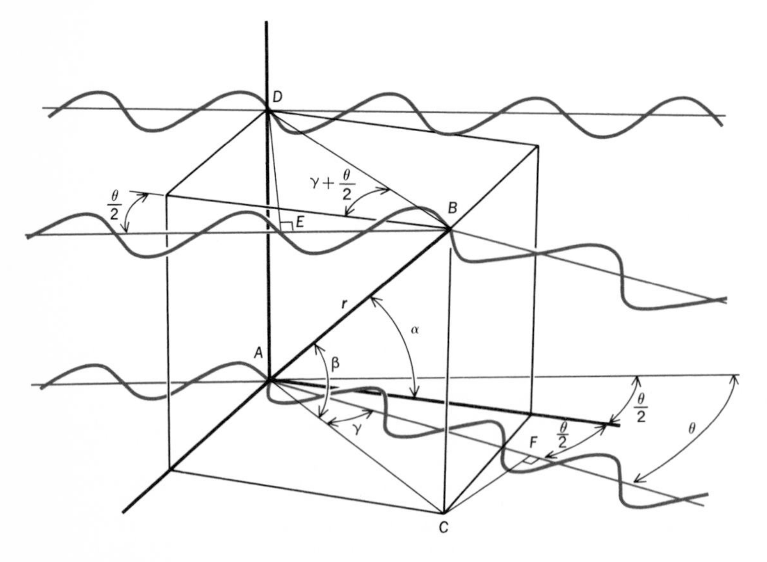

FIGURE 14-4
Diffraction from a diatomic molecule AB. The amount of scattered radiation that comes off at an angle θ to the direction of the incident beam depends on the phase difference imposed by the different path lengths EB and AF.

to follow will be treated as a homonuclear molecule, in the electron beam. We shall deduce an expression for the amount of scattering that comes off at an angle θ to the original beam direction for the molecule with the orientation of that in Fig. 14-4. It will then be necessary to average this expression over all the orientations of the molecule in space. Since only the orientations of the molecules are important in producing different diffraction effects, and not their position in the well-defined electron beam, it is sufficient to keep atom A at the origin and to carry B through all positions that it could adopt and remain a distance r from atom A.

The beam scattered by an angle θ from the original beam arrives at the photographic plate and forms a ring on the plate. The intensity of the beam arriving at a point on the photographic plate at an angle θ will be considered. The result of a rather lengthy derivation will be the expression

$$I(\theta) \propto f_{el}^2 \left[1 + \frac{\sin\left(\dfrac{4\pi r}{\lambda} \sin \dfrac{\theta}{2}\right)}{\dfrac{4\pi r}{\lambda} \sin \dfrac{\theta}{2}} \right] \qquad [9]$$

where f_{el} is the scattering factor for the atoms of the molecule. This is written more conveniently in Eq. [25], after the introduction of s for the term $(4\pi/\lambda) \sin (\theta/2)$. Since the basis and general procedure of the derivation of this important result should be clear, one can proceed directly to that equation if one wishes to pass over the actual derivation.

*The interference between waves coming off in the direction defined by θ from A and B depends on the phase difference that is imparted to the two beams as a result of their different origins. Three shafts of the incoming beam that go through D, E and B, and A are represented. At a cross section perpendicular to such a beam of electromagnetic radiation all the waves will be in phase; otherwise destructive interference would occur and the wave would vanish. Point E is drawn, on the beam that reaches B, on the perpendicular from D to this beam. Thus A, D, and E are on a plane perpendicular to the incoming beam, and at these points the waves must be in phase, as indicated. In a similar manner the directions of the shafts of the scattered beam in the direction θ to the incident beam are shown leaving A and B. Since F is drawn on the perpendicular from C to the beam leaving A, the points B and F are on a line perpendicular to the direction of the scattered beam. The net intensity of the scattered beam depends, therefore, on the extent to which the beams at B and F are in phase. The geometry has been arranged in Fig. 14-4 in such a way that constructive interference would result.

The difference in path length δ of the beams that must combine at B and F is seen from Fig. 14-4 to be given by the difference in the path lengths EB and AF; that is,

$$\delta = EB - AF \qquad [10]$$

The phase difference at B and F depends on the relation of δ to the wavelength of the electron beam. If, for example, δ is an integral multiple of λ, the two scattered beams will arrive in phase and will constructively interfere, as drawn in Fig. 14-4, whereas if δ is a half-integral multiple of λ, the beams will be exactly out of phase and will destructively interfere. The amplitude of the scattered wave from each atom is proportional to a scattering factor for electrons f_{el}, and thus the net amplitude of the combined waves at B will be given by an expression that is based on Eq. [3] or [6], namely,

$$A_R^2 \propto f_{el}^2 \cos^2 \tfrac{1}{2}(\varphi_1 - \varphi_2)$$

This gives, with Eq. [7],

$$A_R^2 \propto f_{el}^2 \cos^2 \frac{\delta\pi}{\lambda}$$

which, since the intensity, or energy, of any classical wave motion is proportional to the square of the amplitude of the wave, can be rewritten

$$I(\theta) \propto f_{el}^2 \cos^2 \frac{\delta\pi}{\lambda} \qquad [11]$$

where $I(\theta)$ denotes the intensity of the beam scattered at an angle θ by the molecule oriented as in Fig. 14-4.

It now remains to relate δ to the geometry of the system, and a considerable amount of geometric manipulation is necessary in order to obtain a convenient expression for δ that will allow integration of Eq. [11] over all orientations that the molecule can adopt.

With the angles labeled as in Fig. 14-4, the distances that contribute to δ can be obtained as

$$DB = AC = r \cos \beta \qquad [12]$$

$$EB = (r \cos \beta) \cos (\gamma + \theta) \qquad [13]$$

$$AF = (r \cos \beta) \cos \gamma \qquad [14]$$

With these values one obtains δ as

$$\begin{aligned} \delta &= EB - AF \\ &= r \cos \beta[\cos (\gamma + \theta) - \cos \gamma] \end{aligned} \qquad [15]$$

The general expression for the difference in two cosines,

$$\cos x - \cos y = 2 \sin \frac{x + y}{2} \sin \frac{x - y}{2}$$

can now be used to convert the expression for δ to the result

$$\delta = 2r \cos \beta \sin \left(\gamma + \frac{\theta}{2}\right) \sin \frac{\theta}{2} \qquad [16]$$

A further manipulation of this result is necessary if the integration of Eq. [11] over all positions of B is to be performed. One needs to recognize that $r \cos \beta$ is the projection of AB on the plane to give AC, and that multiplication by $\sin (\gamma + \theta/2)$ further projects this onto the coordinate axis that lies between the incident and scattered directions. These two projections are equivalent to a direct projection of AB onto this axis. We can therefore make the replacement

$$r \cos \beta \sin \left(\gamma + \frac{\theta}{2}\right) = r \cos \alpha$$

With this simplification, Eq. [16] becomes

$$\delta = 2r \cos \alpha \sin \frac{\theta}{2} \tag{17}$$

The scattered intensity produced by the molecule of Fig. 14-4 is thus given by

$$I(\theta) \propto f_{el}^2 \cos^2 \frac{2\pi r \cos \alpha \sin (\theta/2)}{\lambda} \tag{18}$$

It is customary to make the substitution

$$s = \frac{4\pi}{\lambda} \sin \frac{\theta}{2} \tag{19}$$

where it should be recognized that, for a given electron-beam voltage, a position on the photographic plate could be specified either by θ or, through Eq. [19], by s. With this notation Eq. [18] becomes

$$I(s) \propto f_{el}^2 \cos^2 \frac{sr \cos \alpha}{2} \tag{20}$$

The averaging over all positions of atom B is performed with the spherical-coordinate angular element of $\sin \alpha \, d\alpha \, d\phi$ of Fig. 14-5. Since ϕ does not enter into Eq. [20], the integration of this variable introduces only the constant term 2π, and the total intensity scattered at an angle θ is given by the proportionality equation

$$I(s) \propto f_{el}^2 \int_0^\pi \cos^2 \left(\frac{sr \cos \alpha}{2}\right) \sin \alpha \, d\alpha \tag{21}$$

This troublesome-looking integral turns out to be easily handled. One first writes

$$\sin \alpha \, d\alpha = -d(\cos \alpha) = -\frac{2}{sr} d\left(\frac{sr}{2} \cos \alpha\right) \tag{22}$$

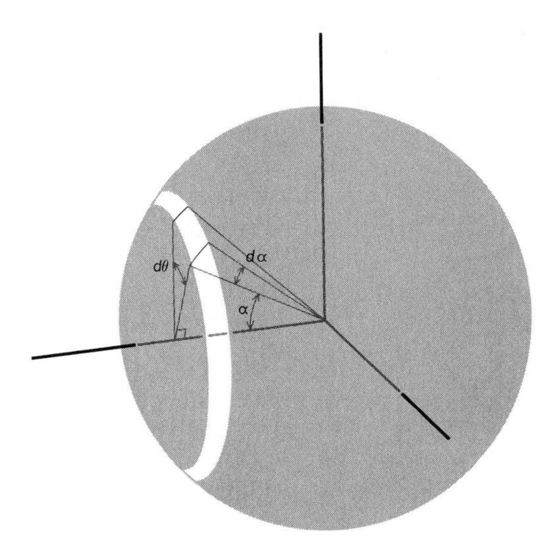

FIGURE 14-5
The angular coordinates that move atom B of Fig. 14-4 through all positions about atom A as θ goes from 0 to 2π and α from 0 to π.

Introduction of y for $(sr/2)\cos\alpha$ puts Eq. [21] in the form

$$I(s) \propto f_{el}^2\left(-\frac{2}{sr}\right)\int_{sr}^{-sr/2}\cos^2 y\,dy \qquad [23]$$

Integration and substitution of the limits give

$$I(s) \propto f_{el}^2\left(-\frac{2}{sr}\right)\frac{1}{2}\Big[y + \sin y\cos y\Big]_{sr/2}^{-sr/2}$$

$$\propto f_{el}^2\left[1 + \frac{2\sin(sr/2)\cos(sr/2)}{sr}\right] \qquad [24]$$

and finally,

$$I(s) \propto f_{el}^2\left(1 + \frac{\sin sr}{sr}\right) \qquad [25]$$

This expression is the Wierl equation for a diatomic, homonuclear molecule and is the desired result of our derivation. It relates the intensity, at a ring at angle θ, of the scattered beam that is involved in the interference effect to the quantity $s = (4\pi/\lambda)\sin(\theta/2)$ and the internuclear distance r.

The Wierl equation for a polyatomic molecule is obtained by recognizing that the net interference of the scattered beams can be deduced by taking the atoms two at a time in a manner like that for the diatomic case. When this is done, the equation applicable to polyatomic molecules is obtained as

$$I(s) \propto \sum_{j,k}(f_{el})_j(f_{el})_k\frac{\sin sr_{jk}}{sr_{jk}} \qquad [26]$$

where the summation is over all atoms j and k of the molecule. It can be easily verified that Eq. [26] reduces to Eq. [25] when a diatomic molecule is treated and it is recalled that $\sin(0/0) = 1$.

The Wierl equation does not, unfortunately, allow the direct calculation of the internuclear distances r_{jk} from measurements of $I(s)$ at various values of s. An indirect method for obtaining these quantities must be resorted to.

In the simplest and earliest of the methods that have been used, the photographic plate is observed visually from the origin outward along any direction. The positions of the darkened rings on the photographic plate are estimated visually, and a plot is sketched for the plate darkening as a function of $s = (4\pi/\lambda)\sin(\theta/2)$. This plot is reported as the *experimental scattering curve*. An example is shown for the CHF_3 molecule in Fig. 14-6. The procedure now requires one to assume a structure for the molecule being studied; for a diatomic molecule this amounts to choosing a value of r. With this assumed structure, Eqs. [25] and [26] can be used, with, in the simplest approach, the assumption that the f_{el} terms are proportional to the atomic numbers, to calcu-

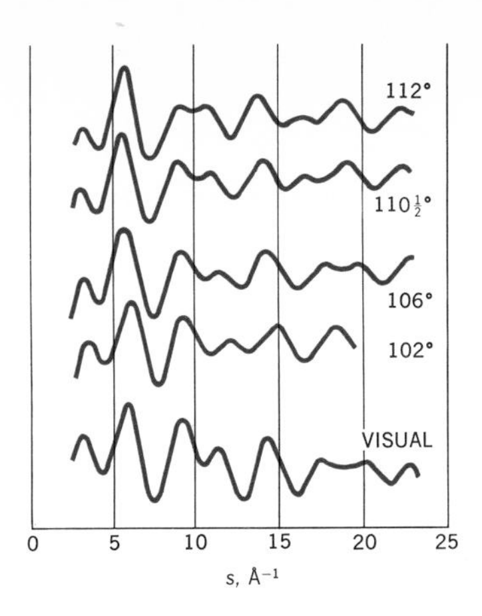

FIGURE 14-6
Comparison of the visual scattering curve for CHF_3 with curves calculated for various F-C-F angles. It is concluded that the angle is between 106 and 110°. (Spectroscopic data give the angle as 108°48′.) (*From L. O. Brockway, in Weissberger, "Physical Methods of Organic Chemistry," 2d ed., pt. II, p. 1123,*)

late $I(s)$ as a function of s. Such calculations can be made for various assumed structures, and plots of $I(s)$ versus s, called *theoretical scattering curves,* can be made. The theoretical curves are then compared, as shown in Fig. 14-6, with the experimental curve, and that most like the experimental curve is taken as being based on the best structure.

For a polyatomic molecule, a number of parameters r_{jk} must be varied to obtain the best structure. One does not now so directly come upon the correct structure. Furthermore, very incorrect structures may happen to lead to a theoretical pattern little different from the experimental one. It is clear, therefore, that a number of difficulties lie in the path of structure determinations by this method. In spite of these difficulties, very many structures have been determined, and Table 14-1 lists a few of these.

*14-4 The Radial-distribution Method

A very great aid to the use of electron-diffraction techniques is the suggestion made by Pauling and Brockway that the data on the photographic plate could be used directly to obtain some information on the structure of the sample molecules. This method, known as the *radial-distribution method,* in effect forms a relation that is the inverse of Eqs. [25] and [26] and gives the distances between scattering centers of the molecule in terms of the data of the photographic plate.

The procedure requires the molecule to be represented by a continuous distribution in space of regions with varying scattering power. In place of the atom B of Fig. 14-4, for example, one would recognize that, as far as the electron-diffraction experiment is concerned, there is merely a region of large scattering power a distance r from the origin. A function $D(r)$, called the *radial distribution function,* can be introduced such that $D(r)$ represents the product of the scattering powers of unit volumes a distance r apart, and $4\pi r^2 D(r)\,dr$ represents the product of the total scattering powers between a distance r and $r + dr$. For a diatomic molecule, therefore, $D(r)$ would have a large value for r near the bond length and would be zero elsewhere. With this description of the scattering effects in a molecule, the

TABLE 14-1 Some molecular-structure results from electron-diffraction studies

Bond angles		Bond lengths (Å)	
$P(CH_3)_3$	C—P—C 98.6 ± 3°	C—P	1.846 ± 0.003
	P—C—H 110.7 ± 5°	C—H	1.091 ± 0.006
CH_3Cl	H—C—H 110 ± 2°	C—Cl	1.784 ± 0.003
		C—H	1.11 ± 0.01
CF_3Cl	F—C—F 108.6 ± 0.4°	C—F	1.328 ± 0.002
		C—Cl	1.751 ± 0.004
CCl_4	Tetrahedral	C—Cl	1.769 ± 0.005
C_2H_4	H—C—H 115.5 ± 0.6°	C═C	1.333 ± 0.002
		C—H	1.084 ± 0.003
CH_3OH	C—O—H 108 ± 3°	C—O	1.427 ± 0.004
	H—C—H 109°28′ (assumed)	C—H	1.095 ± 0.010
		O—H	0.960 ± 0.020
$(CH_3)_2O$	C—O—C 111.5 ± 1.5°	C—O	1.416 ± 0.003
		C—H	1.094 ± 0.006
C_6H_6	C—C—C 120 ± 4°	C—C	1.39 ± 0.03
		C—H	1.08 ± 0.02

Wierl equation becomes the integral equation

$$I(s) \propto \int_0^\infty 4\pi r^2 D(r) \frac{\sin sr}{sr}\, dr \qquad [27]$$

or

$$I(s) = k \int_0^\infty \frac{rD(r) \sin sr}{s}\, dr \qquad [28]$$

where k is a proportionality constant. The factor s in the denominator leads to nothing more than a continuous decrease in scattering from the origin outward, and in the simple visual treatment of the photographic plate this factor cannot be adequately handled and was generally ignored. We can write, therefore,

$$I(s) = k' \int_0^\infty rD(r) \sin sr\, dr \qquad [29]$$

In this form, the integral can be considered to be a Fourier integral representation of the function $I(s)$. The coefficients of the Fourier terms are $rD(r)$, and these are given by the inverted form as

$$rD(r) = k' \int_0^\infty I(s) \sin sr\, ds \qquad [30]$$

or

$$D(r) = k'' \int_0^\infty I(s) \frac{\sin sr}{r}\, ds \qquad [31]$$

This important expression, or ones that treat the smooth fall of intensity with θ or s in somewhat different ways, can be used to convert the experimental data on $I(s)$ as a function of s to a radial distribution function $D(r)$.

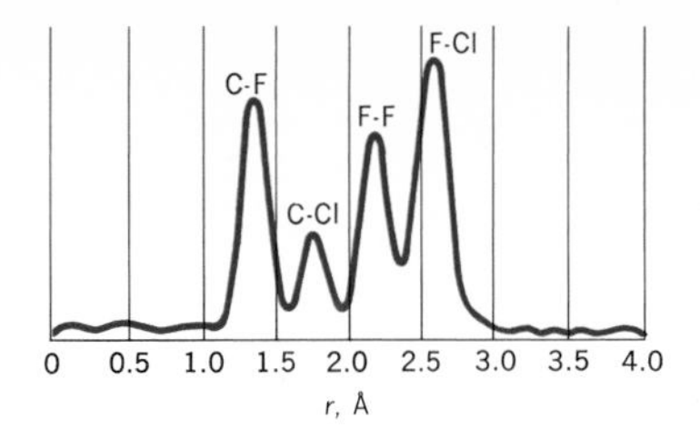

FIGURE 14-7
Radial distribution function for CF_3Cl. [*From L. S. Bartell and L. O. Brockway, J. Chem. Phys.,* **23**:*1860 (1955).*]

Computer integration of the integrals for values of r in the range from r equal to zero to the largest separation of scattering centers expected for the molecule under study gives this function. An example of the resulting radial distribution curve is shown in Fig. 14-7.

One of the chief remaining limitations of the electron-diffraction method is that there are a small number of "wiggles" in the experimental curve. These wiggles are really the data provided by the experiment, and with a small number of data, only a few molecular parameters can be determined. A very large molecule cannot be satisfactorily treated unless, like carbon tetrachloride or benzene, there is some geometric relation which makes several interatomic distances dependent on each other. A second limitation is that the electron beam is not scattered very effectively by hydrogen atoms. It follows that the position of such atoms in a molecule cannot easily be determined by this method. The method of x-ray-diffraction studies of crystals, as we shall see, overcomes the first difficulty by yielding very many scattered beams and therefore very much information. The method of neutron diffraction overcomes the second difficulty since protons have a reasonably large scattering cross section for neutrons.

14-5 Covalent Radii

The accumulation of structural data, such as that of Tables 13-2 and 14-1, by means of spectroscopic studies and both electron and x-ray-diffraction studies, allows one to investigate the possibility of assigning an effective radius for an atom when it is in a covalently bound molecule, i.e., of assigning a *covalent radius* to each atom. One begins by assigning half the length of a homonuclear bond as the covalent radius of the atoms forming the bond. Thus, from the equilibrium bond length of Cl_2 of 1.99 Å, one obtains the value of 1.00 Å for the covalent radius of chlorine; from the carbon-carbon distance of 1.54 Å in ethane, for example, one obtains a value of 0.77 for the covalent radius of carbon and so forth. To proceed one must now establish the extent to which the length of covalent bonds can be treated in terms of the sums of such covalent radii. Some comparisons of calculated and measured bond lengths are shown in Table 14-2.

More extensive treatments of this type show that the bond lengths of many bonds are given within a few hundredths of an angstrom by the sum of assigned atomic covalent radii. This suggests that covalent bonds have lengths that are sufficiently independent of factors other

TABLE 14-2 Some tests of additivity of covalent bond radii,* with values in angstroms

		C—Cl (obs)	
C—C (ethane)	1.54	In CCl_4	1.766
Cl—Cl (Cl_2)	1.99	In $CHCl_3$	1.767
C—Cl (calc)	1.77	In CH_3Cl	1.784
		N—F (obs)	
N—N ($H_2N—NH_2$)	1.46	In NF_3	1.37
F—F (F_2)	1.42	In N_2F_4	1.46
N—F (calc)	1.44		
H—H	0.741	H—F (obs)	
F—F	1.418	In H—F	0.917
H—F (calc)	1.08		
		C—O (obs)	
C—C (ethane)	1.54		
O—O (HO—OH)	1.48	In CH_3OH	1.43
C—O (calc)	1.51	In C_2H_5OH	1.48
		In $CH_3—O—CH_3$	1.42

* Bond-length data given, with references to original sources, in Tables of Interatomic Distances and Configurations on Molecules and Ions, The Chemical Society, Special Publication 11, 1958.

than the fixed radii so that there is some value in assigning radii to the bonded nuclei. Thus covalent radii are tabulated as illustrated in Table 14-3.

Further comparisons of these values with experimental results indicate, as shown in fact by some of the examples of Table 14-2, that serious discrepancies can occur between simply predicted covalent bond lengths and those observed. The C—F bond, for example, is calculated from the data of Table 14-3 to have a length of 1.49 Å, whereas microwave spectral results for CH_3F give it as 1.385 Å and electron-diffraction results for CF_4 give 1.32 Å.

Such discrepancies led Schomaker and Stevenson to suggest that a bond length calculated from covalent radii must be adjusted for the difference in electronegativity of the bonded atoms. They suggested the relation

$$r_{AB} = r_A + r_B - 0.09(x_A - x_B) \qquad [32]$$

Some, but not all, of the interesting violations of simple covalent radii additivity are removed by this empirical expression. In other cases the Stevenson-Schomaker correction makes the agreement with the observed length poorer than is obtained by a simple addition of the covalent radii. Although a number of factors must be operating to affect the length of a bond between a pair of nuclei in any given molecule, the covalent radii of Table 14-3 are often of value in estimating this bond length.

X-RAY DIFFRACTION: DIFFRACTION BY CRYSTALS

When scattering centers exist in an ordered, effectively infinite array, as they do in well-formed crystals, the interference effects between the scattered beams are greatly enhanced compared with the effects that we have been studying so far that depend on interference from different atoms of a single gas-phase molecule. Diffraction from crystals provides, therefore, a powerful method for the deduction of the arrangement and spacing of the atoms or ions that create the scattering centers in a crystal. In the analysis of such diffractions, consideration of the ordered array of atoms or ions is basic to an understanding of the results, just as the analysis of the effect of the random orientations of the molecules was basic in the treatment of electron diffraction by gases. The study of diffraction by crystals will therefore be preceded by treatments of the classification of crystals, the classification of

TABLE 14-3 Covalent radii for atoms involved in single-bonded compounds,* in angstroms

H	C	N	O	F
0.37	0.77	0.74	0.74	0.72
	Si	P	S	Cl
	1.17	1.10	1.04	0.99
	Ge	As	Se	Br
	1.22	1.21	1.17	1.14
	Sn	Sb	Te	I
	1.40	1.41	1.37	1.33

* From A. F. Wells, "Structural Inorganic Chemistry," Oxford University Press, Fair Lawn, N.J., 1962.

ordered arrays of points, or molecules or ions, and the ways in which the arrangement of the elements of such arrays can be described.

14-6 Crystal Shapes

A close look at well-formed single crystals, as occur in natural minerals and occasionally from laboratory crystallizations, shows that crystals are characterized by well-defined and, to some extent, symmetrically arranged planes. A closer look at a number of crystals of a given type shows that, although the shape may depend on the details of the crystallization process, the angles between the planes that are observed remain the same.

The shapes of a number of crystals are shown in Fig. 14-8. These drawings idealize the actual shape that would be observed in a particular specimen, which might be longer and more needlelike or flatter and more platelike, but the crystal faces that would be observed would be oriented with respect to one another as they are in the illustrations. The great variety of crystalline shapes that can occur leads one to attempt to classify crystals in some way. Important classifications result from considerations of the symmetry exhibited by the faces of crystals and by their internal structure. Some of the elements of symmetry will now be introduced so that this concept can be used more effectively than it can with the qualitative idea of an object being either symmetric or not.

14-7 Symmetry Elements and Symmetry Operations

An analysis of what is meant when something is said to be symmetric leads to the introduction of elements of symmetry. Four types of *symmetry elements,* illustrated in Fig. 14-9, can be recognized: the center of symmetry, the plane of symmetry, a rotation axis of symmetry, and a rotation-reflection axis of symmetry. These elements of symmetry are best investigated in terms of operations, called *symmetry operations,* that are associated with them. Thus, to investigate the presence of a center of symmetry, one inverts the object through the suspected center and sees if a result that is indistinguishable from the original is obtained. Likewise, a plane of symmetry exists if the operation of reflection through the plane leads to a result that is indistinguishable from the original. Thus the center and the plane are examples of symmetry elements; the inversion through the center or the reflection through the plane is the symmetry operation.

An axis about which rotation leads to the same configuration as occurred initially must be characterized further by the fraction of a complete revolution that is required. Thus, if a rotation about the axis

FIGURE 14-8
Some idealized crystals. (*From W. E. Ford, Dana's "Textbook of Mineralogy," 4th ed., John Wiley & Sons, Inc., New York, 1942.*)

by 180° leads to a result indistinguishable from the original, two such indistinguishable arrangements will occur in a complete revolution, and the axis is said to be a twofold axis. If rotations of 120 and 240°, i.e., by thirds of a complete revolution, lead to the same result as the original, the axis is said to be a threefold rotation axis, and so forth.

The rotation-reflection axis is described in Fig. 14-9. This symmetry element has associated with it a rotation operation, like that discussed in the preceding paragraph, plus a reflection operation through a plane perpendicular to the axis.

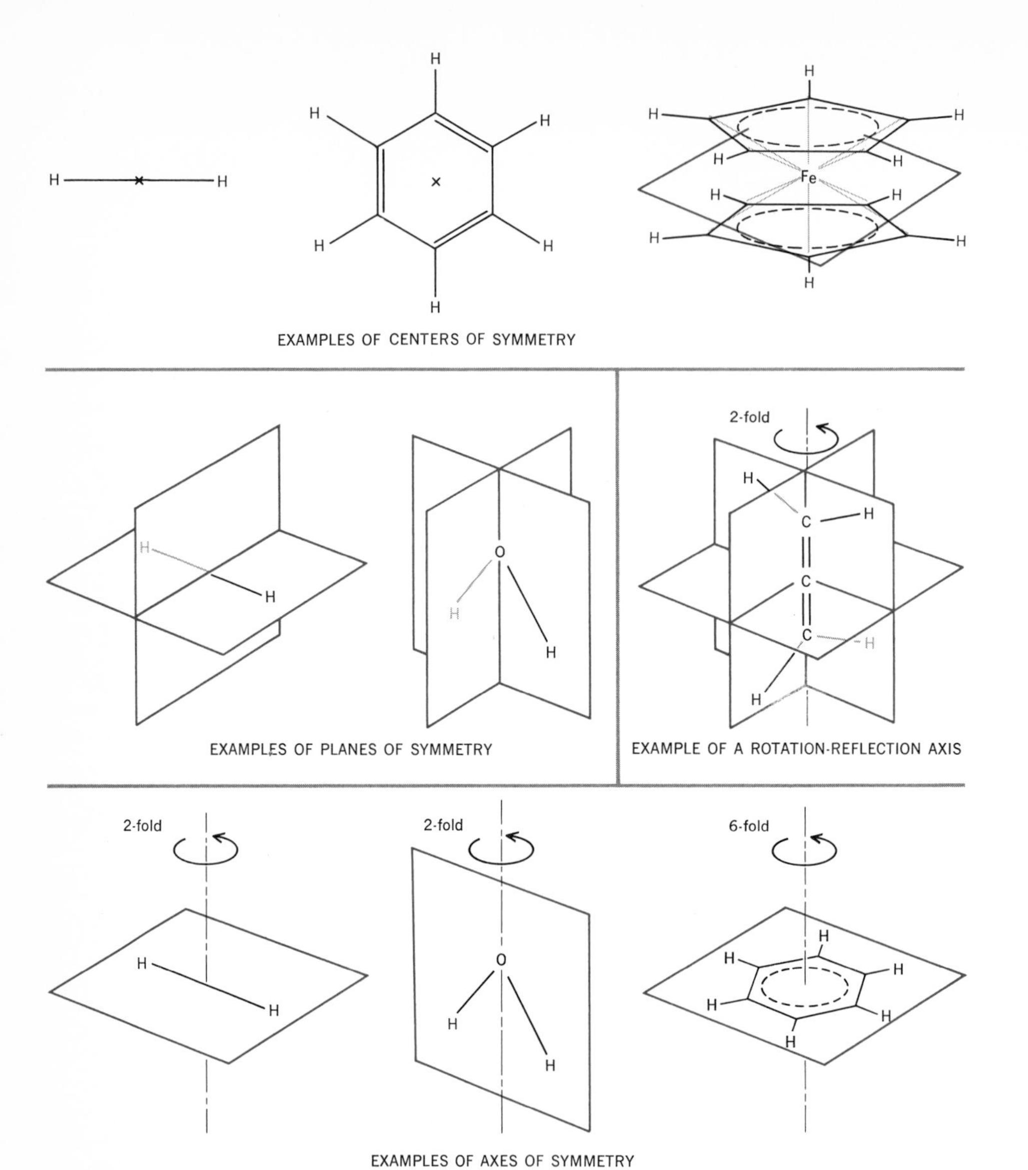

FIGURE 14-9
Molecular examples of elements of symmetry.

Any crystal, or any molecule, will have, as can be most easily verified if crystal or molecular models are available, some collection of these symmetry elements. These combinations of symmetry elements correspond to what are called *point groups*. The word "point" implies that in the symmetry operations one point, perhaps the center of gravity, remains in its position, and the operations are imagined as moving the rest of the object with respect to this fixed point. The word "group" implies more than just a collection of symmetry operations. It is used here in a stricter mathematical sense, since the general theory of groups is applicable to the set of symmetry operations. This aspect of point roups need not be treated here. If one considers symmetric objects hat can be fitted together to form a three-dimensional array, one would nd, in fact, that only 32 different combinations of symmetry elements, e., 32 point groups, can occur.

Detailed studies of the symmetry of the internal as well as external tructure of crystals allow the assignment of a crystal to one of the 2 point groups. The classification procedure can be simplified by rouping together some of these 32 point groups to give just six different categories, known as the six crystal systems. From the external tructure one can deduce some of the symmetry elements of a crystal, nd these symmetry features occur often enough so that the crystal an be assigned to one of six crystal systems. Each crystal system s defined by certain minimum symmetry elements, shown in Table 14-4, nd corresponds to several point groups. Assignment to one of the ix systems will be adequate for our study of crystal structures, and urther assignment to one of the 32 point groups will not be treated.

(It should be mentioned that what is listed in Table 14-4 as the exagonal system is sometimes divided into hexagonal and rhombohedral systems, giving a total of seven systems. We shall not be concerned with the detailed analysis of crystals of this type and need not, herefore, deal with this distinction.)

Further studies of crystal symmetries and classifications would nake extensive use of additional groups of symmetry operations that rise when, as is the case with the effectively infinite array of points vhich corresponds to a crystal, an identical arrangement can be reached

ABLE 14-4 The six crystal systems
In all cases, except for the cubic system, the axes of symmetry may be rotation or rotation-reflection axes.

ystem	No. of lattice types	Minimum symmetry	Unit-cell parameters
ubic	3	Four threefold axes	$a = b = c$ $\alpha = \beta = \gamma = 90°$
exagonal	2	One threefold or one sixfold axis	$a = b \neq c$ $\alpha = \beta = 90°, \gamma = 120°$
etragonal	2	One fourfold axis	$a = b \neq c$ $\alpha = \beta = \gamma = 90°$
rthorhombic	4	Three twofold axes	$a \neq b \neq c$ $\alpha = \beta = \gamma = 90°$
lonoclinic	2	One twofold axis	$a \neq b \neq c$ $\alpha = \gamma = 90°, \beta \neq 90°$
riclinic	1	None	$a \neq b \neq c$ $\alpha \neq \beta \neq \gamma \neq 90°$

by moving a unit of the array with a translational motion. When such symmetry operations are included, one finds that certain groups of operations, called *space groups,* occur. There are 230 different space groups, and the symmetry of the arrangement of ions or molecules of the crystal will be such that the crystal belongs to one of these space groups. Again, it will not be necessary here to show how the space group of a crystal is determined, and it will be adequate to deal with the symmetry operations like those shown in Fig. 14-9.

Assignment of a crystal to a crystal system can sometimes be made, as has been assumed here, on the basis of its external characteristics. The importance for us of the assignment to a crystal system is the fact that the external symmetry of a crystal is dependent on the symmetry of the arrangement of the molecules or ions that make up the crystal. We must now, therefore, investigate the relation between the internal structure of a crystal and the planes and symmetry that constitute its external characteristics.

14-8 Lattices and Unit Cells

It was apparent very early in the study of crystals, particularly as a result of the work of René Just Haüy in 1784, that the shapes of crystals stem from an ordered array of smaller structural units. Although we now know a great deal about the nature of these units, it turns out to be very profitable to consider the ways in which points that are not further characterized can be arranged to give a repeating array.

The limitations on the types of arrangements that can give a repeating pattern in which each point has identical surroundings in the same orientation can best be appreciated from the two-dimensional patterns in Fig. 14-10. Only these five essentially different patterns can be constructed. One can verify that any other two-dimensional pattern that one attempts to draw is identical, except for the relative magnitudes of the spacings a and b and the angle α, with those shown here.

In a similar way there are, as shown by A. Bravais in 1848, only 14 different types of lattices that can be drawn in three dimensions. Units of these lattices, which when repeated in three dimensions produce the lattice, are shown in Fig. 14-11. Any three-dimensional array, such as real crystal, must have an internal structure that corresponds to one of the 14 Bravais lattices.

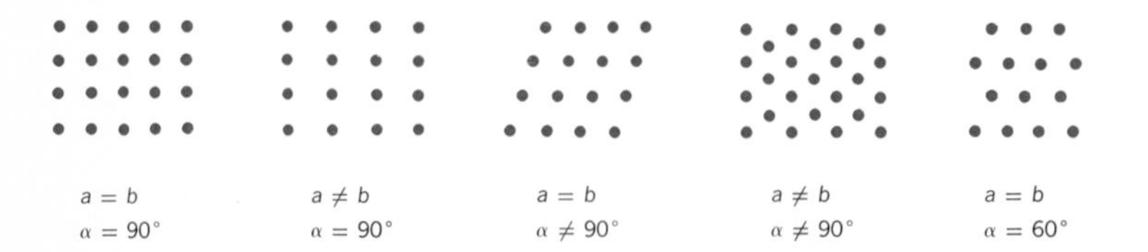

FIGURE 14-10
The five types of repeating arrangements of points that can be drawn in two dimensions with each point identically surrounded.

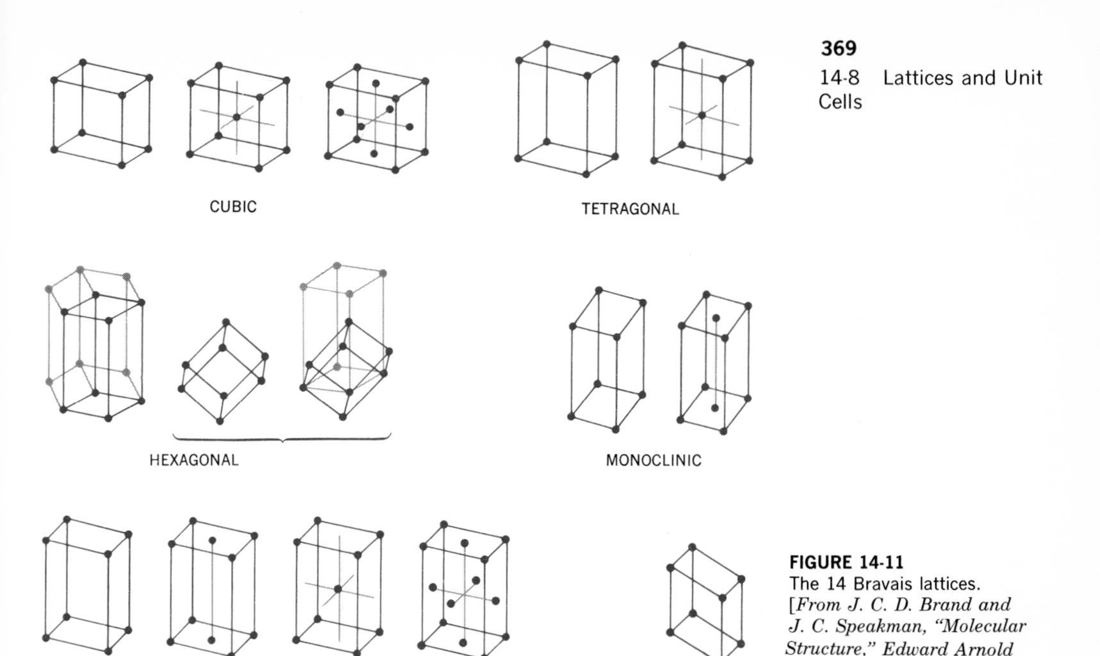

FIGURE 14-11
The 14 Bravais lattices.
[From J. C. D. Brand and J. C. Speakman, "Molecular Structure," Edward Arnold (Publishers) Ltd., London, 1950.]

We have already seen, however, that any crystal can be assigned to one of the six crystal systems on the basis of its symmetry. It follows, therefore, that the repeating units that one constructs have symmetry characteristics that allow them also to be associated with the crystal systems. This has, in fact, been done in Fig. 14-11. There one sees, for example, that the three lattices at the top of the figure have at least four threefold axes of symmetry. They therefore belong to the cubic system. Just as one assigns crystals to crystal systems on the basis of symmetry, so also can one assign the 14 possible lattice arrangements of these systems.

Such assignments, and a number of other features, are most easily done if a unit of the lattice is outlined. Although such outlines are not really pertinent to the nature of the lattice, they are more easily dealt with than the infinite array. The three cubes at the top of the figure, for example, clearly show the cubic symmetry of these three lattices. Such units of the lattice are known as *unit cells.* There is some freedom in the choice of the unit cell for a particular lattice, and the selection is made primarily to exhibit the symmetry of the lattice.

The simplest type of unit cell that can be drawn has lattice points only on the corners. Such unit cells are known as *primitive* unit cells. Since each lattice point is only one-eighth in a particular cell, there is effectively one lattice point per primitive unit cell. Other unit cells, drawn to exhibit the lattice symmetry, have additional lattice points either within the cell, to give *body-centered unit cells,* or on the faces, to give, for example, *face-centered unit cells.* In such cases more than one lattice point is to be assigned to a unit cell.

The concept of unit cells suggests also the characterization of lattices, not only by the symmetry and the primitive, face-centered, or body-centered nature of the unit cell, but also by the type of axes that would most conveniently allow points within the unit cell to be located and planes through various lattice points to be described. Thus, in the cubic system, it is convenient to use the usual orthogonal axes and to measure distances along each axis in terms of the same unit of length. In fact, the unit-cell dimensions provide the most convenient units of length. In the tetragonal system, however, while orthogonal axes are again suitable, lengths in terms of the units $a = b \neq c$ along the x, y, and z axes will be more convenient. Conveniently inclined axes and unit lengths, which can be seen to be related to the unit cells of Fig. 14-11, are tabulated for the different crystal systems in Table 14-4. These lengths and angles do in fact characterize the crystal systems just as the minimum symmetries do.

The concept of lattices, the existence of only 14 types, and the association of these lattices, with the help of unit cells, to the symmetry-based crystal systems provide a valuable connection between internal structure and crystal form. This connection can be further shown by relating planes in the lattices to planes or faces of actual crystals.

14-9 Crystal Planes

The *law of rational indices,* discovered by Abbé René Haüy from observations of the relative orientation of the different faces of a crystal, can be used to show the way in which unit cells and lattices can be related to actual crystals. This empirical law states that the intercepts of the planes of the various faces of a crystal on a suitable set of axes can be expressed by small integral multiples of three unit distances.

Planes that can be similarly described, i.e., by means of intercepts that are integral multiples of unit distances, can be drawn in lattices. In fact, the planes that are drawn to include the largest number of lattice points possible, as are those shown in end view in Fig. 14-12, are naturally described in terms of intercepts that are integral multiples of the unit-cell parameters. The empirical law describing the arrangement of the faces of a crystal is thus understandable if the internal structure of the crystal corresponds to an array of points that constitute a lattice. The important planes of the crystal then correspond to the planes that can be drawn so as to include a relatively large number of lattice points. The orientation of the faces of a crystal, and more importantly the orientation and spacing of the planes that are responsible for x-ray diffraction, can then be compared with the orientation

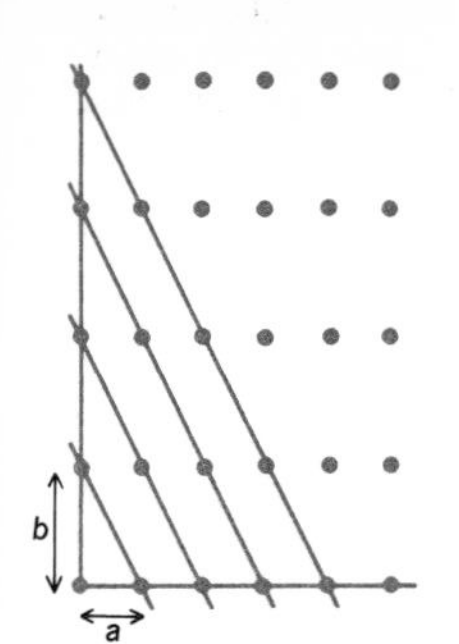

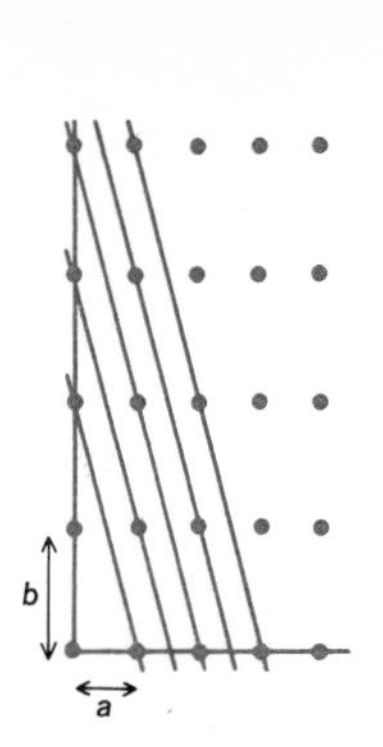

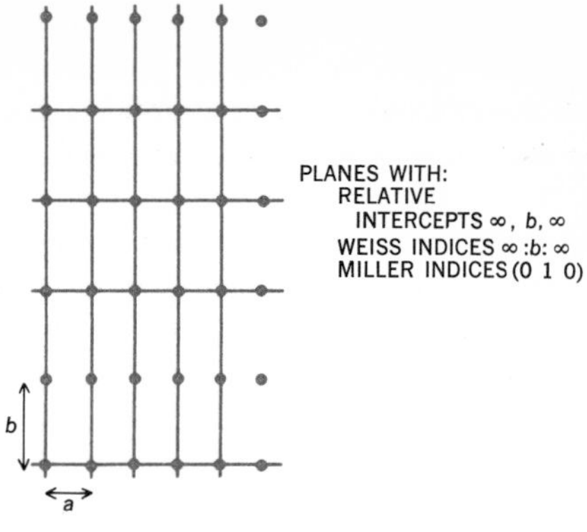

FIGURE 14-12
The end view of a tetragonal lattice showing how planes that pass through relatively large numbers of lattice points can be described in terms of intercepts measured in units of a and b.

and spacing of the planes in the 14 possible Bravais lattices. In this way the internal structure of the crystal can be identified with a particular lattice.

The relative orientation of crystal or lattice planes is of great importance in crystal-structure analysis, and a convenient method for describing these planes is needed. The important planes of a lattice, as Fig. 14-12 suggests, can be described in terms of intercepts that are multiples of the unit-cell dimensions. Since crystal planes will be similarly oriented, they can be similarly described. In this way what are called *Weiss indices* are used, and planes are described by their relative intercepts on the x, y, and z axes as $a:b:c$, or $a:2b:\infty c$, and so forth.

Much more convenient, particularly in the analysis of diffraction data, are sets of numbers called *Miller indices*. These are obtained by taking the reciprocals of the coefficients of the Weiss indices. These three reciprocals are then cleared of fractions and reduced to the smallest set of integers. The plane $a:b:c$ becomes a (111) plane; the $a:2b:c$ plane becomes a (212) plane; the $a:b:\infty c$ plane becomes a (110) plane. The Miller indices, referred to in general by (hkl), describe the relative directions of the crystal planes. Information on the coordinate system and the values of the three unit distances is given separately.

A feeling for these indices can soon be acquired if one remembers their reciprocal nature. A low first number means an intercept at a large distance on the X axis; a low second number means a large intercept on the Y axis; and so forth. A plane parallel to an axis now has a zero term corresponding to the reciprocal of its intercept on that axis.

The Miller indices provide a convenient way of expressing a particular plane, i.e., a particular direction in a crystal. It should be pointed out that as far as direction is concerned, which is sometimes all that is important, the planes (220) and (110) would be the same and the latter notation would be used.

14-10 X-rays and X-ray Diffraction

Studies of the internal structure of crystals depend upon a penetrating radiation that will enter the crystal and will display interference effects as a result of scattering from the ordered array of scattering centers. X-rays have the necessary penetrating power, since they are uncharged and therefore interact to a smaller extent with matter than do electron beams, and they show interference effects since they have wavelengths in the angstrom range. Diffraction effects occur when this radiation is disturbed by the crystal so that some of the scattered waves of the radiation are shifted out of phase with respect to other waves as described in the electron-diffraction treatment earlier in the chapter.

In the Bragg method the phenomenon is observed when nearly monochromatic x-rays are reflected from a crystal. A beam of x-rays is passed into a crystal, which in Fig. 14-13 is represented by layers of particles. The x-rays are scattered by interaction with the electrons of the atoms or ions of the crystal. Since x-rays are known to be quite penetrating, each layer of atoms can be expected to scatter only a small part of the x-ray beam. If the crystal particles did not have a spacing which was of the same order of magnitude as the wavelength of the x-rays, simple reflections and scattering of the x-rays would occur. The reflection is, in fact, not simple, and is greatly disturbed by the interference effect.

The incoming beam of x-rays can be represented as in Fig. 14-13 with all the waves in phase. The nature of the outgoing beams must be investigated. As is seen in Fig. 14-13, the beams scattered from successive layers of crystal particles may show waves that, in a particular direction, are to some extent out of phase with the scattered waves

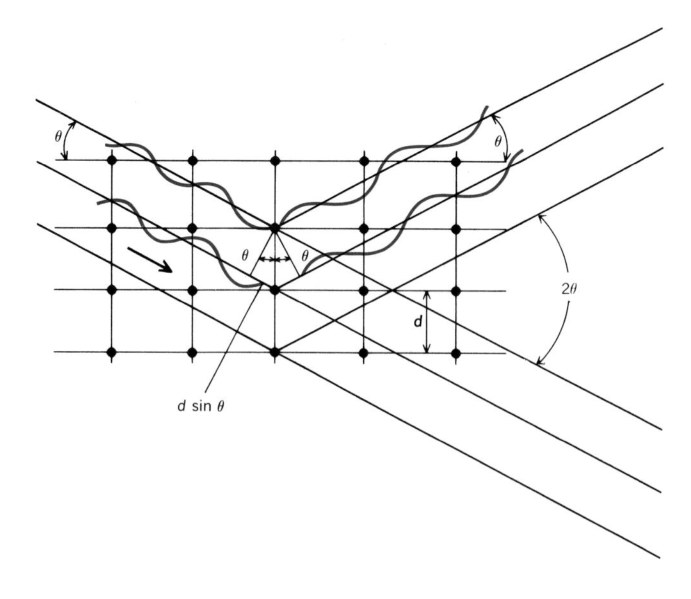

FIGURE 14-13
The Bragg scattering condition.

from the other layers. When this happens and the addition of the waves tends to cancel each other out, *destructive interference* is said to occur. Only if all these scattered beams come out in phase will they add up and contribute to a net scattered beam. This *constructive interference* occurs whenever the phase of beams scattered from successive layers is shifted by an integral multiple of wavelengths. Reference to Fig. 14-13 shows that this happens when the relation

$$n\lambda = 2d \sin\theta \qquad n = 1, 2, 3, \ldots \tag{33}$$

holds. This important equation is known as *Bragg's diffraction law.*

This basic equation shows that for a given value of the x-ray wavelength λ, measurement of the angle θ, or of the term $\sin\theta$, gives the information on the spacing between planes through the scattering centers that make up the crystal. For the crystal arranged as in Fig. 14-14 and subjected to a rotation about the z axis, various planes of the type $(hk0)$ will satisfy Bragg's law and will produce reflections along the xy plane in which the crystal lies. Planes tilted with respect to the z axis will lead to reflections that constitute the "layer lines" that lie above and below this equatorial line.

Arrangement of the Bragg expression to

$$\begin{aligned}\lambda &= 2\frac{d}{n}\sin\theta \\ &= 2d_{(hkl)}\sin\theta\end{aligned} \tag{34}$$

shows that the higher-order reflections from planes with a spacing d can be treated as if they were due to first-order reflections from planes with spacing d/n. Thus each reflection can be labeled with a set of values (hkl), and a spacing $d_{(hkl)}$ can be associated with the reflection. A reflection labeled, or *indexed,* as (200), for example, would have associated with it a spacing half that of the spacing of a (100) plane. This procedure is more satisfactory than treating the reflection as a second-order reflection from planes with the (100) spacing. Instead, therefore, of recognizing reflections of various orders of the $(hk0)$ planes in the equatorial layer line, for example, we shall label, or index, the reflections as (100), (200), (300) and (110), (220), (330), and so forth.

Similar indexing of the reflections in the other layer lines can be done, and all these reflections may need to be considered to deduce the nature of the internal structure of the crystal.

For detailed studies of all but the very simplest crystals one must make use of a single crystal and one of a variety of single-crystal instruments that have evolved from the simple arrangement shown in Fig. 14-14. For some purposes, however, it is possible to make use of a simpler technique that uses a crystalline material ground to a powder as a sample. This method, first used by Debye and Scherrer, is illustrated in Fig. 14-15.

FIGURE 14-14 The single-crystal photographic x-ray technique. One representative point in each of the layers is shown; each layer is made up of a large number of such diffraction spots.

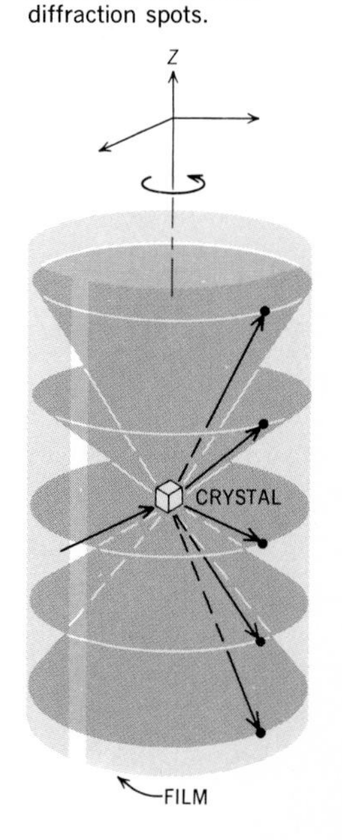

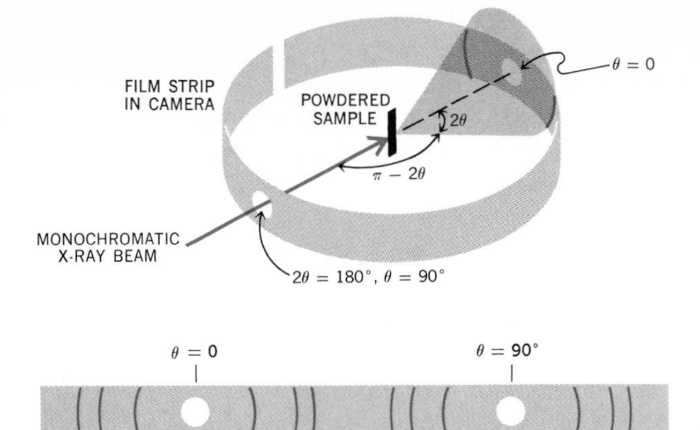

FIGURE 14-15
The x-ray powder method.

The crystals in the sample will present all possible orientations to the x-ray beam. The diffraction obtained will be just like that which would result from mounting a single crystal and turning it through all possible angles. For each crystal plane there will be some one angle at which the Bragg law will be satisfied, and some of the crystals will have this orientation; therefore a diffracted beam will result at the suitable angle, as is depicted in Fig. 14-15. Since there are quite a few crystal planes with a fairly high density of particles, there will be reflections from each of these, and the pattern will show scattering at a large number of angles.

For relatively simple crystals the powder pattern can be used, as will be illustrated in Sec. 14-11, in the study of the crystal structure. For more complex systems, particularly those of low symmetry, many planes in the crystal will happen to have equal or nearly equal spacing, and even if these planes have different directions in the crystal, the powder method will superimpose the reflections from these planes. For this reason, in all but relatively simple cases, one must make use of the more complete display of reflections provided by the single-crystal technique.

Probably the most widespread application of the powder method, which is far more easily used than the single-crystal method, is in the qualitative analysis of solid samples. The spacings between crystal planes, and therefore the position of the line of a powder photograph, are characteristic of a given crystal. The components of a solid mixture can therefore often be rapidly identified from a powder pattern.

Too small a crystal will not, however, give a good diffraction pattern. With any mechanical powdering technique, the crystals remain quite large, on a molecular scale, and a good diffraction pattern is obtained. For crystals that are so small that they do not confront the x-rays with an apparently infinite array of crystal planes, the diffraction pattern becomes poorly defined, and in fact one can use this broadening of the diffraction-pattern lines to determine the size of the crystals. A sample of a polymer like polyethylene consists of small regions of crystalline material embedded in a general amorphous matrix. The x-ray-diffraction pattern shows the broadening expected for small crystalline regions, and as Chap. 25 will point out, the size of the crystalline regions can be deduced from these diffraction results.

14-11 The Determination of the Lattice Type and Unit-cell Dimensions

The use to which such diffraction patterns can be put in the analysis of a crystal structure can be illustrated by seeing the relation of the structure of a cubic crystal to its x-ray powder pattern. The specific example of the NaCl crystal is convenient for our purpose. It can be assumed that, as is suggested by the external characteristics of the crystal, NaCl forms a cubic crystal, and in a way that is analogous to that adopted in a general crystal-structure problem, this assignment to the cubic system will later be verified.

The powder pattern of NaCl is shown in Fig. 14-16. Each reflection corresponds, according to Eq. [34], to a plane (hkl) with a spacing d_{hkl}. The first step in the use of a powder pattern, as it is also for a single-crystal pattern, is the assignment of values of (hkl) for the observed reflections, or as one says, the *indexing* of the observed reflections.

This can be done by recognizing that a general expression for the pattern of spacings between the planes of a particular lattice type can be set up, and therefore the pattern of reflections for this lattice type can be drawn. The observed reflections can then be compared with the pattern deduced, and the desired identification of the reflections can be accomplished.

In Fig. 14-17a, a plane with the general indices (hkl) is shown. In view of the reciprocal nature of the Miller indices and on the basis of a cubic structure, the intercepts can be written, as shown, as a/h, a/k, a/l. It is necessary to calculate d_{hkl} of Fig. 14-17b which is drawn from the origin normal to the plane and which gives, since another (hkl) plane like that shown would pass through the origin, the spacing between these planes. Addition of construction lines from point P to the

FIGURE 14-16
The x-ray powder pattern of NaCl taken with Cr K_α x-rays of wavelength 2.291 Å.

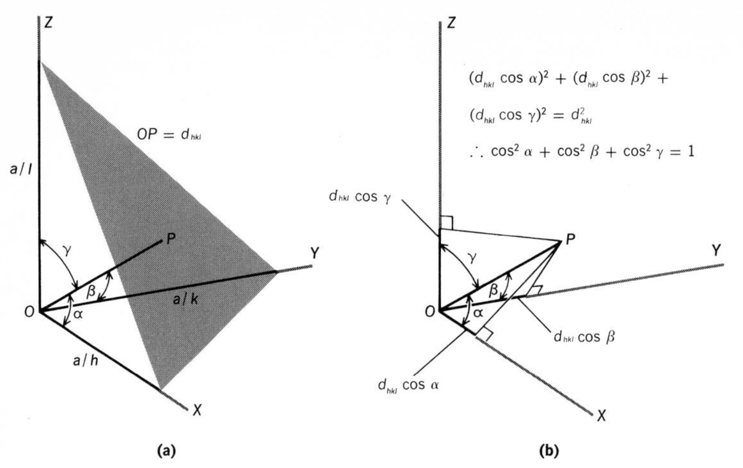

FIGURE 14-17 Steps in the derivation of Eq. [35].

intercepts of the (hkl) plane with the axes produces three right-angled triangles that give the relations

$$\frac{d_{hkl}}{a/h} = \cos \alpha \qquad \frac{d_{hkl}}{a/k} = \cos \beta \qquad \frac{d_{hkl}}{a/l} = \cos \gamma$$

Furthermore, as suggested by Fig. 14-17*b*, the relation

$$\cos^2 \alpha + \cos^2 \beta + \cos^2 \gamma = 1$$

exists for the three angles, and as a result we can obtain the desired expression

$$d_{hkl} = \frac{a}{\sqrt{h^2 + k^2 + l^2}} \qquad [35]$$

or with $\lambda = 2d_{hkl} \sin \theta$ of Eq. [34],

$$\sin \theta = \frac{\lambda}{2} \frac{\sqrt{h^2 + k^2 + l^2}}{a} \qquad [36]$$

or

$$\frac{2a}{\lambda} \sin \theta = \sqrt{h^2 + k^2 + l^2} \qquad [37]$$

Now insertion of all possible combinations of integers for h, k, and l will generate the possible values of $(2a/\lambda) \sin \theta$ for a cubic lattice. The resulting pattern is shown in Fig. 14-18, and you should particularly

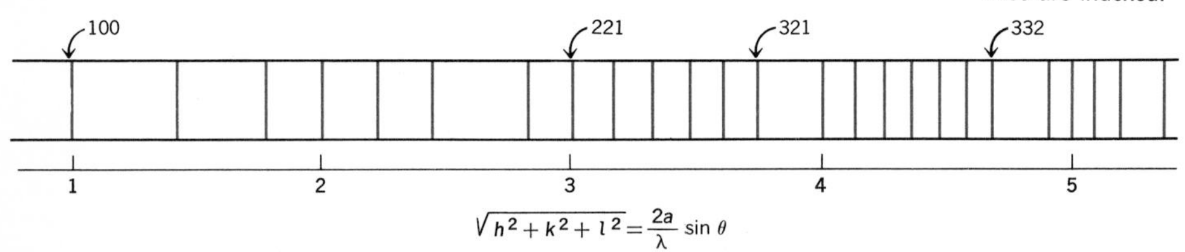

FIGURE 14-18 The pattern calculated for all possible diffraction lines from a cubic crystal. Some of the lines are indexed.

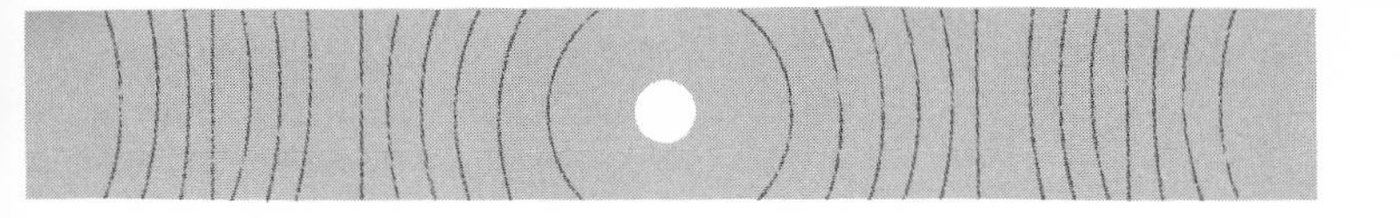

FIGURE 14-19
X-ray diffraction pattern of KCl, which, since K^+ and Cl^- have almost the same scattering effect, acts like a crystal with a simple cubic structure.

notice the gaps that correspond to the failure of sums of the squares of three integers to give values such as 7 or 15.

Some cubic crystals, such as shown in Fig. 14-19, produce a diffraction pattern with the same form as that of Fig. 14-18. This agreement confirms the cubic nature of the crystal, and further allows, if λ for the x-rays is known, a value of the unit-cell dimension a to be deduced.

But more generally, crystals that seem clearly to be cubic produce diffraction patterns that, like Fig. 14-16 for NaCl, appear to be quite different from the calculated pattern of Fig. 14-18. Furthermore, such patterns contain diffraction lines with various intensities. Now, if we consider the intensities with which the possible lines occur, we will see why cubic crystals can produce patterns that are related to, but different from, that of Fig. 14-18.

14-12 The Structure Factor

The crystal symmetry and the unit-cell dimensions are in some cases sufficient information to decide the exact structure of a simple crystal. For most crystals, however, the problem of determining the arrangement of the atoms or ions within each unit cell remains.

Let us now consider a crystal that could be of any of the crystal classes with axes that are orthogonal and let us indicate the unit-cell dimensions along the X, Y, and Z axes as a, b, and c.

To simplify, let us consider a two-dimensional view in which planes are considered, as in Fig. 14-20, to be parallel to the Z axis, that axis being taken to be perpendicular to the plane of the paper. Such planes have indices $(hk0)$, and the edges of the two planes that are shown are for the (210) and (310) planes. Now let us find the effect of an atom B that occupies some general position, at distances x and y, measured in units of the unit-cell dimensions a and b, from the origin of Fig. 14-20, on the intensities of the diffraction lines. After deriving a general expression, we will be able to apply it to the case of a cubic crystal for which the positions of the possible diffraction lines are given by Eq. [36].

Waves scattered by the planes of B atoms will interfere with any waves that are scattered by the $(hk0)$ planes drawn through the chosen origin. The scattering from the successive $(hk0)$ planes shown in Fig. 14-20 is displaced by exactly 2π and leads to the expected scattering as deduced in Sec. 14-11. Now, however, the planes of B atoms interfere by providing scattered beams that are out of phase, by an amount

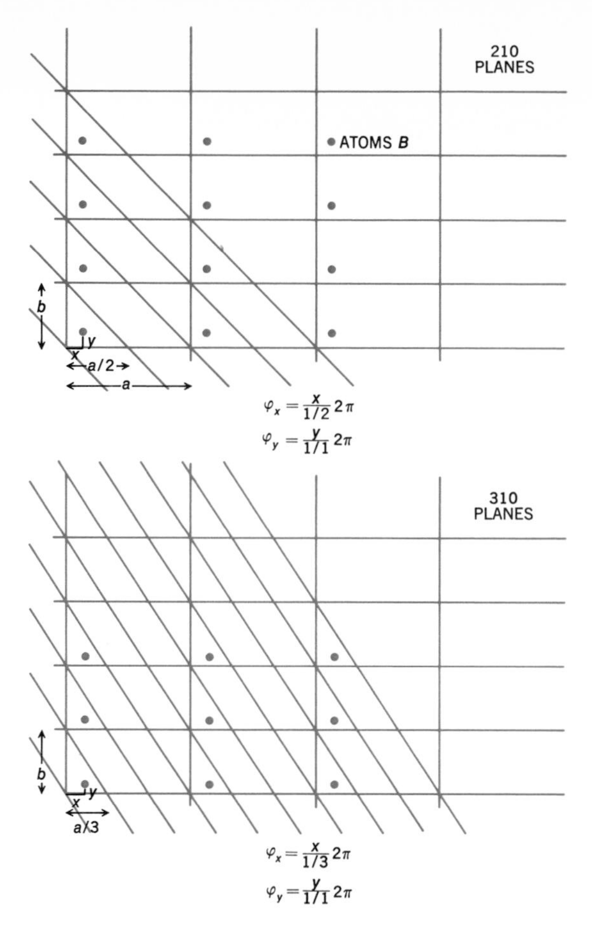

FIGURE 14-20
Phase shifts (φ_x,φ_y) for B atoms when the origins of the unit cells are oriented for constructive interference from (210) and (310) planes.

that depends on the positions of the B atoms. For the (210) reflections the B atoms provide waves that are out of phase by $[x/(1/2)]2\pi$ along the x axis and $[y/(1/1)]2\pi$ along the y axis. Consideration of the (310) reflection shows, as Fig. 14-20 suggests, that for these reflections the B-atom contributions would be out of phase by $[x/(1/3)]2\pi$ and $[y/(1/1)]2\pi$ along the x and y axes. In general, the phase difference of the beam resulting from the displacement of the B atoms from the origin is

$$2\pi(hx + ky)$$

and for a general plane in three dimensions, this phase difference would be

$$2\pi(hx + ky + lz)$$

It should be remembered that h, k, and l are the Miller indices of the reflection plane and diffraction line being considered, and that x, y, and z are the coordinates of the atom B in the unit cell of dimensions a, b, and c.

The net scattering from an (hkl) plane requires the summing up, allowing for the phase differences, of the scattering amplitudes from all atoms in the unit cell.

This sum is properly given, as was shown in Sec. 14-1, by an expression called the *structure factor,* represented by $F(hkl)$. Thus the calculated amplitude of the scattered beam from the (hkl) plane is proportional to $|F(hkl)|$. In view of the form of the amplitude factor of Eq. [5], $F(hkl)$ is written

$$F(hkl) = \sum_{\alpha} f_{\alpha} e^{2\pi i(hx_{\alpha}+ky_{\alpha}+lz_{\alpha})} \qquad [38]$$

where α numbers off the different atoms of the unit cell, and x_{α}, y_{α}, and z_{α} are the coordinates of the α atom of the molecule. The scattering power of the α atom is denoted by f_{α}, and can be taken as approximately proportional to the number of electrons of the atom, i.e., to its atomic number. (In fact, the scattering power of an atom also varies in a known way with the angle of scattering.)

The intensity of an x-ray beam is proportional to the square of the absolute value of the wave amplitude, as mentioned also for an electron beam, and one has the general expression

$$I(hkl) \propto \left| \sum_{\alpha} f_{\alpha} e^{2\pi i(hx_{\alpha}+ky_{\alpha}+lz_{\alpha})} \right|^2 \qquad [39]$$

The trigonometric form of this relation is more easily visualized. The necessary conversion relations are

$$e^{i\theta} = \cos\theta + i\sin\theta$$

and

$$\begin{aligned} |e^{i\theta}|^2 &= (\cos\theta + i\sin\theta)(\cos\theta - i\sin\theta) \\ &= \cos^2\theta + \sin^2\theta \end{aligned} \qquad [40]$$

With these relations, Eq. [39] becomes

$$I(hkl) \propto \left[\sum_{\alpha} f_a \cos 2\pi(hx + ky + lz)\right]^2 + \left[\sum_{\alpha} f_a \sin 2\pi(hx + ky + lz)\right]^2 \qquad [41]$$

Let us investigate the results from this expression for a few special cases. A simple cubic unit cell can be described as accommodating a single atom at the origin, i.e., with coordinates 0, 0, 0. (Note that only the atom at the origin need be specified to produce, as the unit cell is repeated to develop the structure, the atoms that are shown at all the corners of the cube. An alternative but clumsy procedure would consist of specifying one-eighth of an atom at each corner.) The expression for the diffraction intensities is then

$$I(hkl) \propto f^2\cos^2 0 + f^2\sin^2 0 \propto f^2 \qquad [42]$$

Thus all diffractions shown in Fig. 14-18 would be expected, and in this approximation of fixed scattering terms f, would be expected to have the same intensity. Figure 14-19 bears out these expectations.

Now let us consider a body-centered cubic crystal.

The atoms of a body-centered cubic structure can be described by locating one at the origin of the unit cell with coordinates 0, 0, 0, and one in the center of this cell with coordinates $\frac{1}{2}, \frac{1}{2}, \frac{1}{2}$. Now

$$I(hkl) = f^2[1 + \cos \pi(h + k + l)] + f^2[\sin \pi(h + k + l)]$$
$$= f^2[1 + \cos \pi(h + k + l)]^2 \quad [43]$$

The final result follows from the fact that the sine of any integral multiple of π is zero. Now some planes and the intensities of the corresponding diffraction lines can be deduced. Thus

$$I(100) = 0 \qquad I(200) = 4f^2$$
$$I(110) = 4f^2 \qquad I(210) = 0$$
$$I(111) = 0 \qquad I(211) = 4f^2$$

All (hkl) combinations whose sum is odd lead to a diffraction with zero intensity. The expected pattern for a body-centered cubic structure is shown in Fig. 14-21, along with the related simple cubic pattern.

Similar calculations can be made for face-centered cubic crystals which have atoms at 0, 0, 0; $\frac{1}{2}, \frac{1}{2}$, 0; $\frac{1}{2}$, 0, $\frac{1}{2}$; and 0, $\frac{1}{2}, \frac{1}{2}$. The expected patterns for the three basic cubic structures are shown in Fig. 14-21, and you see that each has a characteristic pattern that is found when the intensities, as well as the positions of the lines, are considered.

The structures of some cubic crystals can be determined by comparison of the observed pattern with the three of Fig. 14-21. The pattern for NaCl, for example, is clearly related to the face-centered cubic pattern. So far, however, the analysis has been limited to crystals containing a single type. To interpret the variation in line intensities—beyond the "present" or "absent" deductions made so far—we must work from a structure that locates the atoms of the different types, here Na^+ and Cl^-, in the unit cell. The structure, shown in Fig. 14-22,

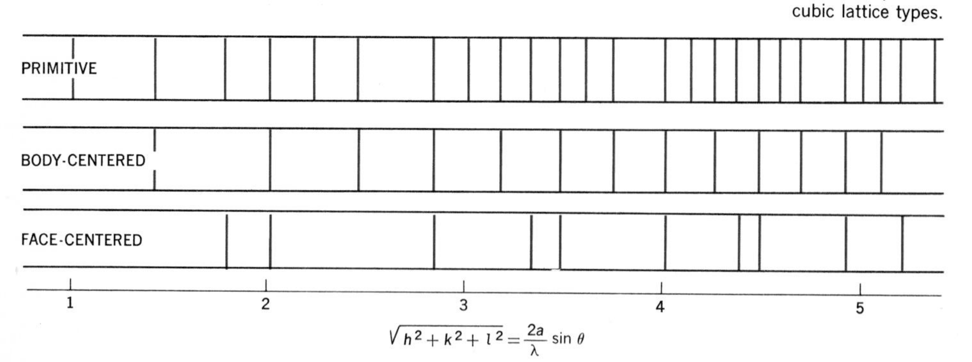

FIGURE 14-21
The line positions for the diffraction patterns for three cubic lattice types.

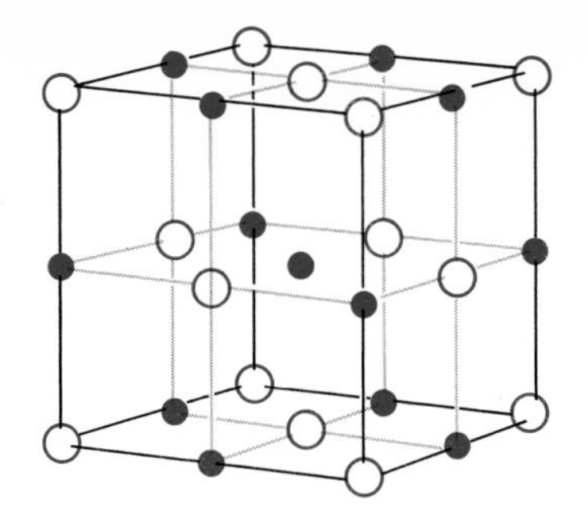

FIGURE 14-22
The unit cell of the NaCl structure.

which accounts for the intensities as well as positions of the diffraction lines is that in which the Na^+ and Cl^- ions form interlocking face-centered cubic arrangements.

If, arbitrarily, the Cl^- ions are given the positions 0, 0, 0; $\frac{1}{2}$, $\frac{1}{2}$, 0; $\frac{1}{2}$, 0, $\frac{1}{2}$; 0, $\frac{1}{2}$, $\frac{1}{2}$; then the Na^+ ions occupy the positions $\frac{1}{2}$, $\frac{1}{2}$, $\frac{1}{2}$; $\frac{1}{2}$, 0, 0; 0, $\frac{1}{2}$, 0; 0, 0, $\frac{1}{2}$. These coordinates can now be entered into Eq. [41]. (If the important simplification that the sine term again will involve only sines of multiples of π is recognized, it can be eliminated at the outset.) The intensity result obtained by summing Eq. [41] over all the atoms of the unit cell is

$$I(hkl) \propto [f_{\text{Cl}} + f_{\text{Cl}} \cos \pi(h + k) + f_{\text{Cl}} \cos \pi(h + l) + f_{\text{Cl}} \cos \pi(k + l) + f_{\text{Na}} \cos \pi(h + k + l) + f_{\text{Na}} \cos \pi h + f_{\text{Na}} \cos \pi k + f_{\text{Na}} \cos \pi l]^2 \quad [44]$$

Now intensity values could be calculated for some of the diffraction lines from NaCl, due allowance being made for the fact that six types of planes contribute to planes indexed as (321), three contribute to those indexed as (200), (220), etc., and only one type contributes to those indexed as (111), (222), etc. Except for the dependence of the f's on scattering angle, the treatment of the intensities would be seen to be in satisfactory agreement with those recorded in Fig. 14-16.

The procedure we have used is mathematical and serves very generally for the calculation of the positions, according to Eq. [36], and the intensities, according to Eq. [41], of the diffraction lines of the crystal system to which the equations correspond. In simpler cases one can see the effect of the constructive or destructive interference that the different types of atoms produce in a given diffraction. Thus the (111) planes of NaCl crystal can be looked on as consisting of planes of Cl^- ions, with planes of Na^+ lying midway between and interfering destructively. The deduction from Eq. [41] leads to an $(f_{\text{Cl}} - f_{\text{Na}})$ factor in the intensity expression. Thus either approach accounts for the weakness of the observed (111) line.

A method has therefore been obtained for going from a crystal structure to a calculated x-ray pattern. One can be content with this result, which shows that, with sufficient trial-and-error steps, a structure for a molecule in a crystal can be obtained. There are, furthermore, very many diffraction spots that can be measured, and a structure which leads to a correct intensity calculation for all the observed spots is almost unquestionably the correct structure. In this regard x-ray-diffraction results are more satisfactory than those of electron diffraction.

It is of course desirable, and for large organic molecules almost necessary, that a method be available for more directly deducing a molecular structure from the observed diffraction pattern without having to resort entirely to this trial-and-error procedure. An indication of the problems that arise when this is attempted and some of the methods that are adopted to overcome these difficulties can now be given.

14-13 The Fourier Synthesis

One of the most important developments in x-ray-diffraction techniques was the recognition by Sir William Bragg that the approach of the preceding section could in fact, to some extent, be turned around. The procedure depends on the suggestion that the crystal can be looked upon not as a set of discrete scattering points, but as a three-dimensional distribution of varying electron densities. X-rays do, in fact, recognize atoms only as regions of high electron density, and therefore high scattering power, compared with that of the surrounding regions. Instead of attempting to deduce the coordinates of the atoms in the crystal, one can look for an electron density function $\rho(xyz)$ that represents the electron density distribution in the crystal. This function, which is like the radial distribution function introduced in electron-diffraction studies, will be quite complicated if the molecules of the crystal are at all large. In view of the periodic nature of the crystal, with periodicities a, b, and c, one can, however, formally represent the function by a Fourier series function with the form

$$\rho(xyz) = \sum_{p,q,r=-\infty}^{+\infty} A(pqr)e^{2\pi i(px+qy+rz)} \qquad [45]$$

where p, q, and r take on all integral values from $-\infty$ to $+\infty$, and $A(pqr)$ are the to-be-determined coefficients of the many Fourier series terms. Determination of the coefficients $A(pqr)$ is the goal of the derivation. Knowledge of these coefficients would allow the electron density to be drawn out and the positions of the atoms to be located.

It is necessary now to recall the amplitude expression of Eq. [38]. The structure factor, which gives the amplitude of the beam diffracted by the (hkl) plane, was found to be

$$F(hkl) = \sum_{\alpha} f_{\alpha} e^{2\pi i(hx_{\alpha}+ky_{\alpha}+lz_{\alpha})}$$

where the summation is over all atoms of the unit cell. In terms of a distribution $\rho(xyz)$ of the electron density throughout the unit cell, this amplitude equation can be written in the integral form

$$F(hkl) = (\text{const}) \int_0^1 \int_0^1 \int_0^1 \rho(xyz) e^{2\pi i(hx+ky+lz)}\, dx\, dy\, dz \qquad [46]$$

where we need not work out the value of the multiplying constant.

Evaluation of the coefficients $A(pqr)$ of Eq. [45] follows from substitution of the Fourier expression for $\rho(xyz)$ into Eq. [46]. This gives

$$\begin{aligned} F(hkl) &= (\text{const}) \int_0^1 \int_0^1 \int_0^1 \sum_{p,q,r} A(pqr) e^{2\pi i(px+qy+rz)} \\ &\qquad\qquad \times e^{2\pi i(hx+ky+lz)}\, dx\, dy\, dz \\ &= (\text{const}) \int_0^1 \int_0^1 \int_0^1 \sum_{p,q,r} A(pqr) e^{2\pi i[(h+p)x+(k+q)y+(l+r)z]} \\ &\qquad\qquad \times dx\, dy\, dz \end{aligned} \qquad [47]$$

Recognition that all terms of the type $\int_0^1 e^{2\pi i\theta}\, d\theta$ are zero, as can be easily seen by expressing the exponential term in trigonometric form, eliminates all terms of the series except those for which $p = -h$, $q = -k$, and $r = -l$. For these, the exponential is $e^{2\pi i(0)} = 1$, and the net result of all the summations and integrations of Eq. [47] is

$$F(hkl) = (\text{const}) A(-h, -k, -l)$$

or

$$A(-h, -k, -l) = \frac{F(hkl)}{\text{const}} \qquad [48]$$

Recognition that the indices p, q, and r are equivalent to $-h$, $-k$, and $-l$ allows substitution of this result in the Fourier series expression [45] for the electron density to give, finally,

$$\rho(xyz) = \frac{1}{\text{const}} \sum_{h,k,l=-\infty}^{+\infty} F(hkl) e^{-2\pi i(hx+ky+lz)} \qquad [49]$$

The power of this elegant result is seen when it is recalled that $F(hkl)$ is the amplitude of the wave scattered by the (hkl) plane. The measured intensity of the beam scattered by the (hkl) plane, furthermore, is proportional to $|F(hkl)|^2$. For centrosymmetric unit cells, for example, measurements lead to relative values of $\pm F(hkl)$. Except for the undetermined sign on the values of $F(hkl)$, the expression of Eq. [49] gives the desired result: a method for using the intensities of the diffraction spots to deduce the crystal structure, as represented by $\rho(xyz)$. If the signs of the amplitudes of the diffracted waves that form the diffraction spots that can be labeled (hkl) were known, the electron density at any point (xyz) in the unit cell could be determined by performing the summation of Eq. [49], for the chosen value of x, y, and z, over all values of (hkl), that is, for all the observed diffraction spots. The more intense diffraction spots correspond to the numerically greater diffraction amplitudes, so that it is at first necessary only to extend the summation to all the more important diffraction spots. Such summations could be performed for various points, specified by values of x, y, and z in the unit cell. By determination of the electron density at sufficient points, an electron-density map could be drawn which would show the positions of the atoms by regions of high electron density.

The undetermined sign, which results because the intensities, and not the amplitudes, of the diffraction beams are obtained, turns out to be very troublesome, and a number of techniques have been suggested to make use of Eq. [46] in spite of the sign difficulty.

These approaches depend, usually, on methods for obtaining the approximate shape of the molecule and its approximate position in the unit cell. Some molecules, for instance, contain a heavy atom, which has a high scattering probability and can be located, whereas the complexities of the remainder of the molecule can initially be ignored. The signs of the most important structure factors can then be guessed at. These few diffraction spots can then be used in Eq. [49] to deduce a crude electron-density map. This first approximation map can lead to the estimate of the position of more of the atoms, and thus, with Eq. [46], to the signs of the structure factors of more of the diffraction results. In this way one refines a structure by working in the information from more and more of the diffraction spots until a structure of the desired detail is obtained.

Typical electron-distribution maps that are obtained as a result of such a study are shown in Fig. 14-23.

The complete deduction of the structure of a complicated molecule by the x-ray-diffraction technique is a major research problem. There is, however, no other method that provides such a wealth of reliable structural information. The number of large molecules whose structure, in the solid state, has been completely worked out continues to grow and to provide valuable basic information for many fields of chemistry.

14-14 Ionic Radii

Just as covalent radii could be assigned, in Sec. 14-5, from the dimensions of covalent molecules, so also ionic radii can be assigned to ions, once the structures and dimensions of simple ionic crystals have been determined. The monatomic ions are represented by spheres, and the radii of these spheres are determined from the dimensions and nature

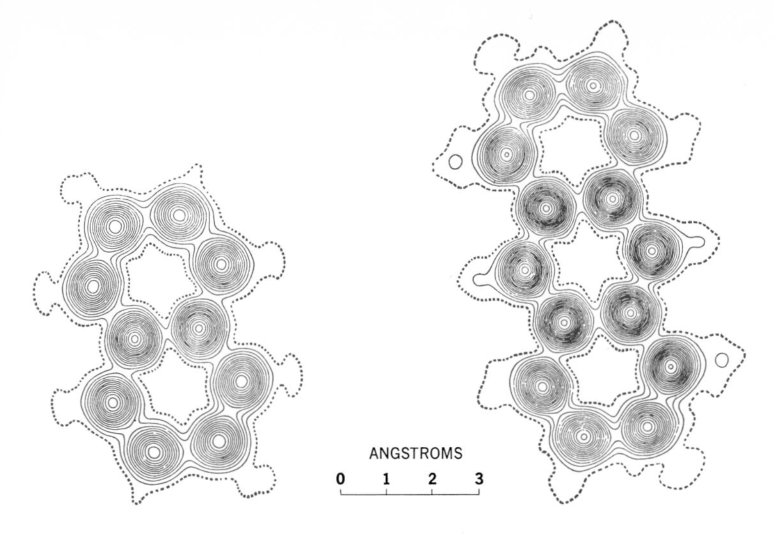

FIGURE 14-23
Electron-density maps of naphthalene and anthracene. [*From x-ray diffraction studies by J. M. Robertson and coworkers, Acta Cryst.,* **2**:*233 (1949) and* **3**:*245 (1950).*] Electron-density contour levels are drawn at intervals of $\frac{1}{2}$ electron per Å^3. That for a density of $\frac{1}{2}$ electron per Å^3 is shown dotted.

of the unit cells of the crystals in which they are found. Again, the effective ionic radius appears to be somewhat dependent on the nature and arrangement of the ions that surround it, and unless different values for different situations are tabulated, the deduced ionic radii that are given are "best" values.

Consideration of unit-cell dimensions of simple ionic crystals suggests that the anions and cations are of such a size as to be in contact with one another, as shown in Fig. 14-24*a*. For such situations one can obtain only the sum of the radii of the two ions which, as Fig. 14-24*b* shows, is equal to half the length of the unit cell.

A number of methods have been suggested to obtain individual ionic radii from these relatively easily obtained sums. Perhaps the simplest depends on the assumption that the crystal with the largest anion, I^-, and the smallest cation, Li^+, will present the situation shown in Fig. 14-24*b*. In this case the cation is so small that there is anion-anion contact and the radius of the anion, the iodide ion, can be ob-

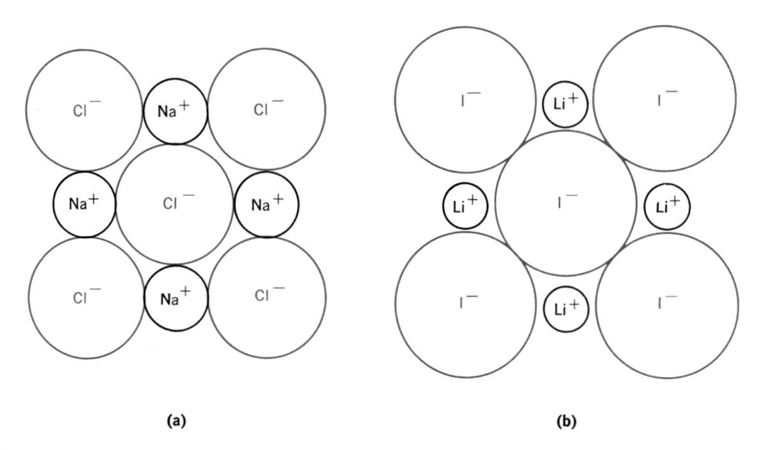

FIGURE 14-24
An example in (*a*) of the unit cell face of NaCl in which the ionic radii cannot be deduced directly from the cell dimensions and (*b*) of the unit cell face of LiI where there is anion-anion contact and a value for the anion radius can be obtained from the length of the diagonal of the unit cell.

TABLE 14-5 Some crystal radii from the compilation by Pauling*

Ion	Radius	Ion	Radius	Ion	Radius	Ion	Radius	Ion	Radius	Ion	Radius	Ion	Radius
												H^-	2.08
Li^+	0.60	Be^{++}	0.31	B^{3+}	0.20	C^{4+}	0.15	N^{5+}	0.11	$O^=$	1.40	F^-	1.36
Na^+	0.95	Mg^{++}	0.65	Al^{3+}	0.50	Si^{4+}	0.41	P^{5+}	0.34	$S^=$	1.84	Cl^-	1.81
K^+	1.33	Ca^{++}	0.99	Se^{3+}	0.81	Ti^{4+}	0.68	V^{5+}	0.59	$Se^=$	1.98	Br^-	1.95
Cu^+	0.96	Zn^{++}	0.74	Ga^{3+}	0.62	Ge^{4+}	0.53	As^{5+}	0.47				
Rb^+	1.48	Sr^{++}	1.13	Y^{3+}	0.93	Zr^{4+}	0.80	Cb^{5+}	0.70	$Te^=$	2.21	I^-	2.16
Ag^+	1.26	Cd^{++}	0.97	In^{3+}	0.81	Sn^{4+}	0.71	Sb^{5+}	0.62				
Cs^+	1.69	Ba^{++}	1.35	La^{3+}	1.15	Ce^{4+}	1.01						
Au^+	1.37	Hg^{++}	1.10	Tl^{3+}	0.95	Pb^{4+}	0.84	Bi^{5+}	0.74				

*L. Pauling, "The Nature of the Chemical Bond," 3d ed., Cornell University Press, Ithaca, N.Y., 1960.

tained from the unit-cell dimensions as suggested in the figure. Once a value for the radius of one ion is obtained, by this or other methods that will not be dealt with here, a table of ionic radii can be constructed from the sums of ionic radii that can be deduced from other crystal structures. A set of values is shown in Table 14-5.

14-15 Van der Waals Radii

Information on distances between covalently bonded atoms, which is obtained by spectroscopic and electron-diffraction studies of gas as well as by x-ray studies of crystals, has already been considered in Sec. 14-5. Here the information, uniquely obtained from diffraction studies of crystals, on the way in which molecules can approach one another and can pack together will be outlined.

Since forces, often treated under the name van der Waals forces, provide the attraction and repulsion between molecules that are responsible for the closeness with which molecules can approach one another, the idea of a van der Waals radius for each covalently bound atom is introduced. Figure 14-25 shows the shapes attributed to molecules as a result of the introduction of van der Waals radii.

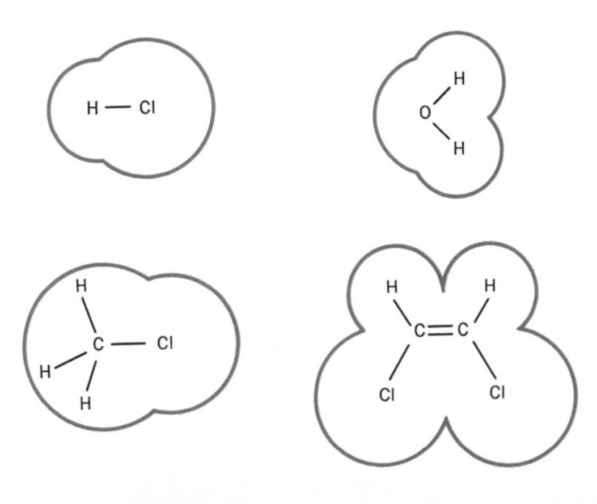

FIGURE 14-25
Examples of molecular structure showing the sizes of the atoms as depicted by van der Waals radii.

ABLE 14-6 **Van der Waals radii of atoms of covalent molecules***

	N	O	F
.2	1.5	1.40	1.35
	P	S	Cl
H3 group 2.0	1.9	1.85	1.80
	As	Se	Br
alf thickness of romatic ring 1.70	2.0	2.00	1.95
	Sb	Te	I
	2.2	2.20	2.15

From L. Pauling, "The Nature of the Chemical Bond," 3d ed., Cornell University Press, Ithaca, N.Y., 1960.

The values of these radii can be deduced from the distances that eparate atoms of different molecules in a crystal lattice. In crystalline 3r$_2$, for example, the shortest distance between a bromine atom of one nolecule and that of an adjacent molecule is 3.90 Å. Half this value, .95 Å, can therefore be assigned as the van der Waals radius of a ovalently bound bromine atom. In similar ways, making use of crystal-tructure data for many organic compounds, the van der Waals radii iven in Table 14-6 can be deduced. These values must be considered eliable to not more than about 0.05 Å, and this uncertainty makes itself evident in the range of values found for a particular element in different compounds and crystals. The values are sufficiently reliable, however, or scale drawings, such as those of Fig. 14-24, to be constructed and ised to see how molecules can fit together. That van der Waals radii an be assigned with some success is attributable to the fact that, as nentioned in Sec. 11-8, the repulsive forces set in very strongly, i.e., he potential-energy curve rises very steeply, as atoms approach one nother. It follows that even when rather different attractive forces perate, the closeness of approach will be little affected. For example, he ionic radius of the chloride ion is very nearly equal to the van der Vaals radius of the covalently bound chlorine atom.

roblems

1 With a slit assembly like that of Fig. 14-2 and a source of essentially monochromatic light of wavelength 5890 Å, constructive interference occurs on a photographic plate to give rings at angles of 10°, 21°, 32°, 44°, and so forth, from the central position on the plate. Deduce the spacing between the slits.

2 Calculate the velocity of electrons, according to Eq. [8], produced by an accelerating potential of 40,000 V. Since the relativistic dependence of mass on velocity is given by the expression $m = m_0(1 - v^2/c^2)^{-1/2}$, calculate the ratio of the rest mass to the mass of the electron accelerated by this voltage.

3 Plot the square of the amplitude of the resultant beam as a function of the phase difference from 0 to 2π (*a*) when waves of equal amplitude interfere, and (*b*) when waves of amplitude A_1 and $A_2 = \frac{1}{4}A_1$ interfere.

4 Plot, underneath one another, the theoretical scattering curves that would be obtained for a CO molecule with assumed bond lengths of 1.00, 1.20, and 1.40 Å. Carry each plot out to a value of s, equal to $(4\pi/\lambda)\sin(\theta/2)$, of about 20.

5 Plot, underneath one another, the theoretical scattering curves for CCl_4 that would be obtained with an assumed tetrahedral structure and carbon-chlorine bond lengths of 1.7 and 1.8. Carry the curves out to $s = 15$. Compare with the reported results of Karle and Karle [*J. Chem. Phys.*, **17**:1052 (1949)] and Bartell, Brockway, and Schwendeman [*J. Chem. Phys.*, **23**:1854 (1955)].

6 The visual appearance of the photograph of the diffracted electron beam, of 40,000 V, from CO_2 shows maxima at $s = 6.7$, 12.2, 17.8, and 23; shoulders on these maxima at about 8.5, 14, and 19; and minima at 4.4, 10.0, 15.4, and 21.

By plotting theoretical scattering curves for assumed values of the bond lengths (assuming a linear symmetric structure), deduce the C═O bond length in CO_2 [see Karle and Karle, *J. Chem. Phys.*, **17**:1052 (1949)].

7 A tetragonal crystal has unit-cell dimensions of 12.04, 12.04, and 19.63 Å. The unique axis is taken as the z axis.

a Prepare sketches showing the planes that have Weiss indices $a{:}b{:}\infty c$, $a{:}b{:}c$, $a{:}\infty b{:}c$, and $a{:}2b{:}2c$.

b What are the Miller indices of these planes?

c Prepare sketches showing the planes that have Miller indices (011), (101), (122), and (021).

8 Tabulate the number of lattice points in each of the unit cells of Fig. 14-11.

9 The density of graphite is 2.25 g/liter^{-1}, and the spacing between the layers is found by x-ray diffraction to be 3.35 Å. What is the carbon-carbon distance in the molecular layers on the assumption of the structure of Fig. 18-1?

Ans. 1.43 Å.

10 A single-crystal diffraction pattern is taken of a graphite crystal, x-rays of wavelength 1.537 Å being used. The crystal is mounted, as in Fig. 14-14, so that the angle θ is taken as zero when the incoming x-ray beam impinges perpendicularly onto the molecular planes of the crystal. Consider the carbon-atom planes as presenting only continuous planes of high scattering. Plot schematically the signal that will be obtained from the detector as a function of θ and the angle through which the crystal is turned, and label the diffraction lines with appropriate Miller indices. The spacing between the molecular planes of graphite is 3.35 Å.

Make a scale drawing showing how the waves from different planes add constructively for the first- and second-order diffractions.

11 Calculate the spacings between (110) planes of KCl viewed as a simple cubic lattice, with the K^+ and Cl^- ions taken as identical and the (100)-plane spacings as 3.152 Å.

At what angles would first- and second-order diffraction from the (100) and the (110) planes be observed if x-rays of wavelength 1.537 Å are used?

Ans. First order, 14°7′ for (100), 20°10′ for (110).

12 By comparison with the patterns of Fig. 14-21, index some of the lines of the NaCl diffraction pattern of Fig. 14-16. Estimate the value of θ for some of these lines, and calculate the unit-cell dimension. Compare with the reported value of 5.627 Å.

13 Consider the crystal of Fig. 14-20 to have atoms A, at the origin, and B such that the scattering factor f_B is one-half of f_A. Furthermore, suppose that the unit cell has $a = 5$ Å and $b = 3$ Å and that the position of B relative to A is given by $x = 0.4$ and $y = 0.3$, in units of the cell dimension.

a Calculate for some assumed wavelength of x-rays the angles at which first-order constructive interference from the A planes will occur for the planes (100), (010), (110), (210), and (120).

b Calculate the relative intensities of the diffraction spots corresponding to each of these planes. (Take $f_A = 1$ and $f_B = 0.5$.)

c Move atom B to a different position, and see that the relative intensities calculated for the different diffractions are changed. Recognize that in this way various positions of atom B could be assumed until a calculated pattern was obtained that matched an observed pattern.

14 Set up the Fourier series expressions according to Eq. [49] for the electron density at atom A, with $x = y = z = 0$, and atom B, with $x = 0.4$, $y = 0.3$, and $z = 0$, as in Prob. 13. Obtain expressions for $\rho(000)$ and $\rho(0.4,0.3,0)$ in terms of the intensities of the reflections calculated in part b of Prob. 13. Recognize that if the signs of the structure factors, which are the square roots of the intensities, were known, the relative electron densities at the two points could be calculated from data that could be obtained experimentally.

15 The angle between the C—Cl bonds in CH_2Cl_2 is determined by both electron-diffraction and microwave-spectroscopic techniques to be 112°, and the C—Cl bond lengths are found to be 1.77 Å. Calculate the Cl···Cl distance in methylene chloride, and discuss this value in view of the van der Waals radius for chlorine given in Table 14-5.

References

WHEATLEY, P. J.: "The Determination of Molecular Structure," chaps. 5–9, Oxford University Press, Fair Lawn, N.J., 1959. An excellent introductory treatment of the determination of structures by x-ray diffraction. After the NaCl example is worked through, a number of additional aspects of the use of symmetry and the treatment of more complex systems are taken up.

BRAND, J. C. D., and J. C. SPEAKMAN: "Molecular Structure," chaps. 2, 8, and 9, Edward Arnold (Publishers) Ltd., London, 1960. A further treatment of the symmetry of molecules and crystals, followed, in chap. 8, by an introduction to the study of crystal structures by x-ray diffraction and, in chap. 9, by a corresponding study of molecular structures by electron diffraction.

WELLS, A. F.: "Structural Inorganic Chemistry," Oxford University Press, Fair Lawn, N.J., 1962. Although devoted primarily to the structure and properties of inorganic compounds, there is included in chaps. 2 to 6 a treatment of the methods by which structures are determined. Of special value are the critical considerations of the way in which structural data are used, as, for example, in the deduction of covalent, ionic, and van der Waals radii.

PHILLIPS, F. C.: "An Introduction to Crystallography," Longmans, Green & Co., Ltd., London, 1963. A rather complete, readable treatment of aspects of crystallography and x-ray diffraction that are important in chemistry. Included are discussions of crystals and crystal symmetry, the use of powder and single-crystal methods, the deduction of unit-cell dimensions, and the determination of atomic positions in simple and complicated systems.

PINSHER, Z. G.: "Electron Diffraction," Butterworth & Co. (Publishers) Ltd., London, 1953. A rather complete treatment of the theory of electron diffraction.

SANDS, D. E.: "Introduction to Crystallography," W. A. Benjamin, Inc., New York, 1969. An excellent extension of the material presented here.

STOUT, G. H., and L. H. JENSEN: "X-ray Structure Determination," The Macmillan Company, New York, 1968. A practical treatment of the principles of x-ray diffraction developed with the objective of crystal-structure analysis.

15

EXPERIMENTAL STUDIES OF THE ELECTRICAL AND MAGNETIC PROPERTIES OF MOLECULES

The properties of molecules that are deduced from spectroscopic or diffraction methods are, in a sense, electrical in that the molecular binding forces result from forces between electrons and the nuclei of the molecules or ions. Molecules and ions do, however, also display electrical and magnetic properties that are more similar to those observed in ordinary-sized systems. It is convenient, therefore, to treat these properties, and to see how they are defined and measured, in a separate chapter.

ELECTRICAL PROPERTIES: MOLECULAR DIPOLE MOMENTS

Since molecules are made up of charged units, electrons and the atomic nuclei, much of the behavior of molecules is understandable in terms of electrical interactions. A detailed theoretical treatment of a molecule by the methods of quantum mechanics would reveal the electron distribution of the molecule and would lead to deductions as to how a particular assembly of charges that make up a molecule would interact with other molecules, with a surrounding medium, or with an electric field. It is possible, also, to learn much about the electric nature of molecules by an experimental approach. In this section the procedure which produces results for the dipole moment of a molecule will be dealt with. The goal, however, is not only to obtain such results, but also to show how some electrical phenomena in chemical systems are treated.

15-1 Dipole Moments of Molecules

The principal characteristic of the charge distribution in a molecule that can be measured is the extent to which the center of the electron distribution of a molecule fails to coincide with the center of the positive nuclear-charge distribution. The charge asymmetry is obtained as the dipole moment of the molecule. This charge asymmetry, as we shall see, may result from an unequal sharing of the bonding electrons, the extreme case of which leads to a molecule with positively and negatively charged ions such as a molecule of NaCl vapor. More subtle electron distributions such as the apparent slight asymmetry of the bonding electrons, as in a CH bond, or the projection of the nonbonding electrons of, for example, the nitrogen atom in ammonia, also lead to molecular dipoles. Dipole-moment results allow such aspects of the electronic configuration of molecules to be discussed.

As Fig. 15-1 indicates, the dipole moment of two equal and opposite charges is defined as the product of the charges and the distance separating them. Thus the dipole moment is given by

$$\mu = qr \qquad [1]$$

The dipole moment has, furthermore, a direction as well as a magnitude; i.e., it is a *vector* quantity. It is frequently convenient diagrammatically to represent a dipole moment by an arrow showing the direction from the positive to the negative charge and the magnitude by the length of the arrow, as is done in Fig. 15-1*a*. In fact, the concept of a dipole arises when, as suggested in Fig. 15-1*b*, the effect of an assembly of charges at some distant point is investigated, and with this approach the dipole moment due to a collection of charges is defined as

$$\boldsymbol{\mu} = \sum_i q_i \mathbf{r}_i \qquad [2]$$

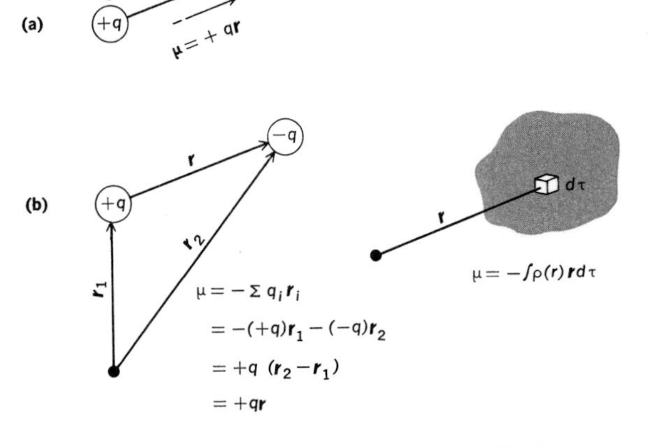

FIGURE 15-1
(*a*) The dipole moment of separated equal but opposite charges. (*b*) The dipole-moment deduction, by means of Eqs. [2] and [3], for the two-particle system and a complex system representing a molecule.

where q_i and $\mathbf{r}_i$ are the charges and vector lengths of the ith charges of the assembly. For a molecule the distribution of the electrons requires an integral form of Eq. [2],

$$\boldsymbol{\mu} = \int \rho(\mathbf{r})\mathbf{r}\, d\tau \qquad [3]$$

where $\rho(\mathbf{r})$ is the charge density at a position defined by the vector $\mathbf{r}$, and $d\tau$ is a volume element. The integration over all the electrons and nuclei required by Eq. [3] can, of course, be carried out only for relatively simple molecules, where the electron distribution can be determined by theoretical methods. In general, the detailed electronic distribution is not known, and information is provided by the measurement of the molecular dipole moment by methods that will be developed in this chapter. The determination of this one quantity does not, of course, allow the charge distribution $\rho(\mathbf{r})$ of Eq. [3] to be deduced. Usually, one makes use of the measured value of the dipole moment of a molecule by depicting the charge asymmetry which it measures in terms of a model, like that of Fig. 15-1a, of separated, opposite charges that have the dipole moment of that measured for the molecule.

The order of magnitude of molecular dipole moments can be deduced by recognizing that these moments result from charges like that of an electron, that is, 0.1602×10^{-18} C, separated by angstrom distances. For one electron separated from an equally charged positive center by a distance of 1 Å, the dipole moment would be

$$\mu = (0.1602 \times 10^{-18})(1 \times 10^{-10}) = 16.02 \times 10^{-30} \text{ m C}$$

In most of the chemical literature, molecular dipole moments are quoted in units of *debyes,* a quantity based on the cgs-esu system. The relation to SI units is

$$1 \text{ debye} = 3.338 \times 10^{-30} \text{ m C}$$

and thus an electronic charge separated from one of opposite sign by a distance of one angstrom has a dipole moment of $16.02 \times 10^{-30}/3.338 \times 10^{-30} = 4.80$ debyes.

In addition to a dipole moment that a molecule can have as a result of its asymmetric charge distribution, there is, for all molecules, the possibility of distorting the electronic distribution in a molecule by applying an electric field. In this way an *induced dipole moment* can be produced. The effectiveness of an applied field in making a molecule polar is determined by the *polarizability* of the molecule. The polarizability, defined as the dipole moment induced by an electric field of unit strength, is the second important electrical property of molecules with which we shall deal. The units and values of this quantity are best left until its determination from measurable quantities is considered. It should be immediately clear, however, that all molecules, symmetric or not, are polarizable and can have an induced dipole moment. On the other hand, symmetric molecules, like H_2, CO_2, and CCl_4, necessarily have zero "permanent" dipole moments.

15-2 Some Basic Electrostatic Ideas

Some basic ideas that enter into the treatment of charged particles in a vacuum (or approximately, in air) must be reviewed. This and the following section, where media other than vacuum are considered, provide the necessary background to the theory of the measurement of molecular dipole moments.

The basic relation in treating the interaction of stationary charges is Coulomb's law, which states that two point charges q_1 and q_2 separated by a distance r in vacuum (or approximately, air) will interact with a force f given by

$$f = \frac{q_1 q_2}{4\pi\epsilon_0 r^2} \qquad [4]$$

where ϵ_0 is the *permittivity constant* and has a value such that

$$4\pi\epsilon_0 = 1.11264 \times 10^{-10}\ \mathrm{C^2\ N^{-1}\ m^{-2}}$$

For q_1 and q_2 of opposite signs the force is one of attraction; with the same signs, one of repulsion.

The interaction of charges at a distance suggests that an electric field exists around each charge. The intensity of the electric field $\mathcal{E}$ at a point is defined as the force which would be exerted on a positive charge placed at that point. Thus the electric field strength about a point charge q is the force on a unit charge, or

$$\mathcal{E} = \frac{f}{q} \qquad [5]$$

The field strength is measured in newtons per coulomb. The force on the unit test charge has a direction as well as a magnitude. The electric field $\mathcal{E}$ is therefore a vector quantity and has the direction as well as the magnitude of the force on a unit positive test charge. It is frequently helpful to draw *lines of force* to represent the intensity and the direction of an electric field.

Consider, for example, the electric field about a point charge $+q$, as in Fig. 15-2. At a distance r from this charge a unit positive charge would be repelled by a force $q/4\pi\epsilon_0 r^2$. The electric field intensity is therefore $q/4\pi\epsilon_0 r^2$ at a distance r from the $+q$ charge. This electric field can be depicted by drawing lines of force emanating from the point charge. As these lines suggest, the *total* field produced by the charge is independent of the distance from the charge and is given, as suggested by Fig. 15-2, as

$$\frac{q}{4\pi\epsilon_0 r^2}(4\pi r^2) = \frac{q}{\epsilon_0} \qquad [6]$$

Another important aspect of an electric field is described by the electrical potential $\mathcal{V}$. This quantity represents the potential energy of a unit positive charge in the electric field. A unit positive charge in

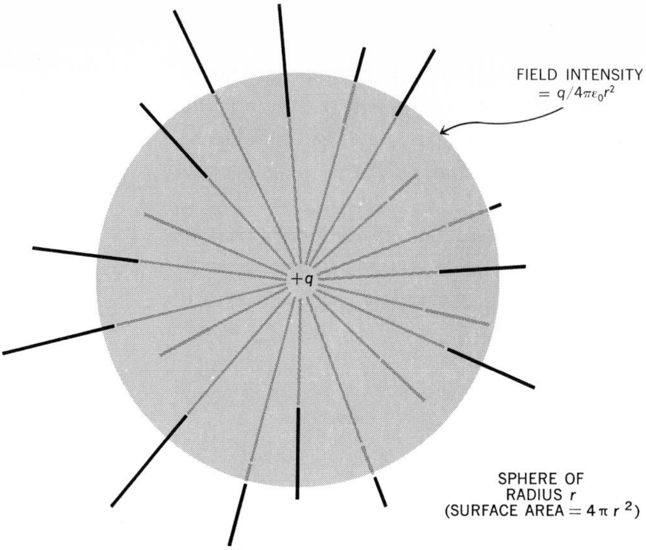

FIELD INTENSITY $= q/4\pi\varepsilon_0 r^2$

SPHERE OF RADIUS r (SURFACE AREA $= 4\pi r^2$)

FIGURE 15-2
The electric field of a charge $+q$ represented by lines of force. A total of q/ϵ_0 lines emanate from a charge of $+q$.

an electric field of intensity $\mathcal{E}$ experiences a force equal to $\mathcal{E}$, and for a given displacement of the unit charge the work involved is the force, or the field strength, times the distance that the charge is moved. Since the potential energy increases as the unit positive test charge is brought closer to the positive charge q which generates the electric field, the change $d\mathcal{V}$ for an infinitesimal change dr is

$$d\mathcal{V} = -\mathcal{E}\,dr \tag{7}$$

Rearrangement of Eq. [7] gives a relation that is useful when electric fields are deduced from applied or known potentials. This expression is

$$\mathcal{E} = -\frac{d\mathcal{V}}{dr} \tag{8}$$

The example of a plane-parallel condenser with air (or more properly, vacuum) between the plates can now be considered. If, as in Fig. 15-3, the condenser is connected to a battery that produces a potential difference of $\mathcal{V}$ volts, there will be a potential drop of $\mathcal{V}$ volts across the condenser. According to Eq. [8], the field between the places of the condenser will be $\mathcal{V}/d$, where d is the distance between the plates.

The electric field in the condenser can also be understood in terms of the charges that the condenser plates acquire when the condenser is connected to a battery. If q is the charge, positive on one plate and negative on the other, and if the area of each condenser plate is A, there will be a charge density of $\sigma = q/A$ charges per unit area. If we now carry over the idea (a procedure which is justified by Gauss' law) that the electric field of a charge $+q$ is q/ϵ_0, as given by Eq. [6], we can write for the field per unit cross section the expression σ/ϵ_0.

FIGURE 15-3
A plane-parallel condenser with no dielectric material between the condenser plates.

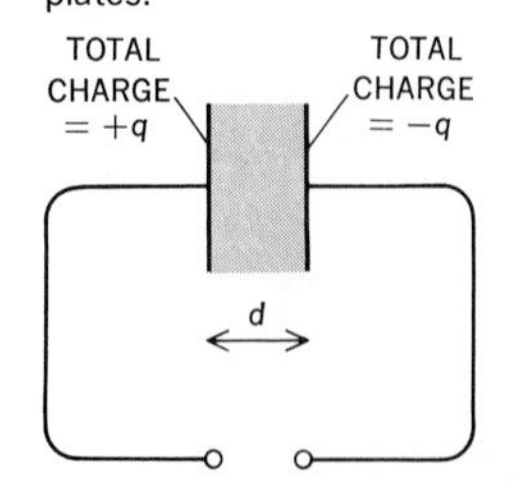

Then, from $\mathcal{E} = \sigma/\epsilon_0$ and $\mathcal{E} = \mathcal{V}/d$, we have the useful relation $\mathcal{V} = \sigma d/\epsilon_0$.

A condenser is most readily characterized experimentally by the ratio of the charge q acquired by the plates to the voltage $\mathcal{V}$ applied across the condenser. The *capacitance* of a condenser is defined by this ratio, and if there is air or a vacuum between the plates, the capacitance is C_0, given by

$$C_0 = \frac{q}{\mathcal{V}} \tag{9}$$

The capacitance is seen to be the charge held by the condenser plates per unit potential.

The capacitance can be related to the geometry of the condenser by substituting the results $q = \sigma A$ and $\mathcal{V} = \sigma d/\epsilon_0$ to give

$$C_0 = \frac{\sigma A}{\sigma d/\epsilon_0} = \frac{\epsilon_0 A}{d} \tag{10}$$

The capacitance is thus a property of the geometry of the condenser and for a plane-parallel condenser is large for large A and small d.

15-3 Electrostatics for Dielectric Media

The presence of a nonconducting, or *dielectric,* material around the charges that have been dealt with alters the relations which have been obtained. The understanding of this dielectric effect will be the basis from which the electrical properties of molecules are deduced.

The starting point is again Coulomb's law. When the charges q_1 and q_2 are immersed in a dielectric material, it is found that the force between them is less at any given value of r than when such a dielectric medium is absent. Coulomb's law can be written to apply to both vacuum and dielectric media by writing it in the form

$$f = \frac{q_1 q_2}{\epsilon r^2} = \frac{q_1 q_2}{(\epsilon/\epsilon_0)(4\pi\epsilon_0 r^2)} \tag{11}$$

where ϵ/ϵ_0 is a constant, at a given temperature, for any medium. It is known as the *dielectric constant,* and represents the effect of the dielectric material in decreasing the force between the charges.

The definition (Eq. [5]) of the electric field as the force exerted on a unit positive charge can now be applied to determine the field of the charge $+q$ when it is immersed in a dielectric material. The force on the unit positive charge at a distance r would be $q/(\epsilon/\epsilon_0)(4\pi\epsilon_0 r^2)$, and the total field for such a surrounded charge would be

$$\frac{q}{(\epsilon/\epsilon_0)(4\pi\epsilon_0 r^2)}(4\pi r^2) = \frac{q}{\epsilon} \tag{12}$$

The electric field is therefore lowered by the presence of the nonconducting materials by a factor equal to the dielectric constant ϵ.

Now we can consider the properties of the condenser of Fig. 15-3 that can be measured and that will allow an understanding of the role of the dielectric material. Suppose the condenser is charged until the same charge q as for vacuum accumulates on the plates. The charge per square centimeter is again $\sigma = q/A$, but in view of Eq. [12] the electric field $\mathcal{E}$ is reduced from σ/ϵ_0 to σ/ϵ. Now $\mathcal{V}$ is given by $\mathcal{E} = -\mathcal{V}/d$, or $\mathcal{V} = \mathcal{E}d$, as

$$\mathcal{V} = \frac{\sigma d}{\epsilon} \qquad [13]$$

and the capacitance is

$$C = \frac{q}{\mathcal{V}} = \frac{\sigma A}{\sigma d/\epsilon} = \epsilon\frac{A}{d} \qquad [14]$$

By comparison with Eq. [10], we see that measured capacitances with and without dielectric material give ϵ/ϵ_0.

$$\frac{C}{C_0} = \frac{\epsilon}{\epsilon_0} \qquad [15]$$

where C_0 is the capacitance of the condenser in vacuum. The dielectric material has therefore increased the capacitance of the condenser. It is this relationship, Eq. [15], that is often used as a more directly operational definition of the dielectric constant, since it is a quite straightforward matter to measure the capacitance of a condenser.

Now it is necessary to investigate the mechanism by which the dielectric material decreases the electric field between the condenser plates when a given charge is placed on the plates. Again consider the condenser to be charged to a suitable potential so that charges $+q$ and $-q$ reside on the two plates. The effect of the dielectric can be understood, without going into the molecular behavior, by supposing that the dielectric material is *polarizable* and that the charges on the plates distort the electric balance within the dielectric material so that it develops an opposing charge arrangement, as shown in Fig. 15-4. This opposing effect is conveniently described by introducing the term *polarization* and symbol p to represent the dipole moment induced in a unit volume of the dielectric material. Considering such a volume of the dielectric, one sees that its dipole moment p can be described as being due to charges of $+p$ on one end and $-p$ on the other. In the interior of the dielectric material between the condenser plates, such charges cancel out, and one is left with only the charges on the surfaces of the dielectric next to the condenser plates. This situation is represented in Fig. 15-4.

The manner in which the dielectric reduces the field between the condenser plates can now be understood. The field $\mathcal{E}_0$ with vacuum

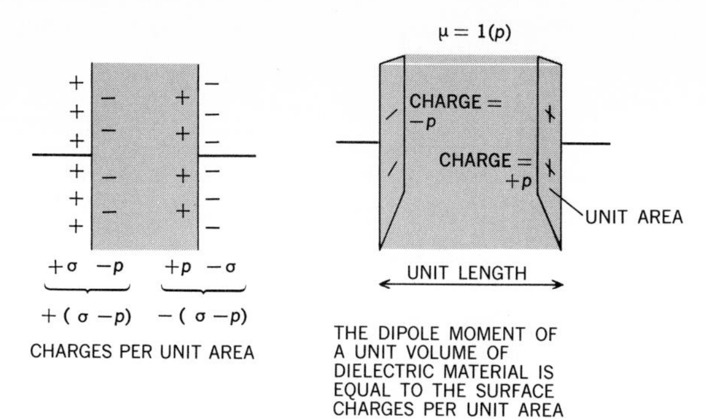

FIGURE 15-4
The electrical behavior of the dielectric material of a condenser.

between the condenser plates is related to the plate charge density by

$$\mathcal{E}_0 = \frac{\sigma}{\epsilon_0} \qquad [16]$$

When a dielectric material is present, the charge q on the plates is partially balanced by the charge p on the surface of the dielectric. This interpretation of the factors affecting $\mathcal{E}$ gives

$$\mathcal{E} = \frac{\sigma - p}{\epsilon_0} \quad \text{or} \quad \mathcal{E} = \mathcal{E}_0 - \frac{p}{\epsilon_0} \qquad [17]$$

Combination of this mechanistic interpretation of the reduction of $\mathcal{E}$ with the expressions $\mathcal{E} = \sigma/\epsilon$ and $\mathcal{E}_0 = \sigma/\epsilon_0$ leads to the desired result

$$\mathcal{E} = \frac{p}{\epsilon - \epsilon_0} \qquad [18]$$

for the electric field in a dielectric medium in terms of the polarization of the dielectric and the dielectric constant.

15-4 The Molecular Basis for Dielectric Behavior

A molecular explanation of the role of dielectric material in effecting electrical phenomena will now be given. Our attention will at first be restricted to the effect of the induced dipole moment that all molecules possess as a result of the electrical distortions of the electron distribution in a molecule by an applied electric field. All molecules are polarizable and therefore will have contributions from this factor.

Let the dipole moment induced in a molecule be denoted by μ_{ind}. If there are n molecules per unit volume of the dielectric material which has a polarization, or dipole moment per unit volume, of p, then

$$p = n\mu_{\text{ind}} \qquad [19]$$

The nature of the dielectric can be understood, therefore, in terms of the molecular property μ_{ind}.

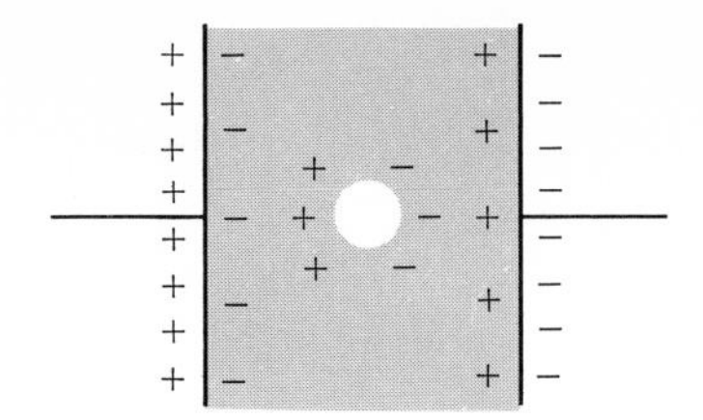

FIGURE 15-5
The charges that contribute to the field $\mathcal{E}$ acting on a molecule which is treated as being in a cavity within the dielectric material.

The simplest relation that would account for the consequences of the induced dipole moment in a molecule is one in which the induced dipole is proportional to the field acting on the molecule. If this field is $\mathcal{E}_{\text{internal}}$ and the proportionality constant is α, the relations

$$\mu_{\text{ind}} = \alpha\epsilon_0\mathcal{E}_{\text{int}} \tag{20}$$

and

$$p = n\alpha\epsilon_0\mathcal{E}_{\text{int}} \tag{21}$$

are obtained when an ϵ_0 is included explicitly rather than incorporated in α. The polarizability α is one of the molecular properties that will be deduced. It represents the ease with which the electron configuration of the molecule can be distorted by an acting electric field.

The field acting on the molecule results from several contributions. With reference to Fig. 15-5, it can be seen that the charges on the plates, the charges at boundaries of the dielectric adjacent to the plates, and the charges on the surface of a small cavity that is supposed to surround the molecule, all contribute to the field on the molecule. The net result of these terms is

$$\mathcal{E}_{\text{int}} = \frac{\sigma}{\epsilon_0} - \frac{p}{\epsilon_0} + \frac{p}{3\epsilon_0} \tag{22}$$

where the last term is the cavity-charge contribution. It is obtained by integrating, over the surface of the sphere, the effects of the surface charge in generating a field in the direction perpendicular to the condenser plates. Finally, with the help of Eq. [17], one eliminates σ and obtains

$$\mathcal{E}_{\text{int}} = \mathcal{E} + \frac{p}{3\epsilon_0} \tag{23}$$

The relation of Eq. [21] allows the not-directly-measurable quantity $\mathcal{E}_{\text{int}}$ to be eliminated and gives

$$p = n\alpha\epsilon_0\left(\mathcal{E} + \frac{p}{3\epsilon_0}\right) \tag{24}$$

Substitution for $\mathcal{E}$ by the relation of Eq. [18] then allows the elimination of $\mathcal{E}$ and p to give, on rearrangement,

$$\tfrac{1}{3}n\alpha = \frac{\epsilon/\epsilon_0 - 1}{\epsilon/\epsilon_0 + 2} \tag{25}$$

The more frequently used form of this result is obtained by writing

$$n = \frac{\rho}{M}\mathfrak{N}$$

where ρ is the density, ρ/M is the number of moles per unit volume, and $(\rho/M)\mathfrak{N}$ is the molecules per unit volume, to get

$$\mathcal{P} = \tfrac{1}{3}\mathfrak{N}\alpha = \frac{\epsilon/\epsilon_0 - 1}{\epsilon/\epsilon_0 + 2}\frac{M}{\rho} \qquad [26]$$

where $\mathcal{P}$, called the molar polarization, has been introduced for the set of quantities $\frac{1}{3}\mathfrak{N}\alpha$. The derivation has now led to an expression, known as the Clausius-Mosotti equation, that allows the calculation of either the polarizability of a molecule α or the molar distortion polarization $\mathcal{P}$ from measurements of the dielectric constant of the dielectric material. Something of the nature and importance of these molecular properties will be mentioned after the more general case, where a molecule having a permanent dipole moment, as well as being polarizable, is considered.

Consider now, as was done by Debye in 1912, the contribution to the dielectric material of any permanent dipole moment μ of the molecules of the material. With no applied field the dipoles will be oriented in all directions and will be ineffective in contributing to the polarization $\mathcal{P}$ of the dielectric. In the presence of a field, however, as Fig. 15-6 illustrates, the molecules will tend to line up with the field so that their dipole moments add to the polarization $\mathcal{P}$. The energy of the dipole varies with the angle with which it is oriented to the acting field direction according to

$$\text{Energy} = -\mu\mathcal{E}_{\text{int}}\cos\theta \qquad [27]$$

The tendency of the molecules to go to the lowest energy position by lining up with the field is opposed, however, by the thermal motions of the molecules. The distribution expression of Boltzmann gives the way in which these two factors operate. Considerable manipulation is necessary for the calculation of the average dipole moment in the direction of the field. It is possible to omit this derivation and proceed with the result shown in Eq. [34].

*According to the Boltzmann distribution law, the number of molecules dN that are lined up at an angle between θ and $\theta + d\theta$ to the field is given

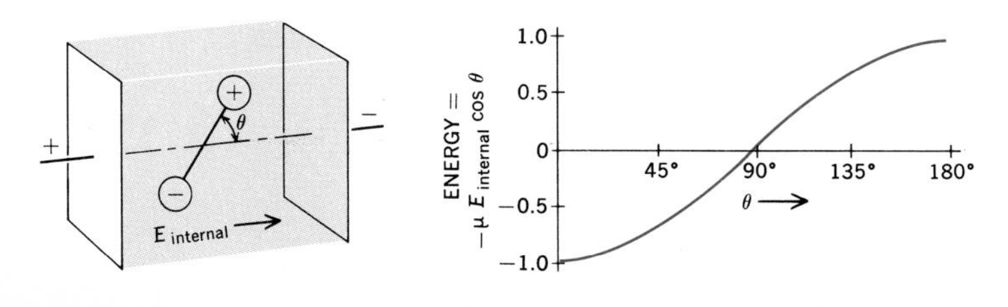

FIGURE 15-6 The energy of a dipole as a function of its orientation in an electric field $\mathcal{E}_{\text{int}}$.

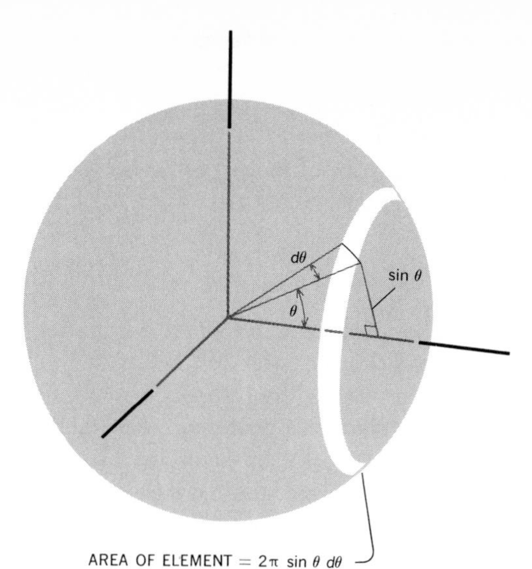

FIGURE 15-7
The solid-angle element for the integration over all molecules with various alignments with respect to the field direction. (Solid angles are equivalent to surface areas on a sphere with unit radius.)

by the expression

$$dN = Ae^{\mu \mathcal{E}_{\text{int}} \cos \theta / kT} (2\pi \sin \theta)\, d\theta \qquad [28]$$

where A is a proportionality constant and, as Fig. 15-7 illustrates, $2\pi \sin \theta \, d\theta$ is the element of solid angle inclined between θ and $\theta + d\theta$ to the field direction. Since the dipole-moment contribution of a molecule so oriented in the direction of the field is $\mu \cos \theta$, the contribution of the molecules in the solid-angle element is

$$\mu \cos \theta \, dN = Ae^{\mu \mathcal{E}_{\text{int}} \cos \theta / kT} (\mu \cos \theta) 2\pi \sin \theta \, d\theta \qquad [29]$$

The total dipole contribution in the field direction will therefore be

$$\int_0^{2\pi} Ae^{\mu \mathcal{E}_{\text{int}} \cos \theta / kT} (\mu \cos \theta) 2\pi \sin \theta \, d\theta \qquad [30]$$

To obtain the average molecular contribution, this total value is divided by the number of molecules and this number is obtained by integrating dN of Eq. [28]. The average dipole contribution μ_{av} is then expressed as

$$\mu_{\text{av}} = \frac{\int_0^{2\pi} 2\pi A \mu e^{\mu \mathcal{E}_{\text{int}} \cos \theta / kT} \cos \theta \sin \theta \, d\theta}{\int_0^{2\pi} 2\pi A e^{\mu \mathcal{E}_{\text{int}} \cos \theta / kT} \sin \theta \, d\theta}$$

$$= \mu \frac{\int_0^{2\pi} e^{\mu \mathcal{E}_{\text{int}} \cos \theta / kT} \cos \theta \sin \theta \, d\theta}{\int_0^{2\pi} e^{\mu \mathcal{E}_{\text{int}} \cos \theta / kT} \sin \theta \, d\theta} \qquad [31]$$

These integrals can be simplified by writing $x = \mu \mathcal{E}_{\text{int}} / kT$ and $y = \cos \theta$. Then, since $dy = -\sin \theta \, d\theta$, one has

$$\mu_{\text{av}} = \frac{\mu \int_{-1}^{+1} y e^{xy} \, dy}{\int_{-1}^{+1} e^{xy} \, dy} \qquad [32]$$

Both integrals are of the exponential type dealt with in Appendix 1. Substitution of the limits shown in Eq. [32] into these results gives

$$\mu_{\text{av}} = \mu\left(\frac{e^x + e^{-x}}{e^x - e^{-x}} - \frac{1}{x}\right) = \mu\left(\coth x - \frac{1}{x}\right) \quad [33]$$

The resulting function, which also occurs in studies of the magnetic effects of molecules with magnetic moments, was developed in this connection by P. Langevin, and is often referred to by his name. In the present connection, the function simplifies because x, which is the ratio of the energy effect of the dipole in the electric field to kT, is much less than unity. The first two terms of the series expansion of tanh x can then be kept to give

$$\coth x = \frac{1}{\tanh x} = \frac{1}{x - x^3/3} = \frac{1}{x}\left(1 - \frac{x^2}{3}\right)^{-1}$$

Furthermore, the binomial expansion of $(1 - x^2/3)^{-1}$ can be used to give

$$\left(1 - \frac{x^2}{3}\right)^{-1} = 1 + \frac{x^2}{3}$$

Appropriate substitutions now give the desired quantity, the average dipole moment in the direction of the field, as

$$\mu_{\text{av}} = \mu(\tfrac{1}{3}x) = \frac{1}{3}\frac{\mu^2 \mathcal{E}_{\text{int}}}{kT}$$

$$= \frac{\mu^2}{3kT}\mathcal{E}_{\text{int}} \quad [34]$$

Comparison of this result with Eq. [20] shows that the factor $\mu^2/3kT$ enters in just the way that the term α did. If the previous derivation were now carried out with the dipole-moment term as well as the polarizability term, it follows that the result, comparable with Eq. [26], would be obtained as

$$\mathcal{P} = \tfrac{1}{3}\mathfrak{N}\left(\alpha + \frac{\mu^2}{3kT}\right) = \frac{\epsilon/\epsilon_0 - 1}{\epsilon/\epsilon_0 + 2}\frac{M}{\rho} \quad [35]$$

This result is known as the *Debye equation*. It shows how the molecular polarizability α and the molecular dipole moment μ contribute to produce a dielectric constant greater than unity in any nonconducting material. Conversely, it suggests that these molecular properties might be deduced from measurement of the dielectric constant.

15-5 Determination of the Dipole Moment and the Molecular Polarizability

The dielectric constant of a material is measured by placing it between the plates of a condenser or, for liquids, filling a cell in which the plates are inserted. This condenser can then be used as one arm of an electrical bridge, like a Wheatstone bridge for measuring resistances, and the capacitance of the sample condenser can be balanced against a variable-reference condenser which has no dielectric between its plates. In principle, the capacitance of the reference condenser can be deduced from its geometry. In this way the capacitance of the sample condenser can be obtained, filled and empty, and the dielectric constant of the sample can be deduced.

When the compound of interest is a liquid or solid material, measurements are generally made on solutions of the material in some inert nonpolar substance such as CCl_4 or benzene. The Debye equation is based on the independent behavior of the polar molecules. Molecules with dipole moments exert considerable interaction on one another, and it is best, therefore, to apply the Debye equation to dilute solutions of polar compounds in nonpolar solvents. For gaseous samples the intermolecular distance is usually sufficiently great so that these interactions present no difficulty.

Measurement of the dielectric constant and use of the Debye equation do not directly lead to the separate determination of α and μ. Two principal ways are available for sorting out these two factors.

The first way consists in measuring ϵ and ρ as functions of temperature and using these data to plot $\frac{\epsilon/\epsilon_0 - 1}{\epsilon/\epsilon_0 + 2}\frac{M}{\rho}$ against $1/T$. The Debye equation leads us to expect such a plot to yield a straight line, and Fig. 15-8 shows that the data for the hydrogen halides do behave in this manner. From such plots the slope of the straight line can be used to obtain the dipole moment μ, and the intercept at $1/T$ can be made to yield the polarizability α. This procedure is quite straightforward and fails only if the molecules are associated to different extents at different temperatures or if the molecular configuration changes with temperature.

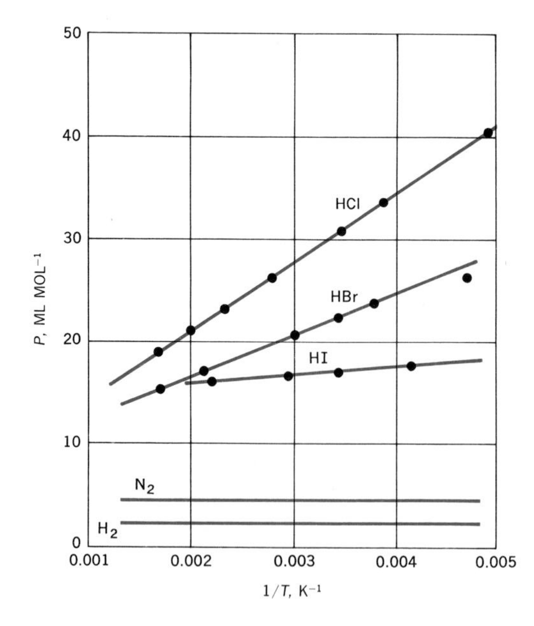

FIGURE 15-8
The molar polarization versus the reciprocal of the absolute temperature. [*From the data of C. T. Zahn, Phys. Rev.,* **24**:*400* (*1924*).]

The second procedure evaluates the polarizability part of the molar polarization $\mathcal{P}$ by an interesting relation between the dielectric constant and the refractive index given by Clerk Maxwell's theory of electromagnetic radiation. The theory cannot be dealt with here, but the basis of the relation can be suggested. The refractive index of a material is the ratio of the velocity of light in a vacuum to the velocity of light in the material. The velocity is always less in dielectric material than it is in vacuum. The slowing down is due to the interaction of the polarizable electrons of the molecules of the medium with the oscillating electric field of the radiation. The permanent dipoles of the molecules would also interfere, but the visible radiation used to measure the refractive index carries such a rapidly alternating electric field, about 10^{15} Hz, that the molecules are unable to orient themselves rapidly enough to keep up with the field. Thus only the polarizability interferes with the passage of the light.

Maxwell's theory shows that for materials composed of molecules with no permanent dipole moment,

$$\epsilon = n_R{}^2 \qquad [36]$$

where n_R is the refractive index of the material. It is apparent, therefore, that even when the molecules of the dielectric do have dipole moments, the polarizability term of the total molar polarization $\mathcal{P}$ can be calculated as

$$\frac{n_R{}^2 - 1}{n_R{}^2 + 2}\frac{M}{\rho} = \tfrac{1}{3}\mathfrak{N}\alpha \qquad [37]$$

Refractive-index data provide, therefore, a value for α that can be used along with the dielectric-constant data to give the molecular dipole moment μ.

Some results for a few simpler molecules are shown in Table 15-1. The information on molecular polarizabilities provided by these experiments finds rather less molecular-structure application than do the dipole-moment results. The data of Table 15-1 should show, however, a qualitative correlation of molecular polarizability with the number of electrons in the molecule and the "looseness" with which they are bound. The polarizability data obtained here will be of value when molecular interactions are treated in the study of liquids.

15-6 Dipole Moments and Ionic Character

In considering at first only diatomic molecules, the measured dipole moment of a molecule gives information on the displacement of the center of negative charge from that of the positive charge. In the past this asymmetry has been interpreted primarily in terms of an unequal sharing of the bonding electrons. The theoretical treatment of hetero-

TABLE 15-1 The dipole moment and the polarizability of some simple molecules

Substance	μ *(debyes)**	*Substance*	μ *(debyes)*	*Substance*	α *(ml)*
HF	1.8	CH_2Cl_2	1.60	He	0.20×10^{-24}
HCl	1.07	CH_3Cl	1.05	Ne	0.39
HBr	0.79	HCN	2.95	A	1.62
HI	0.38	CH_3NO_2	3.50	H_2	0.80
BrClCO	0.13	CH_3OH	1.71	N_2	1.73
H_2O	1.82	CsF†	7.9	H_2O	1.44
NH_3	1.47	CsCl†	10.5	H_2S	3.64
NF_3	0.23	KF†	7.3	CH_4	2.60
PH_3	0.55	KCl†	10.4	CCl_4	10.5
AsH_3	0.18	KBr†	10.5	C_6H_6	25.1
$CHCl_3$	1.94				

*1 debye = 3.338×10^{-30} m C.
† Determined by molecular-beam techniques. N. F. Ramsey, "Molecular Beams," Oxford University Press, Fair Lawn, N.J., 1956.
SOURCE: Dipole moments except as noted are from the compilation by A. L. McClellan, "Tables of Experimental Dipole Moments," H. Freeman and Co., San Francisco, 1963.

nuclear molecules in Chap. 12 anticipated such unequal sharing and introduced *percent ionic character* as a measure of such bonding.

The assumption that a dipole moment results from the location of the bonding electrons leads to a calculation of the ionic character from a dipole moment. The example of HCl, with a dipole moment of 1.07 debyes, or 3.57×10^{-30} m C, and a bond length of 1.275 Å, will illustrate the calculation. If the pair of bonding electrons were completely held by the chlorine atom, the molecule would be represented as positively and negatively charged ions separated by the equilibrium bond length. The dipole moment of such a completely ionic structure would be

$$\begin{aligned} \mu_{\text{ionic}} &= (0.1602 \times 10^{-18})(1.275 \times 10^{-10}) \\ &= 20.42 \times 10^{-30} \text{ m C} \\ &= 6.12 \text{ debyes} \end{aligned} \qquad [38]$$

On the other hand, if the pair of bonding electrons were equally shared, the bonding electrons would be almost symmetrically placed relative to the positive nuclear charges, and the dipole moment would be zero. In fact, the dipole moment is between these two extremes, and the amount of ionic character is calculated as

$$\begin{aligned} \%\text{ ionic character} &= \frac{\mu_{\text{obs}}}{\mu_{\text{ionic}}} \times 100\% \\ &= \frac{1.07}{6.12} \times 100 = 17\% \end{aligned} \qquad [39]$$

The result does not, of course, tell which end of the molecule is positive and which is negative. This aspect must be deduced, or inferred, by other means.

It is necessary to point out, however, that although this procedure for obtaining ionic characters has led to a considerable body of apparently consistent and chemically reasonable data, the method appears to have some inherent uncertainties. The problem arises because there are other electrons in the molecule besides the pair of bonding electrons on which attention has been focused. The inner-shell electrons present little difficulty since even in molecules they undoubtedly remain nearly symmetrically arranged about the nucleus to which they belong. The nonbonding electrons of chlorine in the HCl example, however, cannot be so easily dismissed. The bonding in such a molecule might result from overlap of the $1s$ wave function of the hydrogen atom, with anything from a $3p$ orbital of chlorine to a $3(sp^3)$ orbital of the chlorine. These two extremes are represented in Fig. 15-9, and it is there apparent that, if hybridized orbitals are used for bonding, the nonbonding electrons will project away from the bond and will make a large contribution to the molecular dipole. The role of such *nonbonding* electrons is not easy to evaluate, but it appears difficult to justify the procedure of ignoring them. Thus, although apparently reasonable values of ionic character are obtained from dipole-moment data, the proper interpretation of such results is open to serious question.

In spite of these uncertainties, the data obtained by the electrical measurements described here can be coupled with the electronegativity results developed in Sec. 12-3 to represent the electron-attracting powers of the atoms. The data for the percent ionic character for diatomic molecules can be obtained from measured dipole moments and bond lengths, as shown for HCl, and these data can be plotted against the electronegativity difference, as is done in Fig. 15-10. For these molecules the dipole moment is generally greater, the greater the difference in electronegativity of the bound atoms, as would be expected from a simple interpretation of both factors. Various mathematical expressions have been developed to express the rough relationship which is established by the data for diatomic molecules in Fig. 15-10. These relations find use in the deduction of the ionic character of bonds in polyatomic molecules from the electronegativity difference between the atoms of the bond. In the next section some of the difficulties in the way of a more direct deduction of the ionic character of such bonds from dipole-moment measurements will be treated.

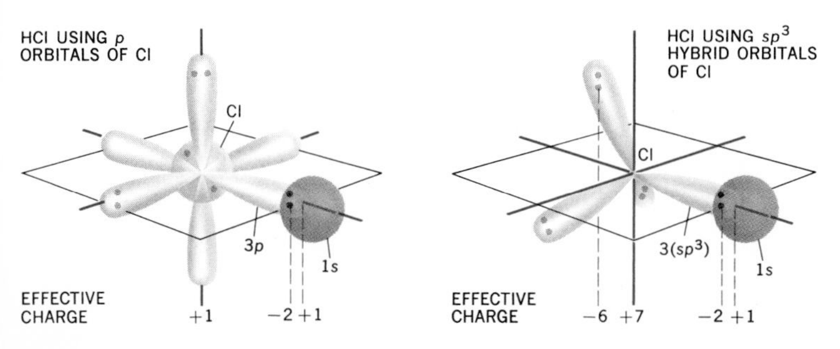

FIGURE 15-9
Descriptions of the bonding in HCl by the use of a p orbital of chlorine and by the use of an sp^3 hybrid orbital of chlorine. With hybrid orbitals the center of the nonbonding electron positions cannot be assumed to coincide with the chlorine nucleus.

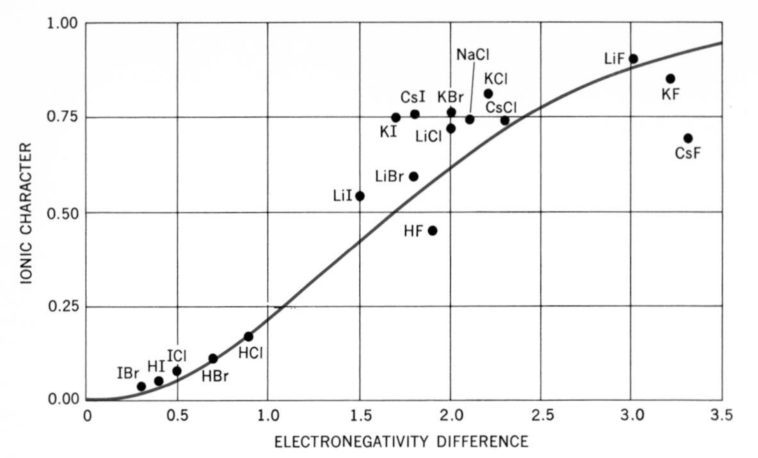

FIGURE 15-10
Curve relating the amount of ionic character of a bond to the electronegativity difference of the two atoms. Experimental points, based upon observed values of the electric dipole moment of diatomic molecules. (*From L. Pauling, "The Nature of the Chemical Bond," 3d ed., Cornell University Press, Ithaca, N.Y., 1960.*)

15-7 Bond Moments

For polyatomic molecules it is customary to try to understand the molecular dipole moment in terms of the contributions of the individual bonds of the molecule in a similar manner to that in which one tried to understand the energy of a molecule in terms of bond energies. With this approach one obtains *bond moments*. For the water molecule, for example, the measured dipole moment is 1.85 debyes, and as Fig. 15-11 shows, this quantity can be resolved into two bond moments which are in the directions of the two bonds and which add together, vectorially, to give the observed molecular moment. The procedure implies that the molecular moment arises from within the separate bonds of the molecule. The contribution of any nonbonding electrons to the total dipole moment will of course be resolved and included in the derived bond moments.

A real difficulty arises when the bond moments of a carbon compound are sought. Methane, for instance, has a zero dipole moment that results from the symmetry of the molecule rather than the electron distribution in the bonds. If such a molecule is drawn, as in Fig. 15-12, it is apparent that one C—H bond is equivalent to the other three CH bonds projecting out at tetrahedral angles from the first bond. No information on the CH bond moment can be deduced from the zero molecular moment of CH_4. In the same manner, CH_3Cl, which is nearly

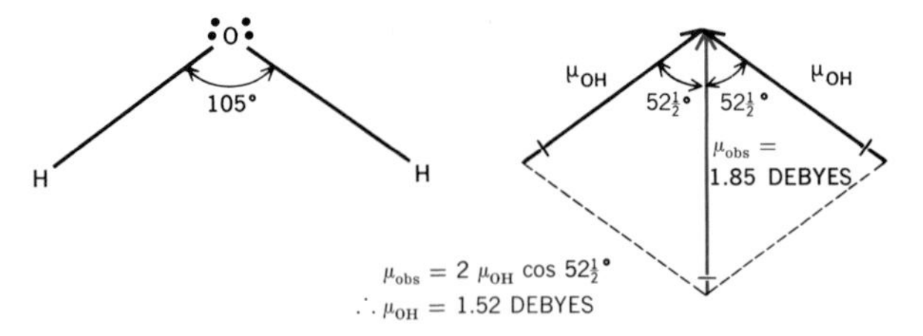

FIGURE 15-11
Resolution of the total dipole moment of the water molecule into OH bond moments. The dipole direction that indicates that H is positive relative to O is assumed.

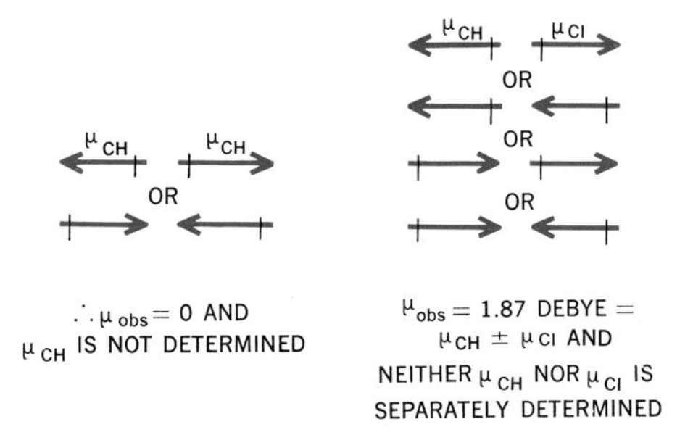

FIGURE 15-12
The problem of deducing bond moments for bonds of carbon compounds.

tetrahedral, is equivalent to opposed CH and CCl bonds. Since the CH bond moment is not known, the CCl bond moment cannot be deduced. No direct method is available by which the individual bond moments of such molecules can be deduced from molecular dipole moments.

Some rather unsatisfactory methods have suggested a CH bond moment of 0.4 debye, the hydrogen being the positive end of the dipole. On the basis of this rather arbitrary value, the bond moments of other atoms joined to a carbon atom can be deduced. Table 15-2 includes some results.

TABLE 15-2 Dipole moments deduced for some chemical bonds

Bond	*Dipole moment (debyes)*	*Bond*	*Dipole moment (debyes)*
H—F	1.9	F—Cl	0.9
H—Cl	1.0	F—Br	1.3
H—Br	0.8	Br—Cl	0.6
H—I	0.4	C—H	(0.4)
H—O	1.5	C—F	1.4
H—N	1.3	C—Cl	1.5
H—P	0.4	C—Br	1.4
P—Cl	0.8	C—I	1.2
P—Br	0.4	C—O	0.7
As—F	2.0	C—N	0.2
As—Cl	1.6	C═O	2.3
As—Br	1.3		

The principal difficulty with the bond-moment analysis is that the electron distribution that results in the molecular dipole moment cannot always be treated in terms of separate noninteracting components. Thus it is rather arbitrary to assign any contribution from the nonbonding electrons to the bonds of the molecule. Likewise, the presence of one bond in a molecule may alter the contribution of a neighboring bond. This interaction occurs through the dipole moment of one bond inducing an opposing dipole moment in a neighboring polarizable bond. In such cases the molecular moment is not understandable in terms of the individual bond-moment contributions.

MAGNETIC PROPERTIES

Magnetic measurements are a tool for molecular studies that have not been of such general applicability as have electric measurements. For certain types of compounds, however, magnetic measurements constitute one of the most powerful approaches to the elucidation of the arrangement of the electrons in the compound. The theory of magnetic studies sufficiently parallels that of electric studies so that a detailed treatment will not be given. Following some mention of the parallels between electric and magnetic phenomena, the applications of magnetic studies will be dealt with.

15-8 Determination of Magnetic Molecular Properties

The effect of magnets on one another at a distance suggests the presence of a magnetic field surrounding a magnet, just as electric fields were suggested by the effects between separated electric charges. The intensity of a magnetic field in vacuum is denoted by $\mathcal{H}$, and the field at a point is defined as the force that would be exerted on a unit magnetic pole placed at that point. The magnetic field strength is expressed in oersteds or gauss. Magnetic fields, like electric fields, are represented in magnitude and direction by lines of force. Figure 15-13 illustrates this description of the magnetic field between two magnetic poles.

When any material is placed between the poles of a magnet, the magnetic field $\mathcal{H}$ is different in the material from what it was in vacuum. The magnetic field in the material is denoted by B. Unlike the case of electric fields and dielectric materials, the magnetic field may be either increased or decreased by the presence of the material. Each of these possibilities is illustrated by the lines-of-force diagrams of Fig. 15-13.

In magnetic measurements the quantity that is obtained experi-

FIGURE 15-13 Magnetic lines of force showing the effect of diamagnetic and paramagnetic materials on the magnetic field.

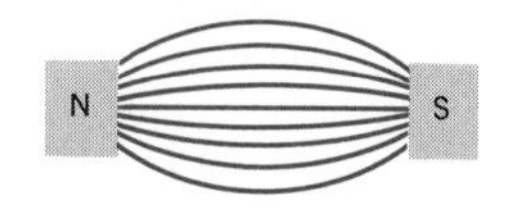

mentally is the intensity of magnetization per unit field strength, that is, $I/\mathcal{H}$. It is convenient, for chemical treatments, to multiply by the molecular volume M/ρ and to deal with the quantity

$$\chi_M = \frac{M}{\rho}\frac{I}{\mathcal{H}} \qquad [40]$$

called the *magnetic susceptibility per mole,* or the *molar magnetic susceptibility.*

The molar magnetic susceptibility, being the magnetization per mole induced by unit field strength, can be recognized as the magnetic counterpart, on a molar basis, of the sum of the electrical polarizability α and the molecular-dipole term $\mu^2/3kT$. This parallel does exist, in fact, and it is therefore convenient to deal with the corresponding terms α_M, the magnetic polarizability, and μ_M, the magnetic moment. Without treating the theory in detail, the relation of these quantities is stated as

$$\chi_M = \mathfrak{N}\left(\alpha_M + \frac{{\mu_M}^2}{3kT}\right) \qquad [41]$$

Several qualitatively different behaviors are recognized. Most organic compounds have only a magnetic-polarizability contribution, and this term acts to reduce the magnetic field in the material; that is, α_M is negative. It follows, since μ_M is zero for these materials, that χ_M and I are also negative. Materials that behave in this manner are said to be *diamagnetic.*

Other materials, which we shall see are characterized by having unpaired electrons, have magnetic moments that correspond to the dipole moments of polar molecules. The magnetic-moment term, when it exists, is almost always much larger than the polarizability term. The magnetic-moment contribution to χ_M is necessarily positive, and this is illustrated by a diagram such as that of Fig. 15-14, in which the microscopic magnets tend to line up to draw the magnetic field into the sample. Materials that behave in this manner are said to be *paramagnetic.*

There are, finally, the important classes of ferromagnetic and antiferromagnetic materials in which the magnetic properties depend on cooperative phenomena among many atoms of the sample. These materials will not be dealt with here.

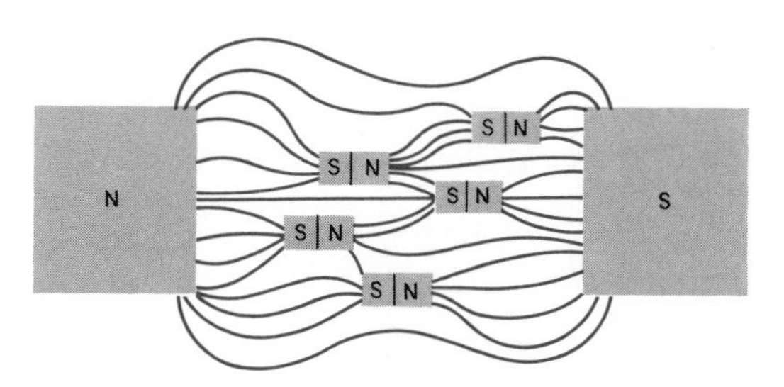

FIGURE 15-14
The microscopic explanation of paramagnetism. The magnetic moments of the sample molecules or ions tend to draw the magnetic field through the sample.

Experimentally, one frequently obtains magnetic-susceptibility data from measurements with a Gouy balance. In this method, as illustrated in Fig. 15-15, a sample is suspended from one arm of a balance in such a way that it is partly in the magnetic field. An electromagnet is ordinarily used, and when the magnet is turned on, the sample is generally repelled by or attracted into the magnetic field. The force required to maintain the position of the sample is measured by the weight that must be added or removed from the balance pan to maintain equilibrium. If the sample is paramagnetic, the magnetic moments will tend to line up with the field and the sample will have lower energy in the magnetic field and will therefore be drawn into the field. If the sample is diamagnetic, the reverse will be the case and the sample will be repelled by the field.

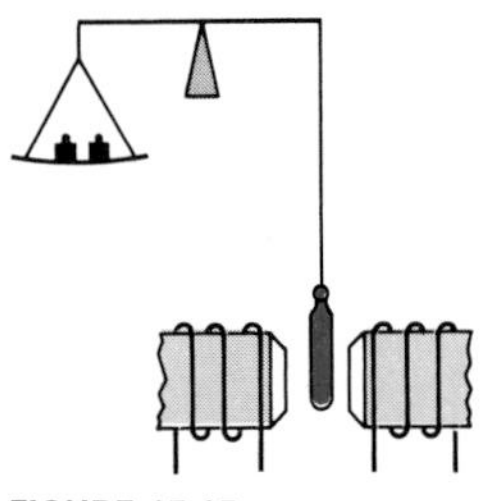
FIGURE 15-15 The Gouy balance for the measurement of magnetic susceptibilities.

The relation between the force exerted on a sample by a nonhomogeneous magnetic field and the magnetic susceptibility of the sample can be determined with reference to Fig. 15-16. It is supposed that the sample is paramagnetic, so that the magnetic field lines up the microscopic magnets of the sample. The magnetic moment per unit volume is $(\rho/M)\chi_M\mathcal{H}$ and the magnetic moment of the section of thickness dz and volume $A\,dz$ in Fig. 15-16 which experiences a magnetic field $\mathcal{H}$ is therefore

$$\frac{\rho}{M}\chi_M\mathcal{H}A\,dz \qquad [42]$$

This induced magnetic moment is in the direction of the applied magnetic field if, as supposed here, the material is paramagnetic. Since the magnetic field increases along the sample toward the center of the magnetic field, the lowering of the potential energy of successive segments will be greater as a result of the lining up of the magnetic moments of expression [42] with the greater magnetic fields. The force corresponding to this varying potential energy is the rate of change of the potential with z. If the magnetic field gradient is $d\mathcal{H}/dz$, the

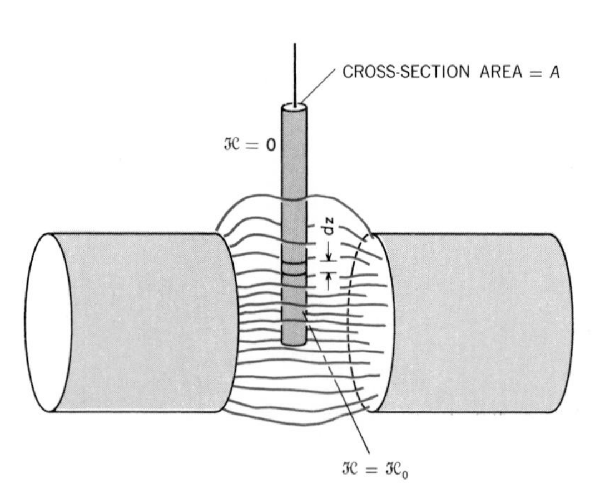

FIGURE 15-16 The effect of an inhomogeneous magnetic field on a sample.

force on a sample segment is

$$\frac{\rho}{M}\chi_M \mathcal{H} A\, dz \frac{d\mathcal{H}}{dz} = \frac{\rho}{M}\chi_M A \mathcal{H}\, d\mathcal{H} \qquad [43]$$

and if the segments of the sample extend from the center of the magnetic field, where the magnetic field value is $\mathcal{H}_0$, to outside the magnetic field, where the value is zero, the total force in the sample is

$$f = \int_0^{\mathcal{H}_0} \frac{\rho}{M}\chi_M A \mathcal{H}\, d\mathcal{H}$$

$$= \frac{1}{2}\frac{\rho}{M}\chi_M A \mathcal{H}_0^2 \qquad [44]$$

This result is the basis of the Gouy-balance method and shows that, for an experimental arrangement such as that of Fig. 15-15, the measurement of the force exerted on the sample by a known maximum field strength $\mathcal{H}_0$ can be used to deduce the magnetic susceptibility of the sample. In practice, however, one frequently compares the force on a standard sample with that on an unknown sample instead of evaluating the cross-section sample area and the magnetic field directly.

A typical experiment with a paramagnetic substance or solution might make use of a sample tube of 1-cm^2 cross section and a magnet with $\mathcal{H}_0$ of 10,000 G. For such an arrangement the force exerted by the magnetic field might be equivalent to a mass of a few tenths of a gram, and for such a value Eq. [44] indicates a susceptibility of about 10^{-4}, or, as susceptibilities are usually reported, 100×10^{-6} cgs unit per gram. Diamagnetic susceptibilities are negative and typically are smaller than this value by a factor of about 100.

15-9 Molecular Interpretation of Diamagnetism

All materials affect a magnetic field in which they are inserted as a result of an induced magnetic moment, which produces a diamagnetic effect, just as all compounds show an induced dipole moment in an electric field. This induced magnetic effect is generally of secondary importance and is almost always greatly overshadowed by the more interesting permanent-magnetic-moment contribution, when one exists. It is necessary, however, to be able to subtract the induced-moment contribution from the total magnetic susceptibility.

The diamagnetic effect is produced by the orbital motion of the electrons of the atoms, ions, or molecules of the sample. It can be understood qualitatively on the basis of Lenz's law, which states that, for an ordinary-sized system of a current flowing in a coil of wire in a magnetic field, the field will induce a current in the coil in such a way as to oppose the applied field. In a similar way, the orbital motion of the electrons is disturbed by the applied field, and the disturbance

TABLE 15-3 Pascal's constants for the determination of the diamagnetic susceptibility,* in cgs units

ATOMIC CONTRIBUTIONS			
H	-2.93×10^{-6}	Cl	-20.1×10^{-6}
C	-6.00	Br	-30.6
N (open chain)	-5.55	I	-44.6
N (ring)	-4.61	S	-15
N (monomide)	-1.54	Se	-23
O (alcohol, ether)	-4.61	B	-7
O (carbonyl)	1.72	Si	-13
O (carboxyl group)	-3.36	P	-10
F	-6.3	As	-21

GROUP CORRECTIONS	
C=C	5.5×10^{-6}
C=C—C=C	10.6
N=N	1.85
C=N	8.15
—C≡N	0.8
Benzene	-1.4
Cyclohexane	-3.0

* From P. W. Selwood, "Magnetochemistry," 2d ed., Interscience Publishers, Inc., New York, 1956.

is such that a magnetic field contribution in opposition to the applied field is produced. The diamagnetic effect is, in fact, temperature-independent and is a function of the quantum numbers of the electrons; i.e., it depends on the nature of the electron orbitals, as these ideas would suggest.

The diamagnetic effect of a given molecule can be estimated from tables that give the contribution of the various atoms within the molecule. Such terms, some of which are shown in Table 15-3, are known as *Pascal's constants*. It is necessary, for good agreement with the observed diamagnetic susceptibilities, to include also terms for any multiple or unusual bonding situation in which the atoms are involved. Some of these terms are also included in Table 15-3. For most purposes, the diamagnetic term can be satisfactorily estimated from such a table of Pascal's constants.

15-10 Molecular Interpretation of Paramagnetism

The paramagnetic effect can be most easily introduced by considering the introductory problem of the magnetic behavior of an electron revolving about a nucleus. A classical treatment is easily made and yields a result which can then be converted to the correct quantum-mechanical result. The motion of an electron in an orbit corresponds, in this connection, to the passage of a current through a coil of wire. It is known that a current in a coil of wire of ordinary dimensions produces a magnetic field perpendicular to the coil. The magnetic field so produced is equal, according to Ampère's law, to that of a magnet with magnetic moment μ_M given by the product of the current and the cross-section area of the loop of wire. For the magnetic moment produced by a current of i amperes circulating in a coil of enclosed area A to have the proper electromagnetic units, it is necessary to divide the Ampère law terms by the factor c, the velocity of light, to obtain

$$\mu_M = \frac{iA}{c} \tag{45}$$

The current corresponding to an electron in orbit is obtained by multiplying the number of times the electron passes any point on the orbit by its electronic charge. Thus, with this classical picture of an electron in an atomic orbit,

$$i = \frac{v}{2\pi r}e \tag{46}$$

where the electron velocity is v, and the orbit has radius r. The cross-section area is

$$A = \pi r^2 \tag{47}$$

Thus the magnetic moment μ_M obtained is

$$\mu_M = \frac{vre}{2c} \tag{48}$$

which can be written

$$\mu_M = \frac{mvre}{2mc} \tag{49}$$

or

$$\frac{\mu_M}{mvr} = \frac{e}{2mc} \tag{50}$$

This final form expresses the result of this classical derivation that can be carried over into quantum-mechanical systems; namely, the ratio of the magnetic moment to the orbital angular momentum is equal to $e/2mc$.

Since, as stated in Sec. 11-2, the orbital angular momentum of an electron of an atom depends on the quantum number l and is given by the expression $\sqrt{l(l+1)}(h/2\pi)$, we can express the magnetic moment due to the orbital motion of the electron as

$$\mu_M = \frac{eh}{4\pi mc}\sqrt{l(l+1)} \tag{51}$$

The constant factor in this equation provides a convenient unit in which to express the magnetic moment of atoms and molecules, and one therefore introduces the symbol μ_0, called the *Bohr magneton,* as

$$\mu_0 = \frac{eh}{4\pi mc}$$

With this unit, the orbital magnetic moment of an electron of an atom is given by

$$\mu_M = \mu_0 \sqrt{l(l + 1)} \qquad [52]$$

When a similar approach is extended to molecules and ions, rather than free atoms, it would seem reasonable to expect that the orbital motions of the electrons would contribute a magnetic moment of the order of a Bohr magneton. This expectation is not generally borne out, and it appears that the orbital motions of the electrons in a polyatomic system are tied into the nuclear configuration of the molecule or the ion so tightly that they are unable to line up with the applied magnetic field and are therefore ineffective. Even for single-atom ions in solution, the interaction of the orbitals of the ion with the solvating molecules is apparently sufficient to prevent the orbitals being oriented so that their magnetic moment contributes in the direction of the field. Thus the orbital-magnetic-moment contribution to the magnetic susceptibility is generally quite small.

We must look to the spinning of the electron about its own axis to explain the larger part of the magnetic moment of those molecules and ions which have magnetic moments. The association of a spin angular momentum of $\sqrt{s(s + 1)}(h/2\pi)$, where s has the value of $\frac{1}{2}$, leads, according to Eq. [50], to the expectation of a spin magnetic moment. Atomic spectral data require, however, a magnetic moment that is twice that expected on the basis of the ratio of the magnetic moment to angular momentum given by Eq. [50]. For the spin magnetic moment due to the electron of an atom or molecule, therefore, we have

$$\mu_M = 2\mu_0 \sqrt{S(S + 1)} \qquad [53]$$

where for one, two, three, . . . unpaired electrons, the spin-angular-momentum quantum number S is $\frac{1}{2}, \frac{2}{2}, \frac{3}{2}, \ldots$. With Eq. [53] and the assumption that the α_M contribution has been taken care of and that the orbital contribution to μ_M is negligible, the magnetic susceptibility of Eq. [41] is related to the total electron spin by the relation

$$\chi_M = \frac{4\mu_0{}^2\mathfrak{N}}{3kT}S(S + 1) \qquad [54]$$

Thus a measurement of χ_M leads to a value of S, and this value can be interpreted in terms of a number of unpaired electrons. Table 15-4

TABLE 15-4 Contributions of unpaired electrons to the magnetic susceptibility

No. of unpaired electrons	Total electron-spin quantum number S	Spin magnetic moment $(2\sqrt{S(S+1)}\mu_0)$	Magnetic susceptibility at 25°C $\chi_M = \frac{4\mu_0{}^2 \mathfrak{N}}{3kT} S(S+1)$
1	$\frac{1}{2}$	$1.73\mu_0$	$1{,}260 \times 10^{-6}$
2	$\frac{2}{2}$	$2.83\mu_0$	$3{,}360 \times 10^{-6}$
3	$\frac{3}{2}$	$3.87\mu_0$	$6{,}290 \times 10^{-6}$
4	$\frac{4}{2}$	$4.90\mu_0$	$10{,}100 \times 10^{-6}$
5	$\frac{5}{2}$	$5.92\mu_0$	$14{,}700 \times 10^{-6}$

shows the results to be expected for χ_M for various numbers of unpaired electrons at 25°C.

Equation [54] implies that, if the magnetic polarizability is not too large, the magnetic susceptibility will vary inversely as the absolute temperature; i.e.,

$$\chi_M = \frac{\text{const}}{T} \qquad [55]$$

This relation, known as *Curie's law,* is in fact found to be valid over not too large a temperature range.

15-11 Magnetic Results for Molecules

The best example of a simple molecule for which the magnetic data bear on the electronic structure is provided by oxygen. Over a considerable temperature range the molar magnetic susceptibility, again in cgs units, as have been used for reporting such results, is found to be represented by

$$\chi_M = \frac{1.00}{T} \qquad [56]$$

giving, at 25°C, the value

$$\chi_M = 3360 \times 10^{-6}$$

The values of Table 15-4 show that this susceptibility is to be interpreted as arising from two unpaired electrons. The bonding in O_2 is therefore unusual in that, although there is an even number of electrons, they are not all paired. The explanation for this, as was seen in Sec. 12-1, can be given in terms of a simple molecular-orbital treatment.

A class of compounds, consisting mostly of highly aromatic substituted systems, shows a tendency to dissociate into fragments, called *free radicals,* that have unpaired electrons. Thus hexaphenyl ethane dissociates according to

$$(C_6H_5)_3C{-}C(C_6H_5)_3 \rightleftharpoons 2\left[(C_6H_5)_3C\cdot\right]$$

Such dissociation is understandable as a result of the number of resonance structures that can be drawn for the free radical. Measurement of the susceptibility for such substances in solution allows the effective number of unpaired electrons and the degree of dissociation to the free radicals to be deduced.

15-12 Magnetic Results for Coordination Compounds

The principal application of magnetic susceptibility in chemistry at present is to the study of coordination complexes of the transition elements. In Sec. 12-9 it was shown that different reasonable bonding arrangements could sometimes be drawn for different ligands on a given metal ion. These bonding descriptions, moreover, were such that different numbers of unpaired electrons could sometimes be expected in the partially filled d orbitals. Magnetic results are of value in assigning an electronic structure to such complexes and, in some cases, in suggesting the geometry of the ligand attachment.

The complexes $Co(NH_3)_6{}^{3+}$ and $CoF_6{}^{3-}$ were used previously to illustrate different bonding descriptions. Magnetic studies of salts containing these ions in aqueous solution yield a magnetic moment of 0 Bohr magnetons for the former and 4.26 for the latter. Reference to Table 15-4 then shows that each $Co(NH_3)_6{}^{3+}$ ion can best be ascribed no unpaired electrons, and each $CoF_6{}^{3-}$ ion, four unpaired electrons. It was these data that were primarily responsible for the different bonding descriptions that were given in Sec. 12-9 to the two ions.

Problems

1 What is the dipole moment of a system of two particles, of charge $+2q$ and $-q$, separated by a distance r?

2 What is the dipole moment of a system of three charges, of $+2e$, $-e$, and $-e$, where e represents the electronic charge at the corners of an equilateral triangle with sides 2 Å long?

3 A potential of 100 V is placed across the plates of a condenser. The area of each plate is 2.4×10^{-4} m, and the distance between the plates is 0.01 m.

a Calculate the electric field between the plates of the condenser and the force that would be exerted on an electron in this region. What would be the charge density on the condenser plates?
Ans. $\mathcal{E}_0 = 10^4$ N C^{-1}; $f = 0.16 \times 10^{-14}$ N; $\epsilon = 8.8 \times 10^{-8}$ C m^{-2}.

b If the condenser is filled with CCl_4, which has $\epsilon/\epsilon_0 = 2.238$, and the charge density on the plates is that in part *a*, what would be the charge density on the carbon tetrachloride adjacent to the plates? What would be the electric field in a cavity in the CCl_4?

c What would be the force exerted on an electron in a cavity in the CCl_4?

4 The maximum voltage that can be applied to dielectric materials, with dielectric constants of about 2.4, is about 10^8 V m^{-1}. Calculate the energy difference between most favorable and most unfavorable orientations of a molecule with a dipole moment of 1 debye in the dielectric subjected to this applied potential. At 25°C what would be the relative population of the two states? Recognize from this the relative unimportance of applied electric fields in the molecular world.

5 The dielectric constant of liquid CCl_4, whose molecules are tetrahedral and have no permanent dipole moment, is 2.238 at 20°C. The density of CCl_4 is 1.595 g ml^{-1}. Calculate the molar polarization $\mathcal{P}$ and the polarizability α.

6 In the deduction of the molar polarizability of a gas at not too high a pressure, the dielectric effect of the medium is small enough so that the final term of Eq. [22], which is due to the charges on the cavity in the dielectric, can be dropped. Follow through the derivation of an expression for α and $\mathcal{P}$ for a low-pressure gas. Notice that this result is also obtained directly from Eq. [25] or [26] when the value of ϵ approaches unity.

7 The following values have been reported for the dielectric constant of BrF_5 vapor at 1 atm pressure [Rogers, Pruett, Thompson, and Speirs, *J. Am. Chem. Soc.*, **78**:44 (1956)]:

T (K)	345.6	362.6	374.9	388.9	402.4	417.2	430.8
ϵ/ϵ_0	1.006320	1.005824	1.005525	1.005180	1.004910	1.004603	1.004378

a Assuming ideal-gas behavior, calculate the molar polarization at each temperature.

b Deduce α and μ for BrF_5 from the plot of these data suggested by Eq. [35]. *Ans.* $\alpha = 7.7 \times 10^{-24}$; $\mu = 1.51$ debyes.

8 The dipole moment of CH_3Cl is 1.86 debyes, and that of $CHCl_3$ is 1.15 debyes. Both molecules are nearly tetrahedral. Assuming that C—H has a bond moment of 0.4 debye with H positive, calculate the C—Cl bond moment for each compound. The disagreement can be understood in terms of the induced dipoles in the three very polarizable chlorine atoms of $CHCl_3$.

9 The dipole moment of NH_3 is found to be 1.46 debyes, and the angle between two N—H bonds is 107°. Calculate the NH bond moment.

10 Compare several of the bond moments given in Table 15-2 with the values that would be obtained using the relation depicted in Fig. 15-10 and other data given earlier in the book.

11 The dipole moment of NF_3 has been measured to be 0.2 debye. The angle between the bonds is 102.5°. Estimate, from the data of Figs. 12-9 and 15-10 and Table 14-2, the expected N—F bond moment, and on this basis deduce a molecular dipole moment for NF_3. Consider the source of the discrepancy.

12 Calculate the ionic character of the five metal halides of Table 15-1 on the basis of point electronic charges separated by the equilibrium bond length. The bond lengths for CsF, CsCl, KF, KCl, and KBr are 2.34, 2.90, 2.55, 2.67, and 2.82 Å, respectively.

13 Compare the frequency of rotation of a representative gas-phase molecule with the frequency of visible light of about 10^{15} Hz. Recognize the implications of this comparison on the ability of a molecule to rotate so as to affect an index-of-refraction measurement.

14 Show that the units ergs per gauss, where ergs = dyne cm, for μ_0 lead to units for χ_M that give the force of Eq. [44] in dynes.

15 The ion $Co(NH_3)_6^{3+}$ has been shown from magnetic measurements to have no unpaired electrons, whereas the ion CoF_6^{3-} has four unpaired electrons. Considering only the effect of unpaired electrons, calculate the change in apparent weight of a 0.1-M solution of salts of these ions in 1-cm-diameter test tubes suspended in a Gouy balance when a magnetic field of 5000 G is turned on. *Ans.* For CoF_6^{3-}, 0.0102 g.

16 Using Table 15-3, calculate the magnetic susceptibility of benzene (C_6H_6). What would be the apparent weight loss of a sample of benzene ($d_{25} = 0.87$ g cm^{-3}) 1 cm^2 in cross section placed in a magnetic field of 3×10^4 G? (Neglect the susceptibility of the glass. Use cgs units throughout.)

17 How sensitive a balance would be needed to measure the magnetic susceptibility of oxygen gas at 1 atm and 25°C? Assume the tube has a cross-section area of 1 cm^2 and that O_2 is an ideal gas. Make use of Eq. [56].

18 Calculate loss of weight for 1 cm^2 of $O_2(l)$ at -180°C, $d = 1.14$ g ml^{-1}, and assume Eq. [56] is valid.

References

HILL, N. E., W. E. VAUGHAN, A. H. PRICE, and M. DAVIES: "Dielectric Properties and Molecular Behavior," Van Nostrand Reinhold Publishing Company, London, 1969. Theory, experiments, and interpretation of the results in terms of the electric and dynamic properties of molecules.

LEFÈVRE, R. W. J.: "Dipole Moments: Their Measurement and Application in Chemistry," Methuen & Co., Ltd., London, 1953. An introductory account of the measurement of dielectric constants and the deduction of molecular properties from the experimental data.

SMYTH, C. P.: "Dielectric Behavior and Structure," McGraw-Hill Book Company, New York, 1955. The deduction of molecular properties from dielectric-constant measurements and the nature and application of dielectric-loss measurements (not dealt with in this text). The last half of the book contains an extensive and detailed discussion of the relation of dipole moments to molecular structure.

SMITH, J. W.: "Electric Dipole Moments," Butterworth & Co. (Publishers), Ltd., London, 1955. Similar to the book by Smyth in scope and level.

PROCH, A., and G. MCCONKEY: "Topics in Chemical Physics: Lectures by P. Debye," chaps. 1 and 3, Elsevier Publishing Company, Amsterdam, 1962. Two very clearly presented treatments are "The Static Electric Field: Dielectric Constant and Polarizability of Gases" and "The Dielectric Properties of Condensed Phases." Reference can also be made to one of the earliest monographs on dipole moments, by P. Debye, entitled "Polar Molecules," Dover Publications, Inc., New York, 1928.

SCHIEBER, M. M.: "Experimental Magnetochemistry," John Wiley & Sons, Inc. (Interscience Publishers Division), New York, 1967. An outline of principles and experimental techniques followed by data, and interpretation, for many types of inorganic compounds.

GOODENOUGH, J. B.: "Magnetism and the Chemical Bond," Interscience Publishers, Inc., New York, 1963. A monograph written for those with some prior knowledge of atomic structure, quantum mechanics, and group theory, dealing with the electronic basis of the magnetic properties of metals and metal-ion systems.

PASS, G., and H. SUTCLIFFE: Measurement of Magnetic Susceptibilities and the Adoption of SI Units, *J. Chem. Educ.*, **48**:180 (1971).

16

THE NATURE OF CHEMICAL REACTIONS: RATES AND MECHANISMS

Our study of the molecular world has so far been concerned with only one of the two broad aspects which interest chemists. Chemical systems at equilibrium have been treated, and the nature of chemical compounds, which may be reactants or products in a chemical reaction, has been studied. Now the actual process of chemical reactions is investigated, and our attention is focused not only on the reactants and the products, but also on the details of the transformation from one set of chemical species to another. That this aspect has previously been neglected, or avoided, is emphasized by recognizing that the time variable has so far been absent and that it will now play a major role.

The question of how reactants are converted to products in some particular chemical reaction calls for an answer in terms of molecular-level happenings. Two aspects of the question and the answer can be recognized.

One focuses on the molecular details of what are called *elementary reactions*. Examples are the collision of two molecules to produce one or more new species, and the decay of some highly energetic molecule. Such elementary processes are considered here only in an introductory way; more detailed treatments are postponed to the following chapter.

The second aspect is the sequence of elementary reactions that constitutes the overall chemical transformation. Answers are found in a variety of experimental results, principally, measurements of the way in which the rate of the reaction depends on the concentrations of any reagents that are effective. Such experimental results are usually expressed analytically by what is known as the *rate equation,* which describes the dependence of the rate of reaction, at a given temperature, on the concentration of the reagents.

The acquisition of data on the rate of a chemical reaction and on its dependence on concentration and the formulation of the rate equation has been a major part of the area of chemistry known as *chemical kinetics.* Such studies provide insights into the sequence of molecular events in a chemical reaction—a topic of obvious interest in a chemical study of transformations. A variety of other experimental results can, and often must, be brought in to support the idea of a particular sequence of reactions.

The sequence of reaction steps for a particular overall reaction is generally known as the *mechanism* of the reaction. Although this term is also sometimes used for the details of an elementary step, it will be used here to imply a sequence of such steps.

It may seem presumptuous, and it is indeed a bold and exciting endeavor, to attempt to learn how molecules come together, how the atoms change their positions, and how the electrons shift so that ultimately new molecules emerge. Much certainly remains to be learned about such intimate details of molecular life, and the unraveling of the details of the reactions of the molecular world is as fascinating a story as is that already introduced of the molecular world in its equilibrium state.

Just as our study of the molecular nature of matter let us more easily understand many of the properties of chemical systems, so also does the study of the molecular details of reactions lead us to a better understanding of chemical reactions. Ideas arising from reaction-mechanism studies are contributing to our classification and understanding of the immense number of individual reactions that occur in organic and inorganic chemistry. It is, in fact, an understanding of reaction mechanisms, rather than of thermodynamics, that leads to most of the qualitative ideas as to what reactions might occur in both organic and inorganic systems of much complexity.

The emphasis on the relation of reaction kinetics to reaction mechanisms should not, however, completely obscure the fact that data on the rate of reactions are often of immediate practical value. It is frequently important to know, for example, under what conditions a particular reaction will proceed rapidly to give a high yield of a product while at the same time the conditions are such that a side reaction that gives an undesirable product is slowed down. Of particular importance, in this regard, is the role of catalysts. The action of catalysts in homogeneous systems will be treated here, but the very important subject of heterogeneous catalysis will be postponed until the phenomenon of adsorption is studied.

Very often, in fact most often, reactions appear not to be the result of a single molecular event, a collision, a rearrangement, and so forth, but rather the result of a series of such steps. Before we tackle the details of the elementary process, let us see how we deduce the sequences of the steps, which is called the *mechanism,* for a particular reaction. The unraveling of these steps is based primarily on the dependence of the rate of the reaction on the concentration of various reactants and products. Therefore we begin by turning our attention to the study of the rates of chemical reactions and the concentration dependence of these rates.

16-1 Measurement of the Rates of Chemical Reactions

Just as there are a great variety of chemical reactions and a correspondingly great range of reaction times, from apparently instantaneous to imperceptibly slow, so there are a great variety of ways of obtaining quantitative data on the rate of reactions.

For a reaction that is slow relative to the time it takes to remove a sample from the reaction mixture and to perform an analysis, the data from which the rate equation can be deduced can be obtained by any analytical method, physical or chemical, that is applicable to the particular system. When the reaction is a relatively rapid one, the time of sampling and analysis is appreciable and the analytical results are not then easily related to any well-defined reaction time. One procedure that is then used is to cool a sample rapidly, a step which generally slows the reaction. In other cases one can quickly dilute a sample of the reaction system. Reactions which proceed rapidly only in the presence of a catalyst can be slowed down by the removal of the catalyst. An example often encountered is an acid-catalyzed reaction that is frozen by dilution or neutralization of the acid.

Frequently more convenient than the withdrawal and analysis of samples is the use of some physical measurement which can be made on the reaction mixture and from which the concentrations of a reactant or a product can be deduced. Any physical property can be made use of as long as the property changes as the reaction mixture changes from reactants to products. Thus one might use the pressure change in a gaseous reaction proceeding in a constant-volume system, or the volume change in a constant-pressure system, or the conductivity change that accompanies many ionic solution reactions, or the optical rotation when the reactants or products are optically active, and so forth.

More satisfactory physical properties, however, are those that are more definitely related to a reactant or a product. One species in the reaction system, for example, may absorb radiation at a particular wavelength in the infrared, visible, or ultraviolet regions, and the amount of absorption at that wavelength can be used as a measure of the concentration of that species. Measurements of such quantities are less susceptible to interference from side reactions than are the physical properties previously listed.

The procedure of mixing the reagents and then measuring some physical property is limited, if a suitably rapid physical method is used, by the time required for the mixing process itself. Much progress has been made by using a flow system in which the two reagents flow together in a T arrangement and a physical property, such as electrical conductivity or ultraviolet absorption, is measured at various positions along the united stream. In this way reactions that proceed appreciably in times as short as a thousandth of a second can be studied.

Even more rapid reactions, however, interest the chemist. The reaction of an acid and a base, for example the combination of H^+ and OH^-, is often said to be instantaneous. A number of methods have been developed for studying such fast reactions. Some of these, known as *relaxation methods*, depend upon the procedure of disturbing a reaction system from equilibrium, as by the sudden imposition of an electric field or a high pressure, and determining the time required for the system to relax, i.e., return to a new equilibrium state. It appears that reactions that occur almost as rapidly as molecules vibrate, i.e., in times down to 10^{-13} s, are not beyond the realm in which the modern kineticist can work.

The way in which the data obtained from chemical or physical measurements on a reacting system at various time intervals are used to deduce the rate of the reaction and rate equation for the reaction will be illustrated by specific examples in the following sections.

16-2 Introduction to Rate Equations

The rate of a reaction is usually expressed in terms of the decrease in the amount of one of the reactants that occurs in some time interval. Alternatively, the increase in the amount of a product can be used. If the reaction system is one of constant or near-constant volume, the change in the amount of reagent will correspond to a change in the concentration of that reagent. For liquid systems one usually states the rate of a reaction in terms of the rate of change of the molar concentration of a reagent. For constant-volume gaseous systems it is generally more convenient to use the partial pressure of the reagent, which, for ideal gases, is seen from $P = (n/V)RT$ to be proportional to a concentration term. The time units that enter into the statement of the rate of a reaction may be seconds, minutes, hours, and so forth.

The rate of a reaction is always considered to be a positive quantity. Thus, for the reaction

$$A + B \rightarrow C + D \quad [1]$$

the rate, which generally is a function of concentrations and temperature, can be given in terms of the decrease in A by writing

$$\text{Rate} = -\frac{d[A]}{dt} \quad [2]$$

where $[A]$ is the molar concentration of A. Alternatively, if one measures the increase in the concentration of C, one might write

$$\text{Rate} = +\frac{d[C]}{dt} \quad [3]$$

Some care must be taken with other reactions such as those of the type

$$A + 2B \rightarrow \text{products} \quad [4]$$

Depending on whether A or B is followed experimentally, one might write the rate as

$$-\frac{d[A]}{dt} \quad \text{or} \quad -\frac{d[B]}{dt} \quad [5]$$

Since two molecules of B are used up for every molecule of A, the latter derivative will be twice as large as the former. A more satisfactory procedure is to define the rate for a reaction

$$aA + bB \rightarrow cC + dD \quad [6]$$

as

$$-\frac{1}{a}\frac{dA}{dt} = -\frac{1}{b}\frac{dB}{dt} = \frac{1}{c}\frac{dC}{dt} = \frac{1}{d}\frac{dD}{dt} \quad [7]$$

All these expressions are equal and no ambiguity arises if such terms are taken to be the rate of the reaction.

It is found that a large number of reactions have rates that, at a given temperature, are proportional to the concentration of one or two of the reactants, with each reactant raised to a small integral power. If reactions are considered in which A and B represent possible reactants, the rate equations for reactions with such concentration dependence would be of the form

$$\begin{aligned} &\text{Rate} = k[A] && \text{(first order)} \\ &\text{Rate} = k[A]^2 \quad \text{or} \quad k[A][B] && \text{(second order)} \end{aligned} \quad [8]$$

Reactions that proceed according to such simple rate equations are said to be reactions of the first or second order as indicated. As we shall see, not all reactions have such simple rate laws. Some involve concentrations raised to nonintegral powers; others consist of more elaborate algebraic expressions. There are, however, enough reactions that, at least under certain conditions, are simple first- or second-order to make the idea of the order of a reaction useful.

It should be pointed out immediately that the rate law has no necessary relation to the form of the equation for the overall reaction. The frequently given, simplified derivation for the form of the equilibrium-constant expression that depends on equating the rate of the forward reaction to the rate of the reverse reaction, each written in terms of the overall equation, is a formalism that should not be taken as implying anything about the form of the actual rate equations.

The units of the rate constants for first- and second-order reactions can be deduced from the form of Eq. [8] and the fact that the rate has units of $-d[A]/dt$. For a first-order reaction, therefore, the rate constant has the units of the reciprocal of time, i.e., reciprocal seconds, reciprocal minutes, and so forth, and is independent of the concentration units. The second-order rate constant, on the other hand, has units such as $(\text{mol liter}^{-1})^{-1}\ \text{s}^{-1}$, or $\text{liter mol}^{-1}\ \text{s}^{-1}$. The units of the rate constants for more complicated rate equations must be determined by inspection of the experimentally deduced rate equation.

Since rate constants always involve reciprocal seconds, reciprocal minutes, and so forth, a quick and approximate idea of how fast a reaction proceeds can be obtained from a reported rate constant by taking the reciprocal of the rate constant. As the units indicate, this gives a quantity that can be interpreted as the time required for the reaction to proceed appreciably when the reagents have about unit concentrations. A more quantitative interpretation will be given when individual reactions are taken up.

It is very important to realize that the order of a reaction and the rate equation are summaries of experimental results. Later, when attempts are made to devise mechanisms that are consistent with the rate equations, it will be necessary to keep in mind that mechanisms are theoretical but that the rate equation is an analytical portrayal of the experimental data.

16-3 The Fitting of Rate Data to First- and Second-order Rate Equations

A first-order reaction is one for which, at a given temperature, the rate of the reaction depends only on the first power of the concentration of a single reacting species. If the concentration of this species is represented by c (for solutions the units of moles per liter are ordinarily used) and if the volume of the system remains essentially constant during the course of the reaction, the first-order rate law can be written

$$-\frac{dc}{dt} = kc \qquad [9]$$

The rate constant k is then a positive quantity and has the units of the reciprocal of time.

The experimental results obtained in a study of the rate of a reaction are usually values of c, or some quantity related to c, at various times. Such data can best be compared with the integrated form of the first-order rate law. If the initial concentration, at time $t = 0$, is c_0 and if at some later time t the concentration has fallen to c, the integration gives

$$-\int_{c_0}^{c} \frac{dc}{c} = k \int_0^t dt$$

and

$$-\ln \frac{c}{c_0} = \ln \frac{c_0}{c} = kt$$

or

$$\log \frac{c_0}{c} = \frac{k}{2.303} t \qquad [10]$$

A sometimes more convenient form is

$$\log c = -\frac{k}{2.303} t + \log c_0 \qquad [11]$$

A reaction can therefore be said to be first-order if a plot of $\log (c_0/c)$, or of $\log c$, against t gives a straight line. If a straight line is obtained, the slope of the line can be used to give the value of the rate constant k. An alternative to this graphical procedure is the calculation of a value of k from the individual measurements of c at the various times t, from Eq. [10], for example. The reaction is classified as first-order if all the data lead to essentially the same values for k, that is, if Eq. [10] is satisfied with k a constant.

These equations can be illustrated by a reaction which is found to be first-order, under certain conditions, and whose mechanism is of some interest. The conversion of *tert*-butyl bromide to *tert*-butyl alcohol in a solvent containing 90 percent acetone and 10 percent, that is, 5 M, water has been studied by Bateman, Hughes, and Ingold. The overall reaction is

$$(CH_3)_3CBr + H_2O \rightarrow (CH_3)_3COH + HBr$$

and the reaction is slow enough so that its progress can be followed by the titration of samples for their HBr content. Some of the data

TABLE 16-1 The concentration of *tert*-butyl bromide as a function of time for the reaction $(CH_3)_3CBr + H_2O \rightarrow (CH_3)_3COH + HBr$ in a 10 percent water–90 percent acetone solvent*

At 25°C		At 50°C	
Time (h)	Conc. of $(CH_3)_3CBr$ (mol liter^{-1})	Time (min)	Conc. of $(CH_3)_3CBr$ (mol liter^{-1})
0	0.1039	0	0.1056
3.15	0.0896	9	0.0961
4.10	0.0859	18	0.0856
6.20	0.0776	27	0.0767
8.20	0.0701	40	0.0645
10.0	0.0639	54	0.0536
13.5	0.0529	72	0.0432
18.3	0.0353	105	0.0270
26.0	0.0270	135	0.0174
30.8	0.0207	180	0.0089
37.3	0.0142		
43.8	0.0101		

* From the data of L. C. Bateman, E. D. Hughes, and C. K. Ingold, *J. Chem. Soc.*, p. 960, 1940.

that were obtained are shown in Table 16-1. Figure 16-1 shows the concentration of the *tert*-butyl bromide plotted against time and also the plot of the logarithm of these concentrations versus time. The linearity of this second plot shows that, in this water-acetone system, the reaction follows first-order kinetics. The rate equation for the reaction is therefore

$$-\frac{d[(CH_3)_3CBr]}{dt} = k[(CH_3)_3CBr]$$

and the slope of the line of Fig. 16-1 leads to a value for k of 1.4×10^{-5} s^{-1}.

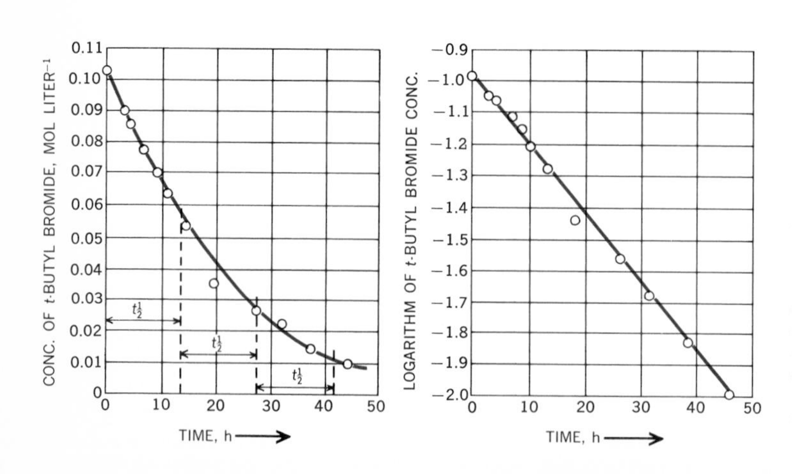

FIGURE 16-1
Graphical representation of the 25°C rate of Table 16-1 showing that, for the conditions used, the reaction $(CH_3)_3CBr + H_2O \rightarrow (CH_3)_3COH + HBr$ is first-order with respect to $(CH_3)_3CBr$. The rate constant is found to be 1.44×10^{-5} s^{-1}, and the half-life 0.48×10^5 s, or 13.4 h.

It is characteristic of first-order reactions that, as Eq. [10] shows, all that need be measured to see whether or not a reaction is first-order is the ratio of the concentrations of the reagent at various times to the concentration at some initial time. The measurement of any quantity that is proportional to the concentration of the reagent can therefore be used, and the actual concentrations need not be calculated. Thus, if some quantity α, perhaps the absorption of some wavelength of light by the reagent, is related to the concentration by the proportionality equations

$$c = (\text{const})\alpha$$

and [12]

$$c_0 = (\text{const})\alpha_0$$

then Eq. [10] becomes

$$\log \frac{\alpha_0}{\alpha} = \frac{k}{2.303} t \qquad [13]$$

From this, or directly from Eq. [11], one gets

$$\log \alpha = -\frac{k}{2.303} t + \log \alpha_0 \qquad [14]$$

A linear plot of $\log (\alpha_0/\alpha)$ or of $\log \alpha$ versus t indicates a first-order reaction. Furthermore, the slope of the straight line gives the same rate constant as would be obtained if the treatment had been in terms of concentrations.

This aspect of first-order reactions can be demonstrated by the gas-phase thermal decomposition of di-*tert*-butyl peroxide, which, in a reaction vessel packed with glass wool, proceeds predominantly according to the overall reaction

$$(CH_3)_3C{-}O{-}O{-}C(CH_3)_3 \rightarrow 2CH_3{-}\overset{\overset{\displaystyle O}{\|}}{C}{-}CH_3 + CH_3CH_3$$

In a constant-volume system the reaction proceeds with a continual increase in pressure due to the occurrence of 3 mol of gaseous products compared with 1 mol of gaseous reactant. If the initial pressure of the di-*tert*-butyl peroxide in some experiment is P_0, the final pressure, when the reaction is complete, will be $3P_0$. A net pressure increase of $2P_0$ occurs, therefore, for the complete reaction. The increase in pressure that occurs from any time t until the end of the reaction is proportional to the amount of di-*tert*-butyl peroxide existing, i.e., that has not yet reacted, at time t. If the pressure of the system at time t is P, the proportionalities that can be set up are

$$c_0 \propto (P_{\text{final}} - P_0) = 3P_0 - P_0 = 2P_0$$

and [15]

$$c \propto (P_{\text{final}} - P) = 3P_0 - P$$

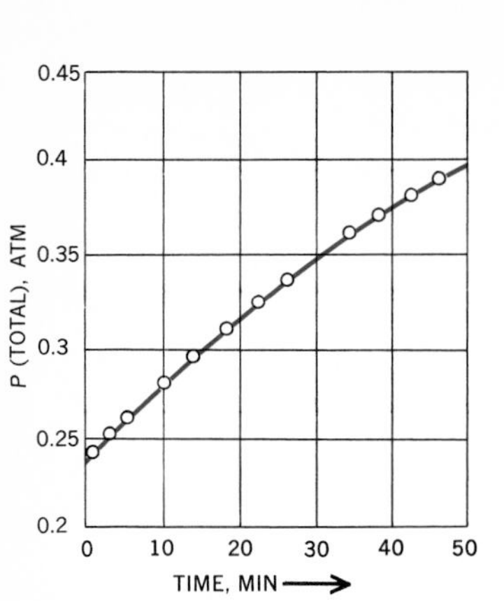

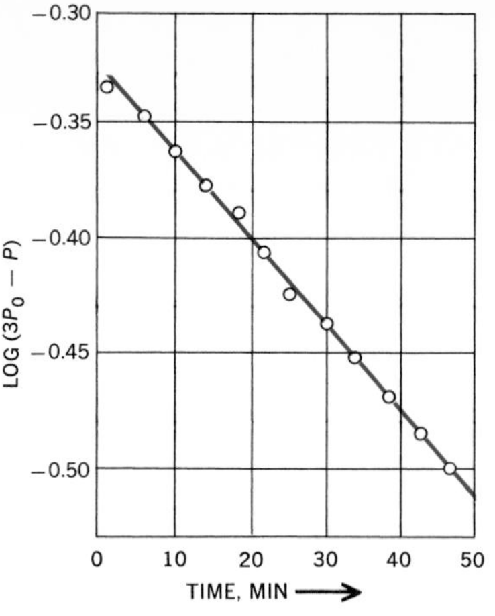

FIGURE 16-2
Graphical representation of the *tert*-butyl peroxide decomposition date of Table 16-2 at 147.2°C. The first-order rate constant is 1.43×10^{-4} s^{-1}.

where c_0 and c are the concentrations of di-*tert*-butyl peroxide at time 0 and t. The plot that tests whether or not the reaction is first-order can be based on Eq. [14], which in this case becomes

$$\log (3P_0 - P) = -\frac{k}{2.303}t + \log 2P_0 \qquad [16]$$

Thus one makes a plot of $\log (3P_0 - P)$ versus t. Figure 16-2 indicates that the data of Table 16-2, which have been calculated from the work of Raley, Rust, and Vaughan, do yield a linear plot. The decomposition, under these reaction conditions, is first-order, and this conclusion is reached without calculation or measurement of the actual concentration of the reacting species.

The use of a quantity such as the total pressure to follow such a reaction is subject to disturbance by any of the side reactions that might occur. A more direct analysis for the reagents is therefore much to be preferred. The data of Table 16-2 have, in fact, been calculated merely to illustrate that the total pressure could be used, from the original mass-spectrometric analyses of Raley, Rust, and Vaughan.

For first-order reactions it is customary to use not only the rate constant k for the reaction but also the related quantity, the *half-life* of the reaction. The half-life is the time required for the concentration, or amount, of the reagent to decrease to half its initial value. For a first-order reaction the relation of the half-life, denoted by $t_{1/2}$, to the rate constant can be found from Eq. [10] by inserting the requirement that at $t = t_{1/2}$ the concentration is $c = \frac{1}{2}c_0$. In this way one obtains

$$\log \frac{c_0}{\frac{1}{2}c_0} = \frac{k}{2.303}t_{1/2}$$

or

$$t_{1/2} = \frac{0.693}{k} \qquad [17]$$

This result shows that, for a first-order reaction, there is a simple reciprocal relation between k and $t_{1/2}$. Furthermore, since the expression involves no term for the concentration, or amount, of material, the time required for half the reactant to be used up is independent of the initial concentration, or amount, of the reactant. This can be seen graphically in Fig. 16-1, where dashed lines have been drawn to show that the time intervals for the amount of *tert*-butyl bromide to decrease to half its value are all equal. It is the simple relation of Eq. [17] that makes the half-life a useful quantity for first-order reactions. Higher-order reactions, as we shall see, have a half-life that is a function of the initial concentration as well as of the rate constant, and the concept of half-life is then of little value.

A type of reaction which is in some ways outside the realm of chemistry but which conforms beautifully to first-order kinetics is that of radioactive decay. It is found that the rate with which a radioactive species decays is proportional to the amount of that species. The decay is therefore first-order, and one can, as is invariably done, use a half-life to characterize the decay rate.

TABLE 16-2 The total pressure* of the gas-phase system in which di-*tert*-butyl peroxide is decomposing, predominantly by the reaction

$$(CH_3)_3C—O—O—C(CH_3)_3 \rightarrow 2(CH_3—\overset{\overset{\displaystyle O}{\|}}{C}—CH_3) + C_2H_6$$

At 147.2°C		At 154.9°C	
Time (min)	Total pressure (atm)	Time (min)	Total pressure (atm)
0	0.2362	0	0.2227
2	0.2466	2	0.2409
6	0.2613	3	0.2489
10	0.2770	5	0.2646
14	0.2910	6	0.2725
18	0.3051	8	0.2877
22	0.3188	9	0.2952
26	0.3322	11	0.3100
30	0.3448	12	0.3160
34	0.3569	14	0.3293
38	0.3686	15	0.3358
42	0.3801	17	0.3480
46	0.3909	18	0.3548
		20	0.3655
		21	0.3744

*Calculated from the results of J. R. Raley, R. F. Rust, and W. E. Vaughan, *J. Am. Chem. Soc.*, **70**:88 (1948).

A reaction is classified as second-order if the rate of the reaction is proportional to the square of the concentration of one of the reagents or to the product of the concentrations of two species of the reagents. The second situation leads to the same equations as the first if the two reactants are used up at the same rate and if their initial concentrations are equal. For these situations, the rate law is

$$-\frac{dc}{dt} = kc^2 \quad [18]$$

where c is the concentration of the single reagent or of one of the two reagents. Again the kinetic data are usually compared with the integrated form of the equation. One has

$$-\int_{c_0}^{c} \frac{dc}{c^2} = k \int_0^t dt \quad [19]$$

and

$$\frac{1}{c} - \frac{1}{c_0} = kt \quad [20]$$

A reaction, of the types considered so far, is second-order, therefore, if a plot of $1/c$ versus t gives a straight line. The slope of the straight line is equal to the rate constant. As Eq. [20] shows, this constant involves the units of concentration and, in this respect, differs from the first-order rate constant that involves only the units of time. Furthermore, the time for the concentration to drop to half its initial value is deduced from Eq. [20] to be

$$t_{1/2} = \frac{1}{kc_0} \quad [21]$$

The half-life depends, therefore, on the initial concentration and is not a convenient way of expressing the rate constant of second-order reactions.

Experimental rate data can be tested for conformity to a second-order rate law in a number of ways. As an illustration, we will consider a case where, instead of working with the concentration of the reacting species as would be done in using Eq. [20], it is more convenient to develop the rate equation by introducing a term for the amount of reaction that has occurred at time t. The overall reaction might, for example, be of the form

$$A + B \rightarrow \text{products}$$

If it is inconvenient to arrange to have the initial concentrations of A and B equal, the analysis that led to Eq. [20] cannot be used. The kinetic data can, however, be treated in terms of the following quantities:

$$
\begin{aligned}
a &= \text{initial concentration of } A \\
b &= \text{initial concentration of } B \\
x &= \text{decrease in } A \text{ or } B \text{ at time } t \\
&= \text{amount of product at time } t \\
a - x &= \text{concentration of } A \text{ at time } t \\
b - x &= \text{concentration of } B \text{ at time } t
\end{aligned}
$$

The differential second-order rate equation would then be

$$
\begin{aligned}
\frac{dx}{dt} &= k[A][B] \\
&= k(a - x)(b - x) \qquad [22]
\end{aligned}
$$

The integration can be performed by using partial fractions. Thus

$$\frac{dx}{(a - x)(b - x)} = k\,dt$$

and

$$\frac{1}{a - b}\int_0^x \left(-\frac{dx}{a - x} + \frac{dx}{b - x}\right) = k\int_0^t dt \qquad [23]$$

which, on integration, gives

$$\frac{1}{a - b}\left[\ln(a - x) - \ln(b - x)\right]_0^x = kt \qquad [24]$$

Insertion of the limits and rearrangement gives, finally,

$$\frac{1}{a - b}\ln\frac{b(a - x)}{a(b - x)} = kt \qquad [25]$$

The data obtained by Dostrovsky and Hughes for the reaction of isobutyl bromide and sodium ethoxide are shown in Table 16-3. Such data can be compared with the second-order rate equation by calculating the values of k for the various times, to get the values listed in Table 16-3, or by making the appropriate plot, as in Fig. 16-3, and observing the linearity of the data. Either test shows that the reaction under these conditions is second-order and has the rate equation

$$
\begin{aligned}
-\frac{d[C_4H_9Br]}{dt} &= -\frac{d[NaOEt]}{dt} \\
&= 5.5 \times 10^{-3}[C_4H_9Br][NaOEt] \qquad [26]
\end{aligned}
$$

where the concentrations are expressed in moles per liter and the time in seconds.

A similar treatment involving the amount of reaction that has occurred at time t can be applied if, for example, the overall reaction follows an equation of the form

$$A + 2B \rightarrow \text{products}$$

The integrated rate equation corresponds to, but is different from, Eq. [25].

TABLE 16-3 The concentrations of isobutyl bromide and sodium ethoxide in ethyl alcohol at 95.15°C*

t (min)	$b - x$ = mol liter^{-1} of C_4H_9Br	$a - x$ = mol liter^{-1} of NaOEt	x = decrease in conc. of C_4H_9Br or NaOEt	k (from Eq. [25])
0	0.0505	0.0762	0.0030	5.6
2.5	0.0475	0.0732	0.0059	5.6
5	0.0446	0.0703	0.0086	5.8
7.5	0.0419	0.0676	0.0107	5.6
10	0.0398	0.0655	0.0135	5.8
13	0.0370	0.0627	0.0166	5.8
17	0.0340	0.0596	0.0182	5.7
20	0.0322	0.0580	0.0230	5.4
30	0.0275	0.0532	0.0277	5.5
40	0.0228	0.0485	0.0311	5.6
50	0.0193	0.0451	0.0335	5.5
60	0.0169	0.0427	0.0355	5.5
70	0.0150	0.0407	0.0386	5.4
90	0.0119	0.0376	0.0421	5.4
120	0.0084	0.0341		

* From the data of I. Dostrovsky and E. D. Hughes, *J. Chem. Soc.*, p. 157, 1946.

16-4 Enzyme Kinetics

Many rate equations that are more complex than the first- and second-order equations of Sec. 16-3 are encountered in chemical-rate studies. But such rate equations can also be illustrated by considering reactions that occur in biological systems, or at least are affected by enzymes that occur in such systems.

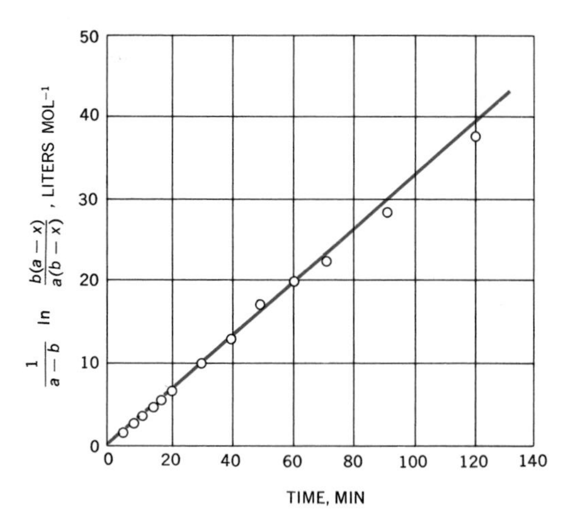

FIGURE 16-3
Plot of Eq. [25] for the data of Table 16-3. The slope of the line gives k as 0.33 liter mol^{-1} min^{-1} or 5.5 × 10^{-3} liter mol^{-1} s^{-1}.

The effect of enzymes on the rate with which chemical reactions move toward their equilibrium position provides one of the most dramatic catalytic effects. Much of the current interest in the subject is centered on the details of the interaction between the *enzyme,* which is the catalyst, and the material, known as the *substrate,* whose reaction it affects. But it is important also to understand the way in which an enzyme-catalyzed reaction proceeds in time and the way, from the measurement of the progress of such reactions, the catalytic activity of the enzyme-substrate pair is evaluated.

The experimental data for enzyme-catalyzed reactions show a variety of forms that depend on the enzyme, the substrate, the temperature, the presence of interfering substances, and so forth. Many of the behaviors that are found can be looked on as variations from the "ideal" curve of Fig. 16-4. It is such rate curves for which we will now develop a rate equation in a form that is conveniently related to the quantities that are measured in enzyme studies.

Inspection of the curves of Fig. 16-4 shows that the rate of the reaction, at high substrate concentrations, is independent of the substrate concentration and is proportional to the total amount of enzyme $[E_t]$ in the system. Furthermore, at low substrate concentrations the rate, as shown by the initial straight-line sections of the curves of Fig. 16-4, is proportional to the enzyme concentration and, as shown by the slopes of these lines, proportional also to the total enzyme concentration. These features can be accounted for by a rate equation, where R denotes the rate of the reaction of the form

$$R = \frac{(\text{const})\,[E_t][S]}{(\text{const}') + [S]}$$

To anticipate the notation that will be introduced when the mechanism of enzyme-catalyzed reactions are dealt with, we will introduce the symbols k_2 and K_M for the two constants and thus write the rate equation in the form

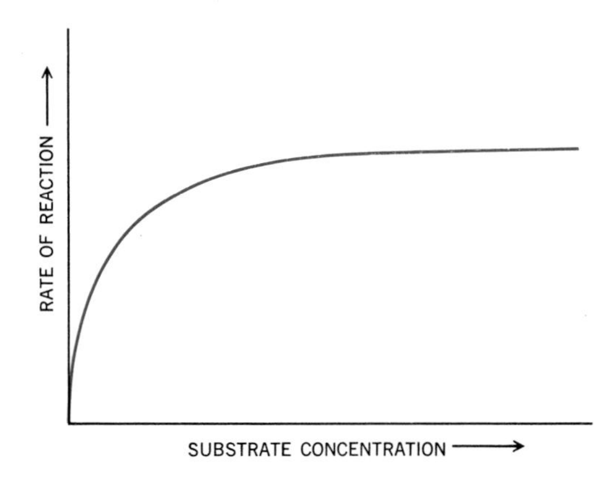

FIGURE 16-4
The form of the rate–versus–substrate-concentration curve for an "ideal" enzyme-catalyzed reaction.

$$R = \frac{k_2[E_T][S]}{K_M + [S]} \qquad [27]$$

Although the parameters k_2 and K_M could be determined so that a function corresponding to the experimental curve on an R versus $[S]$ plot is obtained, a more convenient procedure can be found.

*First, the reciprocal of Eq. [27] is formed as

$$\frac{1}{R} = \frac{1}{k_2[E_T]} + \frac{K_M}{k_2[E_T][S]} \qquad [28]$$

From this form one sees that a plot of $1/R$ versus $1/[S]$ for a fixed amount of enzyme should yield a straight line and that the slope and the intercept of the line will allow K_M and $k_2[E_T]$ to be evaluated. The constant k_2 itself is then obtained if the amount of enzyme is known.

If, however, the available data consist of values for the amount of product formed, or substrate remaining, after various reaction times, an integrated form of Eq. [27] must be used. We first write R as $-d[S]/dt$ and then rearrange Eq. [27] to

$$\left(\frac{K_M}{[S]} + 1\right) d[S] = -k_2[E_T]\,dt \qquad [29]$$

Integration for the range $t = 0$ to t and $[S] = [S_0]$ to $[S]$ leads to

$$[S_0] - [S] + K_M \ln\frac{[S_0]}{[S]} = k_2[E_T]t \qquad [30]$$

This result is difficult to work with except for the initial stages of a reaction where $([S_0] - [S])/[S_0] \ll 1$; then

$$\ln\frac{[S_0]}{[S]} = -\ln\frac{[S]}{[S_0]} = -\ln\left(1 - \frac{[S_0] - [S]}{[S_0]}\right)$$

and since $\ln(1 - x) = -x$ for $x \ll 1$,

$$\ln\frac{[S_0]}{[S]} = \frac{[S_0] - [S]}{[S_0]}$$

Insertion of this result in Eq. [30] gives

$$[S_0] - [S] = \frac{k_2[E_T][S_0]}{K_M + [S_0]}t \qquad [31]$$

Thus the decrease in substrate is seen to be linear with time; for the initial stage of the reaction, the slope $([S_0] - [S])/t$ gives a value for $k_2[E_T][S_0]/(K_M + [S_0])$ for that initial substrate concentration. Measurements of the decrease in substrate concentrations for some time interval at which the reaction is in its initial stages for other values of $[S_0]$ allow values of $k_2[E_T][S_0]/(K_M + [S_0])$ to be deduced for these $[S_0]$ values. Finally, $k_2[E_T]$ and K_M can be extracted, as by forming the reciprocal and plotting, as in the treatment of rate values suggested by Eq. [28].

THE BASIS FOR THE INTERPRETATION OF RATE LAWS IN TERMS OF REACTION MECHANISMS: AN INTRODUCTION TO ELEMENTARY REACTIONS

Mechanism of reactions, i.e., sequences of elementary reactions, can be postulated to explain the form of the rate laws that are deduced from experimental rate data. As a basis for this mechanistic interpretation of reactions, some features of the single-step, elementary reactions must be treated here. More complete probing of the nature of elementary reactions, however, will be postponed until the following chapter.

Both the available experimental approaches and the molecular-level descriptions of elementary processes are so different for the gas phase and the liquid phase that attention will be paid separately to such systems. (Reactions do occur in solids, but these have generally been little noticed by chemists.)

16-5 Introduction to Elementary Processes in the Gas Phase

The molecular model, introduced originally in our kinetic-molecular study of gas behavior, allowed for no molecular transformations and, moreover, insisted that collisions of molecules were "elastic," i.e., that the translational energy of the colliding species was the same after as before the collision. The model must clearly be extended to include *reactive* as well as *elastic* collisions.

The molecular phenomena that constitute elementary reactions can be placed in two or three groups, as shown in Fig. 16-5. In the first group are molecular collisions, which, if they now are to constitute reactions between the colliding species, can be called *reactive collisions*.

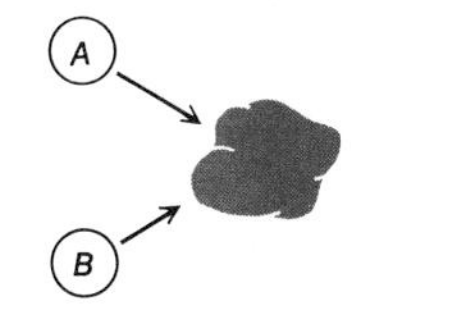

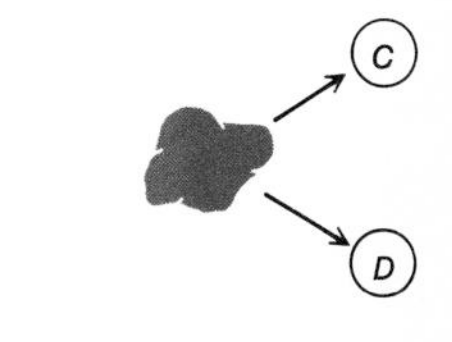

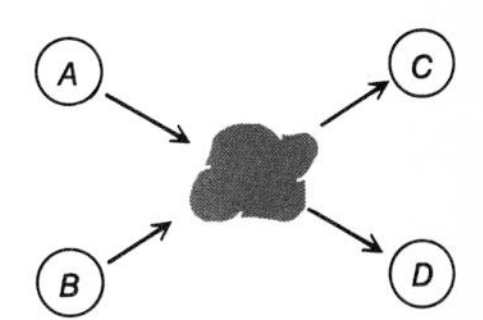

FIGURE 16-5
Types of elementary gas-phase reactions.

The second type of elementary reaction event that can be recognized is the *decay* of a molecular species, the unstable species often being the result of a prior reactive collision.

A third category of such molecular events can be recognized. Here we could place those collisions that lead "instantaneously" to the decay of the collision complex, a process that can be described as a *concerted collision decay*.

Let us now look a little closer at reactive collisions, at the types of species that are formed in such collisions, and at the breakup or decay of such species. This will provide a basis for the study of sequences of elementary reaction steps.

16-6 The Rate of Gas-phase Collisions between Molecules of Different Types

A reaction step in which two different gas-phase molecular types come together and produce one or more new species depends on collisions between the two types of molecules. Here let us see on what the rate of such collisions depends. The collision-number expressions deduced in Sec. 2-8 could again be used, but instead let us develop an expression that will also be useful when a more detailed look at reactive molecular collisions is taken in the next chapter.

Consider a single molecule of A moving through a gas that contains N_B^* molecules of B per unit volume and, to begin with, assume the B molecules are at rest. As shown in Fig. 16-6, the *cross section* for a collision between an A and a B molecule, σ_{AB}, is related to the radii of A and B, if they are assumed to be spherical, by

$$\sigma_{AB} = \pi(r_A + r_B)^2 \qquad [32]$$

The volume in which the A molecule can suffer collisions with B molecules in one second is that swept out by the A molecule, as shown in Fig. 16-6. If the A molecule travels with velocity u_A, the number of collisions by the A molecule with B molecules per second is $N_B^* u_A \sigma_{AB}$.

The probability, or fraction, of A molecules having a velocity in the range u to $u + du$ is given by the Maxwell-Boltzmann distribution, Eq. [31] of Chap. 2,

$$\frac{dN_A}{N_A} = 4\pi \left(\frac{m_A}{2\pi kT}\right)^{3/2} u_A{}^2 e^{-(1/2)mu_A{}^2/kT}\, du_A \qquad [33]$$

Thus, if there are N_A^* molecules of A per unit volume, the number with speeds in the range u and $u + du$ will be

$$N_A^* \left[4\pi \left(\frac{m_A}{2\pi kT}\right)^{3/2} u_A{}^2 e^{-(1/2)mu_A{}^2/kT}\, du_A\right] \qquad [34]$$

The product of the number of collisions an A molecule makes with

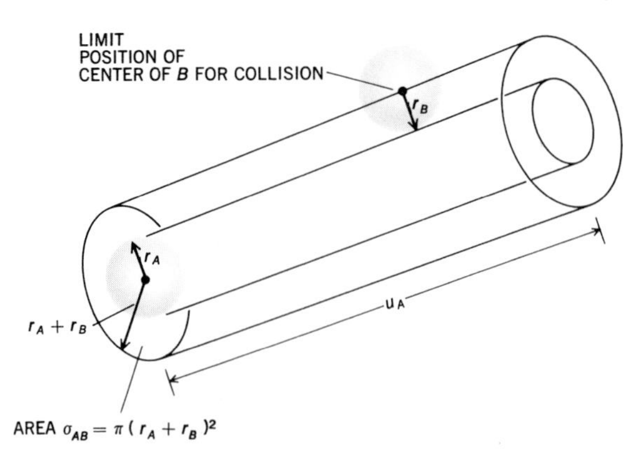

FIGURE 16-6
The collision cross section σ_{AB}. Molecule A will collide with any B molecule whose center is within the volume $u_A\sigma_{AB}$ swept out by the cross-section area σ_{AB}.

B molecules, $N_B^* u_A \sigma_{AB}$, and the number of A molecules per unit volume with a speed in the range u_A to $u_A + du_A$, Eq. [34], gives the number of $A \ldots B$ collisions per unit volume involving A molecules with speeds in the u_A to $u_A + du_A$ range. This result is

$$N_A^* N_B^* (4\pi) \left(\frac{m_A}{2\pi kT}\right)^{3/2} \sigma_{AB}\, u_A{}^3 e^{-(1/2)mu_A{}^2/kT} \qquad [35]$$

This A-focused result will be satisfactory if A molecules are very light and fast-moving relative to B molecules. But in general, we must allow for motions of both A and B. To do this, we must transform to an expression involving the *relative velocity u*, which would be the velocity that an observer on B would attribute to A, or vice versa. When this transformation is carried out, the above expression maintains its form if u_A is replaced by u and m_A is replaced by the reduced mass $\mu = [m_A m_B/(m_A + m_B)]$. We then have, for the number of collisions per unit volume per second that occur with relative velocities in the range u to $u + du$, the expression

$$N_A^* N_B^* (4\pi) \left(\frac{\mu}{2\pi kT}\right)^{3/2} \sigma_{AB}\, u^3 e^{-(1/2)mu^2/kT}\, du \qquad [36]$$

The total number of collisions per unit volume per second, Z_{AB}, is given by

$$Z_{AB} = N_A^* N_B^* (4\pi) \left(\frac{\mu}{2\pi kT}\right)^{3/2} \int_0^\infty \sigma_{AB}\, u^3 e^{-(1/2)mu^2/kT}\, du \qquad [37]$$

If, further, we assume that σ_{AB} is independent of the relative velocity with which the molecules collide, an assumption that will be removed in Sec. 17-3, the definite integrals of Appendix 1 lead to the result

$$Z_{AB} = 2\sigma_{AB} \sqrt{\frac{2kT}{\pi\mu}}\, N_A^* N_B^* \qquad [38]$$

With the substitution $\sigma_{AB} = \pi(d_A + d_B)^2/4$, this result can be easily seen to be related to the collision expressions of Sec. 2-8.

In nonreacting systems this result gives us the number of $A \ldots B$ collisions per unit volume per second. In reacting systems we can expect that only some fraction of these collisions will be effective in producing an elementary reaction step. Even with this limited claim, we do have the important consequence that the number of reactions per unit volume per second should be proportional to the product of the number densities of the species reacting in the elementary collision step. Thus, for a reaction which may constitute a complete chemical transformation, or more likely one step of the transformation, we have the important relation

$$\text{Rate of elementary } A \ldots B \text{ reaction} \propto N_A^* N_B^* \propto [A][B] \qquad [39]$$

It is also informative to calculate a representative value for the maximum rate of a gas-phase collision-produced reaction. If we take, for both the A and B molecules, the reasonable values for small molecules,

$$M = 50$$
$$d = 3\ \text{Å} = 3 \times 10^{-10}\ \text{m}$$

we obtain

$$Z_{AB} = 1.4 \times 10^{-16} N_A^* N_B^* \qquad \text{collision s}^{-1}\ \text{m}^{-3} \qquad [40]$$

If each collision were to lead to a reaction that could be described as

$$A + B \rightarrow AB$$

and if we take a system with $N_A^* = N_B^*$ so that, at all times, $N_A^* = N_B^* = N^*$, we can write for the decrease in the number of A or B molecules

$$-\frac{dN^*}{dt} = 1.4 \times 10^{-16}(N^*)^2 \qquad \text{s}^{-1} \qquad [41]$$

To be specific, let us ask how long, on this basis, it will take for half of an initial amount N_0^* of A and B to be used up. We write

$$-\int_{N_0^*}^{(1/2)N_0^*} \frac{dN^*}{(N^*)^2} = \int_0^{t_{1/2}} 1.4 \times 10^{-16}\, dt$$

or

$$\left.\frac{1}{N^*}\right]_{N_0^*}^{(1/2)N_0^*} = 1.4 \times 10^{-16} t \Big]_0^{t_{1/2}}$$

or

$$t_{1/2} = \frac{1}{1.4 \times 10^{-16}(\frac{1}{2}N_0^*)} \qquad [42]$$

For gases at 1 atm pressure and 25°C,

$$N^* = \frac{6 \times 10^{23}}{0.0224(298/273)} = 2.4 \times 10^{25}\ \text{molecules m}^{-3} \qquad [43]$$

and the time for a half reaction would be

$$t_{1/2} = 0.6 \times 10^{-9}\ \text{s} \qquad [44]$$

Thus we see that if each collision were to be effective in leading to a reaction, the reaction would proceed with explosive rapidity. Most reactions that are studied are, in fact, slower by many powers of 10, and the above calculations represent only an upper limit on the rate of an elementary gas-phase reaction.

16-7 Nature, Lifetime, and Reactions of Gas-phase Free-radical Intermediates

The sequence of elementary reactions that constitute a complete gas-phase reaction often involve intermediate species that are *free radicals,* a term implying a molecular species with an odd, unpaired electron. Although such species are important in some types of solution reactions—photochemical and polymerization reactions—and although there are some quite stable free radicals—triphenyl methane, for example—it is as reactive gas-phase intermediates that free radicals find their most varied roles.

Free radicals can be generated by a variety of means. Absorption of radiation by a molecule leads, for sufficiently high-energy quanta, to the homolytic cleavage of a bond of the absorbing molecule. The products of this primary step of a *photolysis* are thus free radicals. Less selective is the gas-phase *pyrolysis* reaction in which a high temperature is used to produce a similar bond dissociation. The dissociation energy of different bonds, as Table 16-4 shows, varies considerably, and the temperatures necessary for thermal dissociation are likewise varied. The high temperatures of flames lead to considerable free-radical formation, and spectral studies reveal the presence of such species as C_2, CH, OH, NH, and so forth.

Free radicals are generally intermediates and react further to produce the final reaction products. Like other intermediates, however, they can be treated as well-defined molecular species if they have a lifetime long enough so that their identity is established. A species can be treated as a molecular entity if it resists further breakup or fails to undergo further reaction for a time that allows it to exhibit its vibrational character. A typical vibrational frequency has a period of about 10^{-13} s, and at 1 atm pressure and 25°C, as shown in Sec. 2-10, there is an average time of about 10^{-9} s between molecular collisions. Thus even a very reactive intermediate can undergo some 10,000 or so vibrations before it suffers a collision and possible reaction.

It is this basis—the appreciable time available for the energy-rich or newly formed complex to establish its identity—that justifies the treatment of the reaction in terms of the formation, the nature, and the breakup of an elusive, energy-rich species. This species, moreover, can be expected to be longer-lived and better-defined if it is large and complex. Then it is less likely that the energy will concentrate in a vibrational mode that could lead to disruption of the molecule even without a reactive collision.

The final products that emerge from a free-radical reaction depend on the way in which the free-radical intermediates react and produce stable molecules. Several of the types of reactions most frequently attributed to free radicals are listed in Table 16-5. The first two are steps that do not eliminate the free-radical center and do not therefore

TABLE 16-4 Energies for homolytic bond cleavage, in kJ mol^{-1}*

Bond	Energy
CH_3—H	435
C_2H_5—H	410
$(CH_3)_3C$—H	380
H—CN	540
C_6H_5—H	469
CCl_3—H	402
HO—H	498
CH_3O—H	427
CH_3COO—H	469
CH_3—CH_3	368
$C_6H_5CH_2$—CH_3	293
CH_3—NH_2	305
CH_3—OH	381
CH_3—F	452
CH_3—Cl	351
CCl_3—Cl	305
HO—OH	213
t—C_4H_9O—OH	184

* From J. A. Kerr, Bond Dissociation by Kinetic Methods, *Chem. Rev.*, **66**:465 (1966).

TABLE 16-5 Principal types of free-radical reactions

Type of reaction	*Example*
Abstraction	$\cdot CH_3 + CH_3CH_2CH_2CH_3 \rightarrow CH_4 + CH_3\dot{C}HCH_2CH_3$
Addition	$Br\cdot + CH_2{=}CH{-}R \rightarrow BrCH_2{-}\dot{C}H{-}R$
Combination	$2(\cdot CH_3) \rightarrow C_2H_6$

give directly the final products, but constitute only steps in that direction. The abstraction step can lead to *chain reactions,* as in the system of H_2 and Cl_2, that is,

$$H\cdot + Cl_2 \rightarrow HCl + Cl\cdot$$
$$Cl\cdot + H_2 \rightarrow HCl + H\cdot$$
$$H\cdot + Cl_2 \rightarrow \text{etc.}$$

The addition step can lead to *polymerization* reactions, as in the formation of polyethylene,

$$R\cdot + CH_2{=}CH_2 \rightarrow R{-}CH_2{-}\dot{C}H_2$$
$$R{-}CH_2{-}\dot{C}H_2 + CH_2{=}CH_2 \rightarrow R{-}CH_2CH_2{-}CH_2{-}\dot{C}H_2$$
$$R{-}CH_2CH_2CH_2\dot{C}H_2 + CH_2{=}CH_2 \rightarrow \text{etc.}$$

For the destruction of the free-radical centers, which must occur to give the final product, there must be a *radical recombination.* Although a reaction between two highly active species might be expected to proceed easily and rapidly, one must recognize that, for example, two hydrogen atoms coming together will produce something more than 430 J of energy per mole and that, unless this energy is removed, it will be enough to redissociate the molecule. For radical recombination to occur, it is necessary, therefore, to have either a *third body,* usually the wall of the container, present or to be dealing with large molecules, so that many degrees of freedom can spread out the energy evolved in the bond formation.

16-8 Elementary Reactions in Liquid Solution: Diffusion-controlled Reactions

The idea of elementary reaction processes, similar to those recognized for gas-phase reactions, can be carried over to liquid systems. Some significant differences in the molecular model for these steps must, however, be introduced. We begin by deducing the liquid-state counterpart of the gas-phase rate-of-collision expression.

Consider a liquid system consisting of molecular species A and B, which can be identical or different, dissolved in a relatively inert solvent, so as to produce a solution with N_A^* and N_B^* molecules of A and of B per unit volume. What is the rate with which A and B molecules will "encounter" one another?

To proceed, we must borrow some of the concepts and results that will be dealt with in Sec. 25-9. In particular, we must use the concept of a *diffusion coefficient,* which is a measure of the rate with which a material diffuses across a unit cross-section area as a result of a unit concentration gradient. The relation of the liquid-phase reaction process to diffusion can be recognized by, first, focusing on a particular A molecule and asking about the rate with which B molecules would diffuse to it.

The result, which seems reasonable and will not be derived here, is that the rate is proportional to N_B^*, the number of molecules of B per unit volume; D_B, the diffusion coefficient of B; and $r_A + r_B$, the distance to which the center of B must come to that of A for the molecules to be in contact. The derived expression is

$$\text{Rate of } B \text{ molecules diffusing to an } A \text{ molecule} = 4\pi D_B(r_A + r_B)N_B^* \quad [45]$$

For the total rate of $A \ldots B$ encounters per unit volume, on this basis, we would write

$$\text{Rate of encounters per unit volume} = 4\pi D_B(r_A + r_B)N_B^* N_A^* \quad [46]$$

In a similar way, if we focused on a B molecule and asked about the rate with which A molecules diffuse toward it, we would obtain

$$\text{Rate of encounters per unit volume} = 4\pi D_A(r_A + r_B)N_A^* N_B^* \quad [47]$$

The encounter rate that we should expect, since in general both A and B will have appreciable diffusion tendencies, can be written as the mean of these results, namely,

$$\text{Rate of encounters} = 2\pi(D_A + D_B)(r_A + r_B)N_A^* N_B^* \quad [48]$$

We thus have an expression that corresponds to the maximum rate of reaction that could occur in liquid solutions, and we see again the proportionality to the product of the concentration of the reacting species.

To proceed to any absolute value for the rate it is adequate here to anticipate an expression that will be derived in Sec. 25-9 showing that the diffusional coefficient of a sphere in a liquid is related to the coefficient of viscosity η of the liquid and the radius of the sphere by

$$D = \frac{kT}{6\pi r\eta}$$

or

$$D_A = \frac{kT}{6\pi r_A \eta} \quad \text{and} \quad D_B = \frac{kT}{6\pi r_B \eta} \quad [49]$$

where k is Boltzmann's constant. Substitution of these relations in Eq. [48] gives

$$\text{Rate of encounters} = \frac{2kT}{3\eta}\frac{(r_A + r_B)^2}{r_A r_B}N_A^* N_B^* \qquad [50]$$

From this result, a representative value for the maximum rate of a liquid-phase reaction can be calculated by assuming A and B of equal radius and by taking the representative liquid value

$$\eta = 0.01 \text{ poise} = 10^{-3} \text{ kg m}^{-1} \text{ s}^{-1}$$

Calculation gives

$$\text{Rate of encounters} = \frac{8kT}{3\eta}N_A^* N_B^*$$
$$\approx 0.1 \times 10^{-16} N_A^* N_B^* \qquad \text{s}^{-1} \qquad [51]$$

Comparison with the gas-phase counterpart, Eq. [40], shows that, for the same reagent concentrations, the liquid-phase encounter rate is somewhat less than the gas-phase collision rate.

Although the model for the calculation of the encounter rate in liquids is very crude and the derivation introduces further uncertainties, the result, as you will see, does seem to correspond to the rates of some very fast solution reactions. In fact, rates nearly equal to that given by Eq. [51] are said to imply a *diffusion-controlled* reaction.

16-9 Lifetime of Liquid-phase Intermediates: The Cage Effect

The coming together of molecules in the liquid has been called an "encounter," and this usage implies that there is some distinction between an encounter and a collision. The principal distinction, as we will now see, arises from the difficulty with which molecules of a liquid, once adjacent to each other, separate from each other. A similar difficulty stands in the way of the fragments of a decay step escaping from each other. The special character of a liquid system is treated in terms of what is called a *cage effect.*

Consider a pair of molecules, as shown in Fig. 16-7, that have just encountered each other or, what in this regard is equivalent, a pair of molecules that have just arisen from the decay of a parent molecule. The molecules can become separated from each other only as a result of their diffusional motion through the liquid. Since, on the molecular level, diffusion is pictured as the result of repeated jostlings of the neighboring molecules, we can picture the A, B molecular pair as remaining for some time together in a "cage" formed by the surrounding solvent. The whole process of A and B coming together and remaining together for a number of subsequent collisions is known as an *encounter.* Now let us try to estimate the number of collisions (which, in fact, are not easily defined in a liquid system) that occur during an encoun-

FIGURE 16-7
A pair of reacting molecules in a liquid cage.

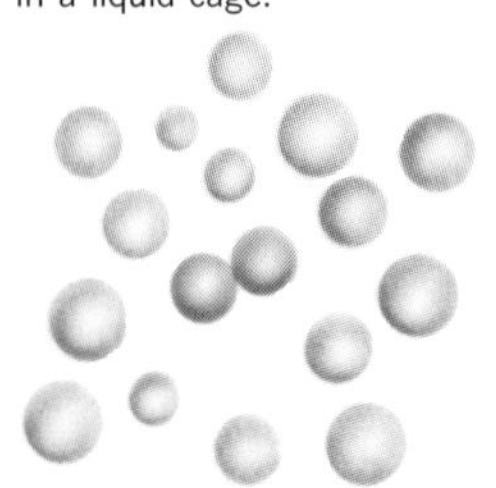

ter. A result from the analysis of the net distance a particle moves as a consequence of a series of random steps taken in some time t must be used. This result, from the "random-walk" problem, is that the average, squared distance is related to the diffusion coefficient by the relation

$$\overline{r^2} = 6Dt \qquad [52]$$

If we say that a pair of colliding molecules have escaped from each other when they are separated by an additional amount $r_A + r_B$, that is, their average diameter, and if we use 3 Å as a representative value for this diameter and 10^{-9} m^2 s^{-1} as a representative diffusion coefficient, we calculate

$$\text{Time duration of an encounter} \approx \frac{\overline{r^2}}{6D}$$

$$= \frac{(3 \times 10^{-10})^2}{6(10^{-9})} = 1.5 \times 10^{-11} \text{ s} \qquad [53]$$

By contrast, the time required for gas-phase molecules subject to no retarding force to become separated by this same 3 Å distance is

$$t = \frac{3 \times 10^{-10}}{\bar{u}} = \frac{3 \times 10^{-10}}{\sqrt{8kT/\pi m}} \qquad [54]$$

With representative values, this is calculated to be

$$t = 0.08 \times 10^{-11} \text{ s} \qquad [55]$$

Thus the encounter lasts some 20 times as long, and other estimates would place this as high as 100 times as long, as it would take the molecules to separate under gas-phase conditions. It follows that in the liquid phase each encounter can be looked upon as consisting of 10 to 100 collisions. This factor makes up for the less frequent liquid-phase encounters than gas-phase collisions and leads us to appreciate the fact that for some of the few simple reactions that have been studied in both phases the rates of the reactions are not greatly different.

The idea of a cage implies also that a decay process cannot be as well defined as in the gas phase. The fragments that are produced have considerable opportunity to react with each other or with the adjacent solvent before they move freely into the bulk of the solution to react with other entities.

16-10 Ionic Intermediates in Liquid-phase Reactions

The most profound distinction between the elementary processes that occur in gas and liquid phases is to be found in the nature of the species that participate or are produced. In the gas phase such species, under most conditions, are neutral atoms, molecules, or free radicals. By contrast, in liquids many reactions proceed with ionic intermediates. These again are often transient reactive species that are not the ionic materials dealt with in equilibrium systems. (Other liquid-phase reactions involve free-radical intermediates, but the discussion of gas-phase free radicals can be carried over satisfactorily, so that no additional features of these intermediates need be pointed out here.)

In organic systems the most important new species that are postulated are those in which a positive or negative charge resides on a carbon atom. Positive ions of this type are called *carbonium ions,* and negative ions are called *carbanions.* A knowledge of the conditions of the formation and the reactions of these intermediates is of great value, not only in understanding reaction kinetics, but also in understanding much of organic chemistry itself.

A few of the more common types of reactions in which carbonium ions and carbanions are postulated to occur as intermediates are shown in Table 16-6.

The carbonium ions and carbanions, being unstable intermediates, must react further, in any complete reaction, to give the final products. Again a large number of possibilities exist, but a few representative types of reactions of carbonium ions and carbanions are shown in Table 16-7.

Some reactions do not occur by the stepwise formation and further reaction of intermediates, but appear to proceed by a single concerted

TABLE 16-6 Some reactions in which carbonium ions and carbanions are postulated as intermediates

SUBSTITUTION REACTIONS	
Reaction	$HO^- + (CH_3)_3CCl \rightarrow HOC(CH_3)_3 + Cl^-$
Mechanism	$(CH_3)_3CCl \rightleftharpoons (CH_3)_3C^+ + Cl^-$
	$(CH_3)_3C^+ + OH^- \rightarrow (CH_3)_3COH$
ADDITION REACTIONS	
Reaction	$HCl + CH_2{=}CH_2 \rightarrow CH_3{-}CH_2Cl$
Mechanism	$H^+ + CH_2{=}CH_2 \rightarrow CH_3{-}\overset{+}{C}H_2$
	$CH_3\overset{+}{C}H_2 + Cl^- \rightarrow CH_3{-}CH_2Cl$
ELIMINATION REACTIONS	
Reaction	$CH_3\underset{}{\overset{Br}{\overset{\vert}{C}}}H{-}CH_3 + HO^- \rightarrow CH_3{-}CH{=}CH_2 + H_2O + Br^-$
Mechanism step	$CH_3\overset{Br}{\overset{\vert}{C}}H{-}CH_3 + HO^- \rightarrow CH_3{-}\overset{Br}{\overset{\vert}{C}}H{-}\overset{\cdot\cdot-}{C}H_2 + H_2O$

TABLE 16-7 Some reactions of carbonium and carbanion intermediates

REARRANGEMENTS (TO LEAD TO A DIFFERENT INTERMEDIATE)

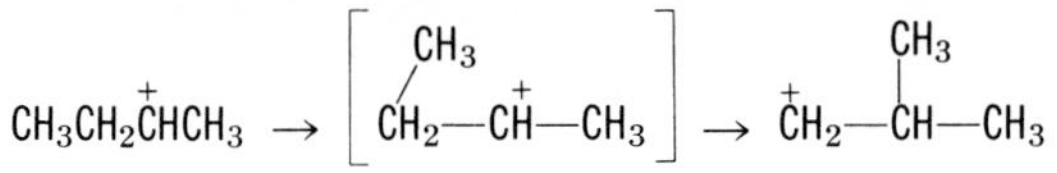

$$CH_3CH_2\overset{+}{C}HCH_3 \rightarrow \left[\overset{CH_3}{\diagup} CH_2\text{—}\overset{+}{C}H\text{—}CH_3 \right] \rightarrow \overset{+}{C}H_2\text{—}\overset{CH_3}{\overset{|}{C}H}\text{—}CH_3$$

REACTIONS TO GIVE STABLE PRODUCTS

Addition of an ionic species:

$$(CH_3)_3C^+ + OH^- \rightarrow (CH_3)_3COH$$

Elimination of an ionic species:

$$CH_3\text{—}\overset{Br}{\overset{|}{C}H}\text{—}\ddot{C}H_2^- \rightarrow \left[CH_3\text{—}\overset{Br}{\overset{|}{C}H}\text{—}\ddot{C}H_2^- \right] \rightarrow CH_3\text{—}CH\text{=}CH_2 + Br^-$$

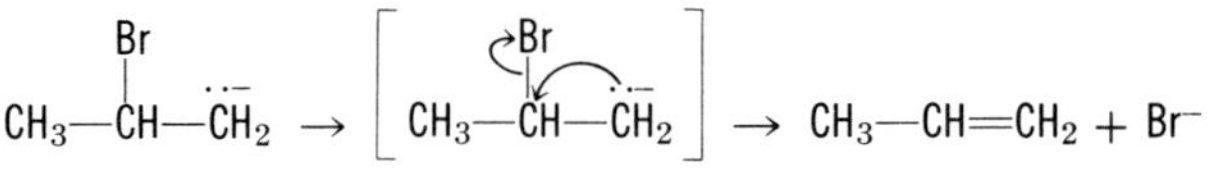

process. These reactions do not involve either free-radical or ionic intermediates. Some concerted reactions, however, can be classified along with ionic, or polar, reactions. Two important reactions of this type are *displacement reactions* and *elimination reactions.*

The first can be illustrated by the example of the displacement of the bromide of an alkyl bromide by an amine.

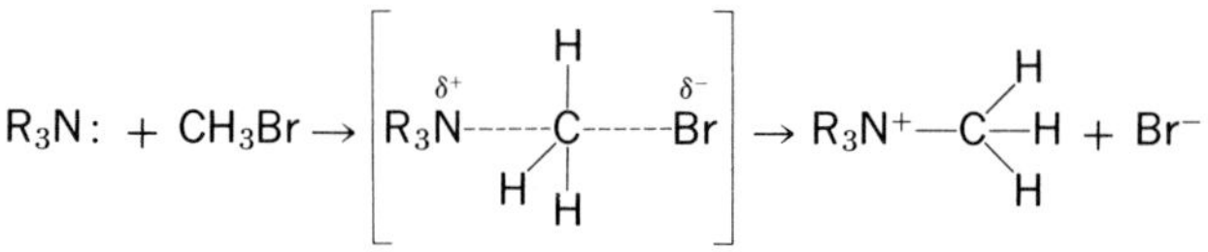

$$R_3N\!: + CH_3Br \rightarrow \left[\overset{\delta^+}{R_3N}\cdots\cdots\overset{H}{\underset{H\;\;H}{C}}\cdots\cdots\overset{\delta^-}{Br} \right] \rightarrow R_3N^+\text{—}C\overset{H}{\underset{H}{\text{—}H}} + Br^-$$

The postulated mechanism consists of the formation of the N—C bond and the simultaneous breaking of the C—Br bond. The intermediate is depicted as the reaction species when this concerted process is partly complete.

A representative concerted elimination reaction is the removal of HBr when an alkyl halide is placed in alkali media. The postulated mechanism is

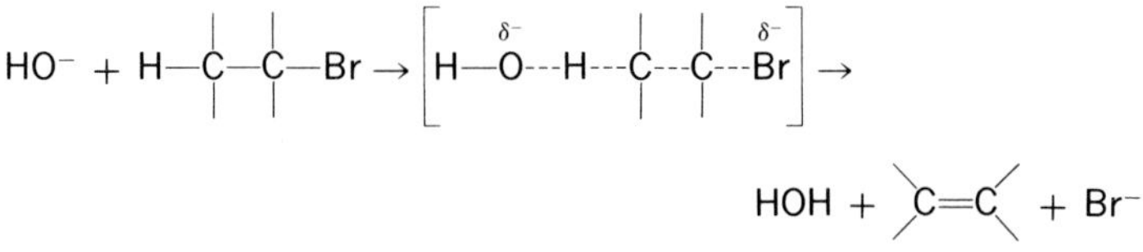

$$HO^- + H\text{—}\underset{|}{\overset{|}{C}}\text{—}\underset{|}{\overset{|}{C}}\text{—}Br \rightarrow \left[H\text{—}\overset{\delta^-}{O}\cdots H\cdots\underset{|}{\overset{|}{C}}\cdots\underset{|}{\overset{|}{C}}\cdots\overset{\delta^-}{Br} \right] \rightarrow$$

$$HOH + \;>C\text{=}C<\; + Br^-$$

in which the net effect is the removal of HBr.

Some reactions will be considered in detail later, and these will further illustrate the formation of charged intermediates and their further reactions to final products.

The physical chemist's role in reaction kinetics has often been considered to end when the experimental measurements of the rate of reaction have been obtained and a rate equation has been formulated. For such efforts to be rewarding, however, it is necessary to continue to the phase of theorizing on the molecular behavior that leads to the observed rate law. In this way one is led to consider reaction mechanisms. Since the study of the details of reaction processes could lead to an investigation of all organic and inorganic chemistry, it is necessary that only a few representative reactions be considered and that the approach to an understanding of chemical reactions that comes from reaction kinetics be indicated.

All the features of some reactions can be accounted for by a mechanism that consists of the single step of a coming together of all the reactant molecules, a rearrangement of the atoms and electrons in the moment of impact, and the flying apart of product molecules. Such a mechanism is said to involve a *concerted* process.

In most reactions, however, such a concerted process would require a complicated and improbable combination of changes to get from reactants to products. It is then more satisfactory to attempt to explain the reaction as proceeding by means of a sequence of steps, each step consisting of simple atomic or electronic moves.

16-11 Reaction Mechanisms and Rate Laws: The Stationary-state Method

Let us here investigate the rate laws that would be expected on the basis of a sequence of elementary reaction steps that appear to enter into the mechanism of many organic and inorganic and ionic and free-radical reactions. We will make use of the conclusions from earlier in this chapter that the rate of an elementary step is proportional to the product of the concentrations of the reactants in that step and that various intermediates may enter into the mechanism even if they are not obtained as products of the overall reaction.

Let us consider some mechanisms in which the first step is a *reversible* reaction that produces one or more intermediates and one or more final-product species. The second step consists in the decay of one of the intermediates, the reaction of one of the intermediates with another intermediate, or the reaction of one of the intermediates with one of the reactant species. The result of the second step may be final products or additional intermediates. In the latter case, for simplicity, we will assume that all further reactions are relatively very rapid. One example of the many variations is

$$\begin{array}{l} A \rightleftharpoons M + C \\ \underline{M + B \rightarrow D} \\ A + B \rightarrow C + D \end{array}$$

where A and B are reactants, M is an intermediate, and C and D are products. Other variations are illustrated by the compilation of suggested mechanisms for reactions in Table 16-8.

The deduction of the rate equation for a reaction that proceeds by one of the mechanisms of this general type can be carried out most easily if we can assume that, for most of the period over which the reaction is studied, the concentrations of the intermediates are small compared with reactants or products. Failure to detect such intermediates lends support to this assumption. If the concentration of the intermediate fails to build up to an appreciable value, it follows that, except at the initial and final stages of the reaction, the rate of change of the intermediate can be set equal to zero. This result, a feature of the *stationary-state* method, is very helpful in the deduction of the rate equation corresponding to a postulated mechanism.

Consider, to be specific,

$$A \underset{k_{-1}}{\overset{k_1}{\rightleftharpoons}} M + C \tag{56}$$

$$M + B \xrightarrow{k_2} \text{products} \tag{57}$$

The value of $[M]$ increases as a result of the elementary reaction associated with k_1 and decreases as a result of the k_{-1} and k_2 reactions.

TABLE 16-8 Some reactions for which mechanisms consisting of an initial reversible step followed by reaction of one of the products of this step have been suggested Additional mechanistic steps beyond the first two are required to account for the overall reaction.

First step	$Br_2 \rightleftharpoons 2Br$	$Co(CN)_5OH_2^{--} \rightleftharpoons Co(CN)_5^{--} + H_2O$
Second step	$Br + H_2 \rightarrow HBr + H$	$Co(CN)_5^{--} + I^- \rightarrow Co(CN)_5\ I^{3-}$
Overall reaction	$Br_2 + H_2 \rightarrow 2HBr$	$Co(CN)_5OH_2^{--} + I^- \rightarrow Co(CN)_5\ I^{3-} + H_2O$
First step	$CHCl_3 + OH^- \rightleftharpoons :CCl_3^- + H_2O$	$CH_3\text{—}\overset{\overset{O}{\|}}{C}\text{—}CH_3 + OH^- \rightleftharpoons CH_3\overset{\overset{O}{\|}}{C}CH_2^- + H_2O$
Second step	$:CCl_3^- \rightarrow :CCl_2 + Cl^-$	$CH_3\text{—}\overset{\overset{O}{\|}}{C}CH_2^- + Br_2 \rightarrow CH_3\overset{\overset{O}{\|}}{C}CH_2Br + Br^-$
Overall reaction	$2CHCl_3 + 7OH \rightarrow CO + HCOO^- + 6Cl^- + 4H_2O$	$CH_3\text{—}\overset{\overset{O}{\|}}{C}\text{—}CH_3 + 3Br_2 + 4OH^- \rightarrow CH_3\overset{\overset{O}{\|}}{C}\text{—}O^- + CHBr_3 + 3Br^- + 3H_2O$

Thus the steady-state assumption lets us write

$$k_1[A] = k_{-1}[M][C] + k_2[M][B]$$

or

$$[M] = \frac{k_1[A]}{k_{-1}[C] + k_2[B]} \qquad [58]$$

According to the mechanism, the rate of formation of products, or the rate of consumption of B, which from the overall stoichiometry is also that for A, is equal to the rate of the second reaction, i.e.,

$$\text{Rate} = -\frac{d[A]}{dt} = -\frac{d[B]}{dt} = k_2[M][B] \qquad [59]$$

The steady-state expression obtained for $[M]$ allows this to be written

$$\text{Rate} = \frac{k_1 k_2[A][B]}{k_{-1}[C] + k_2[B]} \qquad [60]$$

Different, but related, rate equations can be worked out in the same way for the various mechanisms shown in Table 16-8.

That a single mechanism, such as that of Eqs. [56] and [57], can lead to different rate equations can be illustrated by considering two extremes in the values of relative rate constants.

(a) For $k_{-1}[C] \gg k_2[B]$. For this case, Eq. [60] reduces to

$$\text{Rate} = \frac{k_1 k_2}{k_{-1}} \frac{[A][B]}{[C]} \qquad [61]$$

A similar result could have been obtained more directly by assuming that an equilibrium is established by the first reaction and that the second reaction does not use up B fast enough to upset this equilibrium. Then one would have written

$$k_1[A] = k_{-1}[M][C] \qquad \text{or} \qquad \frac{[M][C]}{[A]} = \frac{k_1}{k_{-1}} = K$$

and

$$\text{Rate} = k_2[M][B]$$

$$= \frac{k_1 k_2}{k_{-1}} \frac{[A][B]}{[C]} \qquad [62]$$

An illustration of this situation is the reaction of the iodide ion with hypochlorite according to

$$I^- + OCl^- \rightarrow OI^- + Cl^-$$

The rate of the reaction is influenced by the pH, and the rate law that is deduced from rate data is

$$\text{Rate} = \frac{k[I^-][OCl^-]}{[OH^-]} \qquad [63]$$

A suitable mechanism for this reaction has an initial rapid equilib-ium step

$$OCl^- + H_2O \rightleftharpoons HOCl + OH^-$$

ollowed by a slow step involving the reagents,

$$I^- + HOCl \rightarrow$$

Two, at least, mechanisms can be written beyond this point. Thus we could postulate

$$I^- + HOCl \rightarrow HOI + Cl^-$$

$$OH^- + HOI \rightarrow H_2O + OI^-$$

or

$$I^- + HOCl \rightarrow ICl + OH^-$$

$$ICl + 2OH^- \rightarrow OI^- + Cl^- + H_2O$$

Rate data do not distinguish between these presumed fast, final steps.

(b) For $k_{-1}[C] \ll k_2[B]$. Equation (60) now reduces to

$$\text{Rate} = k_1[A] \tag{64}$$

That is, the rate law would simply be first-order in $[A]$ even though the reaction is still assumed to proceed through the same sequence of steps. One can recognize that the assumed inequality makes the initial dissociation of A the rate-determining step and the later steps in the mechanism of no consequence. The example given for the formation of ethane and acetone from di-*tert*-butyl peroxide in Sec. 16-3 provides a reaction that can be taken as an illustration of this situation.

(c) For $k_{-1}[C]$ and $k_2[B]$ of Comparable Magnitudes. An illustration of this situation is provided by the reaction

$$Co(CN)_5OH_2^{--} + I^- \rightarrow Co(CN)_5I^{3-} + H_2O$$

The observed rate law has the form

$$\text{Rate} = \frac{a[Co(CN)_5OH_2^{--}][I^-]}{b + c[I^-]} \tag{65}$$

A mechanism that is reasonable and leads to an expression of this form is

$$Co(CN)_5OH_2^{--} \underset{k_{-1}}{\overset{k_{-1}}{\rightleftharpoons}} Co(CN)_5^{--} + H_2O$$

$$Co(CN)_5^{--} + I^- \xrightarrow{k_2} Co(CN)_5I^{3-}$$

The analysis that led to Eq. [60] would, for this example, lead to the expected rate equation

$$\text{Rate} = \frac{k_1k_2[Co(CN)_5OH_2^{--}][I^-]}{k_{-1}[H_2O] + k_2[I^-]} \tag{66}$$

Since the reaction is run in aqueous solution with a large and essentially constant concentration of water, we see that the mechanism is consistent with the empirical rate law.

16-12 A Mechanism for Enzyme-catalyzed Reactions

A variety of rate laws is required to portray the rates of enzyme-catalyzed reactions for the great variety of enzymes, substrates, and the rate-influencing reagents and physical conditions that are encountered. The rate law of Sec. 16-4 is, however, a guide to many of these variations and the mechanism of this section, often named after L. Michaelis and M. Menten [*Biochem. Z.*, **49**:333 (1913)], is likewise a base for other variations.

The mechanism that accounts for the rate law of Eq. [27] is similar to those dealt with in the preceding section.

With S representing substrate, E the enzyme, and $E{\cdot}S$ an enzyme-substrate complex, the mechanism of the reaction is presumed to be adequately represented by

$$E + S \rightleftharpoons E{\cdot}S \qquad [67]$$

$$E{\cdot}S \rightarrow E + \text{products} \qquad [68]$$

As in the preceding section, the steady-state assumption, which, however, is not always clearly applicable in these reactions, leads to

$$k_1[E][S] = k_{-1}[E{\cdot}S] + k_2[E{\cdot}S]$$

and

$$[E{\cdot}S] = \frac{k_1}{k_{-1} + k_2}[E][S] \qquad [69]$$

To bring these expressions to a form that can be compared with the empirical rate equation, we must recognize that it is only $[E_T] = [E] + [E{\cdot}S]$, and not $[E]$, that is generally known. Often, in fact, only a quantity proportional to $[E_T]$, and not even values of $[E_T]$, are available.

Replacement of $[E]$ in Eq. [69] leads to

$$[E{\cdot}S] = \frac{k_1[E_T][S]}{(k_{-1} + k_2) + k_1[S]} \qquad [70]$$

Now this expression for the intermediate $E{\cdot}S$ can be inserted into the expression for the rate of the net reaction, which can be based on the formation of products in the second mechanism step, Eq. [68]. We have

$$-\frac{d[S]}{dt} = R = k_2[E \cdot S] = \frac{k_1 k_2[E_T][S]}{k_{-1} + k_2 + k_1[S]}$$

$$= \frac{k_2[E_T][S]}{(k_{-1} + k_2)/k_1 + [S]} \qquad [71]$$

The terms $(k_{-1} + k_2)k_1$ can, as is customary, be indicated by the new symbol K_M, that is,

$$K_M = \frac{k_{-1} + k_2}{k_1} \qquad [72]$$

to give the rate-law result of this mechanism as

$$R = \frac{k_2[E_T][S]}{K_M + [S]}$$

We have come at this stage to the form of the empirical rate law obtained in Sec. 16-4.

We now are in a position to interpret the values of the parameters K_M and $k_2 E_T$ in terms of their roles in the steps of the mechanism.

Reference to Eq. [69] shows that, as the reaction is proceeding,

$$\frac{[E][S]}{[E \cdot S]} = K_M \qquad [73]$$

Thus K_M is related to species concentrations, as is the dissociation constant for the species $[E \cdot S]$. The value of K_M, however, is given by $(k_{-1} + k_2)/k_1$, and this is equal to the value of the dissociation constant for $[E \cdot S]$ only to the extent that k_2 is small and can be neglected compared with k_{-1}. Thus, when the breakup of the $E \cdot S$ complex to form original E and S species dominates the process whereby the complex forms products, the value of K_M approaches the dissociation constant for the $E \cdot S$ complex.

What now is the significance of the term $k_2[E_T]$? One first notes that the rate of the overall reaction is

$$R = k_2[E \cdot S] \qquad [74]$$

It follows that $k_2[E_T]$ is the rate that the reaction would have if all the enzymes were in the form of the enzyme-substrate complex, that is, $k_2[E_T]$ is the maximum rate for a given value of $[E_T]$. The *turnover* rate of an enzyme in a particular enzyme-catalyzed reaction is the rate per mole of enzyme; i.e., the turnover rate is equal to the value of k_2, and this can be calculated from $k_2[E_T]$ if the total enzyme concentration is known.

Problems

1 What are the units of the rate constants of first- and second-order reactions if the concentrations are expressed in moles per liter and the time in seconds? If the rate of a reaction obeyed the rate law $k[A][B]^{2/3}$, what would be the units of k?

2 Using the data of Table 16-1, prepare graphs like those of Fig. 16-1 for the hydrolysis of *tert*-butyl bromide at 50°C.

a Is the reaction first-order at this temperature?

b What are the rate constant and the half-life at 50°C?

Ans. $k = 0.013$ min^{-1}.

c Show the half-lives on the plot of c versus t. See that the concentration does fall to half its value in time $t_{1/2}$ regardless of the concentration considered.

3 The hydration of ethylene oxide in aqueous solution proceeds according to the overall equation

$$\underset{\diagdown O \diagup}{CH_2{-}CH_2} + H_2O \rightarrow CH_2OHCH_2OH$$

The rate of the reaction has been followed by Bronsted, Kilpatrick, and Kilpatrick [*J. Am. Chem. Soc.,* **51**:428 (1929)] by measuring the change in volume of the liquid system. (This is done by observing the height of the liquid in the capillary tube of a dilatometer, a large thermometerlike reaction cell.) They obtained the following results, at 20°C, using 0.12 M ethylene oxide and 0.007574 M $HClO_4$.

t (min)	0	30	60	90	120	240	300	360	390	∞
h (arbitrary units)	18.48	18.05	17.62	17.25	16.89	15.70	15.22	14.80	14.62	12.30

Confirm that the reaction is first-order with respect to ethylene oxide. What is the rate constant? *Ans.* $k = 0.00247\ min^{-1}$.

4 Deduce the pressure of unreacted *tert*-butyl peroxide as a function of time, at 147.2°C, from the total pressure data of Table 16-2. By a suitable graphical treatment, show that these pressure data also indicate a first-order reaction and that the same rate constant as that reported in Fig. 16-2 is obtained.

5 Show that the pressure data for the *tert*-butyl peroxide decomposition at 154.9°C given in Table 16-2 are consistent with a first-order rate equation. What are the rate constant and the half-life of the reaction at this temperature?

6 Calculate a first-order rate constant for each concentration datum at 25°C for the reaction of *tert*-butyl bromide given in Table 16-1. Do these results indicate a first-order reaction?

Assume that the reaction is second-order in *tert*-butyl bromide. Calculate values for the second-order rate constant at each of the reported times. Does the reaction follow a second-order rate law?

7 The gaseous dimerization of butadiene has been followed by measurement of the total gas pressure by Vaughan [*J. Am. Chem. Soc.,* **54**:3863 (1932)]. The data tabulated at the right were obtained at 326°C. What is the order of the reaction, and what is the rate constant? *Ans.* $k = 0.018\ atm^{-1}\ min^{-1}$.

t (min)	*P (atm)*
0	(0.8315)
3.25	0.8138
8.02	0.7886
12.18	0.7686
17.30	0.7464
24.55	0.7194
33.00	0.6944
42.50	0.6701
55.08	0.6450
68.05	0.6244
90.05	0.5964
119.00	0.5694
176.67	0.5332
259.50	0.5013
373	0.4698

8 The displacement of bromide by thiosulfate ion has been studied at 37.50°C, in the reaction

$$n\text{-}C_3H_7Br + S_2O_3^{--} \rightarrow C_3H_7S_2O_3^{-} + Br^{-}$$

by Crowell and Hammett [*J. Am. Chem. Soc.,* **70**:3444 (1948)]. The thiosulfate-ion concentration remaining at various times was determined by titration with iodine. From these data and the known initial concentrations the following data were obtained:

t (s)	Conc. $S_2O_3^{--}$ (mol liter^{-1})	Conc. C_3H_7Br (mol liter^{-1})
0	0.0966	0.0395
1110	0.0904	0.0333
2010	0.0863	0.0292
3192	0.0819	0.0248
5052	0.0766	0.0196
7380	0.0720	0.0149
11,232	0.0668	0.0097
78,840	0.0571	0.0000

Derive the rate equation, including a numerical value for the rate constant, for this reaction. *Ans.* $k = 1.6 \times 10^{-3}$ liter mol^{-1} s^{-1}.

9 Benzene diazonium chloride in aqueous solution decomposes according to the equation

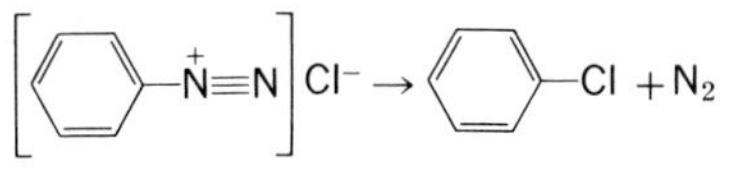

$$\left[C_6H_5{-}\overset{+}{N}{\equiv}N \right] Cl^- \rightarrow C_6H_5{-}Cl + N_2$$

and the reaction can be conveniently followed by the amount of N_2 evolved. Cain and Nicoll [*J. Chem. Soc.*, **81**:1412 (1902)] report the following results for 20°C and 35 ml of a solution containing 10 g of diazobenzene chloride per liter

t (min)	116	192	355	481	1282	1429	∞
ml of N_2 *evolved (measured at 13°C and 0.987 atm)*	9.7	16.2	26.3	33.7	51.4	54.3	60.0

What is the order of the reaction, and what is the rate constant?

10 The radioactive decay of radium, with half-life 1590 years, leads to the formation of the inert gas radon, which, with a half-life of 3.82 days, decays further. If a sample of radium is kept in a sealed vial, the radon gas collects.
a Derive an expression for the number of radon atoms present as a function of time if a 1-g sample of radium is considered. The expression need be valid only for a time interval of less than about 1 year.
b Plot the number of atoms of radon as a function of time in the time interval 0 to 2 weeks.
c The amount of radon is seen to reach a constant value, and radon is said to be in *secular equilibrium* with radium. Why does this differ from ordinary equilibrium?

11 What would be the half-life for a representative gas-phase reaction, as dealt with in Sec. 16-6, if each $A \ldots B$ collision were a reactive collision and the pressures of A and B initially were 10^{-9} atm, a pressure readily obtained in a laboratory vacuum system? What would the half-life be at a pressure of 10^{-11} atm and a temperature of 1500 K as occurs in the earth's atmosphere at an altitude of about 500 km?

12 From the rates of diffusion of hemoglobin in water at 20°C, the value deduced for the diffusion coefficient D is 6.5×10^{-11} m^2 s^{-1}. Using the assumption that the molecule can be treated as a sphere with radius of 30 Å, test the validity of Eq. [49]. The viscosity of water at 20°C is 0.010 poise, a value that corresponds to 0.0010 in the SI viscosity units of kg m^{-1} s^{-1}.

13 The diffusion coefficient of sucrose in water at 20°C is 4.0×10^{-10} m^2 s^{-1}. The viscosity of water is cited in Prob. 12. What value of r does this suggest for the sucrose molecule? Does this seem reasonable?

14 The reaction by which a tertiary chlorine, as in *tert*-butyl chloride, is replaced by a hydroxyl group appears to result from the formation of a carbonium ion which subsequently adds water or hydroxide. The reaction with $(CH_3)_3CCl$ is quite rapid, whereas that with

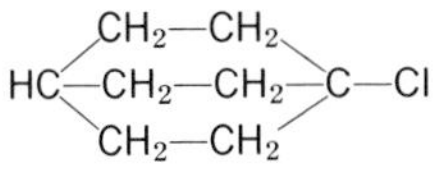

$$HC \begin{matrix} CH_2{-}CH_2 \\ {-}CH_2{-}CH_2{-} \\ CH_2{-}CH_2 \end{matrix} C{-}Cl$$

is imperceptibly slow. What do these relative rates imply about the preferred geometry of the carbonium ion?

15 Consider the schematic hypothetical reaction mechanism

$$A \underset{k_2}{\overset{k_1}{\rightleftharpoons}} M$$

$$M + A \xrightarrow[k_2]{} C$$

a Write down expressions for the net rate of change of each of the species A, M, and C.

b If M is a species that is present in only undetectably small amounts at all times, obtain an expression for the concentration of M in terms of the concentrations of the major reagents A and C.

c Obtain, with the result of part b, rate equations for the disappearance of A and for the formation of C.

d What relative values of the rate constants would result in the reaction being first-order with respect to A? What values would make it second-order?

References

KING, E. L.: "How Chemical Reactions Occur," W. A. Benjamin, Inc., New York, 1963. A short elementary introduction to the study of the rates of chemical reactions and the relationship of such rates to the mechanisms of chemical reactions.

FROST, A. A., and R. G. PEARSON: "Kinetics and Mechanism: A Study of Homogeneous Chemical Reactions," 2d ed., John Wiley & Sons, Inc., New York, 1961. The empirical treatments, experimental studies, and the theories of the rates of chemical reactions are first presented. Then rather detailed treatments of particular reactions or reaction types that serve to illustrate chemical kinetic studies are presented.

BENSON, S. W.: "The Foundations of Chemical Kinetics," McGraw-Hill Book Company, New York, 1960. A rather comprehensive account of the kinetics of chemical reactions. The treatment is somewhat more detailed and more mathematical than that of Frost and Pearson.

GARDENER, W. C.: "Rates and Mechanisms of Chemical Reactions," W. A. Benjamin, Inc., New York, 1969.

DENCE, J. B., H. B. GRAY, and G. S. HAMMOND: "Chemical Dynamics," W. A. Benjamin, Inc., New York, 1968.

HALPERN, J.: Some Aspects of Chemical Dynamics in Solution, *J. Chem. Educ.*, **45**:372 (1968).

Applications of results from kinetic studies are also found in books dealing with the reactions in a particular branch of chemistry. Examples of such books are:

EDWARDS, J. O.: "Inorganic Reaction Mechanisms," W. A. Benjamin, Inc., New York, 1964.

INGOLD, C. K.: "Structure and Mechanism in Organic Chemistry," G. Bell & Sons, Ltd., London, 1963.

LEFFLER, J. E.: "The Reactive Intermediates of Organic Chemistry," Interscience Publishers, Inc., New York, 1958.

A number of monographs on certain topics in the general area are also available, and these often provide a more satisfactory introduction to a particular aspect than do the more general treatments referred to above. Some monographs of interest are:

CALDIN, E. F.: "Fast Reactions in Solution," Blackwell Scientific Publications, Ltd., Oxford, 1964. A treatment of simple classical methods and also relaxation, *nmr*, and *esr* techniques.
MELANDER, L.: "Isotope Effects on Reaction Rates," The Ronald Press Company, New York, 1960.
LAIDLER, K. J.: "The Chemical Kinetics of Excited States," Oxford University Press, Fair Lawn, N.J., 1955.
TROTMAN-DICKENSON, A. F.: "Gas Kinetics," Butterworth Scientific Publications, London, 1955.

17

THE NATURE OF ELEMENTARY REACTIONS

The preceding chapter has shown that the form of the concentration terms of the rate equation is of great value in deducing a sequence of reaction steps that could constitute an overall chemical reaction. It is this aspect of kinetics that leads to much of our present description of the reactions in organic and inorganic chemistry.

To understand these reactions completely, however, we must understand the details of the elementary-reaction steps in which, typically, two molecules come together, rearrange, and depart as species that differ from those that encountered one another. Such elementary processes, until quite recently, were subject to no direct experimental study. Descriptions of some features in such reactions were developed in a theoretical or model way, and the basis for these ideas was primarily the indirect information provided by the temperature dependence of the rates of chemical reactions. Now very direct and revealing experimental results for elementary reactions are becoming available for some relatively simple reactions through the use of molecular-beam techniques. Although such studies reveal the details of only gas-phase elementary reactions, they provide invaluable data for the answer to the question of how elementary chemical reactions occur.

A goal of the detailed study of the elementary reactions can be the theoretical deduction of the numerical value of the rate constants for such reactions. These rate constants, as was shown in the preceding chapter, determine the rate of any overall reaction.

When the numerical values of rate constants are studied, the first striking feature that is noticed is the large variation of all rate constants with temperature. This dependence proves not to be merely a troublesome factor, but rather furnishes the most important clue to the approach that must be used to understand the molecular basis of the rate constants of elementary reaction steps.

17-1 Temperature Dependence of the Rates of Chemical Reactions

The rate equation and the value of the rate constant for a reaction are deduced from measurements of the rate of reaction at a fixed temperature. If experiments are performed at several different temperatures, it is generally found that the concentration dependence exhibited in the rate equation is unchanged but that the value of the rate constant is very temperature-dependent.

In most chemical reactions the temperature dependence of the reaction rate, which shows up according to

$$\text{Rate} = k(T) \times (\text{concentration-dependent term}) \qquad [1]$$

in $k(T)$, the rate "constant," is of the form shown in Fig. 17-1. That is, the rate increases rapidly with increasing temperature—a generalization well illustrated by the photographic developing rule of thumb that the rate of the developing reactions increases by a factor of 2 or 3 for each 10° rise in temperature.

Before dwelling on this typical temperature dependence, we should recognize that other behaviors are observed. Figure 17-2 illustrates that two other types of temperature dependence are found in reactions which reach an explosive stage and in certain enzyme-catalyzed reactions. Other temperature dependences also show up in special circumstances, but the majority of chemical reactions have the temperature dependence illustrated in Fig. 17-1.

In 1889 Arrhenius recognized that this typical temperature depend-

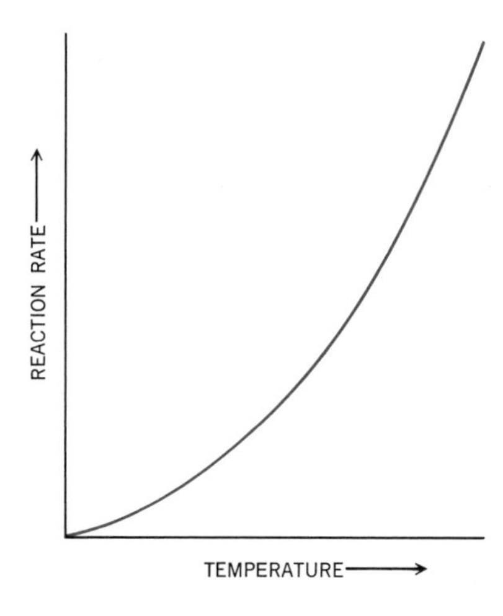

FIGURE 17-1
The shape of the rate-versus-temperature curve followed by most chemical reactions.

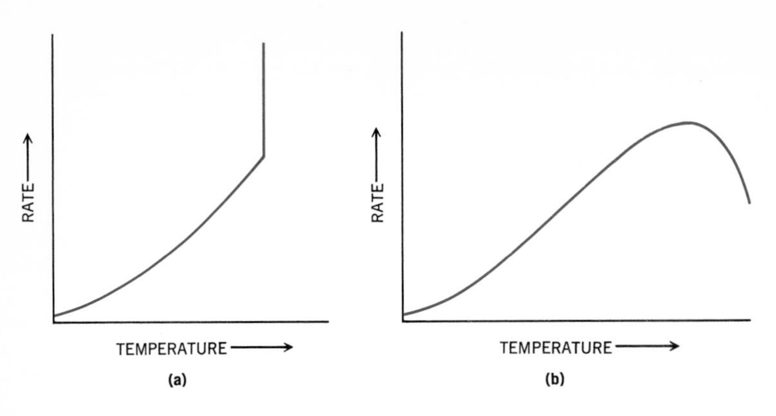

FIGURE 17-2 The rate–versus–temperature curves that are exhibited by some reactions (*a*) that reach an explosive stage and (*b*) that are enzyme-catalyzed.

ence indicates an exponential increase of the rate, or rate constant, with temperature.

It is best to recognize that this can be confirmed empirically by plotting

$$\ln k \text{ versus } \frac{1}{T} \qquad \text{or} \qquad \log k \text{ versus } \frac{1}{T}$$

The data on the rate constant of a particular reaction as a function of temperature, as given in Table 17-1, are shown in such a plot in Fig. 17-3.

Such linear plots imply the relation

$$\log k \propto \frac{1}{T}$$

or

$$\ln k \propto \frac{1}{T} \qquad \text{and} \qquad k \propto e^{\text{const}/T} \qquad [2]$$

In view of interpretations of this temperature dependence to be given later, and in line with the treatment of Arrhenius, this empirical relation can be more conveniently written

$$k = A\, e^{-E_a/RT} \qquad [3]$$

where A is called the *pre-exponential factor,* and E_a is known as the *activation energy*. With this notation one writes the logarithmic form of Eq. [3] as

$$\ln k = -\frac{E_a}{RT} + \ln A$$

or

$$\log k = -\frac{E_a}{2.303R}\frac{1}{T} + \log A \qquad [4]$$

TABLE 17-1 The rate constant as a function of temperature for the reaction

$CH_3I + C_2H_5ONa \rightarrow CH_3OC_2H_5 + NaI$

in ethyl alcohol*

t (°C)	k_2 (mol^{-1} $liter$ s^{-1})
0	5.60×10^{-5}
6	11.8
12	24.5
18	48.8
24	100
30	208

*From W. Hecht and M. Conrad, *Z. Physik. Chem.*, **3**:450 (1889).

The empirical constants E_a and A can therefore be deduced from the slope and intercept of the appropriate plot of the values of k at different temperatures.

Although these expressions are empirical correlations of rate data and the terms E_a and A are, for the present, to be treated as empirical parameters, the form of the expression for the rate constant might have been anticipated from the previously derived relation for the temperature dependence of the equilibrium constant. The thermodynamic equation

$$\frac{d(\ln K)}{dT} = \frac{\Delta H}{RT^2} \qquad [5]$$

can be written

$$K = (\text{const})\, e^{-\Delta H/RT} \qquad [6]$$

and if K is interpreted as k_1/k_{-1}, as can be done for an elementary reaction, the simplest temperature dependence that can be assumed for the rate constants that is consistent with Eq. [6] is

$$k_1 = A_1 e^{-(E_a)_1/RT} \qquad \text{and} \qquad k_{-1} = A_{-1} e^{-(E_a)_{-1}/RT}$$

which gives

$$\frac{k_1}{k_{-1}} = \frac{A_1}{A_{-1}} e^{-[(E_a)_1-(E_a)_{-1}]/RT} \qquad [7]$$

and agreement with the form of Eq. [6].

Most reactions that proceed at a reasonable rate, i.e., that have half-lives of minutes or hours, have values of E_a of 50 to 100 kJ. For such reactions one can use Eq. [4] to verify the photographer's guide that reactions go two or three times as fast when the temperature increases by 10°C.

In Secs. 17-3 and 17-4 we will see that theoretical treatments of elementary reactions lead to equations with temperature dependences

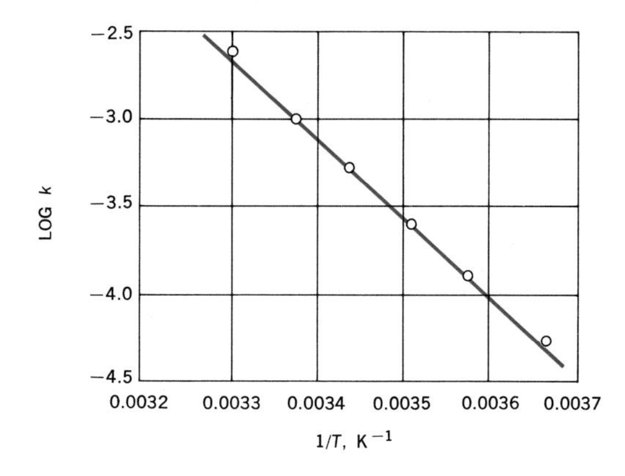

FIGURE 17-3
The Arrhenius plot of Eq. [4] for the data of Table 17-1. The straight line is represented by $\log k = -4250/T + 11.38$, or $\log k = -35{,}300/RT + \log 2.4 \times 10^{11}$, or $\ln k = -81{,}300/RT + \ln 2.4 \times 10^{11}$. Thus $E_a = 81{,}380$ J and $A = 2.4 \times 10^{11}$ mol liter^{-1} s^{-1}.

slightly different from that shown by Eq. [3]. One deduces, for example, expressions like

$$k = A'\sqrt{T}^{-E_a'/RT} \qquad [8]$$

Since the exponential factor, for any appreciable value of E_a, completely dominates the $\sqrt{T}$ factor, this expression is quite consistent with the empirical result of Eq. [3]. One must recognize, however, that if experimental data are fitted to an equation of the form of Eq. [8], values of both A' and E_a' will be somewhat different from A and E_a. Thus the slope of a ln k versus $(1/T)$ plot is $d(\ln k)/d(1/T)$, and this is interpreted according to Eq. [3] as $-E_a/R$, but according to Eq. [8] as $-\frac{1}{2}T - E_a'$. Thus E_a and E_a' are related by

$$E_a = \tfrac{1}{2}RT + E_a' \qquad [9]$$

In most cases the $\frac{1}{2}RT$ term is small but not negligible.

The empirical breakdown of the rate constant according, for example, to Eq. [4] introduces the two quantities A and E_a. Attempts to understand the nature of k become, therefore, attempts to understand the molecular basis and interpretation of A and E_a. A beginning to this was made by Arrhenius himself.

17-2 Introduction to Theories of Elementary-reaction Processes

For reactions whose rate depends on a single elementary reaction, the rate constant for an elementary step has the exponential temperature dependence of the overall reaction, as given, for example, by Eq. [3]. If several steps are involved, a similar dependence for each step is indicated, but now the experimental results will not give directly values for A and E_a for each elementary step. (Special difficulties stand in the way of deducing the parameters for the elementary reactions when, as for example for rate equations of the form of Eq. [27] or [60] of Chap. 16, the rate equation cannot be separated into temperature- and concentration-dependent factors.)

Theories for the value and the temperature dependence of $k(T)$, that is, for the values of A and E_a, will be aimed at elementary reactions, and therefore at overall reactions whose rate depends on one such reaction.

With this qualification we can now turn to an introduction to the theories of the reaction process that occurs in an elementary step, and we can do so in a way that corresponds to the ideas developed by Arrhenius. Two other approaches, which will be presented in the following sections, elaborate on these ideas and thereby produce more quantitative theories.

Either for a decay, or for a collision that is effective in producing product species from reactants, some amount of energy E_a must be available to allow for the necessary bond breakings and rearrangements. The probability of such an "activated complex" occurring will be proportional, among other things, to a Boltzmann factor

$$e^{-E_a/RT}$$

where E_a is the energy necessary for whatever molecular transformation must occur. Thus we can write

$$\frac{\text{Concentration of activated complexes}}{\text{Concentration of reagents}} \propto e^{-E_a/RT} \qquad [10]$$

Then, if it is assumed that the rate of the reaction is proportional to the concentration of activated complexes, we can write, with the introduction of the proportionality factor A,

$$\text{Rate} = Ae^{-E_a/RT} \times (\text{concentration of reagents}) \qquad [11]$$

This, if the reagent-concentration term is treated in more detail, agrees with the observed rate equations since it leads to the relation

$$k = Ae^{-E_a/RT}$$

This approach says that the empirical constant E_a is to be interpreted as the energy of the activated-complex molecules compared with the reagent molecules.

The idea of an activated complex can be presented on a plot of the energy of the system as ordinate versus the *reaction coordinate* as abscissa. The reaction coordinate is not any single internuclear distance, but rather depends on all the internuclear distances that change as the reactant molecules are converted into product molecules. In general, it is impossible, and for the present purpose unnecessary, to give a quantitative description of the reaction coordinate. It consists merely of a qualitative description of the extent of the transformation from reactants to products. The diagram that can be constructed to represent the reaction is shown in Fig. 17-4.

The Arrhenius theory suggests that the energy of the system that is partially transformed from reactants to products is given by the activation-energy term that appears in the empirical expression for the rate constant. This information is added to the elementary-reaction curve of Fig. 17-4, and a smooth, but otherwise undetermined, curve is drawn to pass through the three known energies.

The Arrhenius theory leads to a considerable improvement in our understanding of the reaction process. It is, however, still a very qualitative theory in that it does not show how the pre-exponential factor A depends on the molecular properties of the reaction system, nor does it attempt to predict the value of E_a. Two theories that are now presented make some progress in the interpretation of the factor A. The energy of the activated complex remains, however, too subtle a quantity

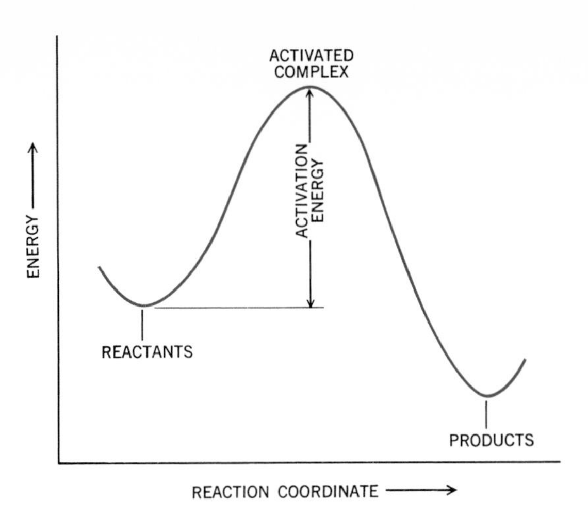

FIGURE 17-4
The energy profile for the transformation of reactants to products.

for evaluation, except in very simple cases where the necessary quantum-mechanical calculations can be carried out.

17-3 The Collision Theory

The collision theory, as its name implies, focuses attention on the idea that the reaction of molecules, considering particularly molecules in the gas phase, can occur only as a result of a coming together, or a collision, of the reactant molecules.

The effectiveness of a collision in producing a particular consequence can be described conveniently by assigning a cross-section area to the colliding molecules that is appropriate to that consequence. Thus the collision diameters introduced in the kinetic-molecular treatment of gases in Chap. 2 would now be said to be related to the *elastic-collision cross section. Reaction* cross sections are, as we will see, generally smaller than elastic-collision cross sections and are more sensitive to the energy involved in the collision.

The simplest kinetic-molecular treatment that assumes hard-sphere molecules with no interaction leads to a total collision cross section that is independent of energy, and can thus be represented as in Fig. 17-5*a*. If attractive interaction between molecules is admitted, the molecules would, at low enough relative velocities, be drawn toward each other, more collisions would occur, and this could be expressed by attributing to the molecules a larger cross section at lower collision energies, as shown in Fig. 17-5*b*.

In a similar way the qualitative dependence of the *reaction* cross section on collision energy can be shown. In the simplest view we might expect no reaction to occur unless some minimum collision energy, known as the *threshold energy,* were available, and that for greater energies, reactions could occur with some constant likelihood. The

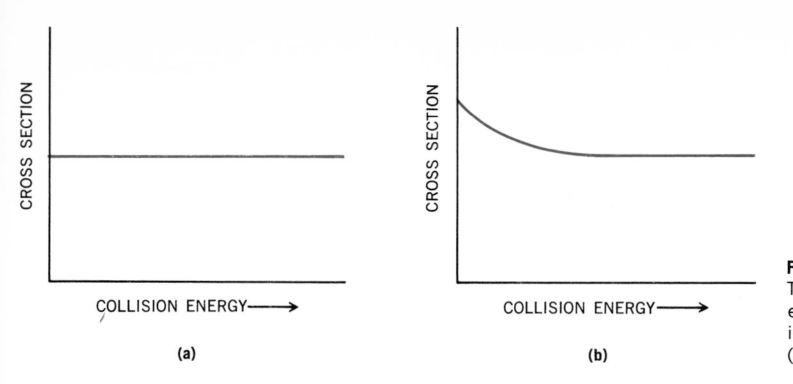

FIGURE 17-5
The cross sections for elastic collisions (*a*) for no intermolecular attraction and (*b*) with such interactions.

graph of Fig. 17-6*a* would express this model. More reasonably, after the threshold-energy requirement has been met, the reactive-collision cross section would increase as more collision energy became available, and a dependence like that shown in Fig. 17-6*b* would be expected.

These reactive-cross-section requirements are to be placed alongside the energy-distribution results that were obtained in Secs. 2-5 and 2-6. It is the product of the probability of molecules having a certain energy times the reactive cross section for that energy that determines the number of collisions that produce elementary reactions. These factors are shown graphically in Fig. 17-7.

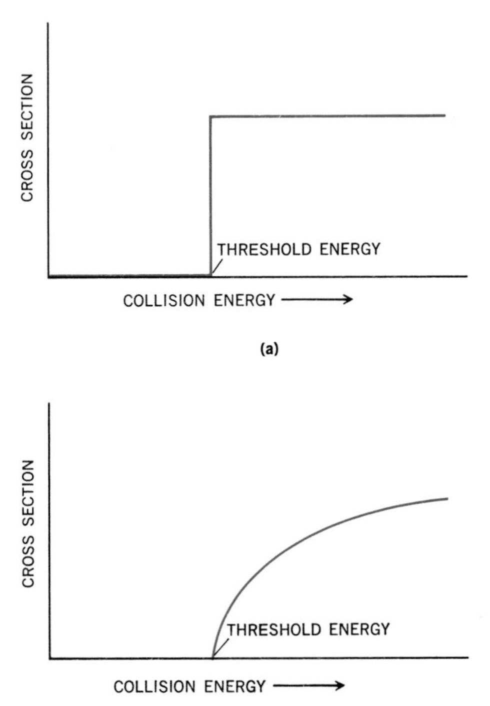

FIGURE 17-6
The cross sections for reactive collisions on the basis of (*a*) a threshold energy and (*b*) a threshold energy and a further dependence of reactive cross section on collision energy.

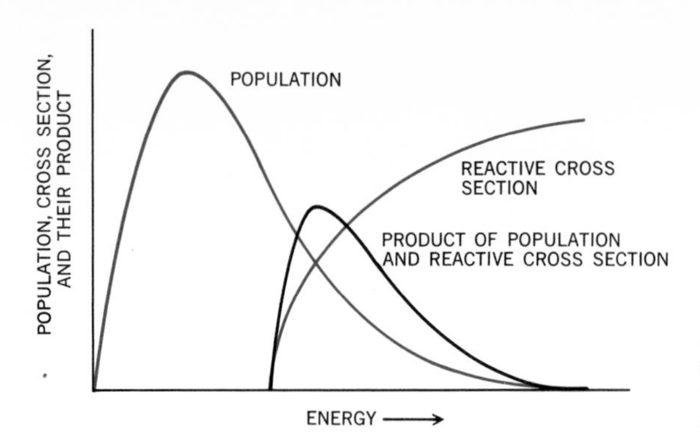

FIGURE 17-7
Schematic representation of the dependence of reactive collisions on energy.

*Now let us develop these ideas analytically. The number of collisions occurring at a particular relative velocity, as given by Eq. [36] of Chap. 16, can be converted to an expression for the number occurring with the possible release of the corresponding translational energy. If only one of the molecular types were moving, this energy would be $\frac{1}{2}m_A u_A{}^2$. To allow, again, for the movement of both molecules, we replace m_A by μ, the reduced mass, and u_A by u, the relative velocity, to obtain

$$\epsilon = \tfrac{1}{2}\mu u^2 \qquad [12]$$

Replacement of u in Eq. [37] of Chap. 16 by ϵ according to this relation now gives the total number of collisions as

$$Z_{AB} = N_A^* N_B^* \left(\frac{1}{\pi\mu}\right)^{1/2} \left(\frac{2}{kT}\right)^{3/2} \int_0^\infty \sigma_{AB}\, \epsilon e^{-\epsilon/kT}\, d\epsilon \qquad [13]$$

(Notice that σ_{AB} has been left inside the integral sign to allow for its dependence on the translational kinetic energy with which the collision occurs.)

The number of these collisions that lead to reaction, which is just the rate of reaction, is obtained by inserting the cross section for reaction $\sigma_{AB,R}$ in place of the collision cross section σ_{AB}. We thus proceed, with this form of the collision theory for a bimolecular reaction, from the equation

$$\text{Rate} = N_A^* N_B^* \left(\frac{1}{\pi\mu}\right)^{1/2} \left(\frac{2}{kT}\right)^{3/2} \int_0^\infty \sigma_{AB,R}\, \epsilon e^{-\epsilon/kT}\, d\epsilon \qquad [14]$$

Comparison with the form of the empirical rate equation

$$\text{Rate} = k'(T) N_A^* N_B^* \qquad [15]$$

then allows us to write, for the rate constant expressed in units of molecules^{-1} m^3 s^{-1}, the result

$$k'(T) = \left(\frac{1}{\pi\mu}\right)^{1/2} \left(\frac{2}{kT}\right)^{3/2} \int_0^\infty \sigma_{AB,R}\, \epsilon e^{-\epsilon/kT}\, d\epsilon \qquad [16]$$

It only remains for us to deal with $\sigma_{AB,R}$, and as will become apparent, a great deal of the problem has been assigned to, or submerged in, this quantity.

The simplest procedure is to assume that the molecules can be treated as hard, noninteracting spheres that collide without reaction for impact energies less than some threshold energy ϵ_0 but that all collisions occurring with energy greater than this amount are effective in producing a reaction. This is the basis for the dependence of cross section on energy as shown

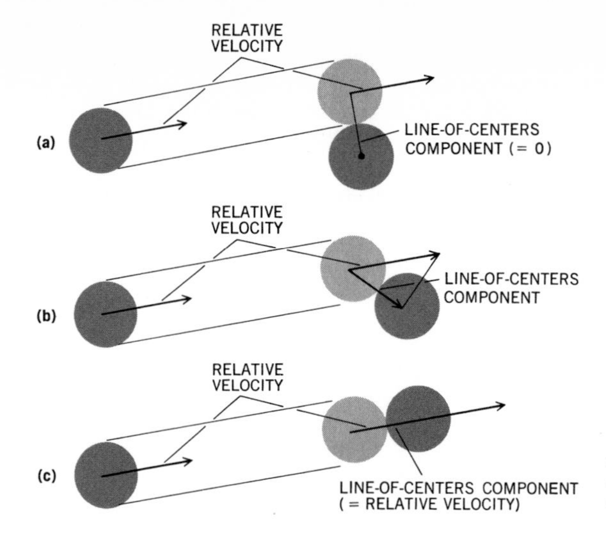

FIGURE 17-8
The dependence of the line-of-centers velocity on the nature of the collision as illustrated by (*a*) a glancing collision, (*b*) an intermediate collision, and (*c*) a direct collision.

in Fig. 17-6*a*. Analytically, this dependence is

$$\sigma_{AB,R} = \begin{cases} 0 & \epsilon < \epsilon_0 \\ \pi(r_A + r_B)^2 & \epsilon \geq \epsilon_0 \end{cases} \qquad [17]$$

The corresponding rate constant is deduced to be

$$k'(T) = \left(\frac{1}{\pi\mu}\right)^{1/2}\left(\frac{2}{kT}\right)^{3/2}\left[\int_0^{\epsilon_0} (0)\epsilon e^{-\epsilon/kT}\,d\epsilon + \int_{\epsilon_0}^{\infty} \pi(r_A + r_B)^2 \epsilon e^{-\epsilon/kT}\,d\epsilon\right]$$

$$= \left(\frac{1}{\pi\mu}\right)^{1/2}\left(\frac{2}{kT}\right)^{3/2}\left[0 + \pi(r_A + r_B)^2(kT)^2\left(1 + \frac{\epsilon_0}{kT}\right)e^{-\epsilon_0/kT}\right]$$

$$= (r_A + r_B)^2\left(\frac{8\pi kT}{\mu}\right)^{1/2}\left(1 + \frac{\epsilon_0}{kT}\right)e^{-\epsilon_0/kT} \qquad [18]$$

Other assumptions regarding the dependence of $\sigma_{AB,R}$ on ϵ can be made, and one that seems more reasonable, and leads to a simple result, assumes again that $\sigma_{AB,R}$ is zero for $\epsilon < \epsilon_0$, but that for $\epsilon \geq \epsilon_0$, $\sigma_{AB,R}$ becomes greater the greater the value of ϵ. One way of developing this idea is to assume that it is not the energy based on the relative velocity of the molecules that matters, but rather the energy component in the direction of the line between the centers. Thus, as Fig. 17-8 suggests, for energies near ϵ_0, only very direct center-to-center collisions can be effective, but for greater values of ϵ, even rather glancing collisions can have a line-of-centers energy greater than ϵ_0. This idea leads to the curve of Fig. 17-6*b*, or to the statement

$$\sigma_{AB,R} = \begin{cases} 0 & \epsilon < \epsilon_0 \\ \pi(r_A + r_B)^2\left(1 - \frac{\epsilon_0}{\epsilon}\right) & \epsilon \geq \epsilon_0 \end{cases} \qquad [19]$$

On this basis

$$k'(T) = \left(\frac{1}{\pi\mu}\right)^{1/2}\left(\frac{2}{kT}\right)^{3/2}\int_0^{\epsilon_0} (0)\epsilon e^{-\epsilon/kT}\,d\epsilon$$

$$+ \int_{\epsilon_0}^{\infty} \pi(r_A + r_B)^2\left(1 - \frac{\epsilon_0}{\epsilon}\right)e^{-\epsilon/kT}\,d\epsilon$$

$$= (r_A + r_B)^2\left(\frac{8\pi kT}{\mu}\right)^{1/2} e^{-\epsilon_0/kT} \qquad [20]$$

This equation, known as the *line-of-centers rate-constant result,* is often used as the representative result of the collision theory.

This result is based on a rate constant expressed in the units molecule^{-1} m^3 s^{-1}. Bimolecular rate constants obtained from the studies of the rates of reactions usually are expressed with concentrations in terms of moles per liter, and thus would have the units mol^{-1} liter s^{-1}.

To convert Eq. [20] to these units, we multiply by the factors $\mathfrak{N}$ molecules mol^{-1} and 10^3 liter m^{-3} and obtain

$$k(T) = 10^3 \mathfrak{N} k'(T)$$

$$= 10^3 (r_A + r_B)^2 \mathfrak{N} \left(\frac{8\pi kT}{\mu} \right)^{1/2} e^{-\epsilon_0/kT} \qquad [21]$$

This is one of the principal results of the collision theory of reaction rates. It is to be compared with empirical results such as those from Eq. [3] or [8].

With the representative results of $M = 50$ g and thus $\mu = 25$ g, or 0.025 kg, and $r_A = r_B = 1.5$ Å, we obtain, for units of mol^{-1} liter s^{-1},

$$k(T) = 5.0 \times 10^9 \sqrt{T} e^{-E_0/RT} \qquad [22]$$

or for temperatures near 25°C,

$$k(T) = 0.8 \times 10^{11} e^{-E_0/RT} \qquad [23]$$

The pre-exponential factors in Table 17-2 are seen mostly to fall around this value. The theory has therefore been successful to some extent. Closer inspection will show that disagreement can become quite appreciable, the empirical values being less than expected on the basis of the collision-theory result by a factor of 10, and sometimes by several powers of 10.

The details that the collision theory has overlooked are readily recognized but are less easily rectified. It seems likely, for example, that for a reaction to occur as a result of a collision, there must be not only enough energy, but also a suitable orientation of the colliding molecules. For larger molecules this *steric factor* can be expected to be much less than unity, because many collisions would not bring the reactive parts of the molecules together. There seems, however, to be no satisfactory quantitative way of allowing for this effect, and all that can be expected is that the observed pre-exponential factor will be less than that calculated when such geometric considerations are important.

An even more serious difficulty arises in the activation-energy term.

TABLE 17-2 Rate constants for second-order gas-phase reactions according to $k = Ae^{-E_a/RT_2}$ (for R in J deg^{-1} mol^{-1})

Reaction	*Rate constant (mol liter^{-1} s^{-1})*
$H_2 + I_2 \rightarrow 2HI$	$1 \times 10^{11} e^{-167,000/RT}$
$2HI \rightarrow H_2 + I_2$	$6 \times 10^{10} e^{-184,000/RT}$
$2NO_2 \rightarrow 2NO + O_2$	$4.5 \times 10^{9} e^{-112,500/RT}$
$2NOCl \rightarrow 2NO + Cl_2$	$9 \times 10^{9} e^{-101,000/RT}$
$NO + Cl_2 \rightarrow NOCl + Cl$	$1.7 \times 10^{9} e^{-83,000/RT}$
$NO + O_3 \rightarrow NO_2 + O_2$	$8 \times 10^{8} e^{-10,500/RT}$
$CH_3I + HI \rightarrow CH_4 + I_2$	$1.6 \times 10^{12} e^{-140,000/RT}$
$2C_2F_4 \rightarrow$ cyclo-C_4F_8	$6.6 \times 10^{7} e^{-108,000/RT}$
$2C_2H_4 \rightarrow$ cyclo-C_4H_8	$7.1 \times 10^{7} e^{-158,000/RT}$

The result that has been obtained is based on the supposition that only the 2 translational degrees of freedom can contribute the energy that goes into surmounting the activated-complex-energy barrier. There is, however, no reason to ignore energy contributions from rotational and vibrational degrees of freedom. If these contribute appreciably, a larger fraction of the collisions will be effective, and the rate of the reaction will be greater than that calculated on the basis of 2 degrees of freedom. Although it is easy to see that corrections of several powers of 10 are in this way possible, there is as yet no way of determining what molecular energies should be included in the calculation.

The chief contribution of the collision theory is that it leads to a definite prediction for the rates of gas-phase reactions. Its chief defects are that it is inflexible in that there is no definite way to allow for steric effects and for participation of the energy from various degrees of freedom and that it is applicable only to gas-phase reactions.

The second theory that attempts to explain reaction rates is now given. Judgment of it, as we shall see, is rather the opposite of that applied to the collision theory. It generally makes no quantitative prediction, and thereby avoids the test that the collision theory could be put to. Its chief merit is that it is very flexible and that the variations found in the rates of different reactions can be discussed in terms of the quantities introduced by the transition-state theory.

17-4 The Transition-state Theory

The collision theory is from the outset tied to the kinetic-molecular theory and, as a consequence, is removed from the realm of thermodynamics. The transition-state theory, on the other hand, is an approach that allows some use to be made of the important thermodynamic concepts. The transition-state theory, like the collision theory, falls far short of the goal of a completely theoretical prediction of rate constants. Nevertheless, the approach of the transition-state theory helps us to understand not only the molecular features of gas-phase reactions, but also some of the molecular features that operate on reactions in solution.

The transition-state theory focuses attention on the species in the reaction process that corresponds to the maximum-energy stage in the reaction process. This species, called the *activated complex,* or *transition state,* is in this theory treated formally as a molecule in spite of its ill-defined nature and transitory existence. More specifically, the theory assumes that this species can be treated as a thermodynamic entity.

Let us, to be specific, consider a bimolecular elementary reaction that leads to products. As such a reaction proceeds, there must at all times be a small fraction of the reacting species A and B that are in the process of undergoing the bond rearrangements that ultimately convert them to product species. These transition-state species will be denoted by the symbol $(AB)^{\ddagger}$, and this species will be treated as though it were a well-defined molecule worthy of thermodynamic consideration. The assumption is satisfactory as long as the reacting system does not lurch so violently from reactants to products that the transition-state species cannot establish an equilibrium with the reactants and products.

When the molecules A and B react to give products, one now can suggest, in an approach similar to that originated by Arrhenius, that A and B establish an equilibrium concentration of the transition-state species and that this species reacts further to form products. Thus

$$A + B \rightleftharpoons (AB)^{\ddagger} \rightarrow \text{products}$$

The rate of the reaction depends on two factors: the concentration of the transition-state species and the rate with which it breaks up to give products.

The concentration of the activated complex can, at least formally, be written in terms of the equilibrium expression

$$K^{\ddagger} = \frac{[(AB)^{\ddagger}]}{[A][B]}$$

or

$$[(AB)^{\ddagger}] = K^{\ddagger}[A][B] \qquad [24]$$

Although no value is given for $K^{\ddagger}$, it will be seen that its thermodynamic interpretation is profitable and justifies its introduction.

The rate with which the complex breaks up can be estimated by recognizing that it can fly apart into product molecules when a suitable vibration happens to have a large enough amplitude to break open the complex. The frequency of such a vibration will therefore be something like the rate with which the complex breaks up.

Let us denote the frequency of vibration along the *reaction coordinate* by ν_{RC}. The rate of the reaction is then given by the transition-state theory as

$$\text{Rate} = -\frac{d[A]}{dt} = -\frac{d[B]}{dt} = \nu_{RC}K^{\ddagger}[A][B] \qquad [25]$$

This expression becomes of value when $K^{\ddagger}$ is given a thermodynamic interpretation.

*Since we have no calorimetric-type thermodynamic data for the species $(AB)^{\ddagger}$, it is best to depend now on our understanding of thermodynamic properties in terms of molecular properties. We can, in fact, in view of Sec. 10-5, go directly to an interpretation of the equilibrium constant $K^{\ddagger}$ in terms of partition functions and the ground-state energy difference of A and B and the species $(AB)^{\ddagger}$. By Eq. [50] of Chap. 10, we write

$$K^{\ddagger} = \frac{[(AB)^{\ddagger}]}{[A][B]} = \frac{\not{q}_{(AB)^{\ddagger}}}{\not{q}_A \not{q}_B} e^{-E_{(AB)^{\ddagger}}/RT} \qquad [26]$$

where $E_{(AB)\ddagger}$ is the energy difference between A and B, on the one hand, and $(AB)^\ddagger$, on the other, and the partition functions are to be modified by having the volume term omitted from the translational components of the partition functions, as in Eq. [50] of Chap. 10. A slash through the partition-function symbol indicates this modification.

The partition function written to show the contributing degrees of freedom is

$$\not{q} = \not{q}_{\text{trans}}\, q_{\text{rot}}\, q_{\text{vib}}\, q_{\text{elec}} \qquad [27]$$

The nature of these partition-function factors is shown again in Table 17-3. It should be clear that, in general, enough data are available for species A and B so that all factors in $\not{q}_A$ and $\not{q}_B$ can be calculated.

For the species $(AB)^\ddagger$ the factors $\not{q}_{\text{trans}}$ and q_{elec} generally cause no difficulty. The rotational factor requires an estimate of the size and shape of the species, and this can be done. Most difficult is an estimate of the vibrational frequencies of the transition-state complex, particularly since, for at least some degrees of freedom, this species will be loosely bound and the corresponding vibrational modes will contribute appreciably to the partition function. One such vibration is that presumed to occur along the reaction coordinate, and it is motion along this coordinate that leads to conversion of the activated complex to product molecules. It is convenient to factor out the term for this degree of freedom from the partition function. If $\not{q}_{(AB)\ddagger}$ denotes the complete partition function for the $(AB)^\ddagger$ species and $\not{q}'_{(AB)\ddagger}$ denotes the partition function with the contribution from the vibrational mode along the reaction coordinate factored out, we can write, in view of Eq. [26] of Chap. 10, for the vibrational-mode partition function,

$$\not{q}_{(AB)\ddagger} = \frac{1}{1 - e^{-h\nu_{RC}/kT}}\not{q}'_{(AB)\ddagger} \qquad [28]$$

Further, if the molecule is only weakly bound along the reaction coordinate, $h\nu_{RC}$ will be much less than kT, and we can write

$$\frac{1}{1 - e^{-h\nu_{RC}/kT}} = \frac{1}{1 - (1 - h\nu_{RC}/kT + \cdots)}$$
$$\approx \frac{kT}{h\nu_{RC}} \qquad [29]$$

TABLE 17-3 Summary of partition-function expressions and representative values for gases at a concentration of 1 mol liter^{-1}

Partition function	Partition function per degree of freedom	Representative value of partition function per degree of freedom
$\not{q}_{\text{trans}} = q_t^{\,3}$	$\not{q}_t = \left(\dfrac{2\pi mkT}{h^2}\right)^{1/2}$	10^{10}
$q_{\text{rot}} = q_r^{\,3}$ (nonlinear) $= q_v^{\,2}$ (linear)	$q_r = \left(\dfrac{8\pi^2 IkT}{h^2}\right)^{1/2}$	10-100
$q_{\text{vib}} = q_v^{\,3n-6}$ (nonlinear) $= q_v^{\,3n-5}$ (linear)	$q_v = \dfrac{1}{1 - e^{-h\nu/kT}}$	1-10

It follows, combining the above equations, that the reaction rate can be expressed as

$$\text{Rate} = \nu_{RC} \frac{kT}{h\nu_{RC}} \frac{q_{AB\ddagger}}{q_A q_B} e^{-E_{(AB)\ddagger}/RT}[A][B]$$

$$= \frac{kT}{h} \frac{q'_{AB\ddagger}}{q_A q_B} e^{-E_{(AB)\ddagger}/RT}[A][B] \qquad [30]$$

Furthermore, comparison with the rate law for a bimolecular step lets us identify the rate constant k_2 as

$$k_2 = \frac{kT}{h} \frac{q'_{AB\ddagger}}{q_A q_B} e^{-E_{AB\ddagger}/RT} \qquad [31]$$

Although for relatively simple species all the partition functions can be calculated and a quantitative result obtained, it seems best to use the transition-state theory to obtain an appreciation of the factors that influence the rate of an elementary-reaction step.

We do this by first recalling the representative values of the various partition-function factors that are listed in Table 17-3.

Then, if we seek an order of magnitude for the deduced rate constant, we can apply such values to both the A and B and the $(AB)^\ddagger$ species. Let us do so for generally shaped, i.e., not monatomic or linear, A and B molecules, with n_A and n_B atoms, respectively, forming a generally shaped $(AB)^\ddagger$ complex containing $n_A + n_B$ atoms. Letting q_t, q_r, and q_v imply partition functions *per degree of freedom* for the type of motion indicated by the subscript, we can write

$$q_A = (q_t)^3(q_r)^3(q_v)^{3n_A-6}$$
$$q_B = (q_t)^3(q_r)^3(q_v)^{3n_B-6} \qquad [32]$$
$$q'_{(AB)\ddagger} = (q_t)^3(q_r)^3(q_v)^{3(n_A+n_B)-7}$$

(Note that the -7 in the vibration exponent for $q'_{(AB)\ddagger}$ results from the prior separation of the contribution from the reaction-coordinate vibration.)

With these representative values, after appropriate cancellations, we have

$$k_2 = \frac{kT}{h} \frac{(q_v)^5}{(q_t)^3(q_r)^3} e^{-E_{(AB)\ddagger}/RT} \qquad [33]$$

Insertion of numerical values gives, for the representative k_2, for the types of molecules being considered,

$$k_2 = 10^{-20} e^{-E_{(AB)\ddagger}/RT} \qquad [34]$$

where k_2 has the units of molecule^{-1} m^3 s^{-1}. Conversion to mol-liter units requires, as shown in Sec. 17-3, multiplication by $10^3\mathfrak{N}$ to give

$$k_2 = 6 \times 10^6 e^{-E_{(AB)\ddagger}/RT} \quad \text{mol liter}^{-1}\ \text{s}^{-1} \qquad [35]$$

More important than this reasonable (cf. Table 17-2) value is the recognition that the formation of the activated complex implies the conversion of translational and rotational degrees of freedom, with large partition functions, to vibrational degrees of freedom with small partition functions, as shown by the ratio of partition functions in Eq. [33]. This loss of freedom that opposes the formation of the activated complex is most important for the generally shaped molecules that have been considered. For reactions

TABLE 17-4 Representative values for the pre-exponential terms of Eq. [31] for various molecular types*

		Representative value at 25°C	
Type of Reaction	Pre-exponential term	molecule^{-1} m^3 s^{-1}	mol^{-1} liter s^{-1}
Atom + atom → diatomic	$\frac{kT}{h}\frac{q_r^2}{q_t^3}$	10^{-15}	10^{12}
Atom + linear → linear	$\frac{kT}{h}\frac{q_v^2}{q_t^3}$	10^{-17}	10^{10}
Atom + linear → nonlinear	$\frac{kT}{h}\frac{q_v q_r}{q_t}$	10^{-16}	10^{11}
Atom + nonlinear → nonlinear	$\frac{kT}{h}\frac{q_v^2}{q_t^3}$	10^{-17}	10^{10}
Linear + linear → nonlinear	$\frac{kT}{h}\frac{q_v^3}{q_t^3 q_r}$	10^{-19}	10^{8}
Nonlinear + nonlinear → nonlinear	$\frac{kT}{h}\frac{q_v^5}{q_t^3 q_r^3}$	10^{-20}	10^{7}

*After A. A. Frost and R. G. Pearson, "Kinetics and Mechanisms," John Wiley & Sons, Inc., New York, 1961.

involving monatomic species, with no rotational degrees of freedom, and linear molecules, with only 2 such degrees of freedom, this effect is less severe. Some representative results, collected in Table 17-4, show the consequence for different types of molecules and transition states.

*17-5 Comparison of the Results of the Collision and the Transition-state Theories

The theories presented in the preceding sections adopt very different approaches, but both succeed in reaching numerical results that are consistent with observed values, at least as regards order of magnitude. Now let us see if a comparison of the analytical form of these results reveals any common ground over which the theories proceed.

Consider the simplest example of monatomic A and B species colliding to form a diatomic-activated complex with internuclear distance $r_A + r_B$.

The transition-state theory requires us to characterize the activated complex by its mass $m_A + m_B$ and its moment of inertia $I = \mu(r_A + r_B)^2$, where μ is the reduced mass $m_A m_B/(m_A + m_B)$. The only vibrational degree of freedom of this species is that along the reaction coordinate, and since this has been given special treatment, there is no remaining contribution to $q'_{AB,v}$. Thus, according to the transition-state theory,

$$k_2 \text{ (transition-state theory)} = \frac{kT}{h}\frac{\not{q}'_{(AB)}}{\not{q}_A \not{q}_B} e^{-E_{(AB)\ddagger}/RT}$$

$$= \frac{kT}{h}\frac{\not{q}_{(AB)\ddagger,\text{trans}}\, q_{(AB)\ddagger,\text{rot}}}{\not{q}_{A,\text{trans}}\, \not{q}_{B,\text{trans}}} e^{-E_{(AB)\ddagger}/RT} \qquad [36]$$

The partition-function expressions that must be used are

$$q_{(AB)\ddagger,\text{trans}} = \left[\frac{2\pi(m_A + m_B)kT}{h^2}\right]^{3/2}$$

$$q_{(AB)\ddagger,\text{rot}} = \frac{8\pi^2(r_A + r_B)^2\mu\, kT}{h^2}$$

$$q_{A,\text{trans}} = \left(\frac{2\pi m_A\, kT}{h^2}\right)^{3/2}$$

$$q_{B,\text{trans}} = \left(\frac{2\pi m_B\, kT}{h^2}\right)^{3/2}$$

Substitution of these relations in Eq. [36], with cancellation of many terms, gives

$$k_2 \text{ (transition-state theory)} = (r_A + r_B)^2 \left(\frac{8\pi kT}{\mu}\right)^{1/2} e^{-E_{(AB)\ddagger}/RT} \qquad [37]$$

The result is virtually identical with Eq. [20], which is the principal result of the collision theory. One need only identify $E_{(AB)\ddagger}$ with the molar equivalent of the molecular threshold energy ϵ_0. Thus one sees that the two theories not only lead to similar numerical results, but also involve similar molecular features.

17-6 Application of the Transition-state Theory to Reactions in Solution

For reactions occurring in the liquid state, one cannot proceed directly with the collision theory or, in the form developed above, the transition-state theory. But the latter theory can be developed so that it gives a theoretical interpretation to the A and E_a parameters of the Arrhenius equation without demanding detailed molecular data. Here, then, one welcomes the thermodynamic functions with their lack of such demands.

We now proceed by treating the formation of the activated complex $(AB)^\ddagger$ and the equilibrium constant $K^\ddagger$ in terms of the *free energy of activation*, the *entropy of activation*, and the *enthalpy of activation.* With such terms $K^\ddagger$ can be interpreted as

$$(\Delta G^\circ)^\ddagger = -RT \ln K^\ddagger$$

or

$$K^\ddagger = e^{-(\Delta G^\circ)^\ddagger/RT} \qquad [38]$$

For the reaction at a given temperature the free energy of activation can be interpreted in terms of an entropy and an enthalpy contribution according to

$$(\Delta G^\circ)^\ddagger = (\Delta H^\circ)^\ddagger - T(\Delta S^\circ)^\ddagger$$

Substitution of this relation in Eq. [38] yields

$$K^\ddagger = e^{+(\Delta S^\circ)^\ddagger/R} e^{-(\Delta H^\circ)^\ddagger/RT} \qquad [39]$$

Insertion of this interpretation in Eq. [25], along with the assumption that the average vibrational energy $h\nu_{RC}$ along the reaction coordinate can be set equal to kT, leads to

$$\text{Rate} = \frac{kT}{h} e^{+(\Delta S^\circ)^\ddagger/R} e^{-(\Delta H^\circ)^\ddagger/RT}[A][B] \qquad [40]$$

Then the rate constant is expressed as

$$k_2 = \frac{kT}{h} e^{+(\Delta S^\circ)^\ddagger/R} e^{-(\Delta H^\circ)^\ddagger/RT} \qquad [41]$$

where, now, the (ΔS°) term is to be calculated for all degrees of freedom except that along the reaction coordinate.

With the recognition that the variation of T is small compared with that in the exponential term, Eq. [41] agrees in form with the empirical Arrhenius expression.

The exponential temperature-dependent term now involves an enthalpy rather than an energy of activation. For liquid systems the difference, as discussed in Sec. 7-3, will be completely negligible, and for gaseous systems will be small, and if necessary, can be calculated from $RT\,\Delta n$.

It is again the theoretical interpretation of the pre-exponential A factor that is of particular interest.

17-7 The Entropy of Activation

Equation [41] can be written to show the exponential temperature-dependent term as an energy rather than enthalpy term, by inserting $(\Delta H^\circ) = (\Delta E^\circ) + RT\,\Delta n$, where Δn is the change in the number of molecules when the complex is formed. When this is done, the transition-state theory interprets the pre-exponential A factor of the Arrhenius equation as

$$A = e^{\Delta n} \frac{kT}{h} e^{(\Delta S^\circ)^\ddagger/R} \qquad [42]$$

Thus, before A can be evaluated, an estimation of the entropy of activation must be made.

Since $(\Delta S^\circ)^\ddagger$ is the entropy change in going from the reagents to the activated complex and since little can be easily said about the properties of the activated complex, the transition-state theory tends to avoid any definite quantitative predictions. In spite of the ill-defined nature of the transition state, a number of conclusions can be drawn concerning $(\Delta S^\circ)^\ddagger$, and these can be compared with the kinetic results.

All gas-phase reactions lead, in the formation of an activated complex from reactant molecules, to the conversion of translational and rotational degrees of freedom of the reactants to vibrational degrees of freedom of the transition-state species. The more widely spaced energy level for this latter type of molecular motion implies a smaller entropy, and thus a negative value, for $(\Delta S)^{\ddagger}$.

For reactions that occur in solution a similar but smaller negative $(\Delta S)^{\ddagger}$ might be expected as the less free translations and rotations are lost and are replaced by vibrations. In addition, however, solvent molecules are involved, and the effect on them can dominate the effect that can be attributed to the reacting molecules themselves. All solutes, and particularly ions, in solvents other than very inert ones, interact with solvent molecules. The solvent molecules are to some extent oriented about the solute, and this orientation imposes a restriction on the motion of some of the solvent molecules. This solvation is an appreciable factor in determining the entropy of the system. Changes in this solvation entropy must therefore be considered in the formation of the activated complex. The uniformly negative value of $\Delta S^{\ddagger}$ for gas-phase reactions, corresponding to a loss of freedom of motion, does not therefore hold for reactions of solvated species in solution.

When oppositely charged ions, for example, react to form a neutral molecule, the extent of solvation is greatly reduced. Even the activated complex, in which the opposite charges will, at least, be close together, can be expected to be formed with a decrease in solvation and a corresponding positive entropy of activation. An example of such a situation is provided by the displacement of water molecules in the reaction

$$Cr(OH_2)_6{}^{3+} + CNS^- \rightarrow [Cr(OH_2)_5CNS]^{++} + H_2O$$

which has a value of $\Delta S^{\ddagger}$ of $+120$ J deg^{-1} mol^{-1}.

On the other hand, the formation of an activated complex that carries charges when the reagents do not will lead to a large negative value for $\Delta S^{\ddagger}$ corresponding to the additional loss of freedom of motion by the solvating molecules. The displacement of bromide in the reaction step

$$C_6H_5\text{—}NH_2 + C_6H_5\text{—}\overset{\displaystyle O}{\overset{\|}{C}}\text{—}CH_2Br \rightarrow \left[H_3\overset{\delta^+}{N}\text{-----}CH_2(C(=O)C_6H_5)\text{-----}\overset{\delta^+}{Br} \right]$$

is an example of such a situation. The entropy of activation for this reaction has been reported as -200 J deg^{-1} mol^{-1}, a very large negative value.

Although not all entropies of activation-of-solution reactions can be so easily rationalized, this type of argument shows how the transition-state theory gives a valuable framework within which observed rate constants can be understood.

RADIATION CHEMISTRY

It has been assumed in our studies of elementary reactions so far that the energy necessary for such reactions is provided by the reservoir of thermal energy in the reacting system. But the necessary energy can be imparted in other ways. Of particular importance is the method of photochemistry, in which electromagnetic radiation in the visible or ultraviolet region initiates a reaction. Much of the remainder of the chapter will be devoted to photochemistry. It seems appropriate, however, to mention also the recent and more specialized process in which molecules are bombarded with high-energy electrons. The molecular ions that are produced are generally detected with a mass spectrometer. Finally, some chemical results due to the interaction of very high energy radiation with matter will be pointed out. This subject tends now to pre-empt the title *radiation chemistry*.

17-8 Light Absorption

In ordinary chemical reactions the energy of activation is supplied by the chance collection in a molecule, or a pair of molecules, of a large amount of thermal energy. An alternative way in which the necessary activation energy can be acquired is through the absorption of quanta of visible or ultraviolet radiation. Reactions which follow as a result of energy so acquired are classified as *photochemical* reactions. With this description, photochemistry appears as a special branch of kinetics. In practice the theory and the experimental arrangements used in photochemistry set it off as a rather special subject. The goal of these reaction studies is, however, still the elucidation of the mechanism of the reaction.

Basic to the understanding of photochemical processes is an appreciation of the energy acquired by a molecule as a result of the absorption of light. It will be recalled that the energy of a light quantum is related to the frequency of the light by Planck's law, $\Delta\epsilon = h\nu$. The energy of an Avogadro's number of light quanta is called an *einstein*. The einstein is therefore the energy acquired when each of an Avogadro's number of molecules absorbs one quantum of radiation. The name of the einstein unit of radiation stems from Einstein's photochemical law, which states that each molecule is activated by the absorption of one light quantum.

The amount of energy in an einstein of radiation can be appreciated from a calculation of this quantity for visible light, say of 6000 Å wavelength. The frequency of such light is given by c/λ, and the energy per quantum by $h\nu = hc/\lambda$. Multiplication by Avogadro's number gives the energy of the einstein. In this way it is found that 1 einstein of yellow light has an energy of 200 kJ. Light of shorter wavelength will have a correspondingly higher energy. In the ultraviolet, for example, at a wavelength of 2000 Å, the energy of 1 einstein is 600 kJ.

From these energy values and the bond energies deduced in the thermodynamic studies of Chap. 7, it is seen that the absorption of light in the visible or ultraviolet can be expected to be sufficient to break a chemical bond or at least to produce a high-energy reactive molecule.

The amount of the chemical reaction that occurs in a photochemical experiment is related to the amount of light that is absorbed. The decrease in the intensity of a given wavelength of light as the light traverses a length dl of a cell containing a light-absorbing compound of concentration c is found to be proportional to I, c, and dl; that is,

$$dI = -\alpha c I \, dl \qquad [43]$$

where α is a proportionality constant, the *absorption coefficient.* This coefficient is generally very dependent on the wavelength and is large for wavelengths at which absorption occurs. If the incident-light intensity on the cell is I_0 and the intensity after traversal of the cell of length l is I, the Beer law expression is obtained from an integration of Eq. [43] as

$$\ln \frac{I_0}{I} = \alpha c l \qquad [44]$$

The experimental arrangement for a photochemical experiment is indicated in Fig. 17-9. The data that are obtained using a setup that has been suitably calibrated allow the determination of the number of light quanta that are absorbed in a particular experiment.

A characteristic, and often limiting, feature of photochemical studies is the fact that only a small fraction of an einstein of radiation can be absorbed in a reasonable time with typical monochromatic light sources and experimental systems. In a typical photolysis of methyl cyclopropyl ketone using 2537-Å radiation, for example, experiments lasting about 6 h resulted in the absorption of about 10^{19} quanta, or 1.6×10^{-5} einstein, of radiation. We shall see that one frequently gets something like one product molecule for each quantum absorbed, and

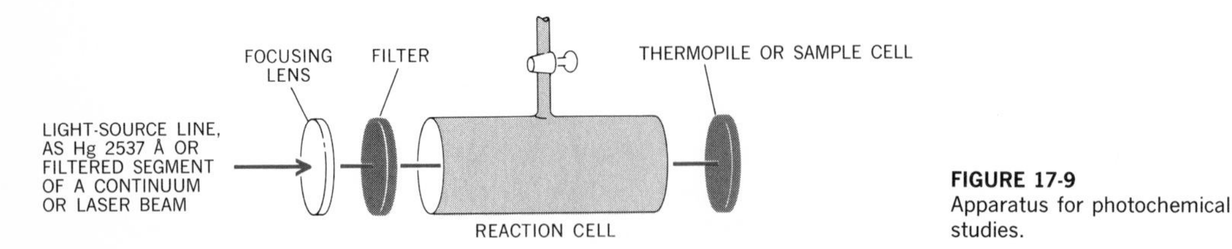

FIGURE 17-9
Apparatus for photochemical studies.

therefore a number of moles of product that is equal to the number of einsteins of radiation absorbed. The analyses of such small fractions of a mole of products, as in this methyl cyclopropyl ketone example, can present considerable experimental difficulties. The use of the mass spectrometer is an invaluable aid in the analysis of these small quantities.

17-9 The Primary Process

A photochemical reaction follows from the absorption of a quantum of radiation by a species in the reaction mixture. This absorption of radiation is known as the *primary process* in the photochemical reaction.

In a system involving atoms, those most commonly studied being mercury atoms, this primary process consists in the formation of an excited atom or, if short-wavelength high-energy radiation is used, of the ionized atom.

In molecular systems the equivalent of these two types of primary processes can be recognized, but a great variety of detail can occur. Some of the possible results of the absorption of a quantum of visible or ultraviolet light were mentioned in Sec. 13-5. Again it is convenient to consider the excitation in terms of potential-energy curves which represent the potential energy of the molecule as a function of the internuclear distance of some bond in the molecule.

The excited electronic state may correspond, in contrast to all the states of Fig. 13-12, to an electronic form in which the electrons of the bond of the absorbing molecule no longer maintain a chemical bond. In such cases the upper-state potential curve shows no minimum, and the absorption of radiation leads to the dissociation of the absorbing molecule. The fragments that are obtained may, furthermore, be in their lowest or in some excited electronic state. In almost all photochemical studies, when a bond breaks as a result of the absorption of a quantum of radiation, a homolytic cleavage of a chemical bond occurs; i.e., dissociation leads to the formation of free radicals, and the earlier discussions of reaction mechanisms lead us to expect chemical reactions to be initiated by these free radicals.

On the other hand, the absorption of radiation may lead to an excited, but bound, state, as in Fig. 13-12. Such species have the necessary activation energy for many chemical reactions. The photochemical consequences of such species depend, however, on the lifetime of this species, or some other related high-energy species, compared with the average time between collisions of this species and a molecule with which it can react. The fate of an excited molecule, or of the directly formed molecular fragments, is treated in Sec. 17-10.

17-10 Secondary Processes

A detailed story of the fate of an excited molecule can seldom be given. A number of courses are generally open to a molecule in a high-energy electronic state; some of these lead to the return of the molecule to the ground state by the emission or dissipation of its excess energy, whereas other paths produce chemical decomposition or reaction.

The most direct path that returns the system to the ground state is, as indicated in Fig. 17-10, the rapid loss of vibrational energy to form the excited state with its lowest vibrational energy and the subsequent emission, which the Franck-Condon principle requires to be vertical, to the ground electronic state. Such a process, known as *fluorescence,* occurs after a relatively short lifetime of about 10^{-9} s of the upper state. The process allows the excited molecule little chance to participate in a photochemical reaction.

A second process that involves the emission of radiation is *phosphorescence.* Phosphorescence is characterized by an emission of radiation at times longer than about 10^{-5} s and up to minutes or hours. It is now quite well established that such long excited-state lifetimes depend on the molecule getting into an excited triplet state, as by the crossing over from one potential-energy curve to another, as shown in Fig. 17-11. Such triplet states cannot readily emit radiation and form the ground singlet state because of the selection rule that transitions can occur only between states of like multiplicity. This rule is particularly effective if the molecule under study is embedded in a rigid glass, which minimizes the intermolecular electric field fluctuations and increases the lifetime of the excited triplet state to times even up to hours. It follows that, if the molecule attains the excited triplet state, it is susceptible to photochemical reaction during its relatively long lifetime.

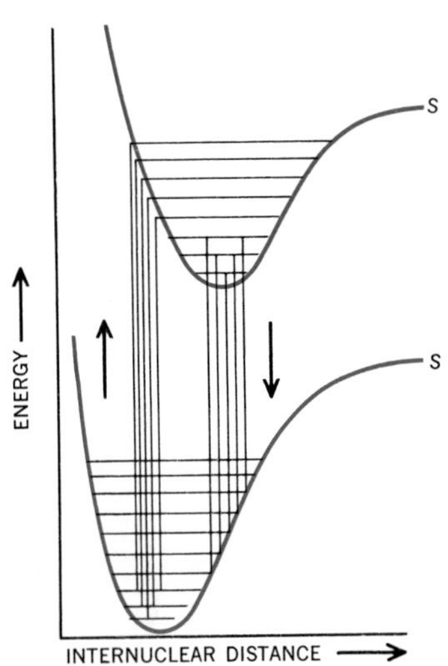

FIGURE 17-10
Electronic excitation and fluorescence in a molecule. Both electronic states are singlets.

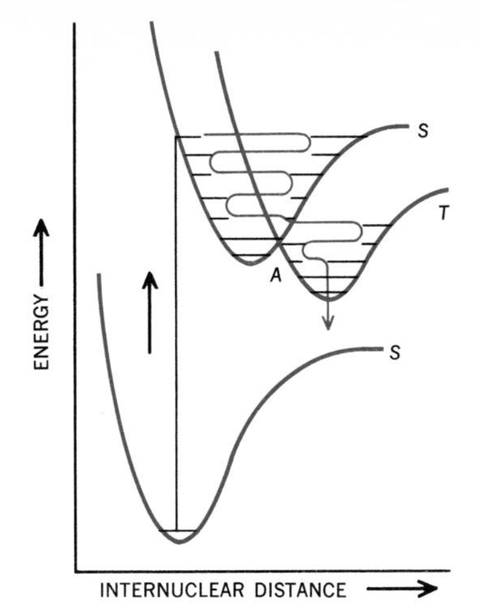

FIGURE 17-11
The formation of an excited triplet state by internal conversion from an excited singlet state. At point A the molecule has the same potential energy, the same internuclear distances, and zero kinetic energy, regardless of whether it has the S or T electronic configuration. Crossing from one to the other is therefore relatively easy.

It may not, however, be necessary for the molecule to emit radiation for the ground state to be reached by the initially excited molecule. For a molecule of any complexity there will be a large number of different excited electronic-energy states, and furthermore, one or more of these may have a potential-energy curve that crosses that of the ground state. Just as the triplet state was reached from the excited singlet state of Fig. 17-11, so also can the ground state be reached by the crossing over from the potential-energy curves of excited states. Such a process, which appears to be very important in polyatomic molecules, is known as *internal conversion.* Since this process can be rapid, the molecule with such potential-energy curves will probably be returned to the ground state before a photochemical reaction occurs.

A photochemical reaction can therefore be expected to occur if the excited state is a repulsive one and free radicals are formed and do not immediately recombine. Furthermore, if the lifetime of a bound excited state is long enough, i.e., if the energy is not lost too rapidly by fluorescence, or internal conversion to the ground state, a collision can occur which forms intermediate or product species and thereby leads to a photochemical reaction.

One can determine the number of quanta absorbed and the number of molecules of reactant used or of product formed in a given experiment. From these data the *quantum yield,* or *quantum efficiency,* defined as the ratio of the number of molecules reacting to the number of quanta absorbed, is determined. A number of reactions have quantum yields close to unity. The extremes, however, go from quantum yields of zero for molecules that absorb visible or ultraviolet radiation and show no photochemical reactions to quantum yields as high as 10^6 for chain reactions such as that which will be discussed for the reaction of H_2 and Cl_2.

17-11 Some Photochemical Reactions

A few representative photochemical reactions will now be discussed to show that detailed reaction mechanisms can be written for such reactions. Although photochemical reactions are studied in both gas and liquid phases, the present examples will be drawn from gas-phase studies.

A particularly clean reaction which can be nicely understood is the photochemical decomposition of HI to give hydrogen and iodine. The absorption of radiation occurs in the ultraviolet at around 2500 Å. The absorption band is a continuum and corresponds to a transition to a state in which the H and I atoms are not bound. The primary process can be written

$$\mathrm{HI} + h\nu \rightarrow \mathrm{H}\cdot + \mathrm{I}\cdot$$

Quantum-yield measurements show that two molecules of HI are decomposed for each quantum absorbed. Secondary reactions that lead to this yield and are energetically reasonable are

$$\mathrm{H}\cdot + \mathrm{HI} \rightarrow \mathrm{H_2} + \mathrm{I}\cdot$$

and

$$\mathrm{M} + \mathrm{I}\cdot + \mathrm{I}\cdot \rightarrow \mathrm{I_2} + \mathrm{M}$$

where M is some third body, perhaps the container wall, that takes up the energy liberated by the bond formation. The sum of the three reaction steps is

$$2\mathrm{HI} + h\nu \rightarrow \mathrm{H_2} + \mathrm{I_2}$$

which agrees with the quantum-yield data.

A number of other steps can be written, but arguments can be given for assuming them to be unimportant contributors. As bond energies indicate, for example the reaction

$$\mathrm{I}\cdot + \mathrm{HI} \rightarrow \mathrm{I_2} + \mathrm{H}\cdot$$

is endothermic to the extent of 148 kJ and would not be expected to be as important as a reaction of H atoms with HI, which is exothermic to the extent of 137 kJ.

The photochemical reaction of H_2 and Cl_2 to form hydrogen chloride is the example par excellence of a photochemical chain reaction. The primary process is again a molecular dissociation, now resulting from the absorption of radiation by Cl_2; that is,

$$\mathrm{Cl_2} + h\nu \rightarrow 2\mathrm{Cl}\cdot$$

The reactions that follow are

$$\mathrm{Cl}\cdot + \mathrm{H_2} \rightarrow \mathrm{HCl} + \mathrm{H}\cdot$$

and

$$H\cdot + Cl_2 \rightarrow HCl + Cl\cdot$$

Again the bond energies of Table 7-4 can be used to show that these reactions are energetically feasible. The pair of reactions proceed with an energy output of

$$2D_{HCl} - D_{H_2} - D_{Cl_2} = 2(432) - 436 - 243 = 185 \text{ kJ mol}^{-1}$$

Repetition of the two steps will not therefore be limited by energy requirements. The chain-terminating steps that will occur are the three-body processes

$$M + 2Cl\cdot \rightarrow Cl_2 + M$$

or

$$M + 2H\cdot \rightarrow H_2 + M$$

or

$$M + H\cdot + Cl\cdot \rightarrow HCl + M$$

At low free-radical concentrations these three-body collisions, or wall reactions, will be highly unlikely, and it appears that as many as a million chain steps can occur before such a chain-terminating step destroys the free radicals. The quantum yield in such chain reactions is dependent on the size of the reaction vessel, and this dependence can be attributed to the chain-terminating surface reactions.

More typical of the photochemical reactions that are being studied is the photolysis of acetone. The presence of the carbonyl group results in the absorption of radiation of about 3000 Å. Such quanta have energy of 400 kJ and are insufficient, therefore, to cause rupture of the carbonyl double bond. Dissociation does occur, however, and it appears that the weaker adjacent C—C bond breaks to give the primary process

$$CH_3-\overset{\overset{\displaystyle O}{\|}}{C}-CH_3 + h\nu \rightarrow CH_3-\overset{\overset{\displaystyle O}{\|}}{C}\cdot + \cdot CH_3$$

Secondary reactions that explain the principal reaction products of some experiments are the further radical breakup

$$CH_3-\overset{\overset{\displaystyle O}{\|}}{C}\cdot \rightarrow CO + \cdot CH_3$$

and the radical combination

$$\cdot CH_3 + \cdot CH_3 \rightarrow C_2H_6$$

Small amounts of the other products that are observed can also be accounted for by the reactions of the species postulated here.

In one of the early studies on the photolysis of acetone, the quantum yield, based on the number of acetone molecules that were decomposed, was measured by Daniels and Damon to be 0.17. Furthermore, a green fluorescence was observed from the reaction cell, and measurements of the quantum output of this radiation showed that it comprised about 3 percent of the quanta lost by the radiation beam in passing through the reaction cell. It follows, therefore, that about 80 percent of the absorbed radiation is in this case degraded without emission or reaction.

The primary bond-breaking process in acetone appears typical of such processes in that the bond cleavage occurs near the group in the molecule which is responsible for the absorption of the radiation. It appears to be generally true that either the absorbing bond, or group, or an immediately adjacent bond breaks when the primary process leads to bond cleavage.

17-12 Flash Photolysis

The deduction of a mechanism to explain the products of a photolysis experiment postulates various intermediates and reaction steps. The use of radiation to generate the reaction intermediates gives us a better idea of some of the high-energy species than can be obtained in ordinary thermally controlled chemical reactions. In both thermal and photochemical reactions these high-energy intermediates, or transition states, are present at any time at only a vanishingly low concentration. It is possible, however, to generate these interesting and elusive species in appreciable amounts by the photochemical method known as *flash photolysis*. The apparatus is indicated schematically in Fig. 17-12.

A flash of light, of duration of about 10^{-5} s, can be generated with as much as 2000 J of light energy. In the visible region this means that about 10^{-3} einstein of radiation is emitted, and therefore, if this flash is allowed to fall on a small sample of millimole size, most of the sample molecules can be brought to an excited state. The opportunity is thereby given for directly studying the initial excited state or those states which follow quickly from it. The opportunity must be quickly grasped, since within a matter of milliseconds all intermediates will have been converted to final products. It is possible, however, to synchronize

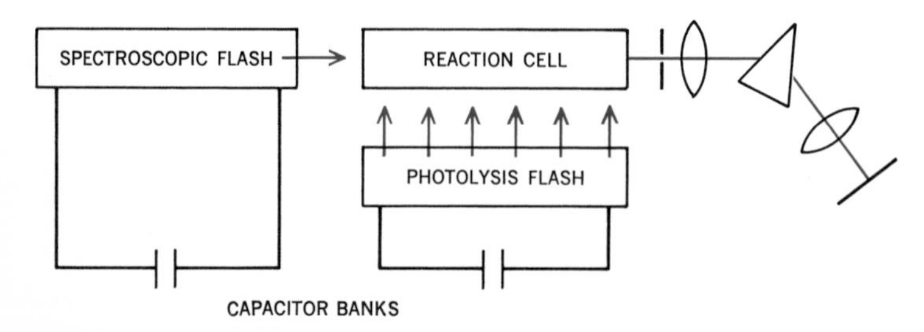

FIGURE 17-12
A flash-photolysis apparatus.

a spectroscopic flash of light to follow the principal flash by some fraction of a millisecond. This spectroscopic flash can be used, in the setup of Fig. 17-12, to obtain an absorption spectrum of the species present at the time of the spectroscopic flash. In this way one can obtain not only indications of what species are generated by the principal flash, but also data as to how these species vary, or give rise to others, in the milliseconds that follow the formation of the primary-process products. A very direct picture of the kinetic behavior of reaction intermediates can thus be obtained.

Many interesting results are being produced by this difficult but powerful technique.

17-13 Mass-spectroscopic Results

Mass spectroscopy provides a rather special method for studying molecular species that do not exist under ordinary conditions but may be important as reaction intermediates. The mass spectrometer has already established itself as an analytical tool of great value, and although the data that are obtained are used primarily for analyses, it is being recognized that interesting chemical features are exhibited by these data.

Mass spectra are obtained in an apparatus shown schematically in Fig. 17-13. Positive ions of the gas sample are produced as a result of the bombarding electron beam which knocks electrons, or negative groups, from the molecule and thereby produces excited and reactive species. Of the ions that are produced, only the singly positively charged

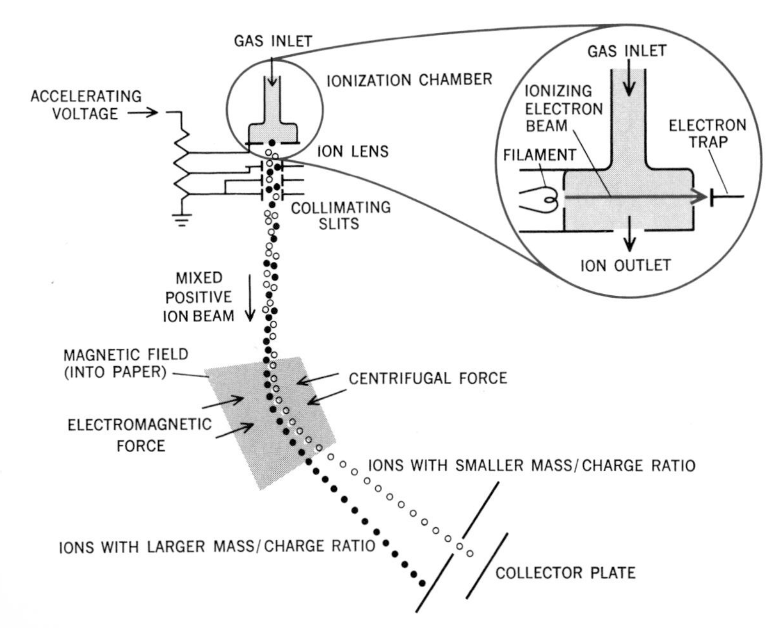

FIGURE 17-13
Schematic diagram of an analytical mass spectrometer. (*General Electric Company.*)

ions are usually detected. They are accelerated, passed through the magnetic field, and detected by an ionization gauge. In the same way as isotopic species are separated by the different curvatures of their paths in the magnetic field, so also are the particles of different mass separated when a molecule is fragmented by an electron beam. Typical mass-spectral fragmentation patterns for some small molecules are shown in Fig. 17-14.

It is such patterns, which are characteristic of the sample molecule, that are of use in analyses. The *parent peak,* which has the same mass number as the molecule under study, may or may not show up prominently. In any case the overall pattern is characteristic of the molecule,

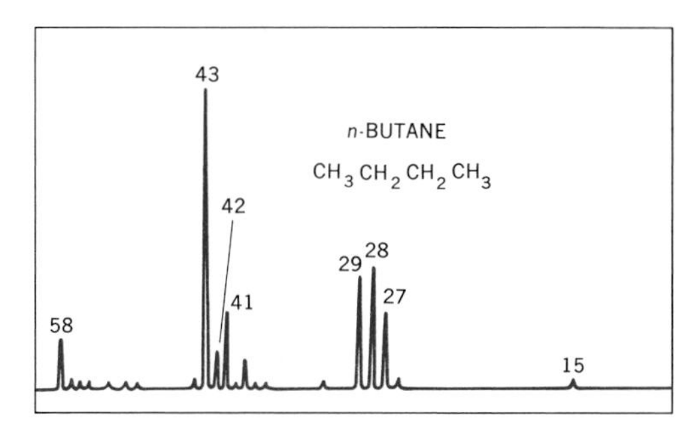

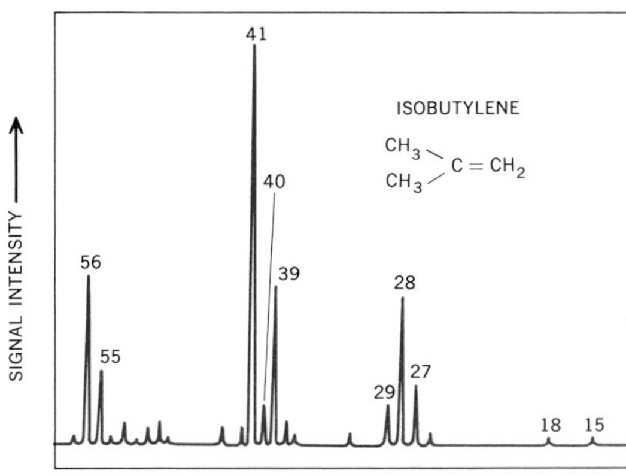

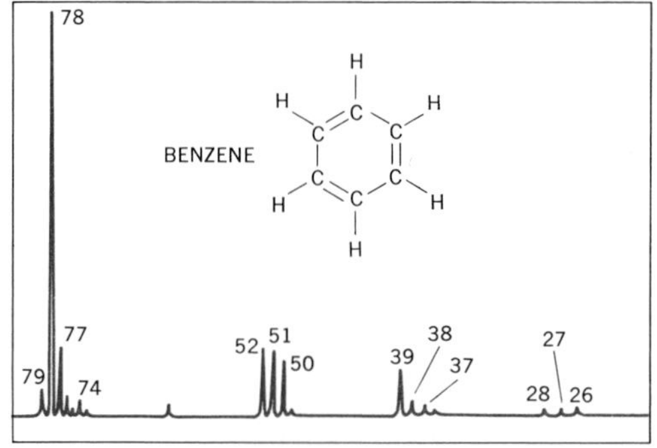

FIGURE 17-14
The mass spectra of several hydrocarbons. Most fragments that are detected have a single positive charge. The abscissa numbers then show the mass numbers of the fragments formed by the electron impact.

and such patterns can be used to distinguish, for example, isomers such as normal butane and isobutane. Clearly, moreover, the data on the fragments that are produced when a gas-phase ion is formed provide a wealth of information on the manner in which these reactive species decompose.

The mass spectrometer can also be used to obtain the minimum energy of the impinging electron beam that is necessary to disrupt the molecule. In this way *appearance potentials* are obtained. For a sample of CH_4 a gradual increase in the electron-beam energy results in the appearance, at about an energy of 12 V, of CH_4^+, and at about 14, 15, and 22 V the species CH_3^+, CH_2^+, and CH^+ appear. The electron-beam energy necessary to form the parent positive ion is known as the appearance potential, and is an indication of the electron-binding power of the molecule, as is the ionization potential of atoms. Some values are shown in Table 17-5.

TABLE 17-5 Data for the ionization of some molecules and free radicals*
1 eV = 96.4905 kJ mol^{-1}

Molecule	Ionization potential (V)	Ionization energy (kJ mol^{-1})	Radical or molecule	Ionization potential (V)	Ionization energy (kJ mol^{-1})
Methane, CH_4	12.70	1225	CH_3	9.84	949
			CH_2	10.40	1003
			CH	11.13	1074
Ethane, C_2H_6	11.52	1112	C_2H_5	8.4	810
Ethylene, C_2H_4	10.45	1008			
Acetylene, C_2H_2	11.40	1100	C_2H	17.22	1663
Benzene, C_6H_6	9.25	892			
Ammonia, NH_3	10.19	983	NH_2	11.4	1100
			NH	13.10	1264
H_2	15.43	1489	F_2	15.7	1520
O_2	12.06	1164	Cl_2	11.48	1108
H_2O	12.61	1217	Br_2	10.53	1016
CO	14.01	1352	CF_4	14.36	1386
CO_2	13.77	1329	CCl_4	11.47	1107
H_2S	10.41	1004	Bf_3	15.5	1496

* From J. L. Franklin, J. G. Dillard, H. M. Rosenstock, J. T. Herron, K. Drasel, and F. H. Field, *Natl. Bur. Std.* NSRDS 26, 1969. Ionization Potentials, Appearance Potentials, and Heats of Formation of Gaseous Positive Ions.

A further indication of the power of mass spectroscopy in providing valuable physical-chemical data is apparent when mention is made that free radicals can be formed and led directly into a mass spectrometer. The mass spectra of these radicals can then be obtained. Not only does the mass spectrometer become a tool for the analysis of these active intermediates, but it also allows the direct study of these species.

It is found, for example, that the appearance potential of CH_3^+ from $\cdot CH_3$ radicals is 9.8 V. This datum can be combined with the appearance potential for CH_3^+ from CH_4 of 14.4 V to give a value for the energy required to break a bond of methane. Thus

$$
\begin{array}{lll}
(1) & \cdot CH_3 \rightarrow CH_3^+ + e^- & +9.8 \text{ eV} \\
(2) & CH_4 \rightarrow CH_3^+ + H\cdot + e^- & +14.4 \text{ eV} \\
(2-1) & CH_4 \rightarrow H\cdot + \cdot CH_3 & 14.4 - 9.8 = 4.6 \text{ eV} \\
 & & = 440 \text{ kJ mol}^{-1}
\end{array}
$$

Such a bond-dissociation energy is given in Table 16-4 and is not the same as the average bond energy obtained from thermochemistry methods in Sec. 7-10. Here one has the more specific information, the energy to break one of the CH_4 bonds. From such applications of mass spectroscopy much detailed information is being obtained on molecular energies.

It is necessary now to point out, as Fig. 17-14 indicates, that the appearance of fragments of a molecule does not generally correspond to the simple pattern exhibited by CH_4. Furthermore, the predominant ion may well be something other than that of the parent molecule. These features suggest that mass-spectral studies can give much information about the interesting chemistry of these gas-phase high-energy carbonium ions. The *cracking* patterns shown in Fig. 17-14 for a number of hydrocarbon molecules indicate that there is, for each molecule, a characteristic pattern of fragments. Although little progress has been made in understanding the basis for these patterns, it is clear that a wealth of information on the disintegrations of gas-phase molecules is given by these mass spectra.

Results such as these suggest some of the many interesting structures that occur in these ion fragments. Their importance as possible reaction intermediates is the purpose for which mass-spectral results were introduced here. However, it should be mentioned that the occurrence of species such as CH_5^+ which are formed presents interesting challenges to our theories of chemical binding.

17-14 High-energy Radiation

The photons used in photochemical studies have energies up to about 600 kJ einstein^{-1}, or about 6.5 eV. The electron beam of a mass spectrometer consists of electrons accelerated to voltages of about 100

V. Radiation with much higher energy is available, however. Electromagnetic radiation with energies up to about 500 eV is usually classified as x-rays; beyond this and up into the million-electron-volt (MeV) range, the electromagnetic radiation, such as is given off by radioactive Co^{60}, is known as *γ-rays.* Also of interest in radiation chemistry are the high-energy electron beams, typically with million-electron-volt energies, produced in a linear accelerator.

The study of the consequences of the passage of such radiation, with energies enormously greater than that required to excite a molecule or break a chemical bond, is relatively recent. It is already evident, however, that this branch of radiation chemistry is exceedingly complex. The selectivity of photochemistry, which allows a particular high-energy electronic state to be populated, is entirely lost. At present few general results can be recognized in the chemical decompositions that follow the passage of very high energy radiation through a material.

Cloud-chamber experiments, in which, for example, cosmic rays are detected as they pass through a supersaturated methyl alcohol vapor by the track of condensed vapor that they leave, are early experiments that can suggest the mechanism of many high-energy-radiation reactions. These cloud-chamber tracks apparently consist of groups of condensed droplets along the path of the high-energy particle. It appears that the high-energy particle knocks out an electron from a molecule and that this electron has sufficient energy, say 100 eV, in turn, to ionize many molecules. The clusters of condensation correspond, therefore, to the positions where the primary electron was generated and reacted further to form many more charged centers. It appears from ionization-chamber experiments that, quite independent of the nature of the material, it takes about 30 to 35 eV to form an ion pair. A 1-MeV particle penetrating a material will therefore form many clusters of ions.

It is generally accepted that, with very high energy radiation, the molecular disruption is such that ionized species are formed, as they certainly are when such radiation passes through a gas-filled ionization chamber.

Thus, in the radiolysis of water, the initial step has been postulated to be

$$H_2O \xrightarrow{h\nu} H_2O^+ + e^-$$

The high-energy H_2O^+ species can be expected to decompose readily according to

$$H_2O^+ \rightarrow H^+ + \cdot OH$$

Likewise, the ejected electron will probably travel away from the site at which it was generated and will be captured by a molecule to give the reaction

$$H_2O + e^- \rightarrow H\cdot + OH^-$$

Thus the expectation is that, along the track of the high-energy radiation, particles H^+ and $\cdot OH$ will be formed, and in the body of the solution there will occur the species $H\cdot$ and OH^-. The small amounts of H_2 and H_2O_2 that are formed can be accounted for by combinations of $H\cdot$ and $\cdot OH$ radicals.

Problems

1 By what factor would the rate of a reaction for which the activation energy is 150 kJ be increased by a temperature rise of 10°C, from 25 to 35°C?

2 The decomposition of nitrogen dioxide according to the equation

$2NO_2 \rightarrow 2NO + O_2$

follows, under certain conditions, the rate law

$\text{Rate} = k[NO_2]^2$

The following values of k as a function of temperature have been reported [M. Bodenstein, *Z. Physik. Chem.*, **100**:106 (1922)]:

T (K)	592	603.2	627	651.5	656
k (mol^{-1} liter s^{-1})	0.522	0.755	1.700	4.020	5.030

a Deduce the values of the parameters in the Arrhenius equation $k = Ae^{-E_a/RT}$.

b Deduce the values of A and E_a that would be required by the equation $k = A\sqrt{T}\,e^{-E_a/RT}$.

3 The rate constant for the combination of methyl radicals CH_3 to form ethane has been deduced to have the value 4.5×10^{10} mol^{-1} liter s^{-1} at 125°C. The activation energy appears to be zero, and it can be expected that there will be little steric hindrance to this radical recombination. Using the expression [21] from the collision-theory derivation and the values of $r_A = r_B = (1.54/2)$ Å from the normal carbon-carbon bond length, calculate a value for k and compare with that deduced from experimental data.

4 A low-frequency molecular vibration would absorb quanta at the far end of the infrared spectral region at a wave number of the order of 100 cm^{-1}. Verify that for such a vibration along the reaction coordinate, we can assume $h\nu_{RC} < kT$ for a room-temperature value of T.

5 Verify, for a representative molecule, the values of the partition functions for each degree of freedom shown in Table 17-3.

6 Show, for one type of molecular reactants and transition-state species found in Table 17-4, other than the example used in Sec. 17-4, that the partition-function factor appearing in the transition-state theory of the rate constant is that listed.

7 It is customary to attribute deviations from simply calculated rate constants to a "steric factor" that depicts the way that molecular size and shape impede the effectiveness with which molecular collision yields products. The transition-state approach can be used to deduce a comparable term by comparing the partition-function terms, as in Table 17-4, with those for the atom + atom reaction for which there can be no steric hindrance. Tabulate the ratio of the partition-function terms for various reaction types to that for the atom + atom reaction. Insert numerical values, and see if these are compatible with the idea that for more complex molecules steric effects can reduce the reaction rate by as much as several powers of 10.

8 Calculate the entropies of activation at 25°C for the reactions of Table 17-2. Recognize that low pre-exponential factors can be explained by either a steric difficulty, as the collision theory would suggest, or a very unfavorable entropy of activation, as the transition-state theory would express it.
Ans. $\Delta S^{\ddagger} = -70$ J deg^{-1} for H_2, I_2.

9 According to Appendix 5, the entropy of a fairly simple gas-phase molecule is in the range 150 to 200 J deg^{-1} mol^{-1} at room temperature. Recognize that the entropies of activation obtained in Prob. 8 do not correspond to the formation of one firmly bound transition-state complex from two reagent molecules. How can the difference between 150 to 200 and the values of Prob. 8 be explained?

10 What is the absorption coefficient of a solute which absorbs 90 percent of a certain wavelength of light when the light beam is passed through a 0.1-m cell containing a 0.25-M solution? *Ans.* $\alpha = 92$ mol^{-1} liter m^{-1}.

11 How many kilojoules of energy would 1 mol of acetone acquire if 1 mol absorbed 1 einstein of ultraviolet radiation of wavelength 2537 Å? How does this compare with the C=O bond energy?

12 Irradiation of HI vapor with ultraviolet radiation of wavelength 2070 Å leads to the formation of H_2 and I_2. It has been observed that for every joule of radiant energy that is absorbed, 0.00044 g of HI is decomposed. How many HI molecules were decomposed per quantum of absorbed radiation?
Ans. Two.

13 A reaction vessel of 1-liter volume contained, at the beginning of a photochemical study, an amount of Cl_2 to give a partial pressure of 0.5 atm at 25°C and an equal pressure of H_2. Irradiation with radiation of 4000-Å wavelength results in the absorption of 6.28 J of radiant energy and the decrease in the partial pressure of Cl_2 from 0.5 to 0.013 atm, the temperature being held at 25°C. How many molecules of HCl are formed for each quantum absorbed?

14 A 100-W sodium-vapor lamp radiates most of its energy in the yellow D line at 5890 Å. How long would such a lamp take to excite more than half the molecules of an absorbing species in a 1-mmol sample if all the radiant energy were absorbed by the sample? *Ans.* 1.0 s.

15 The bond energy of H_2 is 436 kJ mol^{-1}, the ionization energy of a hydrogen atom is 13.60 eV, and its electron affinity is 0.747 eV. What appearance potential might be calculated for H^+ in a mass-spectroscopic analysis of H_2?
The bond energy of H_2^+ can be calculated quantum-mechanically to be 255 kJ mol^{-1}. What would be the appearance potential of H_2^+ in the mass spectrum of H_2?

16 How can the observation that high-energy radiation produces a number of ion pairs that is approximately equal to the energy of the radiation, expressed in electron volts, divided by 30 or 35, be understood when it is known that ionization potentials are much less than 30 V?

References

GREENE, E. F., and A. KUPPERMAN: Chemical Reaction Cross Sections and Rate Constants, *J. Chem. Educ.*, **45**:361 (1968).

SLATER, N. B.: "Theory of Unimolecular Reactions," Methuen & Co., Ltd., London, 1959.

TURRO, N. J.: "Molecular Photochemistry," W. A. Benjamin, Inc., 1967, New York. A short readable account of the nature and chemical consequences of photon-induced excitation.

18

CRYSTALS

The study of many, but not necessarily the most interesting, aspects of crystals is greatly facilitated by the ordered arrangement of the atoms, ions, or molecules that provides the molecular-level characterization of crystals. Although the typical condensed-state feature of interactions between many particles enters, it does not, as it does for liquids, lead to unmanageable complexity. In fact, the structure that results from these interactions locks the particles into the particular structural relationship that, as was shown in Chap. 14, facilitates the study not only of the arrangement of the ions or molecules in the crystal, but also of the arrangements of atoms within these molecules or ions.

Many important properties of crystals depend, however, on minor deviations from perfect regularity in structure and chemical composition. The study here provides a basis from which further investigations of the way in which impurities, dislocation, vacancies, and so forth, contribute to the chemical and electrical properties of crystals.

18-1 Crystal Forces and Crystal Types

The nature of the forces that operate to bind the atoms, molecules, or ions of a crystal together provides a convenient basis for the classification of crystal types.

In a given crystal, although one of these types of forces might predominate, there are contributions from forces of some of the other types. In spite of this overlapping, the classification of crystals in terms of the predominant-force type is very useful.

Ionic Forces The forces operating in ionic crystals are very predominantly the electrostatic, or ionic, forces between the charged ions that were discussed in Sec. 11-8 in terms of pairs of ions. A smaller contribution results from the interaction of the charge of a given ion with the polarizability of the neighboring ions. These electrostatic forces result in a completely nondirectional pulling together of the units of the crystal. We shall see in the following sections that this uncomplicated behavior, together with the well-defined coulombic force law, makes these crystals in many ways the easiest to understand. The structure can be understood, in the case of crystals of simple ions, in terms of nothing more than the efficient packing together of two sets of spheres of different sizes.

Since these electrostatic forces are rather strong, ionic crystals are usually quite hard, brittle, and fairly high melting.

Van der Waals Forces Most organic crystals fall into the crystal type in which van der Waals forces predominate. The forces between the molecules are the same as those we encountered in dealing with gas imperfections. They are therefore called van der Waals forces. The source of these forces can be broken down to dipole-dipole, induction, and London dispersion forces, as will be shown.

These forces are relatively weak, and the typical organic crystal, which is usually held togeher by such forces, is soft and low-melting.

Again the attractive forces are essentially nondirectional, but the simplifying effect of this on the arrangement of the molecules in the crystal is often obscured by the odd shapes of the molecules. The arrangement taken up by a given organic molecule in a crystal can be predicted only with considerable difficulty.

Covalent Bonding A few rather important crystals are made up of atoms joined together throughout the crystal by covalent forces. A number of no less important crystals depend on covalent forces in one or two dimensions and on van der Waals forces in the remaining dimensions.

The example par excellence of the first type is that of diamond. Each carbon atom of the lattice is held to its four tetrahedrally arranged neighbors by covalent bonds. These bonds are little different from those found between carbon atoms in organic molecules. The heat of sublimation of diamond, to form gaseous carbon atoms, now appears to be experimentally established as about 710 kJ mol^{-1}. A diamond crystal with an Avogadro's number $\mathfrak{N}$ of atoms will have $2\mathfrak{N}$ carbon-carbon bonds. The conversion of such a crystal to gaseous atoms will require the expenditure of energy equal to twice the bond energy. The bond energy in diamond is therefore calculated as 355 kJ mol^{-1}. This value is in line with a covalent-bond energy shown in Table 7-4 that seems appropriate to hydrocarbon molecules. The tendency of carbon

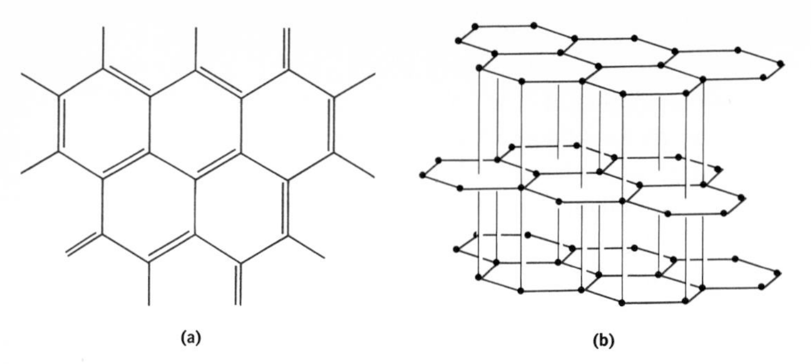

FIGURE 18-1
The structure of the two-dimensionally covalently bonded crystal graphite. (*a*) The bonding in a molecular layer. (*b*) The arrangement of the layers in the crystal.

atoms to form four bonds in tetrahedral directions, rather than the close packing of the carbon atoms, is the determining factor in the crystal structure. The hardness of the crystal also stems from the strength of the covalent bond. A few other crystals have three-dimensional arrangements of covalent bonds. Of these the characteristically tough materials silicon carbide, SiC, and silica, SiO_2, can be mentioned.

The most important representatives of the class of crystals which have a two-dimensional array of covalent bonds are graphite and mica. The structure of graphite is shown in Fig. 18-1, and it is apparent that each carbon atom forms bonds in the plane of the infinite molecules, which are similar to those of benzene and the aromatic molecules. All bonds are equivalent and can be looked on as being single bonds with one-third double-bond character. The layers of graphite are held together by the relatively weak van der Waals forces, and the slippery and flaky nature of graphite is thus readily understood.

The structure of mica is rather more complicated, but again consists of two-dimensional covalent or ionic bonds, with the layers so formed being held together by relatively weak forces.

A number of very different types of compounds form crystals which illustrate the formation of covalent bonds in only one dimension.

Inorganic examples are provided by SiS_2 and $BeCl_2$, the latter being an example of a number of similar and interesting "electron-deficient" compounds. Figure 18-2 illustrates the form taken by the atoms in these crystals.

Organic examples are provided by crystalline polymer molecules, either natural or synthetic. The simplest example is that of polyethylene, $CH_3(CH_2)_nCH_3$, with a very long chain molecule, i.e., a covalently bonded set of atoms. Each molecular chain is more or less oriented to neighboring molecules, and the details of this "more or less" are

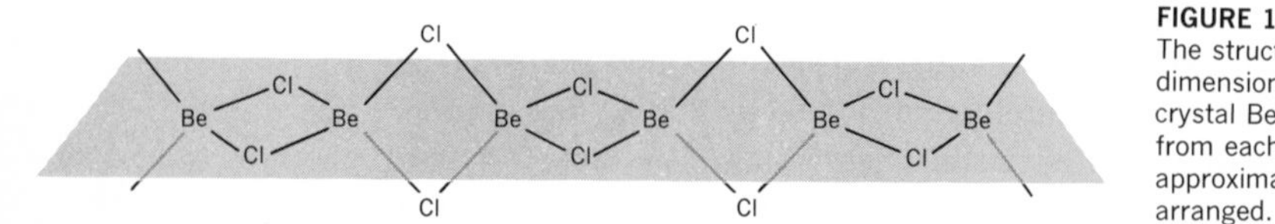

FIGURE 18-2
The structure of the one-dimensionally bonded crystal $BeCl_2$. The bonds from each Be atom are approximately tetrahedrally arranged.

of sufficient interest to be dealt with in our study of macromolecules in Chap. 25.

The final important type of molecule whose crystal forces can be described in this manner is the biologically important molecules, the proteins. These molecules consist of long covalently bonded chains of amino acids held to each other by relatively weak bonds, mainly hydrogen bonds. The structure of this important class of compounds is again of sufficient interest to merit a more complete treatment, and this detailed study will also be deferred to Chap. 25.

Hydrogen Bonds The effect of hydrogen bonds on the form and strength of crystals is both important and widespread. It has been pointed out that some hydrogen atoms, such as those of —O—H bonds, carry a partial positive charge and can form hydrogen bonds to electronegative atoms, such as oxygen and nitrogen, which carry a partial negative charge. The typical hydrogen-bond energy of 20 to 25 $kJ\ mol^{-1}$ puts the strength of this bond between that of the covalent bond and van der Waals effects. The role of the hydrogen bond in crystals is due not only to its strength, but also to the requirement it imposes on the geometry of the molecular arrangement. The maximum strength of the bond is obtained when the hydrogen is located near an electronegative atom and probably positioned near a projecting pair of nonbonding electrons of that atom.

Here again there is a prime example of the effect of this force on the molecular arrangement in a crystal. The crystalline form of ice is unusual in that it is a very open structure, as exhibited, for example, by the fact that ice is less dense than water. The explanation for this fact is that the H_2O molecules arrange themselves in the crystal so that they form good hydrogen bonds to each other rather than so that they are closely packed. The result is a tetrahedral-like arrangement of four hydrogen atoms about each oxygen, two being covalently bonded and the other two hydrogen-bonded. The angle of the water molecule apparently expands from its normal value of 105° to something nearer the tetrahedral angle of $109\frac{1}{2}°$ to accommodate this structure.

Many organic compounds with hydroxyl groups, such as alcohols, phenols, and carboxylic acids, crystallize in a manner at least partly dictated by hydrogen-bond formation. The molecules assume positions in the crystal so that the hydrogen atoms of one molecule can bond to the oxygen or nitrogen atom of a neighboring molecule. Such intermolecular bonds, if they extend throughout the crystal, frequently result in rather high melting and insoluble materials.

The structure of protein crystals, it should be mentioned, probably constitutes the most important consequence of hydrogen bonding. In Chap. 25 it will be shown how the arrangement of adjacent protein molecules to achieve maximum hydrogen bonding is a determining factor in the configuration of protein molecules in a crystal.

Metallic Bonding Although one seldom encounters metals in a form in which their crystallinity is obvious from their external appearance, metals show a ready tendency to crystallize. The nature of the forces which operate can be expected to be different from those forces previously mentioned. The electrical conductivity and metallic luster are two properties which indicate that some unique electronic-structure and bonding phenomena operate. Qualitatively, one thinks of the atoms or, properly, positive ions of the metal being closely packed within a sea of electrons. There are two somewhat quantitative, or at least more detailed, theories which seek to elaborate this view. The qualitative result, however, of relatively free electrons and packed positive ions follows, as we shall see in a later section, from both approaches.

18-2 Cohesive Energy of Ionic Crystals: Thermodynamic Determination

To enter in a more quantitative way into the forces and energies involved in crystals, let us now investigate the experimental results that are available, and useful, for simple ionic crystals. In Sec. 18-3 we shall consider the molecular basis of these crystal energies.

The *lattice energy,* which we shall find to be that directly approached by theoretical considerations of crystal energies, is the energy absorbed in processes such as

$$\text{NaCl(cryst)} \rightarrow \text{Na}^+(g) + \text{Cl}^-(g) \tag{1}$$

Since the energy of such processes cannot be obtained by a direct experimental method, we must resort, as we did in Sec. 7-5, to indirect methods. A procedure based on what is known as the *Born-Haber cycle* accomplishes this in the case of simple ionic crystals.

The process of Eq. [1] can be performed, in the example of NaCl, by the indirect route, indicated by solid arrows:

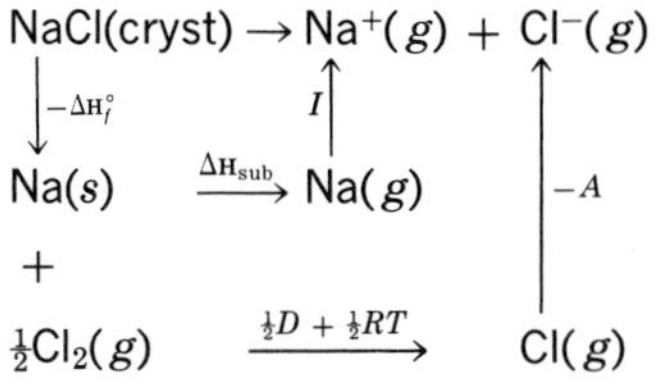

$$\begin{array}{lccc}
\text{NaCl(cryst)} \rightarrow & \text{Na}^+(g) & + & \text{Cl}^-(g) \\
\downarrow -\Delta H_f^\circ & \uparrow I & & \uparrow \\
\text{Na}(s) \xrightarrow{\Delta H_{\text{sub}}} & \text{Na}(g) & & \Big| -A \\
+ & & & \\
\tfrac{1}{2}\text{Cl}_2(g) & \xrightarrow{\tfrac{1}{2}D + \tfrac{1}{2}RT} & & \text{Cl}(g)
\end{array}$$

The enthalpy terms involved in the indirect route of the cycle are

ΔH_f° = standard heat of formation of NaCl
ΔH_{sub} = heat of sublimation of Na
D = dissociation energy of Cl_2
I = ionization potential of Na
A = electron affinity of Cl (usually given as a positive number, but since energy is given out in the process $\text{Cl} + e \rightarrow \text{Cl}^-$, the energy change for the reaction is negative, i.e., is $-A$)

TABLE 18-1 Crystal enthalpies from Born-Haber cycle calculations (Eq. [2]) and from the ionic-model treatment (Eq. [19])
All energies in kJ mol^{-1}

Crystal	ΔH_f°	ΔH_{sub} (metal)	I	$\frac{1}{2}(D + RT)$	A	ΔH (Born-Haber Eq. [2])	ΔH (Eq. [19])	r_0 (10^{-10} m)
LiF	−612	161	520	80	333	1040	1003	2.01
NaF	−569	108	496	80	333	920	896	2.31
KF	−563	89	419	80	333	818	793	2.67
LiCl	−409	161	520	122	350	862	819	2.57
NaCl	−411	108	496	122	350	787	759	2.81
KCl	−436	89	419	122	350	716	689	3.14
NaBr	−376*	108	496	97	330	747	724	2.97
KBr	−408*	89	419	97	330	683	661	3.29
NaI	−319*	108	496	77	300	700	672	3.23
KI	−359*	89	419	77	300	644	621	3.53

*Calculated for gas-phase Br_2 and I_2.

The enthalpy change in the formation of gaseous ions from the ionic crystal, on the basis of the indirect path, can be expressed as

$$\Delta H(\text{cryst}) = -\Delta H_f^\circ + \Delta H_{sub} + (\tfrac{1}{2}D + \tfrac{1}{2}RT) + I - A \qquad [2]$$

One needs, however, for systems involving Br_2 and I_2, where the standard heat of formation ΔH_f° is given in terms of standard states that consist of liquid and solid phases, respectively, to add to $-\Delta H_f$ the amount 16 kJ for half the heat of vaporization of Br_2 and 31 kJ for half the heat of sublimation of I_2. Also, it should be noticed that dissociation energies D are generally energies rather than enthalpies, and the addition of $\frac{1}{2}RT$ is necessary to bring these to enthalpies.

Methods for the determination of all such quantities have already been discussed elsewhere in the text.

The Born-Haber cycle calculation, with the data accumulated in Table 18-1, can now be performed to obtain the experimental values for the lattice energies given in the next-to-the-last column. We can now proceed to a theoretical approach to the lattice energies of such crystals, and we can see if the experimental values obtained here can be accounted for.

18-3 Calculation of the Crystal Energies of Ionic Crystals

The various types of attractive forces, discussed in Sec. 18-1, are not well enough understood to allow, generally, the calculation of the lattice energy of a crystal, i.e., the net energy with which an Avogadro's number of molecules are held together in the lattice.

The well-ordered arrangements of ions in crystals of simple salts, the spherical shape of simple anions and cations, and the predominantly coulombic electrostatic forces that act attractively between the ions, however, allow the energy of such crystals compared with the energy of separated ions to be calculated. The relative ease with which binding energies of ionic crystals can be calculated stems from our ability to express, to a good approximation, the potential-energy term due to the coulombic interactions. For a pair of ions the potential energy is given by

$$U = \frac{(Z_i e)(Z_j e)}{4\pi\epsilon_0 r_{ij}} \tag{3}$$

where i and j identify the ions, which have charges Z_i and Z_j times the electronic charge e, and are separated by a distance r_{ij}. For a gas-phase ionic molecule, with Z_i and Z_j carrying opposite signs, such an expression for the attractive term in the potential energy has already been found, in Sec. 11-8, to be quite satisfactory. The factor $4\pi\epsilon_0$, with the value 0.1113×10^{-9} C^2 m^{-1} J^{-1}, allows the charges to be expressed in coulombs and yields energy values in joules.

The short-range repulsion between ions of opposite charge is less easily understood, but the evidence, as from compressibility studies, that the repulsive energy sets in very sharply as the internuclear distance decreases, makes knowledge of the exact form of the repulsive-potential-energy term less necessary. As already mentioned in Sec. 11-8, it is customary to use expressions with an exponential dependence on the internuclear distance or one which contains this distance raised to some large negative power. Thus

$$U_{\text{rep}} = be^{-r/\rho} \tag{4}$$

as used in Sec. 11-8. Also used are inverse-power equations

$$U_{\text{rep}} = \frac{b'}{r^n} \tag{5}$$

where ρ, b, b', and n are empirical constants, ρ and n typically having values of the order of 0.3×10^{-10} m and between 6 and 12, respectively.

With expressions for the coulombic and short-range repulsive-energy terms for a pair of ions, it is possible to proceed to the calculation of an assembly of an Avogadro's number of ions that constitutes a mole of a crystalline material.

The crystal energy is obtained from the sum of terms given by Eq. [3] and either Eq. [4] or [5] with appropriate summations over all the significant interactions between the ions in the crystal lattice. Let us first consider the summation over all the coulombic terms.

The coulombic potential energy of one cation of the crystal in the field of all the other ions of the crystal can, from Eq. [3], be written

$$U_+ = \frac{Z_+ e}{4\pi\epsilon_0} \sum_i \frac{Z_i e}{r_i^+} \qquad [6]$$

where the summation extends over all ions, other than the selected cation, these ions having charge Z_i and a distance r_i^+ from the reference cation. It turns out to be more convenient to deal with interionic distances measured in terms of crystal dimensions. The length of the unit cell could be chosen, but generally distances are dealt with in terms of a related quantity r, the shortest cation-anion distance in the crystal. To introduce this unit of length, we rewrite Eq. [6] as

$$U_+ = \frac{Z_+ e}{4\pi\epsilon_0 r} \sum_i \frac{Z_i e}{r_i^+/r}$$

$$= \frac{Z_+ e}{4\pi\epsilon_0 r} \sum_i \frac{Z_i e}{R_i^+} \qquad [7]$$

where R_i^+ are the interionic distances in units of r.

One additional manipulation converts Eq. [7] to a form that serves later purposes. Multiplying and dividing by Z_-, the charge of the anion, gives

$$U_+ = \frac{Z_+ Z_- e^2}{4\pi\epsilon_0 r} \sum_i \frac{Z_i/Z_-}{R_i^+} \qquad [8]$$

In a similar way the potential energy of an anion that results from its coulombic interaction with other anions and with the cations of the crystal is given by

$$U_- = \frac{Z_+ Z_- e^2}{4\pi\epsilon_0 r} \sum_i \frac{Z_i/Z_+}{R_i^-} \qquad [9]$$

In general, the environment of the cations and anions of a crystal will not be identical and the summation of Eq. [9] will not be the same, for a given crystal type, as that of Eq. [8].

Nevertheless, the coulombic energy of a crystal, with this model of simple coulombic-ion interactions, with $\mathfrak{N}$ cations and $\mathfrak{N}$ anions, is

$$U = \tfrac{1}{2}\mathfrak{N}(U_+ + U_-) \qquad [10]$$

where the factor of $\frac{1}{2}$ prevents each ion-ion interaction from being counted twice. Then

$$U = \frac{\mathfrak{N} Z_+ Z_- e^2}{4\pi\epsilon_0 r} \left[\frac{1}{2} \sum_i \left(\frac{Z_i/Z_-}{R_i^+} + \frac{Z_i/Z_+}{R_i^-} \right) \right] \qquad [11]$$

The term in square brackets is seen to be completely determined by the charges on the ions and the crystal type. This expression simplifies for crystals like NaCl, for which both ion types have the same magnitude of charge and both are identically dimensionally related to the other ions of the crystal. Then one has

$$U_{\mathrm{NaCl}} = \frac{\mathfrak{N}e^2}{4\pi\epsilon_0 r} \sum_i \frac{\pm 1}{R_i} \qquad [12]$$

where R_i is the distance, in units of the separation between adjacent ions, from either a cation or anion to the other ions of the crystal.

For other crystal types the bracketed term of Eq. [11] must be dealt with, with some care. In particular, for crystals other than those with AB stoichiometry, multiples of $\mathfrak{N}$ cations or anions must be included in the formation of Eq. [10] so that the energy listed is that corresponding to a mole, or formula weight, of the crystal. All such extensions can be included in the bracketed portion of Eq. [11], and this term, which has been calculated and can be tabulated for many crystal types, is known as the *Madelung constant*. Representing this quantity by M allows Eq. [11] for the coulomb potential to be written

$$U_{\mathrm{coul}} = \frac{NZ_+Z_-e^2}{4\pi\epsilon_0 r} M \qquad [13]$$

Values of M are given for several crystal types in Table 18-2.

A similar summation treatment is unnecessary for the repulsive term. The presence of the empirical constants in Eqs. [4] and [5] means that it is adequate to maintain the form of these expressions when a crystal is considered and to introduce new empirical constants. Let us choose to work with Eq. [4] and recognize that the repulsive contribution to the energy of the crystal will be proportional to the factor $e^{-r/\rho}$ for ion pairs. Thus we write, for a mole of a crystal,

$$U_{\mathrm{rep}} = Be^{-r/\rho} \qquad [14]$$

The potential energy of a simple A^+, B^- ionic crystal has now been

TABLE 18-2 Madelung constants defined by Eqs. [11] and [13] for some crystal types

Structure	Stoichiometry	Madelung constant M
Rock salt	AB	1.7476
Cesium chloride	AB	1.7627
Zinc blende	AB	1.6380
Wurtzite	AB	1.6413
Fluorite	AB_2	2.5194
Rutile	AB_2	2.3850

expressed as

$$U_{\text{total}} = -\frac{Ne^2}{4\pi\epsilon_0 r}M + Be^{-r/\rho} \qquad [15]$$

To proceed to the goal of calculating this crystal energy we must now make use of some experimental data in order to evaluate the remaining unknowns, B and ρ.

The empirical factor B is first eliminated by requiring dU/dr to be zero at the equilibrium interionic spacing. From Eq. [15] we obtain

$$\frac{dU}{dr} = \frac{\mathfrak{N}e^2M}{4\pi\epsilon_0 r^2} - \frac{B}{\rho}e^{-r/\rho} \qquad [16]$$

which, when set to zero for r equal to the equilibrium distance r_0, gives

$$B = \frac{\mathfrak{N}Me^2\rho e^{r_0/\rho}}{4\pi\epsilon_0 {r_0}^2} \qquad [17]$$

This result allows Eq. [15] to be written

$$U_{\text{total}} = -\frac{\mathfrak{N}Me^2}{4\pi\epsilon_0 r_0}\left(\frac{r_0}{r} - \frac{\rho}{r_0}e^{(r_0-r)/\rho}\right) \qquad [18]$$

The potential energy of the crystal in its equilibrium structure, i.e., when $r = r_0$, except for small thermal-energy terms, is the negative of the crystal energy ΔE(cryst) which measures the energy required to form free ions from 1 mol of the crystal. We thus calculate

$$\Delta\text{E(cryst)} = \frac{\mathfrak{N}Me^2}{4\pi\epsilon_0 r_0}\left(1 - \frac{\rho}{r_0}\right)$$

and

$$\Delta\text{H(cryst)} = \frac{\mathfrak{N}Me^2}{4\pi\epsilon_0 r_0}\left(1 - \frac{\rho}{r_0}\right) + 2RT \qquad [19]$$

The value of the remaining empirical constant ρ can be deduced from considerations of the relation between the derivative of Eq. [19] with respect to the dimension r_0 and the compressibility of a crystal. This derivation will not be carried through, and we can proceed by stating that when the experimentally determined values of ΔE for alkali halide crystals are used in Eq. [19], values of ρ of about 0.3×10^{-10} m are deduced, and we can therefore proceed, as mentioned earlier, with the assumption of this value.

Results for crystal enthalpies calculated from Eq. [19] with $\rho = 0.34 \times 10^{-10}$ m are included in Table 18-1. This table contains the principal results of a molecular-type treatment for ionic crystals.

The success with which the theoretical values of ΔE follow the experimental values is seen to be quite good and confirms the assumption that the attractive and repulsive terms used in the theory are indeed those of principal importance.

TABLE 18-3 Representative values for c_P and c_V, in J deg^{-1} mol^{-1}, for crystalline materials at 25°C

Crystal	c_P	$c_P - c_V$	c_V
Ag	25.5	1.00	24.5
C (diamond)	6.1	0.002	6.1
Cu	24.5	0.67	23.8
Fe	25.0	0.42	24.6
Pb	26.8	1.76	25.0
Zn	24.6	1.30	23.3

Such calculations, it must be admitted, cannot be carried out with similar success for ionic crystals involving polyatomic ions nor for crystals of polyatomic molecules. Then neither the attractive nor repulsive forces are as simply expressed, and in general, more empirical constants would enter into these expressions than could be evaluated.

18-4 The Heat Capacity of Crystals

Our attention now can be turned to another property of crystals, their heat capacities, for which deductions from the molecular, or ionic, model of crystals can again be compared with the thermodynamic properties.

A very early generalization of the heat capacity of crystals was made by Dulong and Petit in 1819. The data available at that time consisted of heat capacities, measured at atmospheric pressure and room temperature. Dulong and Petit recognized that these data indicated that the heat capacity c_P of solid elements, mostly metals, at room temperature had the same value of about 26 J deg^{-1} mol^{-1}. Furthermore, this generalization holds fairly well also for c_V's since, as Table 18-3 shows, for a few representative solids, the difference $c_P - c_V$, at moderate temperatures, is usually not more than a few tenths of a calorie per degree per gram atom. The basis for these $c_P - c_V$ differences will be investigated when similar data for liquids are encountered.

The kinetic theory of gases provides a background that leads to a ready explanation of this observation. Each vibrational degree of freedom would be expected, on the classical, i.e., non-quantum-mechanical, approach to have an average kinetic energy of $\frac{1}{2}RT$ for an Avogadro's number of particles. The average kinetic energy for the three perpendicular vibration modes of the particles in a crystal lattice would then be $\frac{3}{2}RT$. On the average, also, for a vibrational motion, there is an equal amount of potential energy. We therefore expect

$$\begin{aligned} E_{vib}(\text{classical}) &= E_{kinetic} + E_{potential} \\ &= \tfrac{3}{2}RT + \tfrac{3}{2}RT = 3RT \end{aligned} \qquad [20]$$

The heat capacity is readily calculated by differentiating this result with respect to temperature. Thus one deduces

$$c_V = \left(\frac{\partial E}{\partial T}\right)_V = 3R \qquad [21]$$

This calculation indicates that a heat capacity of about 25 J deg^{-1} mol^{-1} should be expected and that this value should apply to all crystals made up of a lattice of single particles. The indication of the calculation is, furthermore, that the heat capacity should be independent of temperature. The nice, if approximate, success of this derivation was soon disrupted by additional experimental results.

Figure 18-3 shows more extensive data for crystals of elements to which the Dulong and Petit rule presumably applies. One sees that the rule is more or less valid at room temperature if only the heavy metals are considered, but that at lower temperatures the heat capacities show no correlation with this rule. The classical derivation and its prediction of a heat capacity independent of temperature are completely disproved. In the days of classical treatments, before 1900, these results presented a very perplexing problem. There seemed no way to avoid the erroneous deductions of the theory.

With the development of the quantum theory, it was soon recognized that the classical treatment of the vibrations, i.e., with any vibrational energy allowed, was faulty. In 1907 Einstein showed that, when a vibrational frequency is assigned to the vibrations of the crystal particles and the motion is treated quantum-mechanically, curves of the general shape of Fig. 18-3 can be obtained if it is assumed that

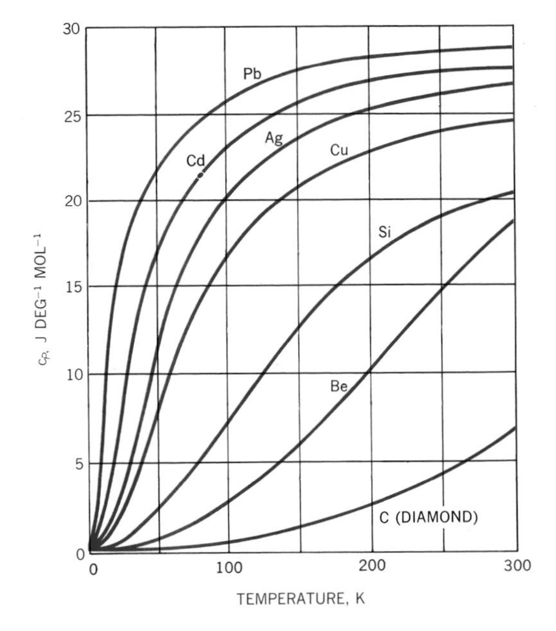

FIGURE 18-3
The heat capacities of simple crystalline solids at low temperatures. (*Adapted from E. B. Millard, "Physical Chemistry for Colleges," McGraw-Hill Book Company, New York, 1953.*)

some energy separation $h\nu$ can be assigned to the energy of the vibration modes of the particles of the crystal.

The calculation of the heat capacity resulting from vibrational modes has already been treated in Sec. 6-13. It is only necessary to note that for a gram atom there are $\mathfrak{N}$ particles, each with 3 vibrational degrees of freedom. The previous formula given in Table 6-4 for molecular vibrations is completely applicable and leads to the heat-capacity prediction for a crystalline solid containing an Avogadro's number of single particles of

$$c_V = 3R\left(\frac{h\nu}{kT}\right)^2 \frac{e^{h\nu/kT}}{(e^{h\nu/kT}-1)^2} \qquad [22]$$

By choosing a value of the vibrational-energy-level spacing $h\nu$ for each crystal, Einstein was able to obtain approximate agreement with each of the curves of Fig. 18-3, as is illustrated by the data for Al in Fig. 18-4.

It should be pointed out that Eq. [22] can be used to form a general curve of c_V versus $h\nu/kT$. To the extent that the Einstein theory is satisfactory in handling the heat-capacity data for simple crystalline materials, it follows that for a given material the measured heat capacities when plotted against $h\nu/kT$, with a suitable value of ν for that material, will fall on the general curve for c_V versus $h\nu/kT$. Such a treatment will be illustrated by the Debye theory, which is rather more successful than that of Einstein.

An improvement on Einstein's theory which gives results in even better agreement with the experimental data was soon forthcoming. It was recognized by Debye that the assumption that all the particles of a crystal vibrate with the same frequency was not entirely sound.

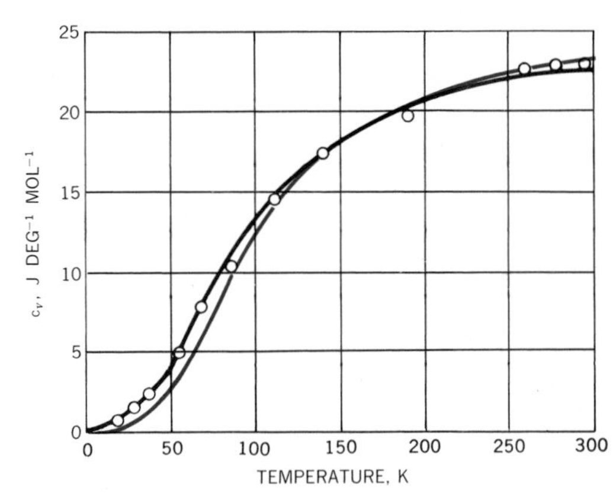

FIGURE 18-4
The experimental heat-capacity data at low temperatures for aluminum. The best fits to the data that can be obtained with the Einstein and Debye theories are indicated.

TABLE 18-4 The Debye characteristic temperatures, the Debye frequencies, and the force-constant factors for some monatomic crystalline materials

Crystal	$\theta_D = h\nu_{max}/k$ (K)	$\bar{\nu}_{max}$ (cm^{-1})	Force-constant factor $M\bar{\nu}^2_{max} \times 10^{-5}$
Ne	63	44	0.4
K	100	70	1.9
Xe	55	38	1.9
Na	150	104	2.5
Li	385	270	5.0
Pb	88	61	7.7
Hg	96	67	9.0
KCl	227	157	9.2
Ca	230	160	10
Al	390	271	20
Ag	215	149	24
Au	170	118	27
Cu	315	218	30
Be	1000	695	43
Fe	420	292	48
W	310	214	84
C (diamond)	1840	1280	197

Although all the particles may be bound by similar forces, the motions of the particles will couple so that vibrations of a wide range of frequencies will result. One finds spectroscopically, for instance, that simple salt crystals show a very broad absorption band in the far-infrared region. On the assumption of a range of vibrational frequencies, Debye derived a heat-capacity equation in a manner similar to that used by Einstein. Debye's expression relates the heat capacity to the term $h\nu_{max}/kT$, where ν_{max} is the maximum of a whole range of frequencies. Like the single frequency of the Einstein theory, the value of ν_{max} must be deduced empirically for each crystal. Figure 18-4 shows how well this theory fits the data of Al, and particularly how the fit at temperatures approaching absolute zero is greatly improved.

Some characteristic frequencies are listed in Table 18-4. It has become customary to introduce the *Debye characteristic temperature* θ_D, defined by $\theta_D = h\nu_{max}/k$, and these numbers are also listed. The values of $h\nu_{max}$ indicate the maximum energy spacing for the vibrational energies of the crystal particles.

In a manner similar to that in which the vibrational frequencies of free molecules indicate the force constants of the bonds between the atoms of the molecule, the Debye or Einstein frequencies indicate the rigidity with which the atoms or ions are held at their crystal-lattice sites. The frequencies depend, however, on both the force constants and the particle masses through a relation of the type $\nu = (1/2\pi)\sqrt{k/m}$, where m is the mass of a single vibrating particle, or a reduced mass if more than one particle moves in the vibration. For comparative purposes, it is sufficient to calculate $M\bar{\nu}^2_{max}$, where M is the atomic mass, for various crystals. This quantity, exhibited in Table 18-4, is a measure of the stiffness of the crystal bonds uncomplicated by the effect of the mass of the crystal particles. The gradation from soft, weakly bonded crystals to hard, strongly bonded materials is evident, and is given in a quantitative way by the data of Table 18-4.

It should be mentioned, however, that it is the vibrational spacings themselves, ν or ν_{max}, that determine the heat-capacity behavior. Those crystals with closely spaced levels, as a result of weak bonds or heavy atoms, as in lead, reach the classical heat-capacity value at relatively low temperatures. Those crystals, on the other hand, with large vibrational spacings, as a result of high force constants or light atoms, reach this limit of $3R$ for c_V only at relatively high temperatures. This latter type, as Fig. 18-3 shows, is illustrated most emphatically by the example of diamond.

It is a matter of considerable practical importance, as was suggested in the discussion of the determination of entropies from the third law of thermodynamics, to be able to predict the way in which the heat capacity of a crystalline solid behaves as the absolute temperature approaches zero. The success of the Debye theory in correlating the measurable heat capacities has led to reliance on its predictions as to the way in which c_V approaches zero. Since the details of the Debye theory have not been given, it is necessary to state simply that the low-temperature limit predicts the behavior

$$c_V = \alpha T^3 \qquad [23]$$

where α is a constant characteristic of each material. With this low-temperature limiting expression it is possible to extrapolate heat-capacity data to absolute zero. The Debye curve for aluminum in Fig. 18-4 shows the form of this predicted behavior. Third-law entropy values depend on this extrapolation of measured heat capacities to absolute zero.

18-5 An Introduction to the Study of Metallic Crystals

Metallic crystals provide systems, like the alkali halide crystals with which we have been primarily concerned, of simple, often cubic, arrangements of like atoms, and discussion of these systems should be included in a treatment of crystals.

Metals are immediately distinguished from the ionic crystals with which we have been dealing by their luster and their electrical conductivity. These characteristics lead directly to the idea of relatively free electrons moving throughout the crystal lattice. The study of metallic crystals becomes, primarily, the study of the way in which these electrons can be described and the way in which their behavior can be investigated.

An approach that is suitable here, in view of the success with which the heat capacities of simple crystals were dealt with in Sec. 18-4, begins with a consideration of the heat capacity of metallic crystals.

Results already presented, in Figs. 18-3 and 18-4, show that the heat capacities of metallic crystals are of a magnitude and a temperature dependence similar to that of simple ionic or covalent crystals. This result should, at first, be surprising since the relatively free electrons of a metal might be expected to lead to an additional heat-capacity contribution of about $3(\frac{1}{2}R)$ if each atom of the metal contributes one electron that is not tightly bound to the atom. A closer look at the heat-capacity data for metals is therefore undertaken.

Data for the heat capacity of silver down to low temperatures are given in Table 18-5. Close inspection of these data shows that, although at the higher temperatures they are in line with the Debye theory, and to a lesser extent with the Einstein theory, they are not, as Prob. 11 suggests, consistent with these theories at the lower temperatures. The data can in fact be fitted, at temperatures low enough for the Debye law to reduce to the T^3 relation, by an empirical equation, which, as will be pointed out, has a basis in theory, of the form

$$c_V = \alpha T^3 + \gamma T \qquad [24]$$

The first term can be looked on as arising from the lattice vibrations, in the manner shown by Debye and dealt with in Sec. 18-4. The linear-temperature term can be attributed to the electronic contribution of the metallic crystal. Heat-capacity data can be compared with Eq. [24] if the rearrangement to

$$\frac{c_V}{T} = \alpha T^2 + \gamma \qquad [25]$$

is first performed. Then, if the additional term does in fact have the proper form, a plot of c_V/T versus T^2 will yield a straight line whose slope will give the value of α and whose intercept will be the value of γ. Such plots are shown in Fig. 18-5, and it is clear that the heat-capacity data for these metals do conform to a heat-capacity equation with a linear term.

Similar behavior is found for other metals, and data in Table 18-6 show the values of γ that are found. The contributions of this electronic term at room temperature are also included, and although, by the graphical treatment of Fig. 18-5, an electronic heat-capacity term for metals has been revealed, it is still clear that it is exceedingly small compared with R or $3R$ at ordinary temperatures.

We now must consider the behavior of the electrons of a metal

TABLE 18-5 The heat capacity of silver at constant volume*

T (K)	c_V (obs) ($J\ deg^{-1}\ mol^{-1}$)
1.35	0.00106
2	0.00262
3	0.00657
4	0.0127
5	0.0213
6	0.0373
7	0.0632
8	0.0987
10	0.199
12	0.347
14	0.559
16	0.845
20	1.671
28.56	4.297
36.16	7.088
47.09	10.80
55.88	13.33
65.19	15.37
74.56	16.90
83.91	18.10
103.14	20.07
124.20	21.27
144.38	22.48
166.78	22.86
190.17	23.34
205.30	23.45

*From C. Kittel, "Introduction to Solid State Physics," John Wiley & Sons, Inc. New York, 1953.

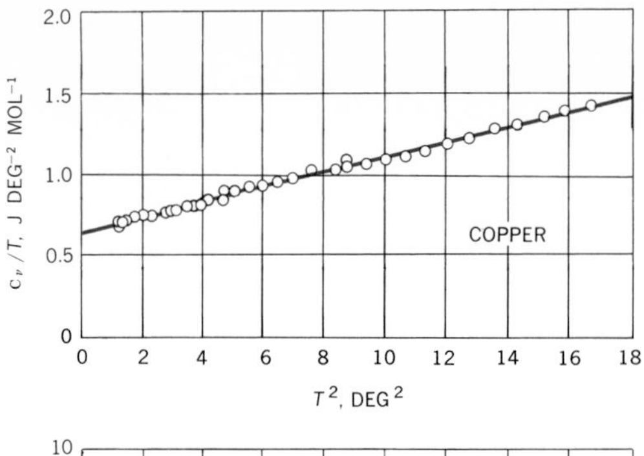

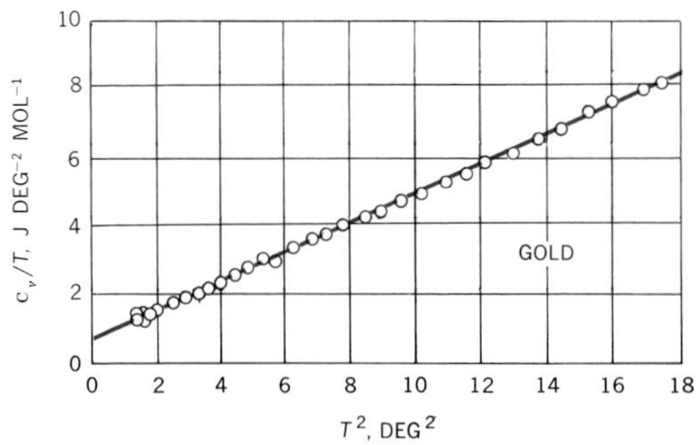

FIGURE 18-5
The c_V/T versus T^2 plots for Cu and Au at low temperatures.

to see if their small heat-capacity contribution can be understood. The simplest model for this purpose, which also has the advantage of leading into more detailed theoretical treatments, is the *free-electron model.* Electrons, perhaps one from each atom, are considered to be free to move throughout a region of uniform potential provided by the metal crystal. These electrons are treated like particles-in-a-box, as already studied in a different connection in Sec. 3-6. This treatment, it will be recalled, yields allowed energy levels according to the formula

$$\epsilon = (n_x^2 + n_y^2 + n_z^2)\frac{h^2}{8ma^2} \qquad \text{where} \begin{cases} n_x = 1, 2, 3, \ldots \\ n_y = 1, 2, 3, \ldots \\ n_z = 1, 2, 3, \ldots \end{cases} \qquad [26]$$

or

$$\epsilon = n^2\frac{h^2}{8ma^2} \qquad \text{where } n^2 = n_x^2 + n_y^2 + n_z^2$$

where m is now the mass of the electron, and a is the dimension of the crystal, here assumed to be a cube.

The N electrons of a gram atom of a metal, at absolute zero, will occupy the lowest-energy-available quantum states. Since electrons obey Fermi-Dirac statistics, two electrons with spins oppositely directed can occupy each state. It follows, as can be seen by considering the states displayed by points as in Fig. 18-6, that to accommodate all N electrons in the lowest available states, all states will be occupied up to those with $n = n_F$, given by

$$2 \times \tfrac{1}{8}(\tfrac{4}{3}\pi n_F^3) = N \qquad [27]$$

TABLE 18-6 The values of γ of Eq. [24] and the electronic contribution γT to c_V at room temperature

Metal	γ (J deg^{-2} mol^{-1})	γT at 25°C (J deg^{-1} mol^{-1})	Metal	γ (J deg^{-2} mol^{-1})	γT at 25°C (J deg^{-1} mol^{-1})
Ag	0.62×10^{-3}	0.18	Na	1.8×10^{-3}	0.54
Al	1.38	0.41	Ni	7.3	2.18
Au	0.71	0.21	Pb	3.3	0.98
Bi	0.04	0.01	Pd	9-13	2.7-3.9
Co	5.0	1.49	Pt	6.8	2.02
Cr	1.5	0.45	Sn	1.8	0.54
Cu	0.7	0.21	Th	5.0	1.49
Fe	5.0	1.49	Ti	3.3	0.98
Mg	1.3	0.39	U	10.8	3.21
Mn	14-18	4.2-5.4	W	0.7-2.0	0.2-0.6

The subscript on n designates that this value of n corresponds to what is known as the *Fermi level.* The energy corresponding to this level, below which, at absolute zero, all states are occupied and above which all states are empty, is designated by ϵ_F and is given by

$$\epsilon_F = \frac{n_F{}^2 h^2}{8ma^2} = \frac{h^2}{8m}\left(\frac{3N}{\pi a^3}\right)^{2/3} \qquad [28]$$

where N/a^3 can be recognized as the electron density which, if one free electron is provided by each atom, is the number of atoms per unit volume.

The way in which the available states are occupied can be shown by a diagram like that of Fig. 18-7. The solid line shows that at absolute zero all states are occupied up to the Fermi energy. For a typical metal, as insertion of numerical values in Eq. [28] will show, the Fermi energy is about 160×10^{-21} J per electron, and this value allows the com-

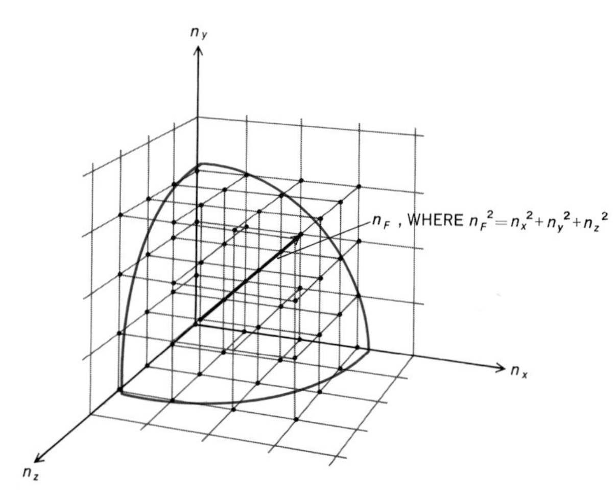

FIGURE 18-6
Graphical representation of the values of the quantum numbers n_x, n_y, and n_z for a particle in a three-dimensional box. All points lying within the spherical segment with radius n_F correspond to energies less than the energy corresponding to n_F.

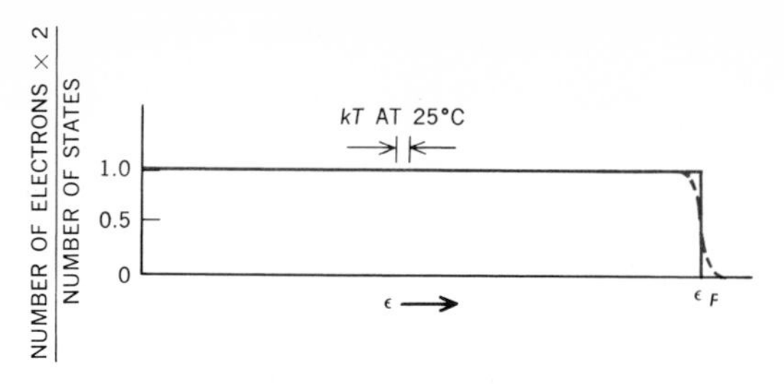

FIGURE 18-7
The occupation of the available states by the free electrons of a metal indicated for absolute zero by the solid line and for room temperature by the broken line.

parison with the room-temperature value of kT of about 4.1×10^{-21} J to be made in Fig. 18-7.

To proceed to a treatment of the occupation of the available states and the total energy of the system at temperatures other than absolute zero, we should have to develop distribution functions, like those obtained in Chap. 5, but now for the case in which states could be occupied by at most two particles. This derivation is somewhat lengthy and will not be carried out. Instead the energy and the heat capacity of the electrons of a metal, according to the model assumed here, will be considered in a very qualitative way. It has been mentioned before that, because of the exponential $e^{-\text{energy}/kT}$ factor that occurs in distribution expressions, we can expect levels that are of about kT energy above some reference level to be appreciably occupied as the temperature is increased from zero to T. This implies that only those states that are within about kT of the Fermi energy will suffer a change in population when the temperature is raised from 0 K to T. We might therefore sketch in the dashed curve of Fig. 18-7 to suggest the curve for the population of the states at room temperature.

We can, furthermore, reach the known linear heat-capacity term by arguing that only a fraction kT/ϵ_F of the electrons gain energy in the temperature interval 0 to T and that on the average they each acquire an energy equal to kT. It follows that the thermal electronic energy would be

$$\mathrm{E}_{\text{thermal, elec}} = \frac{Nk^2}{\epsilon_F} T^2 \qquad [29]$$

and the electronic contribution to the heat capacity would be

$$\mathrm{C}_{V,\text{ elec}} = N\left(\frac{2k^2}{\epsilon_F}\right) T = 2R\frac{kT}{\epsilon_F} \qquad [30]$$

Substitution of numerical values and the representative value of 160×10^{-21} J per electron for ϵ_F leads, in agreement with the experimentally determined coefficient of the linear term, to a value of the order of 10^{-3} J deg^{-2}.

We have come, therefore, by means of a very simple model, to an understanding of the smallness of the electronic-heat-capacity contribution of metals. In a similar way, other properties can be investigated.

FIGURE 18-8
The potential-energy function for the free-electron model for a one-dimensional metallic crystal.

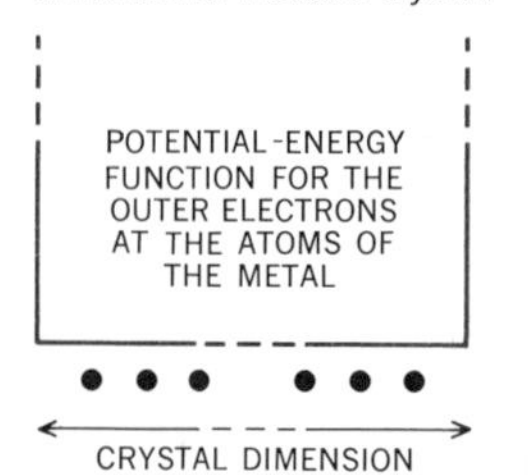

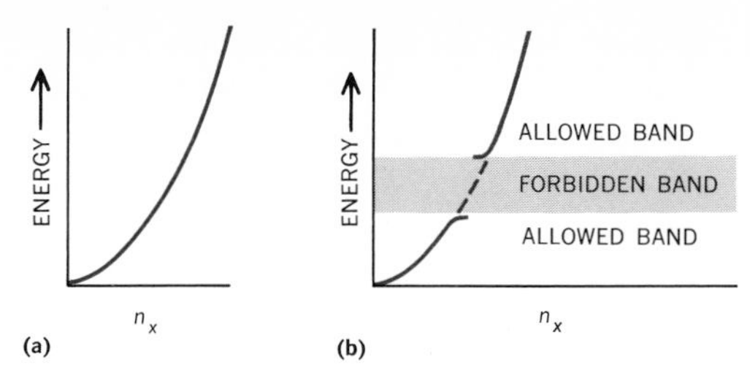

FIGURE 18-9
The allowed energies for the free electrons of a metal as a function of the quantum number (*a*) for the potential function of Fig. 18-7 and (*b*) when the relation between the periodicity of the atomic positions and the maxima in the ψ^2 function for the electrons is taken into account.

One soon finds, it should be pointed out, that the simple free-electron model is inadequate, and in further studies of metallic systems more elaborate models are necessary.

The nature of one of the important modifications that has been made can be indicated by referring to a simple one-dimensional array of metal atoms that forms a one-dimensional crystal, as in Fig. 18-8. The free-electron model would have allowed to the electrons of this system any energy given by the equation

$$\epsilon = \frac{n_x^{\,2}h^2}{8ma^2} \qquad \text{where } n_x = 1, 2, 3, \ldots \tag{31}$$

This relationship between ϵ and the quantum number n_x can be shown graphically, as in Fig. 18-9*a*.

An additional feature enters if we no longer assume a uniform potential energy throughout the crystal but recognize the potential wells that the positive charges at the nuclei produce. The periodicity that now occurs in the potential function has no major effect on the energy of an electron as long, for example, as the electron is in one of the low-lying allowed states where the maxima and minima of the probability function spread out across many nuclei, as suggested in Fig. 18-10*a*. On the other hand, when, for example, the length of the wave that leads to the probability function is equal to the spacing a/N between the nuclei, the positions of maximum electron density can occur anywhere

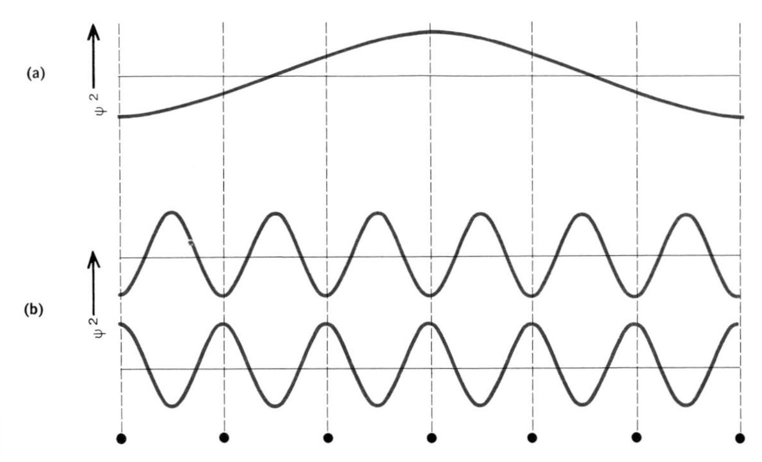

FIGURE 18-10
(*a*) A relatively low energy probability function and (*b*) the two extreme relationships between the probability function and the atomic nuclei arrangement when the periodicity of the function is the same as that of the atomic nuclei of the metallic crystal. Although both probability functions in (*b*) have the same quantum number, they can, because of interactions between the free electrons and the nuclei, correspond to different electron energies.

between two extremes. Either they can occur at the nuclei, as shown in Fig. 18-10*b*, in which case the effect of the periodic potential wells will be a maximum and the energy will be relatively low, or they can occur between the nuclei and, for the same value of the quantum number n_x, lead to a higher energy state. The net effect of the periodicity in the potential function is to produce a diagram like that of Fig. 18-9*b*, with allowed *bands* of energy levels separated by forbidden bands.

Proceeding in this way, one develops the *band theory of solids* that provides an effective approach to many of the properties of metallic conductors and semiconductors. To proceed, however, would take us too far afield from the material normally considered to be included in physical chemistry.

Problems

1 Plot, on the same graph, curves for the attractive and repulsive contributions, according to Eq. [15], to the energy of an NaCl crystal as a function of the distance of closest anion-cation. Use this diagram to explain why the attractive term alone gives nearly the correct lattice energy.

2 Referring to the equations developed for the energy of simple ionic crystals, show why the crystal energy would be about 10 percent too large if the repulsive term were not included in the calculation.

3 Use the form of the repulsive term given in Eq. [5], and eliminate one of the unknown empirical parameters by the procedure used in Sec. 18-3 to eliminate B from the repulsive term that was present in Eq. [14].

4 Assuming an average value of 8 for n, use the expression obtained in Prob. 3 to calculate several values for ΔE(cryst), and compare with the values given in Table 18-1.

5 Deduce an expression for the lattice energy of an ionic crystal on the basis of a repulsion-energy term of the form shown in Eq. [5]. By what fraction is the lattice energy reduced by the repulsion term if n is given the value of 6? What is the reduction if n is 9? Show, by sketching a graph, why the relative repulsion contributions for $n = 6$ and $n = 9$ are those calculated.

6 The heat of sublimation of graphite is 710 J per gram atom. The energy with which the molecular layers are held to one another can be estimated from heats of sublimation of simpler aromatic systems to be 8 J per gram atom. Calculate the carbon-carbon bond energy in graphite. If the bonds of graphite have two-thirds single-bond and one-third double-bond character, what bond energy would be expected from the data of Table 7-4? Comment on the resonance energy of graphite compared with benzene.

7 Explain why KCl can be included in Table 18-4, which otherwise contains calculated values of the force-constant factor only for monatomic substances.

8 Calculate the force-constant factors, as defined in Table 18-4, for CH_4 and H_2O, which have Debye temperatures of 78 and 192°, respectively. Assume that the molecule as a whole vibrates in the lattice vibrations, i.e., that the mass in the force-constant-factor calculation can be taken as that of the whole molecule. Are the values obtained for CH_4 and H_2O in line with what would be expected in view of the data of Table 18-4?

9 Verify, for one of the crystals listed in Table 18-1, the values listed there for ΔH from the Born-Haber cycle and from Eq. [19].

10 Determine a value of the lattice frequency ν for KCl so that the Einstein heat-capacity relation fits, as well as possible, the experimental values for C_V. Compare this value of ν with the value of ν_{max} required by the Debye theory and shown in Table 18-4. Show graphically the fit obtained to the experimental data by the Einstein relation.

Heat-capacity measurements by W. T. Berg and J. A. Morrison [*Proc. Roy. Soc.* (*London*), **A242**:467 (1957)] for KCl give the values tabulated at the right. (*Use the conversion factor* 1 cal = 4.184 J.)

T (*K*)	C_V (*cal deg*$^{-1}$ *mol*$^{-1}$)
10	0.008
20	0.71
30	1.99
40	3.56
60	6.31
80	8.16
100	9.31
140	10.52
180	11.09
220	11.39
260	11.56

11 Fit the data of Table 18-5 to the Einstein and the Debye T^3 functions, and verify that although the fit can be made satisfactory at most temperatures, large discrepancies occur with the very low temperature data.

12 Deduce from the data of Table 18-5 the value of γ for the electronic-heat-capacity term for silver.

13 Assuming each atom contributes one free electron, obtain the necessary density data for several metals, and calculate values of the Fermi energy ϵ_F.

14 If each atom of an array such as that shown in Fig. 18-8 contributes one free electron, what fraction of the total number of electrons that can be accommodated in the first allowed band will in fact be present?

References

KITTEL, C.: "Introduction to Solid State Physics," John Wiley & Sons, Inc., New York, 1953. The classifications of solids, energies of ionic crystals, and heat capacities of ionic and metallic crystals are dealt with in chaps. 2, 6, and 10. The remainder of the text, although not directly related to the material dealt with in this chapter, provides an excellent extension primarily into conducting and semiconducting solids.

PAULING, L.: "The Nature of the Chemical Bond," 3d ed., chaps. 11 and 13, Cornell University Press, Ithaca, N.Y., 1960. Ionic crystals are discussed in chap. 13, with particular emphasis on structure, and metals are treated in chap. 11, from an orbital rather than a free-electron point of view.

19

LIQUIDS

Our study of those physical-chemical phenomena that depend primarily on the mutual effects of a large number of molecules begins in earnest with an investigation into the nature of the liquid state. Neither the free-molecule treatment used for gases nor the structurally fixed arrangement of crystals persists to simplify the molecular treatment of the liquid state. A molecular-level interpretation of liquids and, in succeeding chapters, of solutions will be a prime goal. But the complexities that arise from the molecular interactions that must be considered will often shift, or limit, our studies to the experimentally observed features and the thermodynamic relations that can be established.

From all the preceding physical-chemical studies, you might now expect that when new aspects of the physical world are approached, most of the studies that are made can be placed in the categories of energetics, structure, and kinetics. Such is the case for liquids, the first category being that to which most of the sections can be assigned.

19-1 The Heat of Vaporization and Intermolecular Forces

The energy associated with the liquid state is most directly related to the heat of vaporization, i.e., the heat required to free the molecules of a liquid from the intermolecular interactions. This heat can be measured by direct calorimetric means, or as will be shown in the following chapter, it can be deduced from vapor-pressure measurements. Direct measurement usually depends on the collection and measurement of the amount of liquid converted to vapor by a measured electrical heat input. Table 19-1 shows some heats of vaporization, i.e., the heat required for the conversion of liquid to the equilibrium vapor, for a number of compounds at various temperatures.

TABLE 19-1 The heats of vaporization, in kJ mol^{-1} of liquids, to their equilibrium vapor at various temperatures*

Temperature (°C)	H_2O	C_2H_5OH	$(C_2H_5)_2O$	CCl_4
0	44.8	42.4	28.8	33.5
25	44.01			
40	43.2	41.5	25.7	
80	41.6	39.2	22.3	29.8
100	40.67			
120	39.7	35.1	18.9	27.2
160	37.4	30.0	13.6	24.6
200	34.8	22.1		21.1
240		7.4		16.5
280				6.7

* From International Critical Tables, vol. 5, p. 138, McGraw-Hill Book Company, New York, 1926–1930.

Such data, and Fig. 19-1, emphasize the temperature dependence of the heat of vaporization and show the trend of these heats toward a zero value as the critical point, where the properties of liquid and equilibrium vapor merge, is approached. The dependence of ΔH on temperature is given by the thermodynamic relation previously derived for a chemical reaction,

$$\left(\frac{\partial H}{\partial T}\right)_P = \Delta C_P \qquad [1]$$

It is equally applicable to a physical reaction. For 1 mole of liquid being transformed to vapor,

$$\Delta \mathrm{H} = \Delta \mathrm{H}_{\mathrm{vap}} = \mathrm{H}_v - \mathrm{H}_l \qquad [2]$$

and

$$\Delta \mathrm{C}_P = (\mathrm{C}_P)_v - (\mathrm{C}_P)_l \qquad [3]$$

The nonzero value for $d(\Delta \mathrm{H}_{\mathrm{vap}})/dT$ implies, therefore, that the heat capacity of a liquid is, as would be expected, different from that of the equilibrium vapor. In a later section we shall return to consider a molecular interpretation of the heat capacities of liquids. Now, however, the source of the interactions that are overcome in the vaporization process must be looked into.

Aside from the relatively small zero point and thermal-energy differences that can arise from the different patterns of allowed energy levels for the liquid and vapor molecules, the heat of vaporization can be identified with the potential-energy difference between the intermolecularly bound molecules of the liquid state and, if ideal-gas behavior is assumed, the free molecules of the vapor.

The intermolecular forces of attraction and repulsion due to the interaction of the molecules can be dealt with as a function of the distance between the molecules, and the balancing of these attractive and repulsive forces at some intermolecular distance corresponds to a minimum in the potential energy at that distance. For a liquid the intermolecular distance that must be considered is some average value between a molecule and its nearest neighbors.

It is more convenient to consider the corresponding potential-energy terms. The repulsive contributions to the potential energy form the curve that falls off very steeply with the average molecular separation, as shown in Fig. 19-2. Often-used expressions represent this term, as for the repulsion between ions treated in the preceding chapter, by

$$U_{\text{rep}} = be^{-r/p} \tag{4}$$

b and p being empirical constants, or by

$$U_{\text{rep}} = \frac{B}{r^9} \tag{5}$$

where the B and 9 are empirical constants.

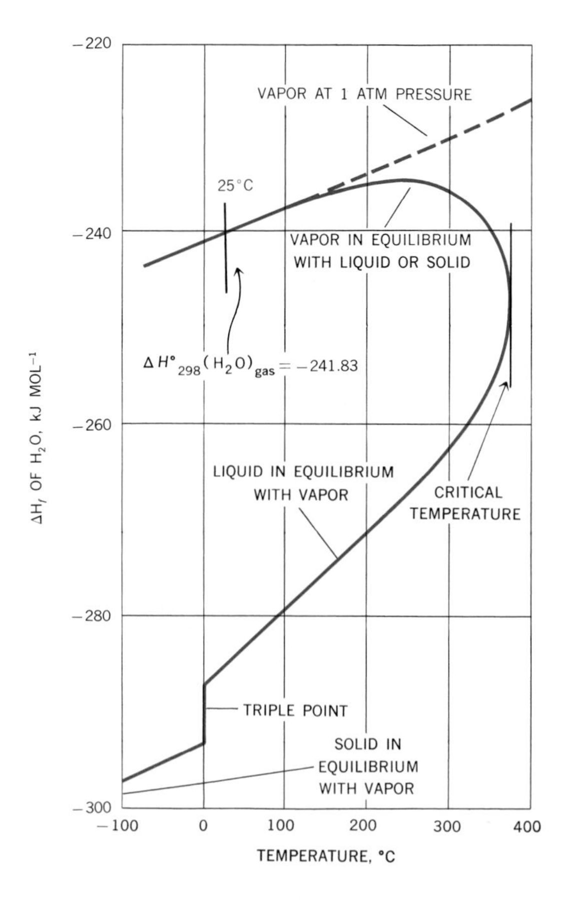

FIGURE 19-1
The heat content of the vapor, liquid, and solid phases of water as a function of temperature.

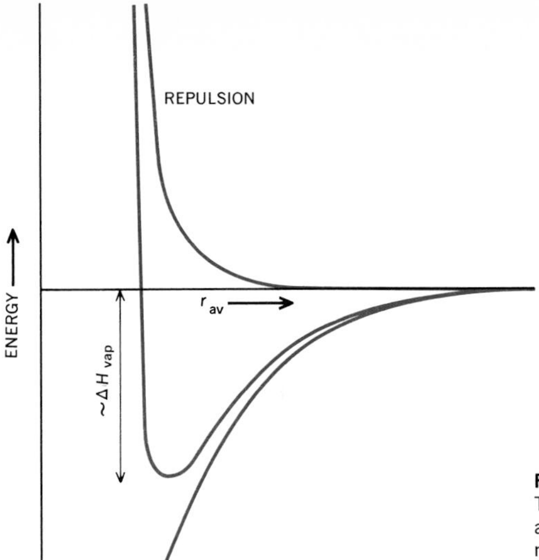

FIGURE 19-2
The net attraction, repulsion, and total energies for the molecules of a liquid. The coordinate r_{av} can be considered to be an average distance from a molecule to its nearest neighbors.

Some understanding of the contributions to the attractive forces can be reached, and it is convenient to consider the interactions that occur between a single pair of molecules. The liquid-state forces will depend on the operation of these forces over many molecules rather than between a pair of molecules. Four types of contributions to intermolecular attractions can be recognized in ordinary liquids.

Dipole-Dipole Attraction Molecules with permanent dipole moments exert a net attraction on each other as a result of the interaction of their dipoles. If the position of one polar molecule is considered to be fixed and if a neighboring molecule is rotated through all possible orientations, the net attraction between the molecules would be zero because of the equal number of repulsive- and attractive-charge orientations. All orientations of the second molecule relative to the first do not, however, occur to the same extent since the attractive, low-potential arrangements are favored over the repulsive, high-energy orientations. A net attraction between two polar molecules therefore results. The average attraction can be worked out, although this will not be done here, by allowing a Boltzmann distribution to yield a higher probability for the attractive orientations. For two polar molecules, with dipole moments μ, separated by a distance r, the dipole-dipole energy of attraction, for values of r large compared with the charge separation in the dipole, can be calculated to be

$$U_{D\text{-}D} = -\frac{2}{3}\frac{\mu^4}{(4\pi\epsilon_0)^2 r^6 kT} \qquad [6]$$

TABLE 19-2 Intermolecular-attraction terms for some simple molecules
An arbitrary intermolecular distance of 5 Å has been used for the comparison.

Molecule	Dipole moment (debyes)	Polarizability (Å^3)	Interaction energy (J per molecule) for two molecules at $r = 5$ Å and $T = 298$ K				ΔH_{vap} (kJ mol^{-1})
			$U_{d\text{-}d} \times 10^{22}$	$U_{ind} \times 10^{22}$	$U_{London} \times 10^{22}$	Total $\times 10^{22}$	
He	0	0.2	0	0	0.05	0.5	0.092
A	0	1.6	0	0	2.9	2.9	6.65
CO	0.12	2.0	0.00021	0.0037	4.6	4.6	6.02
Xe	0	4.0	0	0	18	18	13.0
CCl_4	0	10.5	0	0	116	116	29.9
HCl†	1.07	2.6	1.2	0.36	7.8	9.4	16.1
HBr†	0.79	3.6	0.39	0.28	15	16	17.6
HI†	0.38	5.4	0.021	0.10	33	33	18.2
H_2O†	1.82	1.5	11.9	0.65	2.6	15	39.4
NH_3†	1.47	2.2	5.2	0.63	5.6	11	23.3

† Possibility of hydrogen bonding in addition to listed attractions.

Table 19-2 shows the contribution to the total energy of attraction between a pair of molecules to be expected from this dipole-dipole effect for a few of the simpler molecules. At higher temperatures, as the reciprocal of the temperature in Eq. [6] indicates, the thermal motion of the molecules competes with the tendency toward favorable orientations, and the dipole-dipole attraction becomes less important.

The Induction Effect The presence of a permanent dipole in a molecule always results in an additional contribution which is not included in the previous dipole-dipole term. The dipole of one molecule can interact with and polarize the electrons of the neighboring molecule. The electrons of the second molecule are thereby distorted in such a way that their interaction with the dipole of the first molecule is an attractive one. The induced dipole moments of two nearby polar molecules are sketched in Fig. 19-3, and qualitatively it can be seen that an attraction results. In 1920 P. Debye showed that, for a pair of molecules with dipole moments μ and polarizabilities α, the attractive potential resulting from the dipole-induced-dipole, or induction, effect is

$$U_{ind} = -\frac{2\alpha\mu^2}{4\pi\epsilon_0 r^6} \qquad [7]$$

The polarizability α, introduced in Chap. 15, is defined as the dipole moment induced in a molecule by a unit electric field, and a few results for this quantity were given in Table 15-1. More correctly, one should recognize that the polarizability is different in different directions in he molecule and that again one must be concerned with the relative nolecular orientations. Here, however, it is a very good approximaion to assume all orientations, the induction effect being operative at east to some extent with all molecular orientations.

ondon Dispersion Forces All molecules, including those without a ermanent dipole, attract each other. The liquid state, for example, s exhibited by all compounds and even by the noble gases. The source f the forces that act independently of the molecular dipole, which is ecessary for the attractions discussed under dipole-dipole attraction nd the induction effect, above, was first recognized by F. London in 930. The effect can be treated only quantum-mechanically, being ependent on the detailed motion of the electrons in the neighboring nolecules.

The basis for the attraction, known now as *London,* or *dispersion,* *orces,* can be indicated by reference to a pair of H atoms at a separaon great enough so that the exchange of electrons between the atoms an be ignored. Although a time average would indicate that each atom as a zero-dipole moment, instantaneous-dipole moments can be recgnized. The two instantaneous dipoles of two neighboring atoms annot effectively orient themselves favorably with respect to each other s can the permanent dipoles of molecules. An induction effect, analoous to that treated above, can, however, act between the instantaeous dipole of one atom and the polarizable adjacent atom. Similar ffects, which can be calculated only approximately and only with coniderable difficulty, operate in all molecules, regardless, of course, of hether they have a permanent dipole or not.

Detailed treatments of the quantitative aspects of this interaction ield results that can be variously expressed, but all agree that the otential energy is proportional to the square of the polarizability of ne molecules and inversely as the sixth power of the separation; i.e.,

$$U_{\text{London}} \propto \frac{\alpha^2}{r^6} \qquad [8]$$

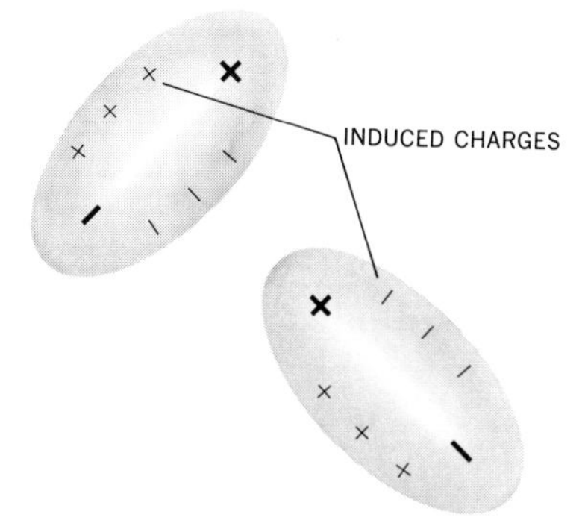

FIGURE 19-3
Mutual attraction of randomly oriented polar molecules through induction. Note that the induced charges are favorably arranged relative to the permanent dipole for attraction to occur.

The proportionality constant turns out to be approximately the same for many molecules and has the value 1.8×10^{-17} J.

Some values that have been calculated for this contribution to the attraction exhibited by a pair of molecules are included in Table 19-2.

Hydrogen Bonding One further, and rather special, interaction must be included. The long-range dipole-dipole interaction calculated from the molecular-dipole moments is not always an adequate treatment of the electrostatic interactions between polar molecules. More localized effects can occur, and the one most often recognized is the *hydrogen bond.* When a somewhat acidic hydrogen atom carrying a partial positive charge, as in the bonds O—H, F—H, N—H, and so forth, can approach an electron-rich basic-type atom, such as the oxygen in water, alcohols, or ethers or the nitrogen in amines, an association occurs that can be represented by

$$\begin{array}{c} \quad\quad\quad\quad\quad\quad\quad\quad\quad\quad\; H \\ \quad\quad\;\; \delta^{+} \quad\;\; \delta^{-} / \\ O - H \cdots\cdots O \\ / \quad\quad\quad\quad\quad\quad\;\; \backslash \\ R \quad\quad\quad\quad\quad\quad\quad\;\; R \end{array}$$

A more revealing diagram emphasizes that it is the details of the charge arrangement in the two molecules that is responsible, and not the molecular dipoles. The lone pair of nonbinding electrons on the oxygen can be exhibited, and this leads to Fig. 19-4. Typical hydrogen-bond distances have been included to emphasize that the hydrogen bond is too long for much covalent character to be expected and is best interpreted as the result of a localized electrostatic attraction.

When hydrogen bonding occurs, it plays a prominent role in determining the intermolecular properties. The high boiling points of H_2O compared with H_2S and of alcohols compared with ethers of the same molecular weight and atomic constitution are primarily the consequence of hydrogen bonding.

A general expression for the potential energy of a hydrogen-bonded pair of molecules as a function of intermolecular distance cannot be given. It can be mentioned, however, for comparison with the other data of Table 19-2, that a pair of alcohol molecules that have approached their equilibrium intermolecular distance are held by a potential-energy contribution of about 25 kJ mol^{-1}, or 400×10^{-22} J per molecular pair.

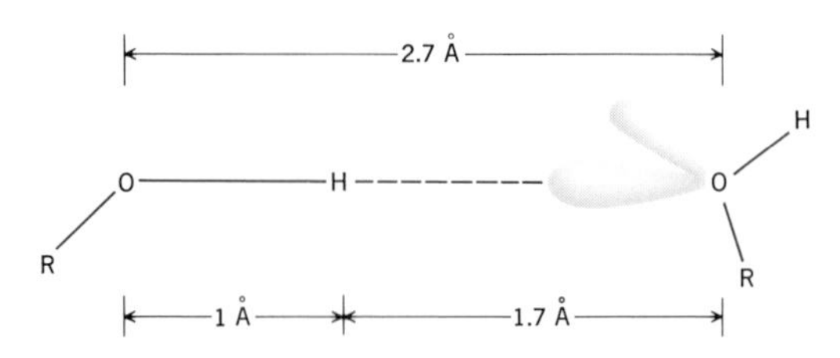

FIGURE 19-4
The hydrogen atom and the electron-pair orbital involved in hydrogen bonding between two water molecules.

The relative importance of the contributions listed in Table 19-2 should be noted. One cannot, however, go easily from the interactions between pairs of molecules to the net effect of a given type of force when the molecules of a liquid are considered. The dipole-dipole forces, for example, require certain favorable orientations, and in a liquid all the neighbors of a molecule must compete for this orientation. London forces, on the other hand, can act as effectively in large collections of molecules as in molecular pairs. Only qualitative relations can therefore be seen in a comparison of the molecular-pair intermolecular forces with the boiling points and heats of vaporization.

19-2 Entropy of Vaporization, Trouton's Rule, and Free-volume Theories of the Liquid State

The heat-of-vaporization data, dealt with in the preceding section, can be used directly to give values for the entropy of vaporization. This quantity can also be investigated from a molecular point of view, and when this is done, one is led to some further ideas that have been developed regarding the behavior of the molecules of a liquid.

An interesting empirical generalization involving heats of vaporization at the normal boiling point was made by F. Trouton in 1884. *Trouton's rule* states that the heat of vaporization in joules per mole divided by the normal boiling point, on the absolute scale, i.e., the entropy of vaporization Δs_{vap}, is approximately 88 J deg^{-1} mol^{-1} for most liquids; i.e.,

$$\Delta s_{vap} = \frac{\Delta H_{vap}}{T_{bp}} \approx 88 \text{ J deg}^{-1} \text{ mol}^{-1} \quad [9]$$

Tests of Trouton's rule are shown for a number of compounds in Table 19-3, and as these data indicate, a very large number of compounds with very different boiling points and heats of vaporization give values of the entropy of vaporization close to 88. Two classes of compounds, however, show serious disagreements, and for both a more consistent set of entropies of vaporization is obtained if they are calculated, not at the normal boiling point, but rather at temperatures at which the vapors of the different compounds have the same molar concentrations. This variation of Trouton's rule, as the results of Prob. 2 will show, improves the overall constancy of the calculated entropies of vaporization, but the abnormalities of the hydrogen-bonded liquids still show up.

Now let us turn to the majority of the compounds that conform to Trouton's rule and show entropies of vaporization near the value 88 J deg^{-1} mol^{-1}. A molecular interpretation of this thermodynamic quantity might be expected to be more easily accomplished than was the corresponding attempt with heats of vaporization because the rela-

TABLE 19-3 Heats and entropies of vaporization of liquids at their normal boiling points

Liquid	Normal bp (°C)	ΔH_{vap} (J mol^{-1})	$\Delta S_{vap} = \Delta H_{vap}/T$ (J deg^{-1} mol^{-1})
Helium, He	−268.9	100	24
Hydrogen, H_2	−252.7	904	44
Acetic acid, CH_3COOH	118.2	24,390	62
Formic acid, HCOOH	100.8	24,100	64
Nitrogen, N_2	−195.5	5,560	72
n-Butane, C_4H_{10}	−1.5	22,260	82
Naphthalene, $C_{10}H_8$	218	40,460	82
Methane, CH_4	−161.4	9,270	83
Ethyl ether, $(C_2H_5)_2O$	34.6	25,980	84
Cyclohexane, (C_6H_{12})	80.7	30,080	85
Carbon tetrachloride, CCl_4	76.7	30,000	86
Stannic chloride, $SnCl_4$	112	33,050	86
Benzene, C_6H_6	80.1	30,760	87
Chloroform, $CHCl_3$	61.5	29,500	88
Hydrogen sulfide, H_2S	−59.6	18,800	88
Mercury, Hg	356.6	59,270	94
Ammonia, NH_3	−33.4	23,260	97
Methyl alcohol, CH_3OH	64.7	35,270	104
Water, H_2O	100.0	40,670	109
Ethyl alcohol, C_2H_5OH	78.5	38,570	110

tive constancy of ΔS_{vap} implies an independence from the individual molecular characteristics.

The entropy of the process

$$\text{Liquid} \rightarrow \text{vapor}$$

can be treated, as were the chemical reactions of Chap. 10, in terms of the energy-level spacing of the product molecules as compared with that of the reactants. The energy difference for a molecule in the two states is not now involved. (It is not, it should be admitted, a straightforward matter to think of molecules in the liquid state. A clear illustration of the difficulty is presented by a liquid metal. The vapor that comes off the liquid is clearly composed of metal atoms, but the properties of the liquid indicate that it is better thought of as metal ions in a sea of electrons. In a similar way any liquid can be thought of as composed of interacting molecules, but if the interactions are very important, the molecular description becomes less satisfactory. For the present entropy calculation it will be assumed that the liquid consists of rather closely packed molecules which compare with similar but less closely packed molecules in the gas phase.)

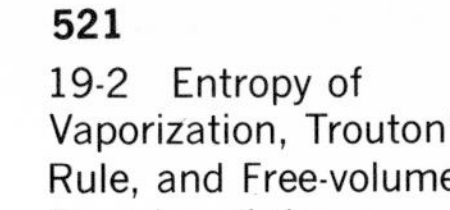

It will be assumed, furthermore, that the rotational- and vibrational-entropy contributions are nearly the same for the molecules of the liquid and the vapor. One finds, for example, that vibrational transitions of a molecule that can be studied in the infrared spectral region are generally only slightly shifted between gas- and liquid-phase spectra. It follows that the allowed vibrational-energy-level pattern is not greatly disturbed by this change of phase. Other infrared spectral results show that small, light molecules do have appreciable rotational freedom in the liquid state. The contours of the infrared absorption bands shown in Fig. 19-5 suggest that, although the discrete, well-defined rotational-energy levels that lead to the rotational structure in

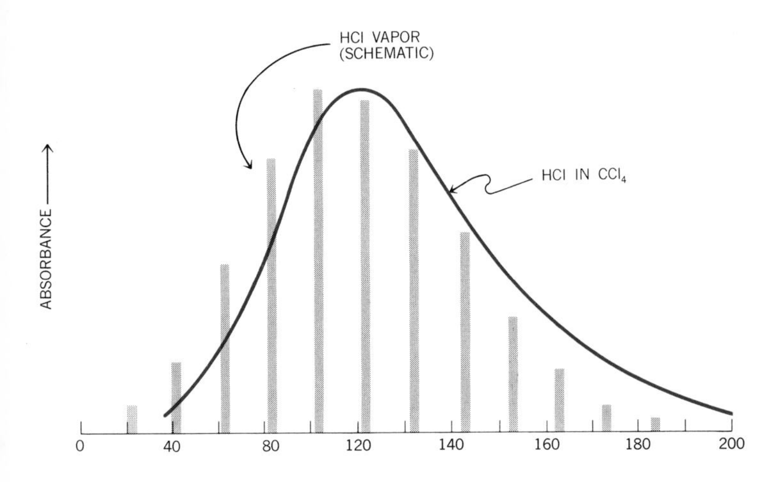

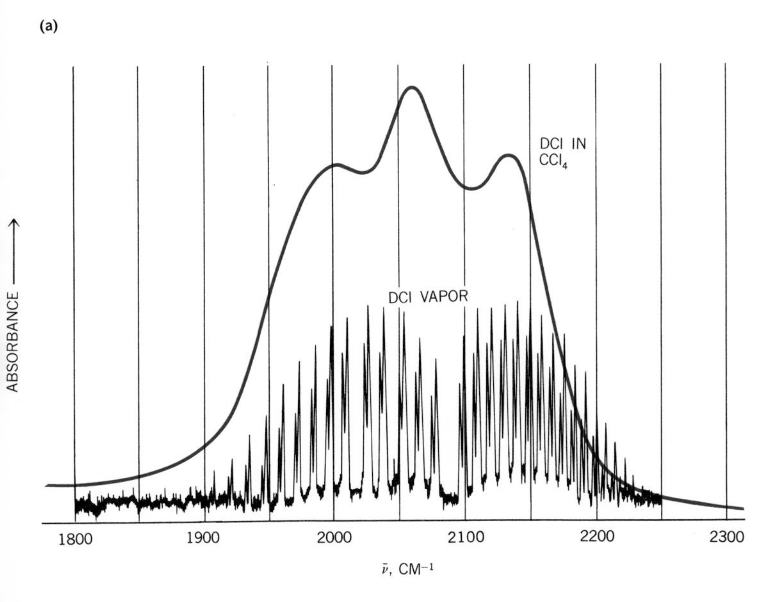

FIGURE 19-5
Vapor and solution spectra of HCl and DCl showing (a) the far infrared region where spectra arise from changes in energy of rotationlike motions, and (b) the infrared absorption bond where such energy changes accompany the vibrational-energy change of the molecules.

a gas-phase rotational band are absent, some rotational effects remain. More empirical justification of the assumption that entropy contributions from vibrational and rotational degrees of freedom are not appreciable in Δs_{vap} comes from the recognition that monatomic and polyatomic molecules, the former having no rotational degrees of freedom and the latter having 2 or 3, conform equally well to Trouton's rule. Furthermore, the near equality of Δs_{vap} for CH_4, which in the vapor phase has a negligible entropy due to rotation and vibration, and CCl_4, in which these degrees of freedom contribute almost half of the total of the vapor-phase entropy, suggests that no appreciable change in these CCl_4 rotational and vibrational contributions occurs. It is necessary, therefore, to look to changes in the translational entropy to account for the principal part of the Trouton rule entropy value.

The translational entropy of vapor molecules is well understood and has been treated in Sec. 10-2. A set of closely spaced translational-energy levels exists, and the spacing of these levels can be calculated by treating each gas molecule as a particle-in-a-box. An actual value for this translation entropy can be calculated, as has been done in Sec. 10-2, but for the present purpose it is enough to recognize that the energy-level spacing decreases with increasing size of the container and that the translational entropy includes the term $R \ln v$.

The counterpart of translation for the molecules of a liquid is less easily defined. A simple assumption that can be made is that a given liquid molecule is free to move around in some small volume, or *cell,* and that it is generally confined to this cell by the neighboring molecules. A similar view led in Sec. 16-9 to the idea of a "cage" which held reacting molecules together for a time that was long compared with the time for a single molecular collision.

The potential energy experienced by a liquid molecule will be a very complicated function of the nature and positions of the neighboring molecules. The effect of these neighbors can, however, be approximated by a simple square-well-type potential which has some low potential-energy value throughout the cell volume but is infinite outside this volume.

The assumption of a simple square-well potential for the molecules of the liquid has made the liquid-to-vapor transformation, as far as the energy-level pattern is concerned, similar to an expansion of a gas from a very small volume and a large volume. The energy-level patterns that are involved are illustrated schematically in Fig. 19-6.

If the volume in which the liquid molecules are free to move, called the *free volume,* is denoted by v_f, the entropy of transformation to the vapor state is given by

$$\Delta s_{vap} = s_v - s_l = R \ln \frac{v_v}{v_f} \qquad [10]$$

Instead of attempting to determine v_f and to obtain thereby a calculated value of Δs_v, it is more satisfactory to insert the Trouton rule result and write

$$88 = R \ln \frac{v_v}{v_f} \qquad [11]$$

which requires

$$\frac{v_v}{v_f} \approx 10{,}000 \qquad [12]$$

Support is given to our explanation of the entropy of vaporization if all normal liquids can be expected to have the ratio of Eq. [12].

First it must be recognized that the entropy of vaporization is very insensitive to the value of v_v/v_f and that it is necessary only that this ratio can be expected to be within a factor of about 2 of the value in Eq. [12] for a reasonable value of the entropy of vaporization to be obtained.

The volume of 1 mol of vapor will be about 20 or 30 liters, depending on the normal boiling point, and such values correspond, according to Eq. [12], to a free volume of 2 or 3 ml. Since a typical molar volume of a liquid is 100 ml, this implies that 2 or 3 percent of the liquid volume is free volume in which the liquid molecules can move. Such a result seems not unreasonable.

A number of other liquid properties can be made to yield values for the free volume of a liquid, and there is general, rough agreement with the idea of a free volume of about 1 percent of the liquid volume.

There are a number of other approaches to the nature of the liquid state. One of these replaces the free volume that is distributed throughout the liquid cells by *holes*, or vacancies, in the liquid structure that can move around and impart to liquids their characteristic properties.

There are some real difficulties connected with improvements of the cell model and also with the simple treatment given here. The assignment of each liquid molecule to a cell is a treatment that differs from that which we adopt for a gas. The restriction of a given molecule

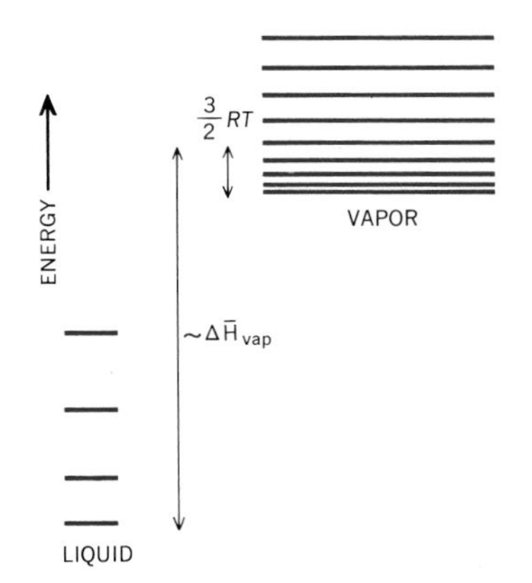

FIGURE 19-6
The liquid-vapor system represented by the energies allowed by square-well potentials.

to a given cell raises a number of complications, mentioned in Sec. 10-1, which, with the less clearly defined cells of the liquid state, cannot easily be taken care of.

We return, finally, to the schematic energy diagram of Fig. 19-6 to understand the equilibrium between the liquid and the vapor. The molecules in the liquid experience a lower potential energy and have a lower enthalpy than do the gas molecules. The close spacing of the energy levels of the vapor, however, gives the vapor a higher entropy. At equilibrium the vapor volume is such that these two factors balance and lead, with the equation

$$\Delta G = \Delta H - T\,\Delta S$$

to a zero free-energy difference.

19-3 The Heat Capacity of Liquids

One remaining thermodynamic property of liquids, the heat capacity, can be dealt with. Again the way in which the thermodynamic data are obtained and treated will be taken up, and then the interpretation, in terms of molecular behavior, will be indicated.

Measurements of the heat capacities of liquids are invariably performed in a constant-pressure apparatus and therefore directly give values of c_P. These constant-pressure results, unfortunately, are not easily approached from a molecular viewpoint because the values for these quantities contain appreciable contributions from the energy absorbed in overcoming the intermolecular attractive forces as expansion with increasing temperature occurs. Values of c_V can be obtained from c_P data and other measurable quantities. We begin with Eq. [38] of Chap. 6,

$$C_P - C_V = \left[P + \left(\frac{\partial E}{\partial V}\right)_T\right]\left(\frac{\partial V}{\partial T}\right)_P \qquad [13]$$

The dependence of internal energy on volume can be deduced from the related quantity, the dependence of enthalpy on pressure, which was obtained in Eq. [45] of Chap. 9 as

$$\left(\frac{\partial H}{\partial P}\right)_T = -T\left(\frac{\partial V}{\partial T}\right)_P + V \qquad [14]$$

With $H = E + PV$, the derivative on the left can be developed to recast this expression to

$$\left(\frac{\partial E}{\partial P}\right)_T + P\left(\frac{\partial V}{\partial P}\right)_T + V = -T\left(\frac{\partial V}{\partial T}\right)_P + V$$

or

$$\left(\frac{\partial E}{\partial P}\right)_T + P\left(\frac{\partial V}{\partial P}\right)_T = -T\left(\frac{\partial V}{\partial T}\right)_P \qquad [15]$$

The interrelation between the derivative of E with respect to V, needed for Eq. [13], and that with respect to P, as appears in Eq. [15], is obtained from the total differential

$$dE = \left(\frac{\partial E}{\partial P}\right)_T dP + \left(\frac{\partial E}{\partial T}\right)_P dT$$

which, after division by dV and specification of constant T, gives

$$\left(\frac{\partial E}{\partial V}\right)_T = \left(\frac{\partial E}{\partial P}\right)_T \left(\frac{\partial P}{\partial V}\right)_T \qquad [16]$$

Substitution of Eqs. [16] and [15] in Eq. [13] now gives

$$C_P - C_V = -T\frac{(\partial V/\partial T)_P{}^2}{(\partial V/\partial P)_T} \qquad [17]$$

With the symbols

$$\alpha = \frac{1}{V}\left(\frac{\partial V}{\partial T}\right)_P$$

for the *coefficient of thermal expansion* and

$$\beta = -\frac{1}{V}\left(\frac{\partial V}{\partial P}\right)_T$$

for the *isothermal compressibility,* we have,

$$C_P - C_V = \frac{\alpha^2 VT}{\beta}$$

or for molar quantities,

$$\mathrm{c}_P - \mathrm{c}_V = \frac{\alpha^2 VT}{\beta} \qquad [18]$$

Data for the various quantities in this expression are given for a number of liquids and solids in Table 19-4. It is clear from these data that although the expansion of a liquid is small compared with that of a gas, and therefore the work against the confining pressure is also relatively small, the difference $\mathrm{c}_P - \mathrm{c}_V$ is appreciable for liquids and cannot be as easily dismissed as in the case of solids.

Now the c_V data can be considered. Most informative, initially, is the value, shown in Table 19-4, of just about $3R$ for mercury. The simplest explanation of this result comes from considering the mercury atoms of the liquid to have, not 3 translational degrees of freedom, which would lead to a heat capacity of only $\frac{3}{2}R$, but rather 3 vibrational degrees of freedom, which, if the atoms are only loosely bound and the vibrational frequencies are low, will lead to each vibrational contribution of R and a total contribution of $3R$. (This assumption regarding the potential-energy function for a molecule of a liquid is not the same, it should be pointed out, as used in the entropy calculation of Sec. 19-2. Thus we can recognize that a better model than the square well used there would be a parabolic potential well. The treatment carried out

TABLE 19-4 Values of c_P and the quantities needed to obtain c_V by means of Eq. [18] for some representative liquids and solids

	c_P (J deg^{-1} mol^{-1})	$\alpha = \frac{1}{V}\left(\frac{\partial V}{\partial T}\right)_P$ (K^{-1})	$\beta = -\frac{1}{V}\left(\frac{\partial V}{\partial P}\right)_T$ (m^2 N^{-1})	V (m^3)	$c_P - c_V$ (J deg^{-1} mol^{-1})	c_V (J deg^{-1} mol^{-1})
LIQUIDS						
H_2O	75.5	2.35×10^{-4}	46×10^{-11}	18×10^{-6}	0.63	74.9
Hg	27.8	1.81	3.4	14.8	4.25	23.6
CS_2	75.7	12.4	96	60	28.6	47.1
CCl_4	131.7	12.5	107	97	42.2	89.5
C_6H_6	134.3	12.5	97	89	42.7	91.6
$CHCl_3$	116.3	13.3	97	80	43.5	72.8
SOLIDS						
C (diamond)	6.07	0.035×10^{-4}	0.68×10^{-11}	3.4	0.0018	6.07
Fe	25.0	0.35	0.59	7.1	0.44	24.6
Cu	24.5	0.48	0.72	7.1	0.68	23.8
Ag	25.5	0.57	0.99	10.2	1.00	24.5
Zn	24.6	0.89	1.69	9.2	1.28	23.3
Pb	26.8	0.86	2.37	18.7	1.74	25.1

on this basis would not, however, be significantly different, with regard to the entropy calculation, from that indicated for the simpler model.)

With the assumption of a contribution of $3R$ from the translational-type motion, we can proceed to polyatomic molecules. The new feature that enters is the counterpart of the molecular rotation. The contribution from the rotational-type motion is, however, not easily handled. If the motion were entirely free, there would be contributions of $\frac{1}{2}R$ per degree of freedom. If, however, the interference of neighboring molecules is such that this motion is more like a low-frequency vibrational motion, the contribution could rise to values equal to about R per degree of freedom. The way to proceed is not now clear, but the examples of Table 19-5 do indicate that the principal contributions to the heat capacity at constant volume of simple liquids have been recognized.

19-4 Surface Tension

One final property of liquids can be included along with the preceding treatments that deal with thermodynamic aspects of liquids. The surface tension of liquids can be looked upon as that property which draws a liquid together and forms a liquid-vapor interface, thereby distinguishing liquids from gases.

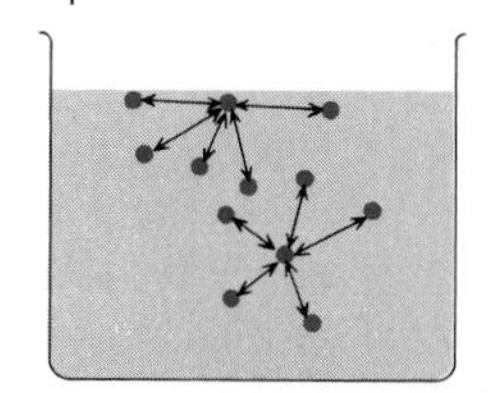

FIGURE 19-7 Interaction forces for a surface molecule compared with one in the body of the liquid.

The molecular basis for this property is indicated in Fig. 19-7, where

TABLE 19-5 Calculations of the heat capacity c_V of some liquids at 25°C
The calculations for free rotation and for a restricted rotation or liberation are shown. Values are in J deg^{-1} mol^{-1}.

	Hg		CS_2		CCl_4	
Motion of center of mass	$3R = 24.9$		$3R = 24.9$		$3R = 24.9$	
Intramolecular vibrational motion (see Sec. 5-7)	0		10.3		45.1	
	Free rot.	Libration	Free rot.	Libration	Free rot.	Libration
Rotational-like motion	0	0	$R = 8.3$	$2R = 16.6$	$\frac{3}{2}R = 12.5$	$3R = 24.9$
Calculated c_V	24.9	24.9	43.5	51.8	82.5	94.9
Observed c_V	23.6		47.1		89.5	

the unbalanced attractions experienced by the surface molecules are shown to lead to a net drawing together of the liquid. On this basis it can be expected that a small amount of free liquid will pull itself together to form a more or less spherical drop. It is also clear that the surface layer will have properties, such as free energy, that will be different from those of the bulk of the liquid.

The surface tension of a liquid can be defined with reference to Fig. 19-8. Most easily pictured is a wire frame, arranged like a piston, used to expand a soap film. The definition also applies, however, to the mechanically more difficult systems where the film would be replaced by a layer of liquid of appreciable thickness. The force required to stretch the film, or liquid layer, is found to be proportional to the length l of the piston. Since there are two surfaces to the film, the total length of the film is $2l$ and the proportionality equation

$$f = \gamma(2l) \qquad [19]$$

can be written. The proportionality constant γ is known as the *surface tension,* and according to Eq. [19], it can be looked upon as the force exerted by a surface of unit length.

Of more general use is the relation between surface tension and surface energy. The work required to expand the surfaces of Fig. 19-8 by moving the piston a distance dx is $f\,dx$, or $2l\gamma\,dx$. Since the area of new surface is $2l\,dx$, the result

$$\frac{\text{Work}}{\text{Change of surface area}} = \frac{2l\gamma\,dx}{2l\,dx} = \gamma \qquad [20]$$

can be obtained. This expression shows that the surface tension can be interpreted as the energy per unit surface area and that it is a mechanical rather than thermal energy. In these terms, the tendency of a surface to reduce its area is just another example of a system tending toward an arrangement of low free energy.

FIGURE 19-8
A wire piston supporting a soap film.

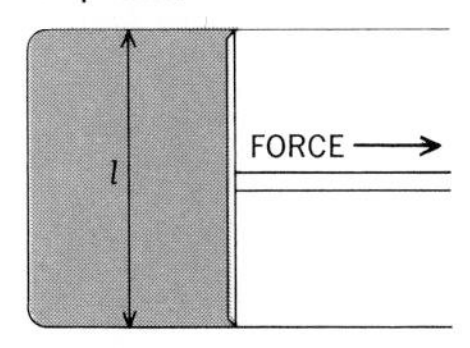

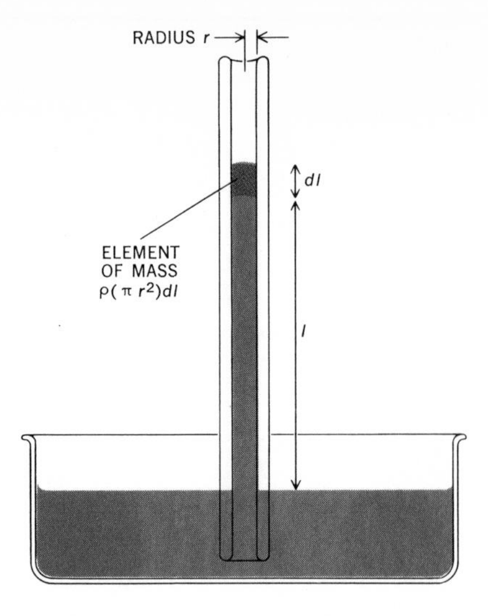

FIGURE 19-9
The capillary rise of a liquid which wets the capillary wall.

A number of methods are available for the measurements of surface tension, but only the *capillary-rise method* will be dealt with here. The arrangement of Fig. 19-9, showing a capillary glass tube inserted into a container of the liquid to be studied, is all that is necessary. All liquids that wet the glass will rise in the tube, and it is this capillary rise that can be used to deduce the surface tension.

The rise of the liquid can be understood if it is assumed that an adsorbed thin film of liquid exists on the wall of the capillary. To reduce its total surface area, the liquid rises in the tube. Equilibrium is reached when the free energy is a minimum; any further rise would expend more free energy in the work to draw up the liquid column than would be saved by the decrease in surface area.

These ideas can be put in more quantitative form by reference to Fig. 19-9. The decrease in surface area that results from a rise in liquid by an amount dl is $2\pi r\, dl$, and the corresponding decrease in surface energy is

$$\begin{aligned} dG_{\text{surface energy}} &= \gamma\, dA \\ &= \gamma(2\pi r)\, dl \end{aligned} \qquad [21]$$

The expenditure of free energy in raising an amount of liquid of volume $\pi r^2\, dl$ and density ρ to a height l is

$$dG_{\text{gravity}} = (\pi r^2\, dl\, \rho)gl \qquad [22]$$

When the column of liquid has risen in the capillary to its equilibrium height, these two free energies balance. For this condition

$$2\pi r\gamma\, dl = \pi r^2 \rho g l\, dl$$

and

$$\gamma = \frac{r\rho g l}{2} \qquad [23]$$

Measurement of the quantities on the right side of Eq. [23] gives a value of γ.

If the liquid does not adhere to the glass, i.e., does not wet it, a capillary rise does not occur and the phenomenon of a depressed liquid in the tube, as observed with mercury and glass, results. Here the mutual attraction between the mercury atoms is greater than that between mercury and glass, and a minimum glass-mercury area is therefore sought by the mercury withdrawing from the inserted tube.

Table 19-6 shows the results obtained for the surface tension of a number of liquids. The only generalization that can be offered is that liquids, like water, which have very strong cohesive forces tend to have high surface tensions. This can be attributed to a greater tendency for the surface molecules to be pulled into the bulk of the liquid. The molten metals and metal salts provide other examples of high surface tension.

19-5 Surface Tension and Vapor Pressure of Small Droplets

Although the surface properties of a liquid are different from those of the bulk liquid, this effect can be ignored except in a few situations. One of these is the case in which a liquid is dispersed into fine droplets and the surface then constitutes a large fraction of the total material. A similar situation occurs with finely divided solid material.

Consider the transfer of dn mol of liquid from a plane surface to a droplet of initial radius r. If the normal vapor pressure of the liquid is P_0 and of the droplet is P, the free-energy change for this process can be written, according to Sec. 9-4, as

$$dG = dn\, RT \ln \frac{P}{P_0} \qquad [24]$$

The free-energy change can also be calculated from the surface-energy change of the droplet that results from the surface-area increase

TABLE 19-6 Surface tensions of some liquids, in N m^{-1}

Liquid	20°C	60°C	100°C
H_2O	0.07275	0.06618	0.05885
C_2H_5OH	0.0223	0.0223	0.0190
C_6H_6	0.0289	0.0237	
$(C_2H_5)_2O$	0.0170		0.0080
Hg	0.480 at 0°C		
Ag	0.800 at 970°C		
NaCl	0.094 at 1080°C		
AgCl	0.125 at 452°C		

due to the addition of dn mol, or volume $M\ dn/\rho$. This volume adds a spherical shell, whose area is $4\pi r^2$. The thickness dr is given by the relation of this spherical shell,

$$\frac{M\,dn}{\rho} = 4\pi r^2\,dr$$

or

$$dr = \frac{M}{4\pi r^2 \rho}\,dn \qquad [25]$$

The increase in surface energy is γ times the increase in surface area resulting from the increase dr in the droplet radius; that is,

$$\begin{aligned} dG &= \gamma\,dA \\ &= \gamma[4\pi(r + dr)^2 - 4\pi r^2] \\ &= 8\pi\gamma r\,dr \end{aligned} \qquad [26]$$

Substitution of Eq. [25] now gives

$$\begin{aligned} dG &= 8\pi\gamma r \frac{M}{4\pi r^2 \rho}\,dn \\ &= \frac{2\gamma M}{\rho r}\,dn \end{aligned} \qquad [27]$$

Equating the two calculations for the free-energy change (Eqs. [24] and [27]) gives

$$dn\,RT \ln \frac{P}{P_0} = \frac{2\gamma M}{\rho r}\,dn$$

and

$$\ln \frac{P}{P_0} = \frac{2\gamma M}{\rho r RT} \qquad [28]$$

If, as is assumed here, SI units are used, care must be taken to express the density in units of kilograms per cubic meter instead of the often-used units of grams per cubic centimeter. The conversion is ρ (kg m^{-3}) = 10^3 ρ (g per cc).

This desired result relates the vapor pressure P of a droplet, or really of a liquid element with highly curved surface, to the vapor pressure P_0 of the bulk liquid. The appearance of r in the denominator implies a dependence of vapor pressure on droplet size that is illustrated in Table 19-7.

These data produce something of a dilemma when the condensation of a vapor to a liquid is considered. The formation of an initial small droplet of liquid would lead to a particle with such a high vapor pressure, according to Eq. [28], that it would evaporate even if the pressure of the vapor were greater than the vapor pressure of the bulk liquid. It is necessary to imagine the condensation to occur on dust

TABLE 19-7 The vapor pressure of water as a function of the radius of curvature of the surface at 25°C (P_0 = 0.03126 atm)

r (m)	r (Å)	P/P_0
10^{-6}	10^4	1.001
10^{-7}	10^3	1.011
10^{-8}	10^2	1.111
10^{-9}	10^1	2.88

particles or other irregularities so that the equilibrium thermodynamic result can be circumvented by some mechanism that avoids an initial slow equilibrium growth of droplets.

Similar considerations are necessary when the reverse process, the boiling of a liquid, which requires the formation of small vapor nuclei, is treated. Chemically, one also encounters this phenomenon in the difficulty with which some precipitates form and in the tendency for liquids to supercool. Likewise, the digestion of a precipitate makes use of the high free energy of the smaller crystals for their conversion to larger particles.

19-6 Structure of Liquids

Let us leave the thermodynamic properties of liquids and consider some other properties whose investigation throws light on the nature of the liquid state. We can follow the approach used earlier in this book and, as for our investigation of molecules, investigate diffraction effects and liquid structure. That there is some organization, or structure, in the liquid state has been assumed in the introduction of the cell theory of the liquid state. In fact, even the concept of nearest neighbors assumes that about any one molecule in a liquid certain nearby positions can be distinguished. Thus one is drawn to the supposition that a certain amount of *short-range order* exists in a liquid and that it is the long-range disorder that gives to a liquid its characteristic properties of fluidity.

This supposition is nicely supported by diffraction studies of simple monatomic liquids. The diffraction principle is essentially that which was used to study molecular structure by electron diffraction in Chap. 14. For the study of liquids, as with that of solids, a more penetrating radiation is necessary so that the interior rather than surface structure can be studied. For this reason liquid structures are investigated by the diffraction of x-rays rather than by electron diffraction. It is found that a diffraction pattern is in fact obtained, as shown in Fig. 19-10, when a beam of x-rays passes through a monatomic-liquid sample, and

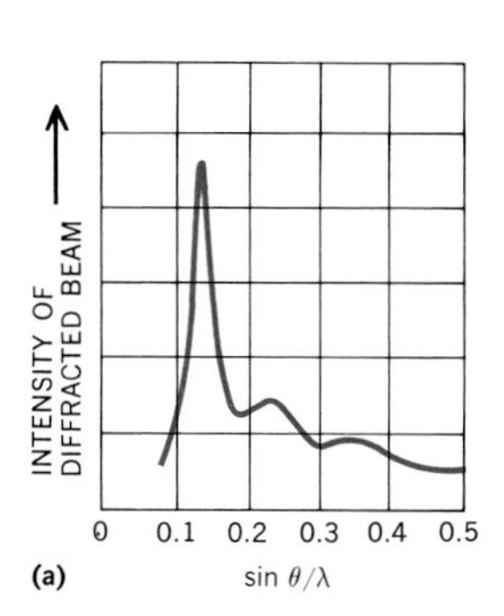

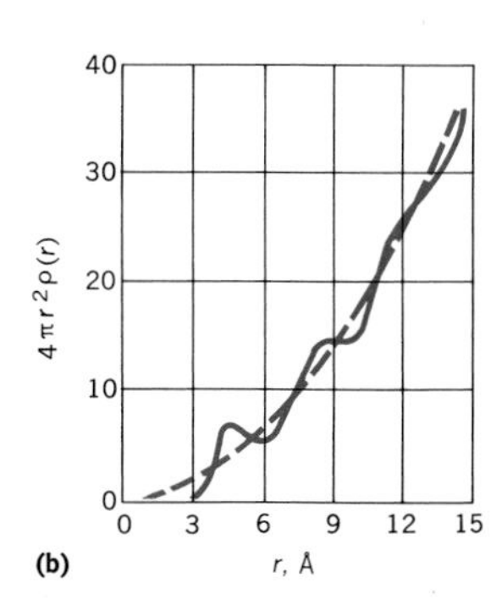

FIGURE 19-10
X-ray-diffraction study of the structure of liquid potassium at 70°C. (*a*) The diffraction intensity with 30-kV x-rays. (*b*) The radial-distribution curve calculated from the data of (*a*); the dashed line indicates the distribution curve that would be expected for no structure in the liquid. [*From C. D. Thomas and N. S. Gingrich, J. Chem. Phys.,* **8**:*411 (1938)*.]

this observation is by itself sufficient for the statement that liquids do exhibit some structure. It will be recalled from the discussion of electron diffraction that the close-in small-angle diffraction depends on small spacings, whereas the wider-angle diffraction is dependent on longer internuclear distances. Liquid-diffraction patterns characteristically show one or two nicely formed small-angle diffraction maxima that correspond to the short-range order, but show rather diffuse diffraction rings corresponding to the farther-removed molecules.

The radial-distribution method introduced in Sec. 14-4 can be used here very effectively. It is only the distance of neighboring molecules from a given central one, and not the angular arrangement, that is of interest. The radial-distribution curve that is calculated from the scattering curve of Fig. 19-10 is also included in that figure. In this way the layer of nearest neighbors clearly shows up, as does the diffuseness of the longer-range structure.

Structures of liquids consisting of molecules of some complexity have been studied, and again a liquid structure, now dependent somewhat on the geometry and intermolecular forces of the molecules, shows up.

All such structure investigations emphasize the similarities between liquids and crystals. We now proceed to another property of liquids that tends to emphasize their likeness to gases.

19-7 The Viscosity of Liquids

The final of the three general categories for our study of liquids, i.e., kinetics or time-dependent phenomena, leads us now to the study of the viscosity of liquids.

The viscosity, or the coefficient of viscosity, has been defined, in connection with our study of gases, by the equation

$$f = \eta A \frac{dv}{dr} \qquad [29]$$

where f is the force necessary to impart, to a fluid with a coefficient of viscosity η, a velocity gradient dv/dr over an area A parallel to the flow direction. Measurements of viscosity are almost always made with a fluid flowing in a tube of circular cross section, and the derivation of the Poiseuille equation (Eq. [27] of Chap. 1), which applies to this case, is now pertinent.

Consider a tube of length l, as in Fig. 19-11. Flow with a uniform velocity results when the pressure drop P along the length l balances the viscous drag of the fluid along this length. The viscous drag stems from the thin film of fluid that, for most liquids, is tightly held to the surface of the tube. This film is effectively stationary, and the fluid must be pushed through the tube against the frictional drag of this

FIGURE 19-11
Flow in a segment of tube of length l for which the pressure drop is P.

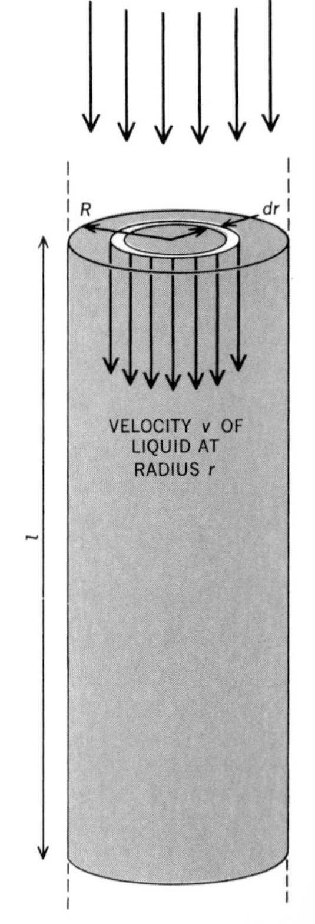

layer. The frictional drag on the cylindrical differential layer of Fig. 19-11 is, according to the definition of Eq. [29],

$$\text{Frictional drag} = \eta(2\pi rl)\left(-\frac{dv}{dr}\right) \qquad [30]$$

where the negative sign has been inserted to make the drag a positive quantity, the velocity gradient being negative.

The pressure acts to drive the central cylinder of fluid down the tube in opposition to the viscous drag at the surface of the cylinder, and for steady flow this driving force of $P(\pi r^2)$ can be equated to the frictional drag to give

$$-\eta(2\pi rl)\frac{dv}{dr} = P(\pi r^2)$$

or

$$dv = -\frac{P}{2\eta l}r\,dr \qquad [31]$$

The velocity gradient across the cylinder is obtained by integration of Eq. [31]. Since the flow velocity is zero along the wall of the tube, i.e., at $r = R$, and since the other integration limit is the velocity v at a radial position r, the integration

$$\int_{v=0}^{v=v} dv = -\frac{P}{2\eta l}\int_{r=R}^{r=r} r\,dr$$

gives

$$v = \frac{P}{4\eta l}(R^2 - r^2) \qquad [32]$$

The velocity contour implied by this equation is shown in Fig. 19-11.

It is at this point that it must be mentioned that fluid flow in a tube does not always conform to the velocity gradient of Eq. [32]. It is found empirically that such an equation governs the flow only for rather small diameter tubes and low flow rates. Flow with a velocity distribution like that of Fig. 19-12a is known as *laminar,* or *viscous,*

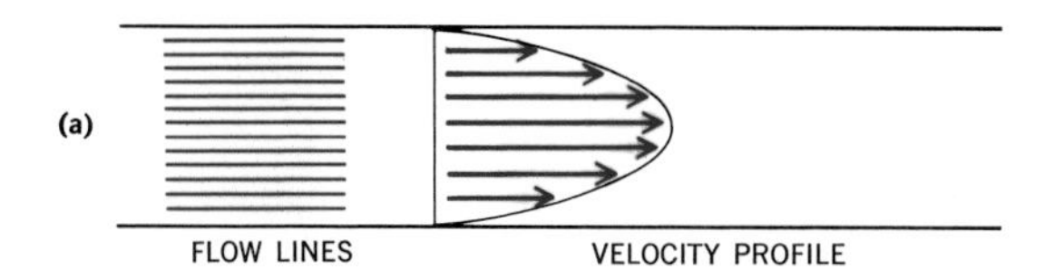

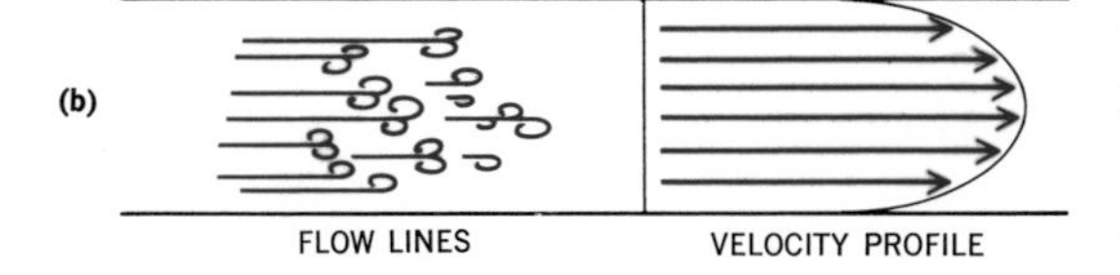

FIGURE 19-12
Nature of laminar flow to which Eq. [32] applies, and turbulent flow, which occurs at high flow rates and for which Eq. [32] does not apply. (a) Laminar, or viscous, flow. (b) Turbulent flow.

flow, and all measurements of viscosity must deal with flow of this type for the derived equations to be valid.

At higher flow rates or for larger-diameter tubing the flow type changes to what is known as *turbulent flow,* which can be described qualitatively in terms of the eddies of Fig. 19-12b. The velocity gradient implied by Eq. [32] is not valid, and it is a matter of considerable difficulty, and considerable practical importance, to treat this flow type. The decision as to which flow type is likely to operate for any set of conditions is usually made in an empirical way on the basis of the dimensionless quantity known as the Reynolds number, which is defined as

$$\text{Reynolds number} = \frac{d\bar{v}\rho}{\eta} \quad [33]$$

where d = tube diameter
v = average velocity of fluid along the tube
ρ = density
η = coefficient of viscosity

Empirically, it has been found that for Reynolds numbers less than about 2100 the flow is laminar, for values greater than 4000 it is turbulent, and for intermediate values of the Reynolds number the type of flow cannot easily be anticipated.

Measurements of viscosity are made, therefore, under conditions of low Reynolds numbers. A suitable and often-used apparatus for many ordinary liquids is the Ostwald viscometer shown in Fig. 19-13. Its use depends on the measurement of the time required for the amount of liquid held between a and b of Fig. 19-13 to flow through the capillary tube. This measurement can be used to obtain a volume rate of flow through the capillary tube.

The volume rate of flow along a cylindrical tube is obtained by integrating the product of the cross-section areas of cylindrical segments and the velocity of flow of the segment. Thus the Poiseuille equation is obtained with the expression for v given by Eq. [32] as

$$\text{Volume rate of flow} = \int_0^R (2\pi r)v\,dr = \frac{\pi P R^4}{8\eta l} \quad [34]$$

It is this equation that can be used to obtain an absolute value for η, all other quantities being measurable.

In practice, one often determines the viscosity by a comparison of the sample with some standard reference sample, using an apparatus like that of Fig. 19-13. For the Ostwald viscometer the Poiseuille equation [34] can be turned around to give

$$\eta = \frac{\pi P R^4}{8l} \times \frac{\text{time}}{\text{unit volume of flow}} \quad [35]$$

Measurement of the time for the same volume of flow of the sample

FIGURE 19-13
An Ostwald viscometer.

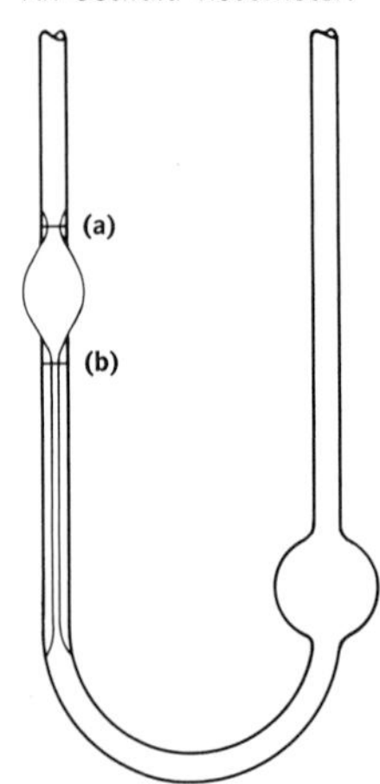

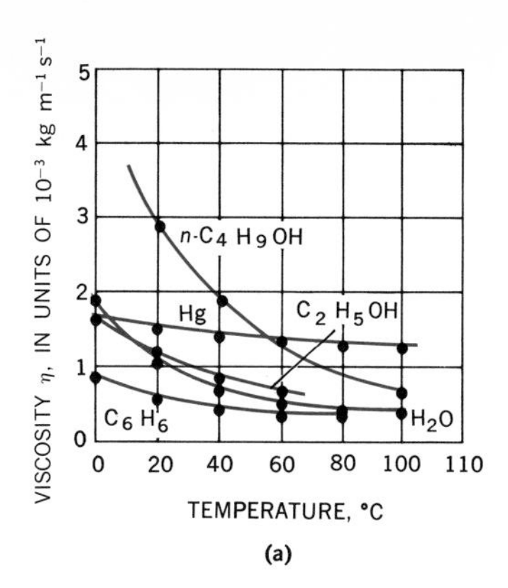

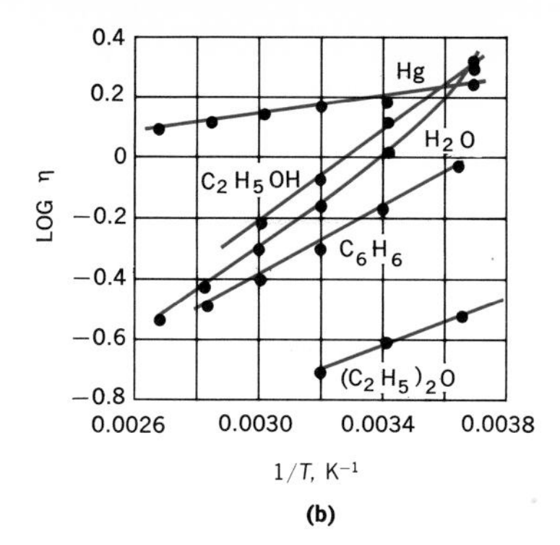

FIGURE 19-14
The dependence of viscosity on temperature for some liquids. (a) η versus T. (b) log η versus $1/T$ for η in units of 10^{-3} kg m^{-1} s^{-1}.

and reference and recognition that the driving force P is proportional to the liquid density ρ allow one to write

$$\frac{\eta_1}{\eta_2} = \frac{\rho_1 t_1}{\rho_2 t_2} \tag{36}$$

Viscosities can be determined, therefore, by comparing the flow time relative to that of a reference liquid in the same viscometer.

Such measurements can be made over a range of temperatures, and as the plots in Fig. 19-14 of the data of Table 19-8 show, the temperature dependence of viscosity conforms to the equation

$$\log \eta = \frac{A}{T} + B \tag{37}$$

It is of interest to try to obtain a theory of liquid viscosities that leads to an expression of the form of this empirical equation. This goal

TABLE 19-8 Coefficient of viscosity of liquids in units of 10^{-3} kg m^{-1} s^{-1}
Viscosities are also often reported in terms of the unit *poise*. The relation of poise to SI units is 1 poise = 10^{-1} kg m^{-1} s^{-1}. Thus the tabulated values must be multiplied by 10^{-1} to obtain coefficients of viscosities in poises.

Liquid	0°C	20°C	40°C	60°C	80°C	100°C
H_2O	1.792	1.005	0.656	0.469	0.356	0.284
C_2H_5OH	1.773	1.200	0.834	0.592		
n-C_4H_3OH	5.186	2.948	1.782			0.540
C_6H_6	0.912	0.652	0.503	0.392	0.329	
$CHCl_3$	0.700	0.563	0.464	0.389		
CCl_4	1.329	0.969	0.739	0.585	0.468	0.384
Hg	1.685	1.554	1.450	1.367	1.298	1.240
$(C_2H_5)_2O$	0.284	0.233	0.197	0.166	0.140	0.118

can be reached, as the following section indicates. The theoretical deduction of the actual values of the viscosities of liquids is, however, an exceedingly difficult theoretical problem.

19-8 Theory of Viscosity

Now an attempt will be made to understand something of the source of viscosity in liquids. Success such as is achieved by the kinetic-molecular theory in its treatment of gas viscosities cannot, however, be attained.

For theoretical purposes it is more satisfactory to think in terms of *fluidity*, i.e., the tendency to flow, rather than viscosity, the resistance to flow. The fluidity is usually represented by ϕ and defined as

$$\phi = \frac{1}{\eta} \qquad [38]$$

The flow of a liquid is a rate process, and we can inquire about the mechanism by which molecules move along a tube in a manner similar to that in which we inquired about the way in which molecules react to form products. Both of the chemical rate theories of Chap. 17 led to the prediction of a rate expression with an exponential term involving an activation energy. A similar form for the rate expression for flow can be anticipated, and the counterpart of the activation energy will be the energy maximum encountered by a molecule as it squeezes past its nearest neighbors to a position farther along the tube. The flow process can be likened to that of passengers pushing down the aisle of a crowded bus, and the activated states correspond to those moments and positions when a person crowds past another to get to a better position farther along the aisle. If the activation energy is designated by ΔE_{visc}, the fraction of the molecules that have energy in excess of this amount, and can therefore move past their neighbors, will be given by a Boltzmann-type expression. The fluidity can therefore be written as

$$\phi \propto e^{-\Delta E_{visc}/RT} \qquad [39]$$

and the viscosity, $\eta = 1/\phi$, as

$$\eta \propto e^{\Delta E_{visc}/RT}$$

or

$$\eta = Ae^{\Delta E_{visc}/RT} \qquad [40]$$

where A is some undetermined constant. The logarithmic form of Eq. [40] is that which was recognized in the plots of Fig. 19-14.

It is of interest to compare the values of ΔE_{visc} listed in Table 19-9 with the heat, or better, the internal-energy change, of vaporization.

TABLE 19-9 The comparison of the activation energy for viscous flow with the internal-energy change for vaporization at the normal boiling point*

Liquid	ΔE_{visc} (kJ mol^{-1})	ΔE_{vap} (kJ mol^{-1})	$\Delta E_{visc}/\Delta E_{vap}$
CCl_4	10.46	27.6	0.38
C_6H_6	10.63	27.9	0.38
CH_4	3.01	7.6	0.40
N_2	1.88	5.1	0.37
O_2	1.66	6.1	0.27
$CHCl_3$	7.36	27.8	0.26
$(C_2H_5)_2O$	6.74	23.8	0.28
Acetone	6.92	26.8	0.26
Hg	2.5	54	0.046
Na (at 500°C)	6.1	98	0.062
Pb (at 700°C)	11.7	178	0.065

*From R. H. Ewell and H. Eyring, *J. Chem. Phys.*, **5**:726 (1937).

The ΔE_{vap} term gives the energy to remove a molecule completely from its neighbors. The flow process seems to require only about a third of this disruption to move a liquid molecule from its initial equilibrium position through a high-energy intermediate position.

This idea can be extended to rationalize the data for liquid metals. Vaporization removes an atom of the metal from the liquid phase. The flow process, on the other hand, can be thought of as involving the movement of positive ions which exist in a sea of electrons. The much smaller size of the positive metal ions compared with the metal atoms can be given as an explanation for the small value $\Delta E_{visc}/\Delta E_{vap}$ for liquid metals.

Problems

1 Calculate the work of expansion done in the vaporization of water at its normal boiling point of 100°C if the vapor is considered to be ideal. What fraction of ΔH_{vap} is due to this work term?

2 Hildebrand's extension of Trouton's rule suggests that the entropy of vaporization be compared for vaporizations from the liquid to a given vapor volume. Assuming ideal-gas behavior, calculate the entropy of vaporization of H_2, N_2, CCl_4, and Hg for vaporization at their normal boiling points to a vapor of volume 22.4 liters. Compare the consistency of these values with those of Trouton's rule given in Table 19-3.

3 Consider two dipoles, each consisting of electronic charges separated by 1 Å. These dipoles are placed with their centers 5 Å apart and with their axes along a straight line joining their centers. They are allowed to take up the low- or the high-energy orientation.

a Using Coulomb's law for the potential energy of separated charges, calculate the energy difference between the two orientations.

Ans. 76.8×10^{-21} J.

b With Boltzmann's distribution deduce the fraction of an Avogadro's number of dipoles that would be in the high-energy orientation at 25°C.

c Repeat the calculations for the centers of the dipoles separated by 10 Å.

d Calculate the energy of the Avogadro's number of dipoles separated by 5 and 10 Å in the case where they are free to adopt either orientation and in the case where both orientations are equally populated. Recognize, as a result of this calculation, the factors that enter into the inverse sixth relation of Eq. [6] that is valid for large distances between the dipoles.

4 Verify that the expressions for the translational contribution to the entropy obtained in Chap. 10 lead to the expression for Δs_{vap} given in Eq. [10].

5 The surface tension of mercury is 0.52 N m^{-1}, and its density is 13.6 g ml^{-1} at 25°C. Mercury does not wet glass.

a Derive an expression for the lowering of the surface of mercury that will occur when a glass tube of internal radius r is inserted into the liquid.

b If r is 1 mm, what will the depression be?

c If r is 5 mm, what will the depression be?

6 A particle of mist has a mass of about 10^{-12} g. What is its vapor pressure compared with that of water if the temperature is 20°C?

Ans. $P = 1.002P_{H_2O}$.

7 In view of Eqs. [23] and [28], consider what will happen as the radius of the capillary tube in which a liquid rises is reduced toward the limit of $r = 0$.

8 Suppose that water in a certain container boils when the liquid is at a temperature at which bubbles of 10^{-7} m diameter can be formed by the equilibrium vapor. Estimate from approximate relations and data previously given:

a The vapor pressure that must be reached by the liquid for these vapor bubbles to be formed.

b The temperature at which boiling will occur at a pressure of 1 atm.

Ans. 100.4°C.

9 The rather viscous material glycerin has the viscosities 134, 12.1, 1.49, 0.63 kg m^{-1} s^{-1} at the temperatures −20, 0, 20, and 30°C, respectively. What is the activation energy for viscous flow for glycerin?

10 The viscosity of chlorobenzene at 20°C is 0.80×10^{-3} kg m^{-1} s^{-1}, and its normal boiling point is 132°C. Assuming that it behaves as a typical liquid, estimate its viscosity at 100°C. Compare with the measured value of 0.37×10^{-3} kg m^{-1} s^{-1}.

11 Water runs out of a tap connected to a $\frac{1}{2}$-in-diameter pipe at a rate of about 1 qt in 1 min. Is the flow of water in the pipe likely to be viscous or turbulent?

Reference

PRYDE, J. A.: "The Liquid State," Hutchinson & Co., Ltd., London, 1966. A short and readable extension into theories of the liquid state.

20

PHASE EQUILIBRIA

PHASE EQUILIBRIA IN ONE-COMPONENT SYSTEMS

A *phase* is defined as that part of a system that is chemically and physically uniform throughout. We have, in fact, been studying the characteristics and the molecular makeup of gas, liquid, and solid phases but have postponed, until now, organized study of systems in which two or more such phases are present together in equilibrium with one another.

Phase-equilibria studies will be introduced with the chemically simplest and least interesting systems, those consisting of a single chemical material, or component. By this it is meant that the composition of each and every phase is completely determined by the specification of a single chemical material for the system. Later, when many-component systems are studied, the implications of the word "component" when used in this technical sense will be explored further.

20-1 Pressure-Temperature Diagrams for One-component Systems

The phase behavior of a one-component system as a function of the pressure and temperature can conveniently be presented on a P versus T plot, as shown for water in Fig. 20-1, for moderate temperatures and pressures.

The meanings of the lines and areas are first considered. Line TC gives the vapor pressure of liquid water up to the critical point C. This line is therefore a plot of the pressures and temperatures at which liquid and vapor exist in equilibrium. At temperatures higher than that of point C, condensation does not occur at any pressure. The areas on

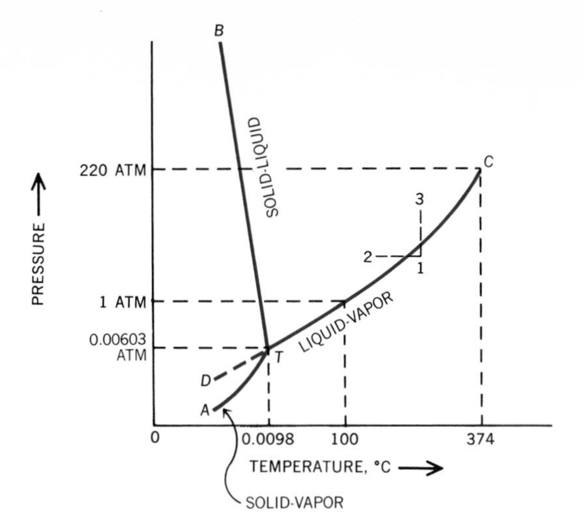

FIGURE 20-1
Phase diagram for water at moderate temperatures (not drawn to scale).

either side of the line TC can be understood by following the changes that occur as a pressure or temperature change results in the system moving across the line. From point 1, for example, the temperature can be lowered to get to point 2, or the pressure can be raised to get to point 3. In either process one crosses the liquid-vapor equilibrium line in the direction of condensation from vapor to liquid. The areas below and above the line TC can therefore be labeled as vapor and liquid, respectively.

The line TA represents the vapor pressure of solid, i.e., the temperatures and pressures at which the solid and vapor are in equilibrium. Again the areas on either side of the line can be labeled, this time as vapor and solid. Finally, line TB gives the melting point of ice as a function of pressure, i.e., the temperatures and pressures at which ice and liquid water are in equilibrium.

Figure 20-1 is a convenient representation of all the available information about the phases of water that occur at moderate pressures and temperatures. It is interesting also to consider Fig. 20-2, which shows the phase behavior of water at very high pressures. Many new solid phases, corresponding to ice with different crystal structures, are encountered. The occurrence of different crystalline forms of a given compound is fairly common and is known as *polymorphism*. It is particularly remarkable that the melting point of ice VII, which exists above about 20,000 atm pressure, is over 100°C.

Mention should be made, furthermore, of the absence of a solid form designated as ice IV. Early studies had been interpreted in terms of an additional phase to those shown in Fig. 20-2. When this phase, which had been labeled as IV, was later shown to be nonexistent, the numbering of the remaining phases was left unaltered.

In Figs. 20-1 and 20-2 the occurrence of a single phase corresponds to an area on a PT diagram; i.e., both these variables can be arbitrarily assigned, within limits, without the appearance of a second phase.

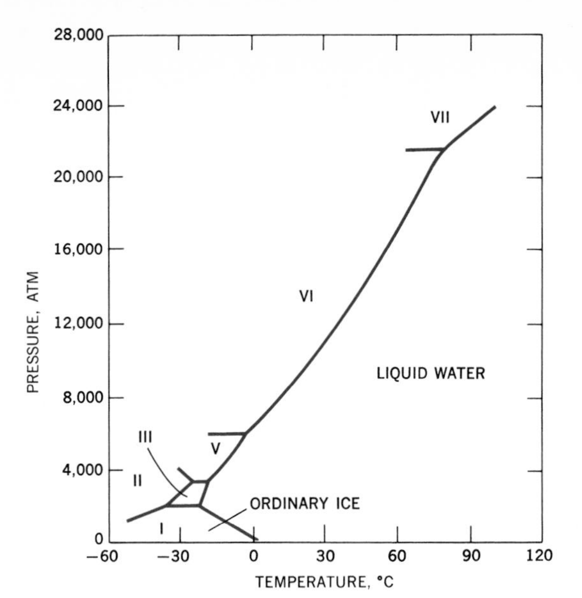

FIGURE 20-2
Phase diagram for water at high pressures.

When two phases are in equilibrium, the diagram shows a line indicating that either P or T may be fixed, but that, when one is fixed, the other must be such that the system is somewhere on the phase-equilibrium line.

Finally, as both Figs. 20-1 and 20-2 show, three phases can exist together, and this occurs at a point on the diagram. Nowhere on the diagram do four phases coexist.

The most important three-phase equilibrium point, called the *triple point,* is that shown by ordinary ice, liquid, and vapor. The temperature and pressure at the triple point are completely determined by the system itself. It is, as a result, a convenient condition on which to base a temperature scale, and as mentioned in Sec. 1-8, it is assigned the value 273.16000 K. The ice point is 0.0098°C lower and thus has the value 273.15 K.

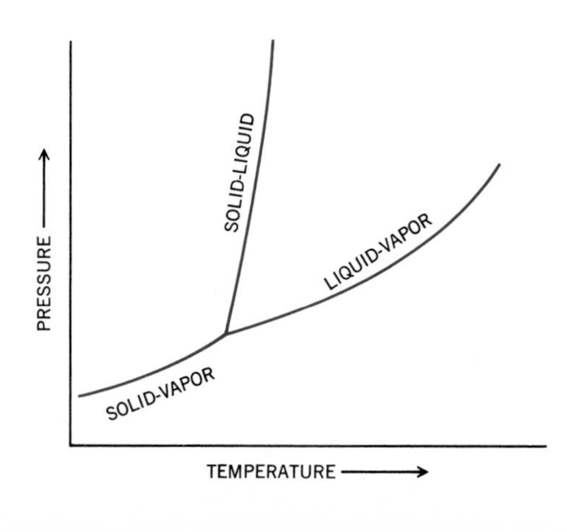

FIGURE 20-3
A representative solid-liquid-vapor phase diagram.

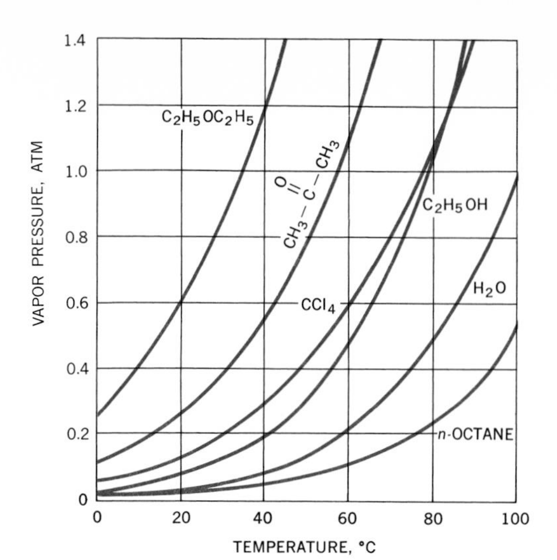

FIGURE 20-4
The dependence of the vapor pressure of some liquids on temperature.

Also, it should be mentioned that liquid water can be cooled below its freezing point to give, as indicated by the dashed line *TD*, *supercooled* water. Supercooled water represents a metastable system because it owes its existence only to the fact that the rate of formation of ice has been interfered with by the use of a very clean sample of water and a smooth container.

The most familiar material, water, that we have used as an illustration of *PT* phase diagrams is, in some ways, not at all representative. More suitable, in this regard, is one in which the solid-liquid equilibrium line, *TB* of Fig. 20-1, has a positive slope, as is the case in Fig. 20-3.

TABLE 20-1 Vapor pressures in atmospheres of liquids as a function of temperature

t°C	H_2O	CCl_4	Acetone	Ethyl ether	Ethyl alcohol	n-Octane
0	0.00603	0.043		0.243	0.016	0.004
10	0.01212	0.074	0.153	0.384	0.032	0.008
20	0.02308	0.120	0.243	0.581	0.058	0.013
30	0.04186	0.188	0.372	0.851	0.104	0.024
40	0.07278	0.284	0.554	1.212	0.178	0.041
50	0.1217	0.417	0.806	1.680	0.292	0.064
60	0.1965	0.593	1.140		0.464	0.103
70	0.3075	0.818	1.579		0.713	0.155
80	0.4672	1.109			1.070	0.230
90	0.6918	1.476			1.562	0.333
100	1.0000	1.925				0.466

Of special value for many studies of liquid solutions are data for the equilibria between the liquid and vapor phases. For pure liquids such data, given by the liquid-vapor curves of Figs. 20-1 and 20-2, are often displayed as *vapor-pressure* curves or as tabular data. Illustrations are given in Fig. 20-4 and Table 20-1. A variety of experimental procedures are used to obtain such data, and the apparatus for two common techniques applicable to the convenient vapor-pressure range from a few hundredths of an atmosphere up to one atmosphere is shown in Fig. 20-5.

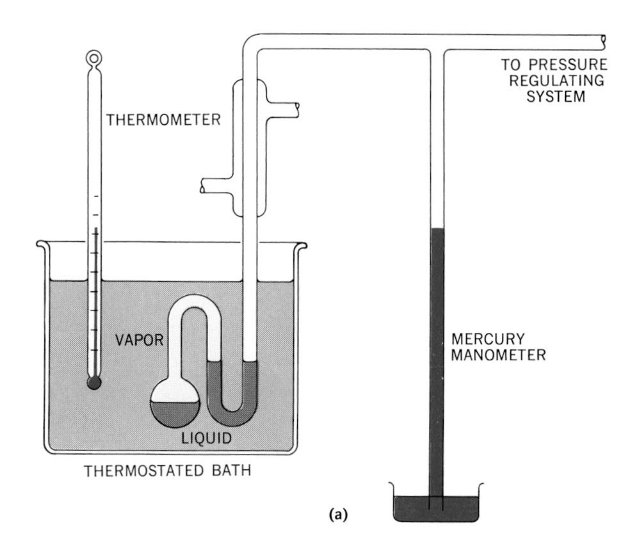

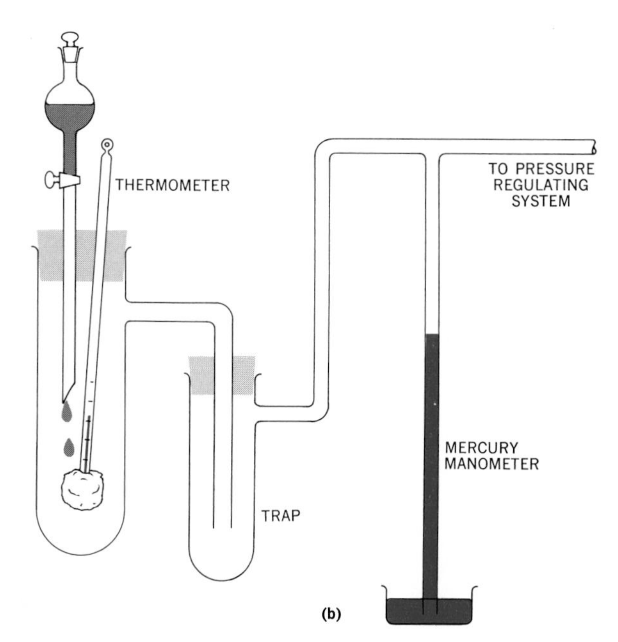

FIGURE 20-5
Methods of vapor-pressure determination. (*a*) The isoteniscope. When the external pressure is equal to the vapor pressure, the sample manometer will show equal heights in the two arms. (*b*) The Ramsey-Young apparatus. The bulb of the thermometer is at the temperature of the liquid, which is in equilibrium with the vapor at the pressure in the system.

20-2 Qualitative Thermodynamic Interpretation of Phase Equilibria of One-component Systems

You will recall from Chap. 9 that the free energy G is the thermodynamic property that is most directly related to the position of equilibrium. It can be used again here to study phase equilibria as portrayed on a pressure-temperature diagram. Important in this regard is the dependence of G on P and T as given by the relations

$$\left(\frac{\partial G}{\partial P}\right)_T = V \qquad \left(\frac{\partial G}{\partial T}\right) = -S$$

or for 1 mol,

$$\left(\frac{\partial \mathrm{G}}{\partial P}\right)_T = \mathrm{v} \qquad \left(\frac{\partial \mathrm{G}}{\partial T}\right) = -\mathrm{s} \qquad [1]$$

From these relations, and values of v and s, the free energy of a phase can be shown as a surface on a graph of G versus P and T.

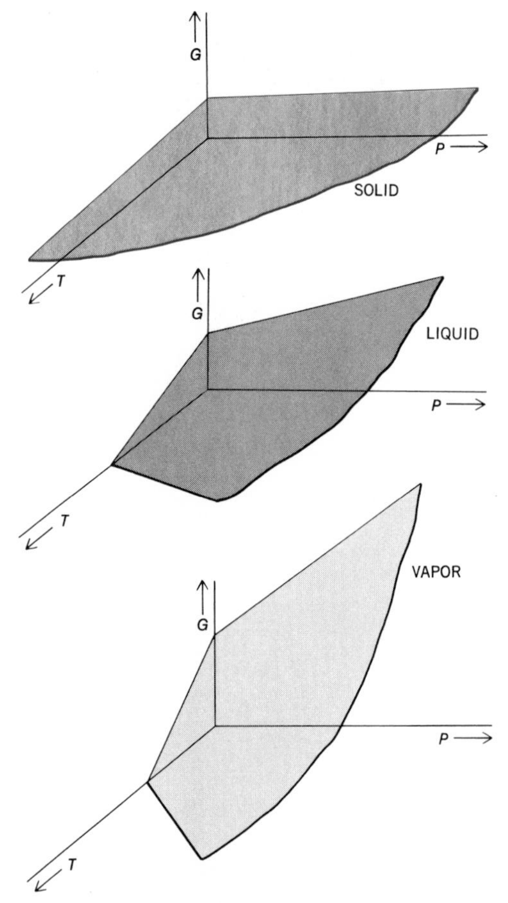

FIGURE 20-6
Representative free energy G versus T and P surfaces for solids, liquids, and gases.

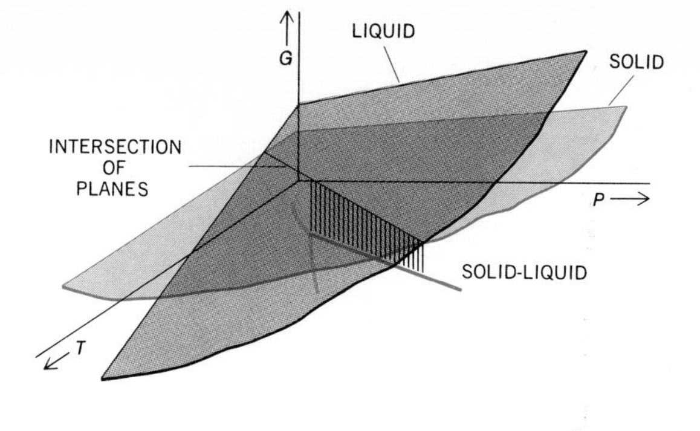

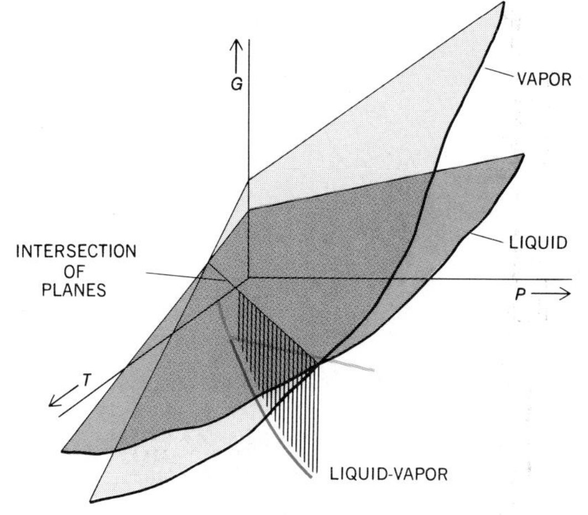

FIGURE 20-7
The intersection of solid-liquid and vapor-liquid free-energy surfaces to give the equilibrium conditions as a function of temperature and pressure.

Such plots are shown for the solid, liquid, and vapor forms in Fig. 20-6. The principal features of such diagrams stem from the relations, applicable to most materials,

$$\mathrm{s} \text{ for gas} > \mathrm{s} \text{ for liquid} > \mathrm{s} \text{ for solid}$$
$$\mathrm{v} \text{ for gas} \gg \mathrm{v} \text{ for liquid} > \mathrm{v} \text{ for solid}$$

Composite diagrams, as in Fig. 20-7, show by the intersection of the planes the pressure and temperature conditions under which different phases give to the material the same molar free energy, and thus equilibrium between the phases.

The exceptional slope of the solid-liquid equilibrium line for water is also seen to be interpretable on the basis of a greater molar volume for ice than for liquid water, in violation of the generalization given above, and the corresponding relation of the slopes of the free-energy surfaces, as shown in Fig. 20-8.

These representations are clearly awkward to use in a quantitative way, even though they clearly reveal the factors determining the lines on PT phase diagrams. To proceed quantitatively, we need analytical expressions for these factors.

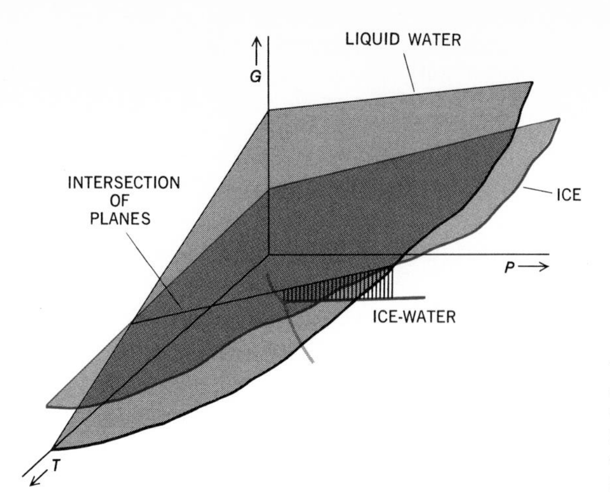

FIGURE 20-8
The intersection of the free-energy surfaces for liquid water and ice and the resulting TP plot for equilibrium between these phases.

20-3 Quantitative Treatment of Phase Equilibria: The Clausius-Clapeyron Equation

Let us now obtain an expression for the pressure-temperature dependence of the state of equilibrium between two phases. To be specific, we will deal with the liquid-vapor equilibrium, but corresponding expressions are applicable to solid-liquid and solid-vapor equilibria.

The free energy of 1 mol of liquid is equal to the free energy of 1 mol of the vapor that is in equilibrium with the liquid. We can write, therefore, with subscript l denoting liquid and v denoting vapor,

$$\mathrm{G}_l = \mathrm{G}_v \qquad [2]$$

and for an infinitesimal change in the system for which equilibrium is maintained, the differential equation

$$d\mathrm{G}_l = d\mathrm{G}_v \qquad [3]$$

can be written.

Such free-energy changes, since only one component is present and the composition is not variable, can be expressed by the total differential

$$d\mathrm{G} = \left(\frac{\partial \mathrm{G}}{\partial P}\right)_T dP + \left(\frac{\partial \mathrm{G}}{\partial T}\right)_P dT \qquad [4]$$

The partial derivatives are related, as used in Sec. 20-2, to the molar volume and entropy, and thus we can write for a molar amount in each phase

$$d\mathrm{G} = \mathrm{v}\, dP - \mathrm{s}\, dT \qquad [5]$$

Application of this equation to the liquid and to the equilibrium vapor, recognizing that although various temperatures and pressures can be considered, both phases are at the same temperature and pressure, gives

$$\mathrm{v}_l\, dP - \mathrm{s}_l\, dT = \mathrm{v}_v\, dP - \mathrm{s}_v\, dT$$

or

$$\frac{dP}{dT} = \frac{\mathrm{s}_v - \mathrm{s}_l}{\mathrm{v}_v - \mathrm{v}_l} = \frac{\Delta\mathrm{s}_{\mathrm{vap}}}{\Delta\mathrm{v}_{\mathrm{vap}}} \tag{6}$$

More generally, we can write

$$\frac{dP}{dT} = \frac{\Delta\mathrm{s}}{\Delta\mathrm{v}} \tag{7}$$

where $\Delta\mathrm{s}$ and $\Delta\mathrm{v}$ signify changes for the two phases being considered.

We thus have an expression for the slope of the phase-equilibrium lines on a PT diagram.

The large value of $\Delta\mathrm{v}$ for solid-vapor or liquid-vapor phases, for example, is related to small values of dP/dT, and thus flatter curves than for solid-liquid phases. Also, curves tend to have positive slopes because the molar entropies and volumes both follow the same order, vapor greater than liquid, and liquid greater than solid. The most notable exception is that for ice-liquid water, where $\Delta\mathrm{s}$ and $\Delta\mathrm{v}$ have opposite signs. This happens also in other cases, as shown by the high-pressure-region diagram for water in Fig. 20-2.

Substitution of

$$\mathrm{s}_v - \mathrm{s}_l = \frac{\Delta\mathrm{H}_{\mathrm{vap}}}{T} \tag{8}$$

in Eq. [6] gives

$$\frac{dP}{dT} = \frac{\Delta\mathrm{H}_{\mathrm{vap}}}{T\,\Delta\mathrm{v}_{\mathrm{vap}}} \tag{9}$$

For liquid-vapor equilibria at temperatures well below the critical temperature, as is often the condition of interest, the liquid volume v_l can be neglected compared with the vapor volume v_v. With this approximation we can write

$$\frac{dP}{dT} = -\frac{\Delta\mathrm{H}_{\mathrm{vap}}}{\mathrm{v}_{\mathrm{vap}}T} \tag{10}$$

This equation is one form of the expression for the temperature–vapor-pressure relation known as the *Clausius-Clapeyron equation.*

If the equilibrium vapor is treated as an ideal gas, the molar vapor volume can be expressed as

$$\mathrm{v}_v = \frac{RT}{P}$$

and substitution of this approximation in Eq. [10] gives

$$\frac{dP}{dT} = \frac{\Delta H_{vap} P}{RT^2} \qquad [11]$$

This result rearranges to the most generally used differential forms of the Clausius-Clapeyron equation,

$$\frac{d(\ln P)}{dT} = \frac{\Delta H_{vap}}{RT^2} \qquad [12]$$

and

$$\frac{d(\ln P)}{d(1/T)} = \frac{\Delta H_{vap}}{R} \qquad [13]$$

The integrated form, with the assumption of a constant value of ΔH_{vap} over the temperature range considered, written in terms of logarithms to the base 10, is

$$\log P = -\frac{\Delta H_{vap}}{2.303R}\frac{1}{T} + \text{const} \qquad [14]$$

The preceding derivation indicates that a plot of log P versus $1/T$ should give a straight line and that the slope of such a line is to be identified with $-(\Delta H_{vap}/2.303R)$. As the curves of Fig. 20-9 show, essentially linear plots are obtained. A more careful look at such results, however, reveals deviations from linearity, and these deviations can be attributed to the approximations that have been introduced in obtaining Eqs. [10], [11], and [14].

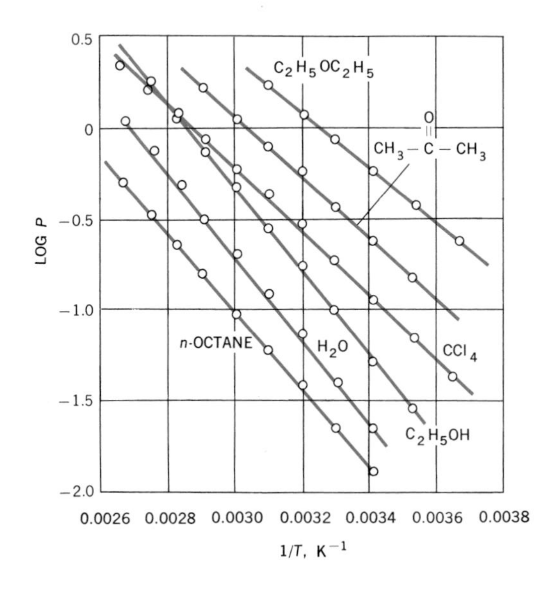

FIGURE 20-9
Log P versus $1/T$ for the vapor pressures of the liquids of Table 20-1.

Before proceeding to the phase equilibria shown by multicomponent systems, a helpful generalization known as the phase rule can be introduced, and it can be illustrated with examples from the one-component systems just studied.

First some terms that are used in the statement of the phase rule must be more carefully defined.

20-4 The Number of Phases

The statement of the thermodynamic rule regarding the phase equilibria that occur in any system requires the prior precise definition of three quantities. The first of these terms is phase. A *phase* is defined as that part of a system which is chemically and physically uniform throughout. The definition is little different from our ordinary use of the word, and only a few points need be made.

A phase may consist of any amount, large or small, of material and may be in one unit or subdivided into a number of smaller units. Thus ice represents a phase whether it is in a single block or subdivided into fine chips. This subdivision must not, however, be carried to molecular dimensions. A solution in which there are two chemical species, for example, is to be considered as one phase, even though subdivision to a molecular scale would reveal that it was not "uniform throughout."

Of particular importance is the *number of phases,* denoted by P, present in a system. Because of the complete mutual solubility of gases, only one gaseous phase can exist in any system. Some liquids are insoluble in one another, and a number of different liquid phases may therefore exist in a system at equilibrium. Different solids, whether they have different chemical composition or the same chemical composition but different crystal structure, constitute different phases.

20-5 The Number of Components

It is necessary now to consider what information must be given to specify the chemical composition of a system. In this connection, the familiar word "components" is used, but a strict definition is attached to it. The *number of components,* denoted by C, is defined as *the least number of independently variable chemical species necessary to describe the composition of each and every phase of the system.*

The composition of a solution of sugar in water, for example, is described by specifying that sugar and water are present. There are two components. If such a solution is cooled, a pure solid sugar phase may begin to separate out. According to the definition, the *system* still has two components even if the solid phase contains only one chemical species.

Some special care is required when the system involves species which are in chemical equilibrium with one another. The number of species that can be arbitrarily varied in a solution of acetic acid in water is two. A number of equilibria are set up in such a system; in particular,

$$CH_3COOH + H_2O \rightarrow CH_3COO^- + H_3O^+$$

Thus there are many chemical species. It should be clear, however, that if the presence of two species is specified, then the presence of the other species is determined by the equilibrium relations that exist. The example should point out that there is no unique set of components among the species in a system. It is only the *number of components* that is unique.

An overstrict attention to the possible equilibria among the species of a system must, however, be avoided. Consider, for example, the gaseous system of water vapor, hydrogen, and oxygen. In the presence of an electric arc or suitable catalyst, the equilibrium

$$2H_2O \rightleftharpoons 2H_2 + O_2$$

is readily established. Under such conditions the system has two components since specification of any two species implies that, at equilibrium, the third will be present. Alternatively, one can say that the concentration of any two species could be arbitrarily set but that the concentration of the third would then be fixed and could, in fact, be calculated from the equilibrium constant of the reaction. At room temperature and in the absence of a catalyst, however, this equilibrium is established so slowly that, for all practical purposes, the reaction connecting the three species can be ignored. Under such conditions the concentrations of all three species can be varied arbitrarily, and the system has three components. A system such as this, which appears stable but is not at the thermodynamic equilibrium position with respect to the reaction, is said to be in *metastable equilibrium.* Many systems have thermodynamically feasible reactions, both chemical and physical, which under certain conditions can be properly ignored.

In a similar way the number of components that one assigns to a system may depend on the conditions that are encountered. For example, the solid-vapor equilibrium system set up by NH_4Cl could, under most circumstances, be said to be a one-component system. The fact that the vapor consists of an equimolar mixture of NH_3 and HCl rather than NH_4Cl molecules would not affect any of the deductions that will be made regarding the behavior of the system. However, if conditions of temperature and pressure could be found so that either the ammonia or the hydrogen chloride separated out from the gas phase, then under these conditions the system would have to be defined as one of two components in just the same way as it would be said to be a two-component system if one could add NH_3 or HCl to the NH_4Cl system.

20-6 The Number of Degrees of Freedom

Some properties of each phase of a system are independent of the amount of the phase present. Thus the temperature, pressure, density, molar heat capacity, and refractive index, for example, of a gas are independent of the amount of gas that one is dealing with. Properties such as these, which are characteristic of the individual phases of the system and are independent of the amounts of the phases, are known as *intensive properties*. Properties such as the weight and volume of a phase, which are dependent on the amount of the phase, are known as *extensive properties*. The latter type of property will not concern us in our study of phase equilibria.

A one-phase system of one component, for example, has a large number of intensive properties. To describe the state of such a simple system, one might measure and report values for many such properties; i.e., one might report the pressure, the temperature, the density, the refractive index, the molar heat capacity, and so forth. We know *from experience*, however, that it is not necessary to specify all these properties to characterize the system completely. All the intensive properties of a sample of pure liquid water, for example, are fixed if the temperature and pressure of the sample are stated. Any two intensive properties, instead of temperature and pressure, might have been fixed, and one would again have found that the sample was completely characterized. Our experience tells us that only a few of the many intensive properties of a sample can be arbitrarily fixed, or as can alternatively be said, need be specified to define the sample.

The number of *degrees of freedom* of a system is defined as *the least number of intensive variables that must be specified to fix the values of all the remaining intensive variables*. The number of degrees of freedom is denoted by Φ. In the previous example of a one-phase system of one component there are 2 degrees of freedom, that is, $\Phi = 2$.

A rearrangement of the statement of the definition, which is sometimes easier to apply, is that the number of degrees of freedom is the number of intensive variables that can be independently varied without changing the number of phases of the system. Examples will bring out the significance of this number and of its definition.

20-7 The Phase Rule for One-component Systems

The identification of an area on a PT diagram with one phase of a one-component system illustrates the two degrees of freedom that exist, these usually being specified as pressure and temperature.

For a two-phase system, the requirement of an equality in the molar free energies of the two phases imposes a relation, such as $dP/dT = \Delta s/\Delta v$, and thus the pressure and temperature cannot both be arbitrarily varied. A two-phase, one-component system thus has a single degree of freedom, as shown by the identification of a line on a PT diagram with two phases in equilibrium.

Finally, for three phases to coexist, the molar free energy of the first pair would have to be equal to that of the additional phase. One more restrictive equation then exists, and thus the last degree of freedom is removed. No arbitrary assignment of variables can be made—the system is entirely self-determined. The one-component PT-diagram feature for three phases is a point.

All this can be summarized by the equation

$$\Phi = 3 - P$$

or in a way that will be shown to be valid for many-component systems, by

$$\Phi = (C + 2) - P = C - P + 2 \qquad [15]$$

where Φ, C, and P are the number of degrees of freedom, components, and phases, respectively.

20-8 The Phase Rule

Rules similar to that given in Sec. 20-7 might be deduced for systems of more than one component. It is possible, however, to proceed more generally and to obtain the *phase rule,* which gives the number of degrees of freedom of a system with C components and P phases. This rule was first obtained by J. Willard Gibbs in 1878, but publication in the rather obscure *Transactions of the Connecticut Academy* resulted in its being overlooked for twenty years.

Consider the C components to be distributed throughout each of the P phases of a system, as schematically indicated in Fig. 20-10. The degrees of freedom of the system can be calculated by first adding up the total number of intensive variables required to describe separately each phase and then subtracting the number of these variables, whose values are fixed by free-energy equilibrium relations among the different phases. To begin with, each component is assumed to be present in every phase.

In each phase there are $C - 1$ concentration terms that will be

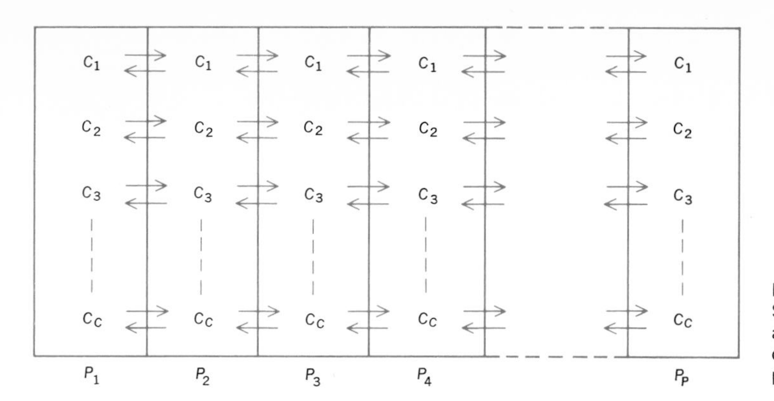

FIGURE 20-10
Schematic representation of a system of C components distributed throughout P phases.

required to define quantitatively the composition of the phase. Thus, if mole fractions are used to measure the concentrations, one needs to specify the mole fraction of all but one of the components, the remaining one being determined because the sum of the mole fractions must be unity. Since there are P phases, there will be a total of $P(C - 1)$ such composition variables. In addition, the pressure and the temperature must be specified, giving a total of $P(C - 1) + 2$ intensive variables if the system is considered phase by phase.

The number of these variables which are fixed by the equilibrium conditions of the system must now be determined. Component C_1, for example, is distributed between phases P_1 and P_2. When equilibrium is established for any one component distributed between any two phases, a distribution relation can be written. Thus, if the concentration of a component in phase P_1 is specified, its concentration in phase P_2 is automatically fixed. Similar equilibria will be set up for each component between the various pairs of phases. For each component there will be $P - 1$ such relations. Thus, for C components a total of $C(P - 1)$ intensive variables will be fixed by the equilibrium conditions.

The number of degrees of freedom, i.e., the net arbitrarily adjustable intensive variables, is therefore

$$\begin{aligned}\Phi &= P(C - 1) + 2 - C(P - 1)\\ &= C - P + 2 \qquad [16]\end{aligned}$$

If a component is not present, or properly, is present to a negligible extent in one of the phases of the system, there will be one fewer intensive variable for that phase since the negligible concentration of the one species is of no interest. There will also be one fewer equilibrium relation. The phase rule applies, therefore, to all systems regardless of whether all phases have the same number of components or not.

This rule is applicable, however, only to what have been termed "ordinary" chemical systems. The properties of some systems might be dependent on the electric or magnetic field throughout the system or the intensity of light shining through the system. If any such additional intensive properties are significant—in "ordinary" chemical systems such intensive variables can be ignored—they must be added into the total number of arbitrarily variable properties and one would then have, for example, $\Phi = C - P + 3$. In practice, we almost always deal with systems for which such additional variables have no noticeable effect on the system, and they can therefore be left out of all consideration.

The phase rule is an important generalization in that, although it tells us nothing that could not be deduced in any given simple system, it is a valuable guide for unraveling phase equilibrium in more complex systems.

PHASE DIAGRAMS FOR MULTICOMPONENT SYSTEMS: CONDENSED PHASES ONLY

A few representative phase diagrams for two- and three-component systems, for which the pressure is high enough so that only the condensed solid and liquid phases occur, will now be discussed. Systems including the vapor phase will be treated separately in the final sections of the chapter.

Here only pictorial presentations will be given. Many quantitative data are implied by these phase diagrams, and some of the interpretations that can be given such data will be pointed out in the following chapter when thermodynamic methods applicable to multicomponent systems are developed.

20-9 Two-component Liquid Systems

The simplest of the two-component phase diagrams are those for liquid systems which may break up into two liquid phases. Such systems are usually treated at some constant pressure, usually atmospheric, high enough so that no vapor can occur in equilibrium with the liquid phases and over a range of temperatures high enough so that no solid phases appear. If the pressure is fixed, the remaining significant variables are the temperature and composition. Diagrams are therefore made showing the phase behavior in terms of these variables. The composition is usually expressed as weight percent of one component or as weight fraction.

Three different types of behavior are recognized. Representatives of these are shown in Fig. 20-11a to c. The heavy line on each of these diagrams bounds the region in which two liquid phases appear. That the line also gives the composition of the liquid layers can be seen by

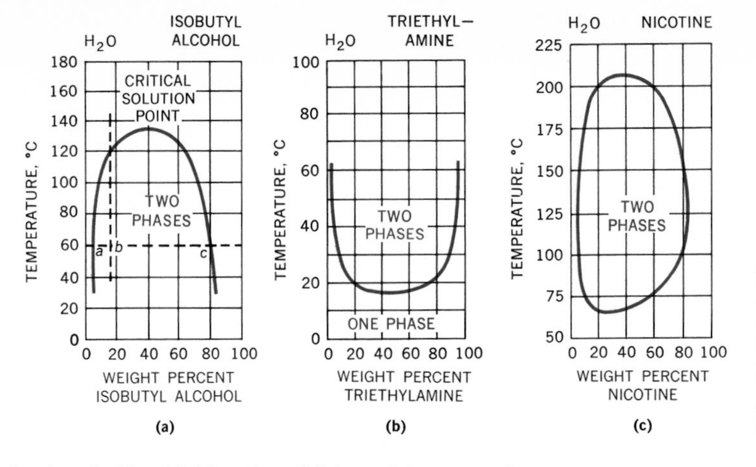

FIGURE 20-11
Partially miscible liquid two-component systems at 1 atm pressure. Measurements were made in a sealed tube in which the pressure was equal to the vapor pressure of the system.

following, in Fig. 20-11a, the addition of the second component, isobutyl alcohol, to an initial quantity of pure water at 60°C. The first additions of butyl alcohol dissolve in the water to form a single phase, and this solubility persists until the total composition of the system corresponds to point a. At this point the solubility of butyl alcohol in water is reached, and further addition produces a second layer of composition c. Thus a total composition of b, at 60°C, results in a two-phase system, the phases having compositions a and c. As the amount of the second component increases, the total composition approaches c, at which point all the water-rich layer has finally dissolved in the butyl alcohol–rich layer to give again a one-phase system.

The relative amounts of the two phases that a system with given total composition gives rise to can be calculated by the procedure indicated in Fig. 20-12. The system has a total weight w and gross weight fraction of component A designated by x. The weights of the two phases that the system breaks up into are w_1 and w_2, and these phases have weight fractions of component A of x_1 and x_2.

The total weight conservation requires

$$w = w_1 + w_2 \qquad [17]$$

and the conservation of component A requires

$$xw = x_1w_1 + x_2w_2 \qquad [18]$$

Substitution of the expression for w from Eq. [17] in Eq. [18] gives

$$x(w_1 + w_2) = x_1w_1 + x_2w_2$$

which rearranges to

$$w_1(x - x_1) = w_2(x_2 - x)$$

and

$$\frac{w_1}{w_2} = \frac{x_2 - x}{x - x_1} \qquad [19]$$

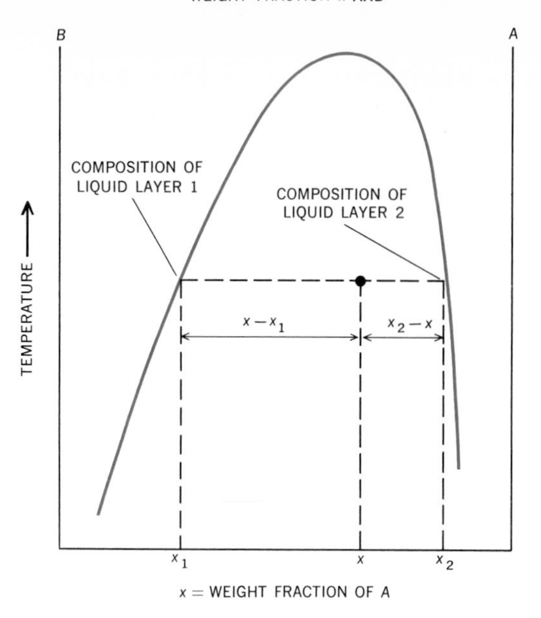

FIGURE 20-12
The relative amounts of the two liquid layers are given in terms of the liquid-layer compositions and the total-system composition by $w_1/w_2 = (x_2 - x)/(x - x_1)$.

Thus the weights of the two phases are in the proportion of the two line lengths indicated in Fig. 20-12.

Application of the phase rule to regions of two liquid phases gives

$$\begin{aligned}\Phi &= C - P + 2 \\ &= 2 - 2 + 2 = 2\end{aligned} \qquad [20]$$

If these 2 degrees of freedom are exercised by specifying T and P, then in the process in which the *total* composition of the system is changed there can be no phase-composition degrees of freedom. The compositions of each of the two phases, which is what the phase rule is concerned with, are fixed, and only the relative amounts of the two phases are varied by changes in the total composition.

It is also of interest to investigate the changes that occur when the temperature of a two-phase system, such as that of Fig. 20-11a, is raised. If the temperature is increased for the system of total composition of b, the system moves up along the dashed line. The fraction of the system composed of the water-rich layer gradually increases until, when the two-phase boundary curve is reached, the last of the butyl alcohol–rich layer appears to dissolve in the water-rich layer. By contrast, if a composition equal to that of the curve maximum is picked, the two layers remain in about equal amounts until, at the two-phase boundary, their composition becomes identical and they form a one-phase system.

The maximum of the curve of Fig. 20-11a is known as the *critical solution temperature,* or the *upper consolute temperature.*

The remaining two diagrams of Fig. 20-11 show that liquid systems can also exhibit a *lower consolute temperature* or, in a few cases, both upper and lower consolute temperatures. This behavior of increased mutual solubility at lower temperatures is certainly not that which is normally expected and must be attributed, very qualitatively, to some interaction between the components that can be effective only at lower temperatures.

20-10 Two-component Solid-Liquid Systems: Formation of a Eutectic Mixture

Consider now a two-component system, at some fixed pressure, where the temperature range treated is such as to include formation of one or more solid phases. A simple behavior is shown by those systems for which the liquids are completely soluble in one another and in which the only solid phases that occur are the pure crystalline forms of the two components. Such phase behavior is shown in Fig. 20-13 for the system benzene-naphthalene. The curved lines AE and BE show the temperatures at which solutions of various compositions are in equilibrium with pure solid benzene and pure solid naphthalene, respectively. The horizontal straight line is the temperature below which no liquid phase exists.

It is instructive to consider what happens when solutions of various concentrations are cooled. The data that are obtained give the temperature of the systems as a function of time. These data are plotted as *cooling curves,* some of which, for concentrations indicated in Fig. 20-13, are shown in Fig. 20-14. It is such cooling curves, in fact, that are used to obtain the data shown in the phase diagram.

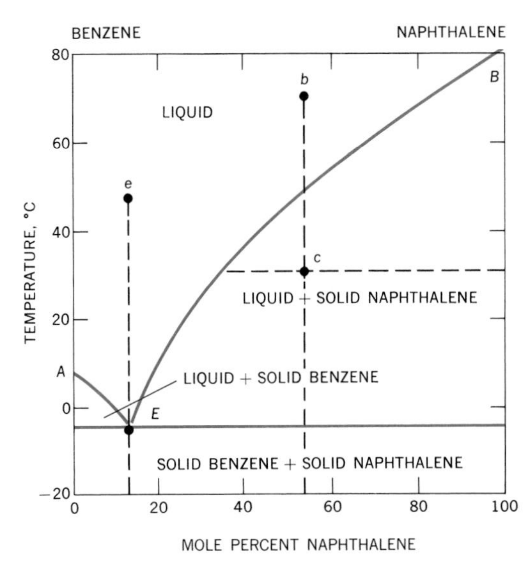

FIGURE 20-13 Freezing-point diagram for the binary system benzene-naphthalene at 1 atm pressure. (*From L. P. Hammett, "Introduction to the Study of Physical Chemistry," McGraw-Hill Book Company, New York, 1952.*)

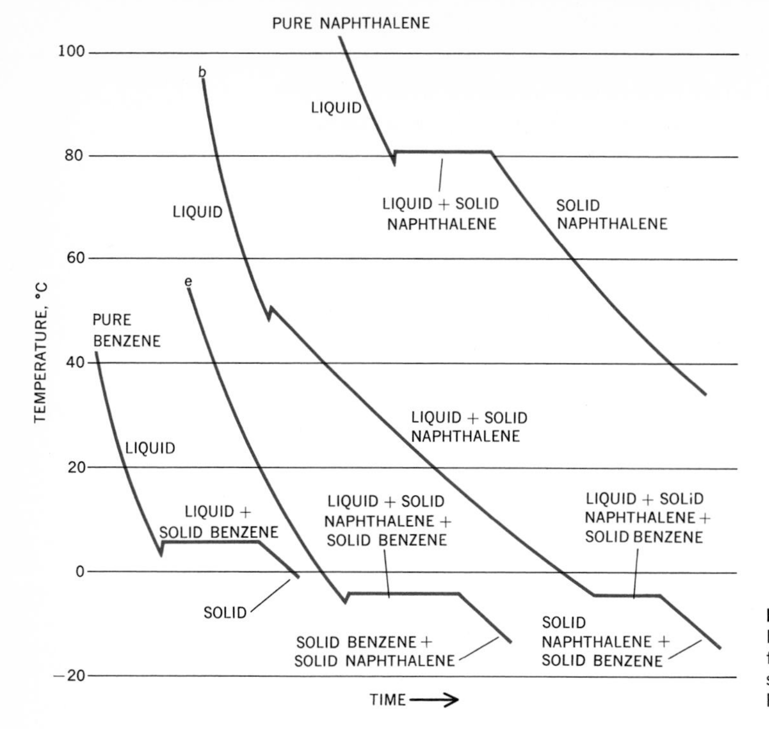

FIGURE 20-14
Examples of cooling curves for pure components and solutions e and b of Fig. 20-13.

The relation between the cooling curves and the information on the phase diagram can be illustrated with one of the cooling curves, b, for example. The liquid system cools until the curve BE is reached, at which point solid naphthalene is in equilibrium with the solution and starts to freeze out. As cooling continues, more naphthalene freezes out, the solution becomes richer in benzene, and its composition and temperature move down along the line BE. This stage is represented on the cooling curve by the slowly falling portion, corresponding to the freezing points of solutions of varying composition. It should be noticed that, although the temperature and overall composition place the system in the area below BE, no phase of such composition exists. Only the two phases, one to the right and the other to the left of the gross-composition point, occur. It is informative, as indicated in Fig. 20-13, to draw a horizontal line through the gross composition, at c for example, to connect or tie together the two phases that are present and are in equilibrium with each other. Such lines can, however, be understood and need not be drawn.

Cooling and freezing out of naphthalene proceeds until the point E is reached by the liquid phase, at which stage the solution becomes in equilibrium with pure solid benzene as well as with pure solid naphthalene. The solution composition and temperature remain constant until the system is entirely converted to the two solids. The point E is called the *eutectic,* from the Greek word meaning "easily melted," and the mixture of solids that separates out is called the *eutectic mixture.*

Application of the phase rule to the system at its eutectic point where there are two solid phases and one liquid phase in equilibrium gives

$$\begin{aligned}\Phi &= C - P + 2\\ &= 2 - 3 + 2 = 1\end{aligned} \qquad [21]$$

Since this 1 degree of freedom is used up by the arbitrarily chosen pressure, we learn that, at a given pressure, the intensive properties of the system at the eutectic point are entirely fixed. That the constant freezing point of a eutectic system does not imply the freezing out of a compound is experimentally verified by the fact that the eutectic mixture has a different composition at different pressures and that microphotographs show the solid to be a mixture of two crystalline forms.

In the following chapters we shall see that solid-liquid equilibrium curves can be subjected very profitably to a thermodynamic treatment. At either end of the composition scale of diagrams like that of Fig. 20-13, the solid-liquid curves show the depression of the freezing point that sets in because of the presence of minor amounts of the second component. The freezing-point depression, as we shall see, will be found to be independent of the nature of the second component for relatively dilute solutions, and this independence will lead to a useful generalization. For the present, we concern ourselves primarily with the diagrams that illustrate various phase situations.

A variation on the formation of a simple eutectic occurs when the solids that separate out can accommodate some of the second component. The system silver-copper is illustrated in Fig. 20-15. The areas at the extreme right and left along the abscissa scale show regions in which there is a solid solution of silver in copper and copper in silver,

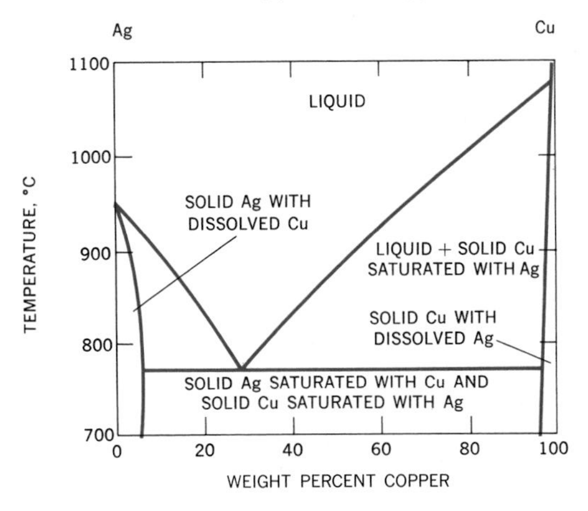

FIGURE 20-15
Freezing-point diagram for the system silver-copper at 1 atm pressure.

respectively. Each region is bordered by a line showing the maximum solubility of the second component in the solid of the first component. Any solution that is cooled will give rise to these solid solutions. The eutectic mixture will of course also be a mixture of saturated solid solutions.

20-11 Two-component Solid-Liquid Systems: Compound Formation

Systems in which the components show some attraction for each other sometimes show the formation of a solid-state compound consisting of a simple mole ratio of the two components. Such a system is that of formic acid and formamide, as shown in Fig. 20-16. Diagrams like this are readily understandable on the basis of the discussion of Sec. 20-10, since each half of Fig. 20-16 corresponds to the simple eutectic diagrams treated there.

Solutions which, on cooling, reach line NM or RQ, of Fig. 20-16, give rise to solid formic acid or formamide, respectively. Solutions which, on cooling, reach line PN or PQ give rise to a solid which is a compound containing equimolar amounts of formic acid and formamide. At point N the solution is in equilibrium with this new compound and with formic acid, and at point Q the solution is in equilibrium with the new compound and formamide. Points N and Q represent two eutectics that generally will have different temperatures.

Compound formation in the solid state is frequently encountered with *hydrates*. Figure 20-17 shows the formation of hydrated compounds of sulfuric acid in the solid state. Again, such diagrams are easily understood as a series of simple eutectic diagrams side by side.

A complication does occur when a solid compound does not have sufficient stability to persist up to the temperature at which it would melt. In such cases the unstable solid breaks down into a solution and

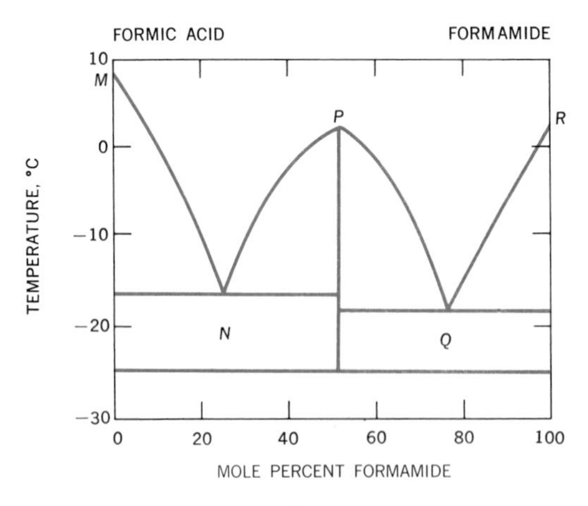

FIGURE 20-16
Freezing-point diagram for the system formic acid–formamide at 1 atm pressure showing the formation of a one-to-one compound in the solid state. (*From L. P. Hammett, "Introduction to the Study of Physical Chemistry," McGraw-Hill Book Company, New York, 1952.*)

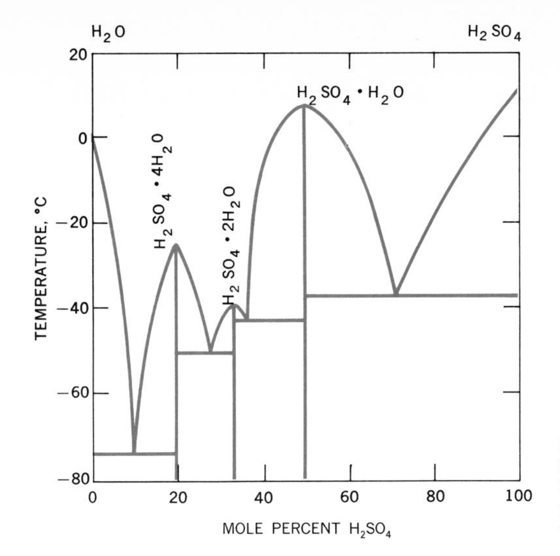

FIGURE 20-17
Freezing-point diagram for $H_2O \cdot H_2SO_4$ showing multiple compound formation in the solid state.

the solid of one or the other of the two components. This is illustrated by the system calcium fluoride–calcium chloride, as shown in Fig. 20-18. The decomposition of such a solid is referred to as a *peritectic reaction,* or an *incongruent melting.* Thus the equimolar crystal $CaF_2 \cdot CaCl_2$ of Fig. 20-18 breaks down at 737°C into a solution of composition B and solid CaF_2. The dashed line shows how the diagram might have looked if the compound had survived to a real or congruent melting point. This line is helpful for visualizing the phase behavior, but has, of course, no real significance.

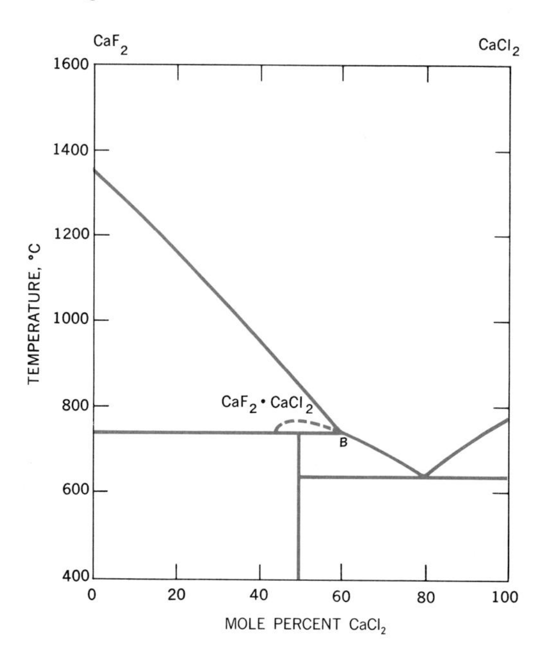

FIGURE 20-18
Freezing-point diagram for the system $CaF_2 \cdot CaCl_2$ showing the incongruent melting of the solid compound CaF_2CaCl_2 at 737°C. (*Data from International Critical Tables, vol. 4, p. 63, McGraw-Hill Book Company, New York, 1927.*)

20-12 Two-component Solid-Liquid System: Miscible Solids

Brief mention can be made, particularly in view of their importance as alloys, of systems forming only one solid phase which is a solid solution. Such behavior is a result of complete mutual solubility of the two solid components. In Sec. 20-10 it was pointed out how partial solubility of the solid phases in each other affected the phase diagram of a system showing a simple eutectic. Such partial solubility frequently occurs when the atoms of one component are small and can fit into the interstices of the lattice of the major component. In this way an *interstitial* alloy is formed. The carbon atoms in a carbon-containing alloy are usually so accommodated.

Complete solubility of two solid phases usually results when the atoms of the two components are about the same size and can substitute for each other in the lattice to form a *substitutional alloy.* The system of copper and nickel, as shown in Fig. 20-19, shows this behavior. The upper of the two curves shows the temperature at which solutions of various compositions start to freeze. The lower curve gives the composition of the solid which separates out at that freezing point. In this system the solid is always richer in the higher-melting component than is the solution from which it separates. The alloy consisting of 60 percent copper and 40 percent nickel is known as *constantan.*

20-13 Three-component Systems

To depict the phase behavior of three-component systems on a two-dimensional diagram, it is necessary to consider both the pressure and the temperature as fixed. The phases of the system as a function of

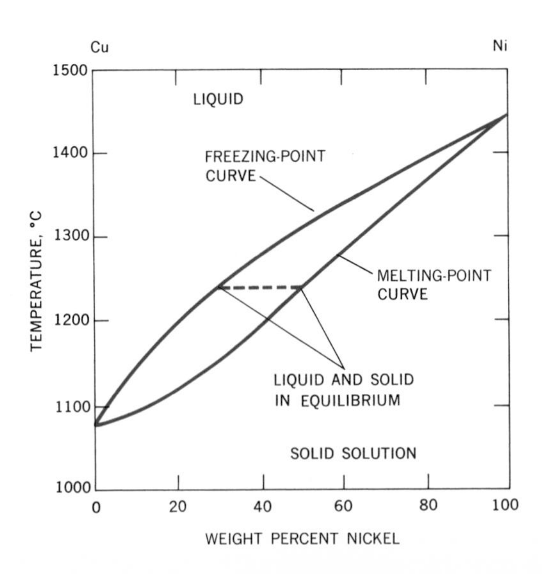

FIGURE 20-19
Solid-liquid phase diagram at 1 atm pressure for a system showing complete liquid and solid miscibility. (*Data from International Critical Tables, vol. 2, p. 433, McGraw-Hill Book Company, New York, 1927.*)

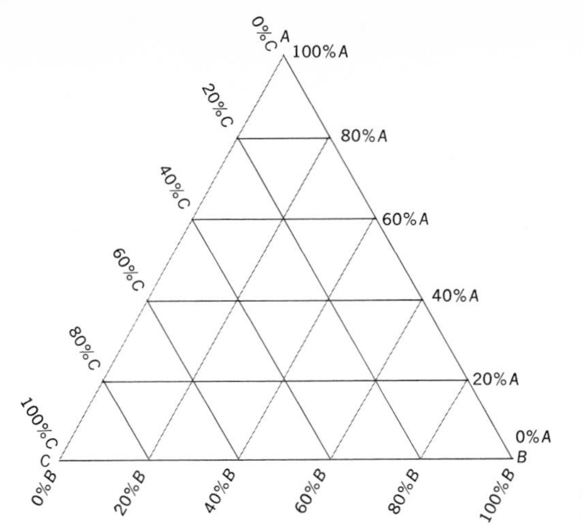

FIGURE 20-20
Diagram for plotting the composition of a three-component system.

the composition can then be shown. The relative amounts of the three components, usually presented as percent by weight, can be shown on a triangular plot, as indicated in Fig. 20-20. The corners of the triangle labeled A, B, and C correspond to the pure components A, B, and C. The side of the triangle opposite the corner labeled A, for example, implies the absence of A. Thus the horizontal lines across the triangle show increasing percentages of A from zero at the base to 100 percent at the apex. In a similar way the percentages of B and C are given by the distances from the other two sides to the remaining two apices. From the three composition scales of the diagram the composition corresponding to any point can be read off. This procedure for handling the composition of three-component systems is possible, and the total composition is always 100 percent, because of the geometric result that the sum of the three perpendicular distances from any point to the three sides of the triangle is equal to the height of the triangle.

As with two-component systems, the simplest three-component systems are those in which a liquid system breaks down into two phases. The system acetic acid–chloroform–water (Fig. 20-21) is such a system, showing, at 18°C, a two-phase region when the amount of acetic acid is small. A necessary part of the diagram are the *tie lines* through the two-phase region joining the compositions of the two phases that are in equilibrium. (In all previous two-component phase diagrams such lines could have been drawn, but since they would have been horizontal constant-temperature lines, it was unnecessary to exhibit them.) Thus a total composition corresponding to point a in the two-phase region gives two phases, one of composition b and the other of composition c. A unique point on the two-phase boundary is that indicated by d. This point, called the *isothermal critical point,* or the *plait point,* is similar to the previously encountered critical-solution temperatures, or consolute points, in that the compositions of the two phases in equilibrium become equal at this point.

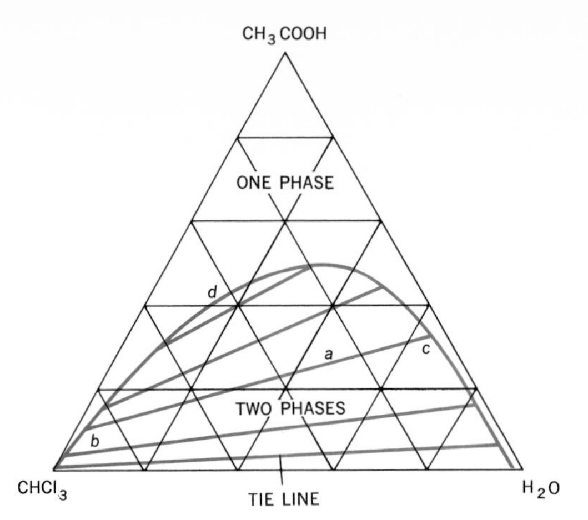

FIGURE 20-21 The liquid system acetic acid–chloroform–water at 1 atm pressure and 18°C. [*From R. H. Perry, C. H. Chilton, and S. D. Kirkpatrick (eds.), "Chemical Engineers' Handbook," 2d ed., McGraw-Hill Book Company, New York, 1941.*]

Application of the phase rule to a system corresponding to a point in the two-phase region gives

$$\begin{aligned}\Phi &= C - P + 2 \\ &= 3 - 2 + 2 = 3\end{aligned} \qquad [22]$$

The 3 degrees of freedom can be accounted for by the pressure, the temperature, and one composition variable. Thus the composition of both phases cannot be arbitrarily fixed. If one is fixed, the tie line from that composition fixes the composition of the second phase.

Three-component systems involving solids and liquids can be introduced by considering systems of two salts and water. The simplest behavior is that shown in Fig. 20-22, where the two salts are somewhat soluble and the diagram gives the curves for the saturated-solution compositions. Such diagrams are perhaps more easily understood if tie lines are also drawn to show that the saturated solutions along DF and EF are in equilibrium with the solid salts B and C, respectively.

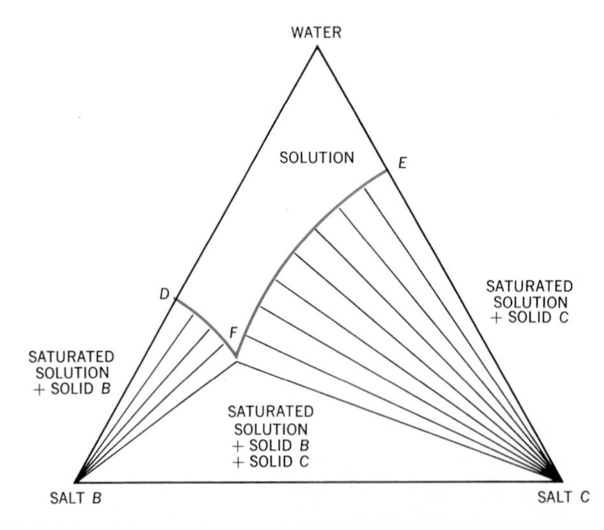

FIGURE 20-22 Phase diagram (schematic) for two salts and water at a fixed temperature and pressure.

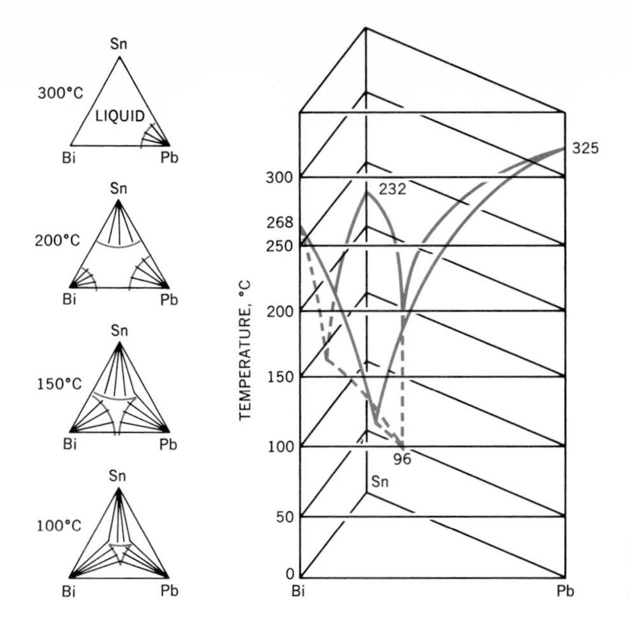

FIGURE 20-23 The three-component system bismuth-tin-lead.

Point F corresponds to a system in which the solution is in equilibrium with both salts. Removal of water from point F moves the total composition toward the base of the triangle. The effect of this is to form more solid salts which remain in equilibrium with the decreasing amount, but constant concentration, of saturated solution.

Finally, three-component systems in which the three components, taken in pairs, form simple eutectics can be illustrated by the system lead-tin-bismuth. A three-dimensional representation (Fig. 20-23) shows, in a descriptive manner, the phase behavior as a function of composition and temperature at the fixed pressure of 1 atm. For quantitative work it is more suitable to express the data at various constant temperatures on triangular plots. Such diagrams for a few temperatures are therefore included. Tie lines are shown to indicate more clearly the solution compositions that are in equilibrium with the solid components.

If a solution containing 32 percent Pb, 15 percent Sn, and 53 percent Bi is cooled, it is found that it remains liquid until, at a temperature of 96°C, all three solid components start separating out. The phase rule indicates that at such a point, called a *ternary eutectic,* there is

$$C - P + 2 = 3 - 4 + 2 = 1 \text{ degree of freedom} \qquad [23]$$

Since this degree has been used up by the fixed pressure, the system has no remaining variables. It is characteristic of such ternary eutectics that the eutectic point is at a low temperature compared with the melting points of the pure components. The ternary eutectic of the metal system of the present example will, for instance, melt in boiling water.

In describing the equilibrium that is set up, at a particular pressure and temperature between a liquid and its vapor, we shall, as for solid-liquid equilibria, place most emphasis on two-component systems. Again phase diagrams will be drawn with a composition variable as the abscissa. Now, however, it will be convenient in some circumstances to consider a fixed temperature and to have pressure as the ordinate and in other situations to have, as was the case in liquid-solid systems, the pressure fixed and the temperature as ordinate.

20-14 Liquid-Vapor Equilibria of Solutions

Mixtures of ideal gases, i.e., gaseous solutions, were treated in Chaps. 1 and 2 and there were found to conform to the simple relation expressed by Dalton's law. This result, that the total pressure of a mixture of ideal gases is equal to the sum of the pressures of the components, was easily understandable on the basis of the kinetic-molecular-theory postulate of noninteracting molecules. Since liquids exist only because of molecular interactions, no such "ideal" liquid solutions can be expected in the same sense as an ideal-gas solution. Some solutions, however, behave in a simple and general enough way to warrant use of the term *ideal solution.*

One might anticipate a simplicity of liquid-solution behavior in solutions of components that are molecularly similar in size and intermolecular interactions. The vapor pressures as functions of composition for solutions, such as carbon tetrachloride–silicon tetrachloride, chlorobenzene-bromobenzene, benzene-toluene, do in fact show similar and simple behavior. Typical results are shown for benzene-toluene in Fig. 20-24 for the total vapor pressure and for the vapor pressure of

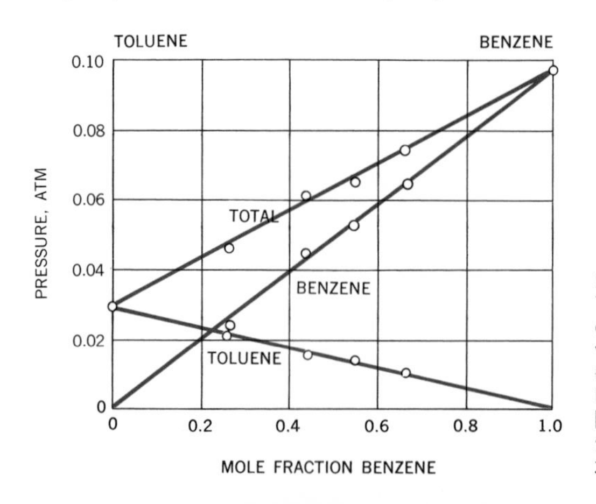

FIGURE 20-24
The vapor pressures of the components and the total vapor pressures for the nearly ideal solution benzene-toluene at 20°C. [*Data from R. Bell and T. Wright, J. Phys. Chem.*, **31:***1884* (*1927*).]

each of the components in equilibrium with the solution as a function of the solution composition. The linear relations of Fig. 20-24, which we shall take as being the primary characteristic of an ideal solution, can be expressed as

$$P_A = x_A P_A^\circ \qquad P_B = x_B P_B^\circ \tag{24}$$

where x_A and x_B are the *mole fractions* of components A and B.

If the vapor behaves ideally, as is being assumed here,

$$P = P_A + P_B \tag{25}$$

where P_A and P_B are the vapor pressure of A and B above a solution of mole fraction x_A and x_B and the vapor pressures of the pure components are P_A° and P_B°. Solutions that obey Eqs. [24] and have vapor-pressure diagrams like that of Fig. 20-24 are said to conform to *Raoult's law*. We shall see in the following chapter that behavior in accordance with Raoult's law merits the designation of the solution as *ideal*. (A more careful statement of an ideal solution, however, includes specification that no volume change should occur and that no heat should be evolved or absorbed on mixing of the two components. This lack of volume and heat effects usually occurs in systems obeying Raoult's law.)

Just as the ideal-gas laws were a useful basis for understanding the specific deviations shown by real gases, so also will the concept of the ideal solution be helpful in understanding the behavior of nonideal solutions.

The vapor-pressure–composition diagrams for nonideal solutions are usually classed as having, as does the chloroform-acetone system of Fig. 20-25, a minimum in the total vapor-pressure–composition curve or as having, as does the carbon tetrachloride–methyl alcohol system of Fig. 20-26, a maximum in this curve. In the following chapter we

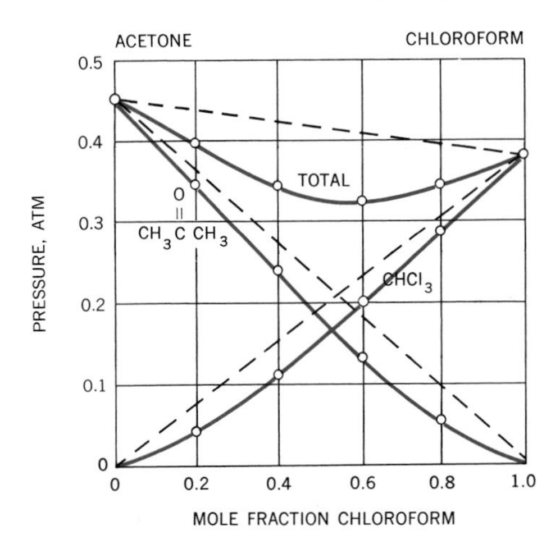

FIGURE 20-25
Vapor-pressure diagram for the system chloroform-acetone at 35°C. [*From data of J. Von Zawidzki, Z. Physik. Chem.*, **35:**129 (1900).]

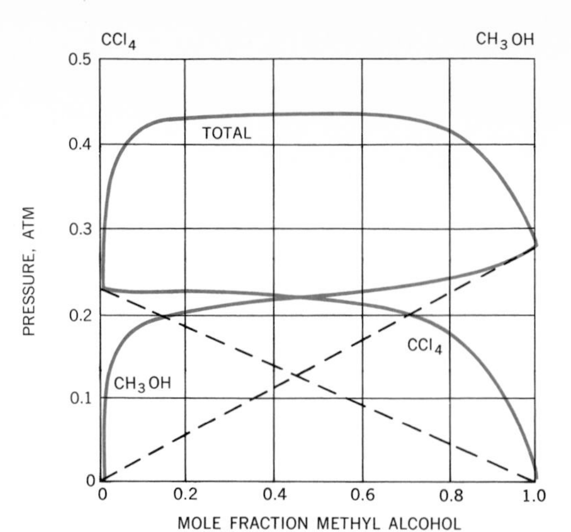

FIGURE 20-26
Vapor-pressure diagram for the system carbon tetrachloride–methyl alcohol at 35°C. (*From J. Timmermans, "Physico-chemical Constants of Binary Systems," vol. 2, Interscience Publishers, Inc., New York, 1959.*)

shall see that some aspects of even these nonideal-system vapor-pressure curves, in particular their behavior near the two composition limits, can be generalized and related to thermodynamic properties of the components.

Here we might make the empirical observation, apparent from Figs. 20-25 and 20-26, that the vapor pressure of the solvent tends to follow the ideal vapor-pressure curve for dilute solutions. This behavior turns out to be quite general, and we state, without specifying ideal behavior, that *the vapor pressure of the solvent of dilute solutions obeys Raoult's law,* $P_A = x_A P_A^\circ$, as the limit of infinite dilution is approached.

A generalization is also possible for volatile solutes of solutions. As can be seen from Figs. 20-25 and 20-26, the vapor pressure of the solute tends to vary linearly with composition for dilute solutions. This also represents a general result, and such behavior is known as *Henry's law.* This law, *obeyed by volatile solutes in dilute solutions,* can be written

$$P_B = kx_B \tag{26}$$

where k, known as the Henry law constant, will depend on the nature of the solvent and the solute and on the units in which the vapor pressure is expressed.

For dilute solutions, which obey Henry's law, it is customary also to use the *molarity,* defined as the number of moles of solute per liter of solution, to describe the solution. The molarity, for a given solvent, is approximately proportional to the mole fraction, and thus Henry's law can also be written

$$P_B = k_M M_B \tag{27}$$

where M implies molarity.

Table 20-2 gives some data showing, more clearly than can be seen from Figs. 20-25 and 20-26, conformity to Henry's law of the solute

HCl in the solvent toluene. Here, since HCl is normally a gas, the Henry law expression has been turned around to

$$x_B = k^{-1}P_B \qquad [28]$$

and in terms of molarity,

$$M_B = k_M^{-1}P_B \qquad [29]$$

This way of writing Henry's law is in line with our tendency to think in terms of the solubility of a gas in a liquid, rather than of the solution giving rise to an equilibrium vapor pressure of a component that is normally a gas.

20-15 Vapor-pressure Diagrams Showing Liquid and Vapor Compositions

Vapor equilibrium data are of use in the study of distillations. In this connection, it is of value to have diagrams showing not only the vapor pressure of a solution of given composition, but also the composition of the vapor that is in equilibrium with the liquid. This additional information can be put on the vapor-pressure–composition diagrams.

For an ideal solution one can calculate the vapor composition that is in equilibrium with a liquid of mole fraction x_A and x_B. Vapor-pressure curves like that of Fig. 20-24 imply the vapor-pressure equations

$$P_A = x_A P_A^\circ \quad \text{and} \quad P_B = x_B P_B^\circ \qquad [30]$$

Since the partial pressure of a gas is proportional to the number of moles of the gas per unit volume, the mole fractions of the vapor can be written

$$x_{A,\text{vap}} = \frac{P_A}{P_A + P_B} \quad \text{and} \quad x_{B,\text{vap}} = \frac{P_B}{P_A + P_B} \qquad [31]$$

or

$$x_{A,\text{vap}} = \frac{x_A P_A^\circ}{P_A + P_B} \quad \text{and} \quad x_{B,\text{vap}} = \frac{x_B P_B^\circ}{P_A + P_B} \qquad [32]$$

TABLE 20-2 Henry law data for HCl in toluene at 25°C*

P(HCl) (atm)	x_B	k^{-1}	M	k_M^{-1}
0.0033	0.00141	0.0427	0.0132	0.400
0.0338	0.00154	0.0456	0.0144	0.425
0.0964	0.00431	0.0446	0.0402	0.409
0.158	0.00702	0.0444	0.0655	0.415
0.282	0.0126	0.0446	0.117	0.415

*Data from S. J. O'Brien and E. G. Bobalek, *J. Am. Chem. Soc.*, **62**:3227 (1940).

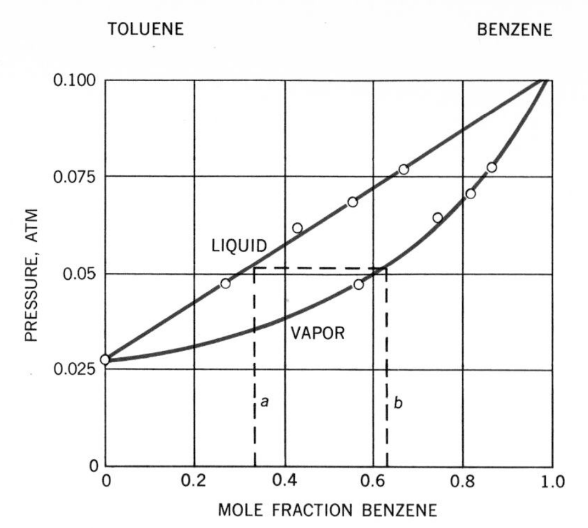

FIGURE 20-27
Vapor-pressure diagram showing liquid and vapor compositions for the nearly ideal system benzene-toluene at 20°C. (Data from Fig. 20-24; curves drawn for ideal behavior.)

The ratio of the mole fractions of the components in the vapor is therefore given as

$$\frac{x_{A,\text{vap}}}{x_{B,\text{vap}}} = \frac{x_A}{x_B}\frac{P_A^\circ}{P_B^\circ} \qquad [33]$$

This expression can be used to calculate the composition of vapor in equilibrium with an ideal solution of any composition. The qualitative result which should be noticed is that the vapor will be relatively richer in A if P_A° is greater than P_B°, that is, if A is the more volatile component.

The vapor-composition information is added to the vapor-pressure–composition diagram by allowing the abscissa to be used for both liquid and vapor compositions, as illustrated for an ideal solution in Fig. 20-27. At a particular vapor pressure one can read, along the horizontal dashed line, for example, the composition of the liquid that gives rise to this vapor pressure and also the composition of the vapor that exists in equilibrium with this liquid. More generally, one uses the diagram by starting with a given liquid composition, a of Fig. 20-27, reading off the vapor pressure of this solution, and also obtaining the composition b of the vapor in equilibrium with the solution.

For nonideal solutions the composition of the vapor in equilibrium with a given solution must be calculated from the experimentally determined vapor pressures of the two components. Typical diagrams for systems showing a minimum and a maximum in their vapor-pressure curves are seen in Figs. 20-28 and 20-29. One finds always that the vapor composition is richer, relative to the liquid, in the more volatile component. This feature can be appreciated by deducing the vapor in equilibrium with various liquid compositions, as shown by the dashed lines of Figs. 20-27 and 20-28.

It is helpful to notice and remember that on vapor-pressure–composition diagrams the liquid-composition curve always lies above

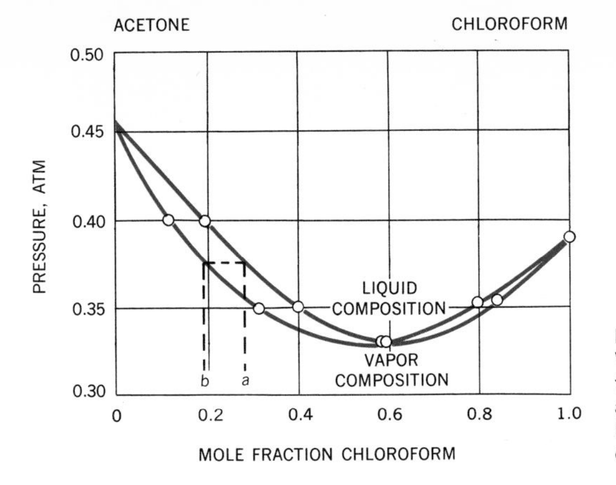

FIGURE 20-28
Vapor-pressure diagram for the system chloroform-acetone at 35°C, showing liquid- and vapor-composition curves.

the vapor-composition curve. Where the curve for the vapor pressure of the liquid shows a maximum or minimum, however, the equilibrium vapor has the same composition as the liquid. Such points will be important when a separation process is considered.

20-16 Boiling-point–composition Diagrams

The discussion of liquid-vapor equilibrium has so far concerned itself with the experimental results that are obtained when the vapor in equilibrium with solutions at some fixed temperature, often 25°C, is studied. Such vapor-pressure data are suited to the theoretical questions that have been discussed. They are not, however, the results that are of primary importance in studies of the more practical aspects of liquid-vapor equilibria. In practice, it is more common to fix the pressure at some constant value, often, but not always, at 1 atm, and to determine the temperature at which liquid and vapor are in equilibrium. In this way, data are obtained from which a *boiling-point–composition dia-*

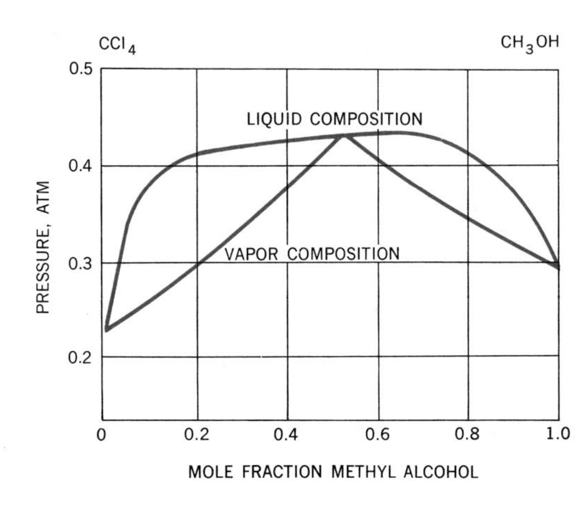

FIGURE 20-29
Vapor-pressure diagram for the system carbon tetrachloride–methyl alcohol at 35°C, showing liquid- and vapor-composition curves.

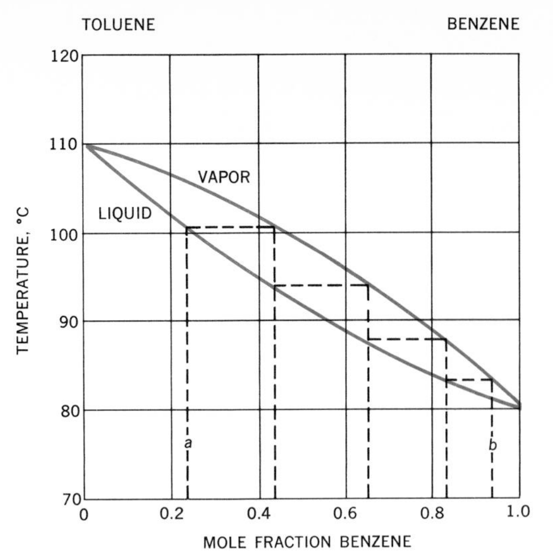

FIGURE 20-30
Boiling-point diagram for the nearly ideal system benzene-toluene at 1 atm pressure.

gram can be constructed. It is again customary to show the composition of the vapor that is in equilibrium with the liquid on the same diagram. Figure 20-30 shows these curves for the ideal system benzene-toluene. Now one notices that, for the vapor to be relatively richer in the more volatile, i.e., *lower*-boiling, component, the liquid-composition curve lies below the vapor-composition curve. The exact shapes of the two curves on the boiling-point diagram, even for an ideal solution, are not so easily deduced as were the curves on the vapor-pressure diagram. The curves depend on the behavior of the system as a function of temperature, and this is less easy to generalize than the constant-temperature behavior of the vapor-pressure diagrams.

The boiling-point curves for the two types of nonideal systems are shown in Figs. 20-31 and 20-32. A minimum in the vapor-pressure–composition curve results in a maximum in the boiling-point–composition curve, and vice versa. Also, as for the ideal solutions, the liquid-composition curve lies lower than the vapor-composition curve on a boiling-point diagram. The significance of these diagrams for any separation process can again be appreciated by following paths, such as those shown dashed, for the conversion of some of a liquid sample of composition a to its equilibrium vapor of composition b.

20-17 Distillation

The important process of distillation can now be investigated. From the boiling-point diagram of Fig. 20-30 one can see that if a small amount of vapor were removed from a solution of composition a, the vapor would have a composition higher in the more volatile component than did the original solution a. Such a single step is, of course, inade-

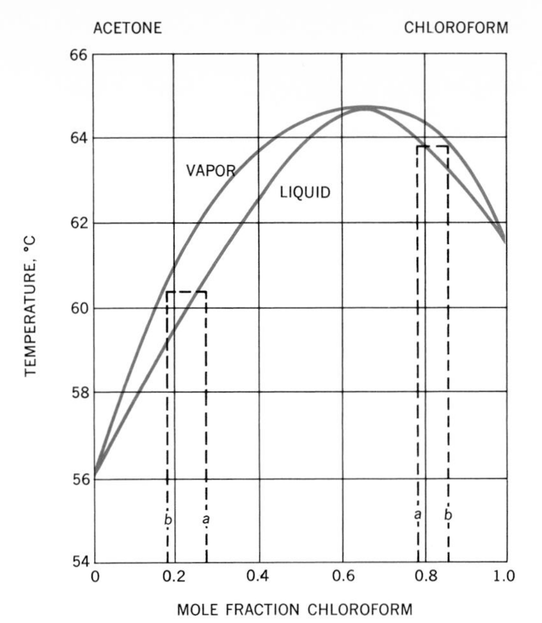

FIGURE 20-31
Boiling-point diagram for the system chloroform-acetone at 1 atm pressure. (*Data from International Critical Tables, McGraw-Hill Book Company, New York, 1926–1930.*)

quate for any appreciable separation of two components unless they have extremely different boiling points. In practice, a process of *fractional distillation* is used in which the separation step just described is, in effect, repeated by condensing some of the vapor, boiling off some vapor from this new solution, collecting and revaporizing this product, and so forth. This procedure has the effect of stepping across the boiling-point diagram, as indicated by the dashed lines of Fig. 20-30. A distillation *column* carries out this stepwise process automatically.

The efficiency of a distillation column is determined by the number

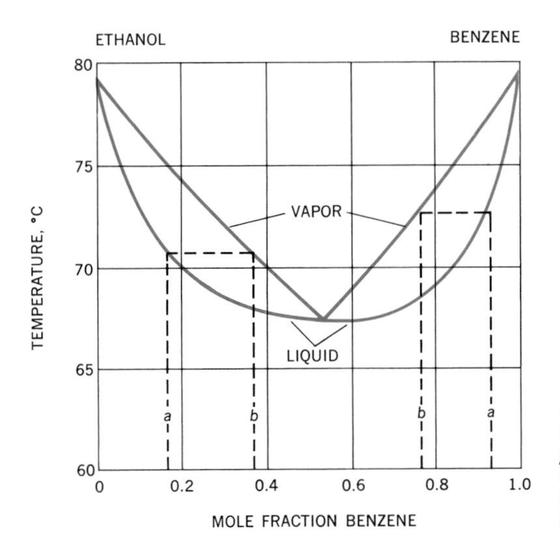

FIGURE 20-32
Boiling-point diagram for the system ethanol-benzene at 0.987 atm pressure. (*Data from International Critical Tables, McGraw-Hill Book Company, New York, 1926–1930.*)

of *theoretical plates* that the separation it performs corresponds to. A column supplied, for example, with a charge of composition a in Fig. 20-30 is operated at total reflux until equilibrium is established. A small sample of distillate is then drawn off and analyzed and, we shall assume, has composition b. The separation that has resulted corresponds to four ideal evaporations and condensations, and the column is said to have four theoretical plates.

For a solution showing a maximum vapor pressure and a minimum boiling point, the distillation process is indicated by the dashed lines of Fig. 20-32. Distillation in a fractional-distillation unit is seen, regardless of the initial solution, to result in a distillate of the composition of the minimum-boiling-point mixture. A separation into one or the other of the pure components could result only by working with the residue. The most important commercial solution that shows this behavior is the ethyl alcohol–water system. Fermentation processes result in an ethyl alcohol concentration of about 10 percent. The object of distillation is to increase this concentration and, possibly, to yield pure ethyl alcohol. The boiling-point diagram of Fig. 20-33 shows that distillation at atmospheric pressure can yield, at best, a distillate of 95 percent ethyl alcohol. It is for this reason that 95 percent ethyl alcohol is a fairly common chemical material. Absolute alcohol can be obtained by a distillation procedure using a three-component system, usually alcohol, water, and benzene.

A different situation arises with solutions that show a maximum in their boiling-point curves, as does the system of Fig. 20-31. If such a solution is merely boiled away, the residue will approach the composi-

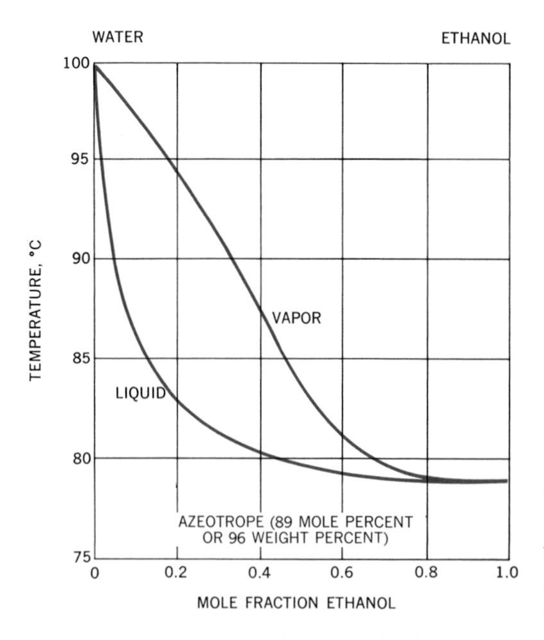

FIGURE 20-33
Boiling-point diagram for the system water-ethanol at 1 atm. (*Data from International Critical Tables, McGraw-Hill Book Company, New York, 1926–1930.*)

tion corresponding to the maximum of the boiling-point curve, and the boiling point will also approach that corresponding to this maximum. Once this solution and boiling point have been reached, the remaining solution will boil at this temperature and will not change its composition. Such a solution is known as a *constant-boiling mixture,* or an *azeotrope.* This same term is applied to a solution having the composition of the minimum of a boiling-point curve.

It is necessary to stress that, although we are dealing, in the case of an azeotrope, with a constant-temperature constant-composition boiling mixture, this mixture is not to be regarded as a compound formed between the two components. A change in the total pressure is usually sufficient to show that the azeotropic composition can be changed.

20-18 Distillation of Immiscible Liquids

It seems desirable to include in our discussion of distillation the distillation of immiscible liquids. The process usually makes use of water or steam and an insoluble organic material and is then referred to as *steam distillation.*

Consider two components which, though somewhat soluble in each other, separate into two layers on mixing. Each layer independently exerts its own vapor pressure, and the pressure measured over such a mixture will be the sum of the two partial pressures. As long as the two layers are present, they will have constant compositions: a layer of A saturated with B and a layer of B saturated with A. It follows that the vapor pressure of each layer (and thus the total vapor pressure) will be independent of the amounts of the two layers. These statements indicate that the boiling point of such a mixture and the equilibrium vapor composition will be constant as long as the two layers are present. Furthermore, the boiling point will be the temperature at which the total pressure is equal to 1 atm, and this will be at a lower temperature than the boiling point of either component. Figure 20-34 shows a boiling-point diagram for immiscible liquids. Any two-phase mixture of the liquids will boil at temperature T_M and produce vapor of composition M. The lines AM and BM correspond to the temperatures at which vapor mixtures would start to condense to give A saturated with B and B saturated with A, respectively.

It is the lowering of the boiling point that makes the process of steam distillation attractive. Many organic materials of high molecular weight boil at such a high temperature that they would decompose. These materials can be conveniently "cleaned up" by steam distillation. A mixture of the organic material and water is heated, usually by the direct addition of steam, and the two liquids distill over at a temperature of less than 100°C. They are collected in amounts that are proportional

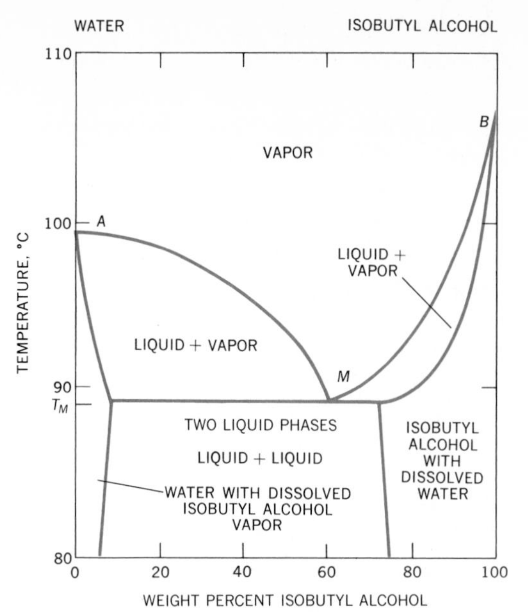

FIGURE 20-34
Boiling-point diagram for a pair of immiscible liquids, water–isobutyl alcohol, at 1 atm pressure. (*From L. P. Hammett, "Introduction to the Study of Physical Chemistry," McGraw-Hill Book Company, New York, 1952.*)

to their vapor pressure. It is this fact that sets a limit to the usefulness of the process since very high boiling materials, having very low vapor pressures, come over in relatively small amounts compared with the amount of steam used. The distillate separates into the two immiscible layers, and the organic layer can be separated and dried. The process has considerable application, but is most useful in the rough purification of organic materials and does not usually handle the same problems as do fractional distillations.

Problems

1 Calculate the slope of the liquid-vapor curve on a pressure-temperature plot for water at 1 atm and 100°C. Estimate the slope of the liquid-solid curve for water, using the heat of fusion of 6.01 kJ mol^{-1} and the density of ice of 1.09 g ml^{-1}, neither being very sensitive to temperature or pressure.

2 The vapor pressure of water at 25°C is 0.0313 atm. Use this datum and the normal boiling point of water to deduce, according to the Clausius-Clapeyron equation, a value for the heat of vaporization of water. (1 atm = 101 325 N m^{-2}.) *Ans.* 42.72 kJ.

3 Normal hexane has a boiling point of 69.0°C. Assuming that Trouton's rule is obeyed, estimate the vapor pressure of n-hexane at 25°C.

4 Assuming that benzene obeys Trouton's rule, calculate dP/dT at the normal boiling point 80.1°C and at 25°C.
Ans. At 25°C, $dP/dT = 5.9 \times 10^{-3}$ atm deg^{-1}.

5 The vapor pressures of neon are reported in the International Critical Tables as a function of temperature as follows:

$t\,(°C)$	−228.7	−233.6	−240.2	−243.7	−245.7	−247.3	−248.5
p (mm Hg)	19,800	10,040	3170	1435	816	486	325

Determine the normal boiling point, the heat of vaporization at the normal boiling point, and the Trouton rule constant. (760 mm Hg = 1 atm = 101 325 N m^{-2}.)

6 At what temperature will water boil when the elevation is such that the barometric pressure is 0.66 atm? *Ans.* 89°C.

7 How many phases are there in:
a A sealed bulb half-filled with liquid water, the other half being air saturated with water vapor?
b A 1-liter bulb containing 2 g of water, and no air, at 100°C? *Ans.* Two.
c A mixture of oil and water which has been dispersed into an emulsion? *Ans.* Two.

8 How many components are there in the systems composed of:
a $N_2(g)$ plus $O_2(g)$?
b What is the answer to part *a* if a catalyst is present that promotes the formation of the many possible oxides of nitrogen? *Ans.* Two.
c NaCl(*s*) and an aqueous solution saturated with NaCl and containing some HCl? *Ans.* Three.
d Any salt plus water?
e The system of part *d* at a high enough temperature so that it consists only of dry solid salt and water vapor?
f The system of part *d* cooled until a solid appears and this solid turns out to be the hydroxide of the metal?

9 A remarkable catalyst is to be imagined that brings a system containing carbon and hydrogen to equilibrium with all hydrocarbons.
a How many components are there if the system is charged with graphite and H_2?
b If the system can be charged only with H_2 and CH_4, how many components will there be?

10 How many remaining degrees of freedom are there in each of the following systems? Suggest variables that could correspond to these degrees of freedom.
a Liquid water and water vapor in equilibrium at a pressure of 1 atm. *Ans.* 0.
b Liquid water and water vapor in equilibrium. *Ans.* 1.
c I_2 dispersed between liquid water and liquid CCl_4 at 1 atm pressure with no solid I_2 present.
d The vapor equilibrium system of NH_3, N_2, and H_2.
e An aqueous solution of H_3PO_4 and NaOH at 1 atm pressure.
f A solution of H_2SO_4 in water in equilibrium with the solid hydrate $H_2SO_4 \cdot 2H_2O$ at 1 atm pressure.

11 Describe, in view of Fig. 20-11*c*, the phase situations that arise when nicotine is gradually added to a small quantity of water at 100°C until the system is transformed to nearly 100 percent nicotine.

12 Describe, in view of Fig. 20-11*c*, the phase situations that arise when an aqueous solution that is 60 percent by weight nicotine is heated from 50 to 250°C.

13 Estimate from Fig. 20-13 the weight of solution and of solid that will be present when 100 g of solution of composition *b* is cooled to point *c*.

14 Describe, with the help of Fig. 20-15, the phases that form and the temperatures at which phase changes occur when a 20 percent copper-in-silver melt at 1000°C is cooled to 700°C.

15 Describe, with the help of Fig. 20-15, the phase changes when solid silver at 900°C is added to a silver-copper melt containing 40 percent copper at 900°C.

16 At about 60 percent by weight, $CaCl_2$ in the system $CaCl_2 \cdot CaF_2$, shown in Fig. 20-18, a peritectic point occurs at 737°C. If the pressure is assumed fixed at 1 atm, apply the phase rule to deduce the number of degrees of freedom at this point.

17 Describe the phases that appear and the temperature of their appearance when the solid compound $CaF_2 \cdot CaCl_2$ is heated from 400 to 1400°C at 1 atm pressure.

18 Verify that the sum of the three perpendicular distances from any point inside an equilateral triangle to the sides of the triangle is equal to the height of the triangle.

19 Estimate from Fig. 20-21 the weights of the three components that are present in 100 g of the one-phase system at point d. Estimate the compositions and the weights of the two phases that occur for 100 g of the system having the total composition of point a of Fig. 20-21.

20 Describe the phases that occur as water is added to an initially anhydrous mixture of 5 percent of the salt B and 95 percent of the salt C in Fig. 20-22.

21 Assuming ideal-solution behavior, calculate the equilibrium vapor pressure and the mole-fraction composition of the vapor in equilibrium, at 40°C, with a solution of carbon tetrachloride–cyclohexane that has 0.4753 mol fraction CCl_4. The vapor pressures at 40°C of pure CCl_4 and cyclohexane are 0.2807 and 0.2429 atm, respectively. Compare with the measured values reported by Scatchard, Wood, and Mochel [*J. Am. Chem. Soc.*, **61**:3206 (1939)] of 0.2677 for the vapor pressure and 0.5116 for the mole fraction of CCl_4 in the vapor.

22 It is found that the boiling point, at 1 atm pressure, of a solution of 0.6589 mol fraction benzene and 0.3411 mol fraction toluene is 88.0°C. At this temperature the vapor pressures of pure benzene and toluene are 1.259 and 0.4993 atm, respectively. What is the vapor composition that boils off this liquid? *Ans.* Mole fraction benzene = 0.830.

23 At 55°C a solution of mol fraction 0.2205 ethanol and 0.7795 cyclohexane has a vapor pressure of 0.4842 atm and a vapor composition of 0.5645 mol fraction ethanol and 0.4355 cyclohexane. The vapor pressures of pure ethanol and cyclohexane at 55°C are 0.3683 and 0.2212 atm, respectively. Plot a possible vapor-pressure diagram having liquid and vapor compositions that are compatible with these data.

24 Data for the total vapor pressure and vapor composition for solutions of carbon tetrachloride and acetonitrile at 45°C have been given by L. Brown and F. Smith [*Aust. J. Chem.*, **7**:269 (1954)]. Some of these data are:

x_{CCl_4} *in liquid*	0.0347	0.1914	0.3752	0.4790	0.6049	0.8069	0.9609
x_{CCl_4} *in vapor*	0.1801	0.4603	0.5429	0.5684	0.5936	0.6470	0.8001
Total vapor pressure (atm)	0.3263	0.4421	0.4797	0.4863	0.4882	0.4773	0.4136

The vapor pressures of pure CCl_4 and pure CH_3CN at this temperature are 0.3405 and 0.2742 atm, respectively.

a Prepare a vapor-pressure–composition diagram, like those of Figs. 20-24 to 20-26, for this system.

b Prepare a vapor-pressure–composition diagram, like those of Figs. 20-27 to 20-29, showing the vapor composition as well as the vapor pressure in terms of the liquid composition.

c Discuss the consequences of distilling solutions that are originally rich in CCl_4 or in CH_3CN.

25 At total reflux, samples from the pot and the still head are withdrawn from an experiment in which ethanol and benzene are refluxing in a packed distillation column operating at 0.987 atm pressure. The pot sample has 0.05 mol fraction benzene, and the still head a composition of about 0.5 mol fraction benzene. According to Fig. 20-32, what would be the temperatures in the pot and head and how many theoretical plates does the column have?

26 Describe with the help of Fig. 20-34 the phase changes, and the temperatures at which they occur, when a 40 percent isobutyl alcohol–60 percent water solution is heated from 50 to 100°C. Similarly, describe the changes that occur when a vapor of this composition is cooled from 100 to 80°C.

References

RICCI, J. E.: "The Phase Rule and Heterogeneous Equilibrium," D. Van Nostrand Company, Inc., Princeton, N.J., 1951.

DENBIGH, K.: "The Principles of Chemical Equilibrium," chaps. 5–8, Cambridge University Press, New York, 1955.

21

THE THERMODYNAMIC TREATMENT OF MULTICOMPONENT SYSTEMS

In almost all the preceding applications of thermodynamics to chemical systems, only systems of one chemical component have been treated. (An exception was the development in Secs. 9-8 and 9-9 of an expression for the equilibrium constant in a system involving various reagents. In that case the simple, and unsubstantiated, assumption was made that the system was *ideal* and each component would have the same thermodynamic properties as it would in the absence of the other reagents.) Now we must develop the thermodynamic apparatus that is necessary for a sound treatment of *real* multicomponent systems. We then will be able to apply the powerful methods of thermodynamics to the many interesting and important solution systems.

First, however, let us treat ideal solutions. The results obtained in these easily visualized cases will provide convenient reference results when the more mathematically based relationships needed for real systems are encountered.

21-1 Introduction to Solutions: The Thermodynamics of Ideal Solutions

Here the formation of ideal-gas or ideal-liquid solutions will be considered and the characterization of ideal solutions will be developed as the various thermodynamic quantities associated with the mixing process are derived.

The Volume of an Ideal Solution The diagram of Fig. 21-1 is intended to suggest the mixing of n_A mol of an ideal gas originally in a volume

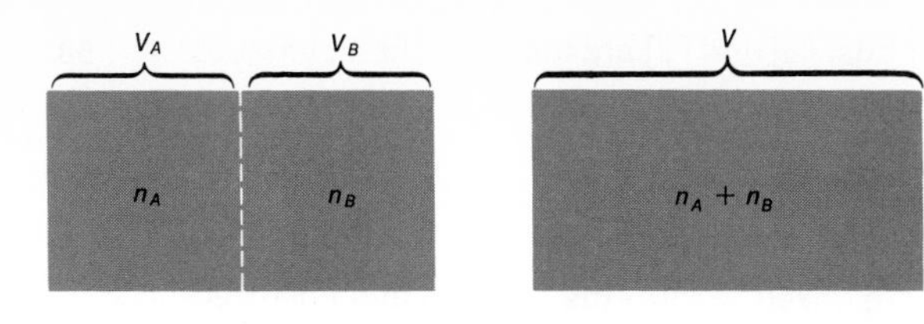

FIGURE 21-1
The mixing of n_A mol of A and n_B mol of B.

V_A at a temperature T and pressure P with n_B mol of another ideal gas originally in a volume V_B, and at the same values of T and P. The two ideal gases are characterized, in part, by obedience to the equations $PV_A = n_A RT$ or $PV_B = n_B RT$. The PVT character of the ideal solution that is presumed to form as a result of the removal of the partition is similarly described. Specifically, we write, for the gas mixture,

$$PV = nRT \qquad n = n_A + n_B$$

The volume V of the mixed-gas system is determined by these relations.

The original, separate gas volumes are

$$V_A = \frac{n_A RT}{P} \qquad \text{and} \qquad V_B = \frac{n_B RT}{P}$$

On mixing, if the ideal-gas solution behaves as an ideal gas and conforms to $PV = nRT$, it follows that

$$\begin{aligned} V &= \frac{nRT}{P} = (n_A + n_B)\frac{RT}{P} \\ &= n_A \frac{RT}{P} + n_B \frac{RT}{P} \\ &= V_A + V_B \end{aligned} \tag{1}$$

Thus an ideal solution is one for which the volume of the solution is equal to the volume of the unmixed components, the temperature and pressure remaining constant. This can be expressed simply as

$$\Delta V_{\text{mix}} = 0 \tag{2}$$

As a consequence of Eq. [2], the volume of the solution is seen to equal that of the total of the separate components. Thus, with the use of small capital letters to represent the volume of 1 mol, the volume $V = V_A + V_B$ of some quantity of an ideal solution can be expressed as

$$V = n_A \text{v}_A + n_B \text{v}_B$$

The volume of a mole of the solution, i.e., one for which the total number of moles is unity, is obtained by dividing by the number of moles $n_A + n_B$ to give

$$\text{v(soln)} = x_A \text{v}_A + x_B \text{v}_B \tag{3}$$

The volume relations of Eqs. [2] and [3] are taken as characteristics of an ideal-liquid solution.

The Enthalpy of an Ideal Solution A further characterization of an ideal solution is that there is no change in the enthalpy of the system when the individual components are mixed to form the solution at some fixed total pressure. The experimental test would be the absence of a temperature change on solution formation in an insulated system. Thus we can write, for an ideal solution,

$$\Delta H_{\mathrm{mix}} = 0 \qquad [4]$$

One finds that the mixing of most gases, at not too high pressure, conforms to this requirement and that the mixing of many liquids with similar properties produces only a small heat effect.

If the molar enthalpies of the individual components are represented by their standard heat-of-formation quantities $\Delta \mathrm{H}_f^\circ(A)$ and $\Delta \mathrm{H}_f^\circ(B)$, and the corresponding standard heat of formation of the solution of $\Delta \mathrm{H}_f^\circ(\text{soln})$, we can write, for an ideal solution,

$$\Delta H_{\mathrm{mix}} = \Delta H_f^\circ(\text{soln}) - [n_A\, \Delta \mathrm{H}_f^\circ(A) + n_B\, \Delta \mathrm{H}_f^\circ(B)] = 0$$

or

$$\Delta H_f^\circ(\text{soln}) = n_A\, \Delta \mathrm{H}_f^\circ(A) + n_B\, \Delta \mathrm{H}_f^\circ(B)$$

The heat of formation of a mole of solution is

$$\Delta \mathrm{H}_f^\circ(\text{soln}) = x_A\, \Delta \mathrm{H}_f^\circ(A) + x_B\, \Delta \mathrm{H}_f^\circ(B) \qquad [5]$$

The Entropy of an Ideal Solution Let us begin again with ideal gases for which we have convenient expressions for the dependence of entropy on pressure or volume. Consider again the solution-formation process of Fig. 21-1.

If the gases are indifferent to each other's presence, the final result could be imagined to be achieved by the isothermal expansion of each gas from its original volume to the final common volume V. The two entropy contributions that would be made are, in view of Example 3 of Sec. 8-2,

$$\Delta S_A = n_A R \ln \frac{V}{V_A}$$

and

$$\Delta S_B = n_B R \ln \frac{V}{V_B}$$

Thus the total entropy change for mixing is

$$\Delta S_{\mathrm{mix}} = \Delta S_A + \Delta S_B = n_A R \ln \frac{V}{V_A} + n_B R \ln \frac{V}{V_B} \qquad [6]$$

Using the ideal-gas relations $V_A = n_A RT/P$, $V_B = n_B RT/P$, and $PV = (n_A + n_B)RT/P$, Eq. [6] becomes

$$\begin{aligned}\Delta S_{\text{mix}} &= n_A R \ln \frac{n_A + n_B}{n_A} + n_B R \ln \frac{n_A + n_B}{n_B} \\ &= -n_A R \ln \frac{n_A}{n_A + n_B} - n_B R \ln \frac{n_B}{n_A + n_B} \\ &= -n_A R \ln x_A - n_B R \ln x_B \qquad [7]\end{aligned}$$

For the formation of 1 mol of a solution we divide this general result by $n_A + n_B$ to obtain

$$\Delta \text{s}_{\text{mix}} = -x_A R \ln x_A - x_B R \ln x_B \qquad [8]$$

This expression is the entropy change for the formation of 1 mol of a solution that we take to characterize an ideal solution, gaseous or liquid.

The entropy of x_A mol of A is $x_A \text{s}_A^\circ$, and that of x_B mol of B is $x_B \text{s}_B^\circ$. For 1 mol of ideal solution, we now can write

$$\Delta \text{s}_{\text{mix}} = \text{s(soln)} - \text{s(components)}$$

or

$$\begin{aligned}\text{s(soln)} &= \Delta \text{s}_{\text{mix}} + \text{s(components)} \\ &= x_A(\text{s}_A^\circ - R \ln x_A) + x_B(\text{s}_B^\circ - R \ln x_B) \qquad [9]\end{aligned}$$

Thus the solution contributions associated with A and with B can be recognized, and we could write

$$\begin{aligned}\text{s}_A\text{(soln)} &= \text{s}_A^\circ - R \ln x_A \\ \text{s}_B\text{(soln)} &= \text{s}_B^\circ - R \ln x_B \qquad [10]\end{aligned}$$

Since the mole fractions x_A and x_B are less than unity, the entropies of each component in an ideal solution are greater than the entropy of the pure material.

The Free Energy of an Ideal Solution Finally, the results of the two preceding parts can be combined to yield a value for the change in free energy that accompanies the formation of an ideal solution. For this constant-temperature, constant-pressure process, we write, in view of Eqs. [4] and [7],

$$\begin{aligned}\Delta G_{\text{mix}} &= \Delta H_{\text{mix}} - T\,\Delta S_{\text{mix}} \\ &= n_A RT \ln x_A + n_B RT \ln x_B \qquad [11]\end{aligned}$$

The free energy of the solution can be expressed by recognizing that the free energy of n_A mol of A and n_B mol of B is given by $n_A\,\Delta \text{G}_f^\circ(A) + n_B\,\Delta \text{G}_f^\circ(B)$. Then, with ΔG_{mix} and a rearrangement as in the derivation of the entropy of the solution,

TABLE 21-1 The entropy and free-energy change at 25°C for the formation of 1 mol of an ideal binary solution

Mole fraction		$x_A R \ln x_A$	$x_B R \ln x_B$	ΔS_{mix}	$T\Delta S_{mix}$	ΔG_{mix}
x_A	x_B	(J)	(J)	(J deg^{-1} mol^{-1})	(J mol^{-1})	(J mol^{-1})
1	0	0	0	0	0	0
0.9	0.1	−0.79	−1.91	2.70	805	−805
0.8	0.2	−1.48	−2.68	4.16	1240	−1240
0.7	0.3	−2.08	−3.00	5.08	1510	−1510
0.6	0.4	−2.55	−3.05	5.60	1670	−1670
0.5	0.5	−2.88	−2.88	5.76	1720	−1720
0.4	0.6	−3.05	−2.55	5.60	1670	−1670
0.3	0.7	−3.00	−2.08	5.08	1510	−1510
0.2	0.8	−2.68	−1.48	4.16	1240	−1240
0.1	0.9	−1.91	−0.79	2.70	805	−805
0	1	0	0	0	0	0

$$\Delta G(\text{soln}) = n_A[\Delta G_f^\circ(A) + RT \ln x_A] + n_B[\Delta G_f^\circ(B) + RT \ln x_B]$$

and

$$\Delta G(\text{soln}) = x_A[\Delta G_f^\circ(A) + RT \ln x_A] + x_B[\Delta G_f^\circ(B) + RT \ln x_B] \quad [12]$$

Again, for this ideal solution, we can associate contributions with the individual components, and write

$$\begin{aligned} \Delta G_A(\text{soln}) &= \Delta G_f^\circ(A) + RT \ln x_A \\ \Delta G_B(\text{soln}) &= \Delta G_f^\circ(B) + RT \ln x_B \end{aligned} \quad [13]$$

In summary, ideal solutions are characterized by $\Delta V_{mix} = 0$, $\Delta H_{mix} = 0$, and by the entropy and free-energy changes associated with the formation of 1 mol of solution given by Eqs. [9] and [12] and shown in Table 21-1 and Fig. 21-2.

21-2 Thermodynamic Properties of Real Solutions

What are the thermodynamic properties of real solutions comparable with those deduced for ideal solutions, and how can they be deduced? ΔV_{mix} and ΔH_{mix} are directly measurable. The quantities ΔS_{mix}, or the more convenient term $T\Delta S_{mix}$, and ΔG_{mix}, as will now be shown, can be determined from the vapor-pressure data as given in Sec. 20-14.

The molar free energies of the equilibrium vapors of the liquids that are to be mixed to form the solution are equal to the molar free energies of the liquids themselves. The molar free energy of the equilibrium vapor of the solution is equal to the molar free energy of the solution. Thus, if we obtain ΔG for the formation of the vapor of the solution from that for the vapors of the components, we can assign this value also to the solutions themselves. The advantage to this procedure is that the vapors can, to a better approximation, be treated as ideal. On this basis we can write:

For x_A mol of component A in the vapor phase:

$$\Delta G_{\text{mix}}(A) = RTx_A \ln \frac{P_A}{P_A^\circ}$$

For x_B mol of component B in the vapor phase:

$$\Delta G_{\text{mix}}(B) = RTx_B \ln \frac{P_B}{P_B^\circ}$$

and thus, for the formation of the solution vapor or the solution itself:

$$\Delta G_{\text{mix}} = RT\left(x_A \ln \frac{P_A}{P_A^\circ} + x_B \ln \frac{P_B}{P_B^\circ}\right) \qquad [14]$$

Here P_A and P_B are the vapor pressures of components A and B over a solution of mole fractions x_A and x_B, and P_A° and P_B° are the vapor pressures of the pure liquids. All these data are provided by vapor-pressure diagrams like those of Figs. 20-25 and 20-26.

Finally, ΔS_{mix}, or $T\,\Delta S_{\text{mix}}$, is obtained from the relation

$$\Delta G_{\text{mix}} = \Delta H_{\text{mix}} - T\,\Delta S_{\text{mix}} \qquad [15]$$

The data for the low-vapor-pressure and high-vapor-pressure examples of Figs. 20-25 and 20-26 can now be used, with additional enthalpy data to give the thermodynamic diagrams of Figs. 21-3 and 21-4.

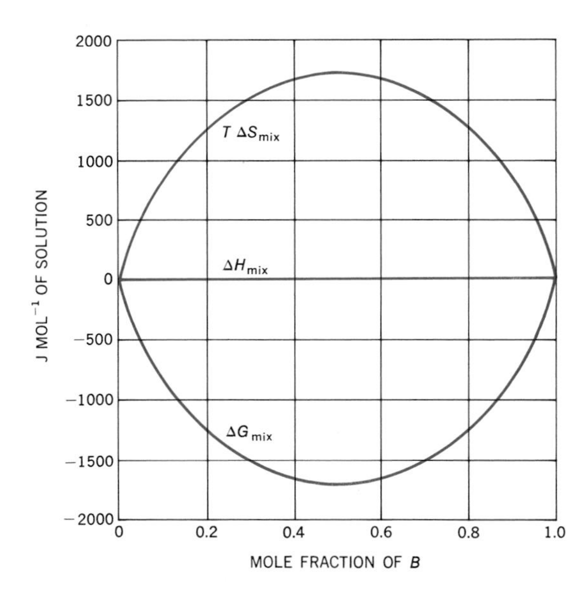

FIGURE 21-2
Changes in the thermodynamic functions for the formation of 1 mol of an ideal solution at 25°C.

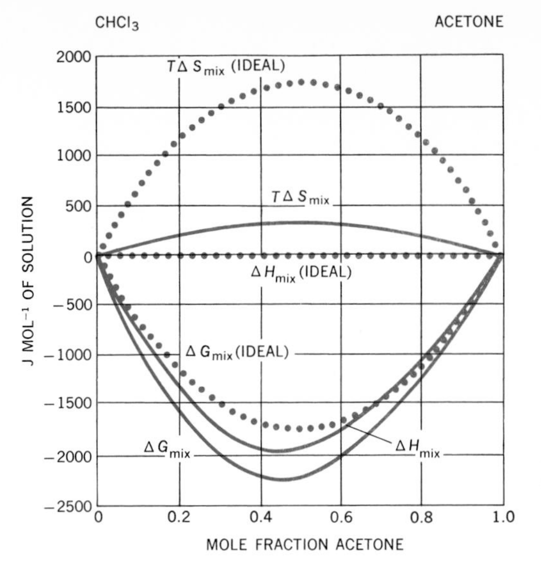

FIGURE 21-3
Changes in the thermodynamic functions for the formation of 1 mol of a chloroform-acetone solution at 25°C. (*Sketched from the data of I. Prigogine and R. Defay, "Thermodynamique chimique," Desoer, Liége, 1950.*) The dotted lines show ideal behavior.

The chloroform-acetone example is representative of solutions in which interactions between the molecules of the different components are different from the interactions between the molecules of an individual component. They consist of mixtures of somewhat acidic and somewhat basic molecules. On mixing, these will interact with one another, and in the chloroform-acetone example, this interaction takes

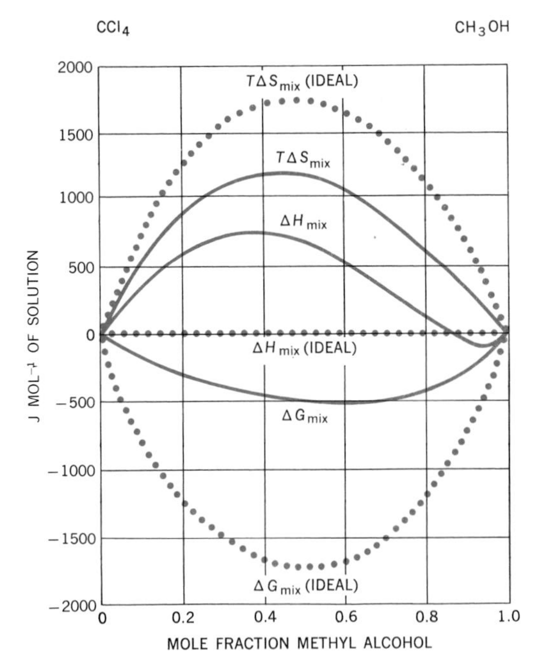

FIGURE 21-4
Thermodynamic changes for the formation of 1 mol of a carbon tetrachloride–methyl alcohol solution at 25°C. The dotted lines show ideal behavior. (*Sketched from the data of I. Prigogine and R. Defay, "Thermodynamique chimique," Desoer, Liége, 1950.*)

the form of hydrogen bonding; i.e.,

$$Cl_3C{-}H\cdots O{=}C(CH_3)_2$$

It can be expected that heat will be evolved when the solution is formed (for chloroform-acetone this can easily be noticed if the two components are mixed in a test tube). The nonideal behavior can be accounted for, in part, by a negative enthalpy of mixing in contrast to the zero enthalpy of mixing of an ideal solution.

The association of the components on mixing will also change the entropy from that calculated for an ideal solution. Very qualitatively, one might expect that the association in the solution would tend to restrict the motion of the molecules, and in view of the general discussion of entropy and freedom in Chap. 10, this would be expected to give the system less entropy than in the ideal case.

Thus our expectations are consistent with the observation that for a system like chloroform-acetone the enthalpy of mixing should be negative and that the entropy of mixing might be less positive than for an ideal solution.

Deviations from ideal behavior in the opposite direction occur and are illustrated by the methyl alcohol–carbon tetrachloride solutions. Systems that show this behavior are frequently those made up of a component that is itself associated, as are water and alcohols, and a more or less inert component. Mixing tends to break up some of the association, and a positive enthalpy term can be expected because of the heat required to break up the associated component. The data for the carbon tetrachloride–methyl alcohol system (Fig. 21-4) indicate that it is primarily this enthalpy effect which produces the free-energy effect. The region of negative heat of mixing and the appreciable entropy term should, however, caution against the use of qualitative arguments, such as those used here, when other than very large effects are being considered.

21-3 Properties of the Components of Real Solutions: Partial Molal Quantities

In addition to treating the total thermodynamic changes that occur when a solution is formed, we find it helpful also to be able to distribute the properties of a solution among the components of which it is composed. (This is particularly clear when one component is present in minor amounts, and is called the solute, and is primarily responsible for the properties or reaction that is being considered.) Now we must see how we can go from measurable properties of the solution to quantities attributable to the components. This was done for ideal solutions in Sec. 21-1. Considerably more development is necessary for real solutions.

In the systems that have been investigated thus far in our studies of thermodynamics, the total amount of reagent in the system being considered has been fixed. Such systems are said to be *closed systems,* and the procedures that have been developed are suitable for finding changes that occur when there is some reaction in such systems or when the temperature or pressure changes and affects such systems.

When one treats the components of solutions it is more convenient to be able to regard a certain amount of the solution as the system and then to investigate what happens when amounts of any of the components are added to the system. Such systems are said to be *open systems,* and we shall see in this and the following sections how we can apply thermodynamic methods to them. To begin with, we will assume that the pressure and temperature are fixed and that only the amounts of the components of the solution are variable.

The volume of a solution is perhaps the easiest of its properties to visualize. Let us therefore first consider how one might treat the volume of a solution and its dependence on the number of moles of the components present. Experiments that would provide the answers might begin with some amount of one component, that designated the *solvent,* for example, and then measurements would be made of the total volume of the system as the other component was added.

A curve showing the results that would be obtained from such measurements for rather dilute solutions of magnesium sulfate in water is shown in Fig. 21-5. The total volumes that can be read off such curves are, of course, dependent on the amounts of solute and solvent present and are therefore not of general use. The slope of the curves, however, gives the rate of change of volume at various points along the abscissa, i.e., at various relative amounts of solute and solvent. These slopes are, if n_A designates the number of moles of solvent and n_B the moles of solute, in calculus notation

$$\left(\frac{\partial V}{\partial n_A}\right)_{n_B} \quad \text{or} \quad \left(\frac{\partial V}{\partial n_B}\right)_{n_A}$$

and it is the value of the second of these partial derivatives that can

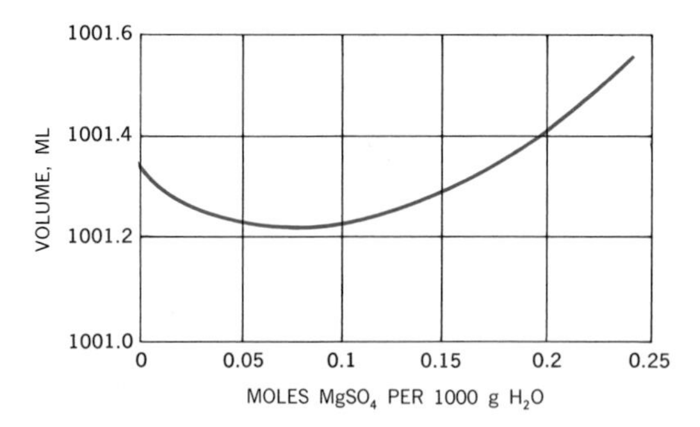

FIGURE 21-5
The volume of solutions containing 1000 g of water and various amounts of magnesium sulfate. [*From G. N. Lewis and M. Randall, "Thermodynamics," 2d ed. (rev. by K. S. Pitzer and L. Brewer), McGraw-Hill Book Company, New York, 1961.*]

be obtained from a curve like that of Fig. 21-5. To be more specific, we should perhaps write these partial derivatives as

$$\left(\frac{\partial V}{\partial n_A}\right)_{n_B,T,P} \qquad \text{and} \qquad \left(\frac{\partial V}{\partial n_B}\right)_{n_A,T,P}$$

to show that they imply the rate of change of volume with respect to the number of moles of one reagent when both the amount of the other reagent and the temperature and pressure are held constant.

Such derivatives, with T,P, and the amounts of the other reagents held constant, are examples of what are called *partial molal quantities*. They play a very important role in the thermodynamic treatment of solutions. Although their derivative nature may make their character and role somewhat obscure at first, one can come to recognize that they do give the contribution of a mole of the material, in the solution specified, to the property being considered. For example, $(\partial V/\partial n_B)_{n_A}$ for a solution containing 0.1 mol of $MgSO_4$ per 1000 g of water is, according to Fig. 21-5, 0.83 liter mol^{-1}. This means that for a system containing 0.1 mol of $MgSO_4$ and 1000 g of water, the rate of change of volume with the number of moles of $MgSO_4$ is 0.83 liter mol^{-1}. More easily pictured is the situation in which one has a very large amount of this solution, so that 1 mol of solute could be added without significantly changing the concentration. Then the change in volume that results from the addition of the 1 mol would itself be $(\partial V/\partial n_B)_{n_A}$, and we see that this partial molal volume is in fact the volume contributed by 1 mol of $MgSO_4$ in a solution of this composition. Thus a partial molal quantity is the contribution of a mole of the reagent to the property being dealt with—in the solution of the specified composition.

Such derivatives also appear if we recognize that, if T and P are considered fixed, a quantity such as the volume of a two-component system is a function of n_A and n_B, and therefore that the exact differential

$$dV = \left(\frac{\partial V}{\partial n_A}\right)_{n_B} dn_A + \left(\frac{\partial V}{\partial n_B}\right)_{n_A} dn_B \qquad [16]$$

can be written. The partial derivative coefficients are seen to be the partial molal volumes, which of course are dependent on the composition of the solution. One can in fact carry over the notation procedure of small capital letters for molar quantities and add a bar over the symbol to designate a partial molal quantity. In this way the nature of the partial molal derivatives can be emphasized, and with

$$\left(\frac{\partial V}{\partial n_A}\right)_{n_B} = (\bar{\mathrm{v}}_A) \qquad \text{and} \qquad \left(\frac{\partial V}{\partial n_B}\right)_{n_A} = (\bar{\mathrm{v}}_B)$$

one can rewrite Eq. [16] as

$$dV = (\bar{\mathrm{v}}_A)\, dn_A + (\bar{\mathrm{v}}_B)\, dn_B \qquad [17]$$

Equation [17] shows how the change in the volume of the solution is related to the partial molal volumes of the components. In a similar way any thermodynamic property can be treated, and partial molar quantities characteristic of each component can be determined and used in the expression for the change in a thermodynamic property of the solution itself.

But even more useful would be an interpretation of the thermodynamic properties of the solution, rather than only *changes* in these properties. For ideal solutions such relations were developed in Sec. 21-1, as for example

$$V = n_A \mathrm{v}_A + n_B \mathrm{v}_B$$

and

$$G(\mathrm{soln}) = n_A[\Delta \mathrm{G}_f^\circ(A) + RT \ln x_A] + n_B[\Delta \mathrm{G}_f^\circ(B) + RT \ln x_B]$$

Now let us begin the deduction of comparable expressions for real solutions.

First a general mathematical expression must be deduced, and this will be done in terms of one representative thermodynamic property, the volume V of the system. Let us investigate the dependence of the volume of a system, any system, on the number of moles $n_A, n_B, \ldots$ of the components present at some fixed temperature and pressure. We know, from experimental results, that if we were to increase the amounts of all the components by some factor a, the volume would increase by this factor. Thus, if the new amounts of the components $n'_A, n'_B, n'_C, \ldots$, related to old amounts n_A, n_B, n_C, by

$$n'_A = an_A \qquad n'_B = an_B \qquad \cdots \tag{18}$$

the new volume $V(n'_A, n'_B, \ldots)$ is related to the old volume $V(n_A, n_B, \ldots)$ by

$$V(n'_A, n'_B, \ldots) = aV(n_A, n_B, \ldots) \tag{19}$$

Now let us differentiate this equation with respect to a. The derivative of $V(n'_A, n'_B, \ldots)$, the left side of Eq. [19], with respect to a can be obtained by writing the total differential and then dividing by da to obtain

$$\begin{aligned}\frac{dV(n'_A,n'_B,\ldots)}{da} &= \frac{\partial V(n'_A,n'_B,\ldots)}{\partial n'_A}\frac{dn'_A}{da} + \frac{\partial V(n'_A,n'_B,\ldots)}{\partial n'_B}\frac{dn'_B}{da} + \cdots \\ &= \frac{\partial V(n'_A,n'_B,\ldots)}{\partial n'_A} n_A + \frac{\partial V(n'_A,n'_B,\ldots)}{\partial n'_B} n_B + \cdots\end{aligned} \tag{20}$$

The derivative of the right side of Eq. [19] with respect to a is, since $V(n_A, n_B, \ldots)$ is independent of a, simply

$$\frac{d}{da}[aV(n_A, n_B, \ldots)] = V(n_A, n_B, \ldots) \tag{21}$$

Thus, by equating the results of Eqs. [20] and [21],

$$\frac{\partial V(n'_A,n'_B,\ldots)}{\partial n'_A}n_A + \frac{\partial V(n'_A,n'_B,\ldots)}{\partial n'_B}n_B + \cdots$$
$$= V(n_A,n_B,\ldots) \quad [22]$$

This result is valid for all values of a, and we find that a useful relation is obtained by setting $a = 1$. Then $n_A = n'_A$, $n_B = n'_B$, etc., and we have

$$n_A\frac{\partial V(n_A,n_B,\ldots)}{\partial n_A} + n_B\frac{\partial V(n_A,n_B,\ldots)}{\partial n_B} + \cdots = V(n_A,n_B,\ldots)$$

or more simply written,

$$n_A\frac{\partial V}{\partial n_A} + n_B\frac{\partial V}{\partial n_B} + \cdots = V \quad [23]$$

The partial derivatives, since all the above has assumed constant T and P, could be written

$$\left(\frac{\partial V}{\partial n_A}\right)_{T,P,n_i} \quad \cdots$$

and they would thereby be identified with the partial molar quantities of Sec. 21-2. Thus, with the small-cap-with-bar notation, we have obtained

$$V = n_A\bar{\text{v}}_A + n_B\bar{\text{v}}_B + \cdots \quad [24]$$

Similar expressions would be obtained for the other thermodynamic properties H, S, and G on which our attention is focused.

We thus see that the experimentally accessible partial molar quantities again can be looked on as the molar value for the component in the solution and that the properties of a solution, as well as changes in these properties, can be deduced from the number of moles of the components and the partial molar quantities of each component by means of relations like

$$dV = \bar{\text{v}}_A\,dn_A + \bar{\text{v}}_B\,dn_B$$

and

$$V = n_A\bar{\text{v}}_A + n_B\bar{\text{v}}_B$$

21-4 The Free Energies of the Components of a Solution: Solvents

The results obtained in Sec. 21-2 for the free energy of solutions, which were based on the vapor pressures of the components above the solution, can now be used to illustrate the way in which the partial molar free energy of a component of a solution can be deduced. The approach is particularly applicable to the solvent, i.e., the major component, and one not too far removed from its pure-liquid state.

The free-energy change that occurs when n_A and n_B mol of two components are combined to give a solution with partial pressures of P_A and P_B was seen, in Sec. 21-2, to be

$$\Delta G_{\text{mix}} = n_A RT \ln \frac{P_A}{P_A^\circ} + n_B RT \ln \frac{P_B}{P_B^\circ} \qquad [25]$$

We also can express ΔG_{mix} in terms of the pure components and the solution as

$$\Delta G_{\text{mix}} = (n_A \overline{G}_A + n_B \overline{G}_B) - (n_A G_A^\circ + n_B G_B^\circ) \qquad [26]$$

Here the G_A° and G_B° terms are the molar free energies of the unmixed components. They can be assigned the standard free-energy-of-formation values, but this is not always the most convenient procedure. Equating and rearranging the above expressions gives

$$n_A \overline{G}_A + n_B \overline{G}_B = n_A G_A^\circ + n_A RT \ln \frac{P_A}{P_A^\circ} + n_B G_B^\circ + n_B RT \ln \frac{P_B}{P_B^\circ}$$

and thus

$$\overline{G}_A = G_A^\circ + RT \ln \frac{P_A}{P_A^\circ} \qquad [27]$$

and

$$\overline{G}_B = G_B^\circ + RT \ln \frac{P_B}{P_B^\circ}$$

From available vapor-pressure data, we can deduce values of the partial molar free energy of a component of a solution. If component A is the solvent it will, at least for dilute solutions, obey Raoult's law. Then $x_A = P_A/P_A^\circ$ and Eq. [27] takes on the form

$$\overline{G}_A = G_A^\circ + RT \ln x_A \qquad [28]$$

This relation, also valid for ideal solutions, as seen in Sec. 21-1, is followed by the solvents of real, dilute solutions even if nonideal behavior sets in as more solute is added. At these higher concentrations it is still convenient to have an expression of the same form as Eq. [28], and this can be done by introducing the *activity a,* defined as $a_A = P_A/P_A^\circ$, so that the partial molar free energy is given, even for nonideal solutions, by

$$\overline{G}_A = G_A^\circ + RT \ln a_A \qquad [29]$$

Since we expect a_A to approach x_A for the solvent of a dilute solution, it is convenient also to introduce the *activity coefficient* γ as

$$\gamma = \frac{a_A}{x_A} \quad \text{or} \quad a_A = \gamma x_A \qquad [30]$$

The activity coefficient shows, by its variation from unity, the nonideality of the solution. Since the solvents of all solutions that are sufficiently dilute obey Raoult's law, the value of γ approaches unity for the solvent of any solution as the solute concentration approaches zero.

The vapor-pressure data of the liquid-vapor equilibrium diagrams of Chap. 20 provide the necessary data for such a calculation of activity and activity coefficients. Some values for the activities and activity coefficients for these systems are shown in Table 21-2. The factors that operate to make the activity coefficients greater or less than unity

TABLE 21-2 The activities and activity coefficients of solvents on the assumption of ideal-gas behavior of the vapor*

Solvent benzene (solute toluene, 20°C)			Solvent toluene (solute benzene, 20°C)		
Mole fraction x	Activity a	Act. coeff. γ	Mole fraction toluene x	Activity a	Act. coeff. γ
1.00	1.00	1.00	1.00	1.00	1.00
0.67	0.65	0.97	0.77	0.78	1.01
0.55	0.54	0.98	0.57	0.55	1.07
0.43	0.46	1.07	0.45	0.47	1.04
Solvent acetone (solute $CHCl_3$, 35°C)			Solvent $CHCl_3$ (solute acetone, 35°C)		
x	a	γ	x	a	γ
1.00	1.00	1.00	1.00	1.00	1.00
0.94	0.94	1.00	0.92	0.91	0.99
0.88	0.87	0.99	0.81	0.76	0.94
0.73	0.70	0.96	0.66	0.55	0.83
0.63	0.57	0.90	0.58	0.48	0.83
0.51	0.42	0.82	0.49	0.38	0.78
Solvent CH_3OH (solute CCl_4, 35°C)			Solvent CCl_4 (solute CH_3OH, 35°C)		
x	a	γ	x	a	γ
1.00	1.00	1.00	1.00	1.00	1.00
0.91	0.95	1.04	0.98	0.99	1.01
0.79	0.88	1.11	0.87	0.97	1.11
0.66	0.84	1.27	0.64	0.94	1.47
0.49	0.80	1.63	0.51	0.92	1.80
0.36	0.78	2.16	0.34	0.87	2.56

*The data are from R. Bell and T. Wright, *J. Phys. Chem.*, **31**:1884 (1927); J. von Zawidski, *Z. Physik. Chem.*, **35**:129 (1900); J. Timmermans, "Physico-chemical Constants of Binary Systems," vol. 2, Interscience Publishers, Inc., New York, 1959.

are of course the same as those mentioned in Sec. 21-3 in connection with the free energy of mixing in the formation of solutions.

It should be noted that the partial molar free energy of a solvent is referred to that of the pure liquid solvent at the temperature and pressure of the solution. This standard state is reached when the value of P_A/P_A°, or of a_A, approaches unity. A rather different standard state is convenient when solutes are treated.

21-5 The Free Energies of the Components of a Solution: Solutes

For some solutions, particularly those made up from liquid components and those with a wide range of the relative amounts of the components, it is not desirable to designate components as solute and solvent. Then both the components can be treated in the manner indicated in Sec. 21-4, where, it will be recalled, the pure component was taken as the standard state. For other solutions, particularly when a component is a solid or a gas and is present in relatively small amounts, it is convenient to designate this component as a solute and to refer its thermodynamic properties to a different standard state.

Let us first recall the general result, shown by the vapor-pressure curves such as those of Figs. 20-25 and 20-26, that the vapor pressure of a solute of a dilute solution conforms to Henry's law; i.e., this vapor pressure is proportional to the mole fraction of the solute. Furthermore, the mole fraction and the molal concentration become proportional to one another as the solution becomes more dilute, and Henry's law for these dilute solutions can be written

$$P_B = kx_B \qquad \text{or} \qquad P_B = k'm \tag{31}$$

where x_B and m are the mole fraction and the molal concentration of the solute, component B, of the solution, and k and k' are Henry's law constants.

The difference in partial molal free energies of a solute at two concentrations can be written in terms of the vapor pressures of the solute over the two solutions as

$$(\overline{G}_B)_2 - (\overline{G}_B)_1 = RT \ln \frac{P_2}{P_1} \tag{32}$$

Furthermore, if the two solutions are dilute enough to conform to Henry's law, one obtains, by insertion of Eq. [31]

$$(\overline{G}_B)_2 - (\overline{G}_B)_1 = RT \ln \frac{x_2}{x_1}$$

or [33]

$$(\overline{G}_B)_2 - (\overline{G}_B)_1 = RT \ln \frac{m_2}{m_1}$$

To proceed to a satisfactory treatment of the thermodynamics of solutes, we must now see how to select a reference state to which the partial molal free energies of solutes can be referred. Now, the pure solute material is not a convenient standard, because the properties of the pure solute material are generally very different from those of the solute in the solution, and furthermore, they are often not pertinent to the problems that arise when the solution is dealt with.

For solutes of solutions that satisfactorily obey Henry's law, and therefore Eqs. [33], up to molal concentrations, it is convenient to choose the standard state as the solution at *unit molal concentration*. Then one writes

$$\overline{G}_B = \overline{G}_B^\circ + RT \ln m \qquad [34]$$

where $\overline{G}_B^\circ$ is the partial molal free energy of the solute in a 1 m solution and where the fact that m is the ratio of the molality to unit molality is implied.

An example of the evaluation of the free energy of this standard state is provided by solutions of oxygen in water.

The solubility of O_2, at a pressure of 1 atm and at 25°C, is 0.00115 mol per 1000 g of water. Since $O_2(g)$ at 1 atm is in equilibrium with $O_2(aq)$ at 0.00115 molal, the free energy of dissolved oxygen at this molality is equal to that of the gaseous oxygen. Standard free energies of formation assign zero to this latter value, and thus the free energy of formation, $\Delta\overline{G}_f$, of oxygen at 0.00115 molal is also zero.

Now, Eq. [34] can be used as

$$0 = \Delta\overline{G}_f^\circ[O_2(aq)] + 2.303(8.314)(298.15) \log 0.00115$$

and

$$\Delta\overline{G}_f^\circ[O_2(aq)] = +16{,}780 \text{ J}$$

Thus the free energy that O_2 would have if it were at 1 molal concentration in water at 25°C would be 16,780 J.

At other concentrations, its molal free energy is given by

$$\Delta\overline{G}_f = 16{,}780 + RT \ln m \quad \text{J} \qquad [35]$$

The result is particularly valuable for the calculation of the free-energy change and equilibrium position of the many biological and nature reactions that occur in aqueous solutions and involve dissolved oxygen.

Solutions whose solutes do not obey Henry's law up to unit molality present a situation very much like that treated in Sec. 9-5 for real gases that do not behave ideally up to a pressure of 1 atm, the standard-state pressure for ideal gases.

As for gases, one could choose some very low concentration where uniformity to the ideal law, in this case Henry's law, was assured, for the standard state. However, such a procedure would not be consistent with that adopted for "ideal" systems. As for gases, the accepted procedure is to choose the standard state as the hypothetical state that the solute would have if Henry's law were obeyed up to 1 m solutions. This procedure is suggested graphically in Fig. 21-6.

In the region of ideal, Henry's law, behavior, $P_B = k'm$, or $m = P_B/k'$. Let us now introduce a solute activity a_B such that even beyond the Henry's law region this activity remains proportional to the partial pressure of B. Further, let $a_B = P_B/k$, and then, as Fig. 21-6 suggests, the activity will be equal to the molarity in the limit of low molality, but at higher molalities will correspond to a hypothetical solution that still maintains conformity with Henry's law.

Now the standard state for the solute can be defined as that which has unit activity, and we can write

$$\overline{G}_B = \overline{G}_B^\circ + RT \ln a_B \qquad [36]$$

where $\overline{G}_B^\circ$ is the partial molal free energy of the standard state, i.e., the value of G_B when $a_B = 1$.

Comparison of Eq. [36] with Eq. [34] suggests, again, the introduction of an activity coefficient defined now, usually, as

$$\gamma = \frac{a_B}{m} \qquad [37]$$

(Since both a_B in Eq. [36] and m in Eq. [34] imply the ratio of these quantities to their unit values, they imply dimensionless quantities. Thus γ is likewise dimensionless.)

With the introduction of the activity coefficient, the molal free energy of the solute of a solution can be written

$$\overline{G}_B = \overline{G}_B^\circ + RT \ln \gamma m \qquad [38]$$

It is, of course, also possible to continue to compare the activity of the solute with the mole fraction rather than the molality of the solute. This procedure is more consistent with that used for solvents, but because of the chemist's tendency to deal with solute concen-

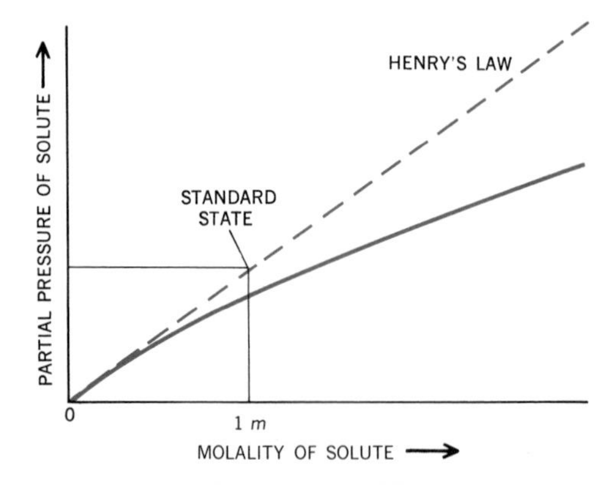

FIGURE 21-6
The standard state of a solute in a solution in which Henry's law is not obeyed up to a concentration of 1 m. (The vapor of the solute is assumed to behave ideally so that its vapor pressure and fugacity can be equated.)

trations in terms of molality, is not generally used. If it is followed, the activities are made equal to mole fractions for dilute solutions, and the activity coefficient is defined as

$$\gamma = \frac{a_B}{x_B} \tag{39}$$

Values of γ so defined will be different from those derived from Eq. [37], and it must be made clear when activity coefficients are used on what concentration term they are based.

To illustrate the use of relations like Eqs. [36] and [38], let us now consider one way of determining the activities and activity coefficients of nonelectrolyte solutes.

21-6 Solute Activities from Solvent Properties: An Application of the Gibbs-Duhem Equation

It is often not feasible to determine the partial molar free energies and the activities of solutes from measurements of the vapor pressure of the solute. For nonvolatile solutes, for example, it is clearly necessary to have an alternative procedure, and here one that depends on the relation between solute and solvent properties as given by the *Gibbs-Duhem* equation will be developed. Specifically, we shall develop the relation between the free energies of a single solute and the solvent of a solution.

One of the relations developed in Sec. 21-3 for partial molal quantities was

$$dG = \bar{G}_A\, dn_A + \bar{G}_B\, dn_B$$

The rate of change of solution free energy with additions of solvent component A is then given by

$$\frac{dG}{dn_A} = \bar{G}_A + \bar{G}_B \frac{dn_B}{dn_A} \tag{40}$$

Alternatively, dG/dn_A can be obtained by differentiating the expression $G = n_A\bar{G}_A + n_B\bar{G}_B$, also obtained in Sec. 21-3, with respect to n_A, and this gives

$$\frac{dG}{dn_A} = n_A \frac{d\bar{G}_A}{dn_A} + \bar{G}_A + n_B \frac{d\bar{G}_B}{dn_A} + \bar{G}_B \frac{dn_B}{dn_A} \tag{41}$$

Equating these two expressions for dG/dn_A leads to the desired Gibbs-Duhem equation

$$n_A \frac{d\bar{G}_A}{dn_A} = -n_B \frac{d\bar{G}_B}{dn_A} \tag{42}$$

An alternative form is obtained by dividing numerators and denominators of both sides by $n_A + n_B$ to obtain mole-fraction terms and the result

$$x_A \frac{d\bar{G}_A}{dx_A} = -x_B \frac{d\bar{G}_B}{dx_A} \qquad [43]$$

Now information on the variation of solvent free energy can be used to deduce the variation of solute free energy.

Expressions for either solvent or solute partial molar free energy have the form

$$\bar{G} = \bar{G}^\circ + RT \ln a$$

Thus

$$\frac{d\bar{G}_A}{dx_A} = RT\frac{d(\ln a_A)}{dx_A} \quad \text{and} \quad \frac{d\bar{G}_B}{dx_A} = RT\frac{d(\ln a_B)}{dx_A} \qquad [44]$$

With these expressions, Eq. [43] becomes

$$x_A \frac{d(\ln a_A)}{dx_A} = -x_B \frac{d(\ln a_B)}{dx_A}$$

or

$$d(\ln a_B) = -\frac{x_A}{x_B} d(\ln a_A) \qquad [45]$$

Integration leads to the information on the solute activity a_B according to

$$\ln \frac{a''_B}{a'_B} = -\int_{a'_B}^{a''_B} \frac{x_A}{x_B} d(\ln a_A)$$

or

$$\log \frac{a'_B}{a''_B} = \int_{a'_B}^{a''_B} \frac{x_A}{x_B} d(\log a_A) \qquad [46]$$

The integration can be performed graphically, as illustrated in Fig. 21-7, if data are available, as by the methods of Sec. 21-4, for the solvent activity as a function of the solution composition. Although some difficulties (which can be avoided by different graphical treatments) enter as infinite dilution is approached because of zero values for x_B and a_B, and therefore an infinity in $\ln a_B$, the integration can be started, not at infinite dilution, but at some solution dilute enough so that Henry's law is obeyed. When this is done, results for $\ln a_B$, and a_B, can be determined for solutions of higher concentrations. Results are shown for solutions of sucrose in water in Table 21-3.

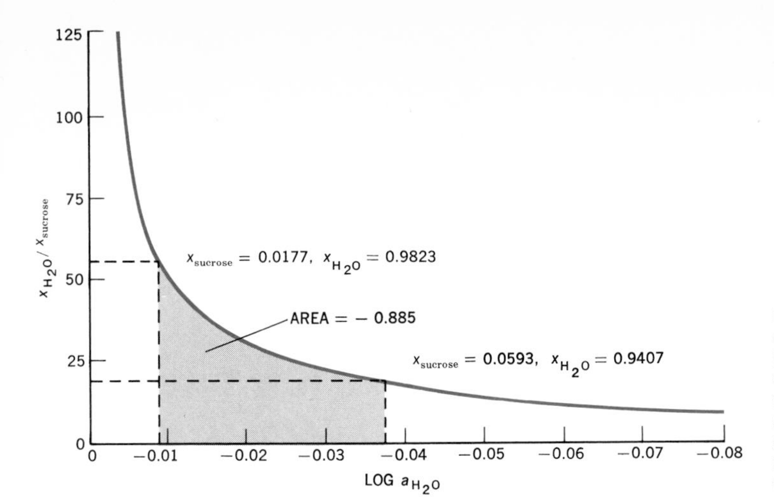

FIGURE 21-7
A plot for the integration of Eq. [46] for the system water-sucrose at 0°C. The shaded area is used to obtain, as an example, the difference in solute activities between the solute mole fractions 0.0177 and 0.0593.

21-7 The Dependence of the Free Energy of the Solvent of an Ideal Solution on Temperature, Pressure, and Composition

The variables that affect the solutions we have been dealing with are pressure, temperature, and composition. The complexities of real solutions are such that in all the developments so far, the pressure and temperature have been presumed to be fixed, so that we could concentrate on the remaining variable, the composition. There are, however, several important properties of dilute solutions, known as *colligative properties,* that require attention to be paid to all three variables. For these dilute solutions, fortunately, the solvent behaves ideally, and the concentration dependence is easily expressed. It then is feasible to

TABLE 21-3 The activities of water and sucrose in water sucrose solutions at 0°C obtained from the vapor pressure of water and the Gibbs-Duhem relation*

Molality of sucrose	Mole fractions: Sucrose	Mole fractions: Water	Vapor pressure of water, mm Hg	a_{water}	$a_{sucrose}$
0	0	1.000	4.579	1.000	0
0.2	0.0036	0.996	4.562	0.996	0.0036
0.5	0.0089	0.991	4.536	0.990	0.0089
1.0	0.0177	0.982	4.489	0.980	0.019
3.5	0.059	0.941	4.195	0.916	0.146
4.5	0.075	0.925	4.064	0.888	0.238
5.0	0.082	0.918	3.994	0.872	0.292
6.0	0.098	0.902	3.867	0.845	0.403

* Data given in International Critical Tables, vol. 3, p. 293, McGraw-Hill Book Company, New York, 1928.

treat the net effect of pressure, temperature, and concentration changes on the free energy of the solvent. Here a general expression that will serve for the interpretation of all four colligative properties will be developed.

The partial molal free energy of the solvent $\overline{\mathrm{G}}_A$ is, generally, a function of T, P, and the number of moles of the components, here assumed to be the solvent A and the solute B. The total differential for G_A can thus be written

$$d\overline{\mathrm{G}}_A = \left(\frac{\partial \overline{\mathrm{G}}_A}{\partial T}\right)_{P,n_A,n_B} dT + \left(\frac{\partial \overline{\mathrm{G}}_A}{\partial P}\right)_{T,n_A,n_B} dP + \left(\frac{\partial \overline{\mathrm{G}}_A}{\partial n_A}\right)_{T,P,n_B} dn_A + \left(\frac{\partial \overline{\mathrm{G}}_A}{\partial n_B}\right)_{T,P,n_A} dn_B \quad [47]$$

To make use of this general expression, the partial derivative coefficients must be determined. The similarity of partial molar quantities to thermodynamic properties of single-component materials has been shown. This leads further to the carryover of thermodynamic relations like those of Eqs. [23] and [24] of Chap. 9, which now give

$$\left(\frac{\partial \overline{\mathrm{G}}_A}{\partial T}\right)_{P,n_A,n_B} = -\overline{\mathrm{S}}_A \quad [48]$$

and

$$\left(\frac{\partial \overline{\mathrm{G}}_A}{\partial P}\right)_{T,n_A,n_B} = \overline{\mathrm{V}}_A \quad [49]$$

For ideal behavior of the solvent, furthermore, we can write

$$\begin{aligned} \overline{\mathrm{G}}_A &= \mathrm{G}_A^\circ + RT \ln x_A \\ &= \mathrm{G}_A^\circ + RT \ln \frac{n_A}{n_A + n_B} \end{aligned} \quad [50]$$

The remaining partial derivatives of Eq. [47] can now be determined, if the derivative formula $d \ln x = (1/x)\, dx$ is recalled, as

$$\begin{aligned} \left(\frac{\partial \overline{\mathrm{G}}_A}{\partial n_A}\right)_{T,P,n_B} &= RT \frac{\partial}{\partial n_A}\left(\ln \frac{n_A}{n_A + n_B}\right)_{n_B} \\ &= RT\left(\frac{n_A + n_B}{n_A} \frac{(n_A + n_B) - n_A}{(n_A + n_B)^2}\right) \\ &= \frac{n_B/n_A}{n_A + n_B} RT \end{aligned} \quad [51]$$

and

$$\begin{aligned} \left(\frac{\partial \mathrm{G}_A}{\partial n_B}\right)_{T,P,n_A} &= RT \frac{\partial}{\partial n_B}\left(\ln \frac{n_A}{n_A + n_B}\right) \\ &= \frac{-1}{n_A + n_B} RT \end{aligned} \quad [52]$$

With these results the complete dependence of the molar free energy of a component of an ideal solution on T, P, and amount of components is given by

$$d\bar{G}_A = -\bar{S}_A\,dT + \bar{V}_A\,dP + \frac{n_B/n_A}{n_A + n_B}RT\,dn_A + \frac{-RT}{n_A + n_B}\,dn_B \qquad [53]$$

With this result we can proceed to the analysis of the experiments that yield the four colligative properties of solutions.

21-8 Vapor-pressure Lowering

What is the effect on the equilibrium vapor pressure of a solution at a fixed temperature when a small amount of a nonvolatile solute is added? Equilibrium is maintained if the change in the molar free energy of the vapor, which is solvent vapor, equals the molar-free-energy change of the solvent in the solution.

The vapor is affected only by the change in pressure, and so we can write, from $(\partial G/\partial P)_T = V$,

$$dG_{A,\text{vap}} = V\,dP$$

If ideal-gas behavior is assumed, this becomes

$$dG_{A,\text{vap}} = \frac{RT}{P}\,dP \qquad [54]$$

If the solution is assumed to be at some total pressure, which is usually 1 atm, the pressure, the temperature, and number of moles of solvent in the solution are fixed. The solvent-free-energy change in the solution results, therefore, only from a change in amount of solute, and according to Eq. [53] this change is

$$d\bar{G}_A = \frac{-RT}{n_A + n_B}\,dn_B \qquad [55]$$

For equilibrium to be maintained,

$$RT\frac{dP}{P} = \frac{-RT}{n_A + n_B}\,dn_B$$

or

$$\frac{dP}{P} = -\frac{dn_B}{n_A + n_B} \qquad [56]$$

For the effect on the initial vapor pressure P_A° of the pure solvent caused by the addition of n_B mol of solute, we write the integral

$$\int_{P_A^\circ}^{P_A} \frac{dP_A}{P_A} = -\int_0^{n_B} \frac{dn_B}{n_A + n_B} \qquad [57]$$

Of principal interest is the result for amounts of n_B that are small compared with n_A and changes in P that are small compared with P°. Then we can write the simpler, approximate expression

$$\frac{1}{P_A^\circ}\int_{P_A^\circ}^{P_A} dP_A \approx \frac{-1}{n_A + n_B}\int_0^{n_B} dn_B \qquad [58]$$

which gives

$$\frac{P_A - P_A^\circ}{P_A^\circ} = -\frac{n_B}{n_A + n_B}$$
$$= -x_B$$

or

$$P_A^\circ - P_A = x_B P_A^\circ \qquad [59]$$

Alternatively, this can be written

$$P_A = P_A^\circ(1 - x_B)$$
$$= P_A^\circ x_A \qquad [60]$$

Thus our analysis of the free-energy changes of solution and vapor have led us to Raoult's law, the empirical result for the vapor pressure of the solvent of a solution.

The lowering, $P_A^\circ - P_A$, of the vapor pressure is seen from Eq. [59] for dilute solutions of nonvolatile solutes to depend only on the mole fraction of the solute and is therefore one of the colligative properties. Use of expression [59] to deduce x_B, and thus the molecular mass of a solute, from vapor-pressure-lowering measurements could be illustrated, but such a procedure will be postponed until the next colligative property, the boiling-point elevation, is dealt with.

21-9 The Boiling-point Elevation

Now consider the change in the boiling point that results from the addition of a small quantity of a nonvolatile solute to a solvent.

The vapor in equilibrium with the solution now suffers a free-energy change only because of a temperature change, the pressure being held constant. Thus, from $(\partial G/\partial T)_P = -S$, we write

$$(d\mathrm{G}_A)_v = -(\mathrm{S}_A)_v \, dT_{\mathrm{bp}} \qquad [61]$$

The solvent in the solution itself is affected both by this temperature change and by the addition of the solute. From Eq. [53] we write

$$d\bar{\mathrm{G}}_A = (-\bar{\mathrm{S}}_A)\, dT_{\mathrm{bp}} + \frac{-1}{n_A + n_B} RT_{\mathrm{bp}}\, dn_B \qquad [62]$$

For equilibrium to be maintained, these free-energy changes are equated.

$$-(\mathrm{s}_A)_v\, dT_{\mathrm{bp}} = -\bar{\mathrm{s}}_A\, dT_{\mathrm{bp}} + \frac{-1}{n_A + n_B} RT_{\mathrm{bp}}\, dn_B$$

or

$$[(\mathrm{s}_A)_v - \bar{\mathrm{s}}_A]\, dT_{\mathrm{bp}} = \frac{RT_{\mathrm{bp}}}{n_A + n_B}\, dn_B \qquad [63]$$

If very dilute solutions are considered, $\bar{\mathrm{s}}_A$ will be little different from $(\mathrm{s}_A)_l$, and with this approximation we can set

$$(\mathrm{s}_A)_v - \bar{\mathrm{s}}_A \approx (\mathrm{s}_A)_v - (\mathrm{s}_A)_l = \frac{\Delta\mathrm{H}_{\mathrm{vap}}}{T_{\mathrm{bp}}} \qquad [64]$$

For small additions of n_B and corresponding small changes in T_{bp} the quantities T_{bp}, $\Delta\mathrm{H}_{\mathrm{vap}}$, and $n_A + n_B$ can be treated as constants, and integration of Eq. [63] to give the change in boiling ΔT_{bp} for the addition of n_B mol of solute yields

$$\Delta T_{\mathrm{bp}} = \frac{RT_{\mathrm{bp}}^2}{\Delta\mathrm{H}_{\mathrm{vap}}} x_B \qquad [65]$$

where x_B has been written for $n_B/(n_A + n_B)$.

It is customary in colligative-property work to use molality rather than mole fraction of the solute. Molality, represented by m, is defined as the moles of solute per 1000 g of solvent, and if n_A is the number of moles of solvent in 1000 g of solvent,

$$x_B = \frac{m}{n_A + m} \approx \frac{m}{n_A} \qquad [66]$$

The final simplification results because for dilute solutions, m is much less than n_A. The quantity n_A, for any solvent A, is readily calculated as 1000 divided by the molar mass A.

In terms of molality, the boiling-point elevation is written

$$\Delta T_{\mathrm{bp}} = \left(\frac{RT_{\mathrm{bp}}^2}{n_A\, \Delta\mathrm{H}_{\mathrm{vap}}}\right) m \qquad [67]$$

The expression in parentheses is called the *boiling-point elevation constant,* or the *ebullioscopic constant,* and is frequently represented by K_{bp}. With this notation we have

$$\Delta T_{\mathrm{bp}} = K_{\mathrm{bp}} m \qquad [68]$$

where

$$K_{\mathrm{bp}} = \frac{RT_{\mathrm{bp}}^2}{n_A\, \Delta\mathrm{H}_{\mathrm{vap}}} \qquad [69]$$

Some results comparing the values of K_{bp} from Eq. [68] and measurements of the boiling-point elevation with the values obtained from

TABLE 21-4 Molal boiling-point-elevation constants at 1 atm pressure

Solvent	BP (°C)	$K_{bp}(\text{obs}) = \dfrac{\Delta T_{bp}}{m}$	$K_{bp}(\text{calc})$ (Eq. [69])
Water	100.0	0.51	0.51
Ethyl alcohol	78.4	1.22	1.20
Benzene	80.1	2.53	2.63
Ethyl ether	34.6	2.02	2.11
Chloroform	61.3	3.63	3.77

Eq. [69] are shown in Table 21-4. This agreement can be expected to be good, however, only for solutes that are neither associated nor dissociated in solution. If either of these processes occurs, the number of solute particles per 1000 g of solvent is not simply the Avogadro's number times the molality.

One type of boiling-point apparatus that is used to measure boiling-point elevations is shown in Fig. 21-8. The principal difficulty in accurately determining the boiling-point elevation stems from the fact that it is the temperature of the boiling liquid and not, as in the case of

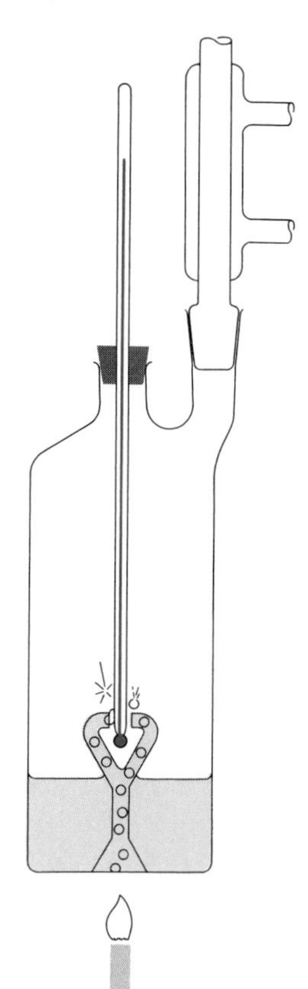

FIGURE 21-8
The Cottrell boiling-point apparatus. (*F. Daniels, J. W. Williams, P. Bender, R. A. Alberty, and C. D. Cornwell, "Experimental Physical Chemistry," 6th ed., McGraw-Hill Book Company, New York, 1962.*)

pure liquids, the temperature of the refluxing vapor that must be measured. The vapor rising from the solution is essentially pure solvent, and this vapor will condense at the boiling point of the pure solvent. It is necessary, therefore, to arrange the apparatus so that the thermometer is drenched with representative samples of the solution and is not merely at the temperature of the condensing vapor. This requirement immediately introduces difficulties due to superheating of the liquid, and it is difficult to obtain boiling-point-elevation results to better than about $\pm 0.01°C$.

The practical application of boiling-point-elevation measurements, and of other colligative properties, follows from their use in the determination of molecular masses. A sample calculation of a molecular mass illustrates this. A sample of unknown molecular mass is found to raise the boiling point of benzene by 1.04°C when 3.63 g of the material is added to 100 g of benzene. From the observed boiling-point elevation and the boiling-point-elevation constant for benzene from Table 21-4, the molality is calculated as

$$m = \frac{\Delta T_{\text{bp}}}{K_{\text{bp}}} = \frac{0.411 \text{ mole}}{1000 \text{ g benzene}}$$

The mass of the solute in 1000 g of the solvent is obtained from the masses of the two components as

$$\frac{3.63}{100} \times 1000 = \frac{36.3 \text{ g}}{1000 \text{ g benzene}}$$

These two results allow the calculation of the molecular mass, i.e., the mass per mole, as

$$\frac{36.3}{0.411} = 88 \text{ g mol}^{-1}$$

The practical difficulties inherent in boiling-point measurements, however, lead to the more frequent use of freezing-point depressions for such molecular-mass determinations.

21-10 The Freezing-point Depression

How is the freezing point of a solvent affected by the addition of a small amount of solute? If we assume, as is often the case, that the freezing process leads to the separation of pure solid solvent, we can proceed as in Secs. 21-8 and 21-9.

The total pressure is fixed, usually at 1 atm, and the free energy of the solid solvent is affected only by the temperature. Thus

$$d(\text{G}_A)_s = -(\text{S}_A)_s \, dT_{\text{fp}} \qquad [70]$$

The solvent in solution is affected both by any temperature change and by the solute addition, and now Eq. [53] becomes

$$d\bar{G}_A = -\bar{S}_A \, dT_{fp} + \frac{-1}{n_A + n_B} RT_{fp} \, dn_B \qquad [71]$$

If these two free-energy changes are required to be equal, and if approximations and rearrangements comparable with those of Sec. 21-9 are made, we arrive at the expression

$$\Delta T_{fp} = -\frac{RT_{fp}^{\ 2}}{\Delta H_{fusion}} x_B \qquad [72]$$

Again this result can be written in terms of molality, and it then becomes

$$\Delta T_{fp} = -K_{fp} m \qquad [73]$$

where

$$K_{fp} = \frac{RT_{fp}^{\ 2}}{n_A \, \Delta H_f} \qquad [74]$$

The constant K_{fp} is known as the *freezing-point-depression constant* and is clearly a function only of the solvent. The freezing-point depression of dilute solutions is therefore a function of the properties of the solvent and is independent of any feature of the solute except its concentration in the solution. The depression of the freezing point is therefore another colligative property.

The freezing-point depression can usually be measured much more easily and accurately than can the boiling-point elevation. A simple, usually suitable, apparatus is shown in Fig. 21-9. The procedure, described as the *Beckmann method,* for determining the freezing point of the pure solvent and of the solution is to fix the bath at some suitable temperature several degrees below the expected freezing point of the solution and to follow the temperature of the sample solution as it cools. The solution is continuously stirred, and particular care is taken to prevent the solvent from freezing out on the wall of the sample tube. The freezing point is determined by the appearance of solid particles or by a change in the slope of the cooling curve. The temperature is usually measured by a differential thermometer that can be set to provide an expanded scale in the region of the freezing point.

For very accurate determinations the temperature is measured with a multijunction thermocouple and an equilibrium method whereby the solid solvent and the solution are agitated until equilibrium is adopted. This procedure prevents the troublesome supercooling that often occurs in the Beckmann method.

Table 21-5 shows the accuracy with which freezing-point depressions can be measured and indicates the degree to which the dilute-solution freezing-point-depression equation is obeyed.

It should be pointed out that freezing-point-depression data are

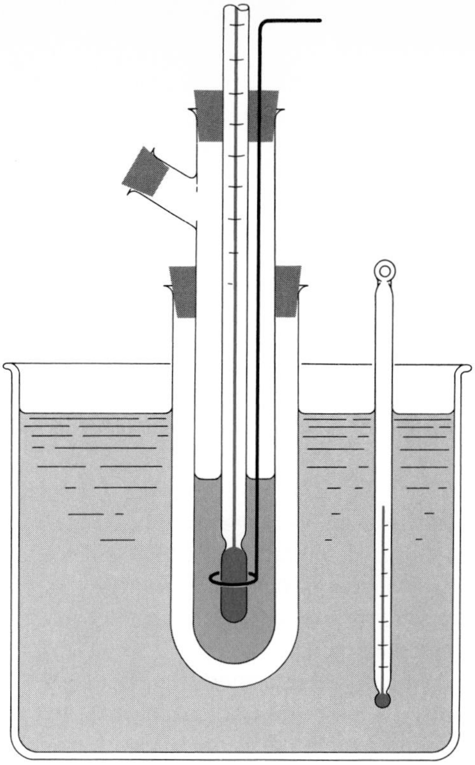

FIGURE 21-9
The Beckmann freezing-point apparatus.

contained in the solid-liquid equilibrium diagrams like those given in Secs. 20-10 and 20-11. The proportionality, for sufficiently dilute solutions, between freezing-point depression and either molality or mole fraction indicated by Eqs. [72] and [73] corresponds to a straight-line portion for the freezing-point curves. This holds in regions of low concentrations of one of the pure components in the other, or of low concentration of a compound that forms as a solid from the solution. The slopes

TABLE 21-5 Freezing-point depressions for solutions of mannitol in water*

Molality	FP (°C)	$-\dfrac{\Delta T_{\text{fp}}}{m}$
0.00402	−0.0075	1.86
0.00842	−0.0157	1.86
0.01404	−0.0260	1.852
0.02829	−0.0525	1.856
0.06259	−0.1162	1.857

* From data of L. H. Adams, *J. Am. Chem. Soc.*, **37**:481 (1915).

TABLE 21-6 Molal freezing-point-depression constants

Solvent	FP (°C)	K_{fp}
Water	0.00	1.86
Acetic acid	16.6	3.90
Benzene	5.5	5.12
Bromoform	7.8	14.4
Cyclohexane	6.5	20
Camphor	173	40

of the straight-line portions or a mole fraction or mole percent graph can, according to Eq. [72], give information on the heat of fusion of the solvent.

The freezing-point-depression constants of a number of solvents that find use in freezing-point-depression work are listed in Table 21-6. The choice of solvent is frequently dictated by the solubility and chemical reactivity of the substance whose molecular mass is to be determined. Increased accuracy results, of course, from the use of a solvent with a large freezing-point-depression constant.

Freezing-point-depression measurements find frequent use in molecular-mass determinations. In the field of organic chemistry it is often very helpful to have a value for the molecular mass of a newly synthesized or isolated material whose structure is being determined.

21-11 Osmotic Pressure

Osmotic pressure needs, perhaps, more of an introduction than did the properties of the preceding sections. The phenomenon of osmosis depends on the existence of *semipermeable membranes*. Such membranes are of a great variety, but they are all characterized by the fact that they allow one component of a solution to pass through them and prevent the passage of another component. Cellophane and a number of animal or protein membranes, for example, are permeable to water but not to higher-molecular-mass compounds. Mention can also be made of the semipermeability of a palladium foil, which is permeable to hydrogen gas but not to nitrogen and other gases. With such a membrane, osmosis can be studied in the vapor phase.

One arrangement that has been used for the quantitative study of osmosis is shown in Fig. 21-10. Any osmosis apparatus depends on the separation of a solution from its pure solvent by means of a membrane, permeable to the solvent but impermeable to the solute. The essential features of the system of Fig. 21-10 are shown schematically in Fig. 21-11. When such an arrangement is made, it is found

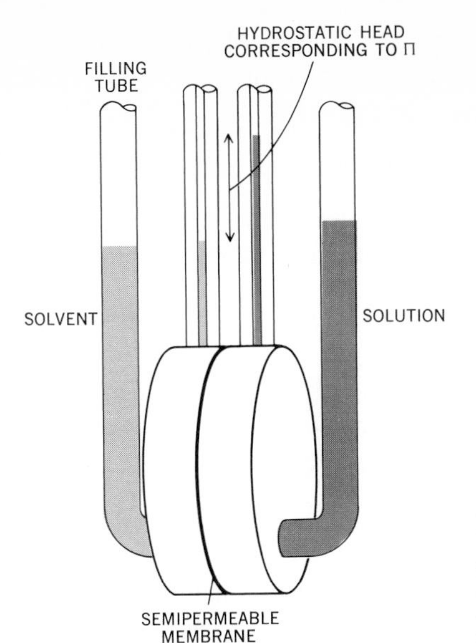

FIGURE 21-10
The osmotic-pressure apparatus (schematic). [*R. M. Fuoss and D. J. Mead, J. Phys. Chem.,* **47**:*59* (*1943*).]

that there is a natural tendency for the solvent to flow from the pure-solvent chamber through the membrane into the solution chamber. This tendency can be opposed by applying pressure to the solution chamber. In the apparatus of Fuoss and Mead this balancing pressure results from the hydrostatic head that is developed. The excess pressure that must be applied to the solution to produce equilibrium is known as the *osmotic pressure* and is denoted by Π. It is through this quantity that the quantitative aspects of osmosis are studied.

The osmotic pressure developed between any dilute solution and its solvent will be shown to be a colligative property. It is therefore dependent only on the concentration of the solution and on the properties of the solvent. It is important to recognize that the nature of the semipermeable membrane and the mechanism by which it allows solvent to pass through it but prevents the passage of solute is of no importance for the study of osmotic pressure as a colligative property.

The thermodynamic basis of the osmotic pressure can now be shown.

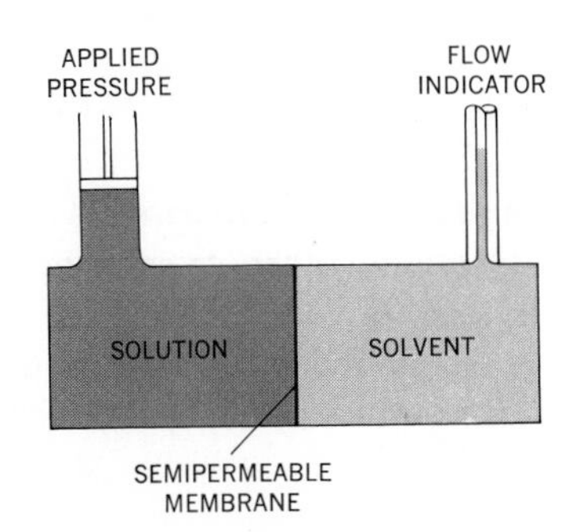

FIGURE 21-11
Schematic representation of an osmotic-pressure apparatus.

Consider the schematic arrangement of Fig. 21-11. The solvent in the pure-solvent compartment will be subject to no pressure or temperature change or solute addition. Thus

$$d(\mathrm{G}_A)_{\text{pure solvent}} = 0 \qquad [75]$$

The solvent in the solution compartment, on the other hand, is subject to a pressure change and the addition of solute. For this solvent, from Eq. [53],

$$d\bar{\mathrm{G}}_A = \bar{\mathrm{v}}_A\,dP - \frac{RT}{n_A + n_B}\,dn_B \qquad [76]$$

For equilibrium to be maintained these two free-energy changes must be equal, that is, $d\bar{\mathrm{G}}_A$ must be zero, or

$$\bar{\mathrm{v}}_A\,dP = \frac{RT}{n_A + n_B}\,dn_B \qquad [77]$$

Integration to show the added pressure that must be applied when n_B mol of solute is added is most easily carried out if it is assumed that n_B is small compared with n_A, and thus $n_A + n_B$ can be taken as constant, and that $\bar{\mathrm{v}}_A$ can be taken as the molar volume of the pure solvent. Then Eq. [77] gives

$$\mathrm{v}_A \int_{P_{\text{init}}}^{P_{\text{final}}} dP = \frac{RT}{n_A + n_B}\int_0^{n_B} dn_B$$

or

$$\mathrm{v}_A(P_{\text{final}} - P_{\text{init}}) = RTx_B \qquad [78]$$

The excess pressure is the osmotic pressure Π, and thus we have the relation

$$\mathrm{v}_A\Pi = RTx_B \qquad [79]$$

An interesting variation of this result is obtained by multiplying by n_A, which is approximately equal to $n_A + n_B$. This gives

$$\begin{aligned} n_A\mathrm{v}_A\Pi &= RTn_Ax_B \\ &\approx n_BRT \end{aligned} \qquad [80]$$

If **V** is introduced to represent $n_A\mathrm{v}_A$, that is, approximately the volume of solution containing n_B mol of solute, we can write

$$\Pi\mathbf{V} = n_BRT \qquad \text{or} \qquad \Pi = MRT \qquad [81]$$

where M is the molarity of the solution. The similarity of this expression to the ideal-gas law led van't Hoff and others to some not very fruitful ideas that view the osmotic pressure as arising from a molecular-bombardment process. It is recognized here that Eq. [81] is merely an approximate form obtained from the thermodynamic-dilute-solution expression (Eq. [79]).

TABLE 21-7 The osmotic pressure of aqueous solutions of sucrose at 20°C*

Molal concentration	Molar concentration	Observed osmotic pressure (atm)	Calculated osmotic pressure	
			From Eq. [79]	From Eq. [81]
0.1	0.098	2.59	2.40	2.36
0.2	0.192	5.06	4.81	4.63
0.3	0.282	7.61	7.21	6.80
0.4	0.370	10.14	9.62	8.90
0.5	0.453	12.75	12.0	10.9
0.6	0.533	15.39	14.4	12.8
0.7	0.610	18.13	16.8	14.7
0.8	0.685	20.91	19.2	16.5
0.9	0.757	23.72	21.6	18.2
1.0	0.825	26.64	24.0	19.8

*Osmotic-pressure data of Morse, reported by A. Findlay, "Osmotic Pressure," Longmans, Green & Co., Inc., New York, 1919.

A comparison of the observed osmotic pressure as a function of concentration with the behavior expected on the basis of the derived expressions is shown in Table 21-7.

To this discussion of the osmotic pressure as a colligative property can be added a few comments on the mechanism by which a semipermeable membrane operates. In some cases the membrane seems to act simply as a mechanical sieve, letting small molecules, like water, pass through and preventing the passage of large molecules. Other membranes do not appear to pass and reject molecules on a simple size basis. In some of these cases a component appears to penetrate the membrane by dissolving in it, whereas another component that is not soluble in the membrane cannot pass through it. Probably the clearest example of this is the passage of hydrogen through palladium. The hydrogen molecules are probably dissociated to atoms on the surface of the palladium. These atoms can penetrate through the solid lattice and on the opposite surface can reunite into hydrogen molecules. Other molecules are not dissociated and cannot pass through the solid.

As previously mentioned, the mechanism of the process at the semipermeable membrane can be studied quite separately from the subject of colligative properties. The expressions for the osmotic pressure derived in this section will apply as long as a membrane is available that will pass solvent and will not pass solute. The procedure by which it accomplishes this is immaterial.

21-12 Osmotic-pressure Determination of Molecular Masses

The principal use of osmotic-pressure measurements is in the determination of the molecular mass of high-molecular-mass compounds. Solutions of high-molecular-mass compounds will have low molal concentrations even though they may be quite concentrated in terms of the mass of solute. The measurable osmotic pressure produced even by solutions of low molality makes it the most suitable of the colligative properties for the study of such compounds.

The physical chemistry of compounds of high molecular mass will be treated in more detail later. For the present it is enough to point out that two types of high-molecular-mass compounds, in which one is interested in the molecular mass, are the synthetic polymers and the naturally occurring materials such as proteins. An example of the first type will illustrate the use of osmotic pressure to obtain the molecular mass of a compound.

The expressions obtained for the osmotic pressure have so far been written in terms of mole fraction or molality. To study the osmotic pressure of a compound of unknown molecular mass, it is convenient to start with Eq. [80] and to write it

$$\Pi = \frac{(n_B)RT}{n_A \mathrm{V}}$$

$$= \frac{(\text{mass of } B)/(\text{mol mass of } B)RT}{\text{vol of soln}} \qquad [82]$$

Introduction of the concentration unit of grams per milliliter for the factor [mass of B)/(vol of soln)], which can be represented by c, allows Eq. [82] to be written

$$\frac{\Pi}{c} = \frac{RT}{\text{mol mass of solute}} \qquad [83]$$

If the osmotic pressure is given in atmospheres and c in grams per

TABLE 21-8 Osmotic-pressure results at 25°C for a polyisobutylene fraction*

	Π (atm)		Π/c (atm ml g^{-1})	
Conc. (g ml^{-1})	In benzene	In cyclohexane	In benzene	In cyclohexane
0.0200	0.00208	0.0117	0.104	0.585
0.0150	0.00152	0.0066	0.101	0.44
0.0100	0.00099	0.0030	0.099	0.30
0.0075		0.00173		0.23
0.0050	0.00049	0.00090	0.098	0.18
0.0025		0.00035		0.14

*Calculated from the data of P. J. Flory, *J. Am. Chem. Soc.*, **65**:372 (1943).

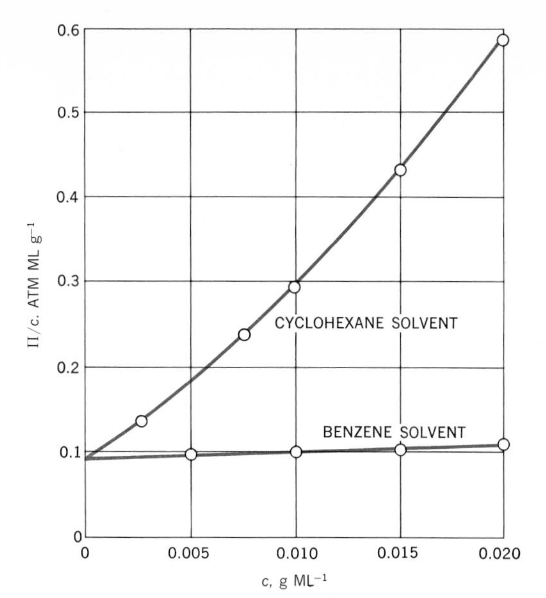

FIGURE 21-12 The extrapolation of the osmotic-pressure–concentration ratio to infinite dilution for a sample of polyisobutylene in cyclohexane and in benzene (data of Table 21-8).

milliliter, the gas constant R must be in milliliter atmospheres per degree.

Equation [83] can be expected to be valid only at infinite dilution. This follows, in addition to the approximations introduced in its derivation, from the fact that large-molecular-mass compounds tend to interact with one another at the concentrations at which the measurements are made. The procedure that must be used, therefore, is to measure Π/c as a function of c and to extrapolate these results to infinite dilution. The intercept of Π/c at zero concentration can then be taken as the value of RT/(mol mass of solute). From this value a valid molecular mass for the solute in solution is obtained.

The data of Table 21-8 show the osmotic pressures and concentrations of solutions of a polyisobutylene fraction in two different solvents. The extrapolations of these data in Fig. 21-12 lead to a value for Π/c of 0.097 atm ml g^{-1} at infinite dilution. With this value and Eq. [83], the molecular mass, or as will be pointed out in Chap. 25, an average molecular mass is calculated as

$$\text{Mol mass polyisobutylene} = \lim_{c\to 0} \frac{RT}{\Pi/c}$$

$$= \frac{(82.05)(298)}{0.097} = 250{,}000$$

It should be noticed that measurements of the osmotic pressure shown in Fig. 21-12 have been made down to molal concentrations of about 10^{-5}. At such concentration the boiling-point elevation and freezing-point depression would be much too small to be readily susceptible to measurement.

Problems

1 Benzene and toluene mix to form nearly ideal solutions. What is the absolute entropy and the standard heat and free energy of formation of 1 mol of a solution at 25°C and 1 atm containing 0.3 mol of benzene and 0.7 mol of toluene? The necessary thermodynamic data for the components are given in Appendix 5.

2 Are the expressions for the thermodynamic changes for ideal-solution formation given in Sec. 21-1 consistent with our ideas on entropy and free-energy effects in spontaneous reactions and with the fact that components that form ideal solutions are completely miscible?

3 What would be the values of Δv_{mix}, ΔH_{mix}, ΔS_{mix}, and ΔG_{mix} for the formation of a solution by the mixing of half a mole of an equimolar two-component ideal solution with half a mole of one of its components?

4 Show that ideal behavior, as described by obedience to Raoult's law, converts the general result for ΔG_{mix} of Sec. 21-2 to that deduced for ideal solutions in Sec. 21-1.

5 The following data for the heat of mixing of carbon tetrachloride and acetonitrile at 45°C have been reported by I. Brown and W. Foch [*Aust. J. Chem.*, **9**:180 (1956)]:

x_{CCl_4}	0.128	0.317	0.407	0.419	0.631	0.821
ΔH_{mix} (J mol^{-1})	414	745	862	858	930	736

From these data and those of Prob. 24 of Chap. 20, prepare a figure showing ΔH_{mix}, ΔG_{mix}, and $T\,\Delta S_{mix}$ for the system CCl_4-CH_3CN at 45°C.

6 Calculate the mole fraction, the activity, and the activity coefficient for water for solutions at 100°C containing:
a 11.8 g NaCl and 100 g water, for which the vapor pressure is 0.931 atm.
b 35.4 g NaCl and 100 g water, for which the vapor pressure is 0.770 atm.
Ans. $\gamma = 0.851$.

7 Deduce the activities and activity coefficients for CCl_4 and for CH_3CN dissolved in one another at 45°C from the data of Prob. 24 of Chap. 20. Use the pure material as the standard state for each component. Plot on a single diagram the activities of each component against the composition of the solution. Add dashed lines to this diagram to show the results that would have been obtained if the solution had behaved ideally.

8 From the standard free energy of O_2 in water obtained in Sec. 21-5, deduce the pressure of oxygen gas in equilibrium with the solution that would be necessary to produce this standard solute state. Assume ideal-gas behavior.

9 Use the activity data for CH_3OH given in Table 21-2 for the system CH_3OH-CCl_4 to check, by the Gibbs-Duhem equation in the form given by Eq. [45], the activity difference for CCl_4 between $x_{CCl_4} = 0.51$ and $x_{CCl_4} = 0.34$.

10 Use the results obtained in Prob. 7 for the activities of CCl_4 and the Gibbs-Duhem equation to deduce the activities of acetonitrile in these solutions. Compare with the results obtained by the use of the acetonitrile vapor pressures.

11 According to Eq. [45], there is a relation between changes in the logarithms of the activities of the two components of a binary solution. Plot for the system CCl_4-CH_3CN, for which the activity data have been calculated in Prob. 7, $\log a_{CCl_4}$ and $\log a_{CH_3CN}$ versus the solution composition, and see if the relationship of Eq. [45] can be recognized in the figure.

12 What is the vapor pressure at 100°C of a solution containing 15.6 g of water and 1.68 g of sucrose, $C_{12}H_{22}O_{11}$? *Ans.* 0.9943 atm.

13 The boiling point of benzene is raised from its normal value of 80.1 to 82.4°C by the addition of 13.76 g of biphenyl, $C_6H_5C_6H_5$, to 100 g of benzene. What are the boiling-point-elevation constant and the heat of vaporization of benzene according to these data? *Ans.* $\Delta H_{vap} = 31$ kJ mol^{-1}.

14 Equation [67] involves the quantity n_A, the moles of solvent in 1000 g of solvent. Explain why, even if the heat to vaporize a given amount of solvent were measured, Eq. [67] could not be used to obtain the molecular mass of the solvent if a measurement of ΔT_{bp} were made with a solute of known molecular mass.

15 A newly synthesized organic compound is analyzed for carbon and hydrogen and found to contain 63.2 percent, by weight, carbon, 8.8 percent hydrogen, and the remainder oxygen. A solution of 0.0702 g of the compound in 0.804 g of camphor is found to freeze 15.3°C lower than the freezing point of the pure camphor. What are the molecular mass and the formula of the new compound? *Ans.* $C_{12}H_{20}O_4$.

16 Henry law constants k_M^{-1} of the relation $M = k_M^{-1}P$ for N_2 and O_2 in water at 0°C are 0.00103 and 0.0022 mol liter^{-1} atm^{-1}, respectively. What will be the difference in freezing point between pure water and water in equilibrium, i.e., saturated with air?

17 The osmotic pressure is measured between water and a solution containing 1 g of glucose, $C_6H_{12}O_6$, and 1 g of sucrose, $C_{12}H_{22}O_{11}$, in 1000 g of water. The temperature is maintained at 25°C.

a What osmotic pressure would be expected?

b If this pressure were measured and it were not known that the solute was a mixture, what molecular mass would have been calculated? *Ans.* 236.

c The measurement of the osmotic pressure gives an average molecular mass of the solute. What kind of average, i.e., number or mass, is this?

18 Verify some of the calculated values in Table 21-7.

19 It is reasonable to expect that a number of water molecules, say five, are bound quite strongly to a sucrose molecule by hydrogen bonding. This would have the effect of decreasing the number of solvent particles relative to the number of solute particles. Calculate what value of the osmotic pressure of the 1-m sucrose solution would be expected with this hydration assumption and Eq. [80]. Compare with the observed value of Table 21-7.

20 A solution of 2.58 g phenol in 100 g bromoform freezes at a temperature 2.374° lower than does pure bromoform.

a What is the apparent molecular mass of phenol at this concentration and temperature in benzene? *Ans.* 157.

b Give a qualitative explanation for this molecular mass.

21 The osmotic pressure of a dilute solution of KNO_3 in water is 0.470 atm when measured against water at 25°C. What would be the vapor pressure at 25°C (the vapor pressure of pure water is 0.03126 atm at this temperature), the freezing point, and the boiling point of the solution?

22 Obtain an expression for the freezing point as a function of pressure, and apply this expression to find the difference in the freezing point of ice at 1 atm pressure and that of ice under its own vapor pressure. The vapor pressure of ice at 0°C is 6.025×10^{-3} atm. The densities of water and ice at 0°C are 1.000 and 0.9168 g ml^{-1}, respectively, and the heat of fusion of ice is 6.01 kJ mol^{-1}.

23 Combine the results of Probs. 16 and 22 to obtain the difference in freezing point of water saturated with air at 1 atm from that of pure water under its own vapor pressure.

References

In addition to the following references, most of the thermodynamics texts referred to in previous chapters include discussions of the treatment of solutions.

HILDEBRAND, J. H., and D. L. SCOTT: "Regular Solutions," Prentice-Hall, Inc., Englewood Cliffs, N.J., 1962. A monograph summarizing Hildebrand's ideas on "regular solutions" that makes use of thermodynamic approaches, and ideas on intermolecular forces and on the liquid state. Excellent material, showing a use and purpose for much of the material that has been presented in this text.

PRIGOGINE, I.: "The Molecular Theory of Solutions," Interscience Publishers, Inc., New York, 1957. The development of a theory of the nature of solutions that builds on the thermodynamic properties of ideal and nonideal systems.

RAWLINSON, J. S.: "Liquids and Liquid Mixtures," Butterworth Scientific Publications, London, 1959. A treatment of many aspects of liquids that brings together the properties and approaches to pure liquids and to liquid solutions. Chapters 4 and 5 contain material on mixtures of simple and complex liquids that supplements the thermodynamic introduction to such systems given in this text.

COVINGTON, A. K., and P. JONES (eds.): "Hydrogen-bonded Solvent Systems," Tylor and Francis, 1968. A collection of symposium articles that show the variety of subjects that can be studied in connection with solute-solvent interactions and reactions.

22

THE NATURE OF ELECTROLYTES IN SOLUTION

In our study of physical chemistry so far, we have learned much about the details of atomic and molecular systems and have used this information to understand better a number of chemical and physical properties. It is a fact, however, that many interesting phenomena such as most that occur in biological systems involve electrolytes in aqueous solutions. Our knowledge of the behavior of free, independent molecules is not a complete basis with which to understand systems in which charged particles are immersed in a medium that is far from inert.

We will begin, in the first part of this chapter, by seeing the experimental basis that led to the now familiar idea that electrolytes—acids, bases, and salts—are more or less dissociated in aqueous solution and that the ions that are present act, to some extent, as free, independent particles. This interpretation of aqueous solutions of electrolytes is akin to the simple kinetic-molecular theory of ideal gases.

Many refinements of this simple theory are necessary if more experimental results are considered and if the finer details of the results on which the simple theory are based are recognized. Thus, in a way that now parallels the van der Waals elaborations of the simple kinetic-molecular theory, two additional aspects of the behavior of ions in solution are introduced. These are concerned with the interaction of the ions with the solvent and the interactions of the ions with one another. The second part of the chapter will be devoted to these refinements of the simple ionic model.

22-1 Electrical Conductivity of Solutions

The fact that aqueous solutions of certain materials, called *electrolytes,* conduct an electric current provides the most direct evidence for the idea that ions capable of independent motion are present. More detailed studies of the electrical conductivity of such solutions provide information on the number and independence of these ions.

Measurements of the conductivity of aqueous solutions are made with a conductivity cell and an electric circuit such as that shown in Fig. 22-1.

When an alternating current is used to prevent buildup of charges of opposite sign near the two electrode surfaces so that there is little metal-solution electric resistance, it is found that the conductivity cell obeys Ohm's law: the current flowing through the cell is proportional to the voltage across the cell. It is possible, therefore, to assign a resistance of so many ohms to such a cell in the same manner as one assigns a resistance to a metallic conductor.

It is more convenient, however, to focus our attention on the *conductance* of an electrolytic solution rather than on its *resistance*. These quantities are reciprocally related, and the conductance, denoted by L, is calculated from the measured resistance as

$$L = \frac{1}{R} \qquad [1]$$

where R is the resistance in ohms, and L has the units, therefore, of

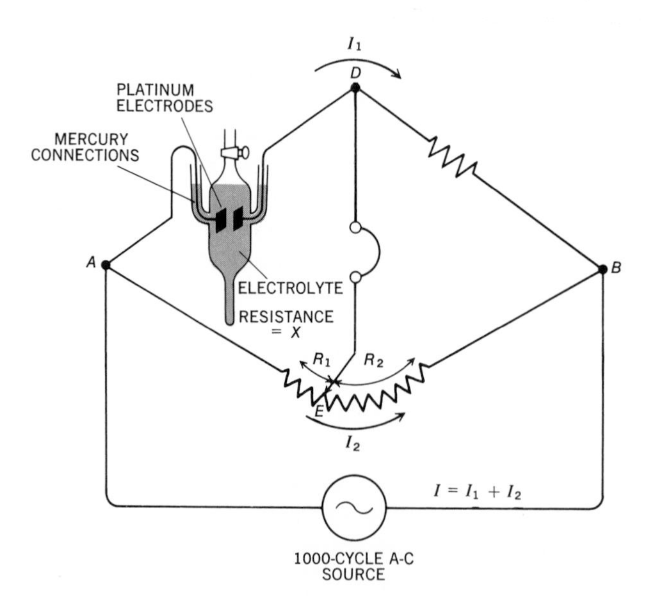

FIGURE 22-1
Schematic diagram of a conductivity cell in a Wheatstone bridge circuit.

TABLE 22-1 Specific conductance of KCl solutions*
Such conductances have generally been reported in the cgs unit of $ohm^{-1}\ cm^{-1}$. Values with these units are obtained by dividing the tabulated values by 100.

Concentration (equiv $liter^{-1}$)	κ ($ohm^{-1}\ m^{-1}$)		
	0°C	18°C	25°C
1	6.543	9.820	11.173
0.1	0.7154	1.1192	1.2886
0.01	0.07751	0.12227	0.14114

* From G. Kortum and J. O'M. Bockris, "Textbook of Electrochemistry," vol. 1, Elsevier Press, Inc., Amsterdam, 1951.

the reciprocal ohm, sometimes called *mho*. As for metallic conductors, the resistance, and therefore the conductance, depend on the cross-section area A and the length l of the conductivity cell, i.e., of the region between the electrodes. Just as for a metallic conductor one has

$$R = \rho \frac{l}{A} \qquad [2]$$

where ρ is the *specific resistance* and is the proportionality factor that corresponds to the resistance of a cell of unit cross-section area and unit length. Thus

$$L = \kappa \frac{A}{l} \qquad [3]$$

where κ, the *specific conductance,* can be thought of as the conductance of a cube of the solution of electrolyte of unit dimensions. (With cgs units the imagined cell is a 1-cm cube, and it is for such units that data have generally been given.)

The specific conductance can, in principle, be obtained from the measured value of R, which gives $L = 1/R$, and of l and A of the cell. In practice, it is more convenient to deduce l and A, or rather the *cell constant* l/A, from a measurement of L when the cell is filled with a solution of known specific conductance. With such a measurement and datum one can use the equation

$$\kappa = \frac{l}{A} L = (\text{cell const})L \qquad [4]$$

to calculate a value for the cell constant. Once this geometric factor has been obtained for a cell, it can be used to deduce κ for an unknown solution from a measured value of L and Eq. [4].

The cell constant is almost always determined by using a solution of KCl. Specific conductances of these reference solutions, shown in Table 22-1, have been determined by measurements with rather elaborately designed electrodes which avoid the uncertainty in the effective current-carrying cross section that exists in ordinary cells. Since the temperature dependence shown in Table 22-1 is characteristic of all conductance results, it is necessary to make measurements of conductances in well-thermostated cells.

Although the specific conductance is a measure of the ease with which a current flows through a unit cube of solution, it is not a convenient quantity for the discussion of the conduction process of solutions of electrolytes. Solutions of different concentrations, for example, will have very different specific conductances simply because a given volume of the different solutions will contain different amounts of electrolyte. Since it will be of particular interest to compare the current-carrying ability of a given number of electrolyte charges at different concentrations, it is helpful to define yet another measure of conductance.

22-2 Equivalent Conductance

An equivalent mass of an electrolyte is that amount which, for complete dissociation, would lead to ions with total positive and negative charges of $+e\mathfrak{N}$ and $-e\mathfrak{N}$, where e is the electronic charge. Thus an equivalent of NaCl gives an Avogadro's number of Na^+ ions and of Cl^- ions; an equivalent of $MgSO_4$ gives half an Avogadro's number of Mg^{++} ions and of SO_4^{--} ions; and so forth. One equivalent of any electrolyte would, on complete dissociation, provide the same effective number of charge-carrying particles. The conductance that would be conveniently dealt with, therefore, is that which would be given by a conductance cell with electrodes a unit distance apart and of large enough cross sections so that the volume of solution containing one equivalent of the electrolyte would be held between the electrodes. This conductance, known as the *equivalent conductance* and designated by Λ, would measure the current-carrying ability of an equivalent of solute.

The volume of solution of concentration c equiv liter^{-1}, i.e., c equiv in 10^{-3} m^3, that holds 1 equiv is $(1/c)$ liters, or $(10^{-3}/c)$ m^3. A cell with this volume and plates separated by 1 m would be equivalent to $(10^{-3}/c)$ unit cells placed alongside one another, and thus the conductance, which is the equivalent conductance, is given by

$$\Lambda = \frac{10^{-3}}{c}\kappa \qquad [5]$$

This relation defines the equivalent conductance in terms of the specific conductance. The concept of the cell holding solution of volume $(10^{-3}/c)$ m^3 is introduced only to suggest the definition of Eq. [5], and one should recognize that, in practice, one uses any convenient conductance cell, measures R, and calculates $L = 1/R$. With this datum one obtains κ = (cell const)L, and finally Λ from Eq. [5].

Many precise measurements of equivalent conductances were

TABLE 22-2 Equivalent conductances Λ, in units of $ohm^{-1}\ m^2$, in aqueous solution at 25°C* Values for $c = 0$ are obtained by extrapolation or, for HAc and NH_4OH, by a combination of extrapolated values.

c	NaCl	KCl	HCl	NaAc	$CuSO_4$	H_2SO_4	HAc	NH_4OH
0.000	0.012645	0.014986	0.042616	0.00910	0.0133	0.04296	0.03907	0.02714
0.0005	0.012450	0.014781	0.042274	0.00892		0.04131	0.00677	0.0047
0.001	0.012374	0.014695	0.042136	0.00885	0.01152	0.03995	0.00492	0.0034
0.010	0.011851	0.014127	0.041200	0.008376	0.00833	0.03364	0.00163	0.00113
0.100	0.010674	0.012896	0.039132	0.007280	0.00505	0.02508		0.00036
1.000		0.01119	0.03328	0.00491	0.00293			

* Data mostly from D. A. MacInnes, "The Principles of Electrochemistry," Reinhold Publishing Corporation, New York, 1939.

made by Kohlrausch and his coworkers between about 1860 and 1880. The data of Table 22-2 are typical of their results.

On the basis of such data and in the absence of any satisfactory theory as to the nature of conduction in these solutions, some valuable empirical relations were deduced. It was recognized that plotting the equivalent conductance of an electrolyte at a fixed temperature against the square root of the concentration led, for some electrolytes, to plots which, at the lower concentrations, conformed very closely to straight lines. Such plots for a few electrolytes are shown in Fig. 22-2. It is clear from this figure that two different types of behavior are exhibited.

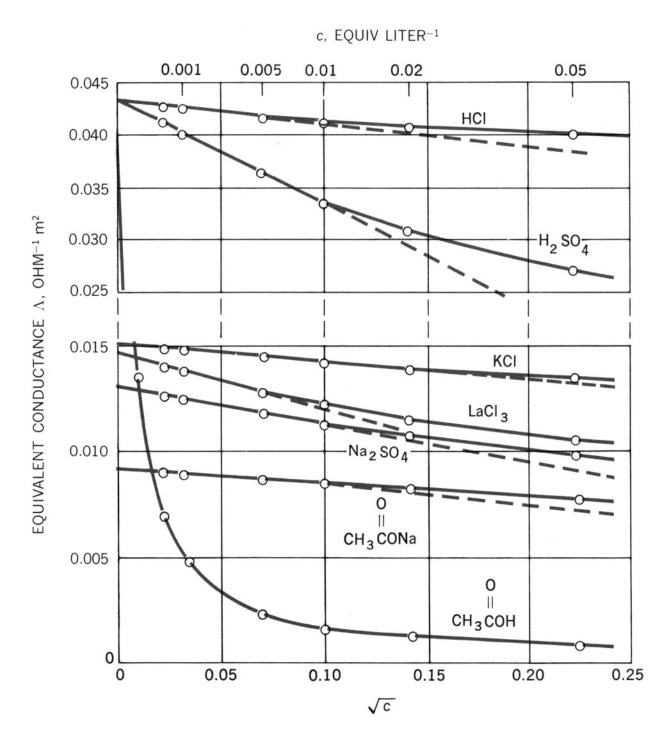

FIGURE 22-2
The equivalent conductance versus $\sqrt{c}$ for some electrolytes in water at 25°C.

Those electrolytes which lead to essentially linear plots are now classed as *strong electrolytes,* and those which seem to approach the dilute-solution limit almost tangentially are classed as *weak electrolytes.*

An important relation can be deduced from the extrapolations of the strong-electrolyte data to infinite dilution to give what are known as *limiting equivalent conductances.* These conductances, denoted by Λ_0, are the basis for *Kohlrausch's law of the independent migration of ions.* The law is more easily stated and understandable if some later ideas are anticipated and the conductance of an electrolyte at infinite dilution is treated as being made up of contributions from the individual ions of the electrolyte. In this way one introduces *equivalent ionic conductances* and writes for the limit of infinite dilution

$$\Lambda_0 = \lambda_0{}^+ + \lambda_0{}^- \qquad [6]$$

The law of Kohlrausch now suggests that at infinite dilution the conductance of an electrolyte, NaCl for example, depends on independent contributions from Na^+ and from Cl^-. The independence of these contributions is judged by a comparison of pairs of electrolytes containing a common ion, as shown in Table 22-3. The differences show the value of $\lambda_0^{K^+} - \lambda_0^{Li^+}$ and $\lambda_0^{Cl^-} - \lambda_0^{NO_3^-}$, and these differences are seen to be independent of the other ionic species present. Kohlrausch's law, it should be noted, gives no way of deducing the contributions of the individual ions.

The immediate practical application of the idea of the independent contribution of the ions at infinite dilution is a method for deducing the limiting equivalent conductance of weak electrolytes. For acetic acid, for example, one can write

$$\Lambda_0(\text{HAc}) = \Lambda_0(\text{NaAc}) + \Lambda_0(\text{HCl}) - \Lambda_0(\text{NaCl})$$

because the right side can be interpreted as

$$\lambda_0^{Na^+} + \lambda_0^{Ac^-} + \lambda_0^{H^+} + \lambda_0^{Cl^-} - \lambda_0^{Na^+} - \lambda_0^{Cl^-} = \lambda_0^{H^+} + \lambda_0^{Ac^-} = \Lambda_0(\text{HAc})$$

TABLE 22-3 The basis for Kohlrausch's law of independent migration of ions* (data for 25°C)
Conductances in units of $ohm^{-1}\ m^2$

	Λ_0	$\Delta(=\lambda_0^{K^+} - \lambda_0^{Li^+})$		Λ_0	$\Delta(=\lambda_0^{Cl^-} - \lambda_0^{NO_3^-})$
KCl	0.014986	34.83×10^{-4}	LiCl	0.011503	4.9×10^{-4}
LiCl	0.011503		$LiNO_3$	0.01101	
$KClO_4$	0.014004	35.06	KCl	0.014986	4.90
$LiClO_4$	0.010598		KNO_3	0.014496	
			HCl	0.042616	4.9
			HNO_3	0.04213	

* Data from H. S. Harned and B. B. Owen, "The Physical Chemistry of Electrolytic Solutions," Reinhold Publishing Corporation, New York, 1950.

In this way, the value at 25°C of

$$\Lambda_0(\mathrm{HAc}) = 0.00910 + 0.04262 - 0.01265 = 0.03907 \qquad [7]$$

is found, and it is clear from Fig. 22-2 that no reliable value could have been determined by a graphical extrapolation.

The availability of limiting equivalent conductances for all types of electrolytes was a prerequisite for the development of the theory for the nature of electrolytes in solution.

22-3 The Arrhenius Theory of Dissociations

Prior to the development of the important theory of Arrhenius, about 1887, a number of suggestions had been made to explain the fact that solutions of electrolytes were conductors of electricity. It is difficult for the modern student, who is brought up with the idea that salts and strong acids and bases are completely ionized in water, to appreciate the early difficulties in recognizing that such solutions of ions could exist. In the nineteenth century chemists were very much impressed with the difficulty of breaking apart stable molecules, and they could not accept the idea that a molecule, like HCl, could be dissociated except at a very high temperature. It must be remembered also that at this time solid salts had not been represented as an array of ions in the crystal lattice.

It was Arrhenius who made the then bold postulate that the dissolving of an electrolyte in an aqueous solution could lead to electrolytic dissociation and, even at ordinary temperatures, the conversion of an appreciable fraction of the electrolyte to free ions. Support for the theory was the explanation offered by Arrhenius for the observed variation of conductance with concentration, as illustrated by the data of Fig. 22-2 and Table 22-2. The increase of equivalent conductance with decreasing concentration, observed in dilute solutions of all electrolytes, was attributed by him to the partial dissociation of the electrolyte. A chemical equilibrium between undissociated electrolyte molecules and the ions that result from dissociation always leads to an increase of the degree of dissociation for more dilute solutions. Since the conductance depends on the presence of charged species, a qualitative explanation for the concentration variation of the equivalent conductance was immediately reached.

These ideas lead to a method for the calculation of the degree of dissociation of an electrolyte from the conductivity data. The supposition of a dissociation equilibrium implies that at infinite dilution all electrolytes are completely dissociated. (This general result depends on the fact that the products contain more particles than do the reactants. Equations [16] and [18], to be obtained in Sec. 22-5, illustrate that, as $c \to 0$, the degree of dissociation α must approach unity.) A

TABLE 22-4 Degrees of dissociation for a strong and a weak electrolyte calculated according to the Arrhenius theory (at temperature 25°C)*

c (equiv liter^{-1})	Λ	$\alpha = \frac{\Lambda}{\Lambda_0}$	$K_\Lambda = \frac{c\alpha^2}{1-\alpha}$	i	$\alpha = \frac{i-1}{2-1}$	$K_i = \frac{c\alpha^2}{1-\alpha}$
HCl						
0	(0.042616)	1.00				
0.001	0.042136	0.99	0.10	1.98	0.98	0.05
0.005	0.041580	0.98	0.24	1.95	0.95	0.09
0.01	0.041200	0.97	0.31	1.94	0.94	0.15
0.05	0.039909	0.94	0.73	1.90	0.90	0.40
0.1	0.039132	0.92	1.1	1.89	0.89	0.72
0.5	0.03592	0.84	2.2			
1.0	0.03328	0.78	2.8	2.12	1.12	−10.4
ACETIC ACID						
0	0.03901	1.000				
0.0000280	0.02103	0.541	1.79×10^{-5}			
0.0001113	0.01277	0.327	1.77			
0.001				1.12	0.12	1.6×10^{-5}
0.001028	0.00481	0.123	1.78			
0.00984	0.001637	0.0419	1.80			
0.0200	0.001156	0.0296	1.80			
0.030				1.07	0.07	1.6
0.0500	0.000736	0.0189	1.81			
0.1000	0.000520	0.0133	1.80	1.04	0.04	1.7
0.2000	0.000365	0.0094	1.78			

*Acetic acid conductivity data from D. A. MacInnes and T. Shedlovsky, *J. Am. Chem. Soc.*, **54**:1429 (1932), and freezing-point-depression data from H. S. Harned and R. W. Ehlers *J. Am. Chem. Soc.*, **55**:652 (1933).

comparison of the equivalent conductance at some finite concentration with that at infinite dilution therefore gives a measure of the fraction of electrolyte dissociated at the higher concentration. One introduces α, the *degree of dissociation,* and writes

$$\alpha = \frac{\Lambda}{\Lambda_0} \qquad [8]$$

In this way results for the degree of dissociation, such as are shown in Table 22-4 for HCl and acetic acid, can be calculated. Arrhenius treated strong and weak electrolytes in essentially the same way, the apparent different behavior revealed in Fig. 22-2 being interpreted merely as a difference in degree of dissociation. It should, however, be mentioned that later ideas tend to substantiate the previous suggestion of two essentially different behaviors being exhibited in Fig. 22-2.

The Arrhenius theory and the deduced degrees of dissociation were received with considerable skepticism, and it was still generally held to be unlikely that the mere solution of an electrolyte could break up the molecules into separate ions. At this stage the measurements of colligative properties of solutions of electrolytes by van't Hoff became available. The interpretation of these results by the electrolytic-dissociation theory of Arrhenius swung support almost completely over to this new theory.

22-4 Colligative Properties of Aqueous Solutions of Electrolytes

In the preceding chapter four properties of dilute solutions were interpreted on the basis of ideal behavior, and the properties, known as the colligative properties, were then seen to be dependent only on the molality of the solution. As the derivations in the preceding chapter showed, thermodynamic analyses of the equilibria involved led directly to information on the dependence of the partial molal free energy of the solvent on solute addition. The assumption of ideal-solution behavior allowed this thermodynamic result to be used to "count" solute particles, as is done when a value of the molality is given.

If the same approach is used with electrolytes, the measured colligative properties can be used to obtain the molality that would be demanded by the ideal-solution model. It was recognized by van't Hoff that this apparent molality could be conveniently represented by im, where m is the molality that would be expected on the basis of the formula of the electrolyte, and i is a factor known as the van't Hoff i factor. In this way, if K_{bp}, K_{fp}, and Π_0 are the values expected for the boiling-point elevation, the freezing-point depression, and the osmotic pressure for molal solutions of nonelectrolytes, we write

$$\begin{aligned} T_{\text{bp}} &= (im)K_{\text{bp}} \\ T_{\text{fp}} &= -(im)K_{\text{fp}} \\ \Pi &= i\Pi_0 \end{aligned} \qquad [9]$$

The van't Hoff i factor shows explicitly the enhancement of the colligative properties encountered with electrolytes. Table 22-6 shows the i factors corresponding to the freezing-point depressions of Table 22-5. These van't Hoff i factors are clearly in qualitative accord with the view that electrolytes are more or less dissociated in solution. Since, for noninteracting solute particles, the colligative measurements give the number of particles in solution, it is apparent that the strong electrolytes behave as though there are about two, three, or four times as many particles as might have been expected. The numbers, furthermore, correspond to the number of ions that would be expected to result from the dissociation of the electrolyte molecule. The colligative-property results provided by van't Hoff gave a virtual proof to the Arrhenius idea of appreciable dissociation of electrolytes in aqueous solution.

TABLE 22-5 Observed freezing-point-depression terms, $iK_{fp} = -\Delta T_{fp}/m$, for electrolytes in water
For nonelectrolytes $K_{fp} = -\Delta T_{fp}/m = 1.86$.

m	NaCl	HCl	$CuSO_4$	$MgSO_4$	H_2SO_4	$Pb(NO_3)_2$	$K_3Fe(CN)_6$
0.001	3.66	3.690		3.38		5.368	7.10
0.01	3.604	3.601	2.703	2.85	4.584	4.898	6.26
0.1	3.478	3.523	2.08	2.252	3.940	3.955	5.30
1.0	3.37	3.94	1.72	2.02	4.04	2.435	

TABLE 22-6 Van't Hoff i factors calculated from the data of Table 22-5 and the freezing-point-depression constant K_{fp} of 1.86 for water

m	NaCl	HCl	$CuSO_4$	$MgSO_4$	H_2SO_4	$Pb(NO_3)_2$	$K_3Fe(CN)_6$
0.001	1.97	1.98		1.82		2.89	3.82
0.01	1.94	1.94	1.45	1.53	2.46	2.63	3.36
0.1	1.87	1.89	1.12	1.21	2.12	2.13	2.85
1.0	1.81	2.12	0.93	1.09	2.17	1.31	

It is necessary, however, to look more closely at the results and to see whether or not the Arrhenius theory can provide an explanation for the fact that the i factors tend to be less than the integer expected for complete dissociation. A quantitative explanation can be attempted in terms of incomplete dissociation, and the degree of dissociation that will be computed from van't Hoff i factors will be compared with the corresponding results from conductivity data.

Consider the general electrolyte A_aB_b, which might undergo complete dissociation to form a positive ions and b negative ions according to the equation

$$A_aB_b \rightleftharpoons aA^{(+)} + bB^{(-)}$$

It is necessary to calculate the net number of particles that result from a degree of dissociation α. If the molality of the electrolyte is m and the degree of dissociation is α, the concentration of undissociated electrolyte will be $m - \alpha m = m(1 - \alpha)$. In addition, the concentration of $A^{(+)}$ and $B^{(-)}$ will be $a\alpha m$ and $b\alpha m$, respectively. The concentration of particles, regardless of their kind, is therefore

$$m(1 - \alpha) + a\alpha m + b\alpha m \qquad [10]$$

It is customary to let ν be the total number of ions yielded by complete dissociation of a molecule; i.e.,

$$\nu = a + b \qquad [11]$$

With this notation the molality of particles for the partially dissociated electrolyte is

$$m(1 - \alpha) + \alpha\nu m \qquad [12]$$

rather than the value of m expected for no dissociation.

The definition of the van't Hoff i factor shows that it is to be identified with the ratio

$$i = \frac{m(1 - \alpha) + \alpha\nu m}{m}$$

$$= 1 - \alpha + \alpha\nu \qquad [13]$$

From this interpretation of i, one obtains

$$\alpha = \frac{i - 1}{\nu - 1} \qquad [14]$$

This important relation provides an alternative way to that given by Eq. [8] for calculating the degree of dissociation of an electrolyte. The results of such calculations for HCl and acetic acid are included in Table 22-4.

For rather dilute solutions the two methods of calculating α lead to fair agreement, and in the early stages of the theory such agreement could be accepted as support for the interpretation of the conductivity and the i-factor data in terms of incomplete dissociation. The modern view, it should be mentioned, does not accept the idea of only partial dissociation of strong electrolytes.

A further test of the Arrhenius theory is possible with the data of Table 22-6. If the equilibria postulated by Arrhenius are, in fact, set up and if this is the only feature of ionic solutions that need be considered, it should be possible to calculate a good equilibrium constant from the deduced degrees of dissociation.

22-5 Dissociation Equilibria

No convenient general expression can be set up for the equilibrium constant of the dissociation reaction in terms of the degree of dissociation for all types of electrolytes. For a particular type of electrolyte such a relation can be easily deduced, as two examples will show.

The treatment of the free energies of solutes in Sec. 21-5 led to an expression for the partial molal free energy of a solute with the same form, but involving the solute activity, as that for the molar free energy, of an ideal gas which involved the pressure of the gas. Since this latter equation led, by the derivation of Sec. 9-8, to the familiar equilibrium-constant expression, it follows that a similar equation can be written involving activities when an equilibrium is set up involving solute species. Here it will be instructive, and in keeping with the assumptions of Arrhenius theory, to assume ideal-solute behavior. We then can replace, in view of the discussion of Sec. 21-5, the activities of the solute species by their molar concentrations.

For a single one-to-one electrolyte AB with a degree of dissociation α at a concentration of c mol liter^{-1}, the concentration of the dissociated species will be $c\alpha$ and that of the remaining undissociated electrolyte will be $c - c\alpha$. Thus one can write the dissociation reaction and the equilibrium concentrations as

$$\begin{array}{lccc} & AB & \rightleftharpoons A^+ & + \; B^- \\ \text{Equil. conc.} & c(1-\alpha) & c\alpha & c\alpha \end{array}$$

The equilibrium-constant expression

$$K = \frac{[A^+][B^-]}{[AB]} \tag{15}$$

can therefore be written

$$K = \frac{(c\alpha)(c\alpha)}{c(1-\alpha)} = \frac{c\alpha^2}{1-\alpha} \tag{16}$$

With this expression, a value of K can be calculated for an electrolyte of the type AB for any concentration for which a value of α is available. Results for such calculations are shown for HCl and acetic acid in Table 22-4.

For an electrolyte of the type A_2B one writes

$$\begin{array}{lccc} & A_2B & \rightleftharpoons 2A^+ & + \; B^= \\ \text{Equil. conc.} & c(1-\alpha) & 2c\alpha & c\alpha \end{array}$$

The equilibrium-constant expression is

$$K = \frac{[A^+]^2[B^=]}{[A_2B]} \tag{17}$$

and substitution of the equilibrium concentrations gives

$$K = \frac{(2c\alpha)^2(c\alpha)}{c(1-\alpha)} = \frac{4c^2\alpha^3}{1-\alpha} \tag{18}$$

Similar relations between K and α can be worked out for any other type of electrolyte.

The calculated equilibrium constants are confusing if one seeks evidence for or against the Arrhenius theory. Those electrolytes previously classified as weak electrolytes give quite constant values for K, whereas strong electrolytes give the unsatisfactory results of which those for HCl are representative.

The modifications to the theory of dissociations that are necessary to bring the calculated results for strong electrolytes reasonably in line with experiment will be postponed until the second part of this chapter.

This is consistent with the historical fact that such unacceptable conclusions from the Arrhenius theory were known but were wisely, if unscientifically, set aside while the acceptable features of the theory were used and developed.

22-6 Electrolysis and the Electrode Process

Much additional information on the properties of the ions in an aqueous solution can be obtained from studies of the passage of a direct current through a cell containing a solution of an electrolyte. Such dc experiments involve chemical reactions at the electrodes, a feature that is avoided in conductivity studies by the use of an alternating current. It is first necessary, therefore, to describe and classify these electrode processes.

When electrodes are inserted in a solution of electrolyte and a sufficient potential, of the order of several volts, is applied, chemical reactions are observed at the electrodes. *Electrolysis* is said to be occurring. The electrode that is charged positively, i.e., that has a deficit of electrons, by the applied potential is called the *anode,* and that charged negatively, i.e., that has an excess of electrons, is called the *cathode.* The electrodes consist of conductors that introduce the source and sink of electrons into the solution. In classifying the reactions that occur as a result of the charged electrodes, it is convenient to distinguish *inert electrodes,* usually a platinum wire, that serve only to transfer electrons to and from the solution, from *reacting electrodes* that enter chemically into the electrode reaction. Most simply, the reacting electrode is a metal that either contributes metal ions to the solution or accepts discharged metal ions from the solution.

The two major categories of electrode reactions that occur in an electrolysis cell are shown in Fig. 22-3. Of course, cells can be constructed that involve various combinations of these reaction types, consistent with the requirement that at the *cathode,* electrons are introduced by the external circuit and *reduction* occurs, whereas at the *anode,* electrons are removed and *oxidation* occurs.

More complicated electrode reactions do occur, but the features of electrolytic solutions that are to be studied in this chapter can be dealt with in terms of electrode reactions of these major types.

Electrolyses of the type illustrated here were extensively and quantitatively studied as early as 1820 by Faraday. He was led to the important conclusion that now can be stated as: *one equivalent of product is produced by the passage of* 96,490 *C of charge.* It can be recognized that 96,490 *C is the charge of an Avogadro's number of electrons.* Faraday's result is understandable because an Avogadro's number of electrons added to or removed from a reagent will produce an equivalent of product.

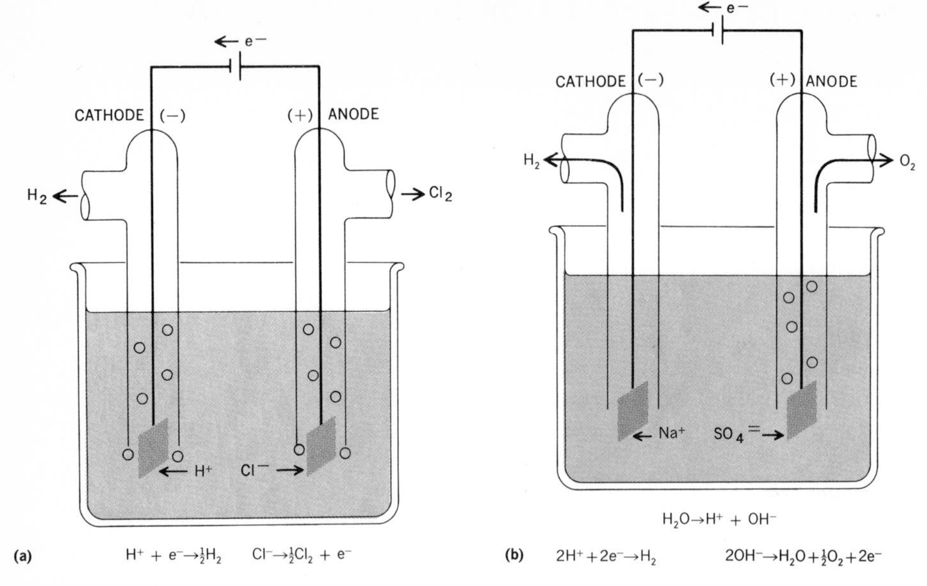

FIGURE 22-3
Types of electrolysis electrode reactions. (*a*) Current-carrying ions are discharged at the electrodes. (*b*) Difficultly discharged cations and anions permit the decomposition of water and the consumption of H^+ and the production of OH^-.

The quantity of charge that corresponds to a chemical equivalent is of enough importance to merit a name, and the unit of a *faraday* is introduced as

$$1\ \mathfrak{F} = 96{,}490\ \text{C}$$

22-7 Transference Numbers

Now that the general features of electrode processes have been mentioned, the details of the passage of the electric current through the body of the solution can be investigated. The flow of either the positive or the negative ions, or both, might be responsible for conduction processes, and the first goal is the determination of the fraction of the current carried by each ion in a given electrolyte. For this purpose the *transference numbers* t_+ and t_- are introduced according to the definitions

$$t_+ = \text{fraction of current carried by cation}$$
$$t_- = \text{fraction of current carried by anion}$$

such that

$$t_+ + t_- = 1 \qquad [19]$$

In metal conductors all the current is carried by the electrons, and for such conductors one could write $t_- = 1$ and $t_+ = 0$. For solutions of electrolytes it is clearly difficult to guess what fraction of the current is carried past some position in the electrolyte by the cations and what fraction by the anions. One method, known as the *Hittorf method,*

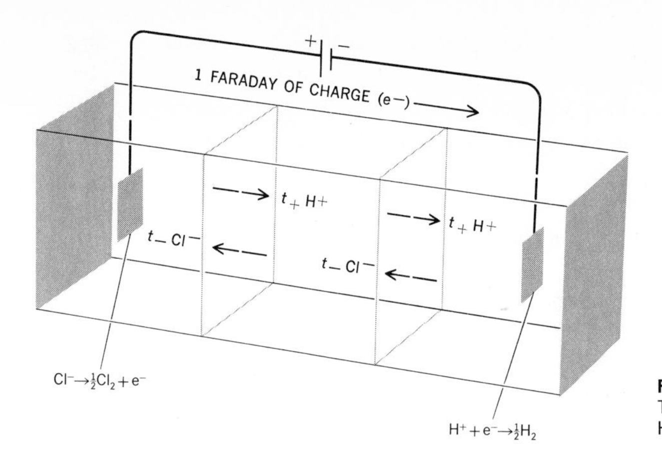

FIGURE 22-4
Transference numbers for HCl by the Hittorf method.

for measuring transference numbers will now be illustrated by two examples.

A schematic diagram of a cell marked off into three compartments is shown in Fig. 22-4. In practice, a cell of the type shown in Fig. 22-5 can be used, and the three compartments that can be drained off correspond to those marked off by the cross-section lines of Fig. 22-4. The following treatment will show that transference numbers can be deduced from the analysis for the amount of electrolyte in the separate compartments following passage of a measured amount of current through the cell.

Consider an experiment in which a cell such as that of Fig. 22-4 or 22-5 is filled with an HCl solution and 1 $\mathfrak{F}$ of charge is passed through the cell. The electrode processes, indicated in Fig. 22-4, were given in

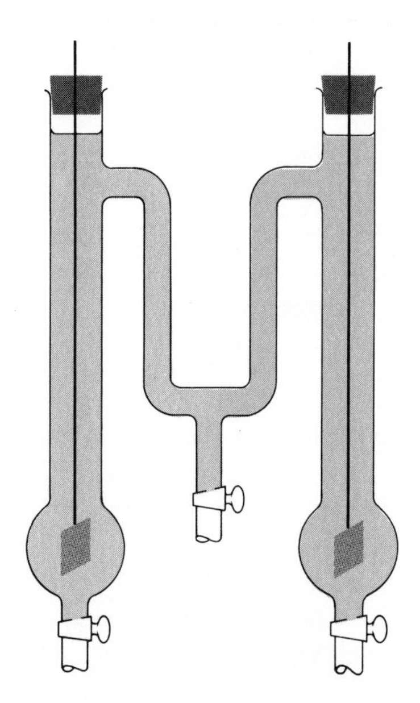

FIGURE 22-5
Hittorf transference-number apparatus.

Fig. 22-3. The current is carried across the cross sections by the flow of ions, and in view of the definitions of t_+ and t_-, the passage of 1 $\mathfrak{F}$ of charge across these sections is accomplished by the flow of t_+ equivalents of H^+ to the right and t_- equivalents of Cl^- to the left. The net flow across these sections is then $t_+ + t_- = 1$ equiv of ions, which corresponds to 1 $\mathfrak{F}$ of charge. It is clear from Fig. 22-4 that the number of equivalents of HCl in the central compartment should not be changed by the passage of current.

Consider now the changes that occur in the cathode portion. The change in equivalents of H^+ and Cl^- due to ion migrations is shown by the transfers across the cross-section line. In addition to migration, there is a removal of 1 equiv of H^+ at the electrode by the reaction

$$H^+ + e^- \rightarrow \tfrac{1}{2}H_2$$

The net cathode-compartment changes for the passage of 1 $\mathfrak{F}$ are calculated as

Change in equivalents of H^+

$$\begin{aligned} &= \text{electrode reaction} + \text{migration effect} \\ &= -1 + t_+ \\ &= t_+ - 1 = -t_- \text{ equiv} \end{aligned}$$

Change in equivalents of Cl^-

$$\begin{aligned} &= \text{electrode reaction} + \text{migration effect} \\ &= 0 - t_- \\ &= -t_- \end{aligned}$$

Passage of 1 $\mathfrak{F}$ of current results, therefore, in the removal of t_- equiv of HCl from the cathode portion.

In a similar manner, the changes in the anode compartment per faraday are calculated as

Change in equivalents of H^+

$$\begin{aligned} &= \text{electrode reaction} + \text{migration effect} \\ &= 0 - t_+ \\ &= -t_+ \end{aligned}$$

Change in equivalents of Cl^-

$$\begin{aligned} &= \text{electrode reaction} + \text{migration effect} \\ &= -1 + t_- \\ &= t_- - 1 = -t_+ \end{aligned}$$

The net effect around the anode is the removal of t_+ equiv of HCl.

It should be noticed that the analysis is in terms of the changes in the number of equivalents, and not concentration. The volume of the compartments, as we shall see, is not critical.

This calculation suggests a method introduced by Hittorf for determining transference numbers. The procedure consists in filling a cell like that of Fig. 22-5 with the HCl solution and, first, without passage of current, draining the compartments and analyzing for the number of equivalents of HCl in each compartment. The number of equivalents would be calculated from the concentrations and compartment volumes. The cell is then refilled with the same solution, and a measured number of coulombs of current is passed. The compartments are then drained and analyzed to give the number of equivalents in each and, from this, the change in equivalents in each compartment.

If not too large an amount of current is passed and if no mixing of the compartment solutions occurs, it will be found, in accordance with the previous treatment, that the number of equivalents in the central compartment will be unchanged. The changes in the number of equivalents in either of the electrode compartments allow the determination of the transference numbers of H^+ and Cl^-.

In practice, of course, an amount of charge much less than 1 $\mathfrak{F}$ is passed through the cell. The observed changes in the electrode compartments can, however, be used to calculate the change expected per faraday of charge passed through the cell. These data can then be used directly with the type of analysis indicated above to give t_+ and t_-.

The treatment is much the same when other electrode processes occur. The electrolysis of a solution of $CuSO_4$ will illustrate this. Again consider the effect of the passage of 1 $\mathfrak{F}$ of charge. The electrode reactions and transfers between the compartments are shown in Fig. 22-6. Again, if the electrolysis is not carried too far, the middle compartment will experience no net change. The net effect on the electrode compartments as a result of the processes of Fig. 22-6 can be shown as in Table 22-7. The overall effect of the passage of a faraday of charge on the solution of the entire cell is the gain of 1 equiv of H^+ and the loss of 1 equiv of Cu^{++} or, in terms of electrolytes, the gain of 1 equiv of H_2SO_4 and the loss of 1 equiv of $CuSO_4$. The changes in the electrode compartments, however, involve the transference number, and thus analyses of these compartments allow values of transference numbers to be obtained.

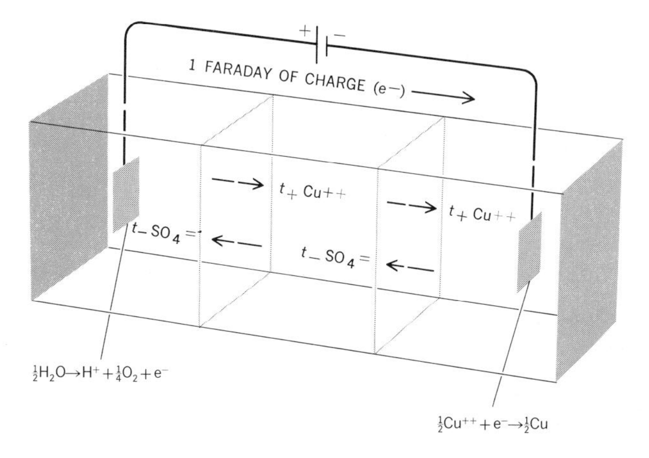

FIGURE 22-6
Transference numbers for $CuSO_4$ by the Hittorf method.

TABLE 22-7 Changes in the electrode compartments of a Hittorf transference-number cell for the passage of 1 $\mathfrak{F}$ of charge through a $CuSO_4$ solution, as in Fig. 22-6

	Cathode compartment	Anode compartment
Change due to electrode reactions	-1 equiv of Cu^{++}	$+1$ equiv of H^+
Change due to migration	$+t_+$ equiv of Cu^{++} $-t_-$ equiv of SO_4^{--}	$-t_+$ equiv of Cu^{++} $+t_-$ equiv of SO_4^{--}
Net electrode compartment changes (using $t_- = 1 - t_+$ and $t_+ = 1 - t_-$)	Loss of t_- equiv of $CuSO_4$	Gain of 1 equiv of H_2SO_4 and loss of t_+ equiv of $CuSO_4$
Net cell change	Gain of 1 equiv of H_2SO_4 and loss of 1 equiv of $CuSO_4$	

Other methods are available for the determination of transference numbers, but the detailed look at the conduction process that the Hittorf method requires makes this method sufficient to illustrate the determination of these quantities.

Table 22-8 shows the transference numbers for several electrolytes at various concentrations. The cation transference numbers t_+ are listed, and the relation $t_+ + t_- = 1$ can be used to give the corresponding anion values. This table shows that both positive and negative ions carry the current through the solution and that they do so to approximately the same extent.

It is important to notice that the transference numbers show some concentration dependence, particularly for electrolytes with highly charged ions. The relative conductance contributions of the ions are therefore a function of concentration.

TABLE 22-8 Transference numbers for positive ions at 25°C and the values obtained by extrapolation to infinite dilution*

Normality	HCl	NaCl	KCl	$CaCl_2$	$LaCl_3$
0	0.821	0.396	0.491	0.438	0.477
0.01	0.825	0.392	0.490	0.426	0.462
0.02	0.827	0.390	0.490	0.422	0.458
0.05	0.829	0.388	0.490	0.414	0.448
0.1	0.831	0.385	0.490	0.406	0.438
0.2	0.834	0.382	0.489	0.395	0.423

* From L. G. Longsworth, *J. Am. Chem. Soc.*, **54**:2741 (1932), and **57**:1185 (1935); L. G. Longsworth and D. A. MacInnes, *J. Am. Chem. Soc.*, **60**:3070 (1938).

Knowledge of the values of transference numbers lets us proceed to a discussion of other ionic properties.

22-8 Ionic Conductances

Values can now be obtained for the contributions that the individual ions of an electrolyte make to the equivalent conductance. The empirical law of Kohlrausch implies that at infinite dilution the equivalent conductance can be interpreted in terms of such ionic contributions and that the contributions of an ion are independent of the other ion of the electrolyte.

At infinite dilution, therefore, it is profitable to write

$$\Lambda_0 = (\lambda_+)_0 + (\lambda_-)_0 \qquad [20]$$

where $(\lambda_+)_0$ and $(\lambda_-)_0$ are the *equivalent ionic conductances at infinite dilution*. Since the transference numbers give the fraction of the total current carried by each ion, i.e., the fraction of the total conductance that each ion contributes, we can write

$$\lambda_0{}^+ = (t_+)_0\Lambda_0 \qquad \text{and} \qquad \lambda_0{}^- = (t_-)_0\Lambda_0 \qquad [21]$$

where $(t_+)_0$ and $(t_-)_0$ are the transference numbers extrapolated to infinite dilution. Table 22-9 shows some values for these limiting equivalent ionic conductances. An immediate use of such tabulated values is the calculation of the limiting equivalent conductance of a weak electrolyte without the addition and subtraction procedures of Sec. 22-2.

In a formal manner one can use the data for the equivalent conductance and the transference numbers at concentrations other than that of infinite dilution to obtain values of λ_+ and λ_- at these higher concentrations. At such concentrations, however, the law of independent migration of the ions fails, and the conductance is really a

TABLE 22-9 Equivalent ionic conductances and ionic mobilities at infinite dilution and 25°C*

Ion	$(\lambda_+)_0$ (ohm^{-1} m^2)	$(v_+)_0$(ms^{-1})/$\mathcal{V}$(V m^{-1})	Ion	$(\lambda_-)_0$ (ohm^{-1} m^2)	$(v_-)_0$(ms^{-1})/$\mathcal{V}$(V m^{-1})
H^+	0.03982	36.3×10^{-8}	OH^-	0.01980	20.5×10^{-8}
Li^+	0.003869	4.01	Cl^-	0.007523	7.91
Na^+	0.005011	5.19	I^-	0.00768	7.95
K^+	0.007352	7.61	CH_3COO^-	0.00409	4.23
Ag^+	0.006192	6.41	$SO_4{}^=$	0.00798	8.27
$NH_4{}^+$	0.00734	7.60			
Ca^{++}	0.005950	6.16			
La^{3+}	0.00696	7.21			

*Data from D. A. MacInnes, "The Principles of Electrochemistry," Reinhold Publishing Corporation, New York, 1939.

property of the electrolyte rather than of the individual ions of the electrolyte. This means that an ionic conductance calculated for a Cl^- ion, for example in a 1-M HCl solution, will be different from that deduced for the Cl^- ion in a 1-M NaCl solution. The concept of ionic conductances is really valuable, therefore, only at infinite dilution.

Rather than try to understand the different current-carrying properties of the ions of Table 22-9 in terms of ionic conductances, we proceed now to obtain an even more fundamental ionic property: the velocity with which the ions travel through the solution under the influence of the applied electric field.

22-9 Ionic Mobilities

Consider a cell of the type used to introduce the concept of equivalent conductance. Such a cell, it will be recalled, consists of two electrodes 1 m apart and of cross-section area A such that an amount of solution that contains 1 equiv of electrolyte is held between the electrodes. A distorted picture of such a cell is shown in Fig. 22-7.

For an applied voltage $\mathcal{V}$, a current I will flow through the cell. These electrical quantities are related, since the conductance of such a cell is the equivalent conductance of the electrolyte, by

$$I = \frac{\mathcal{V}}{R} \qquad \text{or} \qquad I = \Lambda \mathcal{V} \tag{22}$$

At infinite dilution the current can be attributed to the independent flow of positive and negative ions, and one can write

$$\begin{aligned} I &= \Lambda_0 \mathcal{V} = [(\lambda_+)_0 + (\lambda_-)_0]\mathcal{V} \\ &= (\lambda_+)_0 \mathcal{V} + (\lambda_-)_0 \mathcal{V} \\ &= I_+ + I_- \end{aligned} \tag{23}$$

This flow of current through the cell can also be analyzed in terms of the details of the ion movements in the cell. Since the cell contains

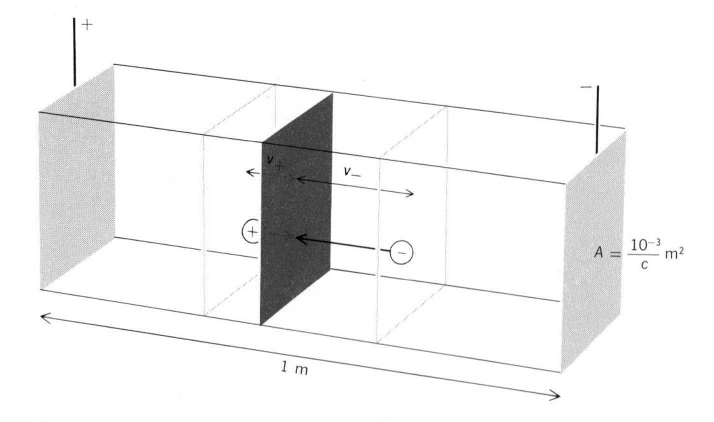

FIGURE 22-7
Diagram for ionic-mobility calculation.

1 equiv of electrolyte, there will be $\mathfrak{N}/Z_+$ positive ions present and $\mathfrak{N}/Z_-$ negative ions present, where $\mathfrak{N}$ is an Avogadro's number and Z_+ and Z_- are the charges of the ions of the electrolyte. The average velocities with which the ions move to their respective electrodes under the influence of the applied voltage are represented by v_+ and v_-. The current passing a cross section of the cell can now be obtained in terms of these ionic quantities and the applied voltage.

For an ion to cross the shaded cross section of Fig. 22-7 within 1 s, it must start within a distance v_+ or v_- m from the cross section. (The ions travel a distance v_+ or v_- in 1 s.) In 1 s, therefore, all the positive ions in the right rectangular compartment and all the negative ions in the left compartment will have crossed the boundary. Since these rectangular volumes have lengths of v_+ and v_- and the cell has a width of 1 m, the volumes will contain fractions $v_+/1$ and $v_-/1$ of the total number of positive and negative ions in the cell. We can write, therefore,

$$\text{No. positive ions crossing boundary per second} = \frac{\mathfrak{N}}{Z_+}\frac{v_+}{1} = \frac{\mathfrak{N}v_+}{Z_+}$$

$$\text{No. negative ions crossing boundary per second} = \frac{\mathfrak{N}}{Z_-}\frac{v_-}{1} = \frac{\mathfrak{N}v_-}{Z_-}$$

The current corresponding to these flow rates is obtained by multiplying by the ion charges eZ_+ and eZ_- to give

$$\text{Positive charge crossing per second} = I_+ = eZ_+\frac{\mathfrak{N}v_+}{Z_+} = e\mathfrak{N}v_+$$

$$\text{Negative charge crossing per second} = I_- = eZ_-\frac{\mathfrak{N}v_-}{Z_-} = e\mathfrak{N}v_-$$

For an infinitely dilute solution, if the average ionic velocities in the electric field direction are represented as $(v_+)_0$ and $(v_-)_0$, comparison with Eq. [23] gives

$$\mathcal{V}(\lambda_+)_0 = e\mathfrak{N}(v_+)_0 \qquad \text{and} \qquad \mathcal{V}(\lambda_-)_0 = e\mathfrak{N}(v_-)_0 \tag{24}$$

Furthermore, the substitution $e\mathfrak{N} = \mathfrak{F}$ can be made to yield, finally,

$$(v_+)_0 = \frac{(\lambda_+)_0\mathcal{V}}{\mathfrak{F}} \qquad \text{and} \qquad (v_-)_0 = \frac{(\lambda_-)_0\mathcal{V}}{\mathfrak{F}} \tag{25}$$

The average velocity with which an ion moves toward an electrode under the influence of a potential of 1 V applied across a 1-m cell is given by $(v_+)_0/\mathcal{V}$ or $(v_-)_0/\mathcal{V}$ and is known as the *ionic mobility*. The ionic mobility is calculated, as Eq. [25] shows, by dividing the ionic conductance, as listed in Table 22-9, by the value of the faraday. These mobility results are also shown in Table 22-9.

The most remarkable feature of the data of Table 22-9 is, perhaps, the high values for the mobilities of the H^+ and OH^- ions compared

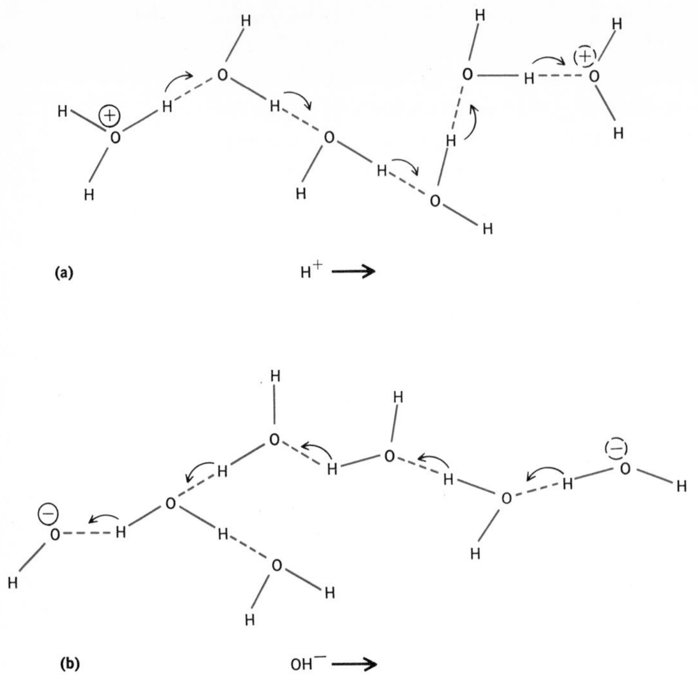

FIGURE 22-8
Movement of (*a*) H^+ and (*b*) OH^- by the Grotthuss mechanism. The charges outlined by dashed circles are formed as a result of the series of proton transfers.

with those of all the other ions. Since the proton is present as an H_3O^+ ion and since both the H_3O^+ and OH^- ions are expected to be highly solvated, an explanation cannot be given in terms of the size of these ions. A mechanism of the type suggested originally by von Grotthuss in 1805 to explain conduction by all electrolytes appears now to be applicable, but only to the H_3O^+ and OH^- ions. Figure 22-8 shows how a series of transfers of protons between neighboring water molecules can have the effect of moving either an H^+ or an OH^- through the solution. The high mobilities of H^+ and OH^- and the fact that they are the dissociation products of the solvent are seen to be related. In other solvents, where such a mechanism could not operate, these ions would show mobilities more in line with those of the other ions.

The values of the mobilities of the other ions in aqueous solution are more difficult to understand. The high degree of solvation expected for the small ions, such as Li^+, and for the highly charged ions, such as La^{3+}, apparently works against the expected dependence of mobility on size and charge. A better understanding of the ionic world than we now have would be necessary before the fairly small variation of mobility such as that shown in Table 22-9, except for H^+ and OH^-, could be interpreted in terms of the properties of the solvated ions.

It is interesting to compare the mobilities of ions with the speeds previously found for molecules of gases. For a reasonable voltage gradient of 10^4 V m^{-1}, for instance, ions would migrate according to the mobility data of Table 22-9 with a velocity of about 5×10^{-4} m

s^{-1}. It would take a typical ion about 30 min to travel 1 m. Comparison can be made with molecular speeds in gases of about 10^2 m s^{-1}, as deduced in Chap. 2, and the corresponding time of 1/100 s to travel 1 m. We conclude, therefore, that the path of an ion under the influence of an electric field is a slow, devious trek of a cumbersome solvated ion through the interfering solvent molecules. This would be explained also by saying that the electric field that is conveniently applied to a solution is not an overwhelming factor in the affairs of ions. The ions are to be thought of as having only a slight directional component imposed on their random motions.

This takes us as far as we can go in our nonthermodynamic study of the behavior and nature of ions in solution. Many facets remain obscure. The problems that existed even before the days of Arrhenius with regard to the extent of dissociation of strong electrolytes are still unsolved for any but rather dilute solutions. The extent of solvation of ions and even an exact interpretation of this term, in view of the rapid coming and going of solvent molecules, remain an open question. Thus, although we have learned a great deal about solutions of electrolytes, a great deal more remains to be discovered.

22-10 Some Applications of Conductance Measurements

Previously, the results of conductance measurements have been used for the investigation of the nature of electrolytes in solution. There are also a number of direct applications of conductance measurements to chemical problems. The usefulness of conductance arises from its dependence on the ionic concentration and from its special sensitivity to the concentration of H^+ and OH^- ions.

Example 1 Conductimetric Titrations An acid-base titration, using HCl and NaOH for example, can be performed in a conductivity cell, and the change of conductance followed as the base is added to the acid. Results such as those shown in Fig. 22-9 are obtained. The net behavior is seen to depend on the high ionic conductances of H^+ and OH^- compared with the ions Na^+ and Cl^- and the equivalence point can then be conveniently taken as the intersection of the two straight lines that can be drawn.

For titrations involving a weak acid or a weak base the behavior is not quite so simple, but the conductance still provides a useful means for following the titration.

Of particular value is the fact that the conductance is derived from the measured resistance of the cell. The change in resistance as the titration proceeds can be used in an instrumental method for following the course of the titration automatically.

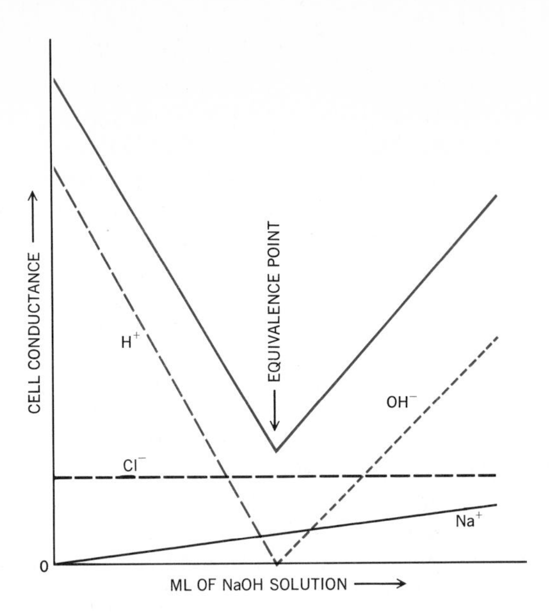

FIGURE 22-9
The conductance curve (solid color line) and the ionic contributions for a titration of HCl with NaOH.

Example 2 Degree of Ionization of Weak Electrolytes For weak electrolytes, for which the ionic concentrations are sometimes very small, the dominant effect on the conductivity is the association of the ions of the weak electrolyte into undissociated molecules. In such cases the Arrhenius relation $\alpha = \Lambda/\Lambda_0$ can be used to deduce the degree of ionization.

A good, but rather special, illustration of a use of conductance measurements is to be found in the determination of the dissociation constant of water,

$$H_2O \rightleftharpoons H^+ + OH^-$$

At 25°C the lowest specific conductance that can be obtained with the most carefully distilled water is 58×10^{-7} ohm^{-1} m^{-1}, and this conductance must be attributed to the equilibrium concentration of H^+ and OH^-. The molarity of pure water is

$$c = \frac{(1000)(0.997)}{18.02} = 55.3 \text{ mol liter}^{-1}$$

which gives, for the equivalent conductance, the result

$$\Lambda = \frac{10^{-3}\kappa}{c}$$
$$= 1.05 \times 10^{-10} \text{ ohm}^{-1} \text{ m}^2$$

The equivalent conductance expected for completely dissociated water is calculated as

$$\Lambda_0 = \lambda_0^{H^+} + \lambda_0^{OH^-}$$
$$= 0.05478$$

The degree of dissociation of water at 25°C is therefore

$$\alpha = \frac{\Lambda}{\Lambda_0} = \frac{1.05 \times 10^{-10}}{0.05478} = 1.9 \times 10^{-9}$$

and the ionic concentrations are

$$[H^+] = [OH^-] = \alpha c = 1.05 \times 10^{-7}$$

The familiar ion product for water is then determined for 25°C as

$$K_w = [H^+][OH^-] = 1.1 \times 10^{-14}$$

The very low concentration of ions makes the Arrhenius relation $\alpha = \Lambda/\Lambda_0$ quite valid, and the value obtained for K_w is therefore reliable.

Example 3 The Ionic Concentrations Produced by Sparingly Soluble Salts A large number of salts are sufficiently insoluble in water to make a chemical analysis of their solubility quite difficult. Information on the concentration of the ions in solution in equilibrium with the solid can be obtained from measurements of the conductance of a saturated solution. For the salt MX, for example, the solubility product $[M^+][X^-]$ can then be deduced. For very insoluble salts the concentration of ions in solution will again be low enough so that the conductance can be taken as a measure of ionic concentrations.

The example, frequently referred to, of an insoluble salt that can be studied in this way is AgCl. The specific conductance of a saturated solution at 25°C is given, after subtraction of the specific conductance of the water itself, as

$$\kappa = 2.28 \times 10^{-4} \text{ ohm}^{-1} \text{ m}^{-1}$$

The conductance of 1 equiv at infinite dilution is obtained from the data of Table 22-9, which are deduced from measurements on more soluble electrolytes. Thus

$$\Lambda_0 = \lambda_0^{Ag^+} + \lambda_0^{Cl^-} = 0.01382 \text{ ohm}^{-1} \text{ m}^2$$

Since the solubility of AgCl is quite low, the equivalent conductance of a saturated solution will be little different from that at infinite dilution. Thus one can use Eq. [5] to write, for the saturated solution,

$$\Lambda = \Lambda_0 = \frac{10^{-3}\kappa}{c}$$

or

$$c = \frac{10^{-3}\kappa}{\Lambda_0} = \frac{10^{-3}(2.28 \times 10^{-4})}{0.01382} = 1.65 \times 10^{-5}$$

The solubility product is then calculated as

$$K_{sol} - [Ag^+][Cl^-] = (1.65 \times 10^{-5})^2 = 2.72 \times 10^{-10}$$

Such a treatment is disturbed if, for example, complex ions such as Ag_2Cl^+ or $AgCl_2^-$ are present, and if any ion pairs or higher-neutral-association species are important, the solubility of the electrolyte would not be simply given by the concentrations of the Ag^+ or Cl^- ions deduced in this way.

REFINEMENTS OF THE MODEL OF IONS IN AQUEOUS SOLUTIONS

As has been pointed out in the preceding sections, the Arrhenius idea of ions in solution in equilibrium with parent molecular species allows many of the properties of ionic solutions to be understood. But a number of difficulties arise, and illustrative of these are the nonconstant equilibrium "constants" that are obtained for the dissociation of strong electrolytes (cf. Table 22-4). Criticisms ultimately were leveled at the Arrhenius theory for postulating, in such cases, molecules instead of ions, a reversal of the initial criticisms of the Arrhenius equation that attacked it for postulating ions in solution instead of molecules!

Refinements to the simplest ideas of the ionic solutions depend on the recognition of the role of the solvent and on the effect of interactions between the ions.

22-11 The Role of the Solvent: Dielectric Effect

A remarkable feature of the Arrhenius theory is that, although it attributes the dissociation process to the solution of the electrolyte, it proceeds to ignore the role of the solvent, or rather treats the solvent as if it were an inert medium. A detailed understanding of the molecular nature of ionic solutions must clearly involve the very important role played by the solvent. It is necessary, for instance, to understand why water is such a unique solvent for ionic systems.

Two aspects of solvent behavior can be recognized. Here the dielectric effect will be pointed out. In the following section the *solvation,* or more particularly *hydration,* of ions by the solvent will be considered.

The electrostatic force of attraction between ions of charge Z^+ and Z^- is given by Coulomb's law:

For vacuum
$$f(r) = \frac{e^2Z^+Z^-}{(4\pi\epsilon_0)r^2}$$

For medium of dielectric ϵ/ϵ_0
$$f(r) = \frac{e^2Z^+Z^-}{(\epsilon/\epsilon_0)(4\pi\epsilon_0)r^2}$$

With the numerical values $e = 0.1602 \times 10^{-18}$ C and $4\pi\epsilon_0 = 1.1126 \times 10^{-10}$ C^2 N^{-1} m^{-2}, the second of these equations is

$$f(r) = 0.2306 \times 10^{-27} \frac{Z^+Z^-}{(\epsilon/\epsilon_0)r^2} \quad \text{N} \qquad [26]$$

For water, the dielectric-constant factor ϵ/ϵ_0 has the very large value of about 80. The force of interaction, and the energy required to overcome coulombic forces, are thus smaller by almost two orders of magnitude in water than in vacuum or materials of very low dielectric. For example, in the gas phase, NaCl molecules exist, and the binding, which holds the particles at an equilibrium separation of 2.36 Å, can be taken as due to coulombic forces. For dissociation, the amount of energy that must be spent to overcome this electrostatic binding is

$$\begin{aligned}
\Delta E &= \int_{r=2.36\times10^{-10}\,\text{m}}^{r=\infty} f(r)\,dr \\
&= 0.2306 \times 10^{-27} \int_{2.36\times10^{-10}\,\text{m}}^{\infty} \frac{1}{r^2}\,dr \\
&= \frac{-0.2306 \times 10^{-27}}{r}\Bigg]_{2.36\times10^{-10}\,\text{m}}^{\infty} \\
&= 0.98 \times 10^{-18} \text{ J per molecule} \\
&= 590 \text{ kJ mol}^{-1} \qquad [27]
\end{aligned}$$

This amount, less the repulsion energy at the equilibrium internuclear distance, as is shown in Sec. 11-8, is approximately the energy for the dissociation of NaCl into ions in the vapor state.

In the presence of a medium with a dielectric constant of 80, this electrostatic energy amounts to only

$$\begin{aligned}
\Delta E &= \frac{0.98 \times 10^{-18}}{80} = 1.22 \times 10^{-20} \text{ J per molecule} \\
&= 7.4 \text{ kJ mol}^{-1} \qquad [28]
\end{aligned}$$

The easy dissociation of electrolytes in aqueous solution as compared with gas-phase or low-dielectric material is therefore readily understandable in terms of the high dielectric constant of water. The initial criticisms raised against the Arrhenius theory for postulating the dissociation of electrolytes in solution remain, however, valid arguments against any theory postulating appreciable dissociation to form free ions in solvents of low dielectric constant. It is not valid to conclude that the dielectric constant is such that all interactions between charged species will be eliminated, an assumption inherent in the Arrhenius theory.

An additional indication of the extent of ionic interaction can be obtained by comparing the energy for complete separation from some arbitrary distance, say 5 Å, for ionic species interacting with electrostatic forces and for neutral molecules interacting with van der Waals forces.

For the ionic case the electrostatic energy to be overcome by singly charged ions would be calculated, as above, to be 4.6×10^{-15} J per molecule for a vacuum environment and 5.8×10^{-17} J per molecule for water as the medium. Typical values for gas-phase molecules can be seen from Table 19-2 to be of the order of 10^{-21} J, much smaller even than values for ions in the high dielectric environment of water.

Thus, although the dielectric effect is a major factor for the formation of ionic species in aqueous solutions, it is not great enough to reduce the intermolecular interaction to the small values found for gas-phase molecules. We must expect, therefore, that for all but extremely dilute solutions, ionic interactions will enter to produce behavior different from the ideal behavior found at infinite dilution.

22-12 The Role of the Solvent: Solvation Energies of Ions

A more specific molecular effect is also the basis for the remarkable solvent features of water. A variety of evidence, some of which will be dealt with here, suggests that the ions present in an aqueous solution have associated with them water molecules that are said to solvate or hydrate the ion. The associations involve various numbers of water molecules and occur with varying energies, and the complexes of ion-plus-water molecules persist for varying lengths of time. Here the energies of hydration will be dealt with.

Results for the heat of solution of electrolytes, which can be obtained calorimetrically, can be listed, and some values are given in Table 22-10. Since these data are the enthalpies of the solutions *compared with* those of the solid salts, they do not directly reflect the relative enthalpies of the ionic solutions. The enthalpy effects of the solution process are most clearly exhibited if, rather, enthalpies for reactions of the type

$$\text{Na}^+(g) + \text{Cl}^-(g) \xrightarrow{(\text{H}_2\text{O})} \text{Na}^+(aq) + \text{Cl}^-(aq)$$

are deduced. To obtain such data it is only necessary to have, in addition to the data of Table 22-10, for the ionic crystals, the enthalpies of processes such as

$$\text{NaCl}(s) \rightarrow \text{Na}^+(g) + \text{Cl}^-(g)$$

Results for such processes, it will be recalled, were obtained in Sec. 18-2 from the Born-Haber cycle and from theoretical calculations. With such results, the combination of equations for the NaCl example

$$\text{NaCl}(s) \xrightarrow{(\text{H}_2\text{O})} \text{Na}^+(aq) + \text{Cl}^-(aq) \qquad \Delta H = 4.3 \text{ kJ}$$

and

$$\text{NaCl}(s) \longrightarrow \text{Na}^+(g) + \text{Cl}^-(g) \qquad \Delta H = 787 \text{ kJ}$$

TABLE 22-10 Values of ΔH, in kJ mol^{-1}, for the solution of crystalline metal halides and gaseous hydrogen halides in water at 25°C

	F^-	Cl^-	Br^-	I^-
H^+	−48.5	−72.8	−83.5	−80.4
Li^+	+4.2	−35.1	−47.1	−61.7
Na^+	+2.5	+4.3	+0.4	−5.2
K^+	−15.1	+17.2	+19.8	+21.8
Rb^+	−24.3	+18.4	+24.9	+27.2
Cs^+	−36.0	+20.0	+28.2	+34.5
Ag^+	−14.2	+66.1	+84.0	+111.8
Tl^+		+42.3	+54.4	+73.3
Mg^{++}	−11.6	−150.3	−181.2	−208.4
Zn^{++}		−65.4	−62.8	−47.3
Cd^{++}		−12.6	−3.2	+4.0
Hg^{++}		+13.2	+6.7	
Cr^{++}		+77.8		
Mn^{++}		+66.9	+66.9	

gives, on subtraction, the desired reaction and enthalpy value

$$Na^+(g) + Cl^-(g) \xrightarrow{(H_2O)} Na^+(aq) + Cl^-(aq) \qquad \Delta H = -783 \text{ kJ}$$

The corresponding calculation for the hydrogen halides requires, instead of the lattice energy, the enthalpy of the reaction

$$HX(g) \rightarrow H^+(g) + X^-(g) \qquad [29]$$

This can be obtained from available data on dissociation energies, ionization potentials, and electron affinities, as illustrated for HCl.

$$\begin{array}{rl} HCl(g) \rightarrow H(g) + Cl(g) & \Delta H = +\ 432 \text{ kJ} \\ H(g) \rightarrow H^+(g) + e^- & \Delta H = +1310 \\ Cl(g) + e^- \rightarrow Cl^-(g) & \Delta H = -\ 347 \\ \hline HCl(g) \rightarrow H^+(g) + Cl^-(g) & \Delta H = +1395 \text{ kJ} \end{array}$$

Combination of this result with the value of −72.8 kJ mol^{-1} for ΔH for the solution of HCl(g) gives, finally,

$$H^+(g) + Cl^-(g) \xrightarrow{(H_2O)} H^+(aq) + Cl^-(aq) \qquad \Delta H = -1468 \text{ kJ}$$

In such ways the data presented for the solution of a variety of gaseous ions of electrolytes in Table 22-11 can be obtained.

This is as far as we can go with a thermodynamic treatment. The fact, however, that the difference between successive values in adjacent columns or rows of Table 22-11 are approximately constant leads us to recognize, in the same way as we did with the data on the limiting equivalent conductances of electrolytes, that the heat of solution of the ions of an electrolyte, at infinite dilution, can be interpreted in terms

TABLE 22-11 Values of ΔH, in kJ mol^{-1}, for the solution of gaseous ions of metal halides and hydrogen halides in water at 25°C [that is, ΔH for the processes of the type $M^+(g) \times X^-(g) \xrightarrow{H_2O} M^+(aq) + X^-(aq)$]

	F^-	Cl^-	Br^-	I^-
H^+	−1594	−1467	−1439	−1397
Li^+	−1036	−897		
Na^+	−918	−783	−747	−705
K^+	−833	−699	−663	−622

of separate contributions made by the ions of the electrolyte. It remains now to see how the values of Table 22-11 can be divided up so that contributions from the separate ions can be obtained.

A variety of attempts to separate the heat of hydration of the ions of an electrolyte into ionic components have been made. Most of these start with the calculation of the relative work required to charge a sphere, which represents the ion, in the gas phase compared with the work required in a dielectric medium.

A result of such a derivation suggests that the hydration energy should be proportional to the square of the charge and inversely proportional to its effective radius in aqueous solution. With such a guide, results such as those of Table 22-11 can be divided into contributions from each of the ions of the electrolyte. One set of values, which can be looked on as based on the assumption of −1090 kJ mol^{-1} for the H^+ ion, is given in Table 22-12.

Some of the trends of the data of Table 22-12 are those expected from the sizes and charges of the ions. (For more extensive data a closer look into the arrangement of the outer electrons of the ion and the way these electrons interact with the adjacent water molecules must be taken.) Of special note, however, is the very large value for the proton. Some understanding of this value is provided by an estimate that the heat of the reaction

$$H^+(g) + H_2O(g) \rightarrow H_3O^+(g) \qquad [30]$$

is −760 kJ mol^{-1}. If this reaction is considered to be a step in the solution of a proton, and if we recall that the heat of vaporization of water is about 40 kJ, the entire heat of solution is accounted for by ascribing 370 kJ mol^{-1} to the heat of solution of the H_3O^+ ion. This value is in line with the values given in Table 22-11 for other cations.

Noteworthy also are the very large values that are commonly to be found for the highly charged ions such as Al^{3+}. These values, if divided by 6, a reasonable value for the number of water molecules that can come in direct contact with the ion, lead to hydration energies for each nearest-neighbor water molecule. Thus, for these ions the

TABLE 22-12 Some values for ΔH, in kJ mol^{-1}, of hydration of gas-phase ions based on the value of −1090 kJ mol^{-1} for $H^+(g) \xrightarrow{H_2O} H^+(aq)$

Ion	ΔH
H^+	(−1090)
Li^+	−520
Na^+	−400
K^+	−320
Ca^{++}	−1580
Zn^{++}	−2040
Al^{3+}	−4680
La^{3+}	−3300
F^-	−500
Cl^-	−380
Br^-	−350
I^-	−310

hydration process must be regarded as involving the formation of "chemical" bonds. For the singly charged ions, on the other hand, the energy per nearest-neighbor water molecule is considerably less, and a very strong hydrogen bond or ion-dipole association is indicated. Thus, although the ion hydration energies cannot be easily understood, at this stage, in a quantitative way, they are clearly significant data in the analysis of the nature of ions in aqueous solution.

*22-13 The Debye-Hückel Theory of Interionic Interactions

Now let us turn to the second of the two features that must be recognized if the simple model of free ions in aqueous solution is to be refined. This feature is the electrostatic interaction between ions that, except at infinite dilution, enters and alters the properties of the solution from those that would be expected from the free-ion model.

The treatment of ion-ion interactions by P. Debye and E. Hückel in 1923 and 1924 led to an explanation of the properties of relatively dilute solutions, less than about 0.01 M. Even this limited success has proved valuable in that a way to extrapolate available experimental data to the limit of infinite dilution was provided, as was a base for more empirical extensions to higher concentrations. It is worthwhile, therefore, to follow through the Debye-Hückel derivation in some detail.

Consider one of the ions, a positive ion to be specific, of an aqueous solution of an electrolyte. It will be affected by coulombic interactions with the other ions of the solution. The potential-energy effect will be greater for the nearby ions and, since the effect varies according to $1/r$, will be less for the more distant ions. But the number of such distant ions increases as the volume of a spherical shell, i.e., as r^2, and thus these more distant ions, and therefore the bulk of the solution, might appear to require our attention in the deduction of the effect on the reference ion. Fortunately, ions of opposite charge can be expected to distribute themselves uniformly at some distance away from a given ion to produce an electrical neutrality well removed from a given ion.

The Debye-Hückel treatment deals with the distribution of ions around a given ion and the net effect these neighboring ions have on the properties of the solution.

It is necessary to present here only a very condensed version that takes in the important features of the theory but ignores many of the subtleties that are involved. Even so, the treatment is somewhat lengthy, due to the use of simple steps rather than some more compact but rather advanced mathematical relationships.

Consider first how the ions in a solution distribute themselves relative to one another. The two factors that play roles are the thermal jostlings and the electrical attractions between oppositely charged particles. Suppose that on the average there are n_i ions of the i type per unit volume. Around any positive ion there will be an increase in the concentration of negative ions and a decrease in the concentration of positive ions. These changes result from the ions moving to the energetically more favored regions, i.e., those in which their potential energy is low, and the tendency for this movement must compete with the random thermal motion.

Boltzmann's equation can be used to give the number of ions that on the average are a distance r from the positive charge. The energy of ions of charge $Z_i e$ in a potential of value $\mathcal{V}$ is $(eZ_i)\mathcal{V}$. If Z_i is positive, the energy is higher near the reference positive charge, and if Z_i is negative, the energy is lower. Boltzmann's equation gives

$$n_i(r) = n_i e^{-(eZ_i\mathcal{V})/kT} \qquad [31]$$

where $n_i(r)$ is the number of ith ions per unit volume at a distance r from the reference positive charge, and n_i is the average number per unit volume in the solution.

This expression cannot be used directly to calculate the density of ions of each type in the neighborhood of the reference positive ion because the neighboring ions, as well as the reference ion, determine the potential $\mathcal{V}$. Some manipulation of expressions for charge densities and potentials is necessary for a solution to this problem.

But to provide a reference for this more complex situation let us first calculate what the density of ions about the reference ion would be if only the reference charge affected the distribution. Then we could write

$$\mathcal{V} = \frac{e}{(\epsilon/\epsilon_0)4\pi\epsilon_0 r} = \frac{e}{4\pi\epsilon r}$$

where e is the charge of the reference ion, and thus

$$n_i(r) = n_i e^{-e^2 Z_i/4\pi\epsilon r kT}$$

or

$$\frac{n_i(r)}{n_i} = e^{-e^2 Z_i/4\pi\epsilon r kT} \qquad [32]$$

This comparison of the number of ions per unit volume at some distance r from the reference charge compared with the average number in the bulk of the solution is shown for various ions in Fig. 22-10.

To correct the data of Fig. 22-10 to allow for the effects of the various types of ions that surround the reference ion on each other, we must first develop an expression for the charge density about a reference ion.

The charge density at a distance r from the unit positive charge can be written as the ion density times the ion charges. This gives the charge density, which is a function of r, as

$$\rho(r) = \sum_i (eZ_i)n_i(r)$$

$$= e\sum_i n_i Z_i e^{-eZ_i\mathcal{V}/kT} \qquad [33]$$

The treatment is mathematically tractable, and the development is physically reasonable only if it is assumed that

$$eZ_i\mathcal{V} \ll kT \qquad [34]$$

i.e., that the energy of the ionic interactions is less than the average thermal energy. Such is the case for dilute solutions. For more concentrated solutions the interionic attractions can more effectively overcome the thermal motion, and associations occur that are not easily treated. The Debye-Hückel treatment applies, therefore, only to solutions in which interionic effects are not too important. For these solutions

$$\frac{eZ_i\mathcal{V}}{kT} \ll 1 \qquad [35]$$

and one can expand the exponential of Eq. [33] to give

$$e^{-eZ_i\mathcal{V}/kT} = 1 - \frac{eZ_i\mathcal{V}}{kT} + \text{higher terms} \qquad [36]$$

If all the higher terms are neglected, Eq. [33] becomes

$$\rho(r) = e\sum_i n_i Z_i - \frac{e^2\mathcal{V}}{kT}\sum_i n_i Z_i^2 \qquad [37]$$

The electrical neutrality of the solution leads to a value of zero for the first summation since it is nothing more than the summation over all the types of ions in the solution of the average number of ith ions per unit volume times the charge of the ith ion. Elimination of this necessarily zero term leaves

$$\rho(r) = -\frac{e^2\mathcal{V}}{kT}\sum_i n_i Z_i^2 \qquad [38]$$

The expression $\sum_i n_i Z_i^2$ is very similar to, and can be related to, the *ionic strength* that was introduced by Lewis and Randall on an empirical basis. This useful quantity is expressed in terms of the molar concentrations of the ions rather than in SI units of ions per cubic meter. Since the number of ions of the ith type per cubic meter is related to the number of moles

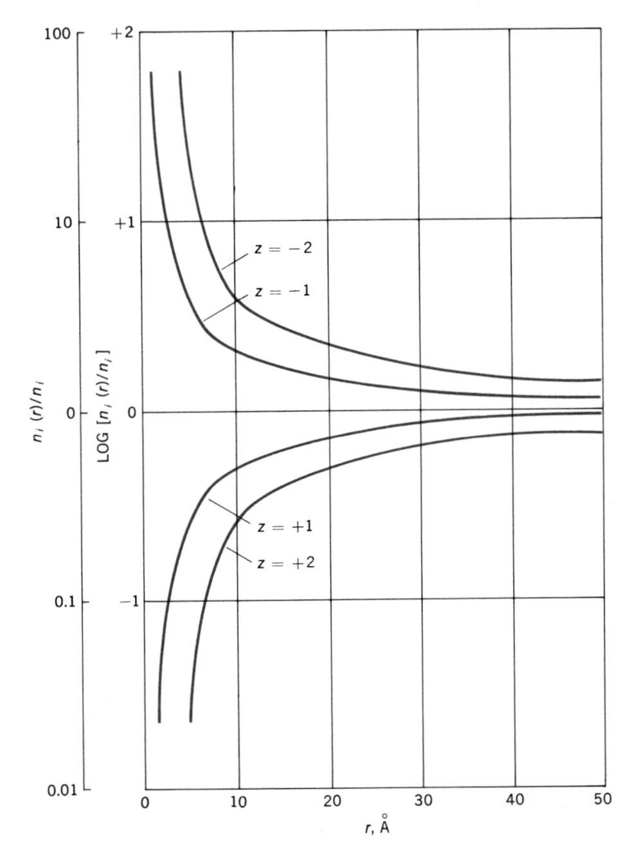

FIGURE 22-10
The variation in the ratio of the density of ions to the average ion density, in the neighborhood of a charge of $+e$, according to Eq. [32], which takes into account only the effect of the reference charge.

per liter c_i by the relation

$$n_i = 10^3 \mathfrak{N} c_i \qquad [39]$$

where $\mathfrak{N}$ is Avogadro's number, we have

$$\sum_i n_i Z_i^2 = 10^3 \mathfrak{N} \sum c_i Z_i^2$$
$$= 2 \times 10^3 \mathfrak{N} \mu \qquad [40]$$

where μ, the customary symbol for the ionic strength, is introduced according to

$$\mu = \frac{1}{2} \sum_i c_i Z_i^2 \qquad [41]$$

With this notation the charge distribution about the central positive-reference ion is written, from Eq. [38], as

$$\rho(r) = \frac{2 \times 10^3 \mathfrak{N} e^2 \mu \mathcal{V}}{kT} \qquad [42]$$

This relation, deduced from the specific considerations of the thermal agitation of ions counteracting the tendency of oppositely charged ions to attract one another, can be introduced into a more general equation that allows the elimination of $\rho(r)$ and gives a relation for $\mathcal{V}$ in terms of the concentrations and ionic strength.

The rate of change, or gradient, of the potential $\mathcal{V}$ is related to the electric field by

$$\mathcal{E} = -\frac{d\mathcal{V}}{dr} \qquad [43]$$

The electric field at the surface of a sphere of radius r that encloses a charge q is thus

$$\mathcal{E} = -\frac{d}{dr}\left(\frac{q}{4\pi\epsilon r}\right) = \frac{q}{4\pi\epsilon r^2} \qquad [44]$$

A more detailed treatment shows that a similar result is obtained for the electric field at the surface of the sphere even if the charge q, instead of being at the center of the sphere, is spread out or dispersed within the sphere of radius r. Therefore, for any spherically symmetric arrangements, one can express the electric field at a distance r from a central point as

$$\mathcal{E} = \frac{\text{enclosed charge}}{4\pi\epsilon r^2} \qquad [45]$$

If a uniform charge density ρ is assumed, the enclosed charge is $\frac{4}{3}\pi r^3 \rho$, and the electric field intensity at r is

$$\mathcal{E} = \frac{\frac{4}{3}\pi r^3 \rho}{4\pi\epsilon r^2} = \frac{1}{3}\frac{r\rho}{\epsilon} \qquad [46]$$

A desired relation is obtained by multiplying both sides by r^2 and differentiating with respect to r to obtain

$$\frac{d}{dr}(r^2 \mathcal{E}) = \frac{d}{dr}\left(\frac{r^3 \rho}{3\epsilon}\right) = \frac{r^2 \rho}{\epsilon}$$

or

$$\frac{1}{r^2}\frac{d}{dr}(r^2 \mathcal{E}) = \frac{\rho}{\epsilon} \qquad [47]$$

Now the relation $\mathcal{E} = -d\mathcal{V}/dr$ can be used again to give the electric potential $\mathcal{V}$ as a function of r and the charge density, i.e.,

$$\frac{1}{r^2}\frac{d}{dr}\left(r^2\frac{d\mathcal{V}}{dr}\right) = -\frac{\rho}{\epsilon} \qquad [48]$$

This expression, which in fact is valid for any charge distribution $\rho(r)$, might be recognized as the form of the Poisson equation appropriate to a spherically symmetric problem.

This result gives the potential in terms of a general charge distribution $\rho(r)$. Elimination of $\rho(r)$ from Eq. [42] by means of Eq. [48] leads to the relation

$$\frac{1}{r^2}\frac{d}{dr}\left(r^2\frac{d\mathcal{V}}{dr}\right) = \frac{2 \times 10^3 \mathfrak{N} e^2 \mu \mathcal{V}}{kT} \qquad [49]$$

or

$$\frac{d}{dr}\left(r^2\frac{d\mathcal{V}}{dr}\right) = \frac{2 \times 10^3 \mathfrak{N} e^2 \mu}{\epsilon kT} r^2 \mathcal{V} \qquad [50]$$

If β is introduced as

$$\beta = \frac{2 \times 10^3 \mathfrak{N} e^2 \mu}{\epsilon kT}$$

Eq. [50] becomes

$$\frac{d}{dr}\left(r^2\frac{d\mathcal{V}}{dr}\right) = \beta r^2 \mathcal{V} \qquad [51]$$

Equation [51], known as the *Poisson-Boltzmann equation,* is a differential equation that can be solved to give $\mathcal{V}$ as a function of r. A certain amount of manipulation is necessary to put Eq. [51] in an easily soluble form. The solution, as can readily be verified by substitution in Eq. [51], is

$$\mathcal{V} = \frac{Ze}{4\pi\epsilon r} e^{-\sqrt{\beta} r} \qquad [52]$$

Thus, in place of the potential function $Ze/4\pi\epsilon r$ that is contributed by a reference charge Ze, we have a modification that depends on β which, among other things, is a function of the ionic strength μ. Illustrations of the effect this factor produces are shown in Fig. 22-11.

We can now return to the question of the ion atmosphere that surrounds

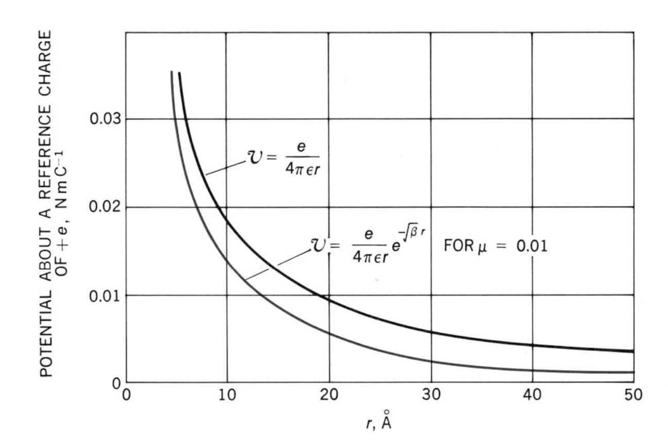

FIGURE 22-11
Illustration of the potential about a reference charge +2 according to Eq. [52] for the case of an aqueous solution at 25°C with ionic strength 0.01.

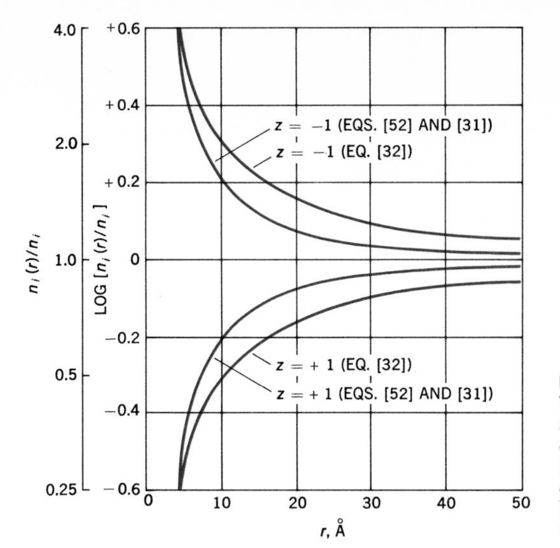

FIGURE 22-12
Variation in ion densities about a reference ion of charge $+e$ according to the Debye-Hückel treatment compared with that which ignores the effect of the neighboring ions.

a reference ion of charge Ze. Although analytically more convenient expressions will be needed later, we can substitute Eq. [52] into Eq. [31] to calculate the ratio $n_i(r)/n_i$ for ions of various types that are contributors to the ionic strength μ of the solution. Some examples are shown in Fig. 22-12.

In a similar way the charge density that exists around the reference ion can now be determined if Eq. [52] is inserted into Eq. [33]. A more convenient expression is obtained if we restrict our attention to solutions and regions for which $\sqrt{\beta}r \ll 1$, which occurs for low ionic strengths and regions not too far removed from the reference ion. Then the exponential of Eq. [52] can be expanded to give

$$\mathcal{V} = \frac{Ze}{4\pi\epsilon r}(1 - \sqrt{\beta}r + \cdots)$$

or

$$\mathcal{V} \approx \frac{Ze}{4\pi\epsilon r} - \frac{\sqrt{\beta}Ze}{4\pi\epsilon} \qquad [53]$$

In this form the difference in the potential energy as a result of the ions in the ion atmosphere is shown explicitly by the second term on the right of Eq. [53]. This term will be of use in our treatment of ion activities in Sec. 23-7. Here it allows the convenient calculation of the charge-density results of Fig. 22-13.

22-14 Interpretation of the Strong Electrolyte Conductance Results

Section 22-13 has shown the way in which the Debye-Hückel theory develops the idea of an ion atmosphere surrounding each ion in an aqueous solution. Although the ions move in a random manner, it was shown that if we focused attention on one ion, that ion would be surrounded more by oppositely charged ions than by like charges. An ion has therefore an oppositely charged *ion atmosphere,* as indicated in Fig. 22-13. The ionic distribution can be looked on as resembling an expanded and loosely held NaCl crystal; for example, the overall arrangement places each ion among nearest neighbors of the opposite charge. This ion atmosphere around each ion is, of course, better formed at higher concentrations.

The application of an electric field, as in a conductance experiment, results initially in the movement of the central ion away from the center of the oppositely charged sphere. The distorted ion atmosphere tends to oppose the applied field, and this decreases the current produced by a given applied electric field. Since the ion atmosphere is more important at higher concentrations, this decrease becomes more important at higher concentrations. The Debye-Hückel evaluation of this factor showed that it contributed to the observed $\sqrt{c}$ dependence of the equivalent conductance. This effect is further enhanced by the tendency of the oppositely charged ions that predominate in the ion atmosphere to move in the opposite direction.

The ionic-atmosphere drag depends on the fact that the atmosphere does not instantaneously adjust itself to the new positions of the central ion. One says that the ionic atmosphere has a *relaxation time;* i.e., when a stress is applied, it takes a finite time for the atmosphere to relax, or to be re-established. The mechanism by which this occurs, as the central ion moves, is better thought of as a building of new ions on the front of the atmosphere and the dropping of some off the back, rather than a maintaining of the same set of ions, which move to keep up with the central ion.

The second factor that acts to decrease the conductance at higher concentrations is an enhanced frictional drag that sets in. When an electric field is applied, the ions set off to the oppositely charged elec-

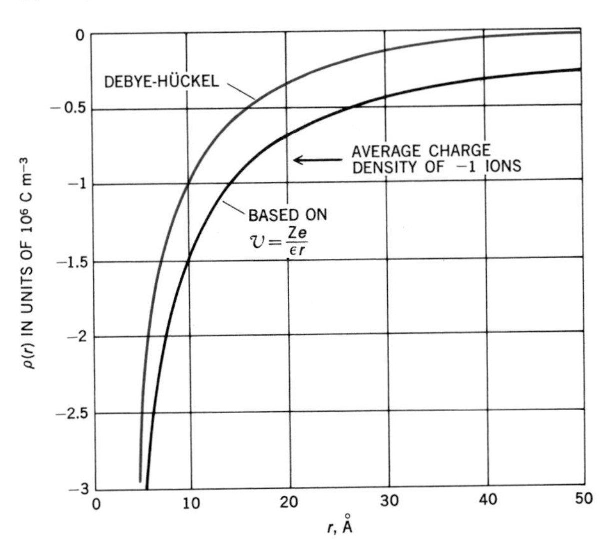

FIGURE 22-13
The charge density in a solution with 0.01 M $+1$ and -1 ions, around a reference ion of charge $+e$ according to the Debye-Hückel theory, Eqs. [38] and [52], and without regard for the effect of the neighboring ions on the potential of the reference ion.

trodes. Each ion moves with a velocity that depends on a balance between the electric force and the viscous drag. The average velocity, and therefore the current, are concentration-dependent because the ions can be thought of as carrying along with them their many solvating molecules, and at higher concentrations an ion seems to swim against the current produced by the oppositely charged, solvated ions moving in the opposite direction.

The theory of Debye and Hückel has been used to draw the slopes of the straight, dashed lines of Fig. 22-2. It is apparent that the effects considered by the theory are adequate to explain the conductance behavior of strong electrolytes up to concentrations of about 0.01 M. Thus the Arrhenius assumption that a decrease in conductance must be interpreted as a decreased number of conducting particles cannot be maintained, and the relation $\alpha = \Lambda/\Lambda_0$ can be applied only when the Debye-Hückel effects are not appreciable or are corrected for. Only for solutions with low ionic concentrations will these effects be small. For weak electrolytes, therefore, one can still rely on Λ/Λ_0 to give a value that can be interpreted primarily in terms of a degree of dissociation.

Higher concentrations, i.e., in the 0.1- and 1-M regions, are not dealt with by the Debye-Hückel theory and are still not susceptible to any satisfactory theoretical treatment. The effects dealt with by the Debye-Hückel theory will still exist, but they are not easily evaluated quantitatively. Furthermore, new features become important, and of these, attention is often directed to the formation of *ion pairs* or *ion triplets*. Primarily as a result of the work of Bjerrum and Fuoss, it has been recognized that a more specific attraction between oppositely charged ions must be recognized at concentrations in the molar range. Bjerrum has calculated that, for a monovalent strong electrolyte with ions of about 2 Å diameter, about 20 percent of the ions, on the average, may be present as ion pairs; i.e., the random motion of the ions in solution will be affected by the electrostatic attractions so that at any time 20 percent of the ions will be making "sticky" collisions with oppositely charged ions. Such ion pairs will affect the conductance and other electrolyte properties. The recognition of these ion pairs is particularly important in solvents of low dielectric constant. In a solvent like chloroform, for example, the neutralization reaction between an acid and a base leads almost exclusively to the formation of ion pairs. One has, for example,

$$\mathrm{HAc} + \mathrm{NEt_3} \rightarrow \mathrm{CH_3{-}\overset{\overset{\displaystyle O}{\|}}{C}{-}O^-} \cdots \mathrm{H{-}\overset{+}{N}Et_3}$$

The recognition of the importance of ion pairs is to some extent a return to the Arrhenius view. For an electrolyte like NaCl, the ion pair that is postulated in the more concentrated solutions might equally well be called an undissociated NaCl molecule. These molecules are present, however, to a much less extent and only in much more concentrated solutions than expected by Arrhenius. The distinction between an ion pair and a molecule becomes more real when one thinks of a strong acid. The ion pair, of HCl for example, would be depicted as

$$\begin{array}{l} H \\ \;\;\diagdown \overset{+}{O}—H\text{-----}Cl^- \\ \;\;\diagup \\ H \end{array}$$

whereas the solvated undissociated molecule would be represented as

$$\begin{array}{l} H \\ \;\;\diagdown O\text{-----}H—Cl \\ \;\;\diagup \\ H \end{array}$$

As with the extent of solvation, the nature of an ion pair, or the formation of undissociated molecules in solutions of strong electrolytes, must be treated with some care. A very short lifetime of an ion pair, for example, is better treated as a collision between the ions than as molecule formation. In these terms the distinction between colliding species and molecules can become poorly defined and will vary with the experimental approach.

Problems

1 The specific conductance of a 0.1-M KCl solution at 25°C is 1.289 ohm^{-1} m^{-1}. What are the conductance and the resistance of a conductance cell for which the electrodes have an effective area of 2.037 cm^2 and are separated by a distance of 0.531 cm? *Ans.* 20.2 ohms.

2 The resistance of a conductance cell filled with a 0.01-M KCl solution is found to be 8.30 ohms at 25°C. What is the cell constant l/A for the cell?

3 A cell whose resistance when filled with 0.1 M KCl is 192.3 ohms is measured to be 6363 ohms when filled with 0.003186 M NaCl solution. All measurements are at 25°C. What are the specific and the equivalent conductance of the NaCl solution? *Ans.* $\Lambda = 0.01224$ ohm^{-1} m^2.

4 The limiting equivalent conductance of NH_4Cl at 25°C is 0.01497; that of NaOH is 0.02478; and that of NaCl is 0.012645. Calculate the limiting equivalent conductance of NH_4OH, and compare with the value reported in Table 22-2.

5 Show from the expression for the equilibrium constant in terms of the concentration of the electrolyte and the degree of dissociation α for an electrolyte AB that, as the concentration approaches zero, the value of α approaches unity, regardless of the value of K.

6 The limiting equivalent conductance of NaOH is 0.02478. With this datum and the results of Table 22-2, calculate what the equivalent conductance of a completely ionized mole of water at infinite dilution in the solvent water at 25°C would be. *Ans.* 0.05476.

7 Calculate, according to the Arrhenius theory, the degree of dissociation and the dissociation equilibrium constants for $CuSO_4$, up to a 1-M concentration at 25°C, from the data of Table 22-2.

8 Calculate according to the Arrhenius theory the degree of dissociation and the dissociation equilibrium constants for the reaction $H_2SO_4 \rightleftharpoons H^+ + HSO_4^-$, up to a 0.1-$M$ concentration, from the data of Table 22-2.

9 Calculate the degree of dissociation and the equilibrium constants for the dissociation of $CuSO_4$ from the van't Hoff i factors of Table 22-6, and compare with the results of Prob. 7.

10 Calculate the degree of dissociation and the equilibrium constants for the dissociation of H_2SO_4 from the van't Hoff i factors of Table 22-6, and compare with the values obtained in Prob. 8.

11 At low concentrations the freezing-point-depression constant $-\Delta T_{\text{fp}}/m$ for one-to-one electrolytes in water approaches the limiting value of 3.72. At higher concentrations the values of $-\Delta T_{\text{fp}}/m$ deviate from this limit. Calculate the value that would be expected for a 1-m solution as a result of the removal of water molecules from the solvent by solvation of each of the ions by six water molecules.

In view of the results shown in Table 22-5 for NaCl, would such an explanation be able to account for the observed variation in $\Delta T_{\text{fp}}/m$?

12 Show by a diagram the process by which current is carried through a solution of NaCl, and show the electrode reactions that carry the current between the solutions and the electrode when an aqueous solution of NaCl is electrolyzed.

13 Show by a diagram the conduction process through an aqueous solution and at the electrodes when an aqueous solution of Na_2SO_4 is electrolyzed between inert electrodes.

Write the electrode reactions.

If 0.342 A is passed through such a cell for 4.80 min, how many equivalents and how many moles of the products of electrolysis will be obtained?

Ans. Equivalents of O_2 = 0.00102; moles of O_2 = 0.00025.

14 What volumes of gases, measured at 1 atm pressure and 0°C, will be obtained at the electrodes when 1000 C of charge is passed through an aqueous NaOH solution?

15 What weight of copper will be deposited at the cathode by the passage of 0.473 A of current through a solution of copper sulfate for 5 min?

Ans. 0.0467 g.

16 A determination of the transference numbers of cadmium and iodide ions by Hittorf gave the following data:

A stock solution of CdI_2 was prepared, and its concentration was determined, by precipitation of iodide as AgI, to be 0.002763 g of CdI_2 per gram of solution.

Another sample of this solution was placed in a Hittorf transference-number cell, and current was passed through the cell. It was found that 0.03462 g of cadmium was deposited at the cathode by the passage of the current. Furthermore, analysis of the anode-compartment solution, which weighed 152.643 g, indicated the presence of 0.3718 g of cadmium iodide.

a What are the electrode reactions?

b Indicate diagrammatically, using t_+ and t_- to represent the transference numbers, the changes in the amounts of the ions in the anode and cathode compartments as a result of migration and electrode reactions.

c How many coulombs of charge were passed through the cell in the experiment?

d What are the transference numbers of Cd^{++} and I^- in this CdI_2 solution?

Ans. $t_+ = 0.443$.

e What was the change in the equivalents of CdI_2 in the cathode compartment?

17 A solution is prepared so that it is 0.01 M in HCl and 0.1 M in NaCl.

a Can the fraction of current carried by the various ions of this solution be rigorously deduced from any of the data given in this chapter?

b Estimate a value for the fraction of the current carried by each of the ions.

18 In aqueous solutions chlorine is hydrolyzed, to some extent, according to the reaction

$$Cl_2 + H_2O \rightleftharpoons H^+ + Cl^- + HOCl$$

The hypochlorous acid is not appreciably dissociated. At 25°C the specific conductance of a 0.0246-M chlorine solution was found to be 0.68 ohm^{-1} m^{-1}. What is the fraction of Cl_2 that has been hydrolyzed? *Ans.* 0.65.

19 The following equivalent conductances of sodium propionate at 25°C have been reported by Belcher [*J. Am. Chem. Soc.*, **60**:2746 (1938)]:

Conc. (mol liter^{-1})	0.002178	0.004180	0.007870	0.01427	0.02597
Λ (*ohm*$^{-1}$ *m*^{2})	0.008253	0.008127	0.007972	0.007788	0.007564

a What is the limiting conductance of sodium propionate?

b What, in view of the data of Table 22-2, is the limiting conductance of propionic acid?

c At a concentration of 1 M the equivalent conductance of propionic acid is 1.4×10^{-4} m^2 ohm^{-1}. What is the degree of dissociation of propionic acid in this solution?

d Estimate from the curves of Fig. 22-2 or the data of Table 22-2 what effect the interionic interaction would have on this conductance measurement.

e Deduce a dissociation constant for propionic acid.

20 Calculate the ionic mobility of the Cl^- ion at an ionic concentration of 0.1 g ion per liter from the transference numbers for the electrolytes HCl, NaCl, and KCl of Table 22-8 and the equivalent conductances of Table 22-2. Recognize that only at infinite dilution is the ionic conductance a property of the ion rather than the electrolyte.

21 The specific conductance, at 18°C, of a saturated silver iodate solution is 1.19×10^{-3} ohm^{-1} m^{-1} more than the water used to prepare the solution. The sum of the limiting equivalent conductances of Ag^+ and IO_3^- is found, from measurements on more soluble salts, to be 0.00873 ohm^{-1} m^2. Calculate the solubility product $[Ag^+][IO_3^-]$ at 18°C. What value of the solubility is obtained if it is assumed that no species other than Ag^+ and IO_3^- are present in the solution?

22 At 18°C the specific conductance of water saturated with CaF_2 is 3.86×10^{-3} and that of the water used in the preparation of this solution is 0.15×10^{-3}. The equivalent ionic conductances at infinite dilution of Ca^{++} and F^- are 0.00510 and 0.00470 ohm^{-1} m^2, respectively. Calculate the solubility product for CaF_2 and, assuming only Ca^{++} and F^- ions in solution, the solubility of CaF_2. *Ans.* Solubility = 0.0148 g $liter^{-1}$.

23 Calculate the energy required to dissociate 1 mol of NaCl in the solvents acetonitrile and benzene, which have dielectric constants 39 and 2.3, respectively. Ignoring the entropy change accompanying dissociation, and the solvation of the ions, which is in fact very important in the polar solvents, use Boltzmann's distribution to deduce the relative amounts of dissociated and undissociated NaCl in benzene, in acetonitrile, and in water at 25°C.

24 Using data given in previous tables and the result that 74 kJ of heat is liberated when 1 mol of HCl is added to a large amount of water, calculate the energy change for the reaction

$$H^+(g) + Cl^-(g) \xrightarrow{H_2O} H^+(aq) + Cl^-(aq)$$

What is the difference in the hydration energy of Na^+ and H^+ in dilute aqueous solutions?

***25** Test the validity of the approximate equation [53] by plotting it and its parent equation [52] on the same graph for some selected values of β.

***26** At what distance from a reference $+e$ ion has the charge density dropped to within 10 percent of the bulk-solution value for the example of Fig. 22-13? What would this distance be if the effect of the reference ion were undiminished by the neighboring ions?

***27** Plot curves showing the charge density about a $+2$ reference ion as a result of $+1$ and -1 ions in the solution, each at a concentration of 0.01 M, according to the Debye-Hückel theory and on the basis of no attenuation of the potential of the reference charge by the surrounding ions.

***28** The charge density about a reference ion, according to the treatment of Sec. 22-13, is dependent only on the charge of the reference ion and the ionic strength of the solution, and not on the number and charges of the ions present.

Using the Debye-Hückel expression for $\mathcal{V}$ given by Eq. [52] or [53], calculate, using Eq. [42], the charge density at 5, 10, 20, and 40 Å from a reference $+1$ charge for:

a A solution containing 4×10^{24} ions of charge $+1$ and 4×10^{24} ions of charge -1 per cubic meter.

b A solution containing 1×10^{24} ions of charge $+2$ and 1×10^{24} ions of charge -2 per cubic meter.

(Both solutions have the same ionic strength.)

References

Some of the references listed here include discussion of electrochemical cells and activities and activity coefficients. These are not dealt with in this text until Chap. 23.

GURNEY, R. W.: "Ionic Processes in Solution," McGraw-Hill Book Company, New York, 1953.

ROBINSON, R. A., and R. H. STOKES: "Electrolyte Solutions," Butterworth & Co. (Publishers), Ltd., London, 1959.

MONK, C. B.: "Electrolytic Dissociation," Academic Press, Inc., New York, 1961.

DAVIES, C. W.: "Ion Association," Butterworth & Co. (Publishers), Ltd., London, 1962.

FUOSS, R. M.: "Electrolytic Conductance," Interscience Publishers, Inc., New York, 1959.

KORTUM, G.: "Treatise on Electrochemistry," 2d ed., Elsevier Publishing Company, Amsterdam, 1965. A revised edition of a comprehensive treatment of electrochemistry.

HARNED, H. S., and B. B. OWEN: "The Physical Chemistry of Electrolytic Solutions," Reinhold Publishing Corporation, New York, 1958.

23

THE ELECTROMOTIVE FORCE OF CHEMICAL CELLS AND THERMODYNAMICS OF ELECTROLYTES

In previous treatments of the thermodynamics of chemical reactions, quantitative use could not be made of the fact that the free-energy change for a reaction is equal to the negative of the useful work which can theoretically be obtained from the reaction. It was necessary to develop more devious ways for determining free-energy changes from measurable thermal quantities.

In this chapter arrangements are considered whereby the work done during reversible chemical change can be determined. The procedure leads to a direct, and frequently very accurate, measurement of the free-energy change of the system. The arrangement consists in allowing the reaction to proceed in an electrochemical cell and determining the work electrically.

As in other thermodynamic studies, we need not be concerned necessarily with theories of the molecular and ionic behavior of the systems being studied. It will be found, however, that the thermodynamic data obtained here provide additional valuable data with which theories of the nature of ionic solutions can be compared.

23-1 Types of Electrodes

Although an electrochemical cell requires two electrodes for an electrochemical reaction to occur and its electrical consequences to be measured, the nature of the cells that can be constructed is best introduced in terms of the individual electrodes. The electrode, including the electrode itself and the reagents that are involved with it, is called the *half cell,* and the component of the total chemical reaction that occurs in the half cell is the *half reaction,* or *electrode reaction.*

Gas Electrodes A gas can be induced to participate in an electrochemical reaction by means of an electrode like that of Fig. 23-1. An example that is important in the development of electrochemical data is the illustrated hydrogen electrode. The electrode is such that on the surface of the inert-metal electrode, the reagents $H_2(g)$, $H^+(aq)$, or $H_3O^+(aq)$ and e^-, the latter in the metallic conductor, can be accommodated. The electrode reaction that can then proceed, written, as is the convention, as a reduction reaction, is

$$H^+ + e^- \rightleftharpoons \tfrac{1}{2}H_2 \qquad [1]$$

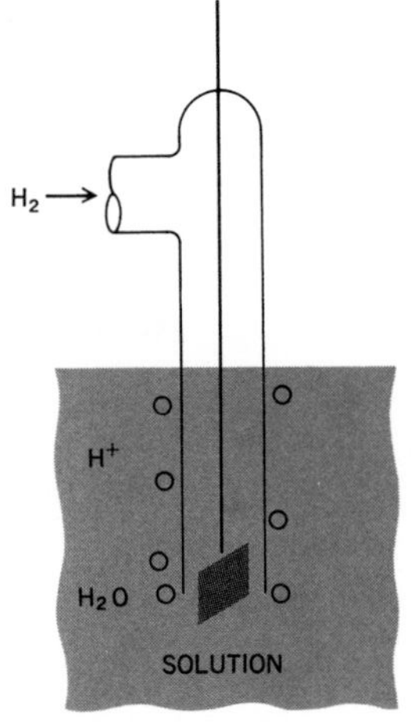

FIGURE 23-1
A gas electrode.

The electrode assembly in which this reaction can proceed is indicated, without producing a diagram like Fig. 23-1, by

$$\mathrm{Pt}|H_2(P \text{ atm})|H^+(c \text{ mol liter}^{-1}) \qquad [2]$$

The vertical lines indicate interfaces between physical states, and the significant features of these states are indicated. The symbol Pt is more restricted than necessary, it being used to imply any inert-metal electrode.

Oxidation-Reduction Electrodes Although all electrodes involve oxidation-reduction reactions in the sense that electrons are being gained or lost at the electrode, this term is generally used for electrodes consisting of an inert metal dipping into a solution containing two different oxidation states of a species. An example consists of a platinum wire dipping into a solution containing ferrous and ferric ions. Such a cell is described by

$$\mathrm{Pt}|Fe^{++}(c_1), Fe^{3+}(c_2)$$

The comma is used to separate the two chemical species which are in the same solution. These electrodes are similar to the gas electrodes except that the two species involved in the electrode reaction are ions. The electrode reaction in the example is

$$Fe^{3+} + e^- \rightleftharpoons Fe^{++} \qquad [3]$$

and there is the possibility of the electrode either donating or accepting electrons.

Oxidation-reduction electrodes can also be made with organic molecules that can exist in two different oxidation states. A generally used material of this type that is related to important biochemical oxidation-reduction reagents is the system of hydroquinone and quinone, which can form the oxidation-reduction system

$$\text{O}=\text{C}_6\text{H}_4=\text{O} + 2\text{H}^+ + 2e^- \rightleftharpoons \text{HO}-\text{C}_6\text{H}_4-\text{OH} \qquad [4]$$

The presence of a platinum electrode in a solution containing these two species again clearly provides an electrode that can donate or accept electrons. One usually represents hydroquinone by QH_2 and quinone by Q, and the cell is then abbreviated as

$$\text{Pt}|\text{QH}_2, \text{Q}, \text{H}^+(c)$$

The occurrence of H^+ as a reagent in Eq. [4] makes it necessary also to state its concentration in the system. This electrode is generally known as the *quinhydrone* electrode, from the name of the crystalline compound $QH_2 \cdot Q$, in which form the material is added to the solution.

Metal–Metal-ion Electrodes The simplest of the electrodes in which the electrode material plays a chemical role is one in which a metal electrode dips into a solution containing ions of the metal. An example is that of a metallic silver electrode in a silver nitrate solution. The electrode is represented as

$$\text{Ag}|\text{Ag}^+(c)$$

and the electrode reaction is

$$\text{Ag}^+ + e^- \rightleftharpoons \text{Ag} \qquad [5]$$

Such an electrode can be set up with any metal that is of intermediate activity. Very active metals react directly with the water itself and cannot be used for such an electrode.

Amalgam Electrodes A variation of the previous electrode is one in which the metal is in the form of an amalgam, i.e., is dissolved in mercury, rather than in the pure form. Electrical contact is made by a platinum wire dipping into the amalgam pool. The reaction is the same as in the metal–metal-ion electrode, the mercury playing no chemical role.

The particular value of amalgam electrodes is that active metals such as sodium can be used in such electrodes. A sodium-amalgam electrode is represented as

$$\text{Na(in mercury at } c_1)|\text{Na}^+(c_2)$$

where the concentration of the sodium metal in the mercury as well as that of the sodium ion in the solution must be given. In addition to allowing the study of active metals, the amalgam electrode is of interest in illustrating some of the thermodynamic relations which will be obtained for electrochemical cells.

Metal–Insoluble-salt Electrodes A more elaborate, but usually satisfactory and frequently used, electrode consists of a metal in contact with an insoluble salt of the metal, which in turn is in contact with a solution containing the anion of the salt. An example is represented as

$$\text{Ag}|\text{AgCl}|\text{Cl}^-(c)$$

The electrode process at such an electrode is

$$\begin{array}{r} \text{Ag}^+ + e^- \rightleftharpoons \text{Ag} \\ \text{AgCl}(s) \rightleftharpoons \text{Ag}^+ + \text{Cl}^- \\ \hline \text{AgCl}(s) + e^- \rightleftharpoons \text{Ag} + \text{Cl}^- \end{array} \quad [6]$$

The electrode reaction involves only the concentration of Cl^- as a variable, in contrast with the Ag|Ag^+ electrode which has the Ag^+ concentration as a variable.

The most frequently used electrode of this type is the *calomel electrode*. This consists of metallic mercury in contact with calomel, Hg_2Cl_2, which is in contact with a chloride solution. Figure 23-2 shows

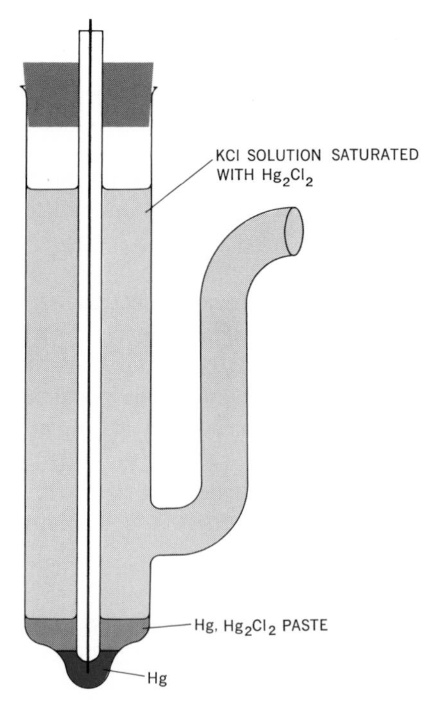

FIGURE 23-2
The calomel electrode.

the usual arrangement of this electrode. The electrode reaction is

$$\begin{array}{l} Hg^+ + e^- \rightleftharpoons Hg \\ \frac{1}{2}Hg_2Cl_2 \rightleftharpoons Hg^+ + Cl^- \\ \hline \frac{1}{2}Hg_2Cl_2 + e^- \rightleftharpoons Hg + Cl^- \end{array} \qquad [7]$$

The calomel electrode is quite easily prepared and is frequently used. The electrode is generally made with a chloride solution of 0.1 *M*, 1.0 *M*, or saturated KCl. The "saturated calomel electrode" is the most common, and for this electrode the calomel, Hg_2Cl_2, is ground up with solid KCl, and the solution is a saturated KCl solution. For this electrode the concentration of the chloride is therefore fixed at a given temperature, and one has an electrode whose emf is completely determined. This is, as we shall see, sometimes convenient.

23-2 Electrochemical Cells, Emfs, and Cell Reactions

When pairs of electrodes such as those of Sec. 23-1 are combined, and are connected by an external electrical conductor, an electric current will flow and chemical reactions will occur in the two half cells. Combining electrodes is done without complication if, as in the example of Fig. 23-3, the electrodes can both operate in the same solution. When this is not tolerated, a connection must be made between the solutions that allow ionic conduction to occur between the half cells but prevents mixing of the two half-cell solutions. A KCl *salt bridge,* illustrated in the assembly of Fig. 23-4, is often used. Such a coupling device, indicated by a double vertical line in a cell diagram, has no net influence on the cell reaction, and although some difficulties are introduced into

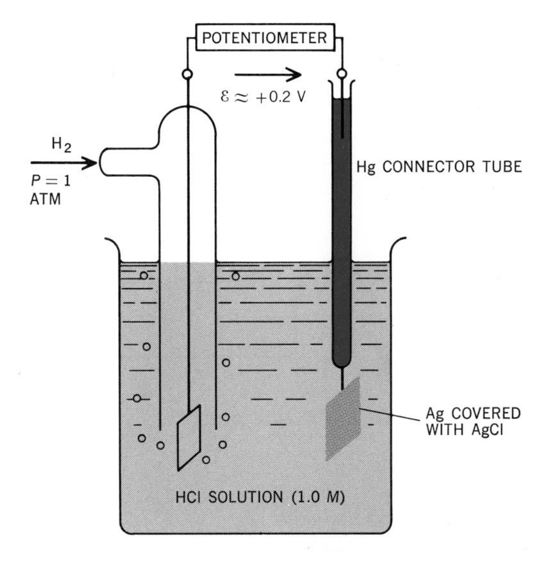

FIGURE 23-3
Electrode arrangement corresponding to the cell $Pt|H_2(1\ atm)|HCl(1.0\ M)|AgCl|Ag$ for a cell in which both electrodes can operate in the same solution.

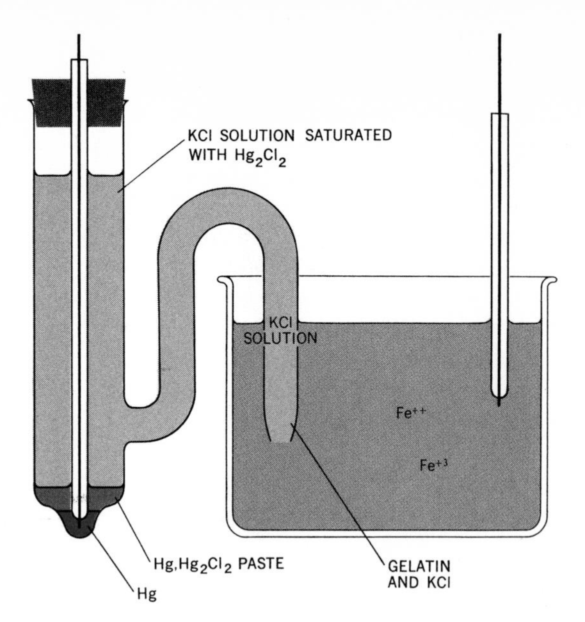

FIGURE 23-4
A cell composed of two electrodes whose solutions cannot be mixed.

the quantitative interpretation of the cell voltage, or *electromotive force* (emf), we will for the time being set these problems aside.

The voltage, or emf, of a cell, like that of Fig. 23-3 or 23-4, can be measured by placing a voltmeter across the terminals. However, it would be found that the measured voltage would depend on the current that is drawn off by the voltmeter. The maximum voltage is found to be that produced in the limit as zero current is drawn. In practice a suitable value for this characteristic of the cell can be obtained with a vacuum-tube voltmeter, or VTVN, which draws a very small current, or more accurately, by a potentiometer arrangement like that shown in Fig. 23-5.

In reporting emf results from laboratory measurements, the direc-

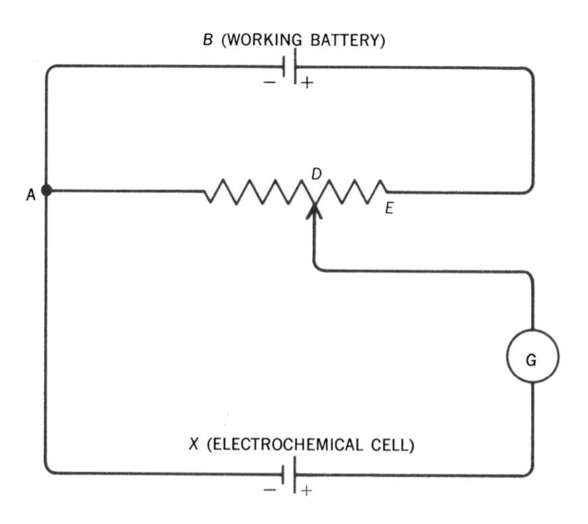

FIGURE 23-5
The potentiometric method for measuring the voltage produced by an electrochemical cell operating at a state of balance. A fraction AD/AE of the working battery can be picked off to produce a voltage that drives a current in the lower circuit in an opposite direction to that in which the electrochemical cell drives the current. When position D is such that no current flows in the lower circuit, as determined by the galvanometer, the voltage of the electrochemical cell at a state of balance is obtained as AD/AE times the voltage of the working battery.

tion of the electron flow would be given, as well as the numerical value of the emf. This direction can be indicated along with the diagram of the electrochemical cell by agreeing that *the emf will be called positive if there is a tendency for electrons to be driven through the external circuit from the electrode written on the left to the electrode written on the right.* For example, the electrochemical cell of Fig. 23-3 is found to drive electrons from the hydrogen electrode to the silver electrode with an emf of about 0.2 V. This could be reported by writing the cell as

$$\mathrm{Pt}|\mathrm{H_2}(1\ \mathrm{atm})|\mathrm{HCl}(1.0\ M)|\mathrm{AgCl}|\mathrm{Ag} \qquad [8]$$

and giving the emf as +0.2 V. If the cell were described as

$$\mathrm{Ag}|\mathrm{AgCl}|\mathrm{HCl}(1.0\ M)|\mathrm{H_2}(1\ \mathrm{atm})|\mathrm{Pt} \qquad [9]$$

the emf would be reported as −0.2 V. In either case the tendency of electrons to be driven through the external circuit from the hydrogen to the silver electrode is shown.

Under balanced, reversible conditions a cell reaction can proceed in either direction, and we must therefore agree on a procedure for writing the cell reaction so that, when a free-energy change is deduced from the electrical work that the cell can perform, the free-energy change will be of appropriate sign for the direction in which the reaction is written.

It is the convention that once a diagram of a cell is written, *the cell reaction will be written so that electrons are accepted from the external circuit by the electrode written on the right and are given up by the electrode on the left.* In the above example, this leads us to write, for Eq. [8],

$$\begin{array}{ll} \text{Right electrode} & \mathrm{AgCl} + e^- \rightleftharpoons \mathrm{Ag} + \mathrm{Cl^-} \\ \text{Left electrode} & \tfrac{1}{2}\mathrm{H_2} \rightleftharpoons \mathrm{H^+} + e^- \\ \hline \text{Cell reaction} & \mathrm{AgCl} + \tfrac{1}{2}\mathrm{H_2} \rightleftharpoons \mathrm{Ag} + \mathrm{H^+} + \mathrm{Cl^-} \end{array} \qquad [10]$$

If the second form, Eq. [9], of the cell diagram had been written, the cell reaction would have to be given as

$$\mathrm{Ag} + \mathrm{H^+} + \mathrm{Cl^-} \rightleftharpoons \mathrm{AgCl} + \tfrac{1}{2}\mathrm{H_2} \qquad [11]$$

The conventions on cell emf and cell reactions together lead to a positive emf being associated with the reaction written in the direction in which it tends to proceed.

23-3 Free-energy Changes for Cell Reactions

When the emf of a cell is measured with the reaction proceeding at a state of balance, or reversibly, as it clearly is when measured with a potentiometer, the work that can be obtained from the cell current is the maximum work, over and above any PV work, that can be delivered to the mechanical reservoir. Thus, in view of the analysis of free energy in Sec. 9-1, this work is equal to the free-energy change accompanying the cell reaction.

The electrical work done by a cell as it drives an Avogadro's number of electrons through an opposing voltage, as imposed, for example, by a potentiometer arrangement, is equal to the product of the total charge of these electrons and the voltage through which this charge is driven. This charge, 96,490 C, of a mole of electrons is frequently encountered and is termed a *faraday,* with symbol $\mathfrak{F}$. With this notation, the work produced as n mol of electrons is driven by the cell emf $\mathcal{E}$ through a nearly equal balancing voltage is

$$\Delta E_{\text{mech res}} = n\mathfrak{F}\mathcal{E} \tag{12}$$

This mechanical energy, being the maximum useful work the reaction can produce, is equal to the *decrease* in the free energy of the reacting system as a result of the process described by the cell reaction. Thus we have

$$\Delta G = -n\mathfrak{F}\mathcal{E} \tag{13}$$

where $\mathcal{E}$ is the reversible emf of the cell. Note that the sign of $\Delta E_{\text{mech res}}$, and thus of ΔG, is such that we do obtain, for the spontaneous direction of reaction, which is associated with a positive emf, the necessary negative ΔG.

Thus, for the cell reaction of Eq. [10] and the $\mathcal{E}$ value of $+0.2$ V, we obtain

$$\begin{aligned} &AgCl + \tfrac{1}{2}H_2 \rightleftharpoons Ag + H^+Cl^- \\ &\Delta G = -(1)(96{,}490)(+0.2) \\ &\quad\;\; = -20 \text{ kJ} \end{aligned} \tag{14}$$

23-4 Standard Emfs and Electrode Potentials

A method for the presentation of the data that are obtained from measurements of the equilibrium emf of electrochemical cells must now be developed. One expects, and a thermodynamic proof of this will be given in Sec. 23-5, that varying the concentration of any reagents that are involved in the electrode process will affect the emf of the cell. In view of this, the emfs of cells with the reagents at standard states are reported. In keeping with the standard states introduced in Chap. 21, these standard-state conditions are again chosen to consist of gases at unit activity, implying for most gases approximately 1 atm pressure, and solutes also at unit activity. For solutes, which for electrochemical cells are frequently ions, the activities, as we saw, can be quite appreciably different from molar concentrations. The way in which activity coefficients can be determined from emf measurements will shortly be presented. For the present it is sufficient to recognize that a standard state is chosen and that it is desired to tabulate the emfs that cells would have for their variable reagents at unit activity.

It is impractical, however, to list the emfs of all possible combinations of electrodes, and it would be much more convenient to have some means of tabulating the relative electron-accepting powers of the individual electrodes.

This can be done by tabulating the emfs of cells composed of a selected reference electrode and a variety of other electrodes. Quite arbitrarily, it is agreed that the hydrogen electrode, with a hydrogen pressure of 1 atm and a unit hydrogen-ion activity in the solution, be used as the reference electrode.

Tabulations such as that of Table 23-1 can then be given of the emfs measured for cells with the standard hydrogen electrode on the left and another electrode on the right. For example,

$$\text{Pt}|\text{H}_2(1\text{ atm})|\text{HCl}(a = 1)|\text{Cl}_2(1\text{ atm})\text{Pt} \qquad [15]$$

The emf of this particular cell is found to be $+1.3595$ V, and it is this value that is given in Table 23-1 for the $Pt|Cl_2|Cl^-$ electrode.

Such tabulated values can also be interpreted as electrode emfs based on a zero emf being assigned to the reference hydrogen electrode. Such electrode emfs are known, if the value is that which would occur for all reagents at unit activity, as *standard electrode potentials* and are indicated by the symbol $\mathcal{V}°$. The hydrogen-chlorine cell data can now be interpreted as

Right electrode	$\frac{1}{2}Cl_2 + e^- \rightleftharpoons Cl^-$	$\mathcal{V}° = +1.3595$ V
Left electrode	$\frac{1}{2}H_2 \rightleftharpoons H^+ + e^-$	$\mathcal{V}° = 0$
Cell reaction	$\frac{1}{2}Cl_2 + \frac{1}{2}H_2 \rightleftharpoons Cl^- + H^+$	$\mathcal{E}° = +1.3595$ V

[16]

The emf of any cell with the reagents in their standard states, not necessarily involving the hydrogen electrode, can be used to obtain data for the standard electrode potentials listed in Table 23-1. Thus the cell

$$\text{Pt}|\text{Cl}_2(1\text{ atm})|\text{HCl}(a = 1)|\text{AgCl}(s)|\text{Ag}$$

is found to have an $\mathcal{E}°$ value of -1.1370 V. This cell emf can be interpreted in terms of the standard *electrode potentials* $\mathcal{V}°$ as

$$\text{Cell emf} = \mathcal{V}°(\text{right electrode}) - \mathcal{V}°(\text{left electrode})$$

or

$$-1.1370 = \mathcal{V}°_{\text{Ag,AgCl}} - (+1.3595)$$

and

$$\mathcal{V}°_{\text{Ag,AgCl}} = 0.2225\text{ V} \qquad [17]$$

TABLE 23-1 Some standard electrode potentials for aqueous solutions at 25°C*

Electrode	Electrode reaction	$\mathcal{V}°$ (volts)
ACID SOLUTION ($a_{H^+} = 1$)		
$Pt \mid F_2 \mid F^-$	$F_2(g) + 2e^- = 2F^-$	+2.87
$Pt \mid H_2O_2 \mid H^+$	$H_2O_2 + 2H^+ + 2e^- = 2H_2O$	+1.77
$Pt \mid Mn^{++}, MnO_4^-$	$MnO_4^- + 8H^+ + 5e^- = Mn^{++} + 4H_2O$	+1.51
$Pt \mid Cl_2 \mid Cl^-$	$Cl_2 + 2e^- = 2Cl^-$	+1.3595
$Pt \mid Tl^+, Tl^{3+}$	$Tl^{3+} + 2e^- = Tl^+$	+1.25
$Pt \mid Br_2 \mid Br^-$	$Br_2 + 2e^- = 2Br^-$	+1.065
$Ag \mid Ag^+$	$Ag^+ + e^- = Ag$	+0.7991
$Pt \mid Fe^{++}, Fe^{3+}$	$Fe^{3+} + e^- = Fe^{++}$	+0.771
$Pt \mid O_2 \mid H_2O_2$	$O_2 + 2H^+ + 2e^- = H_2O_2$	+0.682
$Pt \mid I_2 \mid I^-$	$I_3^- + 2e^- = 3I^-$	+0.536
$Cu \mid Cu^{++}$	$Cu^{++} + 2e^- = Cu$	+0.337
$Pt \mid Hg \mid Hg_2Cl_2 \mid Cl^-$	$Hg_2Cl_2 + 2e^- = 2Cl^- + 2Hg$	+0.2676
$Ag \mid AgCl \mid Cl^-$	$AgCl + e^- = Ag + Cl^-$	+0.2225
$Pt \mid Cu^+, Cu^{++}$	$Cu^{++} + e^- = Cu^+$	+0.153
$Cu \mid CuCl \mid Cl^-$	$CuCl + e^- = CuCl$	+0.137
$Ag \mid AgBr \mid Br^-$	$AgBr + e^- = Ag + Br^-$	+0.0713
$Pt \mid H_2 \mid H^+$	$2H^+ + 2e^- = H_2$	0.0000
$Pb \mid Pb^{++}$	$Pb^{++} + 2e^- = Pb$	−0.126
$Ag \mid AgI \mid I^-$	$AgI + e^- = Ag + I^-$	−0.1518
$Cu \mid CuI \mid I^-$	$CuI + e^- = Cu + I^-$	−0.1852
$Pb \mid PbSO_4 \mid SO_4^{--}$	$PbSO_4 + 2e^- = Pb + SO_4^{--}$	−0.3588
$Pt \mid Ti^{++}, Ti^{3+}$	$Ti^{3+} + e^- = Ti^{++}$	−0.369
$Cd \mid Cd^{++}$	$Cd^{++} + 2e^- = Cd$	−0.403
$Fe \mid Fe^{++}$	$Fe^{++} + 2e^- = Fe$	−0.4402
$Cr \mid Cr^{3+}$	$Cr^{3+} + 3e^- = Cr$	−0.744
$Zn \mid Zn^{++}$	$Zn^{++} + 2e^- = Zn$	−0.7628
$Mn \mid Mn^{++}$	$Mn^{++} + 2e^- = Mn$	−1.180
$Al \mid Al^{3+}$	$Al^{3-} + 3e^- = Al$	−1.662
$Mg \mid Mg^{++}$	$Mg^{++} + 2e^- = Mg$	−2.363
$Na \mid Na^+$	$Na^+ + e^- = Na$	−2.7142
$Ca \mid Ca^{++}$	$Ca^{++} + 2e^- = Ca$	−2.866
$Ba \mid Ba^{++}$	$Ba^{++} + 2e^- = Ba$	−2.906
$K \mid K^+$	$K^+ + e^- = K$	−2.925
$Li \mid Li^+$	$Li^+ + e^- = Li$	−3.045
BASIC SOLUTION ($a_{OH^-} = 1$)		
$Pt \mid MnO_2 \mid MnO_4^-$	$MnO_4^- + 2H_2O + 3e^- = MnO_2 + 4OH^-$	+0.588
$Pt \mid O_2 \mid OH^-$	$O_2 + 2H_2O + 4e^- = 4OH^-$	+0.401
$Pt \mid S \mid S^{--}$	$S + 2e^- = S^{--}$	−0.447
$Pt \mid H_2 \mid OH^-$	$2H_2O + 2e^- = H_2 + 2OH^-$	−0.82806
$Pt \mid SO_3^{--}, SO_4^{--}$	$SO_4^{--} + H_2O + 2e^- = SO_3^- + 2OH^-$	−0.93

*Mostly from a compilation by A. J. deBethune, T. S. Licht, and N. Swendeman, *J. Electrochem. Soc.*, **106:**616 (1959).

By means such as these, convenient pairs of electrodes can be selected and electrode potentials determined. Table 23-1 shows the results that are obtained for the more frequently encountered electrodes. These standard electrode potentials can be called reduction potentials because the listed numbers give the relative tendency for the electrode reaction to proceed with the gain of electrons, i.e., for reduction to occur. The electrodes at the top of the table, with the most positive values of $\mathcal{V}^\circ$, have a high tendency to accept electrons. Those low down in the table have a relatively greater tendency to donate electrons.

The use of this table in calculating a cell emf can be illustrated by the example of the cell

$$\text{Pt}|\text{Tl}^{+}(a = 1),\ \text{Tl}^{3+}(a = 1)\|\text{Cl}^{-}(a = 1)|\text{Hg}_2\text{Cl}_2(s)|\text{Hg}$$

The electrode reactions and the emfs due to the electrodes can be combined to give

Right electrode	$\frac{1}{2}\text{Hg}_2\text{Cl}_2 + e^- \rightleftharpoons \text{Hg} + \text{Cl}^-$	$\text{emf} = \mathcal{V}^\circ_{\text{Hg}\mid\text{Hg}_2\text{Cl}_2\mid\text{Cl}^-} = +0.2676$
Left electrode	$\frac{1}{2}\text{Tl}^+ \rightleftharpoons \frac{1}{2}\text{Tl}^{3+} + e^-$	$\text{emf} = -\mathcal{V}^\circ_{\text{Tl}^{3+},\text{Tl}^+} = -1.25$
Overall reaction	$\frac{1}{2}\text{Tl}^+ + \frac{1}{2}\text{Hg}_2\text{Cl}_2 \rightleftharpoons \frac{1}{2}\text{Tl}^{3+} + \text{Hg} + \text{Cl}^-$	$\text{Cell emf} = \mathcal{E}^\circ = -0.98\ \text{V}$

[18]

Furthermore, for the overall reaction

$$\Delta G^\circ = -n\mathfrak{F}\mathcal{E}^\circ = -(1)(96{,}500)(-0.98) = 94{,}500\ \text{J} \qquad [19]$$

The calculation tells us that the current will tend to flow through the external circuit from the electrode written on the right to that written on the left and that the reaction will tend to proceed spontaneously in the direction opposite to that in which it is written in Eq. [18].

Two features of this calculation should be pointed out. In the first place, the reaction at the left electrode has been written so that it is shown as an oxidation reaction. The sign of the value of the emf of this electrode is therefore the negative of the electrode potential that is found in Table 23-1. The right-electrode equation is a reduction reaction, and the emf is the electrode potential of +0.2676 given in Table 23-1 for that electrode.

The second point to be mentioned is that one can write the reactions, such as those in the example, with one or two electrons exhibited. If two electrons are shown, this corresponds to 2 equiv of reactants, and the free-energy change will be twice as much as for 1 equiv. The emf, of course, will not be altered by a change in the amount of material involved. We might write, therefore,

$\text{Hg}_2\text{Cl}_2 + 2e^- \rightleftharpoons 2\text{Hg} + 2\text{Cl}^-$	$\text{emf} = \mathcal{V}^\circ_{\text{Hg}\mid\text{Hg}_2\text{Cl}_2\mid\text{Cl}^-} = +0.2676$
$\text{Tl}^+ \rightleftharpoons \text{Tl}^{3+} + 2e^-$	$\text{emf} = -\mathcal{V}^\circ_{\text{Tl}^{3+},\text{Tl}^+} = -1.25$
$\text{Tl}^+ + \text{Hg}_2\text{Cl}_2 \rightleftharpoons \text{Tl}^{3+} + 2\text{Hg} + 2\text{Cl}^-$	$\text{Cell emf} = \mathcal{E}^\circ = -0.98\ \text{V}$

[20]

For the overall reaction as written, we calculate

$$\Delta G^\circ = -n\mathfrak{F}\mathcal{E}^\circ = -(2)(96{,}500)(-0.98) = +189{,}000 \text{ J} \qquad [21]$$

Equations [20] and [21] show that the free-energy change as a result of the consumption of 1 mol, that is, 472 g, of mercurous chloride in this way is 189,000 J. Equations [18] and [19] show that for the consumption of half a mole, or 236 g, the free-energy change is 94,500 J. The free-energy change depends, as is seen by the n in the equation, $\Delta G^\circ = -n\mathfrak{F}\mathcal{E}^\circ$, on the amount of reagents being dealt with. The cell emf is independent of the amounts of reagents.

Calculations using the data of Table 23-1 yield only the emfs that electrochemical cells will have if all the reagents are at unit activity. It is now necessary to see how these emfs are related to actual experimental conditions.

23-5 The Concentration and Activity Dependence of the Emf

The dependence of the emf of a cell on the concentration, or more directly on the activity, of the variable reagents can be calculated from our knowledge of the relation between the free energy and the activity.

Consider an electrochemical cell for which the overall chemical reaction is

$$aA + bB \rightleftharpoons cC + dD \qquad [22]$$

Further, suppose that A, B, C, and D are reagents whose concentration can be varied; i.e., they are gases, or ions, or molecules in solution. If, in addition, there are solids involved in the reaction, they will contribute only a constant term to the results, as we shall see.

For species A the free energy of 1 mol can be written

$$\bar{G}_A = G_A^\circ + RT \ln a_A \qquad [23]$$

and for a mol of reagent A, the free energy is

$$\begin{aligned} a\bar{G}_A &= aG_A^\circ + aRT \ln a_A \\ &= aG_A^\circ + RT \ln (a_A)^a \end{aligned} \qquad [24]$$

From expressions like these for all four reagents, the free-energy change for the overall cell reaction can be deduced, as was done in Chap. 9, and it gives

$$\Delta G = \Delta G^\circ + RT \ln \frac{(a_C)^c(a_D)^d}{(a_A)^a(a_B)^b} \qquad [25]$$

where ΔG° is the difference in free energy of the products and the reactants when all reagents are in their standard states. Furthermore, according to the discussion of Sec. 23-3,

$$\Delta G = -n\mathfrak{F}\mathcal{E} \qquad \text{and} \qquad \Delta G^\circ = -n\mathfrak{F}\mathcal{E}^\circ$$

which gives

$$\mathcal{E} = \mathcal{E}^\circ - \frac{RT}{n\mathfrak{F}} \ln \frac{(a_C)^c(a_D)^d}{(a_A)^a(a_B)^b} \qquad [26]$$

This important equation shows how the emf of a cell can be calculated from the emf for all reagents in their standard states and the activities of the reagents. It is clear that when the activities of all the reagents are unity, the logarithmic term drops out and $\mathcal{E} = \mathcal{E}^\circ$, that is, the emf for the standard states.

At 25°C, the temperature at which most standard electrode potentials are reported, the factor before the logarithm term can be explicitly worked out and gives

$$\mathcal{E} = \mathcal{E}^\circ - \frac{0.05915}{n} \log \frac{(a_C)^c(a_D)^d}{(a_A)^a(a_B)^b} \qquad [27]$$

Included in the numerical factor is the term 2.3026 for the conversion to logarithms to the base 10.

It should be recognized that the activity term has the familiar form of the equilibrium-constant expression. It is not, however, the equilibrium constant, since the activities of the reagents have the values that are determined by the solutions used to make up the cell, and these are not generally the equilibrium values.

As in the development of the equilibrium-constant expression, the activities of solid or otherwise fixed-concentration reagents are not explicitly included. The contribution to the free energy and the emf of the cell of such reagents is implicitly included in the ΔG° and $\mathcal{E}^\circ$ terms.

The use of Eq. [27] is illustrated by the calculation of the emf of the cell,

Pt | $H_2(g)$pressure,P | HCl,activity,a | AgCl | Ag

The overall reaction and the standard emf are calculated from the reactions and the data of Table 23-1 as

Right electrode	$AgCl + e^- \rightarrow Ag + Cl^-(a)$	emf $= \mathcal{V}^\circ = +0.2225$
Left electrode	$\frac{1}{2}H_2 \rightarrow H^+(a) + e^-$	emf $= -\mathcal{V}^\circ = 0$
Overall reaction	$\frac{1}{2}H_2 + AgCl \rightarrow Ag + H^+(a) + Cl^-(a)$	Cell emf $= \mathcal{E}^\circ = +0.2225$ V

[28]

Equation [27] now gives

$$\mathcal{E} = \mathcal{E}^\circ - \frac{0.05915}{1} \log \frac{(a_{H^+})(a_{Cl^-})}{(a_{H_2})^{1/2}} \qquad [29]$$

To proceed to a calculated emf we need to know the activities of H^+ and Cl^- for an HCl solution of a given concentration and for H_2 at its given pressure. The methods of Secs. 21-5 and 21-6 yield the necessary solute activities, and those of Sec. 9-5 or 9-6 give the necessary activity-pressure relations for a gas. In practice, however, ionic activities are often determined by the measurement of the cell emf and the deduction, by use of Eq. [29], of the activities. The procedure will be shown in the following section.

23-6 Activities from Emf Measurements

The unknowns in an equation like Eq. [29] for the emf of a cell are really both the activities of the variable concentration species *and* the standard emf $\mathcal{E}°$. Tabulated values of the latter have already been given, but now we must see how these are determined. This determination can be seen to be related to the measurement of activities since $\mathcal{E}°$ values are defined as the cell emf for reagents at unit activity.

Consider again the cell

$Pt|H_2(1\text{ atm})|HCl(c)|AgCl(s)|Ag$

At a pressure of 1 atm, hydrogen will behave ideally, and its activity will be very nearly equal to its pressure. Then Eq. [29] becomes

$$\mathcal{E} = \mathcal{E}° - \frac{0.05915}{1} \log \frac{(a_{H^+})(a_{Cl^-})}{(P_{H_2})^{1/2}} \qquad [30]$$

At a hydrogen pressure of 1 atm this becomes

$$\mathcal{E} = \mathcal{E}° - 0.05915 \log (a_{H^+})(a_{Cl^-}) \qquad [31]$$

In terms of concentrations and activity coefficients the expression can be written

$$\begin{aligned}\mathcal{E} &= \mathcal{E}° - 0.05915 \log (\gamma_+[H^+]\gamma_-[Cl^-]) \\ &= \mathcal{E}° - 0.05915 \log \gamma_+\gamma_- - 0.05915 \log [H^+][Cl^-] \end{aligned} \qquad [32]$$

The brackets imply that the concentrations expressed here are molar concentrations. The directly measurable quantities are now rearranged to the left side of the equation to give

$$\mathcal{E} + 0.05915 \log [H^+][Cl^-] = \mathcal{E}° - 0.05915 \log \gamma_+\gamma_- \qquad [33]$$

Furthermore, for solutions containing appreciable HCl, the concentrations of H^+ and Cl^- are equal and correspond to c, the hydrochloric acid concentration. Let us also introduce the *mean activity coefficient* $\gamma_\pm$, defined as $\sqrt{\gamma_+\gamma_-}$, for each of these activity coefficients. Then

$$\mathcal{E} + 0.05915 \log c^2 = \mathcal{E}° - 0.05915 \log \gamma_\pm^2 \qquad [34]$$

and finally,

$$\mathcal{E} + 0.11830 \log c = \mathcal{E}° - 0.11830 \log \gamma_\pm \qquad [35]$$

The right side of the equation is made up of a constant term and the logarithmic term, which, for solutions in the Debye-Hückel limit-

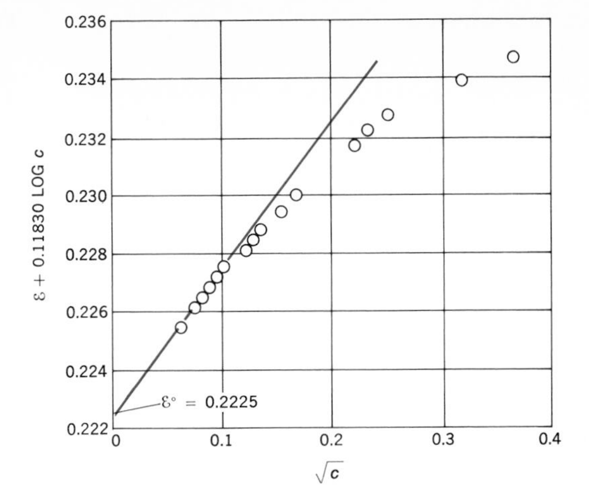

FIGURE 23-6
Determination of $\mathcal{E}°$ for the cell Pt|H_2(1 atm) |HCl(c)|AgCl(s)|Ag from the measured values of $\mathcal{E}$ at various concentrations of HCl. The straight-line extrapolation is drawn with the slope predicted by the Debye-Hückel limiting law. *(Partly from L. P. Hammett, "Introduction to the Study of Physical Chemistry," McGraw-Hill Book Company, New York, 1952.)*

ing-law region, will be shown in Sec. 23-7 to be expected to be proportional to $c^{1/2}$. The left side of the equation, which can be obtained from measurements of $\mathcal{E}$ and c, is therefore plotted against $c^{1/2}$, as shown in Fig. 23-6.

At $c = 0$ the activity coefficient must go to unity, and the logarithmic term on the right side of the equation will vanish. The extrapolation of the data of Fig. 23-6 to $c^{1/2} = 0$ gives, therefore, a value of $\mathcal{E}°$. In this way one finds

$$\mathcal{E}° = +0.2225 \text{ V} \qquad [36]$$

The emf $\mathcal{E}°$ of this cell for the reagents in their standard states can be written

$$\mathcal{E}° = \mathcal{V}^{\circ}_{\text{Ag,AgCl}} - \mathcal{V}^{\circ}_{\text{H}_2}$$

which, with $\mathcal{V}^{\circ}_{\text{H}_2}$ taken as zero, yields

$$\mathcal{V}^{\circ}_{\text{Ag,AgCl}} = 0.2225 \text{ V} \qquad [37]$$

as given in Table 23-1. In a similar way the electrode potentials that other electrodes would have, if all reagents were at unit activity, can be obtained.

Now it can be shown how activity coefficients of HCl can also be determined from the data of Fig. 23-6. Reference to Eq. [35] shows that once a value for $\mathcal{E}°$ is obtained, the measured value of the left side of the equation allows the calculation of the only remaining unknown, the log $\gamma_\pm$ term. One obtains, therefore, as shown in Table 23-2, the mean-activity coefficients of HCl at any concentration for which emf data are available.

In a similar way, the activity coefficient of other electrolytes involved in an electrochemical reaction can be determined. As before, no way is found for the determination of the individual activity coefficients, and therefore again mean-activity coefficients are all that can be considered.

TABLE 23-2 Mean-activity coefficients for HCl from the emf of the cell [Pt | H_2(1 atm) | HCl(c) | AgCl(s) | Ag]

c_{HCl} (mol liter^{-1})	$\mathcal{E}$ (V)	$\gamma^{\pm}$
0.003215	0.52053	0.942
0.005619	0.49257	0.926
0.009138	0.46860	0.909
0.013407	0.44974	0.895
0.02563	0.41824	0.866
0.1238	0.34199	0.788

Extension of this method, or any other thermodynamic method, for the deduction of activities and activity coefficients requires an extension of the idea of mean activities and their coefficients to electrolytes beyond the AB type represented by HCl. An AB_2 electrolyte would dissociate according to

$$AB_2 \rightleftharpoons A^{+2} + 2B^-$$

and the activity term that would appear in all thermodynamic treatments would be of the form

$$(a_{A^+})(a_{B^-})^2$$

The corresponding mean activity $a\pm$ that could be introduced is related by

$$(a_\pm)^3 = (a_A)(a_{B^-})^2 \qquad \text{or} \qquad a_\pm = [(a_A)(a_{B^-})^2]^{1/3}$$

Similarly, the mean-activity coefficient would be related to the ionic activities by

$$\gamma_\pm = (\gamma_{A^+}\gamma_{B^-}{}^2)^{1/3}$$

In general, for an electrolyte A_xB_y,

$$a_\pm = (a^x_{\mathrm{A}(+)}a^y_{B(-)})^{1/(x+y)}$$

and

$$\gamma_\pm = (\gamma^x_{A(+)}\gamma^y_{B(-)})^{1/(x+y)} \qquad [38]$$

Mean activities and activity coefficients are a considerable convenience, and their use loses nothing that is of thermodynamic concern. Furthermore, no thermodynamically exact method has been devised that allows the determination of the individual ion activities or activity coefficients—all, as for the HCl example, lead to terms that can be expressed in terms of these mean quantities.

The deduction of standard-state emfs and the activity coefficients by the method of Sec. 23-6 is aided by, but is not dependent on, the Debye-Hückel theory guide, namely, that the graph of Fig. 23-6 shows a straight-line portion at low concentrations if a $\sqrt{c}$ rather than c coordinate is used. The theory, however, not only provides this aid but also yields values for the activity coefficients of ions in this very dilute solution region. As such it provides an approach based on theories of molecular interactions that supplements the thermodynamic treatments which give, in addition, data for more concentrated solutions.

According to Eq. [38] of Chap. 21, the molar free energy of a solute species can be expressed by the relation

$$\begin{aligned}\bar{G} &= \bar{G}^\circ + RT \ln a \\ &= \bar{G}^\circ + RT \ln \gamma m \\ &= \bar{G}^\circ + RT \ln m + RT \ln \gamma \end{aligned} \quad [39]$$

Since ideal behavior corresponds to a free-energy dependent

$$\bar{G}_{\text{ideal}} = \bar{G}^\circ + RT \ln m \quad [40]$$

the added $RT \ln \gamma$ term can be recognized as the contribution to the free energy as a result of the effects that produce the nonideality of the solution. It is this term that can be interpreted with the results of the Debye-Hückel theory.

*The nonideal contribution to the free energy of ions in solution can be interpreted as the difference in energy required to create the ions in a solution in which there is no interionic interactions and that required where there are such interactions. The energy required to perform the hypothetical creation of a charge depends on the potential that it experiences.

If we imagine the growth of an ionic charge Ze, either by a growth of Z or the value of the electronic charge, we can write

$$\text{Energy}_{\text{ideal}} = \int_0^{Ze} \mathcal{V} d(Ze) \quad [41]$$

With $\mathcal{V} = Ze/4\pi\epsilon r$ for an ion in a dielectric with no interfering ions, this becomes

$$\begin{aligned}\text{Energy}_{\text{ideal}} &= \frac{1}{4\pi\epsilon r}\int_0^{Ze} Ze\, d(ze) \\ &= \frac{1}{4\pi\epsilon r}\frac{(Ze)^2}{2}\end{aligned} \quad [42]$$

If interionic interactions are recognized, as in the Debye-Hückel treatment, the potential is given in Chap. 22 by Eq. [52], or more conveniently by Eq. [53], as

$$\mathcal{V} = \frac{Ze}{4\pi\epsilon r} - \frac{\sqrt{\beta}\, Ze}{4\pi\epsilon}$$

With this expression we obtain

$$\text{Energy}_{\text{nonideal}} = \frac{1}{2}\frac{Ze}{4\pi\epsilon r}Ze - \frac{1}{2}\frac{\sqrt{\beta}\,Ze}{4\pi\epsilon}Ze \qquad [43]$$

It follows that the additional energy term, which is to be identified with the nonideal contribution $RT \ln \gamma$, is the final term of Eq. [43] multiplied by the number of ions $\mathfrak{N}$. Thus we write

$$-\frac{1}{2}\frac{\sqrt{\beta}\,Ze}{4\pi\epsilon}Ze\mathfrak{N} = RT \ln \gamma$$

or

$$\ln \gamma = -\frac{Z^2e^2\sqrt{\beta}\,\mathfrak{N}}{2RT \times 4\pi\epsilon} \qquad [44]$$

On substitution of the expression for β from Sec. 22-13, we obtain

$$\ln \gamma = -\frac{e^3(2 \times 10^3\pi\mathfrak{N})^{1/2}}{(4\pi\epsilon_0 kT)^{3/2}}\frac{Z^2\sqrt{\hat{\mu}}}{(\epsilon/\epsilon_0)^{3/2}} \qquad [45]$$

This is the important limiting law, i.e., for dilute solutions, obtained by Debye and Hückel in 1923. It shows how the activity coefficient of an ion of charge Z can be calculated. It should be mentioned that our derivation has assumed a positive reference charge. The same result is obtained for a negative charge. Thus a positive or a negative value of Z can be used in Eq. [45], and the same calculated value of the activity coefficient of the positive or negative ion will be obtained.

It is convenient to put numerical values in Eq. [45] for the special case of aqueous solutions at 25°C. Substituting values for the constants and 78.54 for ϵ/ϵ_0 for water, and converting to logarithms to the base 10, gives

$$\log \gamma = -0.5091Z^2\sqrt{\mu} \qquad [46]$$

More explicitly, one writes the important results

$$\log \gamma_+ = -0.5091{Z_+}^2\sqrt{\mu}$$

and

$$\log \gamma_- = -0.5091{Z_-}^2\sqrt{\mu} \qquad [47]$$

The thermodynamic expressions have, however, been set up in terms of the mean-activity coefficient, and it is necessary, therefore, to have the Debye-Hückel prediction of the mean-activity coefficient as well as the predictions for individual ions. Some manipulation is necessary.

The definition of a mean coefficient for an electrolyte A_xB_y, where the charge of A is Z_+ and that of B is Z_-, has been written

$$\gamma_\pm = ({\gamma_+}^x{\gamma_-}^y)^{1/(x+y)} \qquad [48]$$

which gives the logarithm of the mean coefficient as

$$\log \gamma_\pm = \frac{1}{x+y} \log (\gamma_+^x \gamma_-^y)$$

$$= \frac{1}{x+y}(x \log \gamma_+ + y \log \gamma_-) \qquad [49]$$

The Debye-Hückel predictions for $\log \gamma_+$ and $\log \gamma_-$, if the charge of A is Z_+ and that of B is Z_-, can be inserted to give

$$\log \gamma_\pm = -\frac{0.5091\sqrt{\mu}}{x+y}(xZ_+^2 + yZ_-^2) \qquad [50]$$

Simplification results from the electrical-neutrality requirement for the electrolyte; i.e.,

$$x(Z_+) + y(Z_-) = 0 \qquad [51]$$

A rearrangement trick consists in multiplying Eq. [51] by Z_+ and, separately, by Z_- to give

$$xZ_+^2 + yZ_+Z_- = 0 \qquad \text{and} \qquad x_+Z_+Z_- + yZ_-^2 = 0 \qquad [52]$$

Addition of these two expressions and rearrangement gives

$$xZ_+^2 + yZ_-^2 = -Z_+Z_-(x+y) \qquad [53]$$

Use of this relation in Eq. [50] leads to the desired Debye-Hückel limiting-law prediction for the mean-activity coefficient for any type of electrolyte in aqueous solution at 25°C as

$$\log \gamma_\pm = 0.5091 Z_+Z_- \sqrt{\mu} \qquad [54]$$

It is this relation, as we now can see, that was used in Sec. 23-6 to guide the extrapolation of Fig. 23-6. The ionic strength was defined by Eq. [41] of the preceding chapter as

$$\mu = \tfrac{1}{2} \sum_i c_i Z_i^2$$

where the summation extends over all the ions of the solution. For a solution containing only singly charged ions of a 1:1 electrolyte like HCl, μ is equal to the electrolyte concentration c, that is,

$$\mu = \tfrac{1}{2}[c(1)^2 + c(1)^2] = c \qquad [55]$$

For this case

$$\log \gamma_\pm = -0.5091\sqrt{c} \qquad [56]$$

Substitution in Eq. [35] leads to the expression, which we expect to be valid as dilute solutions are approached,

$$\mathcal{E} + 0.11830 \log c = \mathcal{E}^\circ - 0.11830(-0.5091)\sqrt{c}$$
$$= \mathcal{E}^\circ + 0.0602\sqrt{c} \qquad [57]$$

Thus we see the basis for the use of a left side versus $\sqrt{c}$ plot for Fig. 23-6, and also the basis for drawing the extrapolation line with a slope of 0.0602.

23-8 Equilibrium Constants and Solubility Products from Emf Data

The data of Table 23-1 constitute a wealth of information on the free energies of inorganic reactions. Although reported as emfs, these data are readily converted to free energies by the expression $\Delta G^\circ = -n\mathfrak{F}\mathcal{E}^\circ$. Such free-energy data are of use in the determination of equilibrium properties and, in particular, of the equilibrium constant for the overall cell reaction.

Consider, for example, the possibility of reducing ferric iron to ferrous iron by the use of metallic zinc as a reducing agent. The reaction in which one would be interested might be performed in the cell

$$\text{Zn}|\text{Zn}^{++}||\text{Fe}^{3+}, \text{Fe}^{++}|\text{Pt}$$

and the reaction would be

$$\begin{array}{ll} \text{Fe}^{3+} + e^- \rightleftharpoons \text{Fe}^{++} & \text{emf} = \mathcal{V}^\circ_{\text{Fe}^{3+},\text{Fe}^{++}} = +0.771 \\ \tfrac{1}{2}\text{Zn} \rightleftharpoons \tfrac{1}{2}\text{Zn}^{++} + e^- & \text{emf} = -\mathcal{V}^\circ_{\text{Zn}|\text{Zn}^{++}} = +0.7628 \\ \hline \text{Fe}^{3+} + \tfrac{1}{2}\text{Zn} \rightleftharpoons \text{Fe}^{++} + \tfrac{1}{2}\text{Zn}^{++} & \mathcal{E}^\circ = +1.534 \text{ V} \end{array} \quad [58]$$

The cell emf is written

$$\mathcal{E} = 1.534 - \frac{0.05915}{1} \log \frac{a_{\text{Fe}^{++}}(a_{\text{Zn}^{++}})^{1/2}}{a_{\text{Fe}^{3+}}} \quad [59]$$

At equilibrium the cell would be able to perform no useful work, and its emf must then be zero. A similar argument based on free energy gives the same result since $\Delta G = -n\mathfrak{F}\mathcal{E}$ and, as was seen in Chap. 9, $\Delta G = 0$ at equilibrium. For equilibrium activities of the variable reagents of Eq. [58] one has, therefore,

$$\mathcal{E} = 0$$

and

$$1.534 = \frac{0.05915}{1} \log \left[\frac{a_{\text{Fe}^{++}}(a_{\text{Zn}^{++}})^{1/2}}{a_{\text{Fe}^{3+}}}\right]_{\text{equilibrium}} \quad [60]$$

The activity expression is now the familiar equilibrium constant, expressed correctly in activities. Thus

$$\left[\frac{a_{\text{Fe}^{++}}(a_{\text{Zn}^{++}})^{1/2}}{a_{\text{Fe}^{3+}}}\right]_{\text{equilibrium}} = K = 8 \times 10^{25} \quad [61]$$

The result shows that essentially all the iron will be reduced to the ferrous state by zinc.

For some cells the overall reaction corresponds to the solution of an insoluble salt. In such cases the equilibrium constant that can be determined is a solubility product. This can be illustrated by the cell

$$\text{Ag}|\text{Ag}^+|\text{Br}^-|\text{AgBr}(s)|\text{Ag}$$

The electrode reactions and the emfs are

$$\begin{array}{rlr} \text{AgBr}(s) + e^- \rightleftharpoons \text{Ag} + \text{Br}^- & \text{emf} = \mathcal{V}^\circ_{\text{AgBr,Ag}} = +0.0713 & \\ \text{Ag} \rightleftharpoons \text{Ag}^+ + e^- & \text{emf} = -\mathcal{V}^\circ_{\text{Ag}^+|\text{Ag}} = -0.7991 & \\ \hline \text{AgBr}(s) \rightleftharpoons \text{Ag}^+ + \text{Br}^- & \mathcal{E}^\circ = -0.7278 \text{ V} & [62] \end{array}$$

The cell emf is written

$$\mathcal{E} = -0.7278 - \frac{0.05915}{1} \log (a_{\text{Ag}^+})(a_{\text{Br}^-}) \qquad [63]$$

The solubility of silver bromide is very small, and the activity coefficients will be sufficiently close to unity to allow, if no other ions are present in appreciable amounts, the activities to be replaced by concentrations. Again at equilibrium the emf of the cell will be zero and

$$0.7278 = -0.05915 \log \{[\text{Ag}^+][\text{Br}^-]\}_{\text{equilibrium}}$$

or

$$\{[\text{Ag}^+][\text{Br}^-]\}_{\text{equilibrium}} = K_{\text{solubility}} = 4.8 \times 10^{-13} \qquad [64]$$

The examples of this section are intended to show that the data of Table 23-1 can be used to determine the equilibrium constant for any reaction which is the overall reaction for a cell made from electrodes included in Table 23-1.

23-9 Electrode-concentration Cells

A particularly simple electrochemical reaction is one that performs the dilution of either the electrode material itself, as studied here, or of the electrolyte, as treated in Sec. 23-10. For the electrode material to be involved in such a process, it must have a variable concentration. Gaseous and amalgam electrodes fall into this classification.

A cell can be constructed from amalgams with two different concentrations of the same metal. The cell

$$\text{Pb-Hg}(a_1)|\text{PbSO}_4(\text{soln})|\text{Hg-Pb}(a_2)$$

allows the electrode reactions

$$\begin{array}{r} \text{Pb}^{++} + 2e^- \rightleftharpoons \text{Pb}(a_2) \\ \text{Pb}(a_1) \rightleftharpoons \text{Pb}^{++} + 2e^- \\ \hline \text{Pb}(a_1) \rightleftharpoons \text{Pb}(a_2) \end{array} \qquad [65]$$

to be written. No chemical change occurs, and the reaction consists

TABLE 23-3 Emfs of cadmium-amalgam electrode-concentration cells at 25°C

Grams Cd per 100 g Hg		Emf (V)	
Left electrode	Right electrode	Observed	Calculated $\left(= \frac{0.05915}{2} \log \frac{10}{1}\right)$
1.000	0.1000	0.02966	0.02957
0.1000	0.01000	0.02960	0.02957
0.01000	0.001000	0.02956	0.02957
0.001000	0.0001000	0.02950	0.02957

of the transfer of lead from an amalgam of one concentration to that of another concentration. The emf of such a cell, which necessarily has $\mathcal{E}^\circ = 0$, is

$$\mathcal{E} = -\frac{0.05915}{2} \log \frac{a_2}{a_1} \qquad [66]$$

The lead will tend to go spontaneously from the high-activity amalgam to that of low activity. For example, if a_1 is greater than a_2, $\mathcal{E}$ is positive and the reaction proceeds in the direction indicated.

One finds that solutions of metals in mercury constitute fairly ideal solutions and that the emfs are almost correctly calculated by using concentrations instead of activities. Table 23-3 shows some data which illustrate this.

The electrode-concentration cells consisting of gas electrodes can be illustrated by the cell

$$\mathrm{Pt}|\mathrm{H_2}(P_1)|\mathrm{HCl}|\mathrm{H_2}(P_2)|\mathrm{Pt}$$

At all ordinary pressures hydrogen behaves very nearly ideally, and the emf corresponding to the overall cell reaction

$$\mathrm{H_2}(P_1) \rightleftharpoons \mathrm{H_2}(P_2) \qquad [67]$$

can be written

$$\mathcal{E} = -\frac{0.05915}{2} \log \frac{P_2}{P_1} \qquad [68]$$

For satisfactorily reversible platinum electrodes one finds experimental results in agreement with this expression.

23-10 Electrolyte-concentration Cells

A second type of cell whose emf is derived only from the free-energy change of a dilution reaction is that in which the electrolyte of the cell is involved in the dilution. If one attempts to construct such a cell by having two solutions of different concentrations in physical contact with each other, complications arise because of the nonequilibrium processes that occur at the liquid-liquid junction. For the present only the simpler *concentration cells without liquid junction* are considered. Such cells can be illustrated with two cells of the type

$$\mathrm{Pt}|\mathrm{H_2}|\mathrm{HCl}(c)|\mathrm{AgCl}(s)|\mathrm{Ag}$$

each of which has a reaction

$$\tfrac{1}{2}\mathrm{H_2} + \mathrm{AgCl} \rightleftharpoons \mathrm{Ag} + \mathrm{H^+}(c) + \mathrm{Cl^-}(c) \qquad [69]$$

Consider two such cells electrically connected through their silver electrodes in the opposed manner,

$$\mathrm{Pt}|\mathrm{H_2}|\mathrm{HCl}(c_1)|\mathrm{AgCl}|\mathrm{Ag}\text{—}\mathrm{Ag}|\mathrm{AgCl}|\mathrm{HCl}(c_2)|\mathrm{H_2}|\mathrm{Pt}$$

The overall reaction is now the sum of the two simple cell reactions. If the pressure of hydrogen gas is the same for both terminal electrodes, the electrode reactions can be combined to give

Right cell	$\mathrm{Ag} + \mathrm{H^+}(c_2) + \mathrm{Cl^-}(c_2) \rightleftharpoons \tfrac{1}{2}\mathrm{H_2} + \mathrm{AgCl}$
Left cell	$\tfrac{1}{2}\mathrm{H_2} + \mathrm{AgCl} \rightleftharpoons \mathrm{Ag} + \mathrm{H^+}(c_1) + \mathrm{Cl^-}(c_1)$
Overall reaction	$\mathrm{H^+}(c_2) + \mathrm{Cl^-}(c_2) \rightleftharpoons \mathrm{H^+}(c_1) + \mathrm{Cl^-}(c_1)$

[70]

The overall reaction involves, therefore, no chemical change and consists only of the transfer of HCl from a concentration c_2 to a concentration c_1. The emf of the complete cell is expressed as

$$\begin{aligned}\mathcal{E} &= -\frac{0.05915}{1}\log\frac{(a_{\mathrm{H^+}})_1(a_{\mathrm{Cl^-}})_1}{(a_{\mathrm{H^+}})_2(a_{\mathrm{Cl^-}})_2} \\ &= -0.05915\log\frac{[a_\pm]_1{}^2}{[a_\pm]_2{}^2} \\ &= -0.11830\log\frac{(a_\pm)_1}{(a_\pm)_2}\end{aligned} \qquad [71]$$

Again one can see that the spontaneous process takes HCl from a higher activity, or concentration, to a lower activity, or concentration. If, for example, c_2 is greater than c_1 and therefore $(a_\pm)_2$ is greater than $(a_\pm)_1$, the emf will be positive and the reaction will proceed in the direction in which it is written.

One can clearly use such concentration cells to determine the activity of an electrolyte at one concentration, or in a solution containing other ions, compared with the activity of an electrolyte in another solution.

The principal purpose in presenting these concentration cells here, however, is to provide a contrast to the situation that arises when concentration cells with liquid junctions are dealt with.

23-11 Electrolyte-concentration Cells with Liquid Junction

The treatment of emfs has so far ignored the problem that arises if one seeks to couple two electrodes which operate in different solutions. If, for instance, one studies the cell consisting of a $Zn|Zn^{++}$ electrode and a $Cu|Cu^{++}$ electrode, one must separate the two solutions, probably $ZnSO_4$ and $CuSO_4$, so that they cannot mix with each other. If the solutions do mix, copper will plate out directly onto the zinc electrode and no emf will be obtained. We are forced, therefore, to form a liquid junction between the solutions, and as we shall see, this gives rise to a *junction potential.* Since the direct contact between solutions of different concentrations is not a balanced state as required for reversible processes, the system is not directly susceptible to thermodynamic analysis. The source of the junction potential will be seen, however, if the cell reaction of a concentration cell is treated in detail.

The dilution of HCl was studied in Sec. 23-10 but can also be accomplished in a cell with a liquid junction, as illustrated in Fig. 23-7. Assume that two HCl solutions of different concentration can be brought together and prevented from mixing. The flowing of two streams of solution together sometimes accomplishes this. One then can set up the cell

$$Pt|H_2|HCl(c_1)|HCl(c_2)|H_2|Pt$$

The emf of the cell can be related to the overall reaction that occurs when 1 $\mathfrak{F}$ of current flows. The reactions that occur at the electrodes and those that occur at the liquid junction can be written separately.

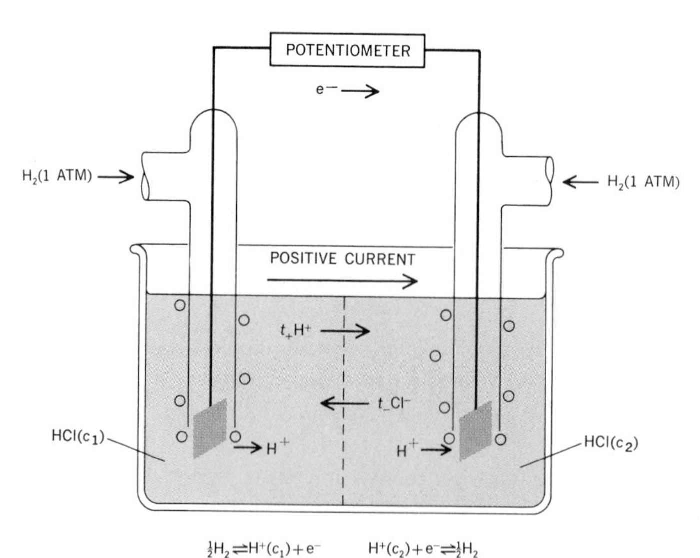

FIGURE 23-7
An electrochemical cell with a liquid junction.

The electrode reactions are

At right electrode	$H^+(c_2) + e^- \rightarrow \frac{1}{2}H_2$	
At left electrode	$\frac{1}{2}H_2 \rightarrow H^+(c_1) + e^-$	
Overall electrode reaction	$H^+(c_2) \rightarrow H^+(c_1)$	[72]

The junction reaction is understood by reference to Fig. 23-7. As the current flows according to our convention, 1 $\mathfrak{F}$ of positive charge must pass through the cell and therefore across the junction. The fraction of current carried by the ions is expressed in terms of their transference numbers, and t_+ equiv of H^+ move to the right whereas t_- equiv of Cl^- move to the left. The junction reactions are therefore

$$t_+H^+(c_1) \rightarrow t_+H^+(c_2)$$

or

$$(1 - t_-)H^+(c_1) \rightarrow (1 - t_-)H^+(c_2)$$

and

$$t_-Cl^-(c_2) \rightarrow t_-Cl^-(c_1) \qquad [73]$$

The electrode reaction and the junction reactions are now combined to give the overall cell reaction. Thus

Electrode reaction	$H^+(c_2) \rightarrow H^+(c_1)$	
Junction reactions	$H^+(c_1) - t_-H^+(c_1) \rightarrow H^+(c_2) - t_-H^+(c_2)$	
	$t_-Cl^-(c_2) \rightarrow t_-Cl^-(c_1)$	
Overall cell reaction	$t_-[H^+(c_2) + Cl^-(c_2)] \rightarrow t_-[H^+(c_1) + Cl^-(c_1)]$	[74]

The emf of this cell, which has $\mathcal{E}° = 0$, can now be written

$$\begin{aligned}\mathcal{E} &= -\frac{0.05915}{1}\log\frac{[(a_{H^+})_1(a_{Cl^-})_1]^{t_-}}{[(a_{H^+})_2(a_{Cl^-})_2]^{t_-}} \\ &= -0.05915t_-\log\frac{[a_\pm]_1^2}{[a_\pm]_2^2} \\ &= -0.11830t_-\log\frac{(a_\pm)_1}{(a_\pm)_2}\end{aligned} \qquad [75]$$

The emf of this cell, unlike that without a liquid junction, depends on the transference numbers. Such cells are frequently described as *concentration cells with transference.*

If the activities of HCl at the two concentrations are known, the measured emf allows the determination of the transference number. The method is satisfactory, and the results compare well with those obtained by the Hittorf method. The difficulties with the method arise through the experimental problem of obtaining a liquid junction that prevents mixing of the two solutions since reproducible and meaningful emfs are not always easily obtained. The assumption has been made, furthermore, that the transference numbers are independent of concentration in the concentration range c_1 to c_2.

23-12 The Salt Bridge

It has already been mentioned that one attempts to circumvent the liquid-junction problem by connecting the different solutions by means of a bridge containing a saturated KCl solution. One then assumes that no junction potential exists. The use of such a device can best be justified by the empirical result that emfs so obtained are generally in satisfactory agreement with results from cells without liquid junctions.

The success of the salt bridge can be attributed to the high concentration of KCl at the solution junction. The effect of the difference between the two electrode solutions is thereby swamped out by the conduction due to the K^+ and Cl^- ions. The fact that the ions of KCl have about equal transference numbers and diffusibilities is also said to be important. The mechanism of the salt bridge, however, is difficult to analyze, and only with a detailed understanding of its operation (it is not a thermodynamic device) can the role of these quantities be understood. The salt bridge is often used as a convenient device for constructing an electrode, as in the assembly of Fig. 23-4. Such an electrode can dip into the solution of another electrode to form an electrochemical cell. When the two electrode solutions could have been mixed and have not been mixed only because of cell-construction convenience, the cell is susceptible to thermodynamic treatment, and thermodynamic functions enter into the cell emf expression; i.e., the activity of an electrolyte will occur, but an individual ion activity will not. The cell

$$\mathrm{Pt}|\mathrm{H_2}(1\text{ atm})|\mathrm{HCl}|\mathrm{KCl}(1\ N)|\mathrm{Hg_2Cl_2}(s)|\mathrm{Hg}$$

might be constructed, for example, and the cell reaction

$$\mathrm{Hg_2Cl_2} + \mathrm{H_2} \rightleftharpoons 2\mathrm{Hg} + 2\mathrm{Cl^-} + 2\mathrm{H^+} \qquad [76]$$

would be written. The cell emf expression is then

$$\mathcal{E} = \mathcal{E}^\circ - \frac{(2)(0.05915)}{2} \log a_{\mathrm{Cl^-}} a_{\mathrm{H^+}} \qquad [77]$$

A typical thermodynamic-activity expression, $a_{\mathrm{Cl^-}} a_{\mathrm{H^+}}$ or $a_\pm^2$, occurs.

The nonthermodynamic nature of the salt bridge is illustrated when the electrode solutions must be separated for an emf to be produced. The cell

$$\mathrm{Pt}|\mathrm{H_2}(1\text{ atm})|\mathrm{HCl}(c = 0.01)\|\mathrm{HCl}(c = 0.1)|\mathrm{H_2}(1\text{ atm})|\mathrm{Pt}$$

can be constructed, and an emf obtained. The electrode reactions are

$$\begin{array}{lll} \text{Right electrode} & \text{H}^+(0.1) + e^- \rightleftharpoons \tfrac{1}{2}\text{H}_2 & \\ \text{Left electrode} & \tfrac{1}{2}\text{H}_2 \rightleftharpoons \text{H}^+(0.01) + e^- & \end{array} \quad [78]$$

The net reaction, on the assumption that no junction reaction need be considered, is

$$\text{H}^+(0.10) \rightleftharpoons \text{H}^+(0.01) \quad [79]$$

Since the $\mathcal{E}^\circ$ value is zero, the cell emf can be written

$$\mathcal{E} = -0.05915 \log \frac{a_{0.01\text{H}^+}}{a_{0.10\text{H}^+}} \quad [80]$$

This apparent approach to the activities of individual ions of an electrolyte, however, is upset by the assumptions made concerning the effectiveness of the salt bridge.

Although a cell making use of a salt bridge is not one that can be analyzed by strict thermodynamic arguments, in a practical way the salt bridge is effective and allows cells to be studied that consist of electrodes requiring different solutions.

23-13 The Glass Electrode

Before proceeding to an important application of emf measurements, brief mention should be made of a component of the most frequently encountered electrochemical instrument. This component is an electrode called the *glass electrode*. Figure 23-8 indicates the electrode assembly of a pH meter, which includes a typical glass electrode. The electrode, not the cell, usually consists of the arrangement

$$\text{Ag}\,|\,\text{AgCl}(s)\,|\,\text{HCl}(c = 1)\,|\,\text{glass}$$

The value of the electrode stems from the fact that, when it is placed in a solution of given acidity and the cell is completed by use of another electrode, the emf of the cell appears to depend primarily on the difference in the concentration or activity of the hydrogen ions on either side of the glass.

The glass membrane of the glass electrode separates two different solutions, as does the KCl salt bridge. Unlike the salt bridge, which provides for electric conduction across the liquid junction, the glass membrane most often used leads to a cell whose emf is primarily responsive to hydrogen ions. Glasses can be made that allow passage of only one type of ion, in this case the hydrogen ion, and thus the electrode can be constructed to be sensitive to this ion only.

Much of the importance of the glass electrode stems from its lack of response to various oxidizing and reducing agents and to a large variety of ionic species. Difficulties may occur, however, if the glass

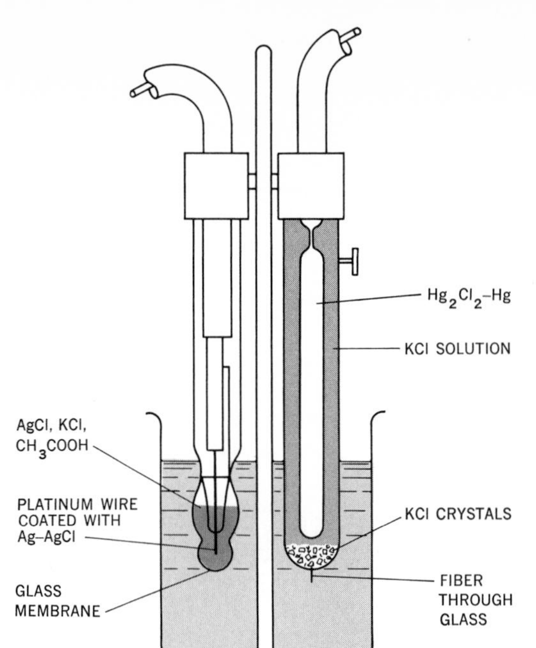

FIGURE 23-8
The glass electrode and calomel electrode of a pH meter. *(From F. Daniels, J. W. Williams, P. Bender, R. A. Alberty, and C. D. Cornwell, "Experimental Physical Chemistry," 6th ed., McGraw-Hill Book Company, New York, 1962.)*

electrode is used in solutions of high sodium-ion concentration or in solution sufficiently alkaline to attack the glass membrane.

23-14 pH Definition and Measurement

In many areas of chemistry it is very convenient to correlate the properties of the system with something related to the hydrogen-ion concentration. Thus, in a titration of an acid with a base, one might try to follow the process and determine the end point by any of a number of physical measurements. Properties that are closely dependent on the hydrogen-ion concentration would be most satisfactory. In a like manner, since some reactions, as was discussed in Chap. 16, are *acid-catalyzed,* the reaction rate of such reactions can be correlated with the concentration of the hydrogen ion.

Such applications are not strictly thermodynamic, and in these applications it is not clearly specified whether it is the hydrogen-ion concentration that is needed or whether it is some effective hydrogen-ion concentration. There seems, therefore, little necessity to try to use the thermodynamically suggested activities or activity coefficients. The fact that the activity of a single ion would be encountered emphasizes the impropriety of inserting this thermodynamic concept.

A convenient form for the expression of hydrogen-ion concentrations was suggested in 1909 by Sörensen. He introduced the term pH, and his original definition gave

$$\mathrm{pH} = -\log\,[\mathrm{H}^+] \qquad [81]$$

Hydrogen-ion concentrations, as was shown in Sec. 22-10, can be deduced for dilute solutions of acids from conductance measurements, and with such data the pH, according to Eq. [81], can be obtained.

Most applications, however, require a measure of something like the hydrogen-ion concentration in solutions that may be concentrated and that may contain a number of other ionic species. Conductance measurements are therefore unsatisfactory, and one is led to consider some electrochemical cell whose emf might give a suitable hydrogen-ion index.

One might consider, for example, the cell

$$\mathrm{Pt}|\mathrm{H_2(1\ atm)}|\mathrm{soln}||\mathrm{Hg_2Cl_2(satd)}|\mathrm{Hg_2Cl_2}(s)|\mathrm{Hg}$$

in which a hydrogen electrode operates in the solution of interest and the cell is completed by a calomel electrode connected through a salt bridge. The electrode reactions are

$$\begin{array}{ll} \text{Right electrode} & \frac{1}{2}\mathrm{Hg_2Cl_2}(s) + e^- \rightleftharpoons \mathrm{Hg} + \mathrm{Cl^-(satd)} \\ \text{Left electrode} & \frac{1}{2}\mathrm{H_2} \rightleftharpoons \mathrm{H^+} + e^- \\ \hline \text{Overall reaction} & \frac{1}{2}\mathrm{H_2} + \frac{1}{2}\mathrm{Hg_2Cl_2} \rightleftharpoons \mathrm{H^+} + \mathrm{Cl^-(satd)} + \mathrm{Hg} \end{array} \qquad [82]$$

If the salt bridge is assumed to be effective and the chloride concentration and activity are fixed and included in a constant emf term, one can write

$$\mathcal{E} = \text{const} - \frac{0.05915}{1} \log a_{\mathrm{H^+}} \qquad [83]$$

The measured emf, after the cell has been standardized by measurements on, for example, a $1\text{-}M$ HCl solution, gives an indication of the hydrogen-ion concentration or activity, the latter being, as pointed out, not well defined. Such a cell does give the information that would be needed, for example, to follow an acid-base titration or to correlate the rate of an acid-catalyzed reaction with the acidity of the solution. The practical difficulties and inconvenience of the hydrogen electrode, however, make the hydrogen-calomel cell unsuitable.

By far the most frequently used electrochemical device is the *pH meter,* which makes use of the combination of a glass electrode and a calomel electrode. The emf of such an assembly is found to depend on the acidity of a solution in much the same way as the hydrogen-calomel electrode. Thus one can formally write for the pH meter the equation

$$\mathcal{E} = \text{const} - 0.05915 \log a_{\mathrm{H^+}}$$

or

$$-\log a_{\mathrm{H^+}} = \frac{\mathcal{E} - \text{const}}{0.05915} \qquad [84]$$

The value of the constant term can be determined from a measurement on a solution of known hydrogen-ion concentration. The measured $\mathcal{E}$ of a pH meter can then be inserted to give a numerical value for the right side of the equation when some test solution is used. The scale of the pH meter can be arranged, moreover, to give directly the right side of Eq. [84] rather than the value of $\mathcal{E}$. Equation [84] suggests that this number will be a suitable hydrogen-ion index. It is convenient, therefore, to drop the original Sörensen pH definition and instead to define pH as

$$\text{pH} = \frac{\mathcal{E} - \text{const}}{0.05915} \qquad [85]$$

where $\mathcal{E}$ is the emf of the pH meter assembly. Other similar pH definitions are also possible. A suitable choice of a value for the constant term allows this pH scale to coincide quite well with that of Eq. [81] based on conductance results.

It is important to recognize that the operational definition of Eq. [85] leads to a hydrogen-ion index that has many important applications. The pH so defined, however, is only loosely related to thermodynamic quantities. A complete and detailed understanding of the glass electrode, or of a salt bridge, would lead to a molecular-type interpretation of pH, and this interpretation would not necessarily involve thermodynamic activities or activity coefficients.

23-15 Activity Coefficients from the Dissociation of a Weak Electrolyte

Other studies of chemical equilibria, besides those that make use of oxidation-reduction reactions in electrochemical cells, can be used to deduce thermodynamic properties of nonideal systems. Acid-base equilibria provide many illustrations. The traditional example is the equilibrium set up by the dissociation of acetic acid, CH_3COOH, here abbreviated HAc.

$$\text{HAc} \rightleftharpoons \text{H}^+ + \text{Ac}^- \qquad [86]$$

The thermodynamic equilibrium constant, denoted by K_{th}, is

$$K_{th} = \frac{(a_{H^+})(a_{Ac^-})}{a_{HAc}} \qquad [87]$$

which can also be written

$$K_{th} = \frac{\gamma_+ \gamma_-}{\gamma_{HAc}} \frac{[\text{H}^+][\text{Ac}^-]}{[\text{HAc}]} \qquad [88]$$

This expression is simplified when it is realized that the electrostatic interactions are primarily responsible for the nonideality which produces activity coefficients different from unity. The uncharged HAc molecule should therefore behave relatively ideally, and we can set $\gamma_{HAc} = 1$.

Introduction of $\gamma_\pm^2$ for $\gamma_+\gamma_-$ and rearrangement gives

$$\log \frac{[H^+][Ac^-]}{[HAc]} = \log K_{th} - 2 \log \gamma_\pm \qquad [89]$$

The concentration expression is that for the equilibrium constant in terms of concentrations, and introduction of the degree of dissociation allows the equation to be written

$$\log \frac{c\alpha^2}{1 - \alpha} = \log K_{th} - 2 \log \gamma_\pm \qquad [90]$$

For solutions that are very dilute in ions one can still use the Arrhenius expression

$$\alpha = \frac{\Lambda}{\Lambda_0} \qquad [91]$$

to obtain the degree of dissociation from the conductivity. More accurately, one can correct for the effect of ion concentration on conductance. In this way, the left side of Eq. [90] is determined for various acetic acid concentrations. The right side consists of a constant term $\log K_{th}$ and a term which the Debye-Hückel theory of Sec. 23-7 suggests will, at low ionic concentrations, be proportional to the square root of the ionic strength. If the solution contains only the H^+ and Ac^- ions from the dissociation of HAc, one has

$$\mu = \tfrac{1}{2}[(c\alpha)(1)^2 + (c\alpha)(-1)^2] = c\alpha \qquad [92]$$

and

$$\sqrt{\mu} = \sqrt{c\alpha}$$

It might be informative, therefore, to plot, as is done in Fig. 23-9, the left side of Eq. [90] against $\sqrt{c\alpha}$. At low concentrations the points do seem to fall along a straight line, in agreement with the prediction of the Debye-Hückel theory. The theory predicts, furthermore, that the last term of Eq. [90] should be

$$-2 \log \gamma_\pm = +(2)(0.5091) \sqrt{c\alpha}$$

and the line of Fig. 23-9, drawn with slope $+1.018$, fits the data satisfactorily.

Extrapolation to zero ionic strength, where $\gamma_\pm = 1$ and $\log \gamma_\pm = 0$, gives

$$\log K_{th} = -4.7565$$

and

$$K_{th} = 1.752 \times 10^{-5} \qquad [93]$$

The quantity K_{th} is that which must be used in the relation

$$\Delta G^\circ = -RT \ln K$$

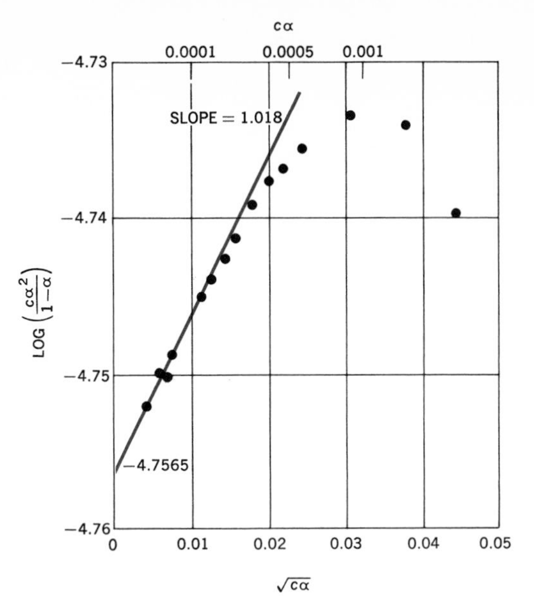

FIGURE 23-9
The extrapolation of the logarithm of the concentration-equilibrium-constant expression $\log [c\alpha^2/1 - \alpha]$, for acetic acid to zero ionic strength, according to the plot suggested by the Debye-Hückel theory.

in order to obtain a $\Delta G°$ value that corresponds to the reagents at unit activity. Here we find for

$$HAc(a = 1) \rightleftharpoons H^+(a = 1) + Ac^-(a = 1)$$

$$\begin{aligned}\Delta G° &= -8.3143(298.15)(2.3026)(-4.7565) \\ &= +27{,}150 \text{ J}\end{aligned} \qquad [94]$$

The value of -399.62 kJ for HAc(aq) and -372.46 kJ for $Ac^-(aq)$, given in Appendixes 5 and 6 for these species, is in accord with this result. Equation [90] can also now be rearranged to give

$$\begin{aligned}\log \gamma_\pm &= \tfrac{1}{2} \log K_{th} - \tfrac{1}{2} \log \frac{c\alpha^2}{1-\alpha} \\ &= -2.3782 - \tfrac{1}{2} \log \frac{c\alpha^2}{1-\alpha}\end{aligned} \qquad [95]$$

From this equation we can determine $\gamma_\pm$ for the dissociation products of HAc at any concentrations for which values of α can be obtained.

23-16 Activity Coefficients from Solubility Measurements

Similar treatments of solubility equilibria yield additional free-energy and activity-coefficient data.

A one-to-one salt AB goes into solution and, possibly among a number of reactions, establishes the equilibrium

$$AB(s) \rightleftharpoons A^+ + B^- \qquad [96]$$

The thermodynamic equilibrium constant, the solubility product, is

$$K_{th} = (a_{A^+})(a_{B^-}) = \gamma_+\gamma_-[A^+][B^-]$$
$$= \gamma_\pm^2[A^+][B^-] \qquad [97]$$

Taking logarithms and rearranging, one gets

$$\log [A^+][B^-] = \log K_{th} - 2 \log \gamma_\pm \qquad [98]$$

The solubility s of such a salt is equal to the moles per liter of the salt that dissolve, which, if no species from the electrolyte AB other than A^+ and B^- exist in solution, gives

$$[A^+] = [B^-] = s$$

Thus

$$\log s^2 = \log K_{th} - 2 \log \gamma_\pm$$

or

$$\log s = \tfrac{1}{2} \log K_{th} - \log \gamma_\pm \qquad [99]$$

Now consider the data that are obtained when a sparingly soluble salt is dissolved in solutions that contain various amounts of non-reacting electrolytes that do not contain the ions A^+ or B^-. The Debye-Hückel theory again suggests that at low ionic strengths the log $\gamma_\pm$ term of the sparingly soluble electrolyte will be proportional to the square root of the ionic strength. One therefore plots the left side of Eq. [99] against $\sqrt{\mu}$ as shown for the complicated but experimentally convenient salt $[Co(NH_3)_4C_2O_4][Co(NH_3)_2(NO_2)_2C_2O_4]$ in Fig. 23-10. The data for solutions with singly charged ions support the Debye-Hückel predictions of a linear relation, and the straight line of Fig. 23-10 has been drawn with the predicted slope of 0.5091. The solutions containing more highly charged ions agree rather less well with the Debye-Hückel prediction.

For this particular set of data one sees that the linear relation is

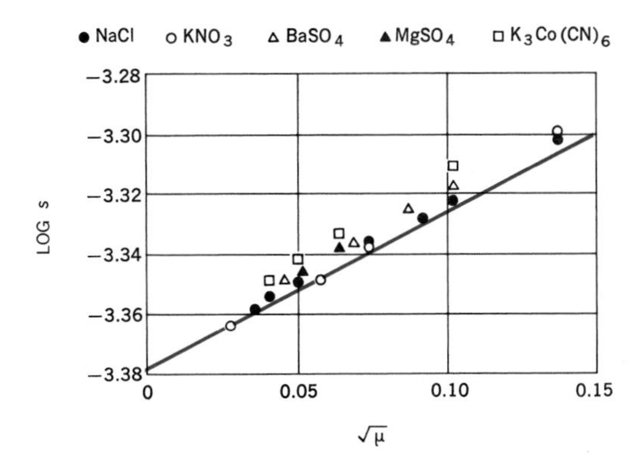

FIGURE 23-10
The effect of added salts on the solubility of a salt $[Co(NH_3)_4C_2O_4][Co(NH_3)_2$-$(NO_2)_2C_2O_4]$ containing a singly charged cation and a singly charged anion. (*From L. P. Hammett, "Introduction to the Study of Physical Chemistry," McGraw-Hill Book Company, New York, 1952.*)

rather rough and that support for the Debye-Hückel theory is given but is not very convincing here. It is just such difficulties that the theory frequently encounters, particularly in solutions other than those of extremely low ionic strength.

An extrapolation can be made in Fig. 23-10, with or without the Debye-Hückel theory, and one obtains for the left side of Eq. [99], at zero ionic strength, the value -3.377. Again the limit of zero ionic strength corresponds to the complete absence of ionic interactions, and therefore to $\gamma_\pm = 1$ and $\log \gamma_\pm = 0$. At this limit Eq. [99] gives, therefore,

$$\tfrac{1}{2} \log K_{th} = -3.377$$

or

$$K_{th} = 1.76 \times 10^{-7} \qquad [100]$$

With this value one is again able to turn Eq. [99] around to give an expression for the activity coefficient of the slightly soluble salt at any concentration as

$$\log \gamma_\pm = -3.377 - \log s \qquad [101]$$

The solubility data used to construct Fig. 23-10 can therefore be made to give $\gamma_\pm$ for the dissolved salt in all the solutions studied. Although the Debye-Hückel theory was used to aid in the extrapolation to obtain K_{th} in Fig. 23-10, it should be recognized that these results do not depend on this theory and are valid for any aqueous solution at this temperature for which the solubility of the salt can be measured.

Salts other than those of the one-to-one type can be handled in a similar manner. For the solubility equilibrium

$$A_2B(s) \rightleftharpoons 2A^+ + B^= \qquad [102]$$

the solubility product is

$$\begin{aligned} K_{th} &= (a_{A^+})^2(a_{B^=}) = \gamma_+{}^2\gamma_-[A^+][B^=] \\ &= \gamma_\pm{}^3[A^+][B^=] \end{aligned} \qquad [103]$$

If the solubility is s mol liter^{-1}, the concentration of A^+ will be $2s$ and of B will be s. In terms of the solubility one has

$$K_{th} = \gamma_\pm{}^3(2s)^2(s) = \gamma_\pm{}^3 4s^3 \qquad [104]$$

Taking logarithms and rearranging now gives

$$\log s = \tfrac{1}{3}(\log K_{th} - \log 4) - \log \gamma_\pm \qquad [105]$$

Data on the solubility of a salt of the type A_2B as a function of the ionic strength of the solution can now be used, in the same manner as for the salt AB, to give the value of K_{th} and values for $\gamma_\pm$.

A collection of activity-coefficient results for a variety of types of electrolytes is shown in Table 23-4. These data correspond to solutions

TABLE 23-4 Mean-activity coefficients for electrolytes in water at 25°C*

Molality	0.001	0.005	0.01	0.05	0.10	0.50	1.00	2.00	4.00
Debye-Hückel theory for $AB \rightarrow A^+ + B^-$	0.965	0.920	0.890	0.770					
HCl	0.965	0.929	0.905	0.830	0.794	0.757	0.809	1.009	1.762
NaCl	0.965	0.927	0.902	0.819	0.778	0.681	0.657	0.668	0.783
NaOH				0.818	0.766	0.693	0.679	0.700	0.890
Debye-Hückel theory for $A_2B \rightarrow 2A^+ + B^=$ $AB_2 \rightarrow A^{++} + 2B^-$	0.880	0.750	0.667						
$CdCl_2$	0.819	0.623	0.524	0.304	0.228	0.100	0.066	0.044	
H_2SO_4	0.830	0.639	0.544	0.340	0.265	0.154	0.130	0.124	0.171
Debye-Hückel theory for $AB \rightleftharpoons A^{++} + B^=$	0.744	0.515							
$ZnSO_4$	0.700	0.477	0.387	0.202	(0.150)	0.063	0.043	0.035	
$CdSO_4$	0.697	0.476	0.383	0.199	(0.150)	0.061	0.041	0.032	

* Data are from H. S. Harned and B. B. Owen, "The Physical Chemistry of Electrolytic Solutions," Reinhold Publishing Corporation, New York, 1958, and B. E. Conway, "Electrochemical Data," Elsevier Publishing Company, Amsterdam, 1962.

in which the only electrolyte present is that indicated. At lower concentrations, particularly for the electrolytes whose ions are singly charged, the Debye-Hückel theory is seen to give values in quite good agreement with the values obtained experimentally.

The Debye-Hückel theory gives, therefore, a molecular interpretation of the behavior of ions that satisfactorily accounts for their thermodynamic behavior in solutions of ionic strength less than about 0.01. The limiting Debye-Hückel law has, moreover, almost the neatness of a law for ideal systems. It characterizes the solution for aqueous solutions at 25°C simply by the dielectric constant of water and the ionic strength of the solution and characterizes the ion only by the absolute value of its charge. In dilute enough solutions, a molecular interpretation needs, apparently, to take into account no other solute or solvent property.

The situation is much more complicated in solutions with higher ionic strength, and although some empirical relations are available, experimental results must often be used.

23-17 Activity Coefficients in More Concentrated Solutions

If attention is given to concentrations that yield ionic strengths greater than about 0.01 but less than about 0.1, a similar pattern of activity-coefficient behavior for different electrolytes is noticed. Figure 23-11 shows some of these data, and such curves can be fitted by an equation that is an extension of the Debye-Hückel limiting law.

$$\frac{\log \gamma_\pm}{-Z_+Z_-} = \frac{-0.5091\sqrt{\mu}}{1+\sqrt{\mu}} + b\mu \qquad [106]$$

where b is an empirical constant. More simply, all the available data are approximately fitted by the expression

$$\frac{\log \gamma_\pm}{-Z_+Z_-} = \frac{-0.5091\sqrt{\mu}}{1+\sqrt{\mu}} \qquad [107]$$

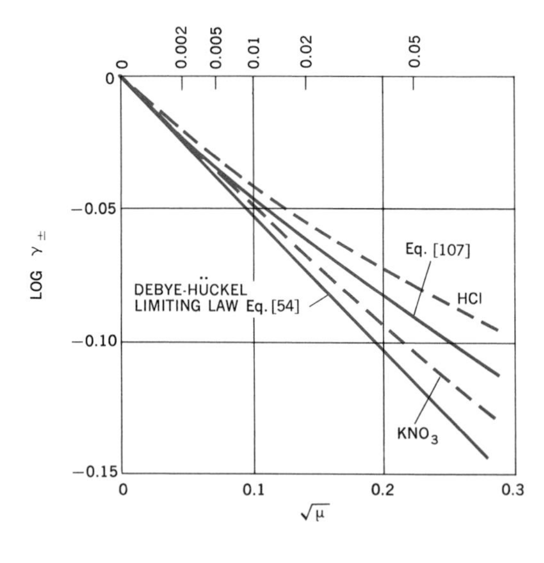

FIGURE 23-11 The activity coefficients of one-to-one electrolytes in the concentration range up to 0.01 *M*. (*From L. P. Hammett, "Introduction to the Study of Physical Chemistry," McGraw-Hill Book Company, New York, 1952.*)

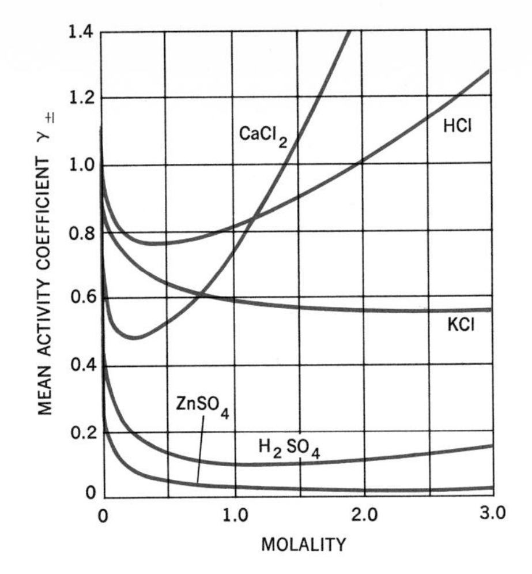

FIGURE 23-12
The mean-activity coefficient of some electrolytes in the concentration range up to 3 *M*. *(Adopted from L. P. Hammett, "Introduction to the Study of Physical Chemistry," McGraw-Hill Book Company, New York, 1952.)*

as shown in Fig. 23-11. Even in the ionic-strength range of less than 0.1, however, it is apparent that the ionic charge is not a sufficient description of an ion to account for its activity coefficient.

At concentrations above those dealt with in Fig. 23-11, nonideality of the solutions, as expressed by the activity coefficients, becomes very striking, as the curves of Fig. 23-12 show. Even electrolytes of the same charge type have very different activity-coefficient curves. No satisfactory theoretical or even semiempirical explanation of these curves is available.

The phenomenon of an ion atmosphere and the resultant stabilization on which the Debye-Hückel theory is based still undoubtedly operates. This effect is much enhanced, even to the extent that ion pairs and ion triplets or other species should be recognized. Such electrostatic associations will necessarily lower the free energies of the ions and produce a value of $\gamma_\pm$ of considerably less than unity.

A qualitative explanation of the activity coefficients that increase with concentration and even become greater than unity can be given in terms of the solvation of the ions. As the ions tie up solvent molecules, the effective concentration, i.e., the moles solute per mole of free solvent, becomes greater than the concentration calculated as moles of solute per mole of solvent. The solute in the apparently more concentrated solution has a higher free energy than would be expected, and this shows up as an increased activity coefficient.

A quantitative explanation for data such as those of Fig. 23-12, however, is as yet nonexistent. In solution chemistry, one of the principal goals of research is that of clearly recognizing the factors that are involved and fitting the data such as are shown in Fig. 23-12 into a quantitative theory.

Since the activity coefficients for ions in solutions of high ionic strengths depart very appreciably from unity, the study of any equilibrium involving ions under these conditions cannot be made simply in terms of concentrations—as one always seems to do in practice problems dealing with ionic equilibria. The activities of the ions should be used, but it is frequently difficult to obtain the activity coefficients that are necessary for the calculation of activities from concentrations.

A frequently encountered approach that attempts to avoid this difficulty arranges, by the addition of a relatively large amount of non-reacting electrolyte, to keep the ionic strength at a high but essentially constant value. Although this procedure makes all the activity coefficients very different from unity, it is intended to keep the activity coefficients of any participating ion constant regardless of the variation in the amount of the other reacting ions. The activity coefficients of the ions in the equilibrium

$$Fe^{3+} + I^- \rightleftharpoons Fe^{++} + \tfrac{1}{2}I_2 \qquad [108]$$

would, for example, be expected to be very sensitive to changes in the ionic strength of the system. That this equilibrium is set up would not be apparent if one attempted to use the concentrations in an equilibrium-constant expression. As the data of Table 23-5 indicate, however, in the presence of 1.65 M KCl the activity coefficients apparently are not changed as the amounts of Fe^{3+}, I^-, and Fe^{++} are varied and a quite constant equilibrium constant based on concentrations results. It must be recognized that this constant may be very different from the thermodynamic equilibrium constant. The experiment really shows only that an equilibrium is established according to the reaction of Eq. [108].

TABLE 23-5 Equilibrium constant at 25°C for the reaction $Fe^{3+} + I^- \rightarrow Fe^{++} + \frac{1}{2}I_2$
Values for K are based on concentrations of the reagents, and a constant ionic strength is maintained by 1.65 M KCl and 0.1 M HCl.

$[Fe^{3+}]$	$[I^-]$	$[Fe^{++}]$	$[I_2]$	K
0.001223	0.00114	0.001257	0.0053	21
0.002644	0.00224	0.003536	0.00129	21.4
0.00483	0.00358	0.007535	0.00238	21.3
0.00900	0.00549	0.01574	0.00415	20.5
0.00436	0.00161	0.000804	0.00032	20.5
0.001104	0.00526	0.003856	0.00104	21.4
0.00043	0.01713	0.005752	0.00076	21.6
0.00192	0.01117	0.01045	0.00185	21.0

SOURCE: From J. N. Brønsted and K. Pedersen, *Z. Physik. Chem.*, **103**:307 (1923).

23-18 Thermodynamic Data from Emf Measurements

The emf data of electrochemical cells provide not only activity-coefficient data, but also much valuable information on the thermodynamic properties of ions in aqueous solution at the standard state of unit activity.

The standard electrode potentials of Table 23-1, for example, give, through the relation $\Delta G^\circ = -n\mathfrak{F}\mathcal{E}^\circ$, values for the free-energy change accompanying the cell reaction. With the added arbitrary assignment of a zero value for the free energy of formation of H^+ at unit activity, values for the standard free energies of individual ionic species, as listed in Appendix 6, are obtained.

Additional information is obtained from measurements of the temperature dependence of the emfs of electrochemical cells.

The thermodynamic relation

$$\left[\frac{\partial(\Delta G)}{\partial T}\right]_P = -\Delta S \qquad \text{or} \qquad \left[\frac{\partial(\Delta G^\circ)}{\partial T}\right]_P = -\Delta S^\circ \tag{109}$$

becomes, with the substitution of

$$\Delta G^\circ = -n\mathfrak{F}\mathcal{E}^\circ$$

the entropy-determining relation

$$-n\mathfrak{F}\left(\frac{\partial \mathcal{E}^\circ}{\partial T}\right)_P = -\Delta S^\circ \qquad \text{or} \qquad \Delta S^\circ = n\mathfrak{F}\left(\frac{\partial \mathcal{E}}{\partial T}\right)_P \tag{110}$$

By this expression the entropy change for the cell reaction can be determined from the temperature coefficient of the cell emf.

This entropy value can, furthermore, be inserted into the expression, for a given temperature,

$$\Delta H = \Delta G + T\,\Delta S$$

to give a value for the enthalpy change for the reaction.

Thus ΔG, ΔH, and ΔS can be evaluated for the cell reaction from measurements of the emf and the temperature coefficient of the emf. This procedure is often a more convenient way of obtaining these thermodynamic properties than direct calorimetric measurements. Electromotive-force studies therefore provide an appreciable amount of thermodynamic information for systems involving ions in aqueous solution.

For example, the cell

$Pt|H_2|HCl(soln)|AgCl|Ag$

has a standard emf, i.e., with variable reagents at unit activity, of 0.2224 V at 25°C and a temperature coefficient of −0.000645 V per degree centigrade.

The cell reaction is

$$\begin{array}{l} \mathrm{AgCl} + e^- \rightleftharpoons \mathrm{Ag} + \mathrm{Cl}^- \\ \tfrac{1}{2}\mathrm{H_2} \rightleftharpoons \mathrm{H}^+ + e^- \\ \hline \mathrm{AgCl} + \tfrac{1}{2}\mathrm{H_2} \rightleftharpoons \mathrm{Ag} + \mathrm{H}^+(a = 1) + \mathrm{Cl}^-(a = 1) \end{array} \qquad [111]$$

and for this reaction one obtains the thermodynamic results

$$\begin{aligned} \Delta G^\circ &= -n\mathfrak{F}\mathcal{E}^\circ \\ &= -(1)(96{,}490)(0.2224) \\ &= -21{,}460 \text{ J} \end{aligned} \qquad [112]$$

$$\begin{aligned} \Delta S^\circ &= n\mathfrak{F}\left(\frac{\partial \mathcal{E}^\circ}{\partial T}\right)_P \\ &= (1)(96{,}490)(-0.000645) \\ &= -62.2 \text{ J deg}^{-1} \end{aligned} \qquad [113]$$

and

$$\begin{aligned} \Delta H^\circ &= \Delta G^\circ + T\Delta S^\circ \\ &= -21{,}460 + 298(-62.2) \\ &= -40{,}000 \text{ J} \end{aligned} \qquad [114]$$

Thus measurements of the emf of cells and the temperature dependence of the emf give thermodynamic data for reactions in which solutions of ions are formed. Such measurements, along with direct calorimetric determinations of the heats of solutions of electrolytes and of the temperature coefficient of the solubilities of electrolytes, provide much of the data of Appendix 6.

23-19 The Effect of Electrostatic Interactions on Reaction Rates

The introduction of the concept of activities and the development of methods for relating activities to concentrations allow the methods of thermodynamics to be applied to solutions of ionic species where the electrostatic interactions are often very considerable. It is true, however, that not only are the equilibrium conditions of a system affected by these interactions, but so also are the rates of chemical reactions proceeding in such systems. Although thermodynamic quantities cannot be carried over with thermodynamic rigor to the treatment of rate effects, some of the ideas previously introduced are valuable in the discussion of reaction rates in ionic systems.

It is particularly profitable, as was shown by Brønsted and Bjerrum, to follow through the transition-state theory with the recognition that activities rather than concentrations should be used in the equilibrium-constant expression for the formation of the activated complex.

The general bimolecular reaction, allowing for charges on the reagents and on the activated complex, can be written

$$A^{Z_A} + B^{Z_B} \rightleftharpoons (AB)^{Z_A+Z_B} \rightarrow \text{products} \qquad [115]$$

where Z_A and Z_B are the charges of the reagents A and B. The equilibrium constant for the formation of the activated complex must now be written

$$K^\ddagger = \frac{a^\ddagger}{a_A a_B} = \frac{c^\ddagger}{c_A c_B} \frac{\gamma^\ddagger}{\gamma_A \gamma_B} \qquad [116]$$

As in the treatment in Sec. 17-4, the transition-state theory attributes the rate of a reaction to the product of the concentration of the activated complex and a frequency factor. Thus the rate

$$-\frac{dc_A}{dt} = k_2 c_A c_B \qquad [117]$$

is interpreted as

$$k_2 c_A c_B = \nu c^\ddagger = \frac{kT}{h} c^\ddagger \qquad [118]$$

Substitution of the expression of Eq. [116] for the activated-complex concentration gives

$$k_2 = \frac{kT}{h} K^\ddagger \frac{\gamma_A \gamma_B}{\gamma^\ddagger} \qquad [119]$$

This treatment leads, therefore, to the appearance of the activity coefficients of the reagents and of the activated complex in the rate constant.

At high ionic concentrations it must be anticipated that the rate constant will behave in some not easily predicted manner if the activated complex carries a charge and therefore has some indeterminate activity coefficient.

In more dilute solutions reliance can be placed on the Debye-Hückel limiting law,

$$\begin{aligned} \log \gamma_+ &= -0.5091 Z_+^2 \sqrt{\mu} \\ \log \gamma_- &= -0.5091 Z_-^2 \sqrt{\mu} \end{aligned} \qquad [120]$$

These expressions for the activity coefficients can be inserted into the logarithm of Eq. [119] to give

$$\begin{aligned} \log k_2 &= \log\left(\frac{kT}{h} K^\ddagger\right) - 0.5091\sqrt{\mu}[Z_A^2 + Z_B^2 - (Z_A + Z_B)^2] \\ &= \log\left(\frac{kT}{h} K^\ddagger\right) + 1.0182\sqrt{\mu} Z_A Z_B \end{aligned} \qquad [121]$$

This result predicts that the rate constant for a reaction depends not only on the $(kT/h)K^\ddagger$ term that appears for noncharged systems but also on the ionic strength of the solution and the charges of the reagents.

The dependence for a given reaction is determined by the product of the charges of the reagents, and the prediction is made that the rate will increase with increasing ionic strength if both reagents have charges of the same sign and will decrease with increasing ionic strength if the reagents are oppositely charged.

It should again be pointed out, however, that in solutions of higher ionic strength, no simple dependence of the rate constant on the ionic strength can be expected. Again one frequently resorts to studies on solutions with a large excess of a nonparticipating electrolyte present to preserve a constant ionic strength. Although this approach is often successful in providing data of some value, the occurrence of specific interactions must be anticipated.

Problems

1 Write the cell reactions for the following cells, and use Table 23-1 to determine the emfs of the cells under standard conditions:

a $Cd|Cd^{++}||KCl|Hg_2Cl_2|Hg$ *Ans.* 0.671 V.

b $Pt|Tl^{+}, Tl^{3+}||Cu^{++}|Cu$

c $Pb|PbSO_4(s)|SO_4^{=}||Cu^{++}|Cu$

2 Deduce, by writing electrochemical cells and calculating their emfs, the standard free-energy changes for the reactions:

a $\frac{1}{2}Br_2 + Ag \rightleftharpoons AgBr(s)$ *Ans.* 0.993 V, −95.8 kJ.

b $H_2 + Cu^{++} \rightleftharpoons 2H^+ + Cu$

c $\frac{1}{2}Cl_2 + Br^- \rightleftharpoons Cl^- + \frac{1}{2}Br_2$

d $Ca^{++} + 2Na \rightleftharpoons Ca + 2Na^+$

e $Hg_2Cl_2 \rightleftharpoons 2Hg + Cl_2$

3 Calculate the emf of the cell and $\Delta G°$ for the cell reaction of

$$Pt|Cl_2(1\ \text{atm})|ZnCl_2(a = 1)|ZnCl_2(a = 1)|Zn$$

using the electrode potentials of Table 23-1.

4 In what direction would the concentrations of the variable reagents in each of the reactions of Prob. 2 be changed to attain cells with zero emf?

5 The emf of the cell $Pt|H_2(1\ \text{atm})|HBr(c)|AgBr(s)|Ag$ has the values at 25°C tabulated at the right according to the measurements of Keston [*J. Am. Chem. Soc.*, **57**:1671 (1935)]. By a suitable graphical method deduce:

a $\mathcal{E}°$ for the cell.

b The activity coefficients for HBr at each of the reported concentrations.

c (*mol liter*$^{-1}$)	$\mathcal{E}$ (*volts*)
0.0003198	0.48469
0.0004042	0.47381
0.0008444	0.43636
0.001355	0.41243
0.001850	0.39667
0.002396	0.38383
0.003719	0.36173

6 The emf of the cell in which the reaction $H_2 + Hg_2Cl_2 \rightarrow 2Hg + 2HCl$ occurs has been studied by Lewis and Randall [*J. Am. Chem. Soc.*, **36**:1969 (1944)] as a function of pressure at 25°C. The pressure was obtained by allowing the hydrogen gas to escape against a hydrostatic head measured in centimeters of water. Their results are:

P (*cm water in excess of 1 atm*)	0	37	63	84
$\mathcal{E}$ (*volts*)	0.40088	0.40137	0.40163	0.40190

Compare the pressure dependence of these results with the thermodynamic predictions.

7 The cell $Zn|ZnCl_2(c)|AgCl(s)|Ag$ has been studied by Lewis and Randall

("Thermodynamics," p. 420, McGraw-Hill Book Company, 1923). They report, for 25°C,

c (*mol liter*$^{-1}$)	0.000772	0.001253	0.001453	0.003112	0.006022	0.01021
$\mathcal{E}$ (*volts*)	1.2475	1.2289	1.2219	1.1953	1.1742	1.1558

a Using a Debye-Hückel limiting-law extrapolation, deduce, as well as possible from these data, the value of $\mathcal{E}°$ and the activity coefficients of $ZnCl_2$ at each concentration.
b Compare the value deduced for $\mathcal{E}°$ from these data with that obtained from the data of Table 23-1.

8 From the data of Table 23-1, calculate the solubility product of $PbSO_4$ at 25°C. *Ans.* 1.35×10^{-8}.

9 Calculate $\Delta G°$ and the equilibrium constant for the reaction $H_2 + O_2 \rightleftharpoons H_2O_2$ at 25°C. At what total pressure would ΔG be equal to zero?

10 If the mean-activity coefficient of the ions formed from the dissociation of Na_3PO_4 is 0.887 in a certain solution, what is the activity coefficient $(\gamma_+{}^3\gamma_-)$ of the salt? *Ans.* $Na_3PO_4 = 0.62$.

11 If the activity coefficient for $CaCl_2$ in a 0.1-M solution is 0.515, what is the mean-activity coefficient for the ions?

12 For 1-M solutions at 25°C the activities a_+a_- of the electrolytes NaCl, $NaNO_3$, HNO_3, and HCl are 0.657, 0.548, 0.724, and 0.809, respectively. Show that these data are such that the activities of electrolytes could not be broken down to contributions from the separate ions, as was done, for example, for conductivities at infinite dilution.

13 Calculate the ionic strengths of solutions that contain:
a 0.30 M $CaCl_2$. *Ans.* 0.9.
b 0.30 M Na_3PO_4.
c 0.10 M Na_2SO_4 plus 0.2 M NaCl. *Ans.* 0.5.
d 0.0078 M acetic acid, which is 4.8 percent dissociated, and 0.5 M dioxane.

14 Calculate, according to the limiting law of the Debye-Hückel theory, the activity coefficient of Ba^{++} and of Cl^- and the mean-activity coefficient of $BaCl_2$ in a 0.0050-M aqueous solution at 25°C.
Ans. $\gamma_+ = 0.56$; $\gamma_- = 0.87$; $\gamma_\pm = 0.75$.

15 Calculate, according to the limiting form of the Debye-Hückel theory, the activity coefficients of each of the ions Na^+, SO_4^{--}, OH^-, and H^+ in a solution that contains both 0.003 M Na_2SO_4 and 0.001 M NaOH.

16 Calculate the activity coefficients for the H^+ and the acetate ion in a solution which is 0.0078 M in acetic acid. The degree of dissociation at this concentration is 4.8 percent.

17 From the conductance data of Table 22-2 and the limiting law of the Debye-Hückel theory, deduce the mean-activity coefficient of the ions of NH_4OH in the concentration range for which data are given.
Ans. At 0.010, $\gamma_\pm = 0.98$.

18 Calculate the concentration, the activity coefficient, and the activity of $[Co(NH_3)_4C_2O_4][Co(NH_3)_2(NO_2)_2C_2O_4]$ when it is present in a saturated solution containing 0.0026 M $K_3Co(CN)_6$ at 15°C. (The necessary data are available from Fig. 23-10.)

19 The solubility of AgCl in water has been deduced from conductivity data to be:

Temperature (°*C*)	1.55	4.68	9.97	17.51	25.86	34.12
Solubility (*g* AgCl/1000 *g* H_2O)	0.0056	0.0066	0.0089	0.0131	0.0194	0.0274

a Calculate the solubility product from these data, and by a suitable graphical treatment, estimate the heat of solution of AgCl at 25°C from the temperature dependence of the solubility product.
b Deduce the free energy of AgCl in its standard, hypothetical state of unit activity in water at 25°C.
c Calculate the entropy of solution of AgCl to its standard unit-activity state in water at 25°C, and with the absolute entropy of AgCl(s) of 96.11 J deg^{-1} mol^{-1}, obtain a value for the entropy of AgCl in aqueous solution.

20 Calculate the entropy change that would be expected for the process

$$Na^+(g) + Cl^-(g) \xrightarrow{H_2O} Na^+(aq) + Cl^-(aq)$$

if the process corresponded to nothing more than the restriction of the ions to a free volume of, say, 2 ml mol^{-1}.

21 Using the thermodynamic data of Appendix 5, deduce the free-energy change accompanying the oxidation of 1 mol of methane, CH_4, to produce CO_2 and H_2O. If this reaction could be performed electrochemically, what emf would be produced? Write balanced oxidation and reduction half reactions for this cell. Verify that the relation $\Delta G° = -n\mathfrak{F}\mathcal{E}°$ recovers the original value of $\Delta G°$.

References

MACINNES, D. A.: "The Principles of Electrochemistry," Dover Publications, Inc., 1961. A new edition of an important earlier work, originally published in 1939, on electrochemistry. Excellent treatment is given not only of the material on emfs, but also on conductance, transference numbers, and so forth.

CONWAY, B. E.: "Electrochemical Data," Elsevier Publishing Company, Amsterdam, 1952. Included along with other data relevant to electrochemical studies are tables of the emfs of a variety of cells and electrodes.

IVES, D. J. G., and G. J. JANZ: "Reference Electrodes," Academic Press, Inc., New York, 1961. Detailed discussions of the theory and use of different types of electrodes, including the hydrogen electrode, the calomel electrode, the silver-silver halide electrodes, the glass electrode, and so forth.

ROBBINS, O., JR.: The Proper Definition of Standard Electromotive Force, *J. Chem. Educ.*, **48**:737 (1971).

EISENMAN, G., R. BATES, G. MATLOCK, and S. M. FRIEDMAN: "The Glass Electrode," Interscience Publishers, Inc., New York, 1965. A reprint volume of articles by the authors on glass electrodes, pH measurements, and sodium- and potassium-sensitive electrodes.

EISENMAN, G. (ed.): "Glass Electrodes for Hydrogen and Other Cations: Principles and Practice," Marcel Dekker, Inc., New York, 1967. Similar in goals and scope to the preceding listing.

DURST, R. A. (ed.): Ion-selective Electrodes, *Nat. Bur. Std.* (*U.S.*) *Spec. Pub.* 314, 1969. A collection of 11 discussions of the theory and application of ion-selective electrodes.

24

ADSORPTION AND HETEROGENEOUS CATALYSIS

It has been mentioned in the discussion of the liquid state that molecules on the surface of a liquid have a different environment and therefore the surface has a different free energy from the bulk of the material. In most chemical systems the fraction of the molecules on a surface and the free-energy difference between the surface and the bulk material are relatively small. Now systems are considered in which the surface effects are dominant. Mention will be made of a liquid film spread out in another liquid, but attention will be devoted principally to systems in which the molecules of a gas are concentrated on the surface of a solid. The molecules are said to be *adsorbed* on the solid surface, and this process is distinguished from the penetration of one component throughout the body of a second, called *absorption*. The chapter will begin with a direct study of the adsorption process and the adsorbed layer.

Such studies are now almost always directed toward an understanding of the chemical reactions that occur at the surface. The surface, it will be shown, enters into reactions as a catalyst. It is interesting that this type of catalysis, called *heterogeneous catalysis,* is understandable only on the basis of some of the information deduced in absorption studies; on the other hand, conclusions that are drawn from the chemical reactions on a surface help to answer some of the problems unsolved by direct adsorption studies.

The goal of modern physical-chemical studies of surface phenomena is the understanding of these phenomena by means of a molecular model. However, systems which have a very thin, often monomolecular layer of gas adsorbed on a complex solid adsorbent resist most of the methods that have already been studied to elucidate the molecular world. The current theories of surface reactions are still very tentative and loosely supported. Furthermore, little can be said about the nature and behavior of surface molecules. Some of the background information and treatments on which the ideas that exist are based will, however, be introduced.

It should be pointed out immediately that heterogeneous catalysis is a procedure of great importance in industrial chemistry. This fact and the challenge of the many unexplained phenomena make the study of the adsorbed state one of the most exciting areas of modern physical-chemical research.

ADSORPTION

Although the distinction between *adsorption* and *absorption* is not always clear-cut and the noncommittal word *sorption* is sometimes used, the processes that will be considered here will be essentially surface effects and the word adsorption will be used.

The most important and interesting type of adsorption is certainly that in which gases are adsorbed on a solid. Before treating this subject, however, the much simpler and more easily treated process of the adsorption, or spreading, of a film of one liquid on the surface of another will be dealt with briefly.

24-1 Liquid Films on Liquids

The most interesting and easily studied liquid films are formed by allowing a small quantity of a *surface-active* material, for example, a long-chain organic acid like stearic acid,

$$CH_3(CH_2)_{16}COOH$$

to spread out on the surface of water. Such molecules are suitable because the acid group shows an attraction for water (short-chain acids are, in fact, soluble in water) that makes the material spread out over the water surface, whereas the long hydrocarbon end prevents the material from dissolving. It can now be shown that such films can be made to form a *monomolecular layer* on the water surface.

Modern studies of such films are made on an apparatus, called a *surface balance,* developed by Langmuir in 1917. The apparatus is shown in Fig. 24-1. The trough is filled with water, and a measured amount of the surface-active material is added. The movable barrier is pushed forward, and measurements of the force exerted on the fixed barrier are read off the delicate tension device. It is customary to plot the results as the force on the fixed barrier versus the surface area per molecule of the surface-active agent, i.e., the surface area covered

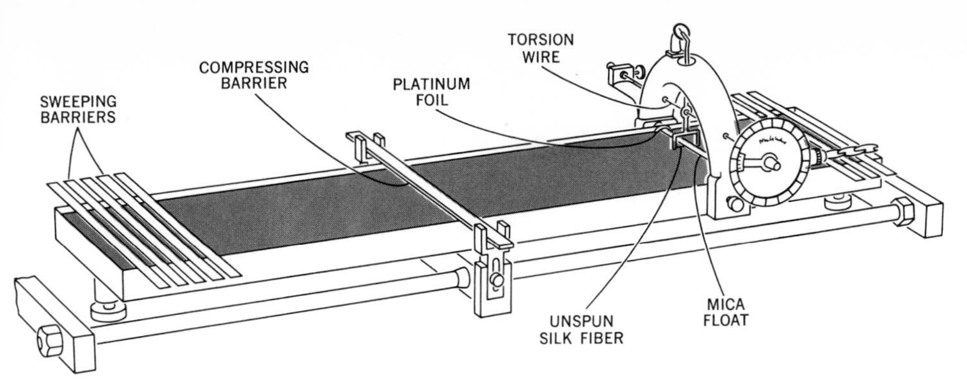

FIGURE 24-1
A Langmuir film-balance apparatus. (*Courtesy of Herman E. Ries, Jr., Standard Oil Company of Indiana, Whiting, Ind.*)

by the surface-active agent divided by the number of molecules in the sample. Typical results for stearic acid are shown in Fig. 24-2.

The initial slow increase in film force with decreasing surface area indicates that the surface is not completely covered by the surface-active film. The beginning of the steep part of the curve is taken to correspond to the completion of the film; further decrease in area must compress the film itself, and a large increase in force is necessary. Finally, the film buckles and folds, and the area can be decreased without any further increase in force.

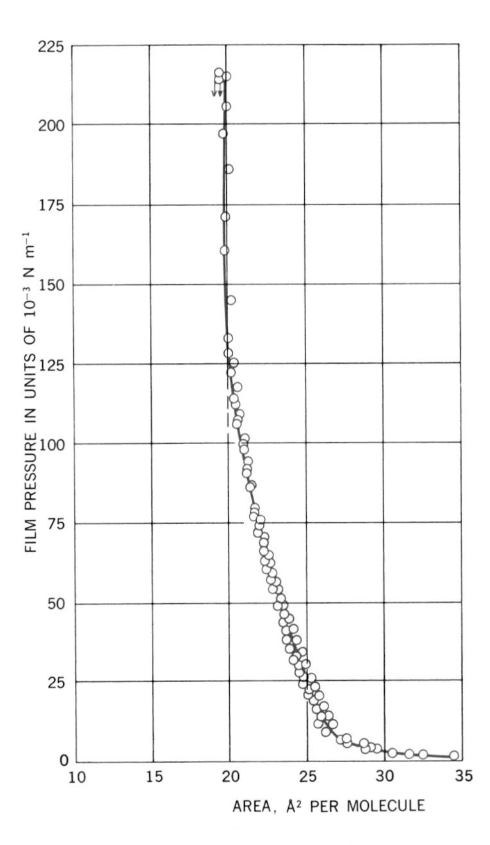

FIGURE 24-2
Pressure-area isotherm and molecular orientation of stearic acid. (*Courtesy of Herman E. Ries, Jr., Standard Oil Company of Indiana, Whiting, Ind.*)

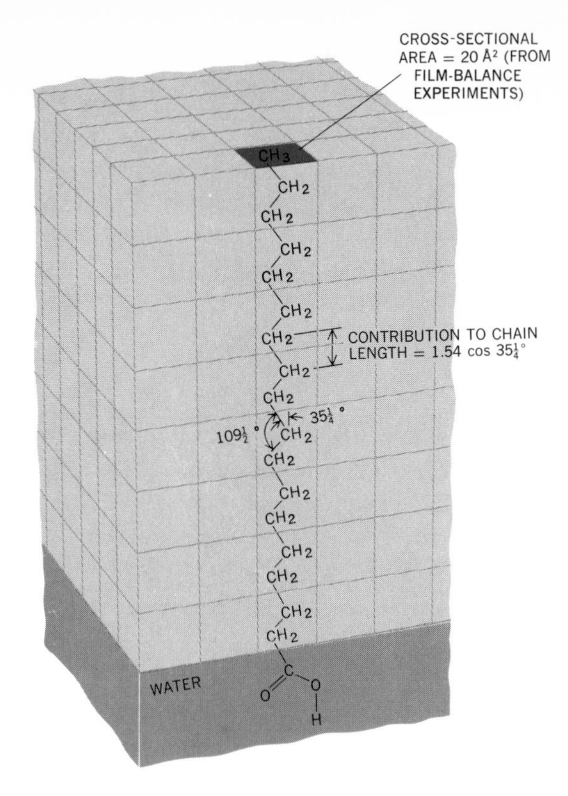

FIGURE 24-3
The dimensions of a monomolecular surface film of stearic acid, $CH_3(CH_2)_{16}COOH$, on water.

A calculation can now be made which supports the idea that the onset of the steep part of the curve corresponds to a monomolecular film. For stearic acid this film has, according to Fig. 24-2, an area of about 20.5 Å^2 per molecule. This value can now be shown to be about that expected for the cross-section area occupied by a stearic acid molecule.

The film is assumed to consist of molecules arranged approximately as depicted in Fig. 24-3. The volume occupied by an Avogadro's number of molecules of stearic acid can be taken to be approximately that of 1 mol of liquid stearic acid, i.e.,

$$\text{Vol of } \mathfrak{N} \text{ molecules} = \frac{M}{\rho} = \frac{284}{0.85} = 330 \text{ ml}$$

and

$$\text{Vol of 1 molecule} = \frac{1}{\mathfrak{N}} = \frac{M}{\rho} = 550 \text{ Å}^3$$

The length of the zigzag chain of carbon atoms can be estimated, as Fig. 24-3 shows, as

$$\text{Length of stearic acid} = (17)(1.54) \cos 35° = 21 \text{ Å}$$

From these estimates one deduces the effective cross-section area as

$$\text{Cross-section area} = 550/21 = 26 \text{ Å}^2$$

This result is in sufficient agreement with the value of 20.5 Å^2 from the film-balance experiment to suggest the existence of a monomolecular layer at the beginning of the steep portion of the force-area curve, such as that of Fig. 24-2.

The nature of these surface layers is nicely shown by microphotographs, as indicated in Fig. 24-4. These surfaces have been *shadow-cast* by a coating of a very thin layer of chromium from a beam directed at an angle to the surface. If the angle of the shadow-casting beam is known, the length of any shadow can be used to deduce the height of any projection. In this way the results for the collapsed layer of Fig. 24-4 can be interpreted in terms of double "sandwich" layers of the fatty acid lying on top of the monomolecular surface film.

Some features of these liquid films will be encountered in the study of the adsorption of gases on solids. The concept of a monomolecular layer will be of great importance, but the transition to multiple layers, for gases on solids, will be less easily detected. Likewise, the surface area of the adsorbent will be talked about but will seldom be as definite a concept as in liquid-film systems. Part of this difficulty stems from the nonhomogeneity that must be anticipated for solid adsorbents. Finally, the nature of the attraction of the surface layer for the adsorbent will be studied and will be found to be rather more complicated than the essentially physical, or van der Waals, attractions that act on the surface film on a liquid. It is all these added complexities which make the study of the nature and reactions of gases adsorbed on solids of special interest.

An extension of studies of liquid-surface films could be made to liquid-liquid interfaces. The molecules of stearic acid, for example,

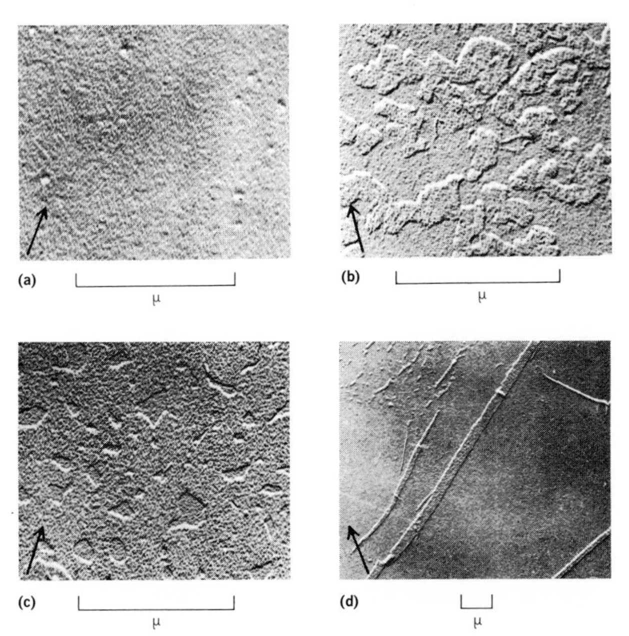

FIGURE 24-4
Electron micrographs of monolayer films of *n*-hexatriacontanoic acid [$CH_3(CH_2)_{34}COOH$]. The arrows give the direction of chromium shadow casting, which was done at an angle of 15° to the surface. The light areas are shadows not covered by chromium. (One micron is 10^4 Å.) (*a*) Blank, no film. (*b*) At 15 dyn cm^{-1}. (*c*) At 25 dyn cm^{-1}. (*d*) After collapse. (*Courtesy of Herman E. Ries, Jr., Standard Oil Company of Indiana, Whiting, Ind.*)

would be expected to concentrate at an oil-water interface just as they do at an air-water interface. All such systems are clearly important in studies of lubrication and in the wetting action of water containing soaps or detergents.

24-2 Classifications of Adsorptions of Gases on Solids

It is very convenient in the study of adsorption to recognize that most adsorptions can be placed in one of two categories. These categories are suggested by the possibilities of having essentially physical forces holding the gas molecules to the solid or of having chemical bonds serve the function. The categories of *physical adsorption* and *chemical adsorption,* or more commonly, *chemisorption,* thus arise. The observed characteristics of any adsorption process usually allow it to be placed in one or the other category. Table 24-1 outlines the experimental features that allow a process to be so categorized. The different behaviors of the two types of adsorption should be recognized as having the features of a physical process, such as condensation, or those of a chemical reaction. It should be pointed out that whether or not more than a monolayer is being formed is not directly observable but, as we shall see, can often be deduced from experimental data.

Most interest in adsorption, and in the closely related field of heterogeneous catalysis, is in chemisorption. Two of the items mentioned in Table 24-1 will therefore be dealt with in more detail and with particular emphasis on chemisorption.

TABLE 24-1 Characteristics of physical adsorption and chemisorption

Physical adsorption	*Chemisorption*
Heat of adsorption is less than about 40 kJ mol^{-1}.	Heat of adsorption is greater than about 80 kJ mol^{-1}.
Adsorption is appreciable only at temperatures below the boiling point of the adsorbate.	Adsorption can occur at high temperatures.
The incremental increase in the amount adsorbed increases with each incremental increase in pressure of the adsorbate.	The incremental increase in the amount adsorbed decreases with each incremental increase in the pressure of the adsorbate.
The amount of adsorption on a surface is more a function of the adsorbate than the adsorbent.	The amount of adsorption is characteristic of both adsorbate and adsorbent.
No appreciable activation energy is involved in the adsorption process.	An activation energy may be involved in the adsorption process.
Multilayer adsorption occurs.	Adsorption leads to, at most, a monolayer.

24-3 Heat of Adsorption

In all adsorptions, heat is given out, and ΔH for the process

$$\text{Gas} \rightarrow \text{adsorbed layer} \tag{1}$$

is negative. Heats of adsorptions are, however, generally listed without sign. The necessity for a negative ΔH, in contrast to chemical reactions in general, arises from the fact that the entropy of the ordered, constrained adsorbed layer is always less than that of the gas; i.e., for the reaction of Eq. [1], ΔS is invariably negative. It follows that, for the process of Eq. [1] to be spontaneous and have a negative value for ΔG, the value of ΔH must be negative and greater than $T\,\Delta S$.

For physical adsorption the heats involved are of the order of heats of vaporization, i.e., generally less than 40 kJ mol^{-1}, and in keeping with the idea that physical adsorption may be leading to the formation of multilayers, these heats are more dependent on the nature of the gas than they are on that of the solid adsorbent.

Adsorptions classed as chemisorptions, on the other hand, have heats of adsorption that compare with those of ordinary chemical reactions; in other words, they have heats of anywhere up to about 630 kJ mol^{-1}.

The approximate values mentioned for physical and chemical adsorptions are not intended to imply a constancy for these heats as a function of the amount of gas adsorbed. Some of the variations in differential heats of adsorption in the chemisorption region are shown in Fig. 24-5. The frequently observed curve over the region of adsorption from low coverages to multilayer formation has a high initial heat that falls off at large amounts of adsorption. This behavior is taken as indicative of an initial chemisorption, to form something like a

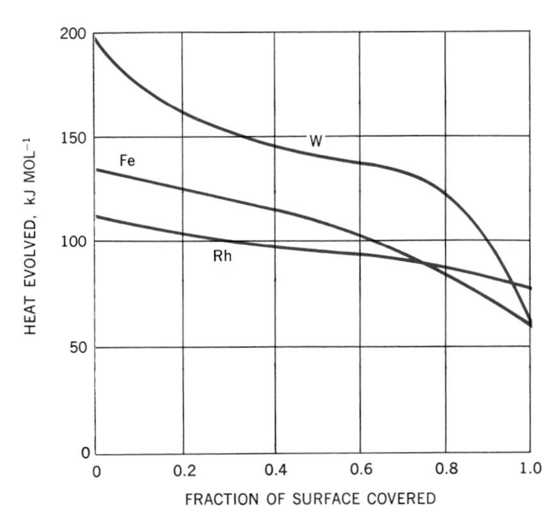

FIGURE 24-5
The heat of adsorption for H_2 on clean metal surfaces. (*From O. Beeck, Discussions Faraday Soc.,* **1950,** p. 118.)

monolayer, followed by the formation of multiple layers that are bound by physical forces.

Even within a range attributed to chemisorption, however, the heat of adsorption is usually found to be a function of the amount adsorbed. A number of molecular explanations have been offered for this variation. Active sites can be assumed to exist on the adsorbent, and as these are occupied by the first additions of gas, the binding of later additions must occur on less active sites and the strength of binding falls off. Alternatively, the binding of some gas to the solid can occur with the giving up of electrons to the solid by the adsorbed molecules or with the withdrawing of electrons from the solid, and as such processes continue, the solid becomes more and more reluctant to gain or lose more electrons. Such an explanation is particularly appropriate to semiconducting and conducting adsorbents. The final important factor that has been suggested is that the mutual repulsion of the adsorbed molecules, especially if they acquire a net charge when they are adsorbed, operates to oppose the addition of further molecules.

24-4 The Adsorption Isotherm

The most frequently encountered adsorption experiment is the measurement of the relation between the amount of gas adsorbed by a given amount of adsorbent and the pressure of the gas. Such measurements are usually made at a constant temperature, and the results are generally presented graphically as an *adsorption isotherm.* Experimentally, one measures either the volume of gas taken up by a given amount of adsorbent or the change in weight of the adsorbent when it is exposed

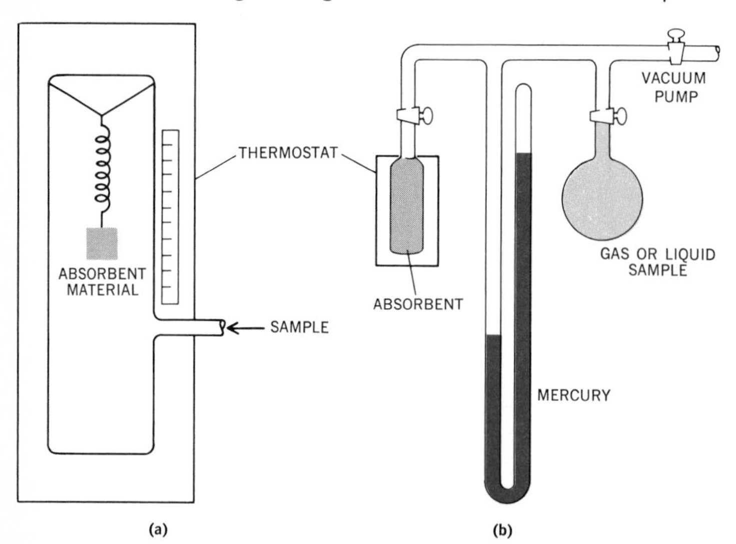

FIGURE 24-6
Adsorption isotherm apparatus. (*a*) Gravimetric. (*b*) Volumetric.

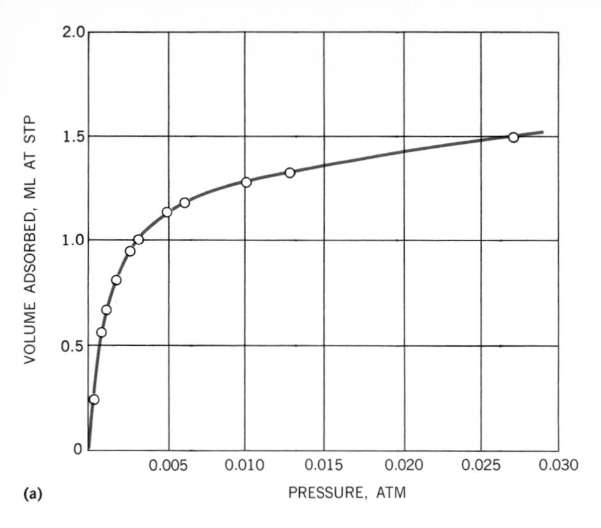

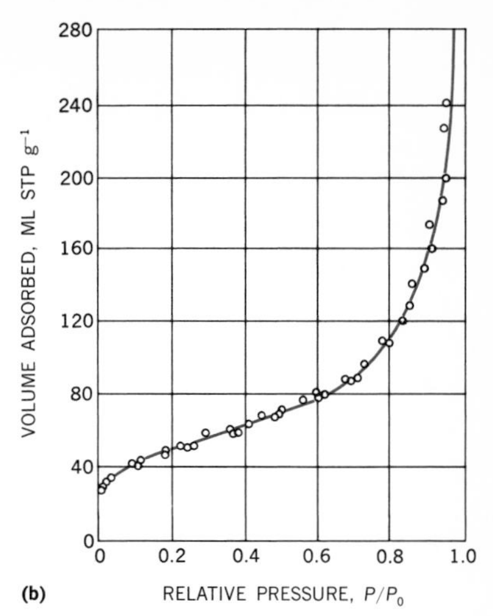

FIGURE 24-7
(*a*) The adsorption isotherm for H_2 on Cu powder at 25°C. [*From A. F. H. Ward, Proc. Roy. Soc. (London),* **A133:***506 (1931)*.] (*b*) The adsorption of N_2 up to the vapor pressure of N_2, on silica. (*From P. Emmett, "Catalysis," vol. 1, Reinhold Publishing Corporation, New York, 1954.*)

to a gas at a given pressure. The apparatus that can be used is shown in Fig. 24-6.

A great variety of adsorption-isotherm shapes are found. Chemisorption is usually accompanied by an initial steeply rising curve that gradually flattens off. The initial rise is taken as corresponding to the strong tendency of the surface to bind the gas molecules, and the leveling off can be attributed to the saturation of these forces, perhaps by one or more of the three mechanisms mentioned in Sec. 24-3. Physical adsorption, on the other hand, is accompanied by an adsorption isotherm that tends to have an increasingly positive slope with increasing gas pressure. Each incremental increase in gas pressure produces a larger increase in the amount of gas adsorbed—up to the limit of a pressure equal to the vapor pressure of the material being adsorbed, at which pressure the adsorption isotherm ascends vertically as condensation occurs.

Some adsorption isotherms, as Fig. 24-7 suggests, can be interpreted as a combination of these chemisorption and physical-adsorption curves. As we shall see, however, no simple, or even complex, explanation can be expected for the detailed shapes of all adsorption isotherms.

24-5 The Langmuir Adsorption Isotherm

A model for the adsorption process, and particularly for the chemisorption process, was presented by Langmuir in 1916 and led him to a simple, but important, theoretical derivation of an adsorption isotherm. The chemisorption process is pictured as leading ultimately to a monomolecular film over the surface of the adsorbent, and the derived adsorption isotherm results from an investigation of the equilibrium that is set up between the gas phase and the partially formed monolayer. When the gas is at a pressure P, the fraction of the surface that is covered is represented by θ. The equilibrium state can be interpreted in terms of the dynamic equilibrium that results from an equal rate of evaporation of the adsorbed material and rate of condensation of the gas-phase molecules.

The Langmuir theory suggests that the rate of evaporation can be taken to be proportional to the fraction of the surface covered and can be written, therefore, as $k_1\theta$, where k_1 is some proportionality constant. This simple proportionality is an assumption that ignores the complications that often make the heat of adsorption dependent on the extent of coverage and that may well be expected to spoil the simple assumption of an evaporation rate proportional to $k_1\theta$. The rate of condensation, furthermore, is taken to be proportional both to the gas pressure P, which according to the kinetic-molecular theory of Chap. 2 determines the number of molecular collisions per unit area per unit time, and to the fraction of the surface not already covered by adsorbed molecules, i.e., to $1 - \theta$. It is assumed that only collisions with this exposed surface can lead to the sticking of a molecule to the surface. The relation between equilibrium surface coverage and gas pressure is then obtained by equating the expressions deduced for the rate of evaporation and the rate of condensation, i.e.,

$$k_1\theta = k_2P(1 - \theta) \qquad [2]$$

where k_2 is another proportionality constant. Rearrangement gives

$$\theta = \frac{k_2P}{k_1 + k_2P} \qquad [3]$$

and introduction of $a = k_1/k_2$ allows this result to be written

$$\theta = \frac{P}{a + P} \qquad [4]$$

Inspection of Eq. [4] shows that a chemisorption-type isotherm is obtained from this theory. At small values of P, where P in the denominator can be neglected compared with a, Eq. [4] reduces to a simple proportionality between θ and P, and this behavior is that corresponding to the initial steep rise of the isotherm curve. At higher pressures the value of P in the denominator contributes appreciably, and the increasing denominator leads to values of θ that do not increase proportionally to the increase in P. For sufficiently large values of P, θ approaches the constant value of unity.

Experimental isotherm data consist of the amount of gas adsorbed by a given weight of adsorbent as a function of the gas pressure. For adsorption, up to a monolayer, the amount of gas y adsorbed at some pressure P and the amount of gas y_m needed to form a monolayer are related to θ according to

$$\frac{y}{y_m} = \theta \qquad [5]$$

and Eq. [4] becomes

$$y = \frac{y_m P}{a + P} \qquad [6]$$

Experimental results can be compared with the Langmuir theory most easily if Eq. [6] is rearranged to

$$\frac{P}{y} = \frac{a}{y_m} + \frac{P}{y_m} \qquad [7]$$

A plot of P/y versus P, if the experimental data are in accord with the Langmuir theory, will yield a straight line. If such a curve is obtained, the intercept can be identified with a/y_m and the slope with the constant $1/y_m$. For many chemisorptions one finds, as Fig. 24-8 shows, a good linear relationship on the Langmuir suggested plot. For physical adsorption isotherms or S-shaped curves, the Langmuir plot does not yield a straight line, and the theory is clearly not applicable to such cases. The success of Eq. [7] in fitting experimental chemisorption-type curves must not, of course, be taken as necessarily confirming the model and assumptions that have been used in the derivation.

Other theories have been developed to explain the more complete adsorption process that leads to multilayer formation. The most important of these treatments is due to Brunauer, Emmett, and Teller. Their theory, like that of Langmuir, leads to an isotherm expression, usually abbreviated as the BET isotherm. Although this expression receives considerable attention as a basis for surface-area determinations, which

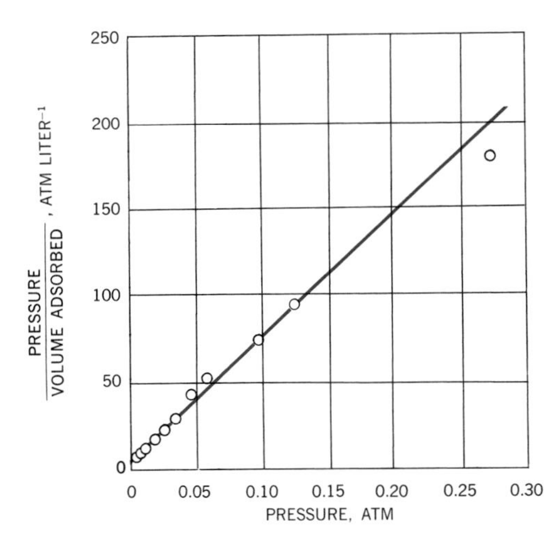

FIGURE 24-8
The Langmuir plot for the adsorption of H_2 on Cu powder at 25°C shown in Fig. 24-7.

will next be dealt with, it will not be necessary for us to investigate here the BET theory.

24-6 Determination of Surface Areas

It is important, if one is to obtain a definite picture of the happenings on a surface, to have some way of estimating the surface area. Since many of the solids that are used in adsorption studies are highly irregular and porous, like charcoal, the area cannot be measured directly and an adsorption method, generally using the BET isotherm, is ordinarily employed. Although the Langmuir isotherm can represent only the chemisorption process, it can be used to show that a surface area can be deduced from adsorption studies.

The specific example of the adsorbent used for the experiments that led to one of the isotherms of Fig. 24-7 can be considered. The Langmuir plot of these data in Fig. 24-8 gives

$$\text{Slope} = \frac{1}{y_m} = 735 \text{ (liter at STP)}^{-1} \tag{8}$$

$$\text{Intercept} = \frac{a}{y_m} = 5.3 \text{ atm (liter at STP)}^{-1} \tag{9}$$

The values of y_m and a are calculated as

$$y_m = 0.00136 \text{ liter at STP} \qquad a = 0.0072 \text{ atm} \tag{10}$$

and the adsorption isotherm is represented by the equation

$$y = \frac{0.00136P}{0.0072 + P} \text{ liters at STP} \qquad \text{for } P\text{, atm} \tag{11}$$

The result is obtained, therefore, that the surface of 1 g of adsorbent would be covered by an amount of H_2 which occupies a volume of 0.00136 liter at STP, i.e., by $0.00136/22.4 \times 6.0 \times 10^{23} = 3.6 \times 10^{19}$ molecules.

The surface area is obtained if the area covered by this much H_2 can be estimated. The easiest, if rather crude, method is to make use of the bulk volume of liquid H_2 and to calculate the effective area per molecule as $(V_{\text{liq}}/\mathfrak{N})^{2/3}$, where V_{liq} is the volume of 1 mol of liquid H_2. In this way one estimates, using the density of 0.070 g ml^{-1} for liquid hydrogen, that the area covered by one molecule is

$$\left(\frac{2 \times 10^6 \text{ ml m}^{-3}}{0.070 \times 6 \times 10^{23}}\right)^{2/3} \qquad \text{or} \qquad 13 \times 10^{-20} \text{ m}^2$$

and the area of 1 g of this charcoal adsorbent is therefore

$$(3.6 \times 10^{19})(13 \times 10^{-20}) = 4.7 \text{ m}^2 \tag{12}$$

In practice, one relies on the BET isotherm and makes use

TABLE 24-2 Estimates of surface area of clean nickel films from the physical adsorption of different gases

Gas	Area molecule^{-1} (Å^2)	Amount adsorbed to give monolayer (on 1 g nickel film)	Surface area of 1 g nickel film (m^2)
Kr	14.6	6.15×10^{19}* molecules	9.0
Kr	14.6	5.85×10^{19}*	8.6
CH_4	15.7	5.40×10^{19}	8.5
n-C_4H_{10}	24.5	3.48×10^{19}	8.5

*On different films.
SOURCE: O. Beeck and A. W. Ritchie, *Discussions Faraday Soc.*, **8**:159 (1950).

of physical-adsorption data rather than the chemisorption data used in this Langmuir example. The calculation procedures are, however, equivalent.

Surface areas estimated from such adsorption studies can often be accepted as generally reliable but approximate. Table 24-2 shows the type of variation in area estimated from different isotherms, and Table 24-3 shows some typical surface areas of adsorbents.

There may, of course, be a number of subtleties connected with the surface that may lead to puzzling area values. The presence of fine pores or capillaries may, for example, be such as to allow one gas to penetrate, whereas another gas with larger molecules finds the pores inaccessible. In this connection it is of interest to note the "molecular-sieve" materials that consist of dehydrated zeolites. They appear to have pores that are of sufficiently uniform size so that an adsorbent can be obtained which can, for example, accept n-paraffin molecules but not branched-chain molecules.

The use of chemisorption data for surface determinations, it should be pointed out, would introduce the very questionable assumption that

TABLE 24-3 Volumes of adsorbed nitrogen to form a monolayer and the surface areas of a number of catalysts*
Area covered by an adsorbed nitrogen molecule taken as 16.2 Å^2.

Material	Monolayer volume (liters at STP g^{-1})	Surface area (m^2 g^{-1})
Fused Cu catalyst	0.09×10^{-3}	0.39
Fe, K_2O catalyst 930	0.14	0.61
Fe, Al_2O_3, K_2O catalyst 931	0.81	3.5
Fe, Al_2O_3 catalyst 954	2.86	12.4
Cr_2O_3 gel	53.3	230
Silica gel	116.2	500

*From P. H. Emmett, "Catalysis," vol. 1, Reinhold Publishing Corporation, New York, 1955.

the "active" surface area is the same both for the gas with which the area is determined and for any other gas that might be studied. In practice, it is much more satisfactory to use the BET-isotherm expression to deduce the surface area since the BET isotherm includes adsorption to form multilayers.

24-7 Adsorption from Solution

A very important, but little understood, process is the adsorption of a solute of a solution onto a solid adsorbent. This procedure is followed, for example, in the decolorizing of solutions using, ordinarily, activated charcoal. The separation technique of chromatography also makes use of the relative adsorption tendencies of the solutes of a solution.

This process of adsorption from solution is even more difficult to treat theoretically than is the corresponding gas-on-solid process. It appears, however, that only a monomolecular layer is formed, any further addition being strongly opposed by the solvating power of the solvent.

A fairly satisfactory empirical isotherm, which can be applied to adsorptions of gases with considerable success but has been used principally for adsorption from solution, has been discussed by Freundlich. If y is the weight of solute adsorbed per gram of adsorbent and c is the concentration of the solute in the solution, this empirical relation is

$$y = kc^{1/n} \tag{13}$$

where k and n are empirical constants. The equation is conveniently used in the logarithmic form

$$\log y = \log k + \frac{1}{n} \log c \tag{14}$$

When applied to gases, y is the amount of gas adsorbed and c is replaced by the pressure of the gas. Experimental results conform to the Freundlich expression if a plot of $\log y$ against $\log c$, or $\log P$, yields a straight line. The constants can then be determined from the slope and intercept. Figure 24-9 shows data treated in terms of the Freundlich expression.

24-8 The Nature of the Adsorbed State

Physical adsorption consists in the binding of molecules to the surface of the adsorbent by essentially van der Waals forces, and the molecules of the adsorbed layers can be expected to be altered only to about the extent that gas molecules are when they are condensed to the liquid state.

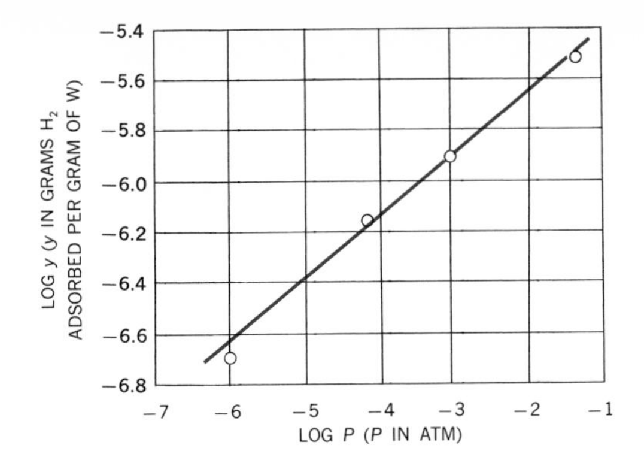

FIGURE 24-9
A Freundlich plot of the isotherm data for H_2 adsorbed on tungsten at 400°C. [*From the more extensive data of W. G. Frankenberg, J. Am. Chem. Soc.,* **66:**1827 (1944).]

Chemisorption, on the other hand, must be expected to produce major changes in the electronic distribution and the bonding in the adsorbed molecules. Much of the current interest in adsorption, and in heterogeneous catalysis, is centered around the description of the molecules that are bound to a solid in chemisorption. Some of the ideas that have been put forward and some of the experiments that have proved to be helpful can now be mentioned.

The use of the word *chemisorption* implies that bonding of the gas molecules to the solid occurs by ordinary chemical bonds, i.e., that the bonds can be described in terms of ionic and covalent character. Such an assumption is profitable for many systems, but one must be prepared, for example, for the sea of electrons of a metal to play a role in the bonding of surface groups that has no direct counterpart in simple chemical systems. It appears from surface dipole moments, however, that the adsorption of Na vapor on tungsten leads to a surface layer of Na^+ ions and a bonding to the tungsten that is comparable with ordinary ionic bonds. Likewise, the infrared spectrum of carbon monoxide adsorbed on platinum indicates the essential invariance of the bonding in the CO bond and suggests a bonding picture such as

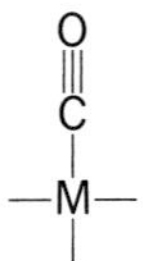

The electron movement in adsorptions such as these can be studied by measuring the magnetic properties of highly dispersed metal particles as a function of the adsorbed gas. The study of coordination compounds by magnetic measurements was seen, in Chap. 15, to reveal the number of unpaired electrons in the metal atom. In a similar way the magnetic properties of a metallic particle depend on the net number of unpaired electrons. If the adsorbed molecules feed electrons into the partially filled d orbitals, or d band, of the metal, the magnetic susceptibility will decrease, and vice versa. From such measurements

Selwood has deduced the following electron shifts for adsorbed species on nickel:

H O C O N H H O N N O N

Such results emphasize that adsorption is at least as complex a phenomenon as chemical-bond formation and that all the variations in electron distribution that we are accustomed to in homogeneous chemical systems will also appear in the bonding of surface molecules to the adsorbent.

Adsorption of saturated hydrocarbons, and hydrogen, presents another problem in that these molecules have no sites for additional chemical bonds. Primarily from exchange studies, such as the passing of a mixture of D_2 and a hydrocarbon over an adsorbent, one deduces that some dissociative mechanism operates. For C_2H_6, for example, one might write the adsorbed state as

CH_3 CH_3

or as

CH_3
CH_2 H

The first is suggested by the fact that ethane can be "cracked" to form methane; the second is suggested by the deuterium exchange occurring in an adsorbed mixture of ethane and D_2.

Of particular interest in this connection are results such as those which show that cyclopentane in an excess of D_2 tends to form, at low temperatures on a suitable adsorbent, the exchange product $C_5H_5D_5$ rather than a more or less deuterium-containing product. A picture for the adsorbed molecule lying on the catalyst with exchange readily occurring only on one side of the approximately planar carbon ring is suggested.

No general principles can be given for the nature of the adsorption bond of any gas on any adsorbent. It appears that a great variety of features must be considered. Not only must the surface of the catalyst be considered, but also the electron reservoir or sink that the solid presents. It is becoming clear that chemisorption encompasses a great variety of chemical reactions and that these reactions cannot be understood in terms of one or a few simple processes. At present, however, many suggestions as to the reactions of surface molecules are being made, and much information will be forthcoming on the nature of chemisorbed species.

24-9 Importance of the Preparation of the Surface

Much of the early work directed toward an understanding of the adsorbed state in chemisorption has recently come under doubt because of uncertainties as to the nature of the surface. It is now recognized that there is an almost infinite variety of surface reactions and that a surface reaction cannot be expected to be correlated with the properties of the bulk material unless the state of the surface is well defined.

The most important developments have been those made in the study of adsorption on freshly prepared metal surfaces. Primarily as a result of Beeck's work it has been shown that, if a metal, such as Ni, W, Pt, etc., is evaporated from an electrically heated filament in an evacuated system, a film several thousand atoms thick can be deposited on the surface of a Pyrex adsorption cell. The film, moreover, is of such a large surface area that it is not appreciably coated by adsorption of the residual gas in the reaction cell. With this technique it has been possible to obtain adsorption isotherms and heats of adsorption that are apparently characteristic of the pure metal. It has been shown, furthermore, that most attempts to clean a surface by prolonged evacuation have been futile, and therefore that the subsequent adsorption studies have been made on surfaces that already are partially covered with a layer of oxygen or nitrogen.

A number of very valuable results have been obtained from studies on these well-defined metal surfaces. On nickel, for example, twice as much carbon monoxide appears to be adsorbed as hydrogen, suggesting that carbon monoxide is attached to a single metal-surface atom, whereas hydrogen, possibly dissociated into two hydrogen atoms, occupies two sites. Adsorption of oxygen proceeds to at least twice the extent that adsorption of carbon monoxide does, and this suggests, as does other evidence, that oxygen diffuses into the metal lattice and forms what are essentially metal oxides.

Of special interest is the observation that the adsorption of small amounts of ethylene leads to the evolution of ethane and the formation on the catalyst surface of a $(CH)_n$ polymeric material. Apparently, the ethylene adsorption occurs with appreciable bond dissociation.

Finally, it should be mentioned that the use of clean metal surfaces has led to some correlation between the adsorption process and the nature of the adsorbent metal. Some correlation has been found, for example, between the available d orbitals of the metal atoms and the heat of adsorption of hydrogen, and this correlation suggests an adsorption mechanism in which the hydrogen molecules dissociate into atoms and the atoms form bonds with the metal atoms on the surface.

The work with clean metal surfaces has emphasized the complexities that undoubtedly occur when metal powders, chemically deposited metal films, oxides of metals, and nonmetals are used as adsorbents. It is these more complicated surfaces, however, that exhibit the many remarkable and industrially important catalytic effects. The study of surface phenomena cannot, therefore, be restricted to clean metal surfaces.

HETEROGENEOUS CATALYSIS

Many chemical processes occur in the presence of certain surfaces that do not proceed at all, or do so very slowly, in the absence of such surfaces. Such reactions are said to be exhibiting *heterogeneous catalysis*. The effect of the surface is often so profound that it may be difficult to keep in mind that this effect is that of hastening the approach to an equilibrium state. A catalyst may, and generally will, provide very different accelerations for the approach of different reactions to their equilibrium state. Full use of this important influence in chemical reactions requires a detailed understanding of the reactions that are occurring on the surface of the catalyst.

The most dramatic surface catalytic effects must be attributed to reactions of chemisorbed species. Physical adsorption is effective in raising the local concentration of the reagents and in supplying a reservoir of thermal energy to these reagents. These factors, however, are probably of minor importance in heterogeneous catalysis. Chemisorption, on the other hand, may result in a rather drastic disruption of the bonding in an adsorbed molecule. It is easily seen that such molecules, or molecular fragments, may enter into reactions in a manner quite different from that in which the unperturbed gas-phase molecules do. In kinetic terms, the molecules on the surface are such that they may react through a state of much lower activation energy than can the normal molecules.

It follows that heterogeneous catalysis can be understood in detail only when the nature of the adsorbed species is so understood. At present, only some features of heterogeneous catalysis can be given a molecular description. It is true, however, that one of the most fruitful approaches to an understanding of the adsorbed state is through studies of the reactions that the molecules of this state undergo.

24-10 Some Experimental Methods and Results

Heterogeneous catalysis is usually studied by passing the gaseous reagent, or reagents, through a tube having a section containing the catalyst. The catalyst can then be held at any desired temperature by an external oven, and the reaction will proceed catalytically at that temperature. On leaving the catalyst chamber the reaction mixture will usually be effectively frozen by the absence of the catalyst, and the products that are collected may be analyzed by any convenient method.

TABLE 24-4 Some representative catalysts and catalytic reactions

Catalyst	*Process*
Silica alumina gel	Cracking of heavy petroleum fractions
Chromic oxide gel, chromia-on-alumina, nickel-aluminum oxide	Hydrogenation, dehydrogenation of hydrocarbons
Phosphoric acid on kieselguhr	Polymerization of alkenes
Co, ThO_2, MgO on kieselguhr (Fischer-Tropsh catalyst)	Synthesis of hydrocarbons from H_2 and CO
Iron	Synthesis of ammonia
Copper	Dehydrogenation of alcohols to aldehydes
Platinum	Isomerization of hydrocarbons
$Al(C_2H_5)_3$; $TiCl_4$	Polymerization of olefins

Many different catalysts are used, and since the exact treatment and mode of preparation are of great importance, many variations in catalytic activity are observed. Some of the more frequently studied and used catalysts are listed in Table 24-4. The important class of *supported catalysts* included there consists of a catalytically active material laid down on some porous support.

One of the areas in which heterogeneous catalysis finds widespread industrial application is petroleum refining. Reactions are desired that convert low-octane hydrocarbons and low- and high-vapor-pressure hydrocarbons to high-octane gasoline, and catalytic reactions are necessary to make such reactions feasible. Table 24-4 summarizes some of the processes that are in use in petroleum refining and in other industrial processes.

Even this brief table shows the variety of reactions that can be stimulated by suitable catalysts. The desire for suitable catalysts for such processes has contributed to the considerable effort that has been directed toward an understanding of the catalytic process.

24-11 Kinetics of Heterogeneous Decompositions

The study of the kinetics of single-phase reactions led, in Chap. 17, to a considerable understanding of the details of reaction mechanisms. A similar study of heterogeneously catalyzed reactions leads only to the more explicit recognition that the catalytic effect is a surface reaction. The study of some relatively simple decompositions that are heterogeneously catalyzed will illustrate this.

The kinetics of decomposition can often be accounted for on the assumption that the rate is proportional to the amount of the reagent on the surface. In line with this assumption it is convenient to treat three situations that are distinguished by the relation between the pressure of the gas and the amount that is adsorbed on the surface.

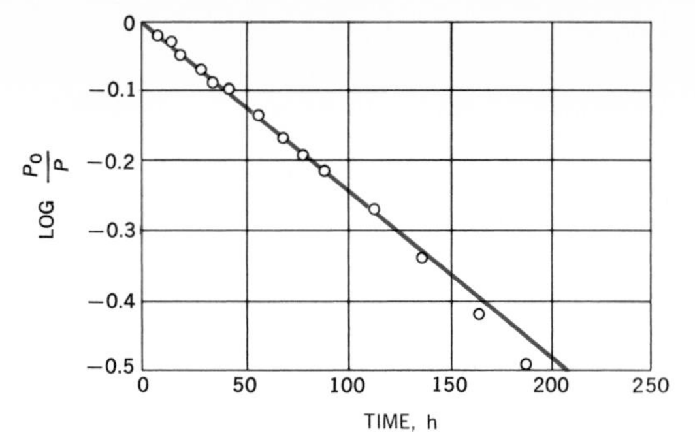

FIGURE 24-10
The decomposition of phosphine on a glass surface as a function of time at 446°C. The partial pressure of phosphine at time t is P, and at zero time is P_0. [*From D. M. Kooij, Z. Physik. Chem.*, **12:**155 *(1893)*.]

An even greater variety exists, however, in the dependence of rate of reaction on reagent pressure, but all the complexities cannot be considered here.

First, for low surface coverages the amount of gas adsorbed is, according to the Langmuir isotherm, approximately proportional to the gas pressure. The rate of decomposition, if decomposition is indeed a surface reaction, would be expected, no other complication occurring, to be proportional to the gas pressure. The rate with which the gas is decomposed, dn/dt mol s^{-1}, would be given by

$$-\frac{dn}{dt} = kP \qquad [15]$$

where k is a proportionality constant. For a constant-volume system, dn can be replaced by $(V/RT)\,dP$, so that the rate law would be

$$-\frac{dP}{dt} = \frac{RT}{V}kP$$

or

$$\ln\frac{P_0}{P} = \frac{RTk}{V}t \qquad [16]$$

where the initial gas pressure is P_0 at time $t = 0$.

The decomposition of phosphine on glass,

$$PH_3 \rightarrow P + \tfrac{3}{2}H_2 \qquad [17]$$

as shown by the data of Fig. 24-10, conforms to this rate law and therefore presumably proceeds by the decomposition of adsorbed molecules. That the surface is involved is readily shown by increasing the surface area, by the addition of glass wool for example, and observing the higher rate constant.

The second decomposition-rate-expression type that will be mentioned here is anticipated for moderate adsorption for which the amount adsorbed can be expected, according to the Langmuir isotherm, to be proportional to the expression $P/(a + P)$. The decomposition might then follow the rate expression

$$-\frac{dP}{dt} = \frac{RT}{V}\frac{kP}{a + P} \qquad [18]$$

where k is a constant. Separation of variables gives

$$-a\frac{dP}{P} - dP = \frac{RTk}{V}dT$$

and integration between the limits P_0 at time $t = 0$ and P at time t gives

$$a \ln\frac{P_0}{P} + (P_0 - P) = \frac{RTk}{V}t \qquad [19]$$

Finally, conversion to base 10 logarithms, with rearrangement so that pressure ratios rather than pressure appear as variables, gives

$$\log\frac{P_0}{P} + \frac{P_0}{2.303a}\left(1 - \frac{P}{P_0}\right) = \frac{RTk}{2.303V}t \qquad [20]$$

The experimental results for the decomposition of stibine on an antimony surface,

$$SbH_3 \rightarrow Sb + \tfrac{3}{2}H_2 \qquad [21]$$

fit this rate law. It is necessary to show that for some value of $P_0/2.303a$, a plot of the left side of Eq. [20] against t yields a straight line, and as Fig. 24-11 shows, such a plot can be obtained.

Finally, for a strongly adsorbed gas, the surface coverage is essentially complete, and the amount of adsorbed material is essentially independent of the pressure. The rate of decomposition would then be expected to be independent of P, and the rate law would be written

$$-\frac{dP}{dt} = k \qquad [22]$$

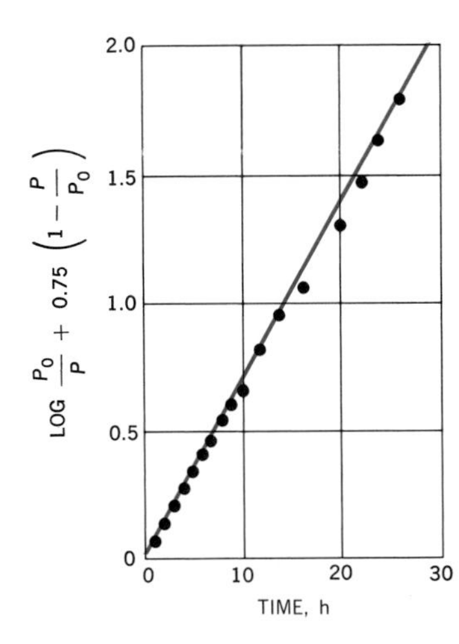

FIGURE 24-11
The decomposition of stibine on an antimony surface as a function of time. The pressure of stibine is P at time t, and P_0 at time zero. [*From A. Stock and M. Bodenstein, Ber.*, **40**:*570 (1907)*.]

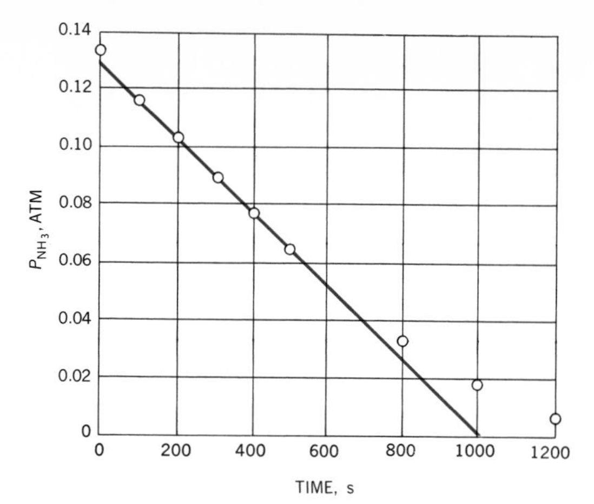

FIGURE 24-12 The decomposition of ammonia on a tungsten surface at 856°C. [*From C. N. Hinshelwood and R. E. Burk, J. Chem. Soc.,* **127:***1105 (1925).*]

and

$$P = -kt + \text{const} \tag{23}$$

The data for the decomposition of ammonia on a tungsten surface are shown plotted in Fig. 24-12 so as to illustrate their conformity to this relation.

Even decompositions of the type used in these examples do not, however, necessarily follow such simply explained rate laws. One or more of the decomposition products may be adsorbed on the catalyst. In such cases the products inhibit the reaction by competing with the reagent for the catalyst surface. An example of this situation is provided by the decomposition of ammonia on a platinum filament. The Langmuir-isotherm derivation leads, for the simultaneous adsorption of NH_3 and H_2, to the coverage expressions

$$\theta_{NH_3} = \frac{b_{NH_3}P_{NH_3}}{1 + b_{NH_3}P_{NH_3} + b_{H_2}P_{H_2}}$$

and [24]

$$\theta_{H_2} = \frac{b_{H_2}P_{H_2}}{1 + b_{NH_3}P_{NH_3} + b_{H_2}P_{H_2}}$$

Adsorption studies of the separate reagents show that hydrogen is adsorbed to a much greater extent than is ammonia, and therefore that $b_{NH_3}P_{NH_3} \ll b_{H_2}P_{H_2}$. For appreciable hydrogen pressures, furthermore, $b_{H_2}P_{H_2}$ will be greater than unity, and the fraction of the surface covered by ammonia becomes

$$\theta_{NH_3} = \frac{b_{NH_3}P_{NH_3}}{b_{H_2}P_{H_2}} \tag{25}$$

If the rate of the decomposition is dependent on the amount of ammonia on the surface, the constant-volume rate expression

$$-\frac{dP_{NH_3}}{dt} = k\frac{P_{NH_3}}{P_{H_2}} \qquad [26]$$

is expected. Such a rate law does, in fact, fit the observed decomposition data of ammonia on a platinum surface.

This type of inhibition is an illustration of the important *catalyst-poison* behavior. The adsorption of a product, or a foreign substance, can compete for the catalyst surface and thereby inhibit the reaction. The effect of a small amount of such adsorption can be very great and leads to the recognition either that some catalysts have only a few active sites at which reactions can occur or that a small amount of adsorbent can alter the electron content of the catalyst to spoil its activity.

This brief discussion of the kinetics of heterogeneously catalyzed reactions should indicate that studies such as those mentioned above can bear out the fact that such reactions are surface reactions. It becomes difficult, however, to study the nature of the surface reaction by kinetic measurements. For a reaction of any complexity the details of the adsorption processes and the details of the surface reaction cannot be deduced from the kinetic data alone.

Problems

1 Sketch the force versus total-surface-area curve that would be expected in a Langmuir surface-film experiment when a 0.1-ml sample of a solution containing 0.1 mol liter^{-1} of palmitic acid, $CH_3(C_{14}H_{28})COOH$, in the volatile solvent ethyl alcohol is placed on the surface of water and compressed on a Langmuir film balance.

2 Estimate, to the extent that you can from Fig. 24-4, the thickness of the almost completed layer of $CH_3(CH_2)_{34}COOH$ indicated in parts *b* and *c*. Similar estimates on the strips in part *d* show them to be two molecules thick.

3 Suggest a mechanism of surface-film collapse that leads to the formation of layers two molecules thick lying on top of the monomolecular layer.

4 The data of Langmuir [*J. Am. Chem. Soc.*, **40**:1361 (1918)] for the adsorption of nitrogen on mica at 90°K are as tabulated at the right.

Show that these data fit a Langmuir-isotherm expression, and evaluate the constants in the expression. *Ans.* $y_m = 1.62 \times 10^{-6}$ mol.

Estimate the area covered by a single nitrogen molecule from the fact that the density of liquid nitrogen is 0.81 g ml^{-1}.

Estimate the surface area of the mica sample in the Langmuir experiment.

P (atm)	*Amt. adsorbed (mm³ at 20°C and 1 atm)*
2.8	12.0
3.4	13.4
4.0	15.1
4.9	17.0
6.0	19.0
7.3	21.6
9.4	23.9
12.8	25.5
17.1	28.2
23.5	30.8
33.5	33.0

5 Acetic acid is adsorbed from solution by activated charcoal. The following data have been reported for the amounts y of acetic acid adsorbed as a function of the concentration c of the equilibrium solution:

c (mol liter⁻¹)	0.018	0.031	0.062	0.126	0.268	0.471	0.882
y (mol)	0.47	0.62	0.80	1.11	1.55	2.04	2.48

Show that these data fit a Freundlich isotherm, and determine the constants in the Freundlich isotherm expression.

6 Show that at low surface coverages the Langmuir isotherm corresponds to the Freundlich expression with $n = 1$. Show also that at high surface coverages the Langmuir equation corresponds to the Freundlich expression with n equal to infinity.

7 Obtain Eqs. [24] for the fraction of surface covered by each of two adsorbents if the adsorptions follow Langmuir's adsorption isotherm.

References

ADAMSON, A. W.: "Physical Chemistry of Surfaces," Interscience Publishers, Inc., New York, 1960. A comprehensive account of the many aspects of studies of surface chemistry.

DREXHAGE, K. H.: Monomolecular Layers and Light, *Sci. Am.,* **222**(3): 108 (1970).

ROSS, S., and J. P. OLIVIER: "On Physical Adsorption," Interscience Publishers, Inc., New York, 1964. A clear exposition of experimental methods, results, and interpretation in the area of physical adsorption.

HAYWARD, D. O., and M. B. W. TRAPNELL: "Chemisorption," 2d ed., Butterworth & Co. (Publishers) Ltd., London, 1964. An excellent summary of the present state of our knowledge of the rates, equilibrium pressures, heats, and mechanisms involved in the chemisorption of gases on solids.

DAVIES, J. T., and E. K. RIDEAL: "Interfacial Phenomena," Academic Press, Inc., New York, 1963. Studies and properties of surfaces and interfaces in liquid systems.

BOND, G. C.: "Catalysis by Metals," Academic Press, Inc., New York, 1962. A complete account of an important area of heterogeneous catalysis that brings together some of the approaches and much of the recent work on chemisorption and catalytic effects of metals.

HAENSEL, V., and R. L. BURWELL, JR.: Catalysis, *Sci. Am.,* **225** (6): 46 (1971).

25

MACROMOLECULES

Many approaches, both theoretical and experimental, to the study of the behavior of chemical systems have already been developed and applied. In these studies a distinction has generally been made between a molecular treatment and a macroscopic one. An important and very interesting class of systems occurs which is, in a way, intermediate between these extremes. These systems often consist of, or contain, molecules that are so large that they can be treated either as large molecules or as small macroscopic particles. Most particles that are of current interest and are in this size range, about 100 to 10,000 Å, are found to be single molecules, and the term *macromolecule* is convenient.

A number of different physical-chemical approaches are required to reveal the nature of these systems. Some of these special techniques have already been mentioned, but find their greatest current application in the study of macromolecule systems. Still other techniques must be introduced for these special systems. The deduction of the details of these large and often very complex molecules is one of the current exciting challenges presented to the physical chemist. Many of the systems, as will be seen, have great biological importance. Some of this area of study is often included in biochemistry, but the term *molecular biology* also seems appropriate. In the short study of macromolecules that can be presented here, it is desirable to discuss both synthetic and naturally occurring macromolecules side by side. Although the areas of plastics and biological materials may seem little related, it will be found that the physical-chemical study of the basic chemical units of these areas has very much in common. It is convenient to distinguish, for these studies, between macromolecules in the solid state and in solution. After a more general introduction to the types of systems that occur or can be produced with particles in this size range, these two principal sections, i.e., solutions and the solid state, will be treated so that information on the structure, shape, and behavior of macromolecules can be obtained.

INTRODUCTION TO MACROMOLECULE MATERIALS

25-1 Types and Sizes of Particles

The existence of particles in the size range that will be dealt with here was suggested by the early observation made by the botanist Robert Brown of the random motion of pollen grains as seen under a microscope. It was later recognized that these particles, though large enough to be seen, were small enough to reveal the effects of random molecular bombardment, the so-called *Brownian motion.* By the end of the nineteenth century, study of small-particle systems, called *colloids,* became an important branch of physical chemistry.

The unique behavior of colloids is now recognized to be exhibited by particles in the size range of about 100 to 10,000 Å. One of the most commonly recognized features of such systems is that they scatter light, as, for example, is observed when a beam of sunlight passes through dusty air or through thin skimmed milk. Furthermore, the particles of a colloidal system do not settle out and, as the chemist has invariably experienced with some silver chloride precipitates, tend to pass through ordinary filter paper. Colloidal systems frequently occur and, as Table 25-1 shows, can be of many different phase types.

A closer look at the chemical world, however, shows that there are particles in this size range of great interest and importance that are not listed in Table 25-1. These stem from the existence of single molecules sufficiently large so that individual molecules have colloidal di-

TABLE 25-1 Some common types of colloids classified as to phase type

Name	Type	Examples
Aerosol	Solid particles in gas	Smoke
Aerosol	Liquid particles in gas	Fog
Sol	Solid particles in liquid	S, Au, AgCl in H_2O
Emulsion	Liquid particles in liquid	Mayonnaise, milk
Foam	Gas bubbles in liquid	Whipped cream
Gels	Liquid in solid matrix	Jellies

TABLE 25-2 Classes of macromolecules

Classes	*Examples*
Synthetic macromolecules:	
Addition polymers	Polyethylene
Condensation polymers	Nylon
Natural macromolecules:	
Proteins:	
Fibrous	Keratin, silk fibroin
Globular	Hemoglobin
Nucleic acids	Deoxyribonucleic acid (DNA)
Polysaccharides	Cellulose
Polyisoprene	Natural rubber

mensions. In view of present interest, these macromolecule systems can be classed as synthetic polymers and as the naturally occurring macromolecules. Most interest in the natural materials is now centered on *proteins* and *nucleic acids,* but natural macromolecules also include the *polysaccharides* and the *polyisoprenes,* the latter being the molecules of natural rubber. These categories of macromolecules are outlined in Table 25-2.

In addition to the many types of highly dispersed systems listed in Table 25-1 and the macromolecules, which will be our principal subject of study in this chapter, listed in Table 25-2, mention should also be made of the colloidal-sized groups known as *micelles.* The turbidity exhibited by soap or detergent solutions is the best-known indication of micelle formation. Since the molecules of soap or detergent are very small compared with colloidal dimensions, the particles causing the turbidity are groups, or micelles, of these molecules. Their formation is closely analogous to that of monomolecular films studied in the preceding chapter. Most soaps and detergents have a long hydrocarbon "tail" and a polar "head." In the soaps the head is the sodium or potassium salt of the carboxylic acid, that is, $RCOO^-K^+$; in the detergents the head is the salt of a sulfonic acid, i.e., of the type $R{-}SO_3^-Na^+$. A micelle can be expected to form in a manner depicted in Fig. 25-1. It will be seen later that the charged layer around the surface of the particles is important for the stability of the individual micelles.

All our studies will now be directed toward an understanding of macromolecule systems. It will be clear that many of the methods dealt with are applicable to all systems with colloidal-sized particles. No specific treatment of the important system of sols, i.e., solids dispersed in liquids, will be given. Such colloids are now studied largely in connection with their role in the precipitation process, and therefore are more suitably treated in a study of analytical chemistry.

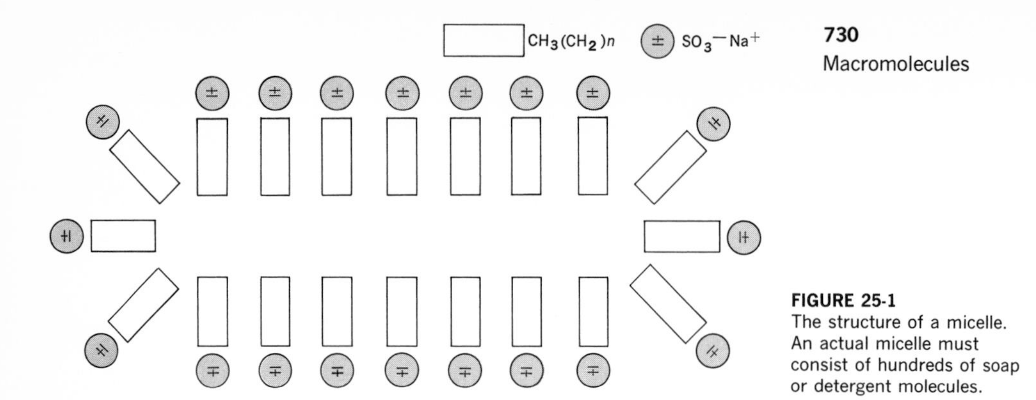

FIGURE 25-1
The structure of a micelle. An actual micelle must consist of hundreds of soap or detergent molecules.

25-2 Synthetic Polymers

A few of the more important synthetic linear polymers are listed in Table 25-3. These *linear polymers* are characterized by covalently bound skeletons that extend throughout the length of the molecule. Probably the nicest example is provided by the synthetic polymer polyethylene,

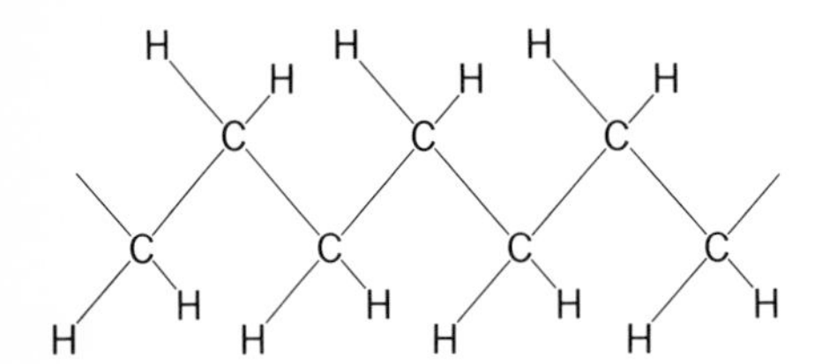

Synthetic polymers are made from the corresponding monomers by one of two general types of reactions. One reaction type is that of *addition polymerization.* The mechanism of this type of polymerization reaction consists in the adding on of monomer units to the growing polymer chain by a free-radical, carbonium-ion (cationic-polymerization), or carbanion (anionic-polymerization) mechanism. The formation of polyethylene from ethylene proceeds, in the high-pressure process, by a free-radical addition which is probably initiated by oxygen mole-

TABLE 25-3 Some synthetic linear polymers

	Chemical unit	Molecular mass (g)
Polyethylene	$[—CH_2—]_n$	5,000–40,000
Polystyrene	$\left[\begin{array}{c} C_6H_5 \\ \vert \\ —CH—CH_2— \end{array}\right]_n$	60,000–1,500,000
Nylon	$\left[\begin{array}{c} \text{H} \quad\quad\quad \text{H} \quad \text{O} \quad\quad\quad\quad \text{O} \\ —N(CH_2)_6N—\overset{\Vert}{C}—(CH_2)_4—\overset{\Vert}{C}— \end{array}\right]_n$	10,000–30,000

cules. Once started, the polymerization proceeds by the process

$$CH_3—(CH_2)_n—\dot{C}H_2 + CH_2{=}CH_2 \rightarrow CH_3—(CH_2)_{n+2}—\dot{C}H_2$$

$$CH_3—(CH_2)_{n+2}—\dot{C}H_2 + CH_2{=}CH_2 \rightarrow \text{etc.}$$

Termination occurs when two radical centers in the system come together and react to pair the electrons of the free radicals.

The second important reaction type for the formation of a polymer is that of *condensation,* in which, usually, water is split out as the monomer units join together. The synthesis of nylon from hexamethylenediamine and adipic acid proceeds by the continuation of the reaction

$$H_2N(CH_2)_6NH_2 + HOOC(CH_2)_4COOH \rightarrow$$

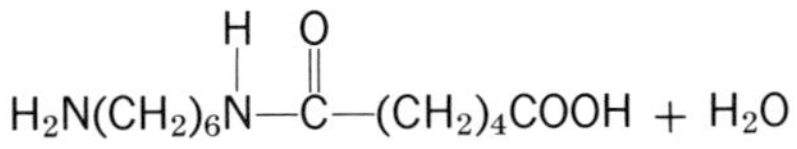

$$H_2N(CH_2)_6\underset{}{\overset{H}{\overset{|}{N}}}—\overset{O}{\overset{\|}{C}}—(CH_2)_4COOH + H_2O$$

The polymers listed in Table 25-3 consist predominantly of linear molecules with minor amounts of branching and cross linking. It should be mentioned, however, that *cross-linked* polymers do exist, or can be formed, and that these constitute a very important type of material. Cross-linked polymers have, at least to some extent, a three-dimensional array of covalent bonds. Common examples are vulcanized rubber and rigid plastics like Bakelite. The vulcanization of rubber can be illustrated by the addition of sulfur monochloride, S_2Cl_2, to rubber. The synthetic-rubber molecule chains

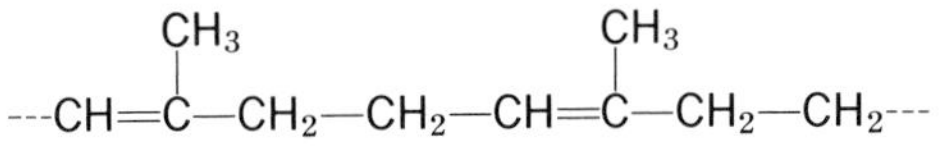

$$\text{---}CH{=}\overset{CH_3}{\overset{|}{C}}—CH_2—CH_2—CH{=}\overset{CH_3}{\overset{|}{C}}—CH_2—CH_2\text{---}$$

are joined together by sulfur cross links in the manner

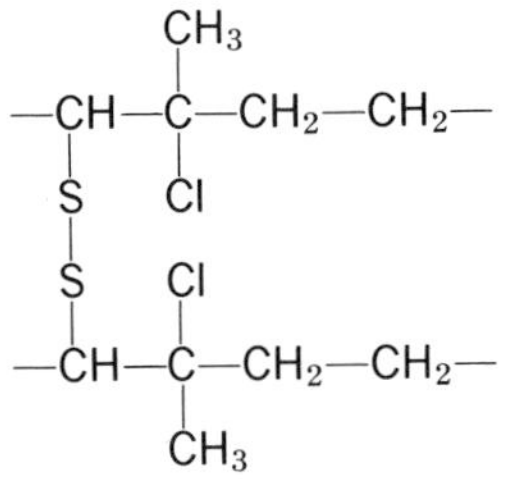

```
     CH3
     |
—CH—C—CH2—CH2—
  |  |
  S  Cl
  |
  S  Cl
  |  |
—CH—C—CH2—CH2—
     |
     CH3
```

Such cross linking produces a characteristic infusible, insoluble material. It is the latter feature that prevents illuminating solution studies from being made on such polymers.

Synthesis of a polymer, such as polyethylene, proceeds principally by the free-radical addition that has been indicated. In fact, one finds that a certain number of side chains are usually produced to give a less perfect linear polymer than that ordinarily shown diagrammatically.

A polyethylene chain may adopt, and does so to some extent in

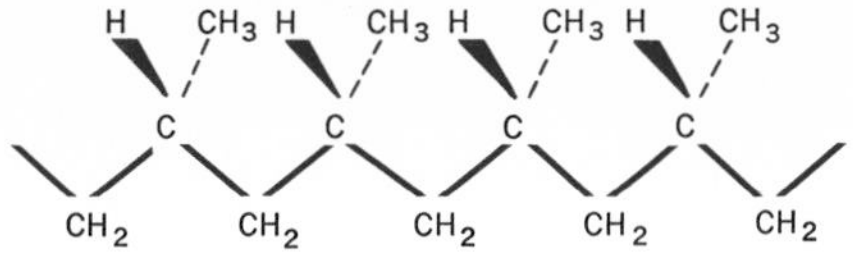

FIGURE 25-2 Segment of an isotactic polymer chain. (*From D. J. Cram and G. S. Hammond, "Organic Chemistry," McGraw-Hill Book Company, New York, 1959.*)

solid polyethylene, the energetically favored, but statistically unlikely, zigzag shape

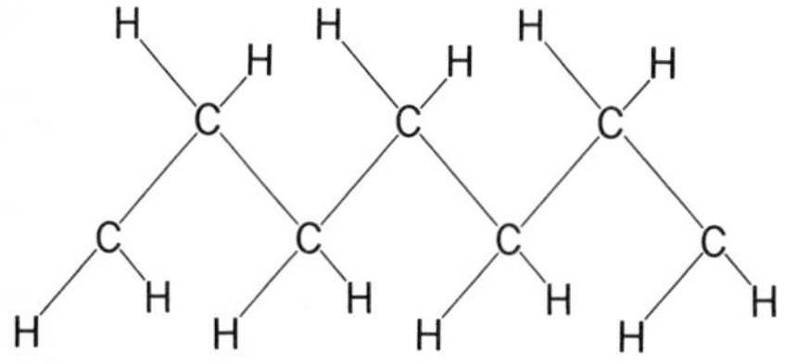

If a similar diagram is attempted for a polymer like polypropylene, one recognizes a number of different possibilities. Polypropylene, as ordinarily prepared, has the methyl groups on every second carbon atom attached at random on either side of the chain, and is known as an *atactic* polymer.

By very carefully selecting the polymerization catalyst it is possible to produce a *stereospecific* form, known as an *isotatic* polymer, with the structure shown in Fig. 25-2. The regular arrangement of the side-chain groups produces considerable changes in the way in which the molecules can pack together, and therefore markedly affects the physical properties of the resulting polymeric material.

Finally, mention should be made of inorganic polymers. The single crystal particles of a finely dispersed ionic or metallic material could be taken as an example. More suitable, however, are the polymers such as occur in the silicones, the metaphosphate glasses, and sulfur. The silicones are synthetic linear polymers of the type

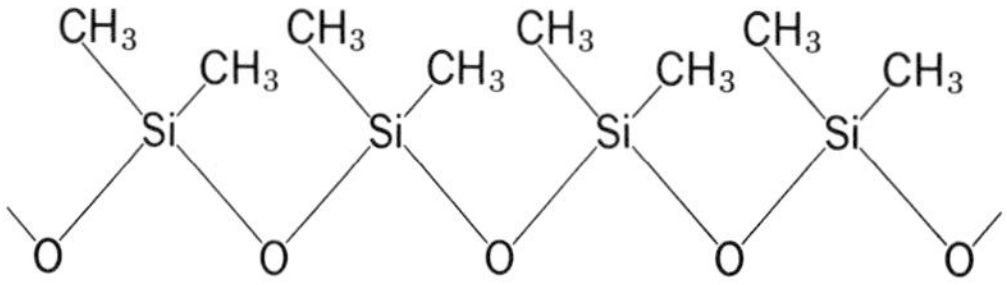

where rotation about the Si—O bonds must be expected to be relatively free. The silicones form liquids and soft plastics or rubbers.

25-3 Proteins

The world of living things is made up of a great variety of chemical substances, many of which fall into the category of macromolecules. In the animal world many of these macromolecules are proteins. These basic units of life have an amazing complexity which, until recently, seemed to be such as to remain forever beyond man's understanding. Much of our present understanding of the nature and biological role of proteins has come about through chemical and biological studies that cannot be dealt with here. In recent years, however, many physical-chemical attacks have been made on the problem of protein structure and behavior. These studies have contributed in great measure to our present remarkable understanding of proteins. It will be clear, nevertheless, that much remains to be done and that the physical-chemical techniques that are now practiced will be refined and new ones will be developed to provide answers to the many questions which still exist.

Very many different proteins can be isolated from the whole variety of living matter. All these proteins can be broken down by hydrolysis to yield organic molecules, all of which are amino acids. These amino acids, furthermore, all have the structure

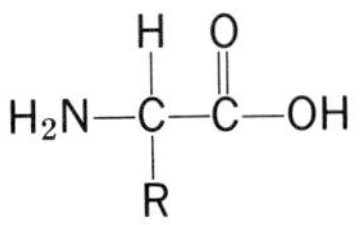

$$H_2N-\underset{\underset{R}{|}}{\overset{\overset{H}{|}}{C}}-\overset{\overset{O}{\|}}{C}-OH$$

i.e., they are α-amino acids, the name signifying that the NH_2 group is attached to the carbon atom adjacent to the acid function. These α-amino acids are distinguished by different R groups. The amino acids that are constituents of proteins are listed in Table 25-4.

Amino acids such as these can be visualized as being linked together through something like a condensation reaction to form *polypeptides*. The attachment of one amino acid to another is through a *peptide bond* similar to that previously illustrated in the formation of nylon. The condensation of two amino acids can be depicted as

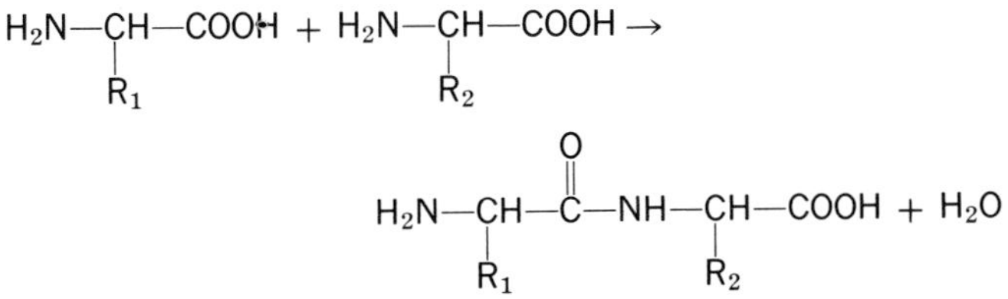

$$H_2N-\underset{\underset{R_1}{|}}{CH}-COOH + H_2N-\underset{\underset{R_2}{|}}{CH}-COOH \rightarrow$$

$$H_2N-\underset{\underset{R_1}{|}}{CH}-\overset{\overset{O}{\|}}{C}-NH-\underset{\underset{R_2}{|}}{CH}-COOH + H_2O$$

By repetition of such a reaction, amino acids can be built up into macromolecules.

True proteins, however, cannot be so easily synthesized. They consist of similar polymers of amino acids, but in a protein there is a definite sequence of a number of different amino acids. If chemically and structurally the exact order of the amino acids seems rather trivial, this is not the case biologically. It is the sequence of amino acids that is an important characteristic of the protein. The sequence, moreover, continues throughout the protein molecule, which, depending on the protein, will have a molecular weight of the order of ten to hundreds

TABLE 25-4 The amino acids obtained from protein hydrolysis*

NEUTRAL AMINO ACIDS			
Glycine	$CH_2(NH_2)CO_2H$	Serine	$HOCH_2CH(NH_2)CO_2H$
Alanine	$CH_3CH(NH_2)CO_2H$	Threonine	$CH_3CH(OH)CH(NH_2)CO_2H$
Valine	$(CH_3)_2CHCH(NH_2)CO_2H$	Methionine	$CH_3SCH_2CH_2CH(NH_2)CO_2H$
Leucine	$(CH_3)_2CHCH_2CH(NH_2)CO_2H$	Cysteine	$HSCH_2CH(NH_2)CO_2H$
Isoleucine	$CH_3CH_2CH(CH_3)CH(NH_2)CO_2H$	Cystine	$SCH_2CH(NH_2)CO_2H$ $SCH_2CH(NH_2)CO_2H$
Phenylalanine	$C_6H_5CH_2CH(NH_2)CO_2H$		
Tyrosine	HO– (benzene ring) –$CH_2CH(NH_2)CO_2H$	Proline	(pyrrolidine ring) N–H, CO_2H
Diiodotyrosine	HO– (ring with I, I) –$CH_2CH(NH_2)CO_2H$	Hydroxyproline	HO (pyrrolidine ring) N–H, CO_2H
Thyroxine	HO– (ring with I, I) –O– (ring with I, I) –$CH_2CH(NH_2)CO_2H$		
Tryptophan	(indole ring, N–H) $CH_2CH(NH_2)CO_2H$		
ACIDIC AMINO ACIDS			
Aspartic acid	$HO_2CCH_2CH_2CH(NH_2)CO_2H$	Glutamic acid	$HO_2CCH_2CH(NH_2)CO_2H$
BASIC AMINO ACIDS			
Lysine	$H_2N(CH_2)_4CH(NH_2)CO_2H$		
Arginine	$H_2NC(=NH)NH(CH_2)_3CH(NH_2)CO_2H$	Histidine	(imidazole ring, N, N–H) $CH_2CH(NH_2)CO_2H$
Ornithine	$H_2N(CH_2)_3CH(NH_2)CO_2H$		

* From D. J. Cram and G. S. Hammond, "Organic Chemistry," McGraw-Hill Book Company, New York, 1959.

of thousands. There are therefore hundreds of amino acids linked in a particular manner in a typical protein.

The chemical constitution, i.e., the percentage of the various amino acids present, has been worked out for many proteins. Furthermore, the remarkable feat of complete analyses of the order of the amino acids of a protein can, in some cases, be accomplished, as was first demonstrated by Sanger. The structure of the relatively simple protein insulin is indicated in Fig. 25-3.

A feature of this protein, which must be anticipated for others, is that it is not one polymeric sequence, but rather two essentially linear chains cross-linked by a disulfide bridge. Any detailed study of the size and shape of such molecules must take into account such cross links.

In this regard it should be mentioned that secondary forces, which are not covalent chemical bonds, can also operate to bind protein chains together. Such forces are usually hydrogen bonds or electrostatic-charge interactions. In aqueous solution, however, *hydrophobic* forces are also recognized. These forces are related to the lowering of the free energy of a protein-water system when the hydrocarbonlike parts of the protein chain are together, and the polar parts of the chain are in contact with the solvent—rather like the soap micelle illustrated in Fig. 25-1. These hydrophobic forces contribute to the overall shape of a protein molecule in an aqueous system.

Hydrogen bonds are important because the protein chain contains a sequence of groups that can engage in hydrogen bonding. Principal of these groups are the NH and C=O groups, which, if suitably positioned relative to each other, can form the hydrogen bond

$$\rangle N{-}H \cdots O{=}C\langle$$

Electrostatic interactions occur because of the presence of acidic and basic centers in some of the amino acids of proteins. In this connection it is important to note that an individual amino acid can be titrated by either acid or base and that the charged species

$$H_3N^+{-}\underset{\displaystyle R}{\underset{|}{C}H}{-}COOH \qquad \text{in acid}$$

and

$$H_2N{-}\underset{\displaystyle R}{\underset{|}{C}H}{-}COO^- \qquad \text{in base}$$

are formed. Furthermore, it appears that the molecule in approximately neutral aqueous solution adopts the zwitterion form

$$H_3N^+{-}\underset{\displaystyle R}{\underset{|}{C}H}{-}COO^-$$

Gly • Ileu • Val • Glu • Glu(NH2) • Cys • Cys • Ala • Ser • Val • Cys • Ser • Leu • Tyr • Glu(NH2) • Leu • Glu • Asp(NH2) • Tyr • Cys • Asp(NH2)

Phe • Val • Asp(NH2) • Glu(NH2) • His • Leu • Cys • Gly • Ser • His • Leu • Val • Glu • Ala • Leu • Tyr • Leu • Val • Cys • Gly • Glu • Arg • Gly • Phe • Phe • Tyr • Thr • Pro • Lys • Ala

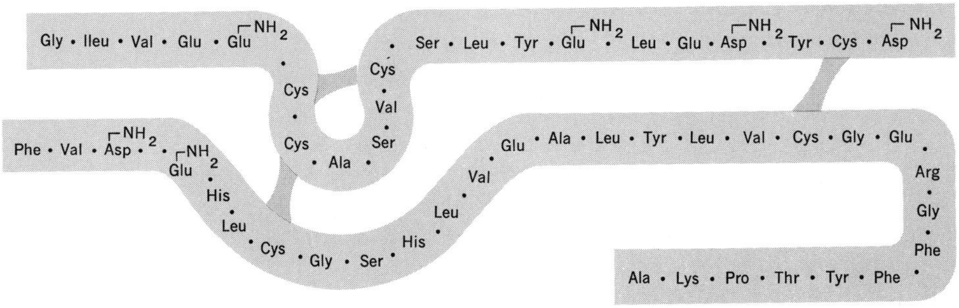

FIGURE 25-3
The arrangement of amino acids in the protein insulin. The shaded connecting links are disulfide bonds. Abbreviations correspond to the amino acids in Table 25-4. (*Courtesy of Prof. Irving M. Klotz, Northwestern University, Evanston, Ill.*)

rather than the noncharged configuration

$$H_2N{-}\underset{\displaystyle R}{\underset{|}{C}H}{-}COOH$$

In the proteins, although most of these amine and carboxylic acid groups are used up in peptide-bond formation, there are free basic and acidic centers which can act in a similar way to give the molecule a net positive or negative charge. Such charges produce secondary forces that have a considerable effect, not only on the electrical properties of proteins but also on the geometric configuration. Like charges can, for example, repel each other to cause the molecule to open up, whereas opposite charges can act to pull the molecule together.

Secondary forces are effective in determining, in solution, the shape that the polymer molecule adopts, i.e., whether it tends to ball up or be extended, and in the solid state both the shape of the protein molecule and the way in which neighboring protein chains are packed together.

In structural studies it is customary to distinguish between *fibrous* and *globular* proteins. The fibrous proteins tend to occur as long chains and are found in structural tissue such as wool, hair, nail, and muscle. The globular proteins, on the other hand, tend to be more or less spherical and, unlike the fibrous proteins, are often water-soluble. Such substances as enzymes, hemoglobin, and egg white fall into this category.

Many features of both types of proteins will be revealed by the physical-chemical studies dealt with in later sections of this chapter.

25-4 Nucleic Acids

The great variety of form that proteins can assume leads one to the question as to how the information for the synthesis of all the protein molecules can be contained in each cell of a living organism. Genetic studies lead to the conclusion that each living cell carries in it sets of information, or codes, for the building up of all the proteins associated with the life of the cell. It appears, furthermore, that the chromosomes, which originate in the nucleus of the cell, are the units in which this vast quantity of information is stored. When a cell divides to form two new cells, the chromosomes (of which man has 48 in each body cell) go through a remarkable series of maneuvers and ultimately split into two parts. One part goes into each of the new cells and continues to function as the information center for protein and other syntheses. The chromosomes contain subdivisions called *genes,* and these control various biochemical syntheses.

The chromosomes are rich in macromolecule species, which since chromosomes occur in the cell nucleus, are called *nucleic acids*. It is these macromolecules that are now recognized as performing the prime function of the genes. Elucidation of the structure and function of nucleic acids, insofar as known at present, will be seen to be at least as great an accomplishment as the corresponding progress that has been made in protein studies.

Nucleic acids occur in at least two main types: deoxyribonucleic acid (DNA) and ribonucleic acid (RNA). In this brief treatment attention will be restricted to DNA, which, as Fig. 25-4*a* indicates, consists of a backbone of alternate sugar residues and phosphate groups. The macromolecule nature of the molecule is best brought out by the more schematic representation of the chain, also shown in Fig. 25-4*a*. Attached to each sugar group of the chain is one of four nitrogen-containing groups. Although we shall not become involved with the detailed behavior of these groups, their geometry is important and they are shown in Fig. 25-4*b*.

Further, it anticipates some of our later structural studies to mention that DNA consists of two macromolecule chains that are associated by secondary forces as indicated in Fig. 25-4. When it is pointed out that there are about 10^{10} side-group positions in a DNA molecule and that the four different groups are attached in some order that apparently represents a code, there is the possibility for storing a vast amount of information. Moreover, the presence of two macromolecule chains anticipates the cell-division process, in which the information unit is divided and new units appear in the two new cells. The biological consequences of a detailed molecular structure for the nucleic acids are fascinating. We shall see that the structural aspects, on which our physical-chemical studies will focus, are equally remarkable.

25-5 The Polysaccharides

The third important type of natural polymers is the polysaccharide molecule. Simple sugar units can be depicted as condensing with the elimination of water to form disaccharides, such as sucrose, and polysaccharides, such as starch and cellulose.

Figure 25-5 shows the structure of glucose and the repeating unit in cellulose.

Cellulose is the chief constituent of wood and cotton. It is perhaps not strictly correct to include such molecules in a study which claims to deal with particles in the range of colloidal sizes. The cellulose molecules can presumably be of macroscopic length, and the average molecular mass of such natural products is not easily determined since any process that breaks the material apart to free the individual molecules may at the same time cleave the molecular chain.

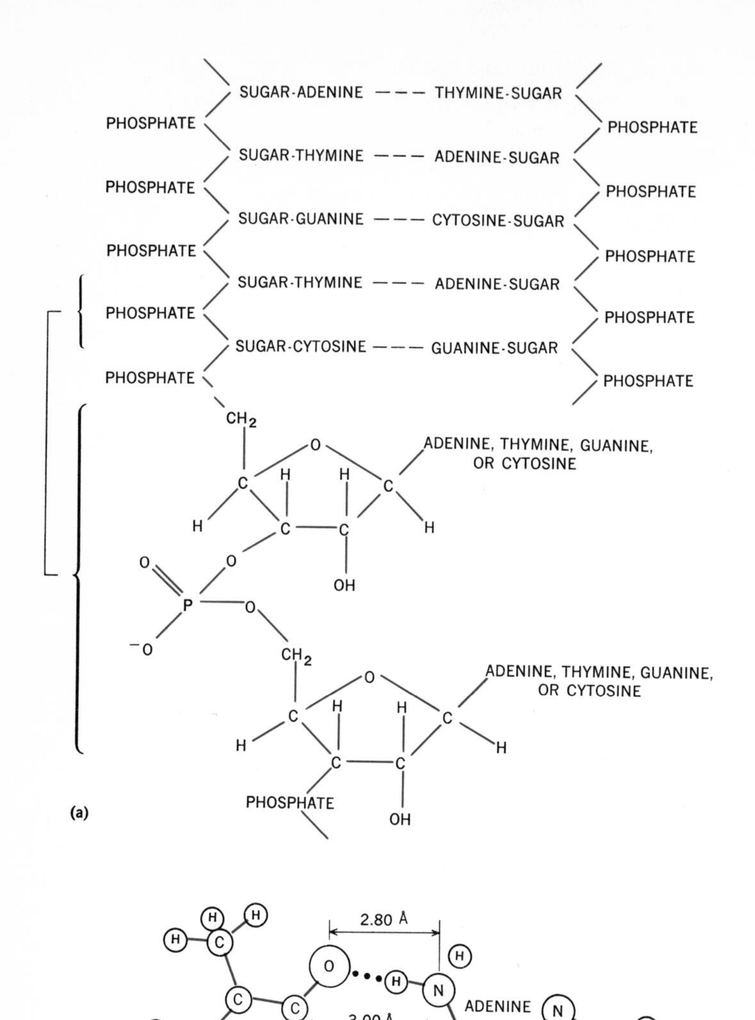

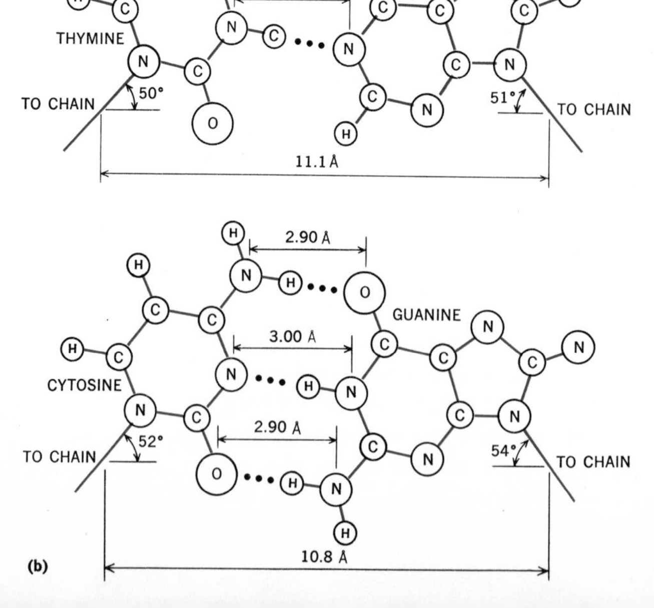

FIGURE 25-4
The structure of deoxyribonucleic acid (DNA). (*a*) Schematic diagram of the DNA chain structure. (*b*) Detail of the hydrogen bonding and geometry of the base pairs of DNA.

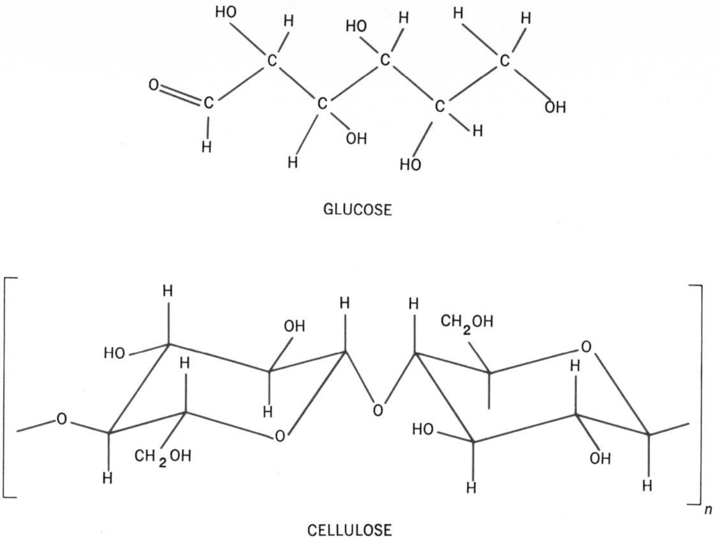

GLUCOSE

CELLULOSE

FIGURE 25-5
The structure of glucose and cellulose.

25-6 The Polyisoprenes

Natural rubber is a polymeric material composed of linear polymers of the isoprene, $CH_2{=}C(CH_3){-}CH{=}CH_2$, molecule. This monomer can in fact be polymerized catalytically to give a synthetic rubber. In naturally occurring rubber the linear polymers have nearly all the remaining double bonds connected in the polymer chain to give a cis configuration, and the polymer can be drawn as

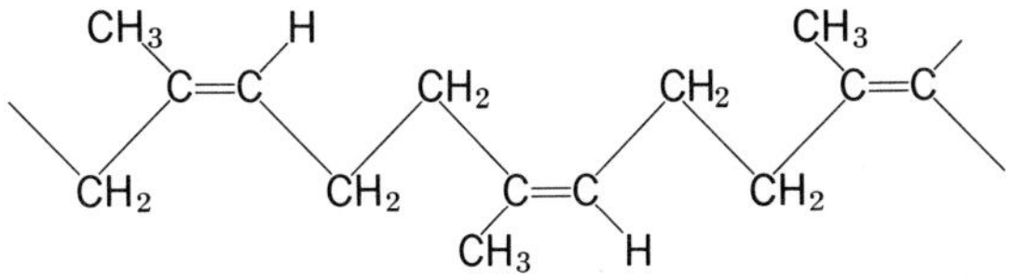

A transpolyisoprene also occurs, and is known as *gutta-percha*. The molecule can be represented as

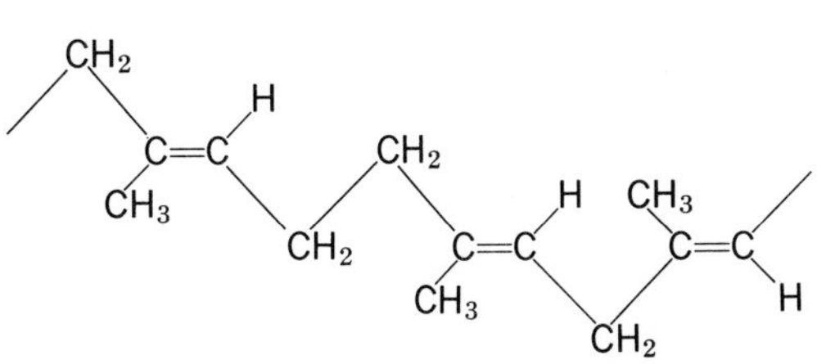

Unlike rubber it is a tough, hard substance.

Much of our information on the size and general shape of macromolecules has been deduced from various properties of solutions containing these molecules. In this section some of the methods used to understand the behavior of solutions of macromolecules will be investigated, and in the process of such studies a number of properties of the macromolecules themselves will be discovered. It is first necessary, however, to discuss the meaning of molecular mass when applied to a polymeric material.

25-7 Molecular Masses of Polymers

Polymerization reactions, both synthetic and natural, can lead to high-molecular-mass compounds. The reaction chain, however, is broken by some termination process that usually occurs in a random manner with respect to the size to which the polymer has already grown. It follows that polymers have a range of molecular masses and that any data for the size or mass of the molecules of a polymer must represent some sort of average value. It will be seen that attempts to deduce molecular masses of polymers lead to *number-average* and *mass-average* molecular masses.

The number average, denoted by M_n, is defined as the mass of sample divided by the total number of moles n in the sample; i.e.,

$$M_n = \frac{\text{mass}}{n} \qquad [1]$$

Any measurement that leads to the number of particles, or molecules, that are present in a given mass of sample will allow the calculation of a number-average molecular mass. If the sample can be considered as made up of fractions consisting of n_1 mol of molecular mass M_1, n_2 mol of molecular mass M_2, and so forth, then

$$M_n = \frac{n_1M_1 + n_2M_2 + \cdots}{n_1 + n_2 + \cdots}$$

$$= \frac{\sum_i n_i M_i}{\sum_i n_i} \qquad [2]$$

In other experiments each particle makes a contribution to the measured result according to its molecular mass. The average molecular mass deduced from such measurements is therefore more dependent on the number of heavier molecules than it is in experiments depending simply on the total number of particles. The appropriate average for such determinations is the *mass average,* defined as

$$M_w = \frac{\sum_i n_i M_i^2}{\sum_i n_i M_i} \qquad [3]$$

For an appreciable distribution of molecular sizes in a polymer sample, these two molecular masses M_n and M_w will be appreciably different.

25-8 Osmotic-pressure Determinations of Molecular Masses

A measurement of any of the colligative properties of a solution of macromolecules leads, essentially, to a value for the number of solute molecules in a given amount of solvent, as was discussed in Chap. 21. There it was pointed out that for the solutions of low molality, such as are always obtained with macromolecules, the only colligative property that is conveniently measured is the osmotic pressure. Such measurements are one of the most important means of molecular-mass determinations. It should be evident from the nature of colligative properties, i.e., their dependence only on the number and not on the nature of the solute molecule, that a number-average molecular mass is obtained.

The high concentration in terms of mass of solute per weight of solvent, even for low molalities, means that solute interactions will occur and nonideal behavior will result. It is almost always necessary, therefore, to extrapolate the measurements to infinite dilution, as was done in Chap. 21. Sensitive osmotic-pressure instruments now allow measurements to be made on very dilute solutions, and molecular masses even up to 500,000 g can be obtained.

25-9 Diffusion

A number of hydrodynamic experiments lead to information about the size and shape of macromolecules in solution. The first of these that will be considered is the diffusion of the macromolecules of a solution across a carefully made, well-defined liquid boundary into pure solvent. Experimentally, one finds that the rate of diffusion is proportional to the concentration difference across the boundary, or more conveniently, to the concentration gradient dc/dx, where c is the macromolecule concentration in mass per unit volume of solution. The diffusion rate is found, furthermore, to be proportional to the cross-section area A. If the diffusion rate is written dw/dt, the mass of macromolecules

TABLE 25-5 Diffusion and sedimentation coefficients and derived molecular masses for some proteins
Values are for aqueous solutions at 20°C.

Protein	Sp vol v ml g^{-1}	liters g^{-1} $= m^3\ kg^{-1}$	Sedimentation coeff. s (s)	Diffusion coeff. D ($m^2\ s^{-1}$)	Mol mass (g) from Eq. [24]
Insulin	0.75	0.75×10^{-3}	3.5×10^{-13}	8.2×10^{-11}	41,000
Hemoglobin	0.75	0.75	4.4	6.3	67,000
Catalase	0.73	0.73	11.3	4.1	250,000
Urease	0.73	0.73	18.6	3.5	470,000
Tobacco mosaic virus	0.73	0.73	185	0.53	31,000,000

transferred across the boundary per second, one has Fick's law of diffusion,

$$\frac{dw}{dt} = -DA\frac{dc}{dx} \qquad [4]$$

The proportionality constant D is called the *diffusion coefficient*, and the negative sign is introduced so that D will have a positive value. The diffusion coefficient can be recognized as the amount of solute that diffuses across a unit area in 1 s under the influence of a unit concentration gradient. The diffusion coefficient is characteristic, therefore, for a given solvent at a given temperature, of the diffusing tendency of the solute. Some measurements of diffusion coefficients for macromolecules are listed in Table 25-5.

It is necessary now to see whether or not these experimental diffusion coefficients can be related to any properties of the system, and particularly of the macromolecule. To do this, a molecular view of the diffusion process is taken.

Consider diffusion across a distance interval dx over which the concentration changes from c to $c - dc$. The force that drives the molecules to the more dilute region can be obtained from the difference in the molar free energy of the solute at concentration c and at concentration $c - dc$. If the corresponding mole fractions of solute are x_B and $x_B - dx_B$ and Henry's law is assumed, the discussion of Sec. 21-5 leads to the free-energy difference *per molecule* of

$$G_{c-dc} - G_c = \frac{RT}{\mathfrak{N}} \ln \frac{x_B - dx_B}{x_B} \qquad [5]$$

where $\mathfrak{N}$ is Avogadro's number. For dilute solutions the concentration in mass per unit volume, such as grams per milliliter or grams per liter, is proportional to mole fraction, and Eq. [5] can be rewritten

$$G_{c-dc} - G_c = \frac{RT}{\mathfrak{N}} \ln \frac{c - dc}{c} \qquad [6]$$

or

$$dG = \frac{RT}{\mathfrak{N}} \ln\left(1 - \frac{dc}{c}\right)$$

$$= -\frac{RT}{\mathfrak{N}}\frac{dc}{c} \qquad [7]$$

where the relation $\ln(1 - y) \approx -y$ for small y has been used.

This free-energy difference corresponds to the work done in the transfer of one macromolecule across the distance dx, and can therefore be written as a force times the distance dx. Thus

$$dG = (\text{Driving force})\, dx$$

or

$$\text{Driving force} = \frac{dG}{dx} = -\frac{RT}{\mathfrak{N}}\frac{1}{c}\frac{dc}{dx} \qquad [8]$$

A frictional force that sets in balances this diffusion force when some constant velocity is reached. The frictional force exerted by a viscous fluid, of viscosity η, has been derived for a macroscopic sphere of radius $\mathbf{r}$ by Stokes as

$$\text{Frictional force} = 6\pi\mathbf{r}\eta v$$

$$= 6\pi\mathbf{r}\eta\frac{dx}{dt} \qquad [9]$$

It appears to be suitable to apply this expression to the motion of reasonably spherical macromolecules. The diffusion velocity increases, therefore, until the force of Eq. [8] just balances that of Eq. [9]. Then

$$6\pi\mathbf{r}\eta\frac{dx}{dt} = -\frac{RT}{\mathfrak{N}}\frac{1}{c}\frac{dc}{dx}$$

or

$$\frac{c\,dx}{dt} = -\frac{RT}{6\pi\mathbf{r}\mathfrak{N}\eta}\frac{dc}{dx} \qquad [10]$$

Comparison with the empirical Fick law expression can be made when it is recognized that $c\,dx$ can be identified with dw since these terms are the mass of solute diffusing across the boundary in time dt. (All the molecules, which have an average diffusion velocity that carries them a distance dx in time dt, will cross the boundary in time dt if they start within a distance dx of the boundary. The mass of these molecules is the volume, dx times the unit cross-section area, times the concentration c in mass per unit volume. Thus $dw = c\,dx$.) Equation [10] becomes

$$\frac{dw}{dt} = -\frac{RT}{6\pi\mathbf{r}\mathfrak{N}\eta}\frac{dc}{dx} \qquad [11]$$

Comparison of this molecularly derived diffusion-rate expression with

Eq. [4] allows the interpretation of the observed diffusion constant D as

$$D = \frac{RT}{6\pi \mathbf{r} \mathfrak{N} \eta} \qquad [12]$$

Measurements of D and η could therefore lead to a value of the radius **r** for the macromolecule. Such a procedure is a little unsatisfactory in that the molecules will not necessarily obey Stokes' law, even for spherical particles, and furthermore, the macromolecules will generally be solvated and in moving through the solution will, to some extent, carry along this solvation layer. The molecular interpretation of D, as given by Eq. [12], is important, however, in determining the effective value of the group of terms $6\pi \mathbf{r} N \eta$ for a given solute and solvent.

25-10 Sedimentation and the Ultracentrifuge

In Sec. 25-9 the tendency of a solute to diffuse across a concentration gradient was treated. Macromolecules in solution can be made to alter their distribution in space by subjecting them to other forces. In the simplest experiment a solution is allowed to stand so that the force of gravity acts. A greater and more easily observed effect can be produced, however, by means of an ultracentrifuge in which a sample of the macromolecule solution rotates at a very high speed, in the neighborhood of 10,000 to 80,000 r min^{-1}. The ultracentrifuge, some features of which are shown in Fig. 25-6, is a very important tool for macromolecule research.

The behavior of solutions of macromolecules on ultracentrifugation will now be investigated. Two essentially different types of experiments can be performed. Either one can centrifuge the sample until an equilibrium distribution is obtained or alternatively, one can observe the rate of movement of the macromolecules during the centrifugation.

The first method, called *sedimentation equilibrium,* allows the process to proceed until an equilibrium distribution of the solute throughout the cell is obtained. Thermodynamics has introduced free energy as a convenient quantity for the study of equilibrium, and it can be used here to deal with the equilibrium concentration gradient

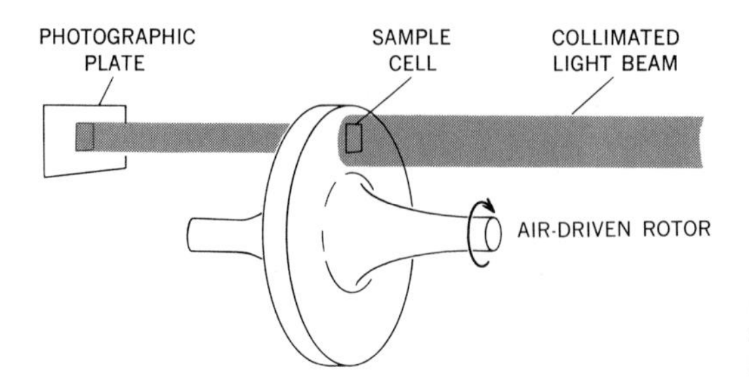

FIGURE 25-6
Rotor and cell for oil-turbine ultracentrifuge. [*T. Svedberg, Endeavour,* **6**:*89 (1947).*]

that develops. In particular, the centrifugal and diffusional contributions to the free energies G_{r_1} and G_{r_2} at the radial positions r_1 and r_2 are calculated. At equilibrium the values of G_{r_1} and G_{r_2} must be equal.

A particle of mass m at a distance r from the center of rotation experiences a force given by

$$f_{\text{centrifugal}} = m'r\omega^2 \qquad [13]$$

where ω is the angular velocity in radians per second, and m' is the effective mass of the particle. The free-energy difference between the particle at r_1 and at r_2 is obtained by finding the work required to move the particle from r_1 to r_2. The centrifugal free-energy difference, which is more negative at larger values of r, is thus

$$\Delta G_{\text{centrifugal}} = -\int_{r_1}^{r_2} (m'r\omega^2)\, dr$$

$$= -\frac{m'\omega^2}{2}(r_2^2 - r_1^2) \qquad [14]$$

This factor tends to concentrate all the particles at large values of r, where the free energy is low.

In solution, however, this centrifugal effect operates only because the macromolecules may be more dense than the solvent, and the quantity m' in Eq. [14] must be interpreted as an effective mass. If v is the specific volume (i.e., the volume per unit mass) of solute and ρ is the density (i.e., the mass per unit volume) of the *solvent,* the mass of a volume v of solvent is ρv. Thus the effective mass m' of the solute is $m - mv\rho = m(1 - v\rho)$, and the corrected centrifugal free-energy contribution is

$$\Delta G_{\text{centrifugal}} = -\frac{m(1 - v\rho)\omega^2(r_2^2 - r_1^2)}{2} \qquad [15]$$

Balance is brought about by the diffusion tendency, which is, according to Eq. [3], per molecule,

$$\Delta G_{\text{diffusion}} = \frac{RT}{\mathfrak{N}} \ln \frac{c_2}{c_1} \qquad [16]$$

where c_2 and c_1 are concentrations at r_2 and r_1.

At equilibrium the decrease in ΔG given by Eq. [15] just balances the increase given by Eq. [16], and for the process of moving solute from r_1 to r_2 one has

$$\Delta G_{\text{centrifugal}} + \Delta G_{\text{diffusion}} = 0$$

which gives, on rearrangement,

$$M = \mathfrak{N}m = \frac{RT \ln (c_2/c_1)}{(1 - v\rho)(\omega^2/2)(r_1^2 - r_2^2)} \qquad [17]$$

Thus, if measurements of the relative concentrations are made at two positions after equilibrium has been obtained, one can use Eq. [17] to calculate a value for the mass of the individual particles or for the mass of an Avogadro's number of particles, i.e., the molecular mass.

The second ultracentrifuge method starts with a well-defined boundary, or layer of solution near the center of rotation, and follows the movement of this layer toward the outside of the cell as a function of time. Such a method is termed a *sedimentation-velocity* experiment.

The force tending to move the macromolecules to the outside of the cell is given by

$$f_{\text{centrifugal}} = m(1 - v\rho)r\omega^2 \qquad [18]$$

This force is balanced for some constant-drift velocity dr/dt by a frictional force that is given by Stokes' law as

$$f_{\text{friction}} = 6\pi\eta\mathbf{r}\frac{dr}{dt} \qquad [19]$$

Equating these forces to find the constant-drift velocity, one obtains

$$m(1 - v\rho)r\omega^2 = 6\pi\eta\mathbf{r}\frac{dr}{dt} \qquad [20]$$

A characteristic of a given macromolecule in a given solution is its *sedimentation coefficient* s, defined as the velocity dr/dt with which the macromolecules move per unit centrifugal field $r\omega^2$. It is therefore often values of

$$s = \frac{dr/dt}{r\omega^2} \qquad [21]$$

that are tabulated to express the results of a sedimentation-velocity experiment. The value of s for many macromolecules comes out to be of the order of 10^{-13} s. A convenient unit having this value has therefore been introduced, and is called a *svedberg*, in honor of T. Svedberg, who did much of the early work with the ultracentrifuge.

According to Eq. [20],

$$s = \frac{dr/dt}{r\omega^2} = \frac{m(1 - v\rho)}{6\pi\eta\mathbf{r}} \qquad [22]$$

Rearrangement and multiplication by an Avogadro's number gives

$$M = \mathfrak{N}m = \frac{6\pi\eta\mathbf{r}\mathfrak{N}s}{1 - v\rho} \qquad [23]$$

Now the troublesome and poorly defined terms involving η and $\mathbf{r}$ can be replaced by their effective values, such as appear in the measurable quantity D of Eq. [12], to give the desired result

$$M = \frac{RTs}{D(1 - v\rho)} \qquad [24]$$

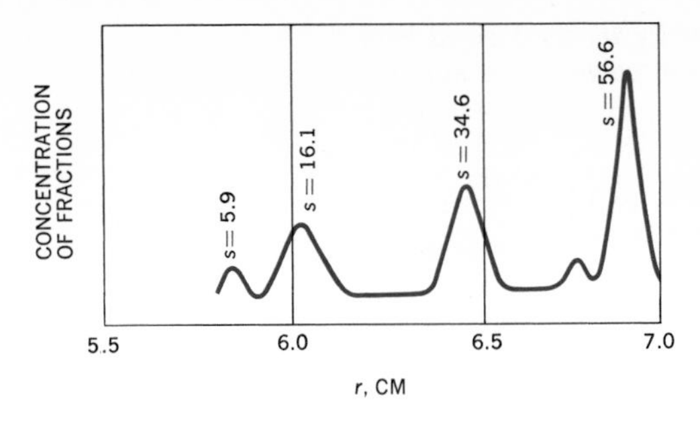

FIGURE 25-7 Separation of Limulus hemocyanin into fractions with different sedimentation constants, given in svedbergs. Centrifugal force is 120,000 times gravity, and time after reaching full speed is 35 min. [*From T. Svedberg, Proc. Roy. Soc.* (*London*), **B127:***1* (*1939*).]

Thus measurements of the sedimentation and diffusion coefficients and of the densities of the solvent and solute allow the deduction of the molecular mass of the macromolecules. The necessary data for such calculations for a few macromolecular materials are given in Table 25-5.

A particular advantage of the sedimentation-velocity technique is that a macromolecule solution containing two or more types of macromolecules is separated according to the molecular masses of the components. Figure 25-7 shows the type of sedimentation diagram that is obtained for a system containing a number of macromolecular species.

25-11 Viscosity

Another hydrodynamic property of solutions of macromolecules which is dependent on the molecular mass is the viscosity of a solution of the macromolecule material. Unlike the previous methods for obtaining molecular masses, however, measurements of viscosity do not yield absolute values. But the measurements are easily made and find wide use in the determination of the molecular mass of a given sample or batch of polymer. Use must be made of calibration measurements of viscosities of solutions containing polymer fractions whose molecular masses have been determined by other methods.

The viscosity, or more properly the coefficient of viscosity, has been treated for gases and liquids in Secs. 2-9 and 19-7 and has been seen to measure the resistance to flow of a fluid. The addition of polymer molecules to a solvent invariably increases the viscosity over that of the pure solvent. In relating this increased viscosity to the properties of the solute, a number of functions of the measured viscosity coefficients η_0 of the pure solvent and η of the solution are used. These are shown in Table 25-6.

Most directly related to the nature of the individual solute molecules is the intrinsic viscosity, which has the effect of macromolecule intermolecular interaction removed by the extrapolation to infinite dilution. It represents the fractional change in the viscosity of a solution per unit concentration of polymer, or macromolecule, at infinite dilution.

Determinations of the intrinsic viscosity for different molecular-

TABLE 25-6 Some viscosity terms derived from the measured solvent and solution viscosities, η_0 and η, respectively
c has usually been used with the units of grams per 100 ml.

Name	*Definition*
Relative viscosity	$\dfrac{\eta}{\eta_0}$
Specific viscosity	$\dfrac{\eta - \eta_0}{\eta_0}$
Reduced specific viscosity	$\dfrac{1}{c}\dfrac{\eta - \eta_0}{\eta_0}$
Intrinsic viscosity	$[\eta] = \lim_{c\to 0}\left(\dfrac{1}{c}\dfrac{\eta - \eta_0}{\eta_0}\right) = \lim_{c\to 0}\left[\dfrac{1}{c}\left(\dfrac{\eta}{\eta_0} - 1\right)\right]$

mass fractions of the same polymer lead to the expression, which is best looked on as being empirical,

$$[\eta] = KM^a \qquad [25]$$

where K and a are empirical constants which depend on the solvent, the polymer, and the temperature. The study of known molecular-mass fractions allows K and a to be evaluated, as is illustrated in the plot of Fig. 25-8. With values for the empirical constants, the molecular mass of any batch of the polymer can be deduced from the easily performed measurements of viscosity.

Some attempts have been made to relate the values of a to the shape of the molecules. The more elongated a molecule is, the more effective are the high-molecular-mass fractions in reducing the viscosity

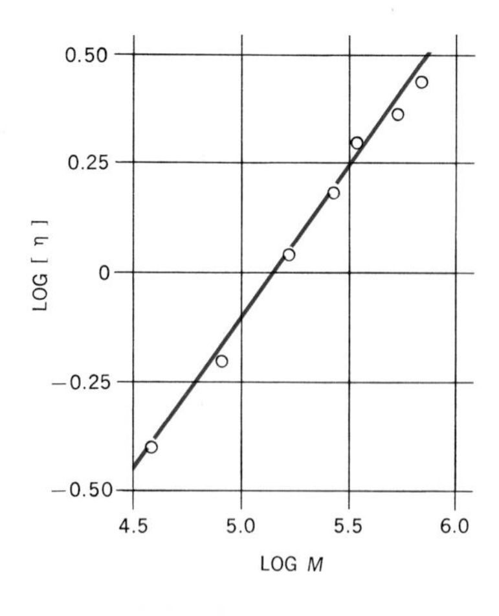

FIGURE 25-8
Relation of $[\eta]$ to M, in grams, for polyisobutylene fractions in cyclohexane at 30°C. [*From W. R. Krigbaum and P. J. Flory, J. Am. Chem. Soc.,* **75:***1775 (1953).*]

TABLE 25-7 Constants for Eq. [25] for various polymer-solvent systems*

Polymer	Solvent	t (°C)	Mol-mass range (g)	K	a
Polystyrene	Benzene	25	32,000–1,300,000	1.03×10^{-4}	0.74
Polystyrene	Methyl ethyl ketone	25	2500–1,700,000	3.9	0.58
Polyisobutylene	Cyclohexane	30	600–3,150,000	2.6	0.70
Polyisobutylene	Benzene	24	1000–3,150,000	8.3	0.50
Natural rubber	Toluene	25	40,000–1,500,000	5.0	0.67

* From P. J. Flory, "Principles of Polymer Chemistry," Cornell University Press, Ithaca, N.Y., 1953.

of the solution, and the values of a are expected to rise from typical average values of 0.6 or 0.7 to 1 or 2. As Table 25-7 shows, many polymer solutions do, however, have values of a that are near 0.6 and 0.7.

The effect of the shape of a polymer molecule is more noticeable in experiments in which the viscosity of the same polymer is studied in different solvents. In a "good" solvent it is expected that the polymer chains will be solvated and will open up; in a "poor" solvent they will tend to remain coiled up. The expectation of a high intrinsic viscosity for polymers in good solvents compared with poor solvents is borne out by the result for polystyrene of 1.20 to 1.30 in good, aromatic solvents and 0.65 to 0.75 in poor, aliphatic solvents.

Again some deductions as to the shape of polymer molecules are possible. The important use of viscosity measurements is, however, in the rapid determination of relative molecular masses.

*25-12 Light Scattering

One of the most distinctive features of a colloidal, or macromolecule, solution is the scattered light, or Tyndall effect, that is observed when a light beam is passed through such a solution. This scattered light can be used in two different ways to help elucidate the nature of colloidal solutions.

The first application is made in the *ultramicroscope,* which is shown schematically in Fig. 25-9. Here the sample is observed through a microscope at right angles to the direction of the entering light beam. Each colloid particle, larger than about 10 Å diameter in very favorable cases, will produce an observable point of scattered light. The individual particles can then be counted, and if the microscope focuses on a definite, known volume of solution, the number of particles per unit volume can be determined. Such data, along with the measurable mass of macromolecule material per unit volume, lead to a value of the average mass of the individual particles.

It should be emphasized that none of the details of the particles can be observed. They merely act as scattering centers, and one observes points of light. Furthermore, unless the refractive index of the colloid particle is very different from that of the solvent, the scattered light is too weak to be seen. The similarity of the refractive index of most macromolecules to

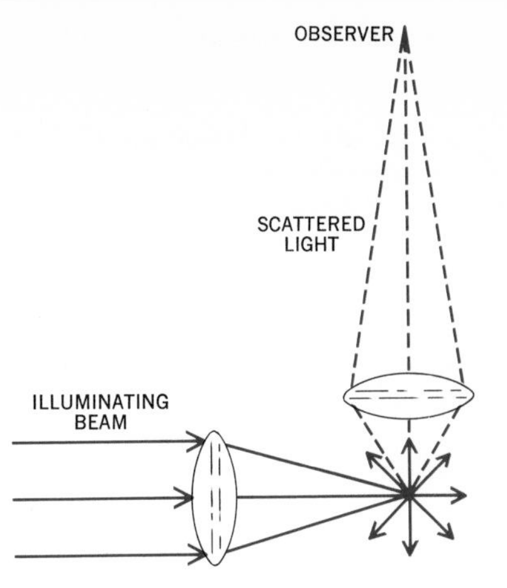

FIGURE 25-9
The detection of particles in the ultramicroscope.

the medium in which they are dispersed means that little scattered light will be given off. The method is therefore mostly applicable to inorganic colloids.

The second application of the scattering of light depends on the measurement and interpretation of the amount of light scattered in various directions as a beam of light passes through a solution of macromolecules. In some experiments the total amount of the scattered light is deduced from the decrease in intensity of the incident beam as it passes through the sample. Just as for Beer's law for the absorption of light (Sec. 17-8), one has the relation

$$I = I_0 e^{-\tau l} \qquad [26]$$

where τ is the measure of the decrease in incident-beam intensity per unit length of a given solution and is known as the *turbidity*. In some experiments, on the other hand, the intensity of light scattered in various directions is measured directly, rather than inferred, from the decrease in intensity of the incident beam.

That the scattered light is related to the particle size and shape can now be shown. We first consider the effect of particles that are small compared with the wavelength of the radiation. Incident plane-polarized radiation imposes, as Fig. 25-10 illustrates, an electric field

$$\mathcal{E} = \mathcal{E}_0 \sin 2\pi\nu t \qquad [27]$$

at the particle. If the particle has a polarizability α, there will be an in-

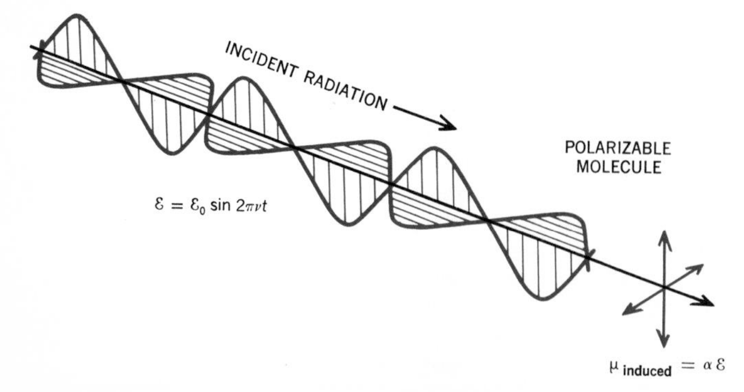

FIGURE 25-10
The induced-dipole moments produced by the two components of incident radiation.

duced-dipole moment given by

$$\mu_{\text{induced}} = \alpha\mathcal{E} = \alpha\mathcal{E}_0 \sin 2\pi\nu t \qquad [28]$$

It is this oscillating-dipole moment that emits secondary radiation and causes the particle to be a scattering center.

It has been mentioned in connection with the difficulties of early atomic theories that in classical electromagnetic theory an accelerated charge must emit electromagnetic radiation. This result can be applied to show what secondary radiation will be emitted by the oscillating induced dipole.

The oscillating-dipole moment of Eq. [28] can be formally written as a charge of value $\alpha\mathcal{E}_0$ oscillating with a unit amplitude relative to an equal and opposite charge. Thus we can write

$$\mu_{\text{induced}} = (\alpha\mathcal{E}_0)x \qquad [29]$$

with

$$x = \sin 2\pi\nu t \qquad [30]$$

In this way a picture is obtained of an induced charge $\alpha\mathcal{E}_0$ in the irradiated particle vibrating with simple harmonic motion. This motion, furthermore, involves an acceleration d^2x/dt^2 which is calculated from Eq. [30] as

$$\frac{d^2x}{dt^2} = 4\pi^2\nu^2 \sin 2\pi\nu t \qquad [31]$$

It is necessary now to quote, without derivation, the very important classical electromagnetic result that the acceleration of a charge q leads to the emission of electromagnetic radiation which produces at a distance r and angle ϕ, as in Fig. 25-11, from the oscillating charge, an electric field e given by the expression

$$e = -\frac{q}{c^2}\frac{d^2x}{dt^2}\frac{\cos\phi}{r} \qquad [32]$$

With this result one calculates that the radiation field of the dipole induced in the particle by one component of the incident radiation is

$$e = -\frac{\alpha\mathcal{E}_0 4\pi^2\nu^2 \sin 2\pi\nu t \cos\phi}{c^2 r} \qquad [33]$$

The propagation of this radiation through space with a velocity c can be represented by including a sinusoidal space dependence to give

$$e = -\frac{4\pi^2\nu^2\mathcal{E}_0 \sin 2\pi\nu(t - x/c)\cos\phi}{c^2 r} \qquad [34]$$

It is, however, not the electric field of the radiation but rather the energy

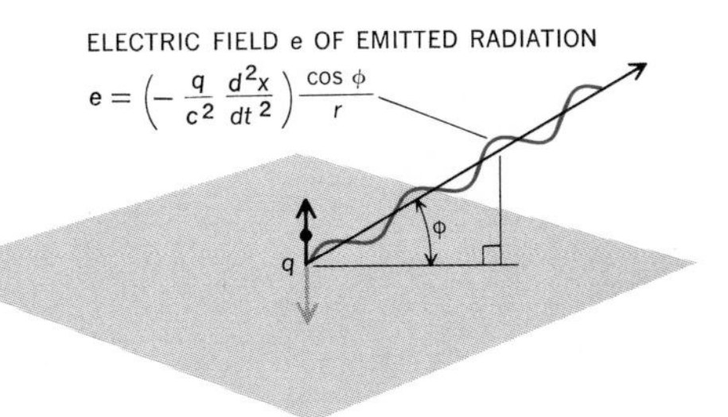

FIGURE 25-11
The angular dependence of the radiation emitted from an oscillating charge.

content that is of interest. This energy is directly related to the square of the field amplitude, and with this relation the intensity, or energy, of the secondary beam is calculated as

$$i = \frac{16\pi^4\nu^4}{c^4r^2}\alpha^2\mathcal{E}_0{}^2 \sin^2 2\pi\nu\left(t - \frac{x}{c}\right)\cos^2\phi \tag{35}$$

What is of importance for comparison with experimental results is the intensity of scattered radiation, at various angles, compared with the intensity of the incident radiation. This incident plane-polarized radiation can be depicted as entering the sample according to the relation

$$I_0 = \mathcal{E}_0{}^2 \sin^2 2\pi\nu\left(t - \frac{x}{c}\right) \tag{36}$$

and this expression can be inserted into Eq. [35] to give

$$\frac{i}{I_0} = \frac{16\pi^4\nu^4}{c^4r^2}\alpha^2\cos^2\phi \tag{37}$$

Introduction of the radiation wavelength by the relation $\nu = c/\lambda$ then gives

$$\frac{i}{I_0} = \frac{16\pi^4}{r^2}\frac{\alpha^2\cos^2\phi}{\lambda^4} \tag{38}$$

When ordinary, nonpolarized radiation is used for the incident beam, the induced-dipole moment in the sample can be considered to have two mutually perpendicular components. The scattered beam consists, then, of two perpendicular components like that of Eq. [38]. This net scattered beam is related to the angle θ of Fig. 25-12 by the equation

$$\frac{i}{I_0} = \frac{8\pi^4\alpha^2}{\lambda^4r^2}(1 + \cos^2\theta) \tag{39}$$

This angular dependence is best verified by checking that it gives the correct summation of the two plane-polarized components in various special directions.

The intensity predicted by Eq. [39] for the scattered beam from small particles is illustrated in Fig. 25-12.

It should be noticed that the forward and backward scattering are equal. Furthermore, the fourth-power dependence of the scattering on the wavelength shown by Eq. [39] should be noticed. It is, for example, to this enhanced scattering of short-wavelength radiation that the blue color of the sky is attributed. The short-wavelength blue end of the visible spectrum is scattered more than the long-wavelength red end, and the "background" color of the sky is therefore blue.

The interpretation of the scattering of radiation that has culminated

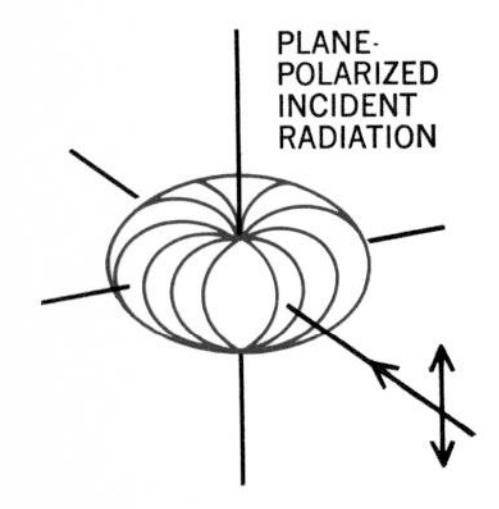

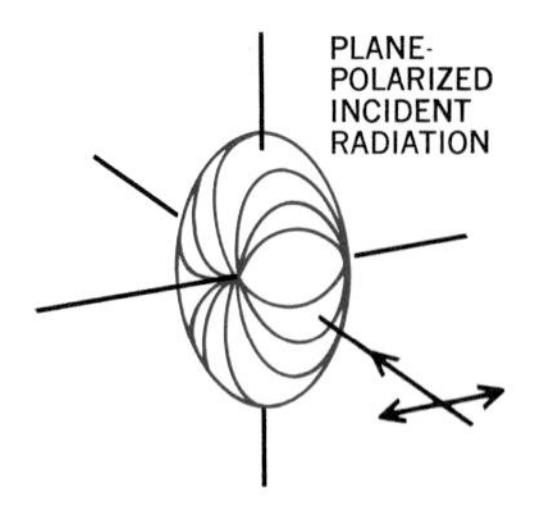

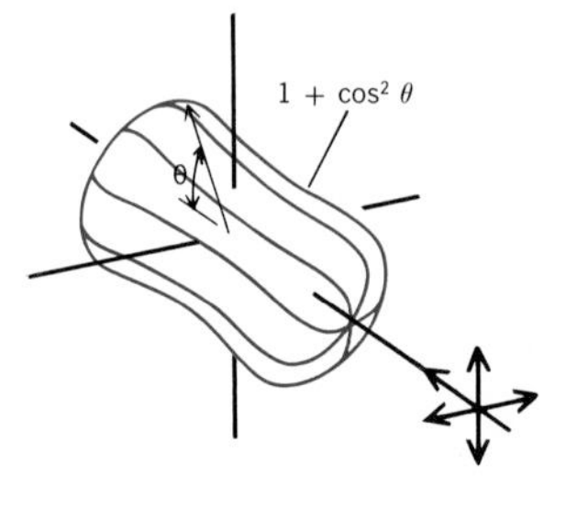

FIGURE 25-12
The angular dependence of secondary, or scattered, radiation from a particle that is small compared with the wavelength of the radiation.

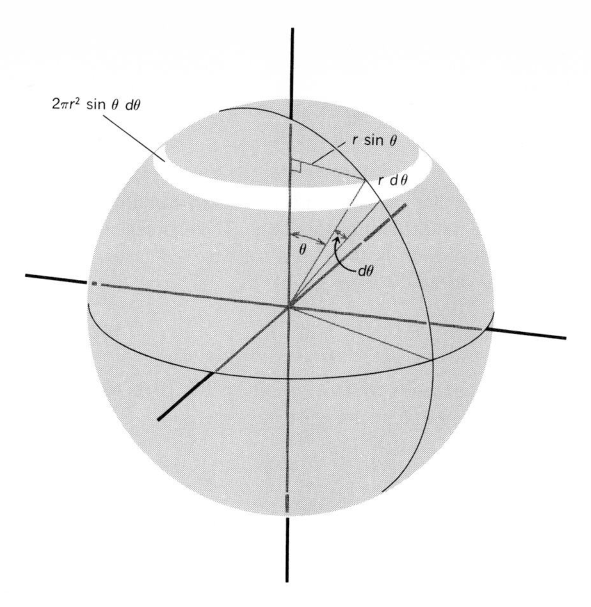

FIGURE 25-13
The surface element for the integration of the scattered radiation.

in Eq. [39] can be brought to a comparison with the experimental quantity, the turbidity. For many samples the amount of scattering is small, and the turbidity expression

$$\frac{I}{I_0} = e^{-\tau l}$$

or

$$\tau l = -\ln \frac{I}{I_0} \qquad [40]$$

can be written, for unit-cell length, as

$$\begin{aligned}\tau &= -\ln\left(\frac{I}{I_0} - \frac{I_0}{I_0} + 1\right)\\ &= -\ln\left(\frac{I - I_0}{I_0} + 1\right)\\ &\approx \frac{I_0 - I}{I_0}\end{aligned} \qquad [41]$$

The term $I_0 - I$ is the intensity removed from the incident beam and is therefore the integral over all angles of the scattered radiation of Eq. [39]. The measure of turbidity can therefore be evaluated, according to the differential surface element of Fig. 25-13, as

$$\tau = \int_0^{\pi} \frac{i}{I_0} 2\pi r^2 \sin\theta \, d\theta \qquad [42]$$

Substitution of the expression for scattered intensity i of Eq. [39] and integration gives

$$\tau = \frac{8\pi}{3}\left(\frac{2\pi}{\lambda}\right)^4 \alpha^2 \qquad [43]$$

For a concentration of c g ml^{-1} and a molecular mass M there will be $(c/M)\mathfrak{N}$ molecules per milliliter, and the turbidity of such a material will be

$$\tau = \frac{8\pi}{3}\left(\frac{2\pi}{\lambda}\right)^4 \alpha^2 \frac{c\mathfrak{N}}{M} \qquad [44]$$

To obtain a calculation of molecular mass from a measured turbidity, it is necessary to have a value of the molecular polarizability. The expression of Eq. [37] of Sec. 15-5 can be reduced for $n_R \approx 1$, as is the case for gases, to the relation between α and the refractive index n_R of

$$2\pi\left(\frac{c}{M}\mathfrak{N}\right)\alpha = n_R - 1 \qquad [45]$$

With this result the turbidity of a gaseous system, in which the particles are small compared with the wavelength, can be written

$$\tau = \frac{32\pi^3 M(n_R - 1)^2}{3\lambda^4 c\mathfrak{N}} \qquad [46]$$

With this expression the measurable turbidity can be related to the molecular mass of the gas-phase particles, the mass of material per unit volume, and the refractive index of the gaseous system. This expression can also be turned around so that a value for Avogadro's number can be obtained from the scattering produced by gas samples.

For the systems of interest here, i.e., macromolecules in a liquid medium, it is necessary to introduce the fact that the scattering depends on the *difference* between the refractive index of the particles and that of the medium. If n_R is the refractive index of the solution and n_R° that of the pure solvent, the appropriate relation comparable with Eq. [46] turns out to be

$$\tau = \frac{32\pi^3 M n_R^\circ}{3\lambda^4 c\mathfrak{N}}(n_R - n_R^\circ)^2 \qquad [47]$$

Now measurements n_R° and n_R for a solution of a given value of c allow the calculation of the molecular mass M. It should be mentioned that, since the polarizability increases with increasing molecular size, the amount scattered by an individual molecule is proportional to its size. The molecular mass that is obtained is therefore a weight-average molecular mass.

In practice, this expression is usually written

$$\tau = \frac{32\pi^3 n_R^\circ}{3\lambda^4\mathfrak{N}}\left(\frac{n_R - n_R^\circ}{c}\right)^2 cM$$

or

$$\tau = HcM \qquad [48]$$

where

$$H = \frac{32\pi^3 n_R^\circ}{3\lambda^4\mathfrak{N}}\left(\frac{n_R - n^\circ}{c}\right)^2 \qquad [49]$$

In dilute solutions, moreover, the term for the change of refractive index with concentration can be written as a differential, and H then is

$$H = \frac{32\pi^3 n_R^\circ}{3\lambda^4\mathfrak{N}}\left(\frac{dn_R}{dc}\right)^2 \qquad [50]$$

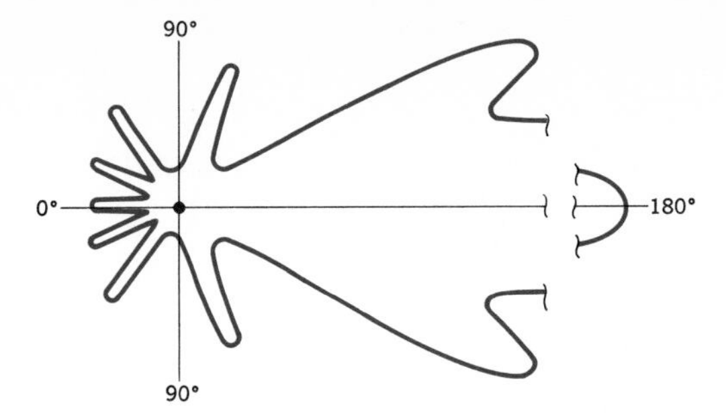

FIGURE 25-14
The scattering of visible light from a spherical particle of radius 5000 Å. [*From V. K. LaMer and M. Kerker, Light Scattered by Particles, Sci. Am.* **188:***69* *(1953)*.]

From measurements of refractive index for the wavelength of light used in the scattering experiments, H can be evaluated and the turbidity can be measured. Equation [48] can then be used to calculate a molecular mass. In practice, an extrapolation to infinite dilution is necessary, and for particles of appreciable size compared with the wavelength of light, so also is recognition of the angular dependence of the scattered light.

A little must now be said about the scattering that results when the molecules are not small with respect to the wavelength of the light. Visible light has wavelengths between about 4000 and 8000 Å, and these lengths are just about the dimensions expected for many macromolecules. As for electron scattering from different atoms of a molecule in an electron-diffraction experiment, the scattering from different parts of the molecule will now interfere with one another. The effect is, in fact, very similar to that studied in detail for electron diffraction. The macromolecule is best thought of as some geometric shape presenting a continuum of scattering centers rather than a few discrete centers. A detailed calculation for the amount of light scattered as a function of angle, for a given wavelength and assumed molecular size and shape, can be performed by integrating the Wierl equation of Sec. 14-3 over all parts of the molecule. The type of light-scattering angular dependence that can result for molecules with dimensions like that of the wavelength of the scattered light is indicated in Fig. 25-14. The details of the pattern are dependent on the shape of the molecule as well as on its overall size. Measurements of light scattering now give promise of being one of the most powerful methods for studying the geometry of macromolecules in solution.

In lieu of a detailed analysis of the molecular shape that would lead to the observed angular dependence of the scattered radiation, it is often sufficient to measure the intensity of the scattered beam at two angles, usually 45 and 135°, to the incident beam. The ratio of these intensities reflects the overall shape of the macromolecule in solution. Calculations have been made, using essentially the Wierl equation for some simple shapes, and these are shown in Fig. 25-15. From observations of the scattered intensity at the angles of 45 and 135°, such curves can be consulted, and lead to some information on the usually unapproachable quantity, the shape of a molecule in solution.

25-13 Electrokinetic Effects

The behavior of colloidal particles dispersed in an aqueous medium is greatly affected by the fact that the particles often carry an electric charge. The presence of acidic and basic groups in proteins, for exam-

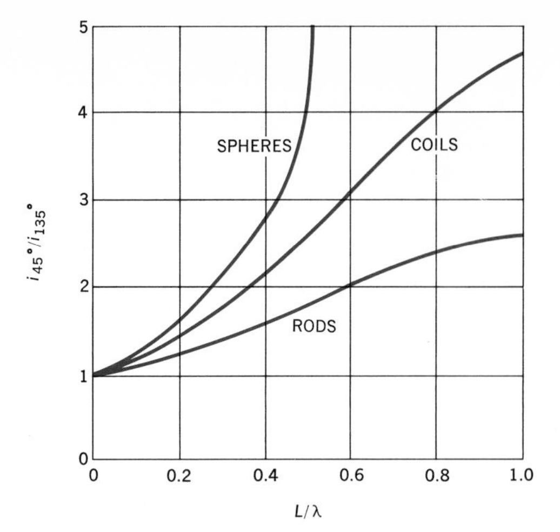

FIGURE 25-15
The ratio of the intensity of the scattered light at 45° to that at 135°. The wavelength of the light is λ, and L is the diameter for spheres, the rms distance between ends for coils, and the length for rods.

ple, means that there will generally be positive or negative charges on the protein molecule. The number and sign of these charges will depend on whether the solution is acidic or basic. Charges are also carried by inorganic colloidal particles, such as AgCl, where the charge can be attributed to a preferential adsorption of Ag^+ or Cl^- ions on the surface of the particles. The nature of the net charge on the particles is clearly important in questions of the tendency of the colloidal particles to come together, or *flocculate,* since this process must overcome the electrostatic repulsion between particles.

The electrical nature of colloidal particles can best be studied in experiments which make the colloid particles and the surrounding medium move relative to one another. Such experiments are said to treat *electrokinetic phenomena.*

The charge distribution around a charged colloid particle might be expected to be that indicated in Fig. 25-16. A protein molecule has

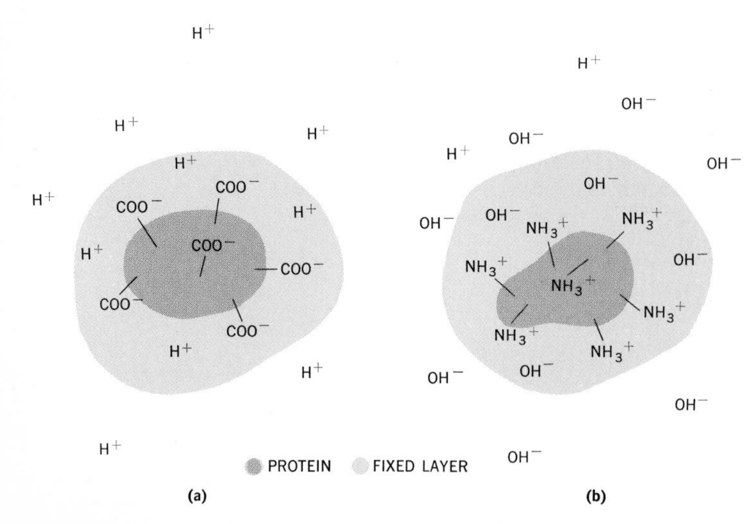

FIGURE 25-16
The fixed charges associated with a protein molecule in (*a*) basic solution and (*b*) acidic solution (schematic).

been used for illustration, but although the charged groups of the protein molecule can be more definitely attributed to the reaction of chemical groups such as —COOH and $—NH_2$, the charge distribution is expected to be similar in inorganic colloids. The effective charge of a particle in electrokinetic experiments is the *fixed charge,* which is made up of the actual charge of the particle and any ionic charges that are held sufficiently close to the particle so that they will remain with the particle as it moves through the solution. The total charge of the system of colloid particles and solution is necessarily electrically neutral, and opposite charges to that of the fixed charge will tend to surround the colloid particle and will form a diffuse layer which will move with the solution rather than the particle. The situation is shown schematically in Fig. 25-16. Electrokinetic phenomena can all be given a qualitative explanation in terms of the concept of the charge fixed to each particle and the surrounding charge which moves with the liquid medium. Attempts at quantitative interpretations introduce the *zeta,* or *electrokinetic, potential,* which expresses the potential drop between the fixed charges and the body of the solution.

The electrokinetic phenomenon of most practical importance is that of *electrophoresis.* In the study of proteins, for example, one forms a boundary between a buffer solution containing a protein sample and the pure buffer solution in a manner similar to that used in studying diffusion. With an electrophoresis instrument one applies a potential difference between electrodes dipping in the two solutions and observes the movement of the boundary between the solutions and thus determines the motion of the macromolecules as a result of a potential gradient. The movement of the boundary is observed optically, the refractive-index gradient at the boundary causing the light to be refracted from its path. In this way the extent to which the macromolecules with different mobilities have migrated from the original boundary can be determined.

The direction of migration is, of course, dependent on the charge of the particles, and one finds that proteins move to the anode for sufficiently basic solutions and to the cathode for sufficiently acidic solutions and show no electrophoretic effect at the isoelectric point.

Different proteins may show different *mobilities.* Electrophoresis is therefore a valuable tool for separating biological fractions into pure components. In this respect, it supplements the ultracentrifuge, which separates according to molecular mass. Electrophoresis experiments can show that, even if a sample is homogeneous with respect to molecular mass, it may contain different components having different electrical properties. Figure 25-17 shows the separation that can be obtained in a complicated natural-product preparation.

The mobilities found for colloidal particles, and particularly for proteins, depend, of course, on the pH of the solution. In general, the mobilities, i.e., the velocity acquired by the particles for a potential gradient of 1 V cm^{-1}, or 100 V m^{-1}, are around 2×10^{-8} m s^{-1} and

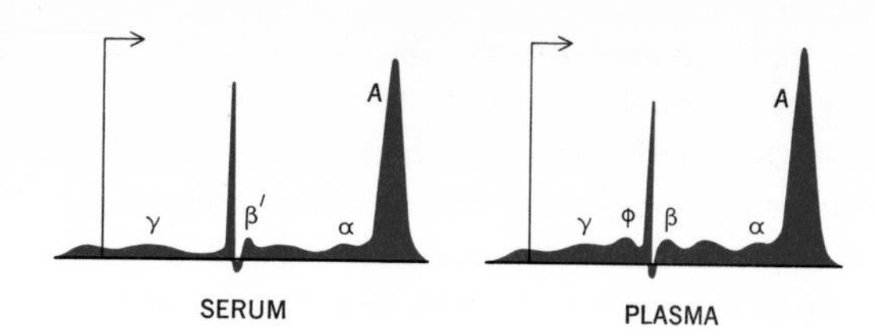

FIGURE 25-17
The electrophoretic patterns obtained with (left) normal human serum and (right) plasma. The main components are identified as A for albumin, ϕ for fibrinogen, and α, β, and γ for the different globulins.

are seen, therefore, to be only a little less than the mobilities found for simple ions, as listed in Table 22-9.

Two interesting electrokinetic effects result from the possibility of holding the position of the charged macromolecules fixed. Thus a membrane or layer of fibrous macromolecules can be introduced into a buffer solution. Again the macromolecules adopt some fixed charge. Unlike any of the systems dealt with in our study of electrochemistry, we now have the movable liquid carrying a *net* charge, the balancing charges being in the fixed macromolecules. The situation is shown diagrammatically in Fig. 25-18. If now an electric potential is applied, the movable ions and their surrounding solvent move to the oppositely charged electrode, where the ions are discharged. The effect of this movement is to carry the solvent, and the movable ions, through the membrane. This process is known as *electroosmosis* and has some practical applications as an adjunct to ordinary osmosis. One should recognize that both electrophoresis and electroosmosis involve the relative flow of the charged macromolecules and the oppositely charged surrounding liquid and are therefore essentially the same phenomenon.

With a similar arrangement as that used for electroosmosis, one can reverse the procedure and produce a potential as a result of forcing the buffer solution to flow through the macromolecule matrix. The potential difference that is developed is known as a *streaming potential.*

The quantitative treatment of all these electrokinetic phenomena depends on the charge fixed to the particle and the movable charges surrounding the particle. Since this charge distribution produces the zeta potential, the phenomena can be analyzed either directly in terms

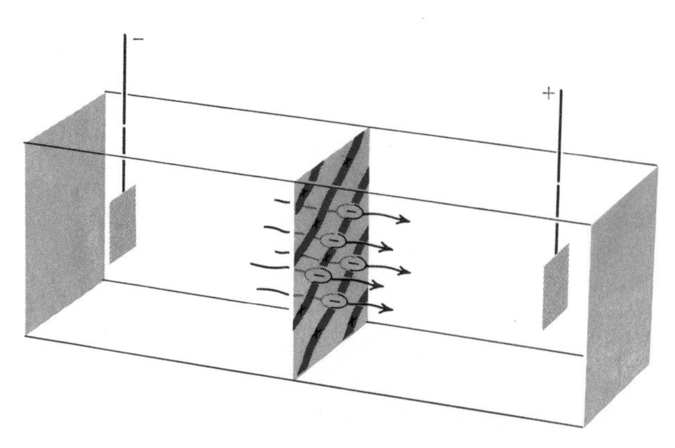

FIGURE 25-18
Electroosmosis through a porous membrane holding a fixed positive charge.

of the charges or, as is more generally done, in terms of the zeta potential.

Before leaving the subject of the charges of colloidal particles, it should be mentioned again that it is these charges which account for much of the stability of colloids. For highly solvated macromolecules in water, i.e., *hydrophilic colloids,* the solvating layers of water help to prevent the individual particles from agglomerating. For nonsolvated particles in water, i.e., hydrophobic colloids, however, such solvation is unimportant, and only the electrical effect operates. The particles of AgCl, for example, carry a net fixed negative charge, due to a preferential adsorption of Cl^- ions surrounded by a diffuse, balancing, positively charged region rich in Ag^+ ions. Because of the electrostatic repulsion between their negative charges, the colloidal particles cannot easily come together. Agglomeration, or flocculation, can be made to occur, however, by adding an electrolyte, particularly one with positive ions of high charge. These added positive charges will surround the colloid, or as might be said, will decrease its zeta potential, and will allow them to approach one another. With inorganic colloids, where solvation is less important than in protein systems, the action of charges in stabilizing or precipitating the colloidal particles is very important.

25-14 The Donnan Membrane Equilibrium and Dialysis

One final electrical phenomenon encountered with macromolecules should now be mentioned. This phenomenon, known as the *Donnan equilibrium,* is not, however, an electrokinetic effect. We shall see, for example, that it does not depend on the zeta potential. The Donnan equilibrium shows up when a colloidal solution in which the particles are charged, most commonly an aqueous protein solution, is separated by a semipermeable membrane from the pure water or from the solution without the colloid. The complications that arise as a result of the charged protein molecules can be seen by reference to Fig. 25-19. It is supposed that the particles P, which can be thought of as protein molecules, carry some negative charge and that an appropriate number of sodium ions balance this charge. Suppose that an osmosis cell is set up, or that dialysis is performed with this solution and a solution

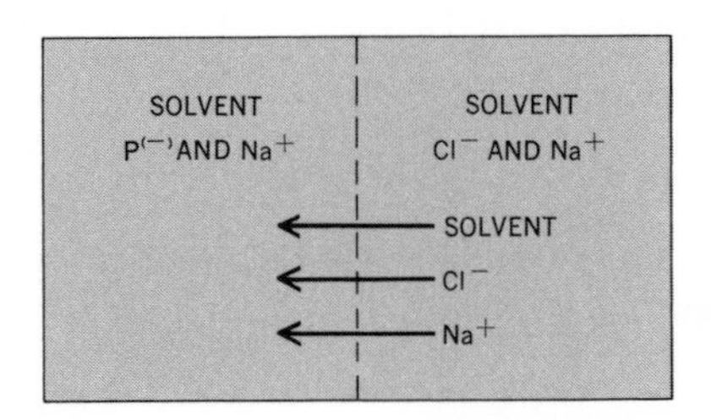

FIGURE 25-19 The passage of solvent and ions through a semipermeable membrane in an osmosis experiment using a solution of macromolecules and electrolyte.

of sodium chloride, as in Fig. 25-19. The macromolecules $P^{(-)}$ cannot pass through the membrane. The Cl^- ions, on the other hand, can, and they will tend to diffuse from the high concentration on the right to the low concentration on the left. To preserve electrical neutrality, an equal number of Na^+ ions will pass from right to left across the membrane. An osmotic-pressure measurement would therefore be complicated by the additional number of particles in the macromolecule side of the membrane.

Suppose that the dialysis or osmosis of the solution containing $P^{(-)}$ and Na^+ is performed against water. Here the Na^+ ions will tend to diffuse to the low-concentration region on the right. Electrical neutrality can now be maintained only by additional dissociation of water to form H^+ ions on the left and OH^- ions on the right. Unless a buffer is used, the pH of the solutions will therefore change during the experiment.

The quantitative nature of these effects can be set down for a system of two equal volumes separated by a semipermeable membrane. If c_1 and c_2 are the initial concentrations and x is the concentration change due to the diffusion of NaCl, the situation is described as

				¦		
Initially		Na^+	$P^{(-)}$	¦	Na	Cl^-
		c_1	c_1	¦	c_2	c_2
At equilibrium	Na^+	$P^{(-)}$	Cl^-	¦	Na^+	Cl^-
	$c_1 + x$	c_1	x	¦	$c_2 - x$	$c_2 - x$

The concentrations are written such that both compartments remain electrically neutral. The transfer of Na^+ ions and Cl^- ions, subject to this neutrality condition, will proceed until the free energy of NaCl in both compartments is equal. More conveniently, this equilibrium stipulation can be expressed as equal activities of NaCl in the two compartments. Thus, in the left compartment,

$$(a_{\mathrm{NaCl}})_l = (a_{\mathrm{Na^+}})_l(a_{\mathrm{Cl^-}})_l = (\gamma_\pm)_l^2[\mathrm{Na^+}]_l[\mathrm{Cl^-}]_l \qquad [51]$$

and in the right compartment

$$(a_{\mathrm{NaCl}})_r = (a_{\mathrm{Na^+}})_r(a_{\mathrm{Cl^-}})_r = (\gamma_\pm)_r^2[\mathrm{Na^+}]_r[\mathrm{Cl^-}]_r \qquad [52]$$

If the mean-activity coefficients in the two compartments can be taken as equal, usually as a result of equal ionic strengths, the Donnan equilibrium relation is obtained as

$$[\mathrm{Na^+}]_l[\mathrm{Cl^-}]_l = [\mathrm{Na^+}]_r[\mathrm{Cl^-}]_r$$

or

$$(c_1 + x)(x) = (c_2 - x)(c_2 - x) \qquad [53]$$

Rearrangement gives the concentration of the NaCl that is transferred to the colloid compartment as

$$x = \frac{c_2^2}{c_1 + 2c_2} \qquad [54]$$

This expression, or a more general one for multivalent ions, is important in any experiment with charged particles in solution that are in any way fixed, or moved, in relation to the other ions in the solution.

MACROMOLECULES IN THE SOLID STATE

As our previous study of the solid state has shown, the existence of an ordered crystalline sample allows, through x-ray diffraction, a rather direct means for the determination of molecular structure. In spite of the complexity of the protein molecules and the frequent lack of simple single crystals of polymers, this technique has been applied and has met with considerable success in unraveling the structures of these macromolecules in the solid state.

Often the x-ray study must be made on a fiberlike material in which the macromolecules are more or less ordered along the fiber axis. From such studies, as Fig. 25-20 indicates, one can often deduce any repeated distances along and perpendicular to the fiber axis. Such results can be of great aid in the elucidation of fiber structures.

These tools and some of the structural properties of proteins, nucleic acids, and solid polymers will now be dealt with.

25-15 The Structure of Proteins

A number of proteins and their simpler analogs, the polymers of amino acids, known as *polypeptides,* exist or can be prepared in fiber or crystalline form. The technique of x-ray diffraction can therefore be applied.

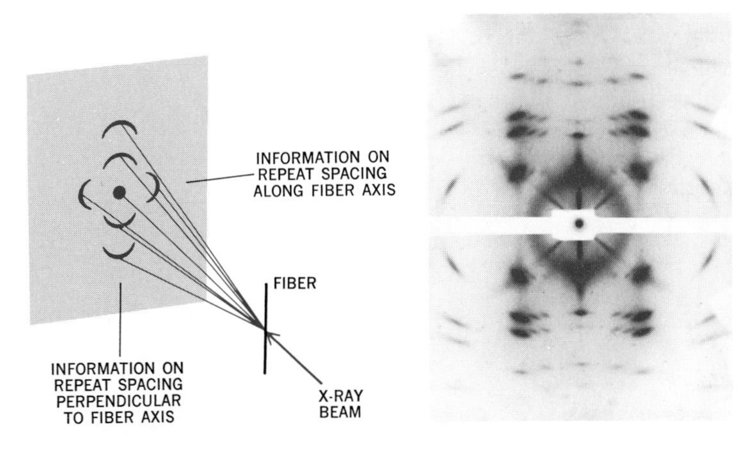

FIGURE 25-20
X-ray patterns of fibers. (*a*) Schematic diagram of x-ray-fiber pattern. (*b*) The x-ray-diffraction pattern of a fiber of polyoxymethylene, CH_2O. (*Courtesy of P. H. Geil, Case Institute of Technology.*)

It is possible to carry over the standard bond lengths and bond angles from simpler molecules and to use these values to construct the geometry of the basic elements of the polypeptide or protein molecules. The idea of resonance of the type

$$\mathrm{C{-}C({=}O){-}N(H){-}C \;\leftrightarrow\; C{-}C(O^{-}){=}N^{+}(H){-}C}$$

suggests, according to the requirement of planarity of a double-bond structure, the planarity of such six-atom groups. Distances carried over from simple molecules now allow the fully extended protein skeleton to be drawn, as in Fig. 25-21. The protein molecule can be looked upon, therefore, as far as structural arrangements are concerned, as a succession of planar groups as in Fig. 25-21, where we have attempted to show that the α-carbon atom, although involved in two planar units, does not fix any necessary relationship between these planes.

Three structural factors, besides the geometry of these planar units and the tetrahedral links, must be considered. They affect both the shape of a single protein chain and the way in which the chains associate with each other.

One factor stems from the size of groups. Adequate space must be provided for any large, bulky side chains, and thus a spatial, or "steric," hindrance to some structural arrangements occurs. This influences both the configuration that the protein chain itself can adopt and the ways in which the neighboring chains can be related to one another. The most general consequence of steric hindrance is the adoption of structures in which the angle φ, of Fig. 25-22, has the value of about 120°. In this way, as Fig. 25-22 tries to show (but models show it much more clearly), the carbonyl group is positioned so that it avoids the sometimes bulky R group on the nitrogen. Even with this general result, which is not always adhered to, various chain structures result from the choices of different values for the angle ψ. (You will also see the effect of the bulkiness of side groups when the structures that result from the packing together of protein molecules are considered.)

The second structure-influencing feature is the tendency for *hydro-*

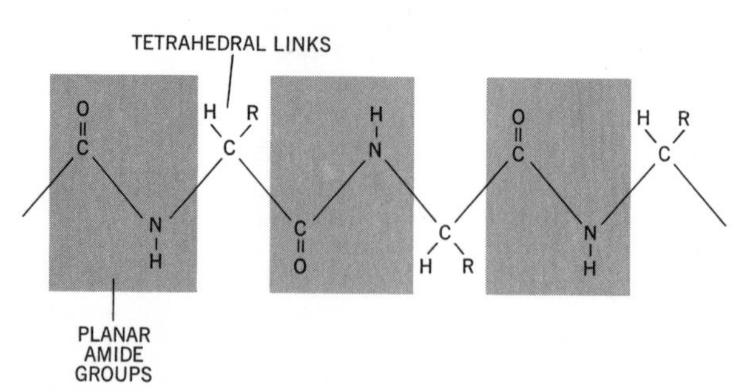

FIGURE 25-21
The planar groups and tetrahedral links that are the basic structural features used to describe a protein structure.

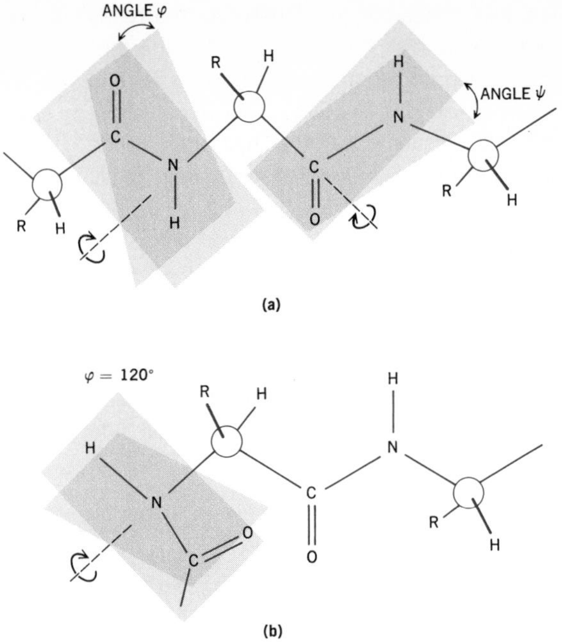

FIGURE 25-22
(a) The description of the rotations of the planar units about the N—C and C—C bonds by the angles φ and ψ, respectively. (b) Rotation to a value of $\varphi = 120°$ places the carbonyl group as far away from the often bulky R group as possible.

gen-bonding interactions to occur, and therefore structures to be adopted so that such bonds are possible. Most prevalent are bonds between N—H and C=O groups of the same or different protein chains. Such bonding is indicated by

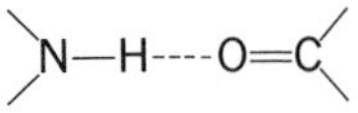

$$\rangle N—H\text{----}O=C\langle$$

Other groups also enter into such bonding, but the amide-carbonyl interactions occur with sufficient frequency to be a principal structure-influencing factor. Again, both the configuration of a single chain and the relationship between planes respond to the influence of hydrogen bonding.

A final factor stems from the same forces that account for the familiar tendency of oil and water to separate from one another. Inspection of the side chains of the amino acids of Table 25-4 shows that there are both polar and nonpolar side chains. Particularly in globular proteins, the protein chain tends to form a ball so that the nonpolar side chains are toward the inside of the ball and the polar groups are to the outside. Because this effect, which is said to be due to *hydrophobic* forces, is in response to the hydrogen bonding and polar character of the surrounding water, these forces must be expected to have a similar importance to hydrogen bonding itself.

Fibrous Proteins: Silk and the β-sheet structure Protein chains in a nearly extended form can, as shown in Fig. 25-23a, be placed side by side in such a way that hydrogen bonds can form between the N—H and C═O groups of adjacent chains. For this to occur, the adjacent chains must point in opposite directions, and as suggested in Fig. 25-23*b* (but clearly seen only by making models), the whole sheet structure is found to be "pleated" in such a way that hydrogen bonds occur. The result is the so-called *antiparallel pleated,* or *β-sheet,* basic protein structure.

These sheets are the basis of the protein structure of silk. The entire material is built by a stacking of such sheets on top of each other. Such packing would be impeded by bulky R groups, which project above and below each sheet. The packing of pleated sheets, and thus the use of this sheet structure, can be adopted by silk, or by silkworms, because of the occurrence of glycine, which has the smallest side chain, H, in every second amino acid position. As a result of the small side

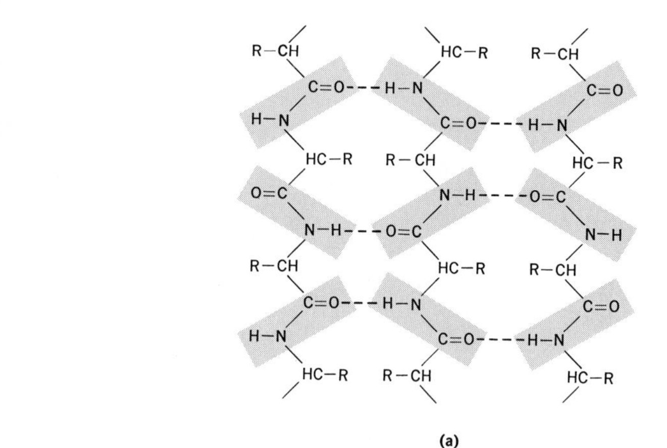

(a)

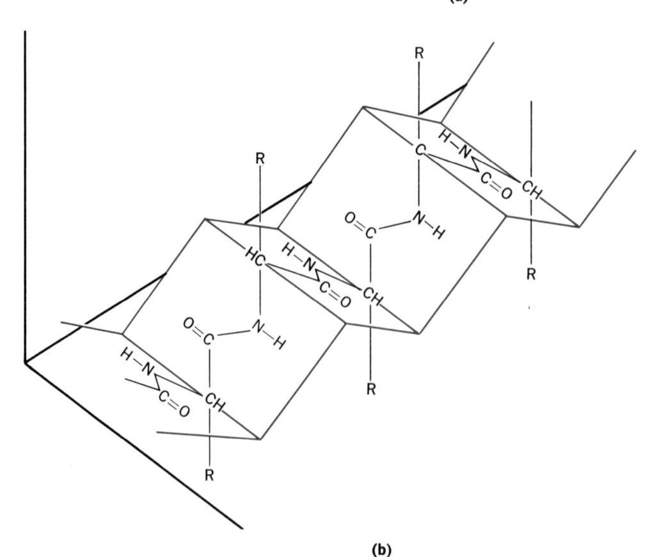

(b)

FIGURE 25-23
(*a*) The "antiparallel" arrangement of two nearly extended protein chains that is part of the pleated-sheet structure. (*b*) The pleated structure of a segment of the pleated sheet.

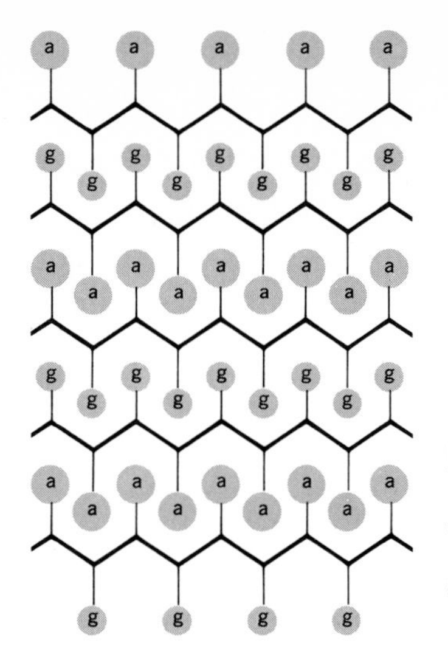

FIGURE 25-24
The packing of pleated sheets in silk, showing the way in which the alternation of the larger alanine and smaller glycine amino acids allow this packing to be very efficient.

chains and the positions of the glycines, a very efficient packing of the β-sheets can occur (Fig. 25-24).

Fibrous Proteins: Hair, Wool, and α-Keratin Many protein structures depend on the formation of hydrogen bonds within a single protein chain rather than, as in the β-sheet structure, between chains. This can be accomplished by forming a helix out of the chain so that the N—H and C=O groups can form hydrogen bonds. A variety of helices are possible; the one shown in Fig. 25-25, which is adopted by a number of proteins, is known as an α *helix*. Its principal features are again best seen by studying a model, but some of these features are indicated in Fig. 25-25. This structure predicts a repeated distance along the fiber axis, say from N to N, of between 1.47 and 1.53 Å. Layer lines indicating a spacing of 1.5 Å have, in fact, been observed in some proteins in which this α helix is expected.

Another repeat distance along the fiber direction that might be expected to show up is the pitch of the helix, i.e., the vertical distance along the axis from one point on the helix to a point on the helix directly above the first point. The α helix has 3.6 amino acid residues per turn, and since each amino acid corresponds to a vertical distance of 1.5 Å, the pitch is expected to be $3.6 \times 1.5 = 5.4$ Å. In fact, a spacing of 5.1 Å is observed. A neat explanation of this discrepancy has been given in terms of a helix with a nonlinear axis, as shown in Fig. 25-26. This shape, moreover, leads nicely into the idea that α-helix protein molecules can form bundles or cables as illustrated in Fig. 25-26. These ideas allow one to see how the detailed structure of large protein groups, of a size that is almost large enough to be seen with an electron microscope, can be deduced.

Studies of fibrous proteins lead, as the above discussion indicates,

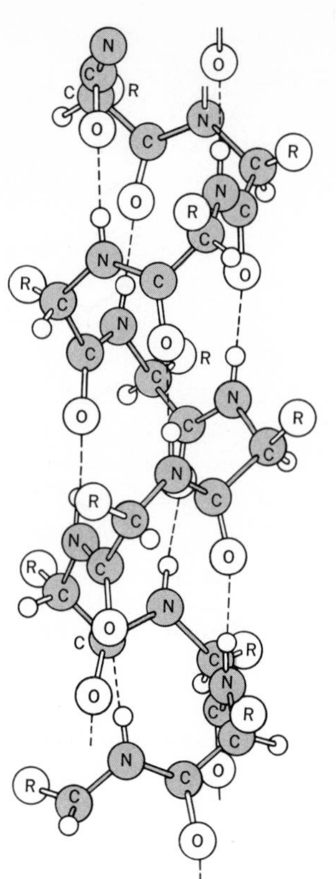

FIGURE 25-25
A portion of the α-helix protein chain structure.

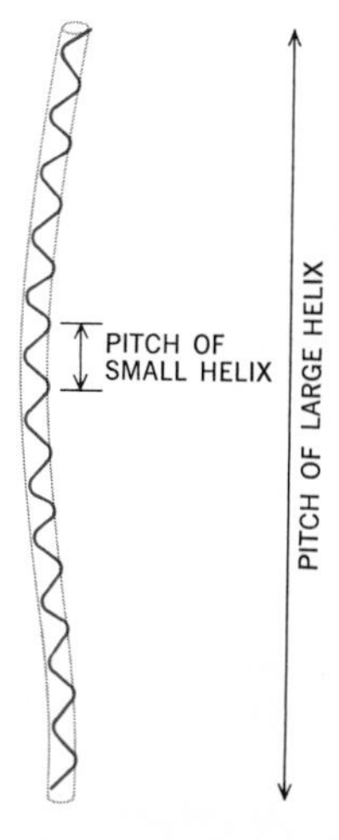

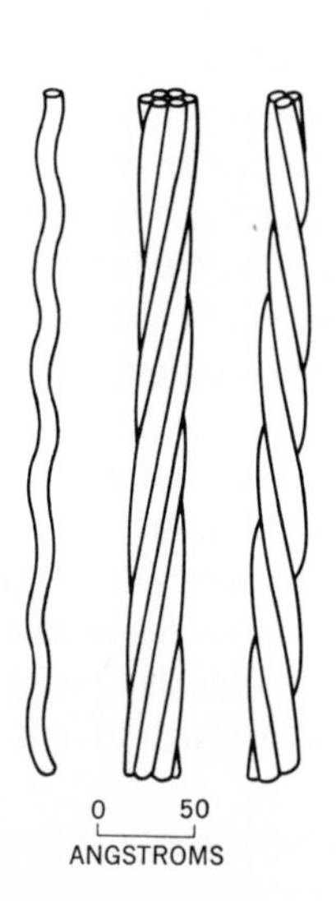

FIGURE 25-26
The helix given a slight coil so that a seven- or three-strand cable can be formed. [*From L. Pauling and R. B. Corey, Nature,* **171:***59* (*1953*).]

to the idea of the pleated sheet and the α helix. Many detailed problems of fibrous protein structure remain: the effect of bulky side groups, the role of disulfide cross links, and so forth. Much progress has been made, and in particular the idea of a helical macromolecule chain has been very fruitful.

Globular Proteins: Myoglobin The second important type of protein has molecules with a spherical or balled-up shape rather than the essentially linear one adopted by fibrous proteins. The tendency at present, however, is to think of the globular protein as consisting partly of an α helix wound in some way into a more or less spherical shape. In spite of the nicely crystalline form that is sometimes adopted by these proteins, the lack of a specific direction of the molecular chain makes the x-ray analysis of their structure very difficult. They become, as it were, huge molecules in a crystal lattice that x-ray-diffraction studies must tackle in a rather direct manner.

Some progress is being made. This progress has followed from the assumption of a protein chain with, perhaps, a helix arrangement and the x-ray-diffraction study of proteins containing a few heavy atoms. These heavy atoms act as strong scattering centers, and the x-ray analysis can treat the spacing between planes containing these relatively few atoms and can, to begin with, ignore the complexity introduced by all the other atoms of the molecule.

A globular protein that has been analyzed in detail is myoglobin. Its composition, structure, and function are related to the more familiar hemoglobin, which consists of four myoglobinlike parts. An early low resolution x-ray study led to the model shown in Fig. 25-27. The convolutions of the protein chain and its relation to the heme group are clearly seen. Detailed structural diagrams showing the locations of all the atoms of the molecule are now available (as in "The Structure and Action of Proteins" by R. E. Dickerson and J. Geis, Harper & Row Publishers, Incorporated, New York, 1969). Only with stereo viewing, however, can these detailed diagrams be fully appreciated.

This section should indicate that the physical-chemical approach to structure problems is making exciting advances even in studies of molecules as complex as proteins.

25-16 The Structure of Nucleic Acids

A number of the general ideas which have been successful in protein-structure studies appear also to be applicable to studies of nucleic acids. The chemical units of the nucleic acid deoxyribonucleic acid (DNA) have been pointed out in Sec. 25-4. It remains to suggest a geometric arrangement for the two molecular chains of Fig. 25-4.

In 1953 Watson and Crick, primarily on the basis of x-ray-diffraction

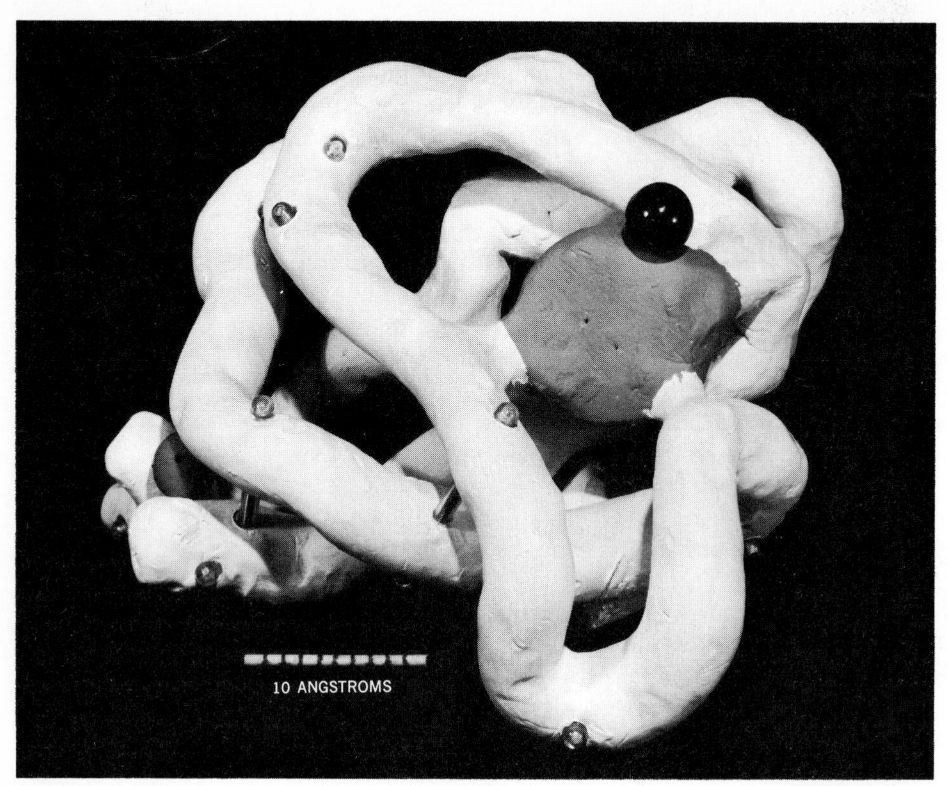

FIGURE 25-27
The configuration of myoglobin. The dark-gray disk is the heme group; the little black ball attached to it is an artificially introduced group required for the x-ray analysis of the macromolecule. The white parts show the polypeptide configuration at a resolution of about 6 Å. (*From M. F. Perutz, Endeavour,* **1958,** *p. 196.*)

studies, proposed that the two molecular chains formed a double helix and that the basic groups of Fig. 25-4 formed hydrogen bonds in a manner which holds the two spirals in position. The proposed structure is shown in Fig. 25-28. It consists of a double helix about 20 Å in diameter and has two residues for every 3.4 Å along the helix. A DNA molecule with a molecular weight of 10 million would have the remarkable length of 50,000 Å, that is, 0.005 mm. Although the double helix would seem to be a rather stiff structure, it appears reasonable that it adopts a coiled or folded shape. The hydrogen-bonding requirements can be satisfied between the chains by the stipulation that a thymine side group lines up opposite an adenine group, and a guanine group opposite a cytosine group. In this way, as Fig. 25-28 shows, the central core of the two-strand helix can maintain a diameter of about 11 Å, and a number of good hydrogen bonds can be formed.

In terms of the double-helix structure of Watson and Crick, the division of a DNA molecule into two daughter molecules, such as occurs in cell division, is pictured as an uncoiling of the double helix. It is in this regard, in addition to its immediate structural interest, that the structural ideas on nucleic acids are of immense interest to biochemists and biologists as well as to physical chemists.

25-17 Crystallinity of High Polymers

The characteristic of most synthetic polymeric materials like polyethylene that sets their x-ray or solid-state study apart from typical studies is the tendency of the polymeric material to be only partly crystalline. Thus one must deal with the *degree of crystallinity* as well as with the structure of the crystalline and amorphous regions. The principal tool for this study remains that of x-ray diffraction.

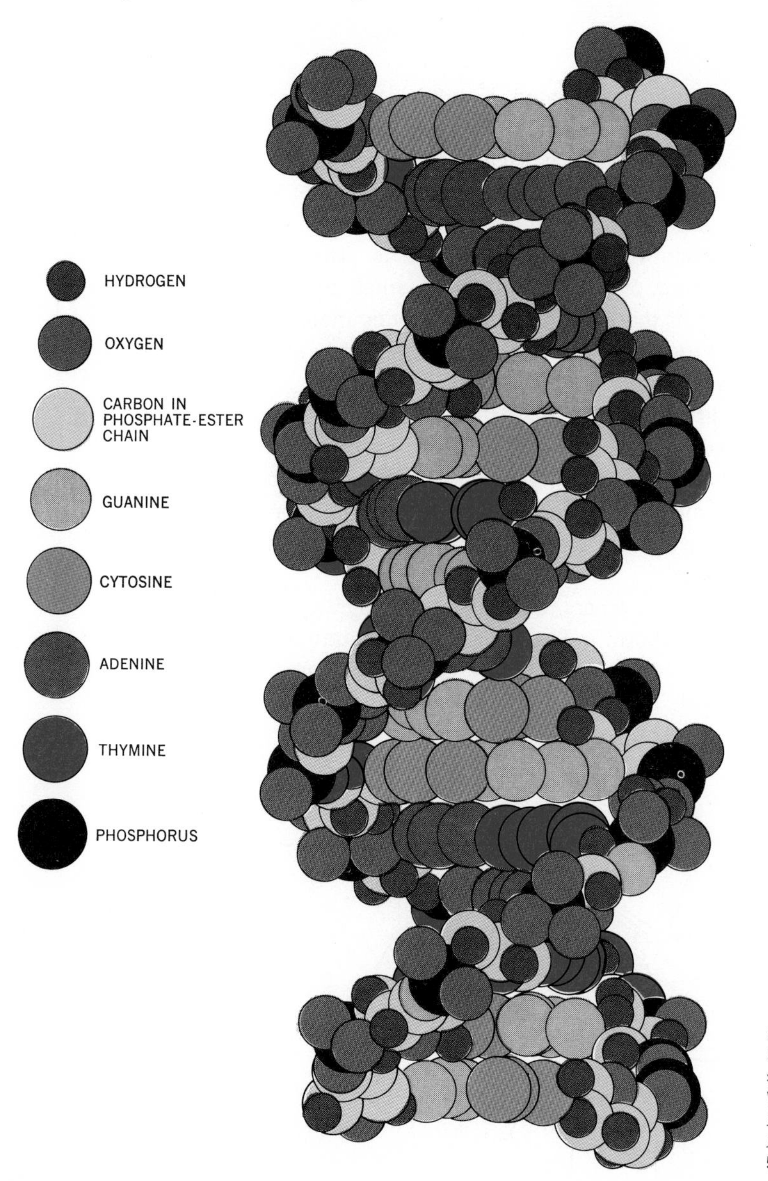

FIGURE 25-28
Double-helix model of the structure of DNA proposed by Watson and Crick. [*From L. D. Hamilton, CA (A Bulletin of Cancer Progress),* **5:**_159_ _(1955)_.]

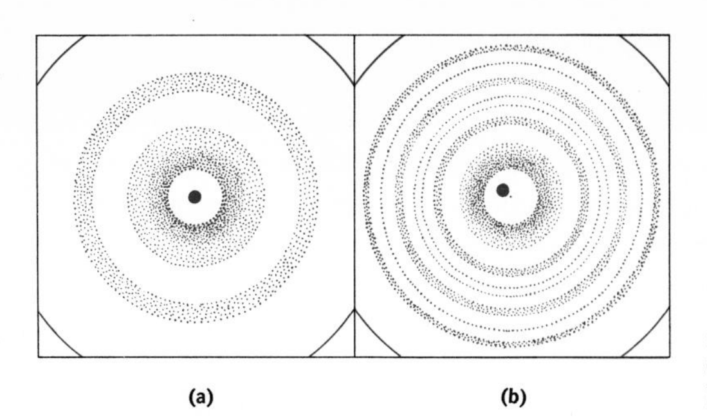

FIGURE 25-29
The x-ray diagram of (*a*) atactic polypropylene, which has only a small amount of crystallinity, and (*b*) highly crystalline isotactic polypropylene.

Differently prepared, or differently treated, samples of polymeric materials show different degrees of crystallinity. A particularly marked difference is shown by isotactic and atactic polymers. The x-ray-diffraction patterns of such types of polypropylene are shown in Fig. 25-29. The isotactic material gives a pattern that corresponds to those mentioned in Chap. 14 in connection with x-ray-powder patterns. The isotactic polymer consists, therefore, of randomly oriented, small crystalline regions. The diffuseness of the pattern from the atactic material indicates that in this form of the polymer there is little crystallinity and the material is essentially amorphous.

The broadening of the diffraction lines that is observed in patterns from partly crystalline polymeric material can be used to estimate the size of the crystalline regions. The broadening results from interference effects that lead to only incomplete constructive and destructive interference. In this way crystallite sizes in the range of tens to hundreds of angstoms have been deduced.

From such measurements, and others, one is led to a diagram for a typical polymeric material, as shown in Fig. 25-30. The amount and size of the crystallites are expected to vary with the particular polymer and with its physical treatment.

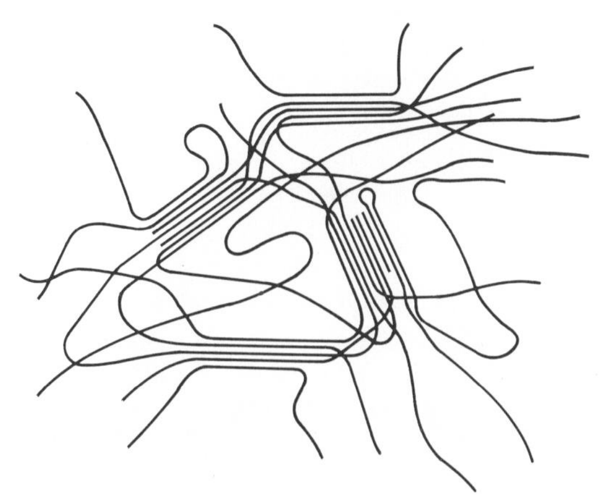

FIGURE 25-30
The arrangement of molecular chains in a partly crystalline linear polymeric material.

Other physical properties can also be used to deduce the degree of crystallinity. The heat of fusion of a polymeric material, for instance, can be compared with that expected for a completely crystalline material. A similar use can be made of the density of the polymer as compared with that expected for crystalline material and for that of completely amorphous material as found in a liquid hydrocarbon.

These ideas on the molecular nature of these synthetic polymer materials are the basis on which the physical properties of the materials are to be understood. The elasticity, for example, can be interpreted in terms of a realignment of the molecular chains in the amorphous regions and of the crystallites, and the x-ray pattern of the stretched material shows, in fact, a similarity to fiber patterns, indicating an ordering of the directions of the molecular chains.

Alterations in the structure of the polymer, such as cross linking produced by irradiation of the material with high-energy radiation, produce molecular changes that lead to marked changes in the physical properties, of which the rigidity imposed by cross linking is probably the most important. The detailed attempts that are being made to understand the physical properties in terms of the molecular configurations cannot, however, be treated here.

25-18 Electron Microscopy

The most straightforward way to investigate the structure of macromolecules would be to make use of a microscope of sufficient resolving power. No progress in this direction can be hoped for by using visible radiation, as does an ordinary microscope, because the wavelength of light, about 6000 Å, is longer than the details of the particles which one hopes to observe. All that one sees is that the molecule acts as a scattering center, a principle used in the ultramicroscope. The suggestion that electrons have a wave nature leads not only to their use in electron-diffraction experiments treated in Chap. 14, but also to their use in place of ordinary electromagnetic radiation in a microscope. According to Eq. [8] of Chap. 14, the wavelength of the electron beam can be made less than 1 Å with fairly readily available voltages of around 10,000 V. It follows, therefore, that such radiation is in principle capable of resolving details of structure almost down to the range at which diffraction methods are effective.

Early developments following the suggestion of the wave nature of electrons showed that a beam of electrons could be focused by electric and magnetic fields. It then became possible to construct a microscope using an electron beam rather than electromagnetic radiation as used in an optical microscope. In view of the previous discussion of electron diffraction, it is apparent that the system must be evacuated so that the electron beam is not diffracted by molecules of air in the system. This requirement and the high energy of the electron beam impose some frequently troublesome restrictions on the nature of the samples that can be studied.

The electron beam passes through the sample, supported on a thin film or screen, and the differences in scattering power of the different parts of the sample lead to a photograph of the transmitted beam that indicates the structure of the sample, insofar as the structure is related to variations in electron scattering power. Under the most favorable conditions, magnifications of 30,000 are possible, giving resolution of structural details down to about 10 Å. In most cases, details with dimensions of 100 or 1000 Å are all that can be satisfactorily recognized. The method, however, is still being developed, and in many macromolecule systems the electron microscope can reveal structural details that are not far removed from structures that can be understood from the molecular results of diffraction experiments.

Problems

1 The molecular masses of a synthetic polymer are distributed according to the expression

$$\frac{1}{N}\frac{dN}{dM} = 0.61 \times 10^{-4} e^{-[(M-10,000)/10,000]^2}$$

where N is the number of molecules with molecular mass M.

Plot the molecular-mass-distribution curve.

Determine graphically the number and mass-average molecular masses.

Ans. M (number av) = 11,000; M (mass av) = 14,500.

2 From the diffusion coefficients of Table 25-5, estimate the radii of the listed macromolecules, using Eq. [12]. The viscosity of water at 25°C is 0.00891 poise, and 1 poise = 10^{-1} kg m^{-1} s^{-1}.

Compare the molecular weights that would be obtained by this method with those listed in Table 25-5. *Ans.* For insulin, M = 97,000 g.

3 A silver chloride–aqueous colloidal solution was examined in an ultramicroscope. In a field of view 0.05 mm diameter and 0.05 mm depth, an average of 8.4 particles were counted.

If the solution had been prepared by the dilution of a solution containing 0.0032 g of AgCl per milliliter by a factor of 1 to 10,000, what was the average mass of the AgCl particles? Is this a number or mass average?

4 The molecular mass of egg albumin is about 40,000 g. What is the freezing-point depression, and what are the vapor-pressure lowering and the osmotic pressure at 25°C of an aqueous solution containing 10 g liter^{-1}?

5 A sample of γ globulin gives the following experimental results at 20°C:

Specific volume v	0.178 ml g^{-1}
Sedimentation coeff. S	7.12×10^{-13} s
Diffusion coeff. D	4.00×10^{-11} m^2 s^{-1}

Calculate the molecular mass of γ globulin.

References

FLORY, P. J.: "Principles of Polymer Chemistry," Cornell University Press, Ithaca, N.Y., 1953. The nature, synthesis, and characterization of polymeric materials. The emphasis is on synthetic polymers and on studies and properties of solutions of these polymeric materials.

GIMBLETT, F. G. R.: "Inorganic Polymer Chemistry," Butterworth & Co. (Publishers), Ltd., London, 1963. The chemical nature and synthesis of inorganic polymers, with some discussions of the properties and structures of these materials.

DICKERSON, R. E., and I. GEIS: "The Structure and Action of Proteins," Harper & Row, Publishers, Incorporated, New York, 1969. A magnificent introduction to the structures of proteins and the way these structures are related to some of the enzyme actions of these molecules.

SHAKMANN, M. A. (ed.): "Polyamino Acids, Polypeptides, and Proteins," University of Wisconsin Press, Madison, Wis., 1962. A summary of recent and current, as of 1961, research on various aspects of polymeric amino acid materials. The collection provides an illustration of the uses to which physical-chemical methods are put in this area of macromolecule studies.

STEINER, R. F., and R. F. BEERS, JR.: "Polynucleotides," Elsevier Publishing Company, Amsterdam, 1961. Included in this monograph are summaries of the physical methods that are of value for these compounds and a discussion of the results of application of these methods. An illustration of the extension of the material of this chapter for one type of macromolecule is thus provided.

APPENDIX 1

EVALUATION OF INTEGRALS OF THE TYPE $\int_0^\infty x^n e^{-ax^2}\,dx$

a. *Reduction of the integrals.* Integration by parts, using the general expression

$$\int u\,dv = uv] - \int v\,du$$

can be carried out with

$$dv = xe^{-ax^2}\,dx \qquad \text{and thus} \qquad v = -\frac{1}{2a}e^{-ax^2}$$

and

$$u = x^{n-1} \qquad \text{and thus} \qquad du = (n-1)x^{n-2}\,dx$$

With these identifications, integration by parts gives

$$\int_0^\infty x^n e^{-ax^2}\,dx = -\frac{x^{n-1}}{2a}e^{-ax^2}\Big]_0^\infty - \int_0^\infty -\frac{1}{2a}(n-1)x^{n-2}e^{-ax^2}\,dx$$

$$= 0 + \frac{n-1}{2a}\int_0^\infty x^{n-2}e^{-ax^2}\,dx$$

In this way integrals of the general type can be reduced either to one involving the integral xe^{-ax^2} or to one involving the integral e^{-ax^2}.

b. *For n odd,* the method of part a leads to the integral

$$\int_0^\infty xe^{-ax^2}\,dx$$

which can be integrated directly to give

$$-\frac{1}{2a}e^{-ax^2}\Big]_0^\infty = \frac{1}{2a}$$

c. *For n even,* the method of part a leads to the integral

$$\int_0^\infty e^{-ax^2}\,dx$$

Evaluation is accomplished by writing the product of two such integrals based on two independent coordinates x and y which can be pictured as being cartesian coordinates. We thus investigate

$$\int_0^\infty e^{-ax^2}\,dx \int_0^\infty e^{-ay^2}\,dy = \int_0^\infty \int_0^\infty e^{-a(x^2+y^2)}\,dx\,dy$$

The coordinates x and y can now be related to new coordinates r and θ by the relations which associate cartesian coordinates with polar coordinates. Thus

$$x^2 + y^2 = r^2$$

and

$$dx\,dy = r\,dr\,d\theta$$

Furthermore, since in the x, y coordinate system the integration is over the first quadrant, the corresponding limits of the integration involving r and θ are 0 to ∞ and 0 to $\pi/2$. We thus have the problem, in r and θ, of

$$\int_0^{\pi/2} \int_0^{\infty} e^{-ar^2} r\,dr\,d\theta$$

Now the two integrations can be performed to give $(\pi/2)(1/2a)$, and therefore the original integral is evaluated as

$$\int_0^{\infty} e^{-ax^2}\,dx = \frac{1}{2}\sqrt{\frac{\pi}{a}}$$

d. *Even and odd character and the limits of integration.* Functions of the type $x^n e^{-ax^2}$ are even if n is even and odd if n is odd. Integration from $-\infty$ to $+\infty$ will therefore give zero for n odd and will give twice the value of the integral from 0 to ∞ for n even.

e. *Values of most often used integrals:*

$$\int_{-\infty}^{+\infty} e^{-ax^2}\,dx = \sqrt{\frac{\pi}{a}} \qquad \int_0^{\infty} x e^{-ax^2}\,dx = \frac{1}{2a}$$

$$\int_{-\infty}^{+\infty} x^2 e^{-ax^2}\,dx = \frac{1}{2a}\sqrt{\frac{\pi}{a}} \qquad \int_0^{\infty} x^3 e^{-ax^2}\,dx = \frac{1}{2a^2}$$

$$\int_{-\infty}^{+\infty} x^4 e^{-ax^2}\,dx = \frac{3}{4a^2}\sqrt{\frac{\pi}{a}}$$

APPENDIX 2

STIRLING'S APPROXIMATION

One form of Stirling's approximation for the value of the factorial of a large number can be derived as follows: One considers the natural logarithm of the factorial and expresses this as a summation. Thus

$$\ln N! = \ln N + \ln (N-1) + \ln (N-2) + \cdots + \ln 2 + \ln 1$$

$$= \sum_{i=1}^{N} \ln i$$

For large N, the argument of Appendix 4 can be used to replace the sum by an integral to obtain

$$\ln N! \approx \int_1^N \ln x\, dx$$

where x is a continuous function.

Integration by parts, with $u = \ln x$, $du = d \ln x = (1/x)\, dx$, and $dv = dx$, $v = x$, leads to the result

$$\ln N! \approx x \ln x \Big]_1^N - \int_1^N x \frac{1}{x}\, dx$$
$$\approx N \ln N - N$$

Although this logarithmic form is often the one used, we can also write

$$N! \approx N^N e^{-N}$$

Finally, it should be mentioned that a better approximation, not derived here, is

$$N! \approx N^N e^{-N} (2N)^{1/2}$$

APPENDIX 3

THE METHOD OF LAGRANGE MULTIPLIERS

An extremum of a function $f(x_1, x_2, \ldots, x_n)$ of a set of n variables $x_1, x_2, \ldots, x_n$ exists when df is zero for variation of any of the x_i's. Since

$$df = \frac{\partial f}{\partial x_1} dx_1 + \frac{\partial f}{\partial x_2} dx_2 + \cdots + \frac{\partial f}{\partial x_n} dx_n$$
$$= \sum \frac{\partial f}{\partial x_i} dx_i \qquad [1]$$

and, for an extremum, df must be zero for all values of $dx_1, dx_2, \ldots, dx_n$, an extremum will occur when all the coefficients of the dx_i are zero. Thus an extremum can be found by writing the n equations

$$\frac{\partial f}{\partial x_i} = 0$$
$$\frac{\partial f}{\partial x_2} = 0$$
$$\vdots \qquad [2]$$
$$\frac{\partial f}{\partial x_n} = 0$$

and solving for the values of $x_1, x_2, \ldots, x_n$ that satisfy this set of n equations.

When, however, the x_i's cannot all be independently varied, the dx_i's cannot take on any value and we cannot require all $\partial f/\partial x_i$'s to be zero. We cannot proceed by the method outlined above to solve for the values of $x_1, x_2, \ldots x_n$ that locate an extremum.

The procedure which locates an extremum in a function such as $f(x_1, x_2, \ldots, x_n)$ when variations in the x_i's are subject to certain constraints depends on the introduction of what are known as *Lagrange multipliers.* If, to be specific, two constraining equations

$$g(x_1, x_2, \ldots, x_n) = 0$$

and [3]

$$h(x_1, x_2, \ldots, x_n) = 0$$

are assumed, the Lagrange method asserts that solution of the set of $n + 2$ equations

$$\frac{\partial f}{\partial x_i} + \alpha\frac{\partial g}{\partial x_i} + \beta\frac{\partial h}{\partial x_i} = 0 \qquad \text{for } i = 1, 2, \ldots, n \qquad [4]$$

$$g(x_1, x_2, \ldots, x_n) = 0 \qquad [5]$$

$$h(x_1, x_2, \ldots, x_n) = 0 \qquad [6]$$

will yield values of $x_1, x_2, \ldots, x_n$ that locate an extremum subject to the constraints. The α and β are *undetermined multipliers* that are constants, not functions of the x_i.

That this procedure locates an extremum for a well-behaved function can now be shown.

Solutions that satisfy Eq. [4] allow the partial derivatives $\partial f/\partial x_i$ to be written

$$\frac{\partial f}{\partial x_i} = -\alpha\frac{\partial g}{\partial x_i} - \beta\frac{\partial h}{\partial x_i} \qquad \text{for all } i$$

These expressions can now be inserted into the total differential expression for df to give

$$df = -\alpha\sum\frac{\partial g}{\partial x_i}dx_i - \beta\sum\frac{\partial h}{\partial x_i}dx_i \qquad [7]$$

Finally, if the constraining equations, Eqs. [5] and [6], are satisfied,

$$\sum\frac{\partial g}{\partial x_i}dx_i = 0 \qquad \text{and} \qquad \sum\frac{\partial h}{\partial x_i}dx_i = 0$$

and it follows from the substitution of these in Eq. [7] that $df = 0$.

Thus solution of Eqs. [4] to [6] for $x_1, x_2, \ldots, x_n$ locates an extremum when the x_i's are subject to constraints such as those of the two expressions $g = 0$ and $h = 0$ used in this illustration. Whether the extremum is a maximum, minimum, or point of inflection is usually determined from physical considerations. The values of the undetermined multipliers, α and β in the example used here, may turn out to be of interest, but their determination is not central to the problem of locating an extremum.

APPENDIX 4

REPLACING A SUM BY AN INTEGRAL

Let us consider the condition under which a summation of terms, each term determined or indexed by the value of an integer, can be replaced by an integration.

Suppose a quantity A is given by a summation of terms according to

$$A = a(1) + a(2) + a(3) + \cdots + a(n)$$

$$= \sum_{i=1}^{n} a(i) \qquad [1]$$

where $a(i)$ implies a term whose value is determined by assigning an integral value to i.

Although only integral values of i are used to generate values for the terms of the summation, we can deal with i as a continuous variable. Then the summation over the range of i's can be written, in a way that turns out to be profitable, as

$$A = a(1) \int_0^1 di + a(2) \int_1^2 di + \cdots + a(n) \int_{n-1}^{n} di$$

$$= \int_0^1 a(1)\, di + \int_1^2 a(2)\, di + \cdots + \int_{n-1}^{n} a(n)\, di \qquad [2]$$

Now, if each $a(i)$ does not vary very much in the unit range of i, that is, in the range of integration of each term, the constant values of $a(1)$, $a(2)$, and so on, obtained by using integers, can be replaced by the varying functions $a(i)$ obtained by using the continuous variable i in each term. When this condition holds, i.e., the variation in each term is small compared with the total variation in the values of the $a(i)$ terms, we can rewrite Eq. [2] as

$$A = \int_0^1 a(i)\, di + \int_1^2 a(i)\, di + \cdots + \int_{n-1}^{n} a(i)\, dn$$

$$= \int_0^n a(i)\, di \qquad [3]$$

By this development, the conditions under which a sum can be calculated by means of the corresponding integral have been displayed.

APPENDIX 5

THERMODYNAMIC PROPERTIES OF SUBSTANCES AT 1 ATM PRESSURE AND 25°C IN THE PHYSICAL STATE INDICATED

Note that S° and C_P° are the entropies and heat capacities of the listed substances. The quantities ΔH_f° and ΔG_f° are the values for the compounds compared with the elements of which they are composed. The ΔG_f° values are related to ΔH_f° and S° values by the relation $\Delta G_f^\circ = \Delta H_f^\circ - T\,\Delta S_f^\circ$, where ΔS_f° can be calculated from the listed entropies of the compound and the elements of which it is composed. Physical states are indicated by (*c*) for crystalline, (*l*) for liquid, and (*g*) for gas.

	Element or compound	ΔH_f° (kJ mol^{-1})	S° (J mol^{-1} deg^{-1})	ΔG_f° (kJ mol^{-1})	C_p° (J mol^{-1} deg^{-1})
	$H_2(g)$	0.0	130.59	0.0	28.84
	$H(g)$	217.94	114.61	203.24	20.79
Group 0	$He(g)$	0.0	126.06	0.0	20.79
	$Ne(g)$	0.0	144.14	0.0	20.79
	$Ar(g)$	0.0	154.72	0.0	20.79
	$Kr(g)$	0.0	163.97	0.0	20.79
	$Xe(g)$	0.0	169.58	0.0	20.79
	$Rn(g)$	0.0	176.15	0.0	20.79
Group 1	$Li(c)$	0.0	28.03	0.0	23.64
	$Li(g)$	155.10	138.67	122.13	20.79
	$Li_2(g)$	199.2	196.90	157.32	35.65
	$Li_2O(c)$	−595.8	37.91	−560.24	
	$LiH(g)$	128.4	170.58	105.4	29.54
	$LiCl(c)$	−408.78	(55.2)	−383.7	
	$Na(c)$	0.0	51.0	0.0	28.41
	$Na(g)$	108.70	153.62	78.11	20.79
	$Na_2(g)$	142.13	230.20	103.97	
	$NaO_2(c)$	−259.0		−194.6	
	$Na_2O(c)$	−415.9	72.8	−376.6	68.2
	$Na_2O_2(c)$	−504.6	(66.9)	−430.1	
	$NaOH(c)$	−426.73	(523)	−377.0	80.3
	$NaCl(c)$	−411.00	72.4	−384.0	49.71
	$NaBr(c)$	−359.95		−347.6	
	$Na_2SO_4(c)$	−1384.49	149.49	−1266.83	127.61

(Continued)

	Element or compound	ΔH_f° (kJ mol^{-1})	S° (J mol^{-1} deg^{-1})	ΔG_f° (kJ mol^{-1})	C_p° (J mol^{-1} deg^{-1})
	$Na_2SO_4 \cdot 10H_2O(c)$	−4324.08	592.87	−3643.97	587.4
	$NaNO_3(c)$	−466.68	116.3	−365.89	93.05
	$Na_2CO_3(c)$	−1130.9	136.0	−1047.7	110.50
	$K(c)$	0.0	63.6	0.0	29.16
	$K(g)$	90.0	160.23	61.17	20.79
	$K_2(g)$	128.9	249.75	92.5	
	$K_2O(c)$	−361.5		−318.8	
	$KOH(c)$	−425.85		−374.5	
	$KCl(c)$	−435.87	82.67	−408.32	51.50
	$KMnO_4(c)$	−813.4	171.71	−713.79	119.2
Group 2	$Be(c)$	0.0	9.54	0.0	17.82
	$Mg(c)$	0.0	32.51	0.0	23.89
	$MgO(c)$	−601.83	26.8	−569.57	37.40
	$Mg(OH)_2(c)$	−924.66	63.14	−833.74	77.03
	$MgCl_2(c)$	−641.82	89.5	−592.32	71.30
	$Ca(c)$	0.0	41.63	0.0	26.27
	$CaO(c)$	−635.09	39.7	−604.2	42.80
	$CaF_2(c)$	−1214.6	68.87	−1161.9	67.02
	$CaCO_3$(c,calcite)	−1206.87	92.9	−1128.76	81.88
	$CaSiO_3(c)$	−1584.1	82.0	−1498.7	85.27
	$CaSO_4$(c,anhydrite)	−1432.68	106.7	−1320.30	99.6
	$CaSO_4 \cdot \frac{1}{2}H_2O(c)$	−1575.15	130.5	−1435.20	119.7
	$CaSO_4 \cdot 2H_2O(c)$	−2021.12	193.97	−1795.73	186.2
	$Ca_3(PO_4)_2(c)$	−4137.5	236.0	−3899.5	227.82
Group 3	$B(c)$	0.0	6.53	0.0	11.97
	$B_2O_3(c)$	−1263.6	54.02	−1184.1	62.26
	$B_2H_6(g)$	31.4	232.88	82.8	56.40
	$B_5H_9(g)$	62.8	275.64	165.7	80
	$Al(c)$	0.0	28.32	0.0	24.34
	$Al_2O_3(c)$	−1669.79	50.99	−1576.41	78.99
Group 4	C(c,diamond)	1.90	2.44	2.87	6.06
	C(c,graphite)	0.0	5.69	0.0	8.64
	$C(g)$	718.38	157.99	672.97	20.84
	$CO(g)$	−110.52	197.91	−137.27	29.14
	$CO_2(g)$	−393.51	213.64	−394.38	37.13
	$CH_4(g)$	−74.85	186.19	−50.79	35.71
	$C_2H_2(g)$	226.75	200.82	209.2	43.93
	$C_2H_4(g)$	52.28	219.45	68.12	43.55
	$C_2H_6(g)$	−84.67	229.49	−32.89	52.65
	$C_6H_6(g)$	82.93	269.20	129.66	81.67

(Continued)

	Element or compound	ΔH_f° (kJ mol^{-1})	S° (J mol^{-1} deg^{-1})	ΔG_f° (kJ mol^{-1})	C_p° (J mol^{-1} deg^{-1})
	$C_6H_6(l)$	49.03	124.50	172.80	
	$CH_3OH(g)$	−201.25	237.6	−161.92	
	$CH_3OH(l)$	−238.64	126.8	−166.31	81.6
	$C_2H_5OH(l)$	−277.63	160.7	−174.76	111.46
	$CH_3CHO(g)$	−166.35	265.7	−133.72	62.8
	$HCOOH(l)$	−409.2	128.95	−346.0	99.04
	$(COOH)_2(c)$	−826.7	120.1	−697.9	109
	$HCN(g)$	130.5	201.79	120.1	35.90
	$CO(NH_2)_2(c)$	−333.19	104.6	−197.15	93.14
	$CS_2(l)$	87.9	151.04	63.6	75.7
	$CCl_4(g)$	−106.69	309.41	−64.22	83.51
	$CCl_4(l)$	−139.49	214.43	−68.74	131.75
	$CH_3Cl(g)$	−81.92	234.18	−58.41	40.79
	$CH_3Br(g)$	−34.3	245.77	−24.69	42.59
	$CHCl_3(g)$	−100	296.48	−67	65.81
	$CHCl_3(l)$	−131.8	202.9	−71.5	116.3
	$Si(c)$	0.0	18.70	0.0	19.87
	$SiO_2(c,\text{quartz})$	−859.4	41.84	−805.0	44.43
Group 5	$N_2(g)$	0.0	191.49	0.0	29.12
	$N(g)$	472.64	153.19	455.51	20.79
	$NO(g)$	90.37	210.62	86.69	29.86
	$NO_2(g)$	33.85	240.45	51.84	37.91
	$N_2O(g)$	81.55	219.99	103.60	
	$N_2O_4(g)$	9.66	304.30	98.29	38.71
	$N_2O_5(c)$	−41.84	113.4	133	79.08
	$NH_3(g)$	−46.19	192.51	−16.63	35.66
	$NH_4Cl(c)$	−315.39	94.6	−203.89	84.1
	$HNO_3(l)$	−173.23	155.60	−79.91	109.87
	$P(c,\text{white})$	0.0	44.0	0.0	23.22
	$P(c,\text{red})$	−18.4	(29.3)	−13.8	
	$P_4(g)$	54.89	279.91	24.35	66.9
	$P_4O_{10}(c)$	−3012.5			
	$PH_3(g)$	9.25	210.0	18.24	
Group 6	$O_2(g)$	0.0	205.03	0.0	29.36
	$O(g)$	247.52	160.95	230.09	21.91
	$O_3(g)$	142.2	237.6	163.43	38.16
	$H_2O(g)$	−241.83	188.72	−228.59	33.58
	$H_2O(l)$	−285.84	69.94	−237.19	75.30
	$H_2O_2(l)$	−187.61	(92)	−113.97	
	$S(c,\text{rhombic})$	0.0	31.88	0.0	22.59
	$S(c,\text{monoclinic})$	0.30	32.55	0.10	23.64

(Continued)

	Element or compound	ΔH_f° (kJ mol^{-1})	S° (J mol^{-1} deg^{-1})	ΔG_f° (kJ mol^{-1})	C_p° (J mol^{-1} deg^{-1})
	$SO(g)$	79.58	221.92	53.47	
	$SO_2(g)$	−296.06	248.52	−300.37	39.79
	$SO_3(g)$	−395.18	256.22	−370.37	50.63
	$H_2S(g)$	−20.15	205.64	−33.02	33.97
	$SF_6(g)$	−1096	290.8	−992	
Group 7	$F_2(g)$	0.0	203.3	0.0	31.46
	$HF(g)$	268.6	173.51	−270.7	29.08
	$Cl_2(g)$	0.0	222.95	0.0	33.93
	$HCl(g)$	−92.31	186.68	−95.26	29.12
	$Br_2(l)$	0.0	152.3	0.0	
	$Br_2(g)$	30.71	245.34	3.14	35.98
	$HBr(g)$	−36.23	198.48	−53.22	29.12
	$I_2(c)$	0.0	116.7	0.0	54.98
	$I_2(g)$	62.24	260.58	19.37	36.86
	$HI(g)$	25.9	206.33	1.30	29.16
Transition metals	$Pb(c)$	0.0	64.89	0.0	26.82
	$Zn(c)$	0.0	41.63	0.0	25.06
	$ZnS(c$,sphalerite)	−202.9	57.74	−198.3	45.2
	$ZnS(c$,wurtzite)	−189.5	(57.74)	−242.5	
	$Hg(l)$	0.0	77.4	0.0	27.82
	$HgO(c$,red)	−90.71	72.0	−58.53	45.73
	$HgO(c$,yellow)	−90.21	73.2	−58.40	
	$HgCl_2(c)$	−230.1	(144.3)	−185.8	
	$Hg_2Cl_2(c)$	−264.93	195.8	−210.66	101.7
	$Cu(c)$	0.0	33.30	0.0	24.47
	$CuO(c)$	−155.2	43.51	−127.2	44.4
	$Cu_2O(c)$	−166.69	100.8	−146.36	69.9
	$CuSO_4(c)$	−769.86	113.4	−661.9	100.8
	$CuSO_4 \cdot 5H_2O(c)$	−2277.98	305.4	−1879.9	281.2
	$Ag(c)$	0.0	42.70	0.0	25.49
	$Ag_2O(c)$	−30.57	121.71	−10.82	65.56
	$AgCl(c)$	−127.03	96.11	−109.72	50.79
	$AgNO_3(c)$	−123.14	140.92	−32.17	93.05
	$Fe(c)$	0.0	27.15	0.0	25.23
	$Fe_2O_3(c$,hematite)	−822.2	90.0	−741.0	104.6
	$Fe_3O_4(c$,magnetite)	−1120.9	146.4	−1014.2	
	$Mn(c)$	0.0	31.76	0.0	26.32
	$MnO_2(c)$	−519.6	53.1	−466.1	54.02

SOURCE: Values mostly from Selected Values of Chemical Thermodynamic Properties, *Natl. Bur. Std. Circ.* 500, 1952.

APPENDIX 6

THERMODYNAMIC PROPERTIES OF SUBSTANCES IN AQUEOUS SOLUTION AT UNIT ACTIVITY AND 25°C

Note that the values are for an effective concentration of 1 molar, that is, unit activities. Particularly for the ionic species, this can be somewhat different from 1-*M* concentration. Ionic properties are based on the assignment of zero value for ΔH_f°, ΔG_f°, and S° for $H^+(aq)$. (aq) = aqueous

	Species in solution	ΔH_f° (kJ mol^{-1})	S° (J mol^{-1} deg^{-1})	ΔG_f° (kJ mol^{-1})
	$H^+(aq)$	0.0	0.0	0.0
	$H_3O^+(aq)$	−285.85	69.96	−237.19
	$OH^-(aq)$	−229.95	−10.54	−157.27
Group 1	$Li^+(aq)$	−278.44	14.2	−293.80
	$Na^+(aq)$	−239.66	60.2	−261.88
	$K^+(aq)$	−251.21	102.5	−282.25
Group 2	$Be^{++}(aq)$	−389		−356.48
	$Mg^{++}(aq)$	−461.95	−118.0	−456.01
	$Ca^{++}(aq)$	−542.96	−55.2	−553.04
Group 3	$H_3BO_3(aq)$	−1067.8	159.8	−963.32
	$H_2BO_3^-(aq)$	−1053.5	30.5	−910.44
Group 4	$CO_2(aq)$	−412.92	121.3	−386.22
	$H_2CO_3(aq)$	−698.7	191.2	−623.42
	$HCO_3^-(aq)$	−691.11	95.0	−587.06
	$CO_3^{--}(aq)$	−676.26	−53.1	−528.10
	$CH_3COOH(aq)$	−488.44		−399.61
	$CH_3COO^-(aq)$	−488.86		−372.46
Group 5	$NH_3(aq)$	−80.83	110.0	−26.61
	$NH_4^+(aq)$	−132.80	112.84	−79.50
	$HNO_3(aq)$	−206.56	146.4	−110.58
	$NO_3^-(aq)$	−206.56	146.4	−110.58
	$H_3PO_4(aq)$	−1289.5	176.1	−1147.2
	$H_2PO_4^-(aq)$	−1302.5	89.1	−1135.1

(Continued)

	Species in solution	ΔH_f° (kJ mol^{-1})	S° (J mol^{-1} deg^{-1})	ΔG_f° (kJ mol^{-1})
	$HPO_4^{--}(aq)$	−1298.7	−36.0	−1094.1
	$PO_4^{3-}(aq)$	−1284.1	−218	−1025.5
Group 6	$H_2S(aq)$	−39.3	122.2	−27.36
	$HS^-(aq)$	−17.66	61.1	12.59
	$S^{--}(aq)$	41.8		83.7
	$H_2SO_4(aq)$	−907.51	17.1	−741.99
	$HSO_4^-(aq)$	−885.75	126.85	−752.86
	$SO_4^{--}(aq)$	−907.51	17.1	−741.99
Group 7	$F^-(aq)$	−329.11	−9.6	−276.48
	$HCl(aq)$	−167.44	55.2	−131.17
	$Cl^-(aq)$	−167.44	55.2	−131.17
	$ClO^-(aq)$		43.1	−37.2
	$ClO_2^-(aq)$	−69.0	100.8	−10.71
	$ClO_3^-(aq)$	−98.3	163	−2.60
	$ClO_4^-(aq)$	−131.42	182.0	−8
	$Br^-(aq)$	−120.92	80.71	−102.80
	$I_2(aq)$	20.9		16.44
	$I_3^-(aq)$	−51.9	173.6	−51.50
	$I^-(aq)$	−55.94	109.36	−51.67
Transition metals	$Cu^+(aq)$	(51.9)	(−26.4)	50.2
	$Cu^{++}(aq)$	64.39	−98.7	64.98
	$Cu(NH_3)_4^{++}(aq)$	(−334.3)	806.7	−256.1
	$Zn^{++}(aq)$	−152.42	−106.48	−147.19
	$Pb^{++}(aq)$	1.63	21.3	−24.31
	$Ag^+(aq)$	105.90	73.93	77.11
	$Ag(NH_3)_2^+(aq)$	−111.80	241.8	−17.40
	$Ni^{++}(aq)$	(−64.0)		−48.24
	$Ni(NH_3)_6^{++}(aq)$			−251.4
	$Ni(CN)_4^{--}(aq)$	363.6	(138.1)	489.9
	$Mn^{++}(aq)$	−218.8	−84	−223.4
	$MnO_4^-(aq)$	−518.4	189.9	−425.1
	$MnO_4^{--}(aq)$			−503.8
	$Cr^{++}(aq)$			−176.1
	$Cr^{3+}(aq)$		−307.5	−215.5
	$Cr_2O_7^{--}(aq)$	−1460.6	213.8	−1257.3
	$CrO_4^{--}(aq)$	−894.33	38.5	−736.8

SOURCE: Values from Selected Values of Chemical Thermodynamic Properties, *Natl. Bur. Std. Circ.* 500, 1953.

APPENDIX 7

SI UNITS

Some aspects of the system of units, known as SI units, (for *Système International d'Unités*) that are likely to be encountered in physical-chemistry studies are given here. For more complete tables and for the rationale of the system, the following references can be consulted.

References

TAYLOR, B. N., D. N. LANGENBERG, and W. H. PARKER: The Fundamental Physical Constants, *Sci. Am.*, **223**(4):62 (1970).

Policy for NBS usage of SI units, *J. Chem. Educ.*, **48**:569 (1971).

NORRIS, A. C.: SI Units in Physico-chemical Calculations, *J. Chem. Educ.*, **48**:797 (1971).

(a) Values of some physical and chemical constants (in SI units and based on the ^{12}C scale)

Avogadro's number	$\mathfrak{N} = 6.0222 \times 10^{23}$ mol^{-1}
Velocity of light	$c = 2.997925 \times 10^{8}$ m s^{-1}
Mass of electron	$m = 0.91096 \times 10^{-30}$ kg
Electronic charge	$e = 0.16022 \times 10^{-18}$ C
Faraday	$\mathfrak{F} = \mathfrak{N}e = 96{,}490$ C mol^{-1}
Planck's constant	$h = 0.66262 \times 10^{-33}$ J s^{-1}
Boltzmann constant	$k = 1.3806 \times 10^{-23}$ J deg^{-1}
Gas constant	$R = \mathfrak{N}k = 8.3143$ J deg^{-1} mol^{-1}
	$= 0.08206$ liter atm deg^{-1} mol^{-1}
Atmospheric pressure	1 atm $= 101{,}325$ N m^{-2}
Standard molar gas volume	22.415 liters
Absolute zero	$-273.15°$C
Permittivity	$4\pi\epsilon_0 = 1.11264 \times 10^{-10}$ C^2 N^{-1} m^{-2}

(b) Basic SI units

Physical quantity	Name of unit	Symbol
Length	meter	m
Mass	kilogram	kg
Time	second	s
Electric current	ampere	A
Thermodynamic temperature	degree Kelvin	K
Amount of substance	mole	mol

(c) Derived SI units

Physical quantity	SI name or special name and symbol	SI symbol
Area	square meter	m^2
Volume	cubic meter	m^3
Density	kilogram per cubic meter	$kg\ m^{-3}$
Velocity	meter per second	$m\ s^{-1}$
Angular velocity	radian per second	$rad\ s^{-1}$
Acceleration	meter per second squared	$m\ s^{-2}$
Force	newton (N)	$kg\ m\ s^{-2} = J\ m^{-1}$
Pressure	newton per square meter	$N\ m^{-2}$
Energy	joule (J)	$kg\ m^2\ s^{-2} = N\ m$
Power	watt (W)	$kg\ m^2\ s^{-3} = J\ s^{-1}$
Electric charge	coulomb (C)	$A\ s$
Electric potential difference	volt (V)	$kg\ m^2\ s^{-3}\ A^{-1} = J\ A^{-1}\ s^{-1}$
Electric field strength	volt per meter	$V\ m^{-1}$
Electric resistance	ohm (Ω)	$kg\ m^2\ s^{-3}\ A^{-2} = V\ A^{-1}$
Electric capacitance	farad (F)	$A^2\ s^4\ kg^{-1}\ m^{-2} = A\ s\ V^{-1}$

(d) Examples of non-SI units

Physical quantity	Name	SI equivalent
Length	angstrom (Å)	10^{-10} m
	inch (in)	0.0254 m
	foot (ft)	0.3048 m
	mile	1,609 m
Volume	liter	$10^{-3}\ m^3$
Mass	pound (lb)	0.4535924 kg
Force	dyne (dyn)	10^{-5} N
	poundal	0.138255 N
Pressure	atmosphere (atm)	101,325 $N\ m^{-2}$
	torr (mm Hg)	133.322 Nm^{-2}
	bar	$10^5\ N\ m^{-2}$
Energy	erg	10^{-7} J
	calorie (cal)	4.1840 J
	electron volt (eV)	0.16021×10^{-18} J
Power	horsepower (hp)	745.700 W
Viscosity	poise	$10^{-1}\ kg\ m^{-1}\ s^{-1}$
Dipole moment	debye	3.338×10^{-30} m C

(e) Fractions and multiples*

	Prefix	Symbol
10^{12}	tera	T
10^{9}	giga	G
10^{6}	mega	M
10^{3}	kilo	k
10^{-3}	milli	m
10^{-6}	micro	μ
10^{-9}	nano	n
10^{-12}	pico	p

*The prefix deci for 10^{-1} and centi for 10^{-2} are also recognized—particularly in the often used centimeter—but their use is being discouraged.

INDEX

Absolute temperature, 4
Absolute zero, 181–184
Absorption coefficient, 476
Absorption spectra:
  rotational, 312–314
  vibrational, 316–321
Acetaldehyde, nmr spectrum of, 337, 340
Acetic acid, degree of dissociation, 624
Acetylene, bonding in, 293
Activation energy, 458
  for viscous flow, 537
Activity coefficient:
  in concentrated solutions, 694–696
  from electrolytic dissociation, 688–690
  from emf measurements, 672–674
  of gases, 203–205
  from solubilities, 690–692
  of solutes, 594–598
  of solvents, 592–594
  table of, for electrolytes, 693
Adiabatic demagnetization, 182–184
Adiabatic process, 130
  expansion of ideal gas, 132–134
Adsorption, 704–719
  of gases on solids, 708
  from solution, 716
Adsorption isotherm, 710–716
Air, pressure-volume relation, 2
Alberty, R. A., 604, 686
Allowed energies:
  from Schrödinger equation, 72–73
  for three-dimensional square well, 77
Alloys, 562
Amalgam electrodes, 661
Amino acids, 733–734
Ammonia, decomposition of, 724
Ampere's law, 413
Angstrom, 49, 61
Angular factor:
  for d orbitals, 300
  of hydrogen-atom wave function, 244–247
Angular momentum:
  of atomic states, 248–250
  and quantum restrictions, 77–80
Angular velocity, 78, 310
Anode, 629
Anthracene, electron density map, 385
Antibonding orbitals, 271
Appearance potential, 485
Argon, entropy of, 223
Aromatic compounds, bonding in, 293–297
Arrhenius, S., 457, 623
Arrhenius theory:
  of dissociation, 623–625
Arrhenius theory:
  of reaction rates, 457–462
Atactic, 732
Atmosphere, model of, 7
Atomic orbitals:
  contour diagrams for, 260–261
  size and energy of, 255–262
Atomic spectra, 63–66
Atomic states, 248–250
Avogadro's number, 31
Azeotrope, 575
Azimuthal quantum number, 241, 244

Bakelite, 731
Balmer, J., 65
Balmer series, 65–67
Band model of metals, 509–510
Bartell, L. S., 361
Bateman, L. C., 426
Beckmann freezing-point apparatus, 607
Beckmann method, 606
Beeck, O., 709, 715
Bell, R., 567, 593
Bender, P., 604, 686
Benzene, bonding in, 294–296
Beryllium chloride, 492
Berzelius, J. J., 262, 263
BET isotherm, 713, 714
Bjerrum, N., 654, 698
Bobalek, E. G., 569

Bockris, J. O'M., 619
Bodenstein, M., 723
Body-centered unit cell, 369
Bohr, Niels, 66, 68, 69, 265
Bohr magneton, 414
Bohr radius, 243
Boiling-point diagrams, 571–572
Boiling-point elevation, 602–605
Boltzmann, L., 27
Boltzmann's constant, 32
Boltzmann's distribution, 34, 94–95
  derivation of, 96–99
  and nuclear states, 334
Bond energies, 159–162
  of diatomic molecules, 280
  table of, 439
Bond enthalpies, 159–162
  table of, 161
Bond lengths of diatomic molecules, 280
Bond moments, 406–407
π-Bonding, 292–293
Bonding with d orbitals, 299–306
Born, M., 264
Born-Haber cycle, 494–495
  and solvation energies, 644
Bose-Einstein statistics, 222
Boyle, R., 2
Boyle's law, 2–4
Bragg, Sir William, 382
Bragg method, 372
Bragg's diffraction law, 370
Bravais, A., 368
Bravais lattices, 368–369
Brewer, L., 588
Brockway, L. O., 359, 361
Brønsted, J. N., 696, 698
Brownian motion, 728
Brunauer, S., 713
Burk, R. E., 724

Cage effect, 442, 522
Calomel electrode, 662
Calorie, 10
Calorimeter, 144
Capacitance, 395
Capillary-rise method, 528
Carbanions, 444–445
Carbon dioxide:
  isotherms for, 15
  molar volume of, 53
  Raman spectrum of, 326
  vibrations of, 326
Carbon disulfide, heat capacity of, 527
Carbon monoxide, microwave spectrum of, 314
Carbon tetrachloride, heat capacity of, 527
Carbonium ions, 444–445
Carbonyl group, electronic energies of, 330
Carnot, S., 174
Carnot cycle, 174–177
β-Carotene, 297–299
Catalysis, heterogeneous, 720–725
Catalyst poison, 725
Catalysts, 721
Catalytic reactions, 721
Cathode, 629
Cell constant, 619
Cell reactions, 665
Cellulose, 737, 739
Celsius scale, 4
Chain reactions, 440
Characteristic frequencies, 321
Charles' law, 5
Chemical shift, 336–339
  table of, 339
Chemisorption, 708, 717
Clausius, R. J. E., 27, 167, 173, 178
Clausius-Clapeyron equation, 546–548
Clausius-Mosotti equation, 399
Clayton, J. O., 186, 187
Coherent scattering, 353
Colligative properties, 601–613
  of electrolyte solutions, 625–627
Collision diameter, 42–50
Collision number, 42–50
Collision properties of gas molecules, 48–50
Collision theory, 462–467, 471–474
Collisions:
  elastic, 435
  rates of, 436–438
  reactive, 435
Colloids:
  classes of, 728
  hydrophobic, 759
Combustion reactions, 144
  enthalpies of, table, 145
Components, number of, 549–550
Compound formation in solid-liquid systems, 560–561
Compressibility factor, 14–15, 17
Condensation of gases, 15–18
Condenser, electrical, 394
Conductance, 618
  applications of, 639–642
Conductance:
  interpretation of, 652–655
Conductimetric titrations, 639–640
Conductivity of solutions, 618–623
Conrad, M., 458
Consolute temperatures, 556, 557
Constant-boiling mixture, 575
Constantan, 562
Conway, B. E., 693
Cooling curves, 558
Coordination compounds:
  bonding in, 301–306
  magnetic properties of, 416
Copper, heat capacity of, 506
Copper sulfate, transference numbers in, 633–634
Cornwell, C. D., 604, 686
Correlation diagram for diatomic molecules, 281
Corresponding states, law of, 15–18, 55–56
Cottrell boiling-point apparatus, 604
Coulomb's law, 395
Covalent bond, 265–274
Covalent bonding in crystals, 491
Covalent radii, 361–363
Cracking pattern, 486
Cram, D. J., 732, 734
Crick, F. H. C., 767–769
Critical point, 15–18
  and van der Waals' equation, 53–55
Critical solution temperature, 556
Crystal energies, 494–500
  force constant factors, 503
  shapes, 364–365
  systems, 367
  types, 490–494
Crystal-field theory, 301–302
Crystal forces, 490–494
Crystals, heat capacities of, 500–504
Curie's law, 415

d orbitals:
  equivalent, 300
  use in bonding, 299–306
Dalton, J., 262
Dalton's law, 10–12
Damon, G. H., 482
Daniels, F., 482, 604
Davy, H., 262
de Bethune, A. J., 668

de Broglie, L., 67, 68
de Broglie waves, 68, 352
Debye, P., 373, 399, 502, 516, 647
Debye (unit), 392
Debye characteristic temperature, 503
Debye equation, 401–402
Debye-Hückel theory, 647–654
  and activity coefficients, 675–678
Defay, R., 586
Degenerate states, 77
Degree of ionization from conductance, 640–641
Degrees of freedom, 34, 85
  in phase equilibria, 551
Deoxyribonucleic acid, 737
  structure of, 767–769
Dialysis, 759–760
Diamagnetism, 408
  determination of, 411–412
Diameter:
  collision, 42–50
  of ions, 384–386
  from van der Waals' equation, 55
Diamond, 149, 491
Diatomic molecules, bonding in, 277–281
Di-*tert*-butyl peroxide, decomposition of, 427–429
Dickerson, R. E., 767
Dielectric, 395
  effect of solvent, 642–644
  molecular basis of, 397–401
Dielectric constant, 395
Diffraction methods, 347–389
Diffusion:
  of gases, 19
  of macromolecules, 741–744
Diffusion coefficient, 742
  in solution reactions, 441
Diffusion-controlled reactions, 442
Dillard, J. G., 485
Dipole-dipole attractions, 515–516
Dipole moment:
  determination of, 401–403
  and ionic character, 403–406
Dipole moments:
  induced, 392
  of molecules, 390–408
Dispersion forces, 517
Displacement reactions, 445
Dissociation:
  degree of, 624
  energy, 162
  equilibria, 627–628
Distillation, 572–576
  steam, 575
Distinguishable particles, effect on probability, 220
DNA (deoxyribonucleic acid), 737
  structure of, 767–769
Donnan membrane equilibrium, 759–760
Dostrovsky, I., 431, 432
Drasel, K., 485
Driving force, 168
Droplets, vapor pressure of, 529–530
Dulong and Petit, law of, 500
Dushman, S., 21

Eastman, E. D., 15
Ebullioscopic constant, 603
  table of, 604
Eddington, A., 173
Efficiency of heat-work transformation, 178–179
Effusion, 19, 41–42
  Graham's law of, 19–20
Ehlers, R. W., 624
Einstein, A., 62, 501, 502
Einstein (unit), 475–476
Einstein relation, 67
Elastic collisions, 435
  cross section for, 462–463
Electrochemical cells, 663–665
  electrode-concentration cells, 679–680
  electrolyte-concentration cells, 680–683
Electrode-concentration cells, 679–680
Electrode potential, 666–670
Electrode processes, 629–630
Electrode reaction, 660
Electrodes, 629
  types of, 659–663
Electrokinetic effects, 755–760
Electrokinetic potential, 757
Electrolysis, 629–630
Electrolyte-concentration cells, 680–683
Electrolytes, 617–654
Electromagnetic radiation, 61–63
Electromotive force, 664–665
  activity determination from, 672–674
  concentration dependence, 670–672
  equilibrium constant from, 678
  solubility product from, 679
Electron affinity, 250–251
  and electronegativity, 285
  table of, for atoms, 251
Electron configuration:
  of atoms, 254
  table of, 256–257
Electron diffraction, 352–361
Electron micrograph, 707
Electron microscope, 771–772
Electron-pair bond, 270–272
Electronegativities, 285–288
  table of, 288
Electronic spectra, 326–331
Electronic states of molecules, 92–93
Electrons, wave nature of, 352
Electroosmosis, 758
Electrophoresis, 757
Electrostatics, 393–397
Elementary reactions, 419, 456–487
  in gas phase, 435
  in solution, 440–442
Elimination reactions, 445
Emf, 663–665
  activity determination from, 672–674
  concentration dependence of, 670–672
  equilibrium constant from, 678
  solubility product from, 679
Emmett, P., 711, 713, 715
Encounter, 442
Endothermic, 146
Energies:
  of collections of molecules, 94–110
  of molecules, 83–93
Energy, 115
Energy level diagram for the Na atom, 249
Enthalpy, 122–124
  of activation, 472
  changes of, in reactions, 145–146
  dependence on pressure, 205–206
  dependence on temperature, 153–156
  function, table of, 136
  in ideal gas expansions, 127–128
  of ideal solutions, 582
  relation to internal energy, 146–147
  standard, of formation, 779–784
Entropy:
  of activation, 472–474

Entropy:
dependence on pressure, 205–206
of ideal solutions, 582–584
molecular interpretation of, 219–228
of nitrogen, 187
of real solutions, 585–586
rotational, 224–227
and second law, 168–174
standard, 779–784
as a state function, 179–181
and third law, 184–187
translational, 222–224
of vaporization, 169
vibrational, 227–228
Enzyme, 433
Enzyme kinetics, 432–434
Equation of state, 13
Equilibrium, molecular interpretation of, 228–235
Equilibrium constant:
for ammonia synthesis, 211
from emf data, 678
at high ionic strengths, 696
for real-gas systems, 210–211
relation to free energy, 206–211
temperature dependence of, 211–215
Equivalent conductance, 620–623
Equivalent ionic conductance, 622
ESCA, 331
Esr (electron-spin resonance) spectroscopy, 341–343
Ethyl chloroformate, photoelectron spectrum of, 331
Ethylene, bonding in, 292
Eutectic, 559
ternary, 565
Ewell, R. H., 537
Exchange of electrons in H_2, 272
Excluded volume in gases, 50–51
Exothermic, 146
Expansions:
of ideal gas, 127–128
work of, 119–122
Extensive properties, 124, 551
Eyring, H., 537

Face-centered unit cell, 370
Faraday, M., 629
Faraday (unit), 630, 666
Fermi-Dirac statistics, 222
and atomic states, 253
and metallic crystals, 506
Fermi energy, 507
Fermi level, 507
Fiber, x-ray pattern of, 761
Fibrous protein, 736, 763–767
Fick's law, 742
Field, F. H., 485
Films, 704–707
Findlay, A., 611
First law of thermodynamics, 115–118
First-order rate equations, 423–429
Flash photolysis, 482–483
Flory, P. J., 612, 748, 749
Fluidity, 536
Fluorescence, 478
Force constants, 89, 316
determination of, 319
table of, 320
Forces:
centrifugal, 745
crystal, 490–494
hydrophobic, 735
intermolecular, 512–519
London dispersion, 517
van der Waals, 50–55
Ford, W. E., 365
Fourier synthesis, 382–384
Franck-Condon principle, 327
Frankenberg, W. G., 717
Franklin, J. L., 485
Free-electron model of metals, 506–510
Free-electron molecular orbitals, 297–299
Free energy, 190–193
of activation, 472
calculation of, 193–196
dependence on temperature and pressure, 197–198
of electrochemical reactions, 665–666
of ideal solutions, 583–584
molecular interpretation of, 228–235
of nonideal gases, 198–202
of real solutions, 585–586
relation to equilibrium constant, 206–211
of solution components, 591–598
of solvents, 599–601
standard, of formation, 196, 779–784
Free-energy function, 234–235
Free-radical intermediates, 439–440
Free-radical reactions, 440
Free radicals, magnetic properties of, 416
Free volume in liquids, 522
Freedom, degrees of, 34
Freezing-point depression, 605–608
of electrolyte solutions, 626
Frequencies, characteristic, 321
Frequency, 61
Freundlich isotherm, 716–717
Frost, A. A., 471
Fugacity:
of methane, 200
of nonideal gases, 200–202
Fuoss, R. M., 609, 654

Gaddy, V. L., 14, 15
Gas constant:
energy units of, 9–10
R, 8
Gas electrode, 660
Gas laws, 2–8
Gas mixtures, 10–12
Gases:
condensation of, 15–18
kinetic-molecular theory of, 25–56
nonideal, 12–18
properties of, 1–21
viscosity of, 20–21, 45–47
Gay-Lussac, J., 4
Gay-Lussac's law, 5
Geil, P. H., 761
Geis, J., 767
Gerade, 279
Giauque, W. F., 186, 187
Gibbs, J., 552
Gibbs-Duhem equation, 597–599
Gibbs free energy, 193
Gingrich, N. S., 531
Glass electrode, 685–686
Globular protein, 736, 767
Glucose, 739
Gold, heat capacity of, 507
Gouy balance, 410
Graham, T., 19
Graham's law of effusion, 19–20
Graphite, 149, 492
Gray, H. B., 251
Grotthuss mechanism, 638
Ground state, 84
of atoms, 253
Gutta-percha, 739

Hair, 765
Half cell, 660
Half life, 428
Half reaction, 660
Hamilton, L. D., 769
Hamiltonian, 258
Hammett, L. P., 13, 213, 557, 560, 576, 673, 691, 694, 695

Hammond, G. S., 732-734
Harmonic-oscillator model, 317
Harned, H. S., 623, 624, 693
Hauy, R. J., 368
Heat, 118
  of absorption, 709-710
  of reaction: measurement of, 143-145
    molecular interpretation of, 157-159
    temperature dependence of, 153-157
  of vaporization, 512-519
    table of, 513, 520
Heat capacity:
  of crystals, 500-504
  dependence on pressure, 205-206
  of ideal gases, 129-130
  of liquids, 524-527
  molecular interpretation of, 136-139
  of solids and liquids, 526
  standard, 779-784
  table of, 156, 527
    for diatomic molecules, 139
  vibrational contribution to, 137-138
Hecht, W., 458
Heisenberg, W., 69
Heitler, W., 271-272
Helium, 182
Helmholtz free energy, 193
Henry's law, 568, 569, 594-597
Herron, J. T., 485
Herzberg, G., 64, 327
Hess' law, 148
Heterogeneous catalysis, 720-725
Heteronuclear bonds, 281-285
High-energy radiation, 486-488
Hinshelwood, C. N., 724
Hittorf method, 630-634
Hooke's law, 89-90, 317
Hückel, E., 647
Hughes, E. D., 426, 431, 432
Hybrid orbitals, 289-293
Hybridization, 288-293
Hydrates, 560
Hydration, 642
  enthalpies of, 646
Hydrochloric acid:
  activity coefficients from emf measurements table, 674
  degree of dissociation, 624
  transference numbers in, 632
Hydrogen atom:
  energy level diagram for, 67
  Schrödinger equation for, 239-242
Hydrogen atom:
  wave functions for, 242-247
Hydrogen-atom spectrum, 64-67
Hydrogen bonding:
  in crystals, 493
  in liquids, 518-519
  and protein structures, 762-763
  in real solutions, 587
Hydrogen chloride:
  infrared spectra of, 521
  rotation-vibration spectrum of, 323
  rotational spectrum of, 314
Hydrogen fluoride, bonding in, 291
Hydrogen molecule:
  bonding in, 270-272
  virial theorem and, 273-274
Hydrogen-molecule ion, 267-270
Hydrophobic forces, 735

Ice calorimeter, 113-114
Ideal gas, 4
  adiabatic expansion of, 132-134
  heat capacities of, 129-130
  pressure-volume relation, 1-3
  temperature-volume relation, 4-6
  thermometer, 18
Ideal solutions:
  liquid-vapor diagrams of, 566-567
  thermodynamics of, 580-584
Incoherent scattering, 353
Incongruent melting, 561
Independent variables, 125
Indices:
  Miller, 371
  Weiss, 371
Indistinguishable particles:
  electrons in H_2, 272
  and probability, 220
Induction effect, 516-517
Infrared:
  of liquids, 521
  region, 63
  spectra, 322
Ingold, C. K., 426
Insulin, 735
Integrals, evaluation of, 774-775
Intensive properties, 124, 551
Interference phenomena, 348-352
Intermolecular forces:
  in liquids, 512-519
  table of, 516
Internal conversion, 479
Internal energy, 117-118
  molecular interpretation of, 134-136
Internal energy changes:
  in chemical reactions, 145-146
  in ideal gas expansions, 127-128
  relation to enthalpy changes, 146-147
  in reversible and irreversible processes, 118-122
Interstitial alloy, 562
Intrinsic viscosity, 748
Inversion temperature, 131
Ion atmosphere, 653
Ion pairs, 654
Ionic bond, 263-265
Ionic character, 281-285
  and dipole moment, 404
  and electronegativity, 406
Ionic conductance, 635
Ionic crystals, cohesive energy of, 494-500
Ionic forces in crystals, 491
Ionic intermediates, 443-445
Ionic mobilities, 635-639
Ionic radii, 384-386
Ionic strength, 649
  in Debye-Hückel theory, 677
Ionization energy:
  of molecules, 485
  table of, for atoms, 250
Ionization potential:
  and electronegativity, 285
  of molecules, 485
Irreversible processes, 118-119
Isobutyl bromide, 431-432
Isotactic, 732
Isoteniscope, 543
Isotherm, 3
  critical, 16
Isothermal compressibility, coefficient of, 525
Isothermal critical point, 563
Isothermal expansion, entropy change in, 171-173
Isotropic, 34

James, T. L., 331
Joule, J., 127-128, 130
Joule-Thomson coefficient, 130-132, 206
  table of, 131
Joule-Thomson expansion, 182
Junction potential, 682

Kasha, M., 281
Kekule, F. A., 262

Kelvin, Lord (W. Thomson), 5, 130, 167, 178
Kelvin temperature scale, 4
α-Keratin, 765
Kerker, M., 755
Kerr, J. A., 439
Kinetic energy of gas molecules, 30–34
Kinetic-molecular theory, 26–56
Kinetics:
  of chemical reactions, 419–487
  of heterogeneous decompositions, 721–725
Kittel, C., 505
Klotz, Irving M., 735
Kohlrausch's law, 622
Kooij, D. M., 722
Kortum, G., 619
Kossel, W., 263
Krigbaum, W. R., 748
Kvalnes, H. M., 14, 15

Lagrange multipliers, 98, 776–777
LaMer, V. K., 755
Laminar flow, 533
Langmuir, I., 704, 711
Langmuir adsorption isotherm, 711–714, 722
Langmuir film balance, 705
Lattice energy, 494–495
Lattices, 368–369
LCAO (linear combination of atomic orbitals) method, 270
LeBel, J. A., 262, 288
LeChatelier's principle, 209
Lewis, G. N., 185, 265, 588
Lewis diagrams, 266–267
Licht, T. S., 668
Ligands, 301
Light, 61–63
  corpuscular model of, 61–63
  wave model of, 61–63
Light scattering, 749–755
Line-of-centers velocity, 465
Lines of force, 393
Liquid films, 704–707
Liquid junction, 682
Liquids, 512–537
  free volume theory of, 519–524
  heat capacity of, 524–526
  structure of, 531–532
  surface tension of, 526–531
  two-component systems, 554–567
  viscosity of, 532–537
London, F., 271, 272, 517
London dispersion forces, 516–518
Longsworth, L. G., 634

McClellan, A. L., 404
MacInnes, D. A., 624, 634, 635
Macromolecules, 727
  classes of, 729
  molecular mass of, 740–741
  in solid state, 761–772
  in solution, 740–761
Madelung constant, 498–499
Magnesium sulfate, partial molal volume of, 589
Magnetic field, effect on nuclear energies, 332–334
Magnetic properties:
  of coordination compounds, 416
  determination of, 408–415
  of free radicals, 416
  of oxygen, 415
Magnetic quantum number, 241, 244
Magnetic susceptibility, 409
  and unpaired electrons, 415
Mannitol, 607
Mass spectra, 484
Mass spectrometer, 483–486
Maxwell, J. C., 27, 403
Maxwell-Boltzmann distribution, 37–40
Maxwell's relations, 205
Mayer, J. E., 264
Mead, D. J., 609
Mean free path, 42–50
  table of, 49
Mechanical energy, relation to thermal energy, 113–115
Mechanical reservoir, 114–116
Mechanism, 420
Mechanisms:
  of enzyme reactions, 450–451
  of reactions, 446–449
Menten, M., 450
Mercury, heat capacity of, 527
Metallic bonding, 494
Metallic crystals, 504–510
Metastable equilibrium, 550
Mica, 492
Micelle, 729–730
Michaelis, L., 450
Microwave spectra, 313–314
Microwaves, 63
Millard, E. B., 501
Miller indices, 371
Miscible solids, 562
MO-LCAO method, 277
Mobilities:
  of ions, 636–639
Mobilities:
  of ions, table of, 635
  of proteins, 757
Model, definition of, 26
Mokacsi, E., 339
Mole, 8
Molecular beam, 41–42
Molecular dimensions from rotational spectra, 315
Molecular interpretation:
  of entropy, 219–228
  of equilibria, 228–235
  of free energy, 228–235
  of third law, 235–236
Molecular mass:
  from gas densities, 8–9
  of gas mixtures, 11
  from osmotic pressures, 741
  of polymers, 740–741
Molecular orbitals:
  for aromatic molecules, 296–297
  for conjugated systems, 297–299
  contour diagrams for, 277–282
  for diatomic molecules, 282–283
Molecular speeds, 32–34
  table of, 33
Molecular structures from electron diffraction, 360
Molecular velocities:
  distribution in one dimension, 34–37
  distribution in three dimensions, 37–40
Moment of inertia, 78, 310–311
Momentum, relation to wave length, 68
Monomolecular layer, 704–707
Moore, C. E., 250
Mulliken, R. S., 287
Multicomponent systems, thermodynamic treatment of, 580–616

Naphthalene, electron density map, 385
Newton, I., 62
Newton, R. H., 202
Nickel, surface area of, 715
Nitrogen, adsorbed, 715
nmr (nuclear magnetic resonance) spectrometer, 335
nmr spectroscopy, 335–336
Nonideal gases, 12–18
  and van der Waals' equation, 50–56

Normal vibrations, 320–321
Normalized wave functions, 72
Nuclear charge, effective, 252
Nuclear energy levels in
magnetic fields, 332–334
Nuclear magnetic resonance,
332–340
(*See also under* nmr)
Nucleic acids, 729, 736–738
structures of, 767–769

O'Brien, S. J., 569
Octahedral field, 302
One-component systems:
phase rule for, 552
pressure-temperature
diagrams for, 539–543
Open systems, 588
Operator, 255, 258
Orbital energy diagram, 253
Orbitals:
for electrons of many-electron
atoms, 251–253
for hydrogen atom, 243–247
Osmotic pressure, 608–613
apparatus for, 609
of macromolecule solutions,
741
and molecular mass, 612–613
of sucrose solutions, 611
Ostwald viscometer, 534
Owen, B. B., 622, 693
Oxidation-reduction electrodes,
660–661
Oxygen:
electronic states of, 327
free energy of dissolved, 595
magnetic properties of, 415

Paramagnetic salts, 182–183
Paramagnetism, 408, 412–415
Parent peak, 484
Partial molal quantities,
587–591
Partial pressure, 10–12
Particle-in-a-box example:
one-dimensional, 71–74
three-dimensional, 75–77
Partition function, 101–103
rotational, 105–106
in transition-state theory,
469–472
translational, 104
vibrational, 106–107
Pascal's constants, 412
Pauli, W., 253
Pauli exclusion principle, 226
for atomic states, 253
Pauling, L., 161, 286–289, 303,
386, 387, 406, 766
Pearson, R. G., 471
Pedersen, K., 696
Peptide bond, 733
Periodic table, 254
Peritectic reaction, 561
Permittivity constant, 393
Perry, R. H., 200, 202, 564
Perutz, M. F., 768
pH, 686–688
pH meter, 687
Phase, 549
Phase diagrams:
for liquid-vapor systems,
566–574
for multicomponent systems,
554–576
Phase equilibria, 539–576
in one-component systems,
539–548
thermodynamics of,
544–548
Phase rule, 552–554
Phosphine, decomposition of,
722
Phosphorescence, 478
Photochemical reactions,
475–482
Photoelectron spectroscopy, 331
Photolysis, 439
flash, 482
Physical adsorption, 708
Pitzer, K. S., 274, 588
Plait point, 563
Planck, M., 62
Planck's constant, 62
Planck's relation, 74
Plane, R. A., 257
Point groups, 366–367
Poise, 21
Poiseuille equation, 534
Poisson-Boltzmann equation,
651
Polar coordinates, 239–240
Polarizability:
of molecules, 392
and Raman spectra, 325
Polarization, 396
determination of, 401–403
molar, 399
temperature dependence of,
402
Polyethylene, 492, 731–732
Polyisobutylene, 612–613
viscosity of, 748
Polyisoprene, 729, 739
Polymerization, 440
addition, 730
condensation, 731
Polymers:
crystallinity of, 769
synthetic, 730–732
Polymorphism, 540
Polypeptides, 733, 761
Polypropylene, crystallinity of,
770
Polysaccharide, 729, 737–739
Potential energy:
in Schrödinger equation, 70
for square wells, 71–77
Potential energy curve:
for H_2, 271
for H_2^+, 268
for NaCl, 264, 265
Potentiometer, 664
Powell, R. E., 300
Pre-exponential factor, 458
values of, 471
Pressure:
critical, 16
and free energy, 197–198
and kinetic-molecular theory,
27–30
partial, 10–12
reduced, 17–56
and van der Waals' equation,
50–55
and virial equation, 13
and volume of gases, 2–4
Pressure fraction, 11–12
Pressure-temperature diagrams
in one-component systems,
539–543
Prigogine, I., 586
Primary process in
photochemical reactions,
477
Primitive unit cell, 369
Principal quantum number, 241
Probability, 96–97, 219–222
Probability functions:
for one-dimensional square
well, 73
for three-dimensional square
well, 76–77
Promotional energy, 289
Proteins, 729, 732–736
hydrolysis products from, 734
molecular mass of, table, 742
structure of, 761–767
Pyrolysis, 439

Quantum efficiency, 479
Quantum numbers:
from angular momenta, 79
and de Broglie waves, 68
for diatomic molecule orbitals,
281
for electrons in atoms, 241,
249, 256–257
and molecular rotations, 105,
310

Quantum numbers:
and molecular vibrations, 106, 318
Quantum yield, 479
Quinhydrone electrode, 661

Radar, 63
Radial distribution:
in electron diffraction, 359–361
in liquids, 531
Radial-distribution function, 243–244
Radiation, high-energy, 486–488
Radiation chemistry, 475–483
Radii of ions, 384–386
Radio waves, 63
Radioactive decay, 429
Radiolysis, 487
Raley, J. R., 429
Raman spectroscopy, 325–326
Ramsey, N. F., 404
Ramsey-Young apparatus, 543
Randall, M., 185, 588
Raoult's law, 567, 602
Rate of reaction, 419–488
for decompositions, 721–725
effect of ions on, 698–700
first- and second-order equations for, 424–432
measurement of, 421–422
temperature dependence of, 457–460
Rate constants, table of, 466
Rate equation, 420, 422–424
Rational indices, law of, 370
Reaction coordinate, 461, 462
vibration along, 468
Reaction cross section, 462–463
Reactive collisions, 435
Reduced mass, 312
Reduced variables, 17
Refractive index and molar polarization, 403
Relaxation time of ionic atmosphere, 653
Resistance of solutions of electrolytes, 618
Resonance, 293–296
Resonance energy, 296
Reversible processes, 118–119
Reynolds number, 534
Ribonucleic acid (RNA), 737
Ries, E., Jr., 705, 707
Ritchie, A. W., 715
RNA (ribonucleic acid), 737
Robertson, J. M., 385
Rollefson, G. K., 15
Root-mean-square speed, 33
Rosenstock, H. M., 485
Rossini, F. A., 139
Rotation, hindered, 139
Rotation-vibration spectra, 321–325
Rotational energy:
and angular momentum, 80, 87
levels for, 88
of a mole of a gas, 105–106
of molecules of a gas, 87–88
and rotational spectra, 310
Rotational entropy, 224–227
Rotational spectra, 310–316
Rubber, 739
Rust, R. F., 429
Rutherford, E., 263
Rydberg, J. R., 65
Rydberg formula, 65–66

Sackur-Tetrode equation, 223
Salt bridge, 663, 684–685
Sanger, F., 734
Scattering of electrons, 353
Scherrer, P., 373
Schomaker, V., 362
Schrödinger, E., 69
Schrödinger equation:
applied to hydrogen atom, 239–242
in one dimension, 69–74
in polar coordinates, 240
in three dimensions, 75–77
Second law of thermodynamics, 167–174
Second-order rate equations, 423, 430–431
Secondary process in photochemical reactions, 478–479
Sedimentation, 744–747
Sedimentation coefficient, 746
table of, 742
Selection rule:
for rotational transitions, 312
for vibrational transitions, 318
Self-consistent field, 255–262
Selwood, P. W., 412
Semiempirical equation, 52
Semipermeable membrane, 608
Shedlovsky, T., 624
β-Sheet, 763–764
Short-range order in liquids, 531
SI (Système International) units, 9, 785–787
Siemens, W., 182
Sienko, M. J., 257
Silicon carbide, 492
Silk, 763–765
Silver heat capacity of, 505
Singlet state, 328, 479
Sodium atom:
effective nuclear charge, 252
energy levels of, 249
Sodium chloride:
lattice energy of, 494–495
unit cell of, 381
x-ray powder pattern of, 375
Solubility:
activity coefficients from, 690–694
from conductance measurements, 641
Solubility product:
from emf data, 679
of sparingly soluble salts, 641
Solutes:
activity of, 594–598
activity coefficient of, 594–598
free energies of, 594–598
standard state of, 596
Solvation, 642
energies of, 644–647
Solvents:
activities of, 592–594
activity coefficient of, 592–594
effect on solute properties, 642–655
free energies of, 591–594
Sørensen, S. P. L., 686
Sorption, 704
Space groups, 368
Speakman, J. C., 369
Specific conductance, 619
Specific resistance, 619
Spectra:
atomic, 64
esr, 342, 343
far infrared, of HCl, 314
hydrogen atom, 64
infrared, 322
of HCl, 323
nmr, 337
photoelectron, 331
Raman, 326
Spectrograph, 63
Spectroscopy:
electronic, 326–331
esr, 341–343
molecular, 309–343
nmr, 335–336
photoelectron, 331
Raman, 325–326
rotational, 310–316
vibrational, 316–321
Speed of molecules, 40
Spin quantum number, 241
Spontaneity, relation to free energy, 192–193
Standard electrode potential, 666–670
Standard emf, 666

Standard free energies, 196
Standard heats of formation, 149–151
of aqueous ions, 151–153
Standard state:
for free energy of real gases, 203–205
for heats of formation, 150
for solutes, 595
for solvents, 593
Starch, 737
State function, 117
properties of, 124–127
Stationary-state method, 446–451
Steam distillation, 575
Stearic acid, 706
Stevenson, D. P., 362
Stibine, decomposition of, 723
Stirling's approximation, 775–776
Stock, A., 723
Stokes, R. H., 743
Stokes' law, 743–744
Streaming potential, 758
Strong electrolytes, 622
Structure factor, 377–382
Su, Goug-Jen, 17
Substitutional alloy, 562
Substrate, 433
Sucrose, 611, 737
Sum, relation to integral, 778
Supercooled water, 541
Surface, preparation of, 719
Surface-active, 704
Surface area, 714–716
Surface balance, 704–705
Surface energy, 527
Surface tension, 526–531
table of, 529
Svedberg, T., 744, 746, 747
Svedberg (unit), 746
Swendeman, N., 668
Symmetry:
of crystals, 365
of molecules, 366
of vibrations, 320–321, 326
Symmetry elements, 364–368
Symmetry number, 226
Symmetry operations, 364–368
System, 115–116
Système International d'Unites, 9, 785–787

Teller, E., 713
Temperature:
absolute, 4, 18–19
absolute zero of, 181–184
and Boltzmann equation, 94–101
Temperature:
critical, 15–17, 53
and kinetic energy, 30–31
reduced, 17, 56
Temperature dependence:
of emf, 697
of equilibrium constants, 211–215
of free energy, 211–215
of heat capacities, 156
of heats of reaction, 153–157
of reaction rates, 457–460
of vapor pressure, 542
of viscosity, 535–537
Ternary eutectic, 565
tert-butyl bromide, hydrolysis of, 425–426
Tetrahedron, 288
Tetramethyl silane, 337, 339
Thermal energy:
calculation of, for NO_2, 135
categories of, 83–85
electronic, 92–93
relation to mechanical energy, 113–115
rotational, 87–88, 105–106
summary of, 109–110
translational, 86–87, 99–101, 103–104
vibrational, 89–92, 106–109
Thermal enthalpy terms, 136, 158
Thermal expansion, coefficient of, 525
Thermal reservoir, 114–116
Thermochemical equations, 147–148
Thermochemistry, 143–162
Thermodynamic data, 779–784
from emf measurements, 697–698
Thermodynamics:
first law of, 115–118
nature of, 112–115
second law of, 167–174
third law of, 184–187
Thermometer, ideal gas, 18–19
Third law of thermodynamics, 184–187
molecular interpretation of, 235–236
Thomas, C. D., 531
Thomson, J. J., 263
Thomson, W. (*see* Kelvin, Lord)
Three-component systems, 562–565
Threshold energy, 462–463
Tie lines, 563
Timmermans, J., 568, 593
Titration, conductometric, 639–640
Torsional motion, 139
Total differential, 126
Transference numbers, 630–634
table of, 634
Transition-state theory, 467–472
application to solutions, 472–474
Translational energies, 84–85
of a mole of gas, 103–104
of molecules of a gas, 86–87, 99–101
Translational entropy, 222–224
Trigonal orbitals, 292–293
Triple point, 541
of water, 19, 541
Triplet state, 329, 478
Trouton, F., 519
Trouton's rule, 519–520, 522
Turbidity, 750
Turbulent flow, 533
Turnover rate, 451
Two-component systems:
liquids, 554–556
solid-liquid, 557–562
Tyndall effect, 749

Ultracentrifuge, 744
Ultramicroscope, 749–750
Ultraviolet, 63
Undetermined multipliers, 777
Ungerade, 279
Unit cell:
dimensions of, 375–377
parameters for, 367
two-dimensional, 368
types of, 369–370
Units, SI, 785–787
Universe of a process, 115–116

Valence bond, 288–293
for conjugated systems, 293–296
van der Waals, J. C., 50
van der Waals' equation, 50–56
van der Waals' forces in crystals, 491
van der Waals' radii, 386–387
van't Hoff, J. H., 262, 288, 610
van't Hoff *i* factor, 625–627
Vapor-liquid systems, 566–576
Vapor pressure:
determination of, 543
as function of temperature, 542
and Gibbs-Duhem equation, 598–599
lowering by solutes, 601
of small droplets, 529–531
of two-component systems, 566–571

Vaporization:
  entropy of, 169, 519–524
  heat of, 512–520
  of water, 121
Variation theorem, 259
  and electronegativities, 286
Vaughan, W. E., 429
Vector diagram for atomic
    states, 249
Velocities of molecules:
  in one dimension, 34–37
  in three dimensions, 37–40
Vibrational energies, 84–85
  of a mole of gas, 106–109
  of molecules of a gas, 89–92
Vibrational entropy, 227–228
Vibrational heat capacity,
    137–138
Vibrational spectra, 316–321
Virial coefficients, 14
Virial equation, 13
Virial theorem, 272–274
Viscosity:
  of gases, 20–21
  of liquids, 532–537
  of solutions of macromolecules,
    747–749
Viscous flow, activation energy
    for, 537
Visible radiation, 63
Volume:
  critical, 15–17
  excluded, 51
  of ideal solutions, 580–582
  partial molal, 588–590
  reduced, 17, 56
Volume fraction, 11–12
von Laue, M., 351
von Zawidzki, J., 567, 593

Wahl, A. C., 260, 282
Ward, A. F. H., 711
Water:
  heat content of, 514
  hydrogen bonding in, 518
  phase diagram for, 540, 541
  radiolysis of, 487
Water droplets, vapor pressure
    of, 530
Water molecule:
  bonding in, 291
  vibrations of, 321
Watson, J. D., 767–769
Wave function, 70
Wave length, 61
Wave mechanics, 69–70
Wave nature:
  of electrons, 352
  of particles, 66–69
Wave number, 65, 319
Weak electrolytes, 622
Weiss indices, 371
Wells, A. F., 363
Werner, A., 263
Wheatstone bridge, 618
Wierl equation, 354–359
Williams, J. W., 604, 686
Wool, 765
Work, 118
  of expansions, 119–122
Wright, T., 567, 593

X-ray diffraction, 363–384
  from fibers, 761
  powder method, 374
X-rays, 63

Zahn, C. T., 433
Zeta potential, 757

LIST OF THE ATOMIC MASSES OF THE ELEMENTS

Element	Symbol	Atomic Number	Atomic Mass*
Actinium	Ac	89	(227)
Aluminum	Al	13	26.98
Americium	Am	95	(243)
Antimony	Sb	51	121.75
Argon	Ar	18	39.95
Arsenic	As	33	74.92
Astatine	At	85	(210)
Barium	Ba	56	137.34
Berkelium	Bk	97	(249)
Beryllium	Be	4	9.012
Bismuth	Bi	83	208.98
Boron	B	5	10.81
Bromine	Br	35	79.91
Cadmium	Cd	48	112.40
Calcium	Ca	20	40.08
Californium	Cf	98	(251)
Carbon	C	6	12.011
Cerium	Ce	58	140.12
Cesium	Cs	55	132.91
Chlorine	Cl	17	35.45
Chromium	Cr	24	52.00
Cobalt	Co	27	58.93
Copper	Cu	29	63.54
Curium	Cm	96	(247)
Dysprosium	Dy	66	162.50
Einsteinium	Es	99	(254)
Erbium	Er	68	167.26
Europium	Eu	63	151.96
Fermium	Fm	100	(253)
Fluorine	F	9	19.00
Francium	Fr	87	(223)
Gadolinium	Gd	64	157.25
Gallium	Ga	31	69.72
Germanium	Ge	32	72.59
Gold	Au	79	196.97
Hafnium	Hf	72	178.49
Hahnium	Ha	105	(260)
Helium	He	2	4.003
Holmium	Ho	67	164.93
Hydrogen	H	1	1.0080
Indium	In	49	114.82
Iodine	I	53	126.90
Iridium	Ir	77	192.2
Iron	Fe	26	55.85

*Based on mass of ^{12}C at 12.000. Values in parentheses represent the most stable known isotopes.